INTEGRATED DESIGN AND OPERATION OF WATER TREATMENT FACILITIES

New Mokawk water treatment plant (100 mgd), Tulsa, Oklahoma.

INTEGRATED DESIGN AND OPERATION OF WATER TREATMENT FACILITIES

SECOND EDITION

Susumu Kawamura

JOHN WILEY & SONS, INC.

New York • Chichester • Weinheim • Brisbane • Singapore • Toronto

This book is printed on acid-free paper.♾

Published simultaneously in Canada.

Library of Congress Cataloging-in-Publication Data:
Kawamura, Susumu.
Integrated design and operation of water treatment facilities / Susumu Kawamura.—2nd ed.
p. cm.
Includes bibliographical reference and index.
ISBN 0-471-35093-1 (cloth:alk. paper)
1. Water treatment plants—Design and construction. I. Title.
TH4538.K38 2000
628.1′62—dc21

99-089640

Printed in the United States of America

10 9 8 7 6 5 4 3 2

CONTENTS

PREFACE

My book entitled *Integrated Design of Water Treatment Facilities* was published in 1991. It has gained a good reputation among professional water treatment engineers, plant operators, as well as many college students because it covers a broad spectrum of water treatment process design (in detail) and important operational issues. I have tried to make this book unique by incorporating many practical ideas that are not covered in many design books. During the past 49 years I have been involved in all facets of water treatment issues—pilot studies, design work, construction management, and four years of actual plant operation— in various regions (United States, Asia, South America, and Australia) and wish to share my knowledge and experience with the readers. As a testimony to its value and usefulness, the book has been translated into Japanese and Korean, and a Spanish version is currently in preparation.

Treatment process technology has advanced rapidly during the past 10 years, due to various recent treatment issues and more stringent drinking water quality standards. For this reason I have decided to create a second edition of the original book. The title has been modified to *Integrated Design and Operation of Water Treatment Facilities* to reflect the many important plant operational issues that can benefit plant operators. The book has only one author, thus retaining a uniform style for easy readability. Moreover, I have tried to indicate clearly how to select, from many alternatives, the proper unit process and equipment in order to maximize overall process efficiency, thereby minimizing operation and maintenance efforts and costs.

This new edition covers several new subjects: value engineering, watershed management, dissolved air flotation process, filtered water reservoir (clearwell) design, electrical system design, ultraviolet ray disinfection, dichlorination, and four new appendices of useful information. There has also been a significant degree of modification and addition for certain subjects: privatization delivery process objectives for finished water quality, treatment process selection, enhanced coagulation, high-speed microsand settling process, basic hydraulics of filtration, taste-and-odor control, instrumentation and control, disinfection process, ozonation, disinfection by-product control, GAC process, and membrane filtration process.

The history of modern water treatment for the public water supply began approximately during the mid-nineteenth century in Europe with control of pathogenic bacteria by chlorination. Several decades later, the use of chemical flocculation, sedimentation, and the rapid sand filtration process train (prior to chlorination) were established in the United States early in the twentieth century to make public waters safe from pathogenic bacteria. No significant changes occurred until the early 1970s; the Safe Drinking Water Act (1974) ushered in a new era for the field of water treatment in the United States by controlling disinfection by-products (DBPs) created through the use of chlorine disinfection as well as pathogens. Since the mid-1970s new methods, such as ozonation, the enhanced coagulation process to control precursors of DBPs, the $C \times t$ concept for good final disinfection, and high rate filtration with dual media and coarse deep media beds with synthetic polymers, have been developed to meet the requirements. However, another new water treatment issue emerged around 1990 due to outbreaks of *Cryptosporidosis* in both the United States and other major developed countries. Since the *Cryptoparvum* oocyst cannot be inactivated by regular chlorination, there must be a high degree of removal of suspended solids, including the oocysts, prior to final chlorination or even ozonation. Moreover, the filter wash waste also requires a high degree of treatment prior to recycling.

Membrane filtration, which has been practiced in both the medical field and certain food industries, has now gained recognition as a new water treatment process, due to the effective removal of almost all suspended solids, including all pathogens, such as *Cryptosporidium* oocysts, without the treatment processes of coagulation and sedimentation. Membrane filtration is also a highly automated system that saves operator time and attention and allows for easy disposal of residuals. Despite the few limitations of the membrane process, it will undoubtedly become an important as well as practical water treatment process into the new millennium.

Susumu Kawamura

San Gabriel, CA
June 2000

ACKNOWLEDGMENTS

This book is dedicated to my parents, for their guidance, exceptional caring, and devotion to providing all their children an education to the highest level.

I would like to express my deepest appreciation to my daughter, Mika Kawamura, for her devoted efforts in editing and organizing the manuscript. I am also grateful to Mrs. Debbie Montesanti for typing the manuscript. Particular appreciation is given to many of my colleagues, especially to Murli Tolaney, Chairman of the Board of Montgomery Watson, Inc., for his encouragement. Finally, I wish to express my sincere tribute to my wife, Reiko, and my elder brother, Dr. Shigeru Tanaka, for their spiritual support and understanding.

INTEGRATED DESIGN AND OPERATION OF WATER TREATMENT FACILITIES

CHAPTER 1

Management of a Design Project

1.1 ROLE OF THE PROFESSIONAL ENGINEER

For projects that are not privatized, the owner must prepare a budget to finance the operation and a time schedule for completion. The design phase of most municipal water treatment plants typically costs 5 to 10% of the estimated construction cost. The time required to design and construct a treatment plant is approximately 3 years, including a 9- to 12-month design period as practiced in the United States. If the owner of the water treatment plant wishes to restrict both the cost and time required to complete the project to these figures, he must hire well-qualified consulting engineers or equipment manufacturers and depend on their experience, judgment, and ingenuity.

Most European and Asian countries generally rely on either equipment manufacturers or contractors offering packaged proprietary processes to design, construct, and start operation of a water treatment plant. The owner or hired consultant often establishes the plant site, the hydraulic profile across the unit processes, and a few basic design criteria, such as the hydraulic loading rates for the basins and the filters and the finished water quality standard. Based on the criteria established, the successful bidder then furnishes the major plant process facilities and all ancillary facilities required to complete the plant.

In the European and Asian method, tenders are invited to bid on the project after only a few design criteria have been established. The tender has the freedom to choose the process configurations, plant layout, type of process units, number and dimensions of the unit process, mode of the control systems, and building materials. The hydraulic loadings of the treatment processes are occasionally upgraded from those dictated by the owners by submitting the results of pilot studies that meet the criteria established for the finished water quality. The proposals are then evaluated by the owner and judged according to their quality and adequacy and the estimated design and construction costs.

Few large public water purveyors in Europe or Asia are capable of performing all phases of a project, including detailed design, construction management, and plant startup, using their own staff. In many cases, detailed predesign studies such as evaluation of alternative processes and site selection tend to be less thorough than work performed by consultants in the United States. In regard to municipal and publically owned water treatment plants, standard practice in the United States is generally quite different from that in Europe, except in the following cases: small plants processing less than 10 million gallons per day (mgd) (0.4 m^3/s) of municipal water, any size plant used in the industrial field, or water treatment plants owned by investors such as land developers. In the United States it is customary for professional engineers to complete 80 to 85% of the detailed engineering design prior to bidding. In contrast, the European and Asian approach completes the bulk of the detailed design (70 to 80%) after bidding by the contractor.

In the United States, professional engineers are required to prepare documents that are complete with a series of drawings and a set of specifications prior to being invited to bid on a project. The drawings alone may total 250 to 500 sheets of civil, architectural, structural, mechanical, and electrical design, as well as drawings of the instrumentation and landscaping. The printed specifications may be 1 to 3 in. thick prior to bidding and must present the various design criteria established by the professional engineers: the hydraulic profile across the plant, the various process trains, and the site plan. This document should also include a detailed design that defines the exact number, configuration, and size of each unit process, yard piping scheme, civil works, and necessary ancillary facilities. Once the bid has been awarded, the engineer merely reviews the shop drawings submitted by the manufacturers and manages the general construction work.

In the United States a consulting engineer is a professional who is experienced in applying scientific principles to engineering problems. As a professional, the consulting engineer has a duty to both the public and the client. Thus the engineer should be registered, that is, pass the required state examinations. The American Society of Civil Engineers (ASCE) publishes a guide on how to engage the engineering services of consulting engineers; refer to Manual 45 (1981 and 1996).

1.2 SELECTION OF CONSULTANTS

In principle, a consultant does not competitively bid his or her engineering services. If consultants are selected on the basis of price, the owner risks hiring incompetent and inadequate services. It is therefore important for the owner to hire the best professional service that is available by paying an appropriate fee. The fee structure is discussed in Section 1.3.

Owners of water treatment plants should follow the normal procedure of selecting a professional consultant in the following order:

1. Issue a *request for qualification* (RFQ) for the project and review the qualifications submitted by each firm.
2. Compile a short list of three to five firms based on experience, knowledge, and ability to undertake the project.
3. Issue a *request for proposal* (RFP) to each firm selected. Each company should be asked to submit a detailed presentation of its qualifications and ability to undertake the project. The proposal must include information on the size of the firm, number of staff members, availability of qualified personnel to be

assigned to the project, and experience of these engineers in similar lines of work.

4. Select the most qualified firm and a backup in case contract negotiations fail with the first choice.
5. Notify the firm of its selection and begin negotiating the fee for the project and a detailed scope of the professional services that are to be rendered.

Several important points should be evaluated during selection of the consultant: technical qualifications of the firm; personality and administrative skills of key engineers, such as the project manager and project engineer; existing workload of the firm (i.e., the ability of the firm to absorb the additional workload in relation to its capacity); financial stability of the firm; and experience, reputation, and past accomplishments of the firm in similar lines of work.

1.3 FEE STRUCTURE FOR SERVICES

Although most engineers assigned to a project are not involved in negotiating the fee structure of a project, as this is in the realm of the project engineer or his or her superiors, it is still important for all engineers to be familiar with this subject. After all, design engineers are responsible for both the fiscal and technical aspects of the project. The fees charged by the firm are based on acceptable compensation curves or yardsticks and the prevailing standards of the industry. For more detail on this subject, refer to the ASCE *Manual and Report on Engineering Practice, Manual 45*, and the American Consulting Engineers Council's *Compilation of Fees*.

There are four basic methods by which fees are determined: (1) salary cost × a multiplier, plus other direct costs; (2) cost plus a fixed fee; (3) fixed lump sum; and (4) a percentage of the construction cost. Most civil engineering contracts are based on the first two methods.

1. *Salary cost × a multiplier, plus other direct costs.* This type of contract is commonly used when the scope of work cannot be defined accurately. In this method, reimbursement is calculated by multiplying salary cost, payroll factor, and a multiplier. Salary cost is defined as the direct payroll plus fringe benefits given to the project team members; total benefits are approximately 130 to 140% of actual salary cost. The multiplier applied to the salary cost is a factor that compensates for overhead and supplies, a reasonable margin for contingencies, interest on invested capital, insurance, and profit. This multiplier varies with many factors, including geographic area, and is negotiable. The multiplier is approximately 2.3 to 2.5 times the salary cost but is subject to periodic reevaluation. For instance, if the average salary rate of a project team is \$25 per hour, the payroll factor is 130%, and the multiplier is 2.4, the payment provision in the contract should stipulate that the consultant be reimbursed \$78 per hour for the designer's time:

$$(\$25 \times 1.3) \times 2.4 = \$78 \text{ or } \$25 \times 3.12 = \$78$$

Other identifiable costs or direct nonsalary expenses include travel, computer time, telephone calls, subcontractors' charges, printing and binding, and a processing fee for executing the work. These expenses must be reimbursed by the client at actual cost plus a service charge of 10 to 15%.

This particular method is not limited to projects with an undefined scope of work but may also be applied in situations where the scope is fairly well defined. With this type of contract, consultants stand a good chance of not losing money.

2. *Cost plus a fixed fee.* This type of contract is used in cases where the scope of the work cannot readily be defined. Yet it is important to define the scope of the work as completely as possible, for only then will the owner agree to reimburse the consultant for direct cost plus a fee. Reimbursable costs include the technical payroll and actual expenditures for the project that are incurred directly. The fixed fee contains profit, nonallowable costs, and allowable costs. The *nonallowable costs* include contingencies, interest on invested capital, and the availability of the consulting team. *Allowable costs* are defined as the direct labor costs, direct project costs, and indirect costs incurred by the labor base.

The fixed fee varies with the size of the project as well as the complexity and scope of the job. It is traditionally calculated as a percentage of engineering costs and varies from 10 to 25%. Due to the nature of this type of contract, it is important to eliminate any increase in the engineering fee unless the scope of the engineering services is revised.

3. *Fixed lump-sum fee.* This method is frequently used for investigations and studies or design projects with a well-defined scope and complexity. The fixed fee is computed by estimating the work-hours required to execute the project and the anticipated costs for rendering the service. For water treatment plant design projects, any experienced professional engineer can estimate the cost for services by analyzing the demands of the project and the total number of required drawings.

The fixed fee may also be calculated as a percentage of the estimated construction cost. However, this particular method is not generally recommended because the figures tend to be inaccurate, especially with respect to plant expansion projects. Unless otherwise specified, the fixed lump-sum fee contract has the client pay the firm on a monthly basis during execution of the work. It is therefore important that the contract include a time limit for the professional services and a provision for adjusting the fee in case the project is delayed by factors that are beyond the control of the engineer.

4. *Percentage of construction cost.* This method is a variation of the fixed lump-sum fee. It was very popular when the economy was stable and the available technology was not changing rapidly. Yet this type of contract is still used for some design projects. This method of assessing fees relies heavily on the engineering reputation of the consulting firm and the customary percentage by the industry; refer to the compensation curves published by ASCE (Figure 1.3-1).

This method is not recommended for plant expansion projects because the design work is generally more detailed and therefore more costly than for new plants.

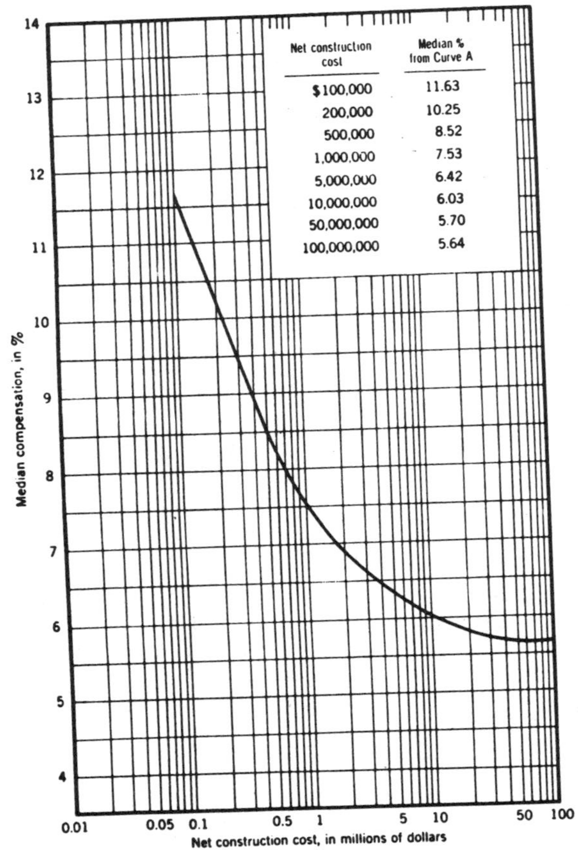

CURVE A. MEDIAN COMPENSATION FOR BASIC SERVICES EXPRESSED AS A PERCENTAGE OF NET CONSTRUCTION COST FOR PROJECTS OF ABOVE-AVERAGE COMPLEXITY (1980)

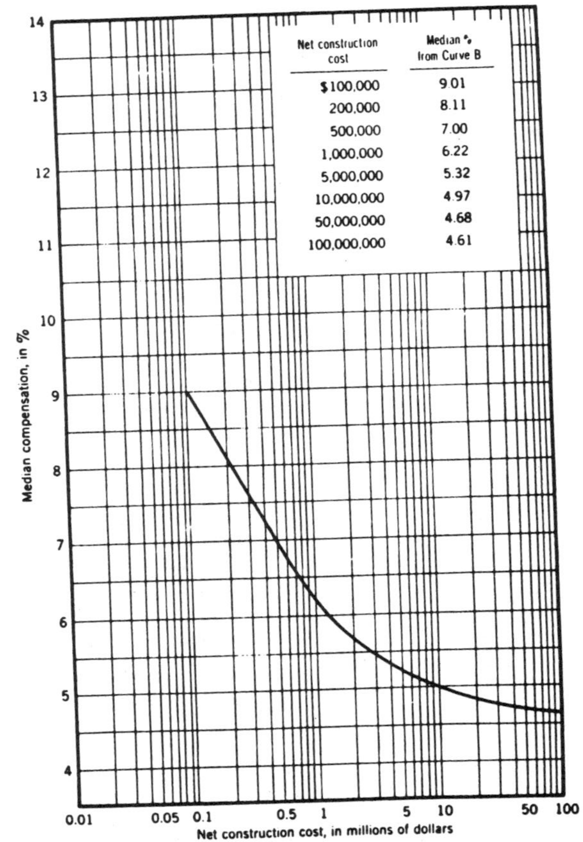

CURVE B. MEDIAN COMPENSATION FOR BASIC SERVICES EXPRESSED AS A PERCENTAGE OF NET CONSTRUCTION COST FOR PROJECTS OF AVERAGE COMPLEXITY (1980)

Figure 1.3-1 Compensation curves. (From ASCE Manual 45, 1981).

TABLE 1.4-1 Contract Brief and Billing Summary

CONTRACT BRIEF

Client Name ______________________________

Job Number ______________ Profit Center ______________ Date Opened ________

A. Contract Information

C __________ For contracts already in our database, provide contract number, and proceed to Section B.
Complete remainder of Section A for new contracts which are not in our database

Contract Description ______________________________

Profit Center __________ Proposal Number __________ Contract Amount $__________

Contract Manager Name ______________________________

Client Address ______________________________ Employee Number __________

____

City ______________ State ______________ Zip__________

Client Contact: First Name ______________ Middle Initial____ Last Name__________

Title ______________________________ Phone Number (____) __________

Department ______________________________ Division __________

Referred by Client Number ______________ Parent Client Number __________

___ Notice to proceed required?____ Standard Fees & Conditions attached? Form No. __________

B. Contract Financial Element Information

E __________ For elements already in our database, provide financial element number, and proceed to Section C.
Complete remainder of Section B for new elements which are not in our database

Element Description ______________________________

Project Manager Name ______________________________ Employee Number__________

Profit Center ______________ Contract Type ________ Element Amount: $ __________

Client Authorization:

In Accordance with:

____ Our proposal countersigned by Client dated__________

____ Our agreement dated __________

____ Letter of authorization dated __________

____ Client's contract No. __________ dated __________

____ Client's P.O. No.__________ dated __________

Source of Funds:
First Source__________
Second Source__________
Third Source__________

C. Job Information

Computer Job Description ______________________________

Job Description ______________________________

Project Engineer Name ______________________________ Employee Number __________

Business Element ______________ Project Control Element__________

Job Site Location: City ______________ State ______________ Zip__________

____ QA/QC Job Percent of fees ____ % CCM scheduled: date __________ location__________

Requested CCM members: Name ______________ Emp. No.:________ Name__________ Emp. No.:________

Name ______________ Emp. No.: ________ Name ______________ Emp. No.:________

(attach additional sheet if nessary)

Subcontractors: ______________________________

____

(attach additional sheet if nessary)

D. Approvals

Project Manager ______________ Date__________ Contract Manager ______________ Date__________

Dept. Mgr. or BOM ______________ Date__________ Group Manager ______________ Date__________

Verified by Billing______________ Date__________ Verified by Corp. Counsel______________ Date__________

CA-12(3/89)

TABLE 1.4-1 *(Continued)*

BILLING SUMMARY

E. **Job Upper Limit**

Fees $ ______ ___ Firm ___ Estimated Budgeted Cost $______ Profit ___ %

(Budget Worksheets are not required on Contracts with less than $10,000 in fees)

F. **Invoicing Information**

Bill on same invoice as job no. ______ ___ Show jobs separately on invoice

When to bill: ___ monthly ___ in stages ___ upon completion

G. **Contract Type (select one)**

	G-1. Salary Cost overhead/mark-up %	G-2. Cost + Fixed Fee overhead/mark-up %	G-3. Hourly Rate mark-up %	G-4. Lump Sum ___ check here if fixed lump sum Period Amt. $____
Billing Group				
Office Labor	____ %	____ %	____ %	
Field Labor	____ %	____ %	____ %	
Overseas Labor	____ %	____ %	____ %	
Rented Labor	____ %	____ %		
Outside Services	____ %	____ %		
Other Charges (ODC)	____ %	____ %		
		___ Prov. Overhead rate	___ Current Std.- code ____	
	Salary Cost ___ %	Fixed Fee $____	___ Historical Std.- code ____	
			___ Other (attach rate sch.)- code ____	

H. **Other Information**

Retention ______ % Mileage Rate: ___ JMM Standard or ____ ¢

Interest Rate ______ % After No. of Days ______

Backup (only if required by client) ___ Labor ___ ODC

I. **Special Instructions**

______

______

______

J. **Billing Contact Information**
(complete only if different from address on reverse side)

Street Address ______ State ______ Zip ______

City ______ Country ______

County ______ Middle Initial ______ Last Name ______

First Name ______ Phone Number (___) ______

Title ______

K. **Invoice Description**

Professional Services Rendered during the Period of: Start Date ______ To ______

______

______

______

CA-12(3/89)

The existing plant must be evaluated extensively, and all new parts must be compatible with existing parts. This detailed analysis does not allow freedom in design and the engineer must spend more time on the project.

It is very important to note that final agreement on the fee (by both parties) should always be negotiated after a detailed discussion of both the scope of the work and the elements of the engineering costs.

1.4 PROJECT CONTROL

Professional engineers manage a broad range of activities, including technical investigation and analysis, environmental and preliminary studies, planning, design, construction, compilation of the operation and maintenance manual, and plant startup, as well as providing advice and general consultation. Regardless of the type of service, each project must be completed on schedule, within budget, and with a minimal number of errors and omissions—ideally, error free. A rigorous control measure is therefore vital for any project.

To execute effective control over a project, a solid system of control must be established. However, if the firm already has a standard procedure, all projects should follow this method. An example of a design project control system is as follows:

1. Upon receipt of a signed contract, assign a job number to the project, including the year if possible. This number should be used on all work, including calculations, drawings, and correspondence. All costs and charges pertaining to the project should be identified by this number.
2. Prepare a contract brief, billing summary, and a budget worksheet for the project. Refer to Tables 1.4-1 and 1.4-2 for examples of the forms.
3. Create project files.
4. Prepare a control schedule for the project, including the period of activity, the budget for each discipline of the project team, meeting dates for coordinating each section of the project, a final check date, a date for obtaining the final authorized signature, and a date for printing and binding the proposal. Although several methods are used in scheduling the workforce requirements during the design phase, the simplest and most common are the bar chart, a graphic representation of each discipline of the job plotted against time, the critical path method (CPM), the program evaluation and review technique (PERT), and the integrated budget and schedule monitoring technique (IBSM; Figure 1.4-1).
5. Investigate all requirements established by local, state, and federal agencies.
6. Organize a project team.
7. Arrange for all necessary outside services, such as soil analysis and site survey.
8. Select members for the technical advisory committee and the value engineering team.

TABLE 1.4-2 Project Budget Worksheet

Job Number ☐☐☐☐.☐☐☐☐

Client ____________________

Description ____________________

Fee Estimate (see reverse):

1. Fee Basis: ________ Upper Limit $________
   FF = Fixed Fee ________ SC+ = Salary Cost Plus ________ %
   HR = Hourly Rate (Schedule) ________ %C = Percent of Construction, ASCE Curve ________
   HP+ = Payroll Cost Plus ________ O = Other ________
2. Estimate of Fee (show calcs on back)........ $________

Cost Estimate: Item		Class	Actual Salary ($/hr.)	Hours	Amount ($)	Remarks
3	Senior Company Officer	011				
4	Principal Professional	012				
5	Supervising Professional	013				
6	Senior Professional	014				
7	Professional	015				
8	Associate Professional	016				
9	Assistant Professional	017				
10	Senior Designer	021				
11	Designer	022				
12	Drafter	023				
13	Temporary Labor	098				
14	Supervising Administrator	151				
15	Senior Administrator	152				
16	Administrator	153				
17	Secretarial	154				
18	Typist	155				
19	Data Processor	156				
20	Reproduction Technician	157				
21	Clerk	158				
22	SUBTOTAL (3 to 21) Office Direct Labor[2]	X	X	X	$	
23	Senior Resident Engineer	031				
24	Resident Engineer-Inspector	032				
25	Senior Surveyor	041				
26	Surveyor	042				
27	SUBTOTAL (23 to 26) Field Direct Labor	X	X	X	$	
28	TOTAL DIRECT LABOR (22 and 27)	X	X	X	$	
29	Payroll Cost (% of Total 28)	201	X	X		
30	Office Burden (% of 22)	202	X	X		
31	Domestic Field Burden (% of 27)	202	X	X		
32	Overseas Burden (% of O/S 28)	202	X	X		
33	TOTAL of 29 to 32	X	X	X	$	
34	Outside Professional Services	203	X	X		
35	Services Division (see reverse)	204	X	X		
36	Miscellaneous ODC's	205	X	X		
37	Computer	207	X	X		
38	Travel and Subsistence	208	X	X		
39	Laboratory	210	X	X		
40	Overseas Differential (%of salary)	213	X	X		
41	Equipment and Supplies	216	X	X		
42[3]			X	X		
43[3]			X	X		
44[3]			X	X		
45	SUBTOTAL of 34 to 44, ODC's	X	X	X	$	
46	TOTAL ESTIMATE OF COST (28 + 33 + 45)	X	X	X	$	

Profit:

47 Estimate of Profit (2 less total 46) $________

48 Profit as Percent of Fee (47 divided by 2) ________ %

Check One: Original Budget ________ Revised Budget ________ Revision No. ________

Submitted: ____________________ Project Manager ________ Date

Approved: ____________________ Group Manager ________ Date

Recommended: ____________________ Department Manager ________ Date

Circulated by Contracts Administration: ____________________ ________ Date

9. Determine the number of technical advisory meetings that will be held during the project. The first meeting should take place after approximately 5 to 10% of the project is completed.
10. Prepare a memorandum after each meeting.
11. Encourage active input from the client. Keep the client informed on the progress of the project by holding periodic meetings and exchange ideas and experiences through workshops.

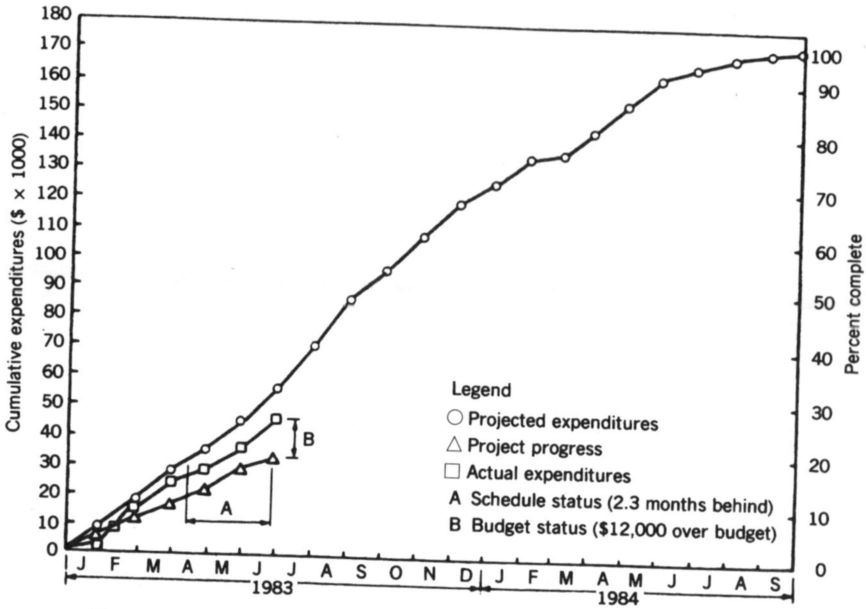

Figure 1.4-1 Integrated budget and schedule monitoring (IBSM).

12. Review the cost of the project at the end of each month.
13. Prepare construction specifications based on the standard specifications and edit them to make the job more specific.
14. Have the completed drawings and specifications checked by the project engineer and an independent checker. Each sheet should be red and yellow, lined.
15. Edit the bid documents and submit the preliminary drawings and specifications to the client for review.
16. Arrange an estimate for the construction costs.
17. Schedule the production of construction documents.
18. Obtain signatures from the company officer and the client on the final tracings.
19. Present the final drawings and specifications to the client and appropriate governmental agency (local, state, or federal) as required.
20. Arrange for advertising of bids and bid openings.

It is imperative that the calculations, drawings, and specifications be checked prior to finalizing the bid documents. Failure to do so may result in errors that could potentially have disastrous effects on plant construction and operation or even result in a lawsuit. Since many specifications are similar among various projects, in both outline and technical provisions, it is strongly advised that firms standardize their specifications as a means of project control.

Finally, it is essential to schedule the total workload of a firm as well as those for individual projects. Without proper scheduling, a firm will operate inefficiently and will have problems in effectively controlling ongoing projects.

1.5 VALUE ENGINEERING

Value analysis was developed by staff members at General Electric in the late 1940s to improve product efficiency. This process intentionally seeks substitute materials and methods to replace more costly components. Until the early 1970s, the engineering and construction industries had limited interest in the process of value analysis or value engineering. However, value engineering (VE) became an important part of design and construction in March 1972 when the Public Building Service of the General Service Administration (PBS/GSA) made VE a requirement for most new architectural, engineering, and construction management contracts. Both the Society of American Value Engineers (SAVE) and the defense industry have made great contributions to the development of value engineering.

One early study identified seven reasons/opportunities for the application of VE:

1. Advances in technology
2. Evaluation of specifications for worth (value)
3. Introduction of alternative design concepts, skills, and information
4. Recognition of a change in the needs of the client
5. Recognition of a design with added value
6. Feedback from pilot and field test studies, in addition to actual operations
7. Recognition of current design deficiencies

The value engineering review team should consist of a minimum of one or two licensed VE engineers and several experts in all technical disciplines being analyzed. It is recommended that the inclusion of professors and scientists be limited only to those who possess both theoretical and actual experience in design and construction. The VE review process is time consuming; the duration of a small project is at least 1 week. The focus of the process is to evaluate the functional capability of the facility and secondary considerations such as aesthetics and specific preferences relating to design and operations may be excluded from the VE review. A well-run VE review generally results in recommendations producing a cost savings of 15 to 30% per project.

1.5.1 Scope and Methodology of Value Engineering

Value engineering is both a systematic and creative method of identifying unnecessary cost within a project. The goal is to achieve savings in construction costs, as well as operation and maintenance costs, without sacrificing the effectiveness and reliability of the project. The parameters are tight: Maintain necessary functional

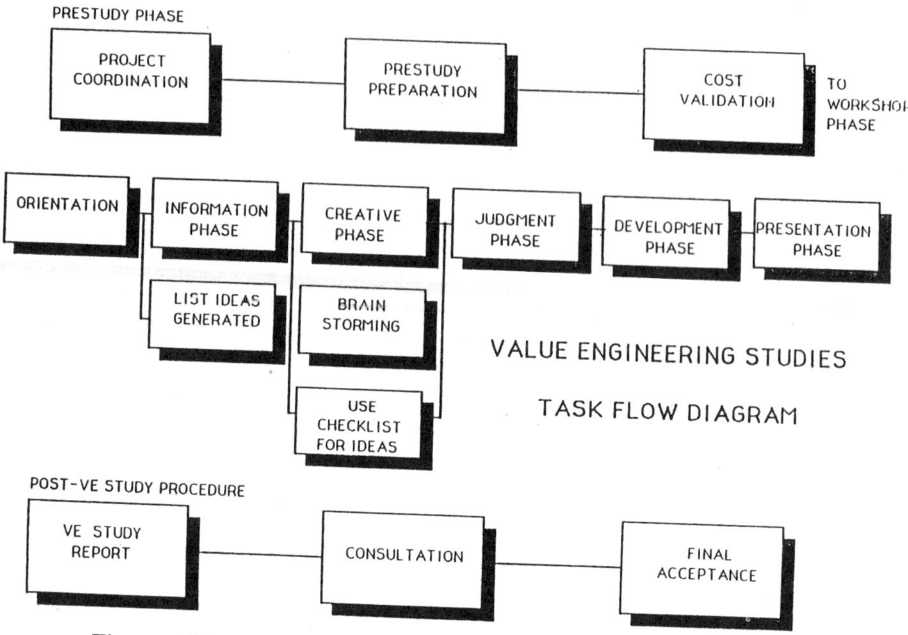

Figure 1.5-1 Typical task flow diagram of value engineering studies.

capacity without waste. Not all VE reviews result in cost reductions; however, additional "value" may still be derived if the recommendations increase the capabilities of the project. It is important to note that VE is not a detailed review of all project features. Rather, VE evaluates selected high-cost items; areas in which either costs may be reduced without sacrificing functional capabilities or areas in which the capabilities may be enhanced without an increase in cost.

The VE team only provides recommendations. The design engineer and the client must conduct a thorough analysis of the applicability, feasibility, and viability of each recommendation before deciding to adopt or reject them. The VE study is composed of two segments: workshop and presentation. The work plan is set in the first segment (workshop) and provides an organized structure for identifying potential review areas for alternative design and development. The work plan should contain the following five phases: (1) information, (2) creative, (3) judgment, (4) development, and (5) recommendation. Figure 1.5.1 is a task flow diagram of a VE study.

1.5.2 Information Phase

The project engineer must prepare handouts containing all essential project information. These should be distributed to each member of the VE review team 1 week prior to the session. The review session begins with an oral overview presented by the project engineer. It is essential that the entire VE review team understand both the background and decisions that influence the project design: site conditions, rationale for the design, project constraints, design considerations, raw water quality and treatment goals, treatability studies, factors affecting decision making, and the rationale for selection of the processes, hydraulic profile, and layout of the treatment plant.

After oral presentation, the VE team members should be more knowledgeable about the design and begin to critically analyze the various costs and effectiveness of the proposed treatment process train, each process unit, equipment, and control systems. After evaluating the treatment facilities, each team member prepares a functional analysis. It is important for the team to review and assess the preliminary cost estimates of the project, as this figure serves as a benchmark for the various alternatives generated during the review workshop.

1.5.3 Creative Phase

This second step of the VE review is a brainstorming session which lists areas that could potentially be improved while incurring less cost. The purpose of this phase is to generate a number of new ideas (alternatives); those worthy of further evaluation are identified in a later phase.

1.5.4 Judgment Phase

The various design alternatives are now evaluated and refined. Each is assigned an associated cost, then ranked in order of their potential costs savings. Secondary considerations such as aesthetics, efficiency, reliability, and durability may be evaluated and applied to the ranked alternatives.

1.5.5 Development Phase

Design alternatives receiving high ranking are further developed during this phase. The team should study the most promising alternatives in detail and conduct a cost comparison in relation to the original design. A life-cycle cost analysis should be conducted in cases where the capital cost differential is not significant, but the operation and maintenance cost differential is notable.

1.5.6 Recommendation Phase

The recommendation phase is the last step of the workshop segment. After the VE review team has carefully restudied and rereviewed all feasible alternatives, a written recommendation is composed. An oral presentation of these recommendations is then delivered to the project engineer and the owner/representatives. There should not be any debates or arguments during this presentation. As mentioned previously, these recommendations are not final decisions, only suggestions. It is therefore essential that the VE team present its finding and recommendations in an intelligent, convincing, and credible manner. This can be achieved only if the VE team has a thorough understanding of the project goals, the function of all unit processes, including the performance of all equipment and control systems, and the project constraints. Delivery of an elaborate VE report must follow the oral presentation. Preparation of this final report may last 1 to 2 weeks, depending on the magnitude of the report.

1.6 PRIVATIZATION DELIVERY PROCESS

Although the private sector utilized items such as design-build-operated (DBO), design-build-maintain (DBM), public-private partnerships (PPPs), and long-term contract operations, these were not practiced in the domestic water industry until the early 1990s. Some 35% of U.S. utilities now consider using these approaches at the end of the twentieth century. These project delivery processes have distinct advantages over the traditional processes of design, bid, build—having RFQ, FRP, evaluations, negotiations, bidding, building, as well as protests and claims. The traditional process is both time consuming and costly.

The privatized processes have three basic advantages over the traditional processes: lower cost, lower risk, and an accelerated schedule. The expected savings in these three categories are approximately 15–30%. Yet, the disadvantages are severe penalties if the project does not meet contracted requirements, concerns regarding quality control and the potential reduction in plant operational flexibility due to the low cost of the project.

The recent emergence of the privatized delivery process in the domestic water industry is the result of internal and external competition. Factors such as regulatory requirements governing the upgrade of existing plants or the building of new plants, the consequences of deferred maintenance, public resistance to rate increases, and communities embroiled in financial crisis, force many public utilities to consider the options provided by the privatization processes so that they can meet their goals. This trend toward privatization is likely to become more popular on a global scale.

BIBLIOGRAPHY

Abdallah, E. T., *Preparing Specifications for Design-Bid-Build Projects,* ASCE Manual, January 2000.

"DBO and DBM Gain Popularity," Roundtable, *J.AWWA,* 91:4:18(April 1999).

Dysad, J. A., "Trends in Privatization," *J.AWWA,* 91:11:44 (November 1999).

General Services Administration, *Value Engineering,* Public Building Service/GSA Handbook P8000.0, U.S. Government Printing Office, Washington, DC, 1972.

Haarmeyer, D., "Privatization Infrastructure: Option for Municipal Water Systems," *J.AWWA,* 86:3:42(March 1994).

O'Brien, J. J., *Value Analysis in Design and Construction,* McGraw-Hill, New York, 1976.

"Public-Private Partnerships," *J.AWWA,* 88:4:26 (April 1996).

CHAPTER 2

Preliminary Studies

2.1 MASTER PLANNING

The first step in the design of a regional water supply system is master planning. The following areas must be covered: study of the water treatment facilities, evaluation of the system conveying raw water, evaluation of the finished water distribution system, and evaluation of a few basic infrastructures of the region (sewer system and sewage treatment facilities).

The number, size, and location of the local water distribution reservoirs depend on the anticipated pressure zone of the service area. Since a large portion of potable water is converted into sewage and various other wastewater, the conveyance and treatment of wastewater must also be evaluated. Other infrastructures that should be considered are new access roads, power supply, and utility systems near the plant site. It is important to remember that a water treatment plant is a "factory" that handles a large amount of chemicals, including hazardous chemicals such as chlorine and caustic soda, in addition to producing wastes. Thus the project engineer must consider whether it is necessary to reclassify the existing zoning or modify existing codes. The master development plan should also study the influence of and connection to existing waterworks and sewage handling facilities. In some cases, issues such as water rights and sludge disposal may become difficult to solve when they affect neighboring communities.

During the early stages of planning, the project engineer must evaluate potential sources of water and assess the quality and safe yield of each. Water may be obtained from a river, lake, reservoir, or groundwater. In general, groundwater is preferred over surface water due to its superior and steady water quality. Although these characteristics make groundwater easier and less costly to treat, and results in a smaller amount of waste, the available safe yield from groundwater wells is generally limited. Nonetheless, the master plan should evaluate groundwater sources and issues pertaining to groundwater contamination, ground subsidence, and seawater intrusion due to excessive withdrawal of groundwater. Groundwater contamination may result from human, animal, or industrial waste, agricultural activity, or geologic conditions. Recently, the analysis of radon levels in groundwater has become an important issue, due to its classification as a carcinogen.

The master plan must also include the following items for study: proposed watershed management, a brief evaluation of site conditions, including topography, geologic conditions, history of flooding and earthquakes, site accessibility, land use of the site, and preliminary archeological and environmental assessments. One of the most difficult items in the master plan is the prediction of the region's future water demand. Although there are basic mathematical equations, variables unique to the region (changes in socioeconomic conditions, significant change in weather pattern) and global variables may greatly fluctuate over a period of 10 to 15 years. It is unfortunate that these variables are not easily predicted, as demonstrated by many recent case histories throughout the world. Consequently, prediction of water demand 50 years into the future is inaccurate and has little value.

In recent years, other unique concerns have found their way into the master plan. The first is the anticipated enforcement of the more stringent Drinking Water Quality Standards beyond the year 2000. A second is emerging water treatment technologies such as the membrane filtration process. In the past, master development plans demonstrated the accommodation of future increases in plant capacity merely as additions to the original conventional treatment process module. This concept is no longer valid, due to the emergence of new water treatment technologies and the need for more sophisticated processes that may require a different (larger or smaller) overall footprint. For this reason, estimating the required acreage for the proposed water treatment plant site and the selection of future water treatment processes has become more difficult. Yet it is safe to allocate an extra area of the plant site exclusively for the addition of future water treatment processes on the master plan. The importance of the master development plan must not be overlooked. The master plan provides an early rough estimate of the overall capital expenditure, duration of the project, and regional planning scheme.

2.2 FEASIBILITY STUDY

The feasibility study is the starting point for all water treatment design projects. It is often conducted as part of the master development plan for the region or regions. Since this subject is not the focal point of this book, it is discussed only briefly.

2.2.1 Planning Period

A span of 10 to 20 years is considered to be a reasonable period for planning a water supply system, including the capacity of the water treatment plant. It is important to note that some regions, such as southern California and the Bay area of northern California, have already exceeded the water demand predicted in the early 1980s. Yet some cities in the United States are well below the predicted levels, due to declining industrial activities and severe weather conditions. However, a span of 15 years is generally considered to be an acceptable planning period for immediate system design and construction.

2.2.2 Water Supply Areas

When determining the areas to be served by the water supply system, engineers should limit their evaluation to areas considered to receive water during the design period. However, adjacent areas and the potential amalgamation of neighboring districts should also be included in the overall design.

2.2.3 Future Population

There are several methods to predict the growth or decline of the future population. Since most college textbooks cover this subject in great detail, it is not discussed here.

2.2.4 Maximum Daily Water Demand

The first task in the feasibility study is the estimation of the maximum daily water demand per capita. For a new system, this estimation is generally derived from data acquired from other communities sharing similar backgrounds, characteristics, and developmental trends. In the case of plant expansion, the best estimation is provided by the trends of the existing water supply system. The maximum daily demand per capita is the basis for evaluating the water treatment capacity. Once this figure is established, the maximum daily demand may be calculated as the product of the maximum daily per capita demand and the estimated population to be served within the planning period. An average annual rate of use in gallons per capita per day (gpcd) of 180 (with a range of 150 to 210) or 680 liters per capita per day (with a range of 570 to 795) is common in the United States, and the maximum daily demand is usually 150% of the average annual rate.

2.2.5 Evaluation and Selection of the Water Source

The source of the raw water may be a river, lake, artificial reservoir, groundwater, and in some cases, reclaimed sewage or seawater. The evaluation and selection of the proper water source should be based on the following issues: (1) quantity of water required, (2) quality of the raw water, (3) climatic conditions (i.e., icing), (4) potential difficulties in constructing the intake, (5) operator safety, (6) providing minimal operations and maintenance costs for the treatment plant, (7) possibility of future contamination of the water source, and (8) ease of enlarging the intake if required at a future date.

Underground water, including water collected by an infiltration gallery, should be considered seriously as a water source because it can potentially save a significant amount of capital as well as operations and maintenance costs for small communities. The quality of underground water is generally superior to that of surface water, thus resulting in minimal waste (sludge) and no problems associated with icing during the winter months—an added benefit for cold-weather regions.

2.2.6 Size of the Water Treatment Plant

The capacity of a water treatment plant is determined by the maximum daily demand placed on the system. However, future water demand of the area being served should also be taken into account when calculating the size of the plant and the site area. Additional issues that affect plant size are the reliability of the water supply and the cost-effectiveness of supplying water from one large plant versus two or even three medium-sized plants at different locations and different elevations. As a rule of thumb, the site area required for a conventional plant may be estimated by the formula

$$A \geq Q^{0.7}$$

where A is area in acres (1 acre = 0.405 ha) and Q is the ultimate plant capacity in millions of gallons per day [mgd; 1 mgd = 3.784 cubic meters per day (m^3/d)].

2.2.7 Treatment Plant Site

Evaluation of the treatment plant site is based primarily on the distance from the intake, the layout of the treatment process units, the environmental impact of the treatment plant, and the method of water distribution (gravity versus pumping). Moreover, the following items must be included in evaluation of the treatment plant site: (1) geographical location, (2) information obtained from the geological study, (3) availability of electric power and utilities, (4) accessibility to major highways, (5) history of flooding or the presence of earthquake faults, (6) construction costs, (7) site maintenance costs, (8) operator safety and the safety of neighboring homes, (9) provisions for future plant expansion, and (10) environmental impact study, including an archeological investigation.

2.2.8 Financing

The preliminary site evaluation must present various means by which to fund the project. Several methods are used to obtain necessary funding for constructing water treatment plants: revenue bonds, general obligation bonds, special assessment bonds, state and federal aid funds, operating revenues, contributions by customers, and private funds. The most common method is the issuance of revenue bonds, since water treatment plants are revenue-producing facilities.

The first step in obtaining financing requires the engineer to provide an estimate of the plant construction costs. A team of consultants (engineering, financial, and legal) must then study the proposal. It is imperative that the team include a financial specialist and legal expert so that the utility can decide if it can issue revenue bonds or if the community should be asked to issue general obligations bonds. A bond attorney is required to draw up the proceedings and to issue the approving opinion, without which the bonds cannot be marketed. Finally, the preliminary engineering, financial, and legal studies must be submitted to the city council and should disclose whether the project can be accomplished at a reasonable cost through acceptable financing methods within the existing statutory limitations. These studies often save the governing board of the utility from unnecessary embarrassment and keeps the project from becoming a political football.

In recent years, *privatization,* which involves private financing and turnkey projects, has emerged. In the public water-utility industry the term *turnkey construction* refers to having a single entity design, construct, and deliver a properly functioning facility that is ready for service. This type of contract has the advantages of maintaining engineering continuity and quality control. Privatization of the water treatment industry has allowed organizations from the public sector to enter into a market that has traditionally been dominated by public funding. Private companies may now finance, own, construct, and operate the facilities. Since these facilities are run as a business enterprise, their design, construction, and operation are generally more cost-efficient than their public counterparts. However, local governments have the option of purchasing these types of facilities at fair market value at the end of the lease terms. Despite the tax reform bill passed by the U.S. House of Representatives in December 1986—which removed a large part of the tax exemption benefits for the investor and owners—the tax-exempt interest of this income, for the investor or lessor, should still attract private investors.

BIBLIOGRAPHY

Alla, P. M. J., and Manzi, D., "Sydney Water's Public–Private Partnership," *J. AWWA,* 88:4:108 (April 1996).

AWWA Joint Discussion, "Pros and Cons of Turn-Key Construction," *J. AWWA,* 65:12:766 (December 1973).

AWWA Round Table Discussions, "Gaining Access to Capital Markets," *J. AWWA,* 87:2:23 (February 1995).

AWWA Round Table Discussions, "Privatization," *J. AWWA,* 79:11:12 (November 1987).

Ayres, D. W., and Thorpe, S., "Financing Capital Improvements," *J. AWWA,* 83:8:50 (August 1991).

Benson, R. B., "Effective Revenue Bond Marketing: Put Your Best Foot Forward," *J. AWWA,* 83:8:40 (August 1991).

Booker, A., "British Privatization: Balancing Needs," *J. AWWA,* 86:3:56 (March 1994).

Bray, F. A., "Municipal Utility Bond Financing," *J. AWWA,* 62:8:468 (August 1970).

Brice, R. L., and Unangst, E. R., "Long-Range Financial Planning for Water Utilities," *J. AWWA,* 81:5:48 (May 1989).

Casasus, C., "Privatizing the Mexican Water Industry," *J. AWWA,* 86:3:69 (March 1994).

Doctor, R. D., "Private Sector Financing for Water Systems," *J. AWWA,* 78:2:47 (February 1986).

Earl, T. E., and Greenstein, S. A., "Economic Considerations in Facility Planning," *J. AWWA,* 62:11:705 (November 1970).

Gullet, B. M., and Bean, D. O., "The Charlotte Model for Competitions," *J. AWWA,* 89:4:100 (April 1997).

Haarmeyer, D., "Privatizing Infrastructure: Options for Municipal Water Systems," *J. AWWA,* 86:3:42 (March 1994).

Jeffrey, J., "Privatization in England and Wales," *J. AWWA,* 86:3:64 (March 1994).

Pontek, R., and Wehmeyer, M., "Competition: Strategies for a Public Utility," *J. AWWA,* 89:4:119 (April 1997).

Sagraves, B. R., Peterson, J. H., and Williams, P. C., "Financing Strategies for Small Systems," *J. AWWA,* 80:8:40 (August 1988).

Seader, D., "Privatization: An Emerging Management and Financing Trend," *Water Eng. Manage.,* p. 44 (March 1984).

Shanker, A., and Rodman, L., "Public–Private Partnerships," *J. AWWA,* 88:4:102 (April 1996).

2.3 BENCH-SCALE AND PILOT-PLANT STUDIES

The predesign efforts must include bench-scale and pilot-plant studies whenever possible. These studies allow the engineer to obtain valuable basic design data and either to optimize or confirm the proposed raw water and filter waste-wash treatment processes. The bench-scale study may be conducted in a short period of time and with a limited budget. However, it requires a minimum of 6 months to 1 year for a pilot study to yield meaningful and reliable data. Consequently, these studies are quite expensive.

2.3.1 Bench-Scale Studies

A proper bench-scale study addresses the following objectives: optimization of chemical coagulants, the chemical application sequence, confirmation of the proper mixing conditions for flocculation, estimation of the hydraulic surface loading for the sedimentation process through the measurement of floc settling velocities, the total level of trihalomethane potentially produced by the water, and the control of taste- and odor-producing compounds through the use of oxidants or activated carbon. The Phipps–Bird jar tester is the instrument employed primarily for these studies, and a minimum of 200 work-hours is generally required to complete the bench-scale studies, which includes the time required to write the report.

The coagulant optimization study involves selecting the most effective type of coagulant and the appropriate dosage that produces good settled water quality, as well as optimal conditions for the filtration process. Engineers should realize that a conventional water treatment process is comprised of many unit processes: coagulation, flocculation, sedimentation, and filtration, and that these subsystems must be interrelated and compatible if the sequence is to produce good-quality water. It is recommended that the effectiveness of coagulants, such as alum, ferric salts, and cationic polymers, be evaluated both individually and as a combination of alum and polymer. The recommended jar testing procedure is presented in Example 1 of Section 3.2.7 and in Appendix 10. The suggested coagulant dosages for the first raw water jar test run are as follows:

- *Alum or ferric chloride dosages* (mg/L): 3, 6, 12, 20, 40, 60
- *Cationic polymer alone* (mg/L): 0.25, 0.5, 1, 2, 4, 8
- *Alum with cationic polymer* (mg/L): 6 to 0.25, 12 to 0.25, 20 to 0.25, 6 to 0.5, 12 to 0.5, and 20 to 0.5

See Table A10-1 of Appendix 10 for sample jar test results.

The flocculant optimization test should be designed to determine the most effective type of polymer or additives (such as bentonite clay), optimum chemical dosage, and proper application sequence for both coagulant and chemical flocculation aids. The most effective flocculant aids are usually anionic polymers or non-ionic polymers because they can greatly increase both floc size and settling velocity. However, the timing of their application is very important; they should be fed when pinpoint floc masses have formed, which is approximately 5 min after alum flocculation begins. Table 3.2.3-1 illustrates this issue clearly. The sequence of chemical application is a critical consideration for other water treatment chemicals, such as lime and alum, chlorine and polymer, and chlorine and powdered activated carbon (PAC).

The mixing conditions for the flocculation process may be optimized by a jar test, and data obtained in this manner are useful for the plant design. The jar test allows the engineer to determine the optimum level of the mixing energy input (G value), the mixing time, and the energy input pathway (tapered mixing requirement). Although most existing conventional treatment plants use an average G value of about 35 with a mixing time of 30 min, a wide range of numbers should be tested at optimum coagulant and flocculant dosages to maximize the conditions. Tapered mixing—the use of high-intensity mixing at the beginning of flocculation, followed by a gradual reduction toward the end—helps promote the formation of large floc masses and should therefore be studied as well.

Four major parameters that significantly influence the optimum mixing conditions are water temperature, raw water quality, type of coagulant, and coagulant dosage. Thus it is recommended that bench-scale tests be performed during the winter, summer, and either spring or autumn to better comprehend the need to tailor coagulant dosages to the changing raw water characteristics and to estimate the seasonal sludge production rate. The fourth objective of bench-scale studies—floc settling velocities—must be evaluated at optimal levels of coagulant and mixing conditions. The test should be conducted during both cold- and hot-weather months, for reasons stated previously.

It is relatively simple to produce a settling velocity distribution curve from a jar test. The turbidity of the settled water is measured for samples drawn from 10 cm below the surface of a 2-L beaker at 1, 2, 4, 8, and 16 min after the jar test has been initiated. The samples collected at these times represent the respective settling velocities of 10, 5, 2.5, 1.25, and 0.625 cm/min. A settling velocity of 4 cm/min is equivalent to a surface loading of 1 gpm/ft^2. The measurement of floc settling velocities is discussed in detail by Camp (1946) and Hudson (1981) (see the Bibliography at the end of this section). Engineers must note that bench-scale test results may not be relevant if the turbidity is produced artificially or if the water is not fresh, more than 1 to 2 days old, since the test results do not accurately portray actual water treatment conditions. Refer to Section 7.4 for a discussion of the total potential THM formation and Section 7.3 for taste and odor control by oxidants or activated carbon.

2.3.2 Pilot-Plant Studies

In recent years, engineers have been forced to evaluate nonconventional treatment processes due to new EPA drinking water regulations, public concern over the safety of drinking water, the ever-increasing costs of plant construction, and the emergence of new water treatment technologies. Consequently, the use of pilot-plant studies to evaluate proposed treatment processes is increasingly becoming a necessary part of the design process.

Although pilot studies can potentially result in significant cost savings in plant construction and operations, the cost of the pilot studies themselves is rather large: A good pilot study may cost anywhere from \$100,000 to \$1,000,000. Thus if meaningful results are to be obtained from a pilot study, the most appropriate type and best manufactured equipment must be selected. Moreover, the pilot plant must be operated by highly qualified personnel for a period of at least 6 to 12 months.

Yet, a good pilot study can save millions of dollars in plant construction cost and result in a significant reduction in plant operation and maintenance costs by providing a simple, but effective high-rate unit process with a small foot-print for the plant site; as supported by a few cases.

Pilot-plant studies may be necessary to achieve some or all of the following objectives:

1. Obtain permits or variances from regulatory agencies for nonconventional processes.
2. Evaluate the practicability of a new treatment process.
3. Compare the effectiveness of alternative processes.
4. Obtain a guide for process design criteria and operating costs.
5. Improve existing processes.
6. Investigate the cause of problems.
7. Confirm the effectiveness of the proposed treatment process.
8. Discover unforeseen problems resulting from the implementation of the proposed treatment process.

Although pilot-plant studies are a valuable guide in establishing both process design criteria and operational parameters of a plant, they also have their limitations. The major problems are (1) difficulty in testing the raw water on a year-round basis; (2) use of an improper type of clay when simulating abnormal raw water conditions (high turbidity); (3) using raw water that has been stored for over 1 to 2 days; (4) differences in operational conditions, including the technical and operational knowledge of the plant operators—that is, pilot plant versus actual treatment plant; (5) problems encountered in scaling up; (6) failure to foresee long-term effects of the new process; and (7) arriving at conclusions biased by personal expectations.

Pilot-plant studies may be classified into two basic categories: study of hydraulic characteristics and study of unit process performance. The first includes the study of flow characteristics in clearwells or sedimentation tanks, and the second includes pilot studies on corrosion and the processes of filtration, flocculation, dissolved air floatation (DAF), GAC adsorption, and ozonation.

Once a pilot study is planned, the engineer must determine the variables to be studied and what their characteristics should be. Since flocculation, sedimentation, DAF, and preozonation are filtration pretreatment processes, it is very important to select a combination of these unit processes that yields both maximum efficiency and optimum economic benefits for the overall treatment system. From a practical standpoint, a pilot study usually cannot examine all the relevant factors due to time and budgetary constraints. Thus it is necessary to reduce the number of variables by selecting only a few very important items. However, engineers must first establish the following issues during the initial stages of the study: (1) the purpose of the study, (2) the duration and cost of the experiment, (3) the availability of equipment and technical staff, and (4) the important variables of the study.

A basic pilot plant has a coagulation, flocculation, clarification, filtration, and final disinfection process. The study should first optimize the dosage of the coagulant and other required chemicals, that is, determine the dosage that yields the best filter performance for the filter bed configuration selected. The physical variables of the pretreatment process, such as mixing intensity, detention time, and hydraulic loading, should then be evaluated using the optimum coagulant dosage. The optimum dosage is based on results obtained from jar test studies. However, the use of ordinary pilot plants is virtually useless in predicting settled water quality or in optimizing horizontal flow sedimentation tanks. This subject is discussed later.

The piloting filter, including the GAC adsorption process, yields meaningful data and is relatively easy to monitor. Filter performance may be effectively evaluated through the use of small-diameter columns [2 to 4 in. (5 to 10 cm)]. However, if filter washing is conducted for study purposes, large columns should be used to minimize the "sidewall" effect. A rule of thumb is to use a column diameter that is 1000 times the average size of the granular filter medium. For example, if the average size of the filter sand is 0.8 mm, the size of the filter column should be 80 cm (30 in.) in diameter. In the case of filtration studies, the size of the filter column should be 100 times the effective size of the filter medium.

The pilot filter study should address the following issues: optimization of the filter bed and filter rate, evaluation of the filter washing conditions, and control of initial and terminal turbidity breakthrough. In some cases, declining rate versus constant rate filtration may be included as a study item. To achieve these objectives, a minimum of three filter columns is absolutely necessary; five to six columns are preferred, with one of the units as a control column. Media optimization involves the evaluation of different types of medium materials, determination of the size-to-depth relationship, and comparison of the effectiveness of monomedium versus multimedia. Basic information on size, depth, and material of filter beds is presented in Table 3.2.7-1.

One of the most popular filter beds in the last half of the twentieth century is a dual-media filter bed with a depth of 30 in. (0.75 m). The bed is composed of a layer of anthracite coal that is 20 in. (0.5 m) thick (1 mm effective size) and 10 in.

(0.25 m) of filter sand (0.5 mm effective size). This bed should be used as a control for comparing the performance of other types of bed.

Engineers are advised to use a constant-rate filtration process for the pilot filter study. Filtration rates of 3, 6, and 9 gpm/ft^2 (7.5, 15, and 22.5 m/h) should be evaluated since a filtration rate of 6 gpm/ft^2 (15 m/h) is generally considered to be an acceptable filtration rate for high-rate filtration using dual-media beds. Figure 2.3.2-1 is a schematic diagram of the recommended standard pilot-plant configuration.

The quality of the filter effluent obtained from pilot filters is often significantly below that produced by actual operational filters. Under identical conditions, the turbidity of the pilot filter effluent may be twice as high as the actual filter: 0.2 nepherometric turbidity unit (ntu) for pilot-plant filters versus 0.1 ntu for actual filters. Engineers should also be aware that the initial turbidity breakthrough, at the beginning of the filtration cycle, is usually quite severe for pilot plants. This may be attributed to the degree of ripening or maturation of the filter bed. Unlike operational treatment plants, the filters of the pilot plant are always thoroughly washed. Consequently, the filter beds are seldom allowed to become well ripened. The reasons for the frequent washings are (1) minimizing the conditions present in previous pilot filter tests, and (2) lack of operational knowledge by the pilot filter operators, who are usually engineers or scientists. Scientists and engineers generally wash the filter beds until the waste wash water becomes clear, in contrast to the practice of actual plant operators, who allow a turbidity of 5 to 10 ntu to remain in the filter wash waste.

As mentioned previously, pilot tests for horizontal sedimentation tanks have limitations. If the purpose of the piloting is to study the hydraulic characteristics of the tank, a hydraulic-scale model that applies Froude's similarity law should be adopted.

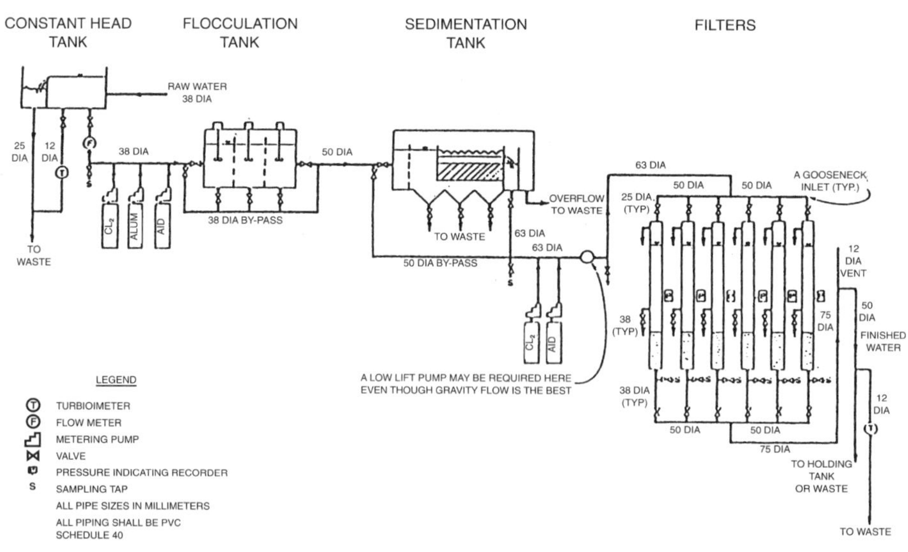

Figure 2.3.2-1 Pilot-plant setup.

However, if the object of the study is to evaluate the ability of the tank to remove turbidity or other constituents, it is difficult to scale down the horizontal sedimentation tanks due to problems associated with adjusting the physical characteristics of the floc, even if the hydraulic flow characteristics have been properly established. Consequently, traditional design criteria, such as the hydraulic overflow rate and detention time, are used to size the pilot units based on a manageable (small) flow rate.

Pilot sedimentation tanks whose designs are based on actual basin parameters always have geometrically distorted shapes, for example, very deep and short. Furthermore, unlike their prototypes these basins fail to produce effluent of acceptable quality and are incapable of producing a good sludge deposition profile. The way to produce acceptable quality effluent is to improve floc settleability through the application of an appropriate type of polymer (as a flocculant aid) in conjunction with alum, in addition to installing a high-rate settler module, such as a tube settler in the pilot tank. The addition of the high-rate settler module provides an overflow rate of approximately 1 to 2 gpm/ft^2 (2.5 to 5.0 m/h) for the surface area of the tank covered by the settler module.

Based on data obtained from numerous pilot-scale upflow sludge blanket clarifiers, it may be stated that the overflow rate of the prototype pilot scale unit may be used, to an acceptable degree, to model the upflow sludge blanket clarifiers and the high-rate settler modules. The relationship between the upward flow velocity and the floc settling velocity is the major clarification mechanism for both upflow reactor-clarifiers and the high-rate settlers. This simple relationship enables the pilot study to be possible, provided that there is no distinct flow short-circuiting within the unit.

Hydraulic-scale model studies are not frequently conducted even for large-scale projects because they are always tailored for specific projects and therefore cannot easily be used for other projects, thus making them very costly. However, the hydraulic-scale model study is the only positive means of studying the flow characteristics of a sedimentation tank, including the selection of an effective flow diffuser and baffle walls. A few academic and research institutes conduct hydraulic-scale model studies purely for academic interests or for large and costly projects.

Froude's similarity law should be applied to all scale-model studies:

$$Fr = \frac{v^2}{Lg}$$

where v is the characteristic velocity of the system, L the characteristic linear dimension, and g the acceleration of gravity. Froude's law expresses the condition for similarity between the forces of gravity and inertia. In other words, when the Froude number is the same for both the model and the prototype, the gravity force to inertial force ratios are the same; the paths of flow are similar. The fundamental relationships for models of structure, based on Froude's law, are presented in Table 2.3.2-1.

TABLE 2.3.2-1 Relationships Between Hydraulic Scale Model and Prototype

Scale	Relation[a]	Example (1/25 Scale Model)
Length	$1{:}\alpha$	1:1/25
Area	$1{:}\alpha^2$	1:1/625
Volume	$1{:}\alpha^3$	1:1/15,625
Time	$1{:}\alpha^{0.5}$	1:1/5
Velocity	$1{:}\alpha^{0.5}$	1:1/5
Discharge	$1{:}\alpha^{2.5}$	1:3,125
Acceleration	$1{:}\alpha^0 = 1$	1:1
Force	$1{:}\alpha^3$	1:1/15,625
Work	$1{:}\alpha^4$	1:1/390,625

[a]α is the size reduction ratio for the hydraulic scale model.

BIBLIOGRAPHY

AWWA, "Design of Pilot-Plant Studies," *AWWA Seminar Proceedings,* No. 20164, American Water Works Association, Denver, CO, May 1982.

Camp, T. R., "Sedimentation and the Design of Settling Tanks," *ASCE Trans.,* 3:985 (1946).

Hickox, G. H., "Hydraulic Models," in *Handbook of Applied Hydraulics,* 2nd ed., C. V. Davic, ed.-in-chief, McGraw-Hill, New York, 1952.

Hudson, H. E., Jr., *Water Clarification Processes,* Van Nostrand Reinhold, New York, 1981.

Kawamura, S., "Hydraulic Scale-Model Simulation in the Sedimentation Process," *J. AWWA,* 73:7:417 (July 1981) and *Proceedings of the 3rd Annual Conference of the Japan Water Works Association,* May 1952.

2.4 PRELIMINARY ENGINEERING STUDY

Once the owner has decided to proceed with the construction phase of a project based on the feasibility study, a preliminary engineering study, including planning, must be initiated. This study should address the following issues:

1. Quality and treatability of the raw water
2. Objectives for the finished water quality
3. Additional goals and objectives
4. Restrictions and constraints on plant design
5. Alternative treatment processes
6. Preliminary design criteria for treatment unit processes, including chemical feed systems
7. Hydraulic grade across the plant
8. Geotechnical considerations
9. Structural design conditions and criteria

10. Plant waste handling and disposal
11. Instrumentation and control systems
12. Preliminary cost estimates
13. Recommended treatment processes as well as instrumentation and control, and waste handling and disposal
14. Plant layout and architectural design
15. Management planning of the design phase
16. Environmental impact statement (EIS) for the new plant

The preliminary study must conclude with concrete recommendations with respect to the proper treatment process train, hydraulic loading values for each unit process, and all major equipment and appurtenant facilities. At this stage of the project, proper procedure dictates that each of the recommendations be discussed with the appropriate regulatory agencies to solicit their opinions.

2.4.1 Quality and Treatability of the Raw Water

Data on the surface water quality, taken over a sufficient period of time (5 to 10 years), should be both reviewed and evaluated to assess the physical, chemical, microbiological, and radiological characteristics of the raw water. A risk assessment must also be made in regard to possible contamination of the water supply by chemical spills or radioactive wastes. Moreover, the degree of present and future land development in the watershed must be studied.

If groundwater is selected as the source of the process water, the same considerations associated with surface water apply. Groundwater as a raw water source necessitates additional studies, such as the geological conditions, water table, the drawdown of the water table as the result of pumping, problems associated with seawater intrusion, and the potential leaching of industrial wastes, domestic wastes, agricultural chemicals, and fertilizers into the groundwater.

For the reasons just mentioned, data analysis is a very important aspect of evaluating the quality of a water source. Although arithmetic averages are often presented, a statistical presentation is generally more meaningful. The statistical presentation may be constructed by plotting data on an arithmetic or probability scale. The two most important aspects of data analysis are that it is a means of describing the central tendency of the data and it allows engineers to determine the degree of variability in the data. Components in the raw water whose maximum concentration levels are limited by the Drinking Water Quality Standards must be analyzed and evaluated carefully.

The treatability of the raw water may be evaluated through the use of bench-scale tests and a pilot study. These issues are discussed in Section 2.3. If there is an existing water treatment plant in the vicinity of the proposed plant site, the design engineers should consult the operational data of the existing plant because it will provide valuable information on the treatability of the raw water.

The 1986 Amendment to the Safe Drinking Water Act (SDWA-PL-99-339) required the U.S. Environmental Protection Agency (EPA) to promulgate a National Primary Drinking Water Regulation (NPDWR) by December 19, 1987. This amendment specified the criteria by which filtration was ordered as a standard treatment technique for all public water systems supplied by any source of surface water. The most significant regulations which have the greatest impact on the treatment process design are the Surface Water Treatment Rule (SWTR) and the Enhanced Surface Treatment Rule (ESWTR).

Specific parameters of importance include the range of water temperature, certain microbiological organisms, total organic carbon (TOC), and either toxic or suspected cancer-causing synthetic organic compounds (SOCs) and inorganic compounds (IOCs). Microbiological characteristics of raw water that must be monitored are viruses, *Giardia* cysts, and *Cryptosporidium* oocysts.

2.4.2 Objectives for the Finished Water Quality

The objective of a public water supply water system is to provide safe and aesthetically appealing water to consumers without interruption and at a reasonable cost. The required quality is measured against the standards set by state and federal governments; both regulate drinking water quality. Since only the national regulations are discussed in this section, engineers are strongly advised to contact local regulatory agencies for further restrictions. In 1974 the EPA established the National Interim Primary Drinking Water Regulations (NIPDWR) and thereby set the maximum contaminant levels (MCLs) for a variety of inorganic and organic chemicals, as well as physical, microbiological, and radioactive contaminants. These primary standards have been effect since June 1977 and are designed to protect the public health. Compliance with these standards is therefore mandatory. The National Interim Secondary Drinking Water Regulations generally relate to the aesthetic quality of a water supply and thus are recommended goals.

After the enactment of the 1986 SDWA amendments, the EPA promulgated many regulations. They include the following:

1. *Phase 1:* criteria for eight volatile organic compounds (VOCs) and fluoride
2. *Total Coliform Rule:* MCLG = 0
3. *Surface Water Treatment Rule* (SWTR)
4. *Phase 2:* 38 VOCs, synthetic organic chemicals (SOCs), inorganic chemicals (IOCs), and microbial contaminants
5. *Lead and Copper Rule*
6. *Phase 5:* 23 VOCs, SOCs, and IOCs
7. *Information Collection Rule:* monitoring for microbial pathogens and disinfection by-products (Certain large systems are required to conduct bench- or pilot studies of GAC or membrane filtration to evaluate the formation of DBP precursor and TOC removal.)

In 1996, Congress reauthorized the Safe Drinking Water Act. These amendments made several modifications to the EPA's drinking water program; the EPA must evaluate five contaminants for possible regulation from a list of drinking water contaminant candidates. Sulfate, radon, and arsenic are the major items.

Surface Water Treatment Rule (SWTR) The SWTR applies to all public water systems using surface water sources or groundwater sources "under the direct influence of surface water." The SWTR establishes a Maximum Contaminant Level Goal (MCLG) for the following microbiological organisms:

- *Giardia lamblia:* 0
- *Viruses:* 0
- *Legionella:* 0

Since these microorganisms are neither easily monitored nor detected, the SWTR specifies a treatment technique (disinfection and/or filtration) instead of MCLs. Criteria for the removal or inactivation of microbiological organisms for public water systems are established in the basic requirements of the SWTR.

- 3 log (99.9%) removal and/or inactivation of *Giardia lamblia* cysts
- 4 log (99.99%) removal and/or inactivation of viruses

In systems utilizing filtration, compliance to a specified treatment level is judged by the type of filtration process and $C \cdot t$ values, where C is the residual concentration of disinfectant (mg/L) and t is the effective contact time t_{10} (min). The effective contact time is determined by tracer test or by using the baffling factors presented in the SWTR guidance manual. The removal/inactivation of microbiological contaminants, as recommended by the EPA, are presented in Table 2.4.2-1 (June 1989) as a function of treatment process.

Under the SWTR, a public water system using filtration will achieve compliance with the SWTR by meeting the following performance standards:

1. The combined filter effluent must be less than 0.5 ntu in 95% of monthly samples and must never exceed 5 ntu for conventional and direct filtration treatment plants.
2. The system must maintain a 0.2 mg/L disinfectant residual entering the distribution system and must never fall below 0.2 mg/L for more than 4 h. Heterotrophic plate counts (HPCs) can be measured in lieu of disinfectant residual. For compliance purposes, a sample with a HPC of less than 500 colonies/mL is considered to have detectable residual.

Disinfectants/Disinfection By-products Rule (D/DBP Rule) The D/DBP Rule (published in the *Federal Register* December 16, 1998) is shown in Table 2.4.2-2. The use of best available technology (BAT) was proposed to reduce or control all four contaminants. Bromate and chlorite could be controlled by a specific mode of the ozonation method. However, enhanced coagulation or enhanced lime softening would be required to reduce/control TTHM and HAA precursors.

TABLE 2.4.2-1 Assumed Log Removal/Inactivation by Disinfection

	Assumed Log Removal by Filtration		Recommended Minimum Log Inactivation by Disinfection	
Filtration Type	Giardia	Viruses	Giardia	Viruses
Conventional	2.5	2.0	0.5	2.0
Direct	2.0	1.0	1.0	3.0
Slow sand	2.0	2.0	1.0	2.0
Diatomaceous earth	2.0	1.0	1.0	3.0

The limits on disinfectants as dictated by the D/DBP rule are as follows:

- *Chlorine:* 4.0 mg/L
- *Chloramines:* 4.0 mg/L as total chlorine
- *Chlorine dioxide:* 0.8 mg/L

Enhanced Coagulation Requirements Enhanced coagulation is defined as the addition of excess coagulant to improve the removal of disinfection by-product precursors through the use of conventional water treatment processes. Specific treatment systems must be utilized to achieve the percent reduction of TOC between the raw water and treated water prior to final continuous disinfection. A system using ozone and biologically active filters (BAFs) must also achieve the percentage reduction of TOC prior to the addition of a residual disinfectant (chlorine) unless the avoidance criteria are met. Table 2.4.2-3 (August 1997) illustrates how the required TOC removal targets are linked to source water TOC and the alkalinity level of the source water.

The avoidance criteria for enhanced coagulation are as follows:

1. Raw water SUVA must be less than 2.0 L/mg per meter, where SUVA is equal to (UV254/TOC) $\times$ 100. It should be noted that preozonated raw water may be considered as the raw water for subsequent conventional treatment process.

TABLE 2.4.2-2 Maximum Contaminant Levels of DBPs (mg/L)

Contaminant	Stage 1	Stage 2 (Proposed)
Total THMs (TTHMs)	0.080	0.040
Five haloacetic acids (HAA_5s)[a]	0.060	0.030
Bromate ion	0.010	
Chlorite ion	1.0	

[a] HAA_5 includes trichloroacetic acid, dichloroacetic acid, monochloroacetic acid, dibromoacetic acid, and monobromoacetic acid.

TABLE 2.4.2-3 Required TOC Removal

	Source Water Alkalinity (%) for:		
Source Water TOC (mg/L)[a]	0–60 mg/L	> 60–120 mg/L	> 120 mg/L
2–4	35	25	15
> 4–8	45	35	25
> 8	50	40	30

[a]No action when TOC is less than 2.0 mg/L.

2. The TOC for treated water must be less than 2 mg/L.
3. Raw water TOC must be less than 4 mg/L, alkalinity must be greater than 60 mg/L, and the distribution system TTHM/HAA levels must be less than 0.04/0.03 mg/L.
4. Distribution system TTHM/HAA levels must be less than 0.04/0.03 mg/L, with free chlorine as both primary and secondary disinfectant (not chlorine/chloramines)

It should be noted that the enhanced coagulation requirement does not apply to treatment systems using direct filtration or systems obtaining a waiver from the filtration requirements of SWTR.

Enhanced Surface Water Treatment Rule (ESWTR) The final Interim ESWTR was scheduled to be published in December 1998 and the rule will become effective in the year 2001. The rule will apply to surface water and groundwater that are under the direct influence of surface water systems serving over 10,000 people. The EPA is developing a long-term ESWTR that will be composed of two rules, Long Term (1) and Long Term (2). Currently, the requirements of Long Term (1) will be applicable to systems serving less than 10,000 people, with the possibility of future modifications. Long Term (2) requirements will apply to systems serving over 10,000 people and will incorporate the results of the Information Collection Rule. Long Term standards are expected to be published in the year 2002 and become effective in the year 2005.

The main features of the ESWTR are as follows:

1. The minimum removal of *Cryptosporidium* is 2 log.
2. The combined filtered water turbidity must be less than 0.3 ntu in at least 95% of the measurements taken each month and must never exceed 1.0 ntu.
3. Systems that employ rapid granular filtration, slow sand filtration, and diatomaceous earth filtration will be granted a 2 log reduction credit of *Cryptosporidium* if they meet the turbidity requirements presented above.
4. All rapid granular filtration surface water systems serving over 10,000 people must monitor the turbidity of each filter continuously and furnish an exceptions report (violations of the turbidity criteria) to the state on a monthly basis.

Other Water Quality–Related Items Design engineers must also pay particular attention to several water quality issues. These are listed below.

Microbial Regrowth Biologically stable finished water is particularly important for any treatment alternatives utilizing ozone. Biological stability requires a reduction in either assimilable organic carbon (AOC) or biodegradable dissolved organic carbon (BDOC) to raw water levels.

MX MX is a synonym for 3-chloro-4(dichloromethyl)-5-hydroxy-2[5H]-furanone, which is a by-product of chlorine disinfection. MX has been identified as a strong mutagen in the TA100 strain of *Salmonella typhimurium.* Additionally, MX studies have indicated that it can induce chromosomal aberrations in hamsters, however, it is not currently being considered for regulation in ESWTR.

Methyl Tertiary Butyl Ether (MTBE) MTBE is a gasoline additive used in the United States. Leakage of underground storage tanks has resulted in MTBE contamination of groundwaters in certain areas. Currently, there is no MCL enforced for MTBE; however, studies have suggested a maximum drinking water level ranging from 40 to 200 μg/L. Some states have set more stringent limits, ranging from 50 to 100 μg/L. Three treatment processes are currently being considered for the treatment of MTBE: air stripping, GAC adsorption, and advanced oxidation. Some studies indicate that conventional air stripping is very inefficient for MTBE removal. GAC adsorption is an effective process but a costly method of MTBE removal. Advanced oxidation processes include the UV/H_2O_2 and O_3/H_2O_2 oxidation processes. Although oxidation processes are equally effective and have high capitalization and operation costs, the O_3/H_2O_2 process is considered to be more cost-effective.

Perchlorate Perchlorate (ClO_4) is an electrochemical oxidant (explosive) that is in solid state. It is used in solid rocket fuel (NH_4C1O_4), fireworks, and various pyrotechnics. Perchlorate is extremely stable when dissolved in water at concentrations as high as 1000 mg/L. In very high doses, it has been used in medicine to treat Grave's disease and therefore has great biological impact; when injected, perchlorate inhibits the ability of the thyroid gland to perform its normal function. Based on 1995 evaluations of perchlorate by the EPA and the States' Drinking Water Program, a maximum level of 0.018 mg/L has been set to protect public health. It is certainly possible that the action level could be lowered in the future. Currently, the issue of perchlorate is limited primarily to groundwater contamination of a few regions in the United States. There are four possible treatment processes: anaerobic biodegradation, reverse osmosis, GAC adsorption, and ion exchange.

Arsenic The current MCL for arsenic is 0.05 mg/L. However, it could be reduced to 0.005 mg/L. The final rule is to be released in January 2001.

Filter Wash Wastewater Recycling Rule In July 1998 the EPA held a public meeting to discuss this rule. The following components are considered to be potential issues for future regulations: (1) requirement to recycle filter wash wastewater to the plant headworks, (2) requirement to practice flow equalization of the recycle stream, (3) a ban on the recycling of sludge, and (4) a requirement to provide effective treatment prior to the recycling of filter wash wastewater. This regulation is scheduled to be published in the summer of 1999 and the final rule published by August 2000.

Algal Toxins Certain blue-green algae produce hepatotoxin and neurotoxins. Hepatotoxins are primarily peptides. Commonly identified compounds are microcystins released by *Microcystis aerginosa* and nodularin produced by *Noduralia damage.* These toxins can cause death or liver cancer. Neurotoxins are alkaloids such as the anatoxins released by *Anabaena flos aquae.* Neurotoxins can cause rapid death through paralysis of the respiratory and cardiac muscles.

Number of Particles in Filtered Water (2 to 10 μm Range) *Cryptosporidium* oocysts (2 to 4 μm) and *Giardia* cysts (4 to 10 μm), are not easily disinfected by conventional chlorination. Although ozonation and membrane filtration are very effective against these protozoans, many existing plants have not adopted either of these processes, for many reasons, among them high capital cost, tight hydraulic conditions, and high plant output. Recent pilot studies, however, indicate that a regular conventional treatment process is capable of removing particles in the range 2 to 10 μm range at a 3 to 4 log removal rate provided that the operation of each process is optimized. The current goal of filtered water turbidity of less than 0.1 ntu is achieved primarily because of the high level of particle removal. However, certain states are starting to adopt more stringent guidelines based on a number of particles per milliliter. The regulatory agency of New York has set a criterion of less than 50 particles/mL for particles in the range 2 to 10 μm. Major consulting engineers are setting a goal of 25 particles/mL for the same range of particles for their pilot filter studies.

Federal Drinking Water Regulations The current Treatment Rules and Drinking Water Quality Standards are subject to be modified upon the publication of various studies. For instance, the ICR studies conducted by the EPA are anticipated to be published early in the twenty-first century; some of its conclusions may alter a few of the current MCLs for certain contaminants. Under these circumstances, engineers should be aware of current federal and state regulations. An excellent summary of updated Rules and Drinking Water Quality Standards, with BAT information, is published on an annual basis by the *Journal of the American Water Works Association* (JAWWA).

The currently enforceable USEPA Primary Drinking Water Standards are shown in Table 2.4.2-4 and the Secondary Standards in Table 2.4.2-5. Table 2.4.2-4 also presents European Community (EU) Drinking Water Guide levels in addition to the World Health Organization (WHO) guidelines for projects outside the United States.

TABLE 2.4.2-4 Primary Drinking Water Standards of U.S. EPA (1997) and International Drinking Water Guidelines (1989)

Parameter	EU Guide levels	World Health Organization Guidelines	U.S. Environmental Protection Agency Standards
	Microbials		
Bacteria	<10/ml at 37°C <100/ml at 22°C		
Coliform Group	Not detected in 100 ml	Not detected in 100 ml	Not present
E. coli			Not present
Giardia			99.9% removal/ inactivation
Viruses			99.99% removal/ inactivation
	Inorganics		
Lead	<0.05 mg/L*	<0.01 mg/L	<0.015 & treatment standard
Fluoride	<1.5 mg/L* 8-12°C <0.7 mg/L* 25-30°C	<1.5 mg/L	<2.0 mg/L**
Arsenic	<0.05 mg/L*	<0.01 mg/L (P)	<0.05 mg/L
Selenium	<0.01 mg/L*	<0.01 mg/L	<0.05 mg/L
Mercury	<0.001 mg/L*	<0.001 mg/L	<0.002 mg/L
Cyanide	<0.05 mg/L*	<0.07 mg/L	<0.2 mg/L
Chromium (Cr^{6+})	<0.05 mg/L*	<0.05 mg/L (P)	<0.1 mg/L***
NH_3^-N	<0.05 mg/L as NH_4	<1.5 mg/L as NH_3	
NO_2^-N			<1 mg/L
NO_3^-N	<25 mg/L as NO_3^-	<50 mg/L as NO_3^-	<10 mg/L
Total Nitrate + Nitrite			<10 mg/L
Cadmium	<0.005 mg/L*	<0.003 mg/L	<0.005 mg/L
Fluoride			<4.0 mg/L
Antimony			<0.006 mg/L
Asbestos			<7 MFL
Barium			<2 mg/L
Beryllium			<0.004 mg/L
Nickel			<0.1 mg/L
Thallium			<0.002 mg/L
Turbidity[2]	<0.4 JU	<5 NTU	<=0.5 NTU in 95% of samples per month

TABLE 2.4.2-4 *(Continued)*

Parameter	EU Guide levels	World Health Organization Guidelines	U.S. Environmental Protection Agency Standards
	Radionuclides		
Radium-226 + Radium-228			<5 pCi/L
Gross Alpha			<15 pCi/L
Beta particle & photon activity			4 mRem/yr exposure
	Organics		
Phenol	<0.0005 mg/L*		
Total Trihalomethanes			<0.1 mg/L
1,1,1-Trichloroethane		2 mg/L (P)	<0.2 mg/L
Tetrachloroethylene		<0.04 mg/L	<0.005 mg/L
Trichloroethylene		<0.07 mg/L	<0.005 mg/L
Dichloromethane		<0.02 mg/L	<0.005 mg/L
Benzene		<0.01 mg/L	<0.005 mg/L
Toluene		<0.7 mg/L	<1 mg/L
Ethylbenzene		<0.3 mg/L	<0.7 mg/L
Xylenes		<0.5 mg/L	<10 mg/L
1,1-Dichloroethylene		<0.03 mg/L	<0.007 mg/L
Carbon tetrachloride		<0.002 mg/L	<0.005 mg/L
Aldrin and Dieldrin		<0.00003 mg/L	
Benzo(a)pyrene		<0.0007 mg/L	<0.0002 mg/L
Chloroform		<0.2 mg/L	See total trihalomethanes
2,4-D		<0.03 mg/L	<0.07 mg/L
DDT		<0.002 mg/L	
1,1-Dichloroethene		<0.003 mg/L	
Heptachlor epoxide and Heptachlor		<0.00003 mg/L	<0.0002 mg/L & <0.0004 mg/L, respectively
Hexachlorobenzene		<0.001 mg/L	<0.001 mg/L
Lindane		<0.002 mg/L	<0.0002 mg/L
Methoxychlor		<0.02 mg/L	<0.04 mg/L
Pentachlorophenol		<0.009 mg/L (P)	<0.001 mg/L
2,4,6-Trichlorophenol		<0.2 mg/L	
1,1,2-Trichloroethane			<0.005 mg/L
1,2-Dichloroethane			<0.005 mg/L
1,2-Dichloropropane			<0.005 mg/L
1,2,4-Trichlorobenzene			<0.07 mg/L
cis-1,2-Dichloroethylene			<0.07 mg/L
trans-1,2-Dichloroethylene			<0.1 mg/L
Monochlorobenzene			<0.1 mg/L

TABLE 2.4.2-4 ***(Continued)***

Parameter	EU Guide levels	World Health Organization Guidelines	U.S. Environmental Protection Agency Standards
o-Dichlorobenzene			<0.6 mg/L
p-Dichlorobenzene			<0.075 mg/L
Styrene			<0.1 mg/L
Vinyl chloride			<0.002 mg/L
2,3,7,8-TCDD (Dioxin)			$<3 \times 10^{-8}$ mg/L
2,4,5-TP (Silvex)			<0.05 mg/L
Alachlor			<0.002 mg/L
Atrazine			<0.003 mg/L
Carbofuran			<0.04 mg/L
Chlordane			<0.002 mg/L
Dalapon			<0.2 mg/L
Di(2-ethylhexyl)adipate			<0.5 mg/L
Di(2-ethylhexyl)phthalate			<0.006 mg/L
Dibromochloropropane (DBCP)			<0.0002 mg/L
Dinoseb			<0.007 mg/L
Diquat			<0.02 mg/L
Endothall			<0.1 mg/L
Endrin			<0.002 mg/L
Ethylene dibromide			<0.00005 mg/L
Glyphosphate			<0.7 mg/L
Hexachlorocyclopentadiene			<0.05 mg/L
Oxamyl (vydate)			<0.2 mg/L
Picloram			<0.5 mg/L
Polychlorinated biphenyl (PCB)			<0.0005 mg/L
Simazine			<0.004 mg/L
Toxaphene			<0.003 mg/L
Hexachlorobenzene			<0.001 mg/L

(P) Provisional value
*Maximum Admissible Concentration
**Secondary standard
***Total for all species

2.4.3 Additional Goals and Objectives

Beyond the finished water quality objectives, the owner's goals for the water treatment plant must be identified and established. The goals of the overall treatment plant should state what the plant is to accomplish and how these goals affect the water distribution systems and the environment. Once this is established, the required function of each unit process (of the treatment process train) must be identified and the objectives of each of these units should be defined.

TABLE 2.4.2-5 EPA Secondary Drinking Water Standards

Contaminant	Maximum contaminant level
Aluminum	0.05 to 0.2 mg/L[a]
Chloride	250 mg/L
Color	15 color units
Copper	1 mg/L
Corrosivity	Neither corrosive nor scale forming
Fluoride	2.0 mg/L
Foaming agents	0.5 mg/L
Iron	0.3 mg/L
Manganese	0.05 mg/L
Odor	Three threshold odor numbers
pH	6.5 to 8.5
Silver	0.1 mg/L
Sulfate	250 mg/L
Total dissolved solids	500 mg/L
Zinc	5 mg/L

[a]Level selected for aluminum depends on the discretion of the state primacy agency.

Design engineers have traditionally considered each unit process as an individual item and have optimized each unit separately on the premise that this would result in the most efficient overall plant design. However, such suboptimization usually does not result in optimization of the total plant design. Additionally, objectives such as optimization of initial costs, life-cycle costs, and construction time should be evaluated in the preliminary engineering study.

During the early stages of the project, engineers should establish basic design parameters for major finished water quality and a few plant operational issues. Suggested goals are:

1. The range of plant flow rate should be no more than 4 : 1.
2. The clarifier effluent turbidity must be less than 2 ntu.
3. The frequency of filter washing should be limited to once a day per filter at the designed filtration rate.
4. Filtered water turbidity of less than 0.1 ntu.
5. Particles in the filtered water should be less than 50 particles/mL (those larger than 2 μm).
6. After backwash, discard filtered water with turbidity above 0.25 ntu.
7. Aluminum in finished water should be less than 0.05 mg/L.
8. TOC of filtered water should be less than 2.0 mg/L (or less than 2.0 SUVA).
9. TTHMs should be less than 40 μg/L.
10. HAA_5 should be less than 30 μg/L.
11. Bromate should be less than 10 μg/L.

2.4.4 Restrictions and Constraints on Plant Design

Project engineers are rarely, if at all, given complete freedom in their designs. Thus restrictions and constraints must be clearly defined at the beginning of the design phase. Restrictions may be due to economic, physical, chemical, temporal, climatic, geological, sociological, legal, or aesthetic considerations and may be imposed by local, state, or federal agencies.

Constraints are defined as restrictions placed on the acceptable values of design variables (properties of the system that are controlled by the designers). Examples of constraints are building codes, zoning laws, and OSHA regulations and standards. For instance, a 33-in. (84 cm)-diameter steel pipe is not standard size, nor is a 70-hp (53-kW) electric motor a standard item. These restrictions and constraints may fix the component properties or establish a range in which they must fall. Although these restrictions are generally made for reasons of standardization and safety, serious design problems often arise when designing treatment plants for developing countries, due to the limited number of available components, materials, technology, and qualified personnel, in addition to the bureaucratic red tape associated with the importation of goods.

2.4.5 Alternative Treatment Processes

The basis for selecting treatment process alternatives is established by the characteristics of the raw water and the finished water quality goals. Consideration must be given to future implementation of more stringent EPA drinking water quality standards and to possible changes and variability in the raw water quality. Thus the goals and objectives, as well as the restrictions and constraints defined in the preceding section, all bear upon the selection of alternative processes. Furthermore, the availability of major equipment, postinstallation services, and the capability of operators and maintenance personnel, as well as the waste handling requirements and the availability and cost of water treatment chemicals, all greatly affect the selection of the water treatment process, especially in remote regions and developing countries.

Final selection of the most appropriate treatment process scheme must be based on reliability, constructability, ease of operation, simple maintenance, and most important, cost. Value analysis may be practiced during the selection process, with respect to the whole system and the individual unit processes. A cost analysis of the various alternatives should be based on two criteria: present worth and useful life.

Bench-scale studies and pilot studies, conducted during either the feasibility study or preliminary engineering study phase, are quite valuable in confirming the effectiveness of the alternative processes selected if the engineer is unable to draw on previous experiences in treating the raw water. Experience acquired from existing water treatment plants (treating the same source of water) and plant-scale simulation tests provide important guides in selecting both the treatment process scheme and the hydraulic loading of each unit process.

There are three basic water purification processes with six modified schemes: (1) micro membrane filtration, (2) direct filtration, and (3) conventional complete treatment. Modified versions are ultra membrane filtration, nano membrane filtration, reverse osmosis membrane filtration, in-line (contact) filtration, two-stage filtration, and lime softening process. The first three are modified microfiltration processes using smaller pore membranes. In-line filtration is attained by deleting the flocculation stage from the direct-line filtration process. Two-stage filtration uses a coarse media roughing filter in lieu of the regular flocculation process of direct filtration. The lime softening process is similar to a conventional complete treatment process except that the primary coagulant is lime, recarbonation is achieved through the use of carbon dioxide, and clarification is conducted before filtration (standard scheme). Table 2.4.5-1 presents the applicable raw water qualities for some practical water treatment processes. Figure 2.4.5-1 illustrates six water treatment process trains. It is vital that the project engineer provide both space and extra hydraulic head across the proposed treatment process train during the final design stage, as the rules and regulations governing water quality will be modified in the future.

2.4.6 Preliminary Design Criteria for Treatment Unit Processes

The preliminary size of all major treatment unit processes must be established in preparation for the cost comparisons and site plans used in evaluating site adequacy. This preliminary report should consider the coagulation, flocculation, sedimentation, and filtration processes; the disinfection process, including type of disinfectant, disinfecting condition, and by-products; the filter washing, waste-wash handling, sludge handling, and chemical feed and storage facilities; the clearwell; and the control building. The ozonation and GAC adsorption processes should also be included in the preliminary design report if they are integrated into the treatment plant. Table 2.4.6-1 is an example of preliminary design criteria.

2.4.7 Hydraulic Grade Across the Plant

It is important to establish the hydraulic grade line across the plant early in the preliminary engineering study because both the proper selection of the plant site elevation and the suitability of the site (for executing balanced cuts and fills to accommodate all process units requiring specific water elevations and depths of structures) depend on this consideration.

Most conventional water treatment plants require 16 to 17 ft (4.9 to 5.2 m) of headloss across the plant. This means that a difference of 16 to 17 ft must exist between the water level at the head of the plant and the high water level in the clearwell, which is the tail end of the treatment plant process train. Modern treatment plants using preozonation, as well as postozonation and granular activated carbon adsorption processes, require almost 25 ft (7.6 m) of available head across the plant. Under these circumstances, if the plant site is flat, the following criteria must be met; the high water level in the clearwell must be set at ground level because of the groundwater table; the water level at the head of the process train must be 25 ft (7.6 m) above the ground level; and the majority of the unit processes in the first half of the process train must necessarily be in the air unless a pumping station is

TABLE 2.4.5-1 Suggested Raw Water Quality for Practical Treatment Processes

	Micro Filtration	Direct Filtration	DAF Filtration	Direct Filtration with Ozone	Conventional Complete	Conventional Complete with Ozone
Turbidity (NTU)	<100	<20	<100	<30	$<3,000$	$<3,000$
Color (apparent)	<15	<20	<100	<100	$<1,000$	$<1,500$
Alkalinity (mg/L)	<150	<200	<300	<200	<500	<500
Hardness (mg/L)	<150	<150	<200	<200	<700	<700
Iron (mg/L)	<0.5	<0.5	<0.6	<0.8	<2.0	<2.5
Manganese (mg/L)	<0.1	<0.1	<0.1	<0.2	<0.5	<1.0
TOC (mg/L)	<2	<2.5	<5.0	<7.0	<7.0	<10.0
Taste and odor (TON)	<3	<4.5	<5.0	<7.0	<10.0	<15.0
Algae (ASU/mL)	$<10^3$	$<10^3$	$<10^4$	$<10^3$	$<10^4$	$<10^4$
Giardia (100 L)	<100	<3	<5	<10	<20	<50
Cryptosporidium (100 L)	<100	<1	<3	<10	<10	<20
Coliform (no./mL)	$<10^4$	$<10^3$	$<10^4$	$<10^6$	$<10^6$	$<10^7$

Notes: (1) The criteria shown are a general condition thus exceptions are possible.

(2) The conventional complete treatment process includes a lime softening process.

(3) If the raw water turbidity is expected to exceed 1,000 ntu, it is recommended that a pre-sedimentation process be added to the conventional complete treatment process. The inclusion of a pre-sedimentation process will allow the conventional process to treat raw water with turbidity greater than 3,000 ntu.

(4) Under certain conditions, slow sand filtration is still considered to be a basic treatment process. In general, the raw water quality is the same as the direct filtration process.

(5) Membrane filtration may be used in situations where the raw water quality requires the use of a conventional complete process, provided that proper pre-treatment is conducted to control TOC, color, taste and odor, and excessive suspended solids prior to the membrane filtration process.

included in the process train. It is therefore obvious that a flat and level site is not the best choice for this type of treatment plant. The ideal plant site will have a 3 to 5% one-way slope and a ground elevation that satisfies the necessary elevations. Figure 6.1-2 is an example of the hydraulic grade line across a plant.

2.4.8 Geotechnical Considerations

Ground considerations of the proposed plant site greatly influence the entire construction cost, the duration of construction, and plant maintenance. Items of major importance are the availability of information necessary to design the foundations of the treatment facilities, ground characteristics that could affect construction, and soil characteristics that allow engineers to provide measures against cut-and-fill slopes as well as corrosion of pipes and concrete structures.

The following items can be only obtained from a geotechnical investigation.

1. *Soil pressure:* allowable bearing pressures under footings, lateral earth pressure against cantilever walls, lateral earth pressure against fully restrained walls, passive resistance force, friction coefficient between footing bases and the soil, and anticipated degree of settlement
2. *Data on excavation and fill:* cut slope for excavation, permanent cut-and-fill slopes, shoring cut angle for pipe trenches, type of equipment required for excavation, and compaction requirements under structures, around structures, and under roadways
3. *Groundwater level:* expected high groundwater level and seasonal fluctuations
4. *Site seismicity:* peak ground acceleration force and the seismic surcharge earth load against the walls of buried structures; must be used in the structural design

2.4.9 Structural Design Conditions and Criteria

The preliminary engineering study includes both general and specific structural design considerations and criteria for the proposed plant site. Generally, all structures must be structurally sound, that is, capable of withstanding dead weight, live loads, water pressure, earth pressure, various forces resulting from earthquakes, vibrations produced by moving mechanical equipment, loads arising from the snow pack, ice pressure, wind pressure, and the anticipated loading and impact during construction. Since a majority of the structures are water bearing (tanks and basins), they must be leakproof. Moreover, structures holding filtered and finished water must not be polluted by untreated water or wastewater. Finally, structures situated in areas with high underground water levels or where the underground water levels become high during the rainy seasons should be designed to resist buoyancy, especially when the tanks are empty. When all these requirements are considered, the minimum thickness of the reinforced concrete walls and slabs of water-bearing structures should be 8 in. (200 mm), regardless of the loading conditions.

(*text continues on page 49*)

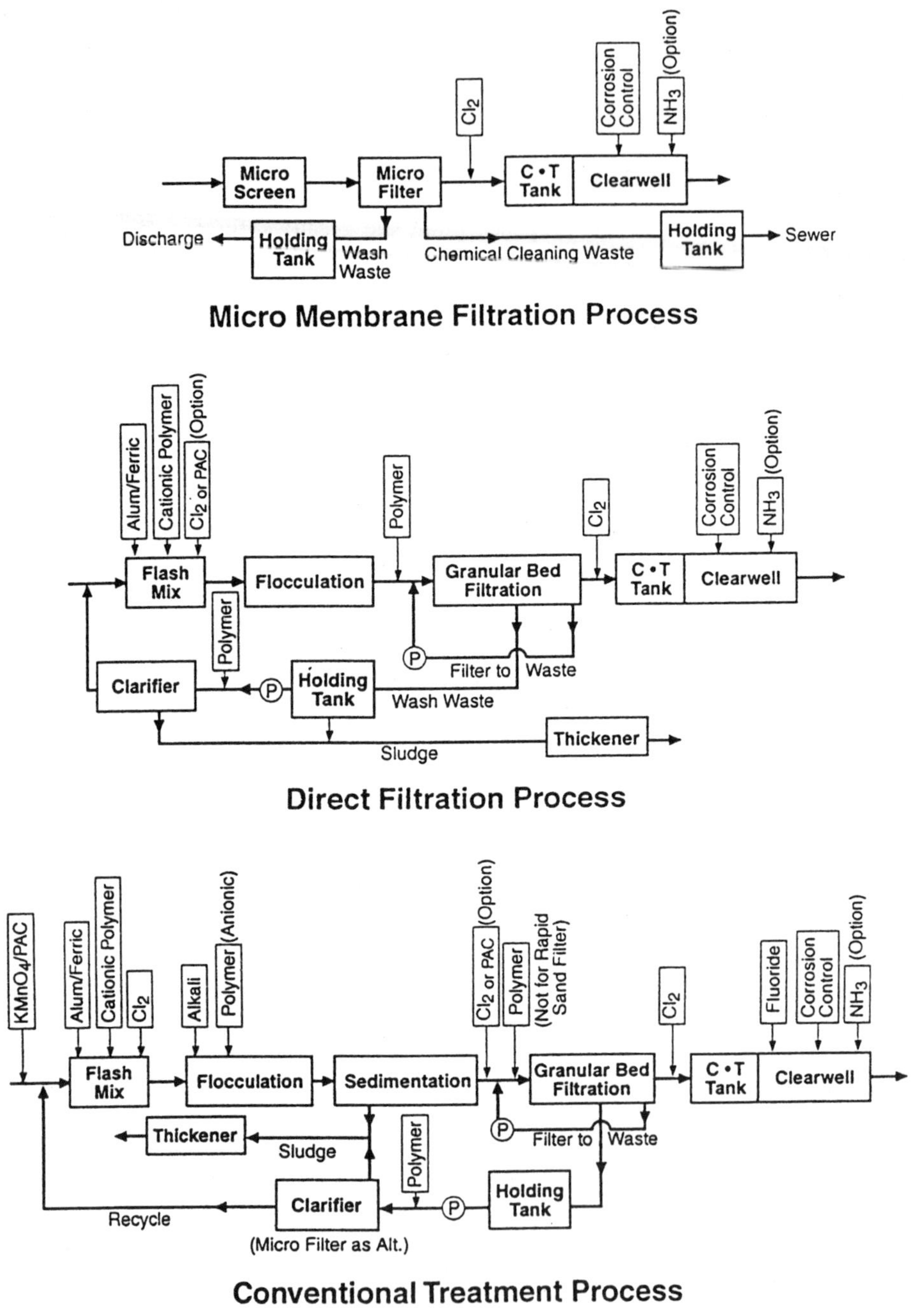

(*a*)

Figure 2.4.5-1 (*a*) Three basic water treatment processes; (*b*) modified processes.

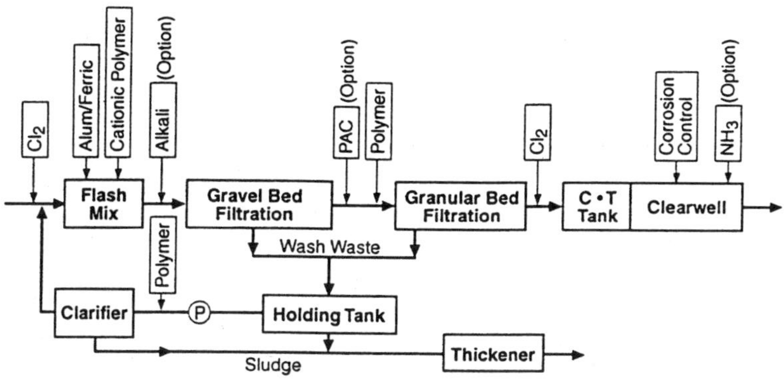

Two-Stage Filtration Process

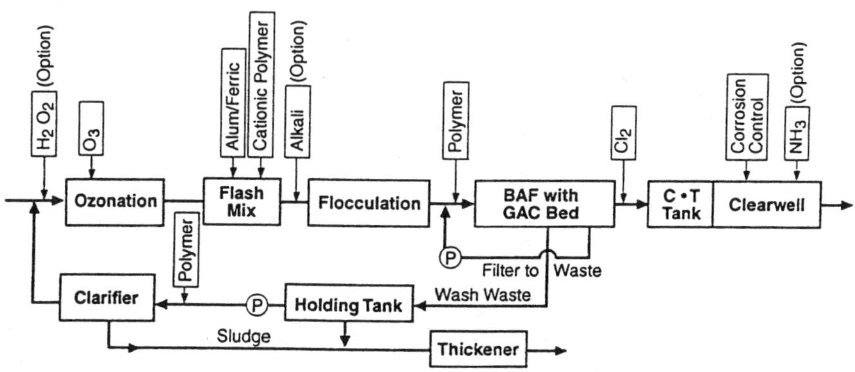

Direct Filtration Process with Preozonation

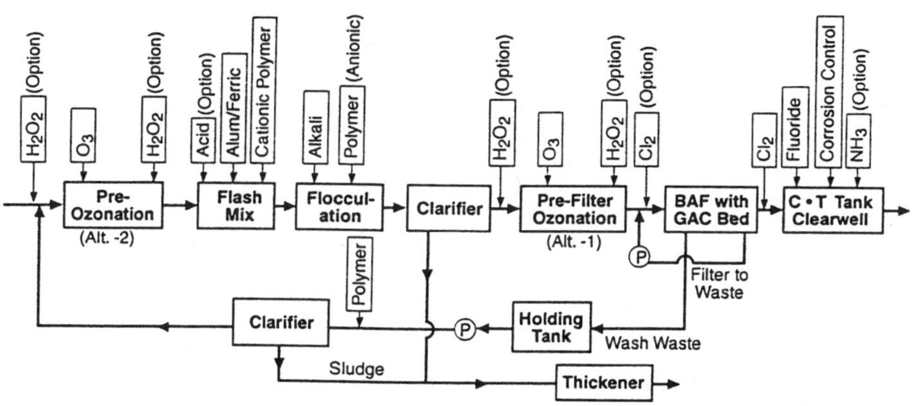

Conventional Treatment Process with Ozonation and GAC Filters

(*b*)

Figure 2.4.5-1 *(Continued)*

TABLE 2.4.6-1 Example of Design Criteria

FAIRFAX COUNTY WATER AUTHORITY
FREDERICK P. GRIFFITH, JR.
WATER TREATMENT PLANT

	Value	Units
Plant capacity		
Design capacity	126	mgd
	87,500	gpm
	195	cfs
	11,697	cfm
Average daily flow	75	mgd
	52,083	gpm
	116	cfs
	6,963	cfm
Minimum flow	40	mgd
	27,778	gpm
	62	cfs
	3,713	cfm
Influent flow metering		
Number	2	ea
Type: Magnetic flowmeter		
Size	54	in
Range	20-85	mgd
Flash mixing		
Primary system		
Number	1	ea
Type: Hydraulic pump diffusion		
Size	72	in
Mixing energy (G)	730	sec^{-1}
Mixing zone length	9.0	ft
Mixing zone diameter	6.0	ft
Mixing zone volume	254.3	cu ft
Mixing time		
@126 mgd	1.3	sec
@40 mgd	4.1	sec
Mixing pump		
Number	1	ea
Type: Horizontal end suction		
Capacity	2,300	gpm
TDH	70	ft
Nominal motor size	60	hp
Secondary system		
Number	2	ea
Type: Hydraulic pump diffusion		
Size	54	in
Mixing energy (G)	690	sec^{-1}
Mixing zone length	6.8	ft
Mixing zone diameter	4.5	ft
Mixing zone volume	107.0	cu ft

TABLE 2.4.6-1 ***(Continued)***

	Value	Units
Mixing time		
@126 mgd	1.1	sec
@40 mgd	3.5	sec
Mixing pump		
Number	2	ea
Type: Horizontal end suction		
Capacity	1,200	gpm
TDH	48	ft
Nominal motor size	20	hp
Flocculation basins		
Number	4	ea
Stages/basin	3	ea
Water depth	15	ft
Basin volume (each)	105,300	cu ft
Total volume	421,200	cu ft
Detention time		
@126 mgd	36	min
@40 mgd	113	min
Stages 1 & 2		
Type: Hydraulic flocculation (baffled horizontal serpentine flow)		
Baffles/stage	24	ea
Energy input (G × t)		
1st stage		
@126 mgd	41,150	
@40 mgd	24,400	
2nd stage		
@126 mgd	28,800	
@40 mgd	16,600	
Stage 3		
Type: Mechanical flocculation (variable speed, vertical hydrofoil impellers)		
Number (per basin)	6	ea
Compartment size	20 × 19.5	ft × ft
Compartments/basin	6	ea
Compartment volume	5,850	cu ft
Depth to equivalent diameter ratio	0.77	
Energy input (G × t)		
@126 mgd	variable	
@40 mgd	variable	
Motor size	1	hp
Sedimentation basins		
Type: rectangular		
Number	4	ea
Size	295 × 120	ft × ft
Water depth	16	ft
Basin volume (each)	566,400	cu ft
Total volume	2,285,600	cu ft

TABLE 2.4.6-1 ***(Continued)***

	Value	Units
Detention Time		
@126 mgd	194	min
@40 mgd	610	min
Horizontal flow velocity		
@126 mgd	1.5	fpm
@40 mgd	0.5	fpm
Surface loading		
@126 mgd	0.6	gpm/sq ft
@40 mgd	0.2	gpm/sq ft
Basin length to width ratio	4.9:1	
Weir loading		
Effluent weir length (each)	100	ft
Effluent weir length (total)	400	ft
@126 mgd	219	gpm/ft
@49 mgd	69	gpm/ft
Sludge withdrawal		
Type: chain and flight		
Number of longitudinal collectors per basin	12	ea
Number of cross collectors per basin	1	ea
Filters		
Type: deep bed gravity		
Number	14	ea
Number of cells/filter	2	ea
Inside dimentions (each cell)	34 × 17	ft × ft
Media surface area (each filter)	1,156	sq ft
Media surface area (total)	16,184	sq ft
Filtration rates (@126 mgd)		
Two filters off line	6.3	gpm/sq ft
All filters on line	5.4	gpm/sq ft
Filter media		
Type: granular activated carbon/sand		
Depth	72/10	in
Effective size	1.4/0.54	mm
Uniformity coefficent	<1.4/<1.4	
Specific gravity	1.35/2.65	
Underdrain type	Dual lateral	
Filter backwash		
Maximum rate	22	gpm/sq ft
Normal rate	16-18	gpm/sq ft
Duration (total)	15	min
Backwash pumps		
Type of pump: vertical centrifugal		
Number	3	ea
Capacity (each)	12,500	gpm
TDH	30	ft
Motor size	125	hp
Air scour blowers		
Type: positive displacment		

TABLE 2.4.6-1 ***(Continued)***

	Value	Units
Number	3	ea
Air scour rate	3	scfm/sq ft
Capacity	2,000	scfm
Pressure	8.5	psi
Motor size	125	hp
Clearwell		
Type: rectangular, buried cast-in-place conc		
Number of cells	2	ea
Capacity (each)	6	mg
Total capacity	12	mg
Dimensions	208 × 352	ft × ft
Water depth (maximum)	20	ft
Finished Water Pumping Station		
High service pumps		
Type: vertical diffusion vane (turbine)		
Number	3	ea
Capacity (each)	21,600 (31)	gpm (mgd)
Total capacity	64,800 (93)	gpm (mgd)
TDH	270	ft
Motor size	1,800	hp
Main service pumps		
Type: vertical diffusion vane (turbine)		
Number	3	ea
Capacity (each)	16,000 (23)	gpm (mgd)
Total capacity	48,000 (69)	gpm (mgd)
TDH	95	ft
Motor size	500	hp

Ozone System

General

Condition	Plant Flow (mgd)	Applied Ozone Dosage (mg/L)	Ozone Concen. (%)	Ozone Production (lb/day)	Total Gas Flow (scfm)	Gas Flow per Contactor (scfm)	No. of Diffusers	Gas flow per Diffuser (scfm)	Oxygen Required (tpd)
Design	126	3.8	12	4,000	300	75	70	1.1	17
Average	75	3.8	7	2,400	300	75	70	1.1	17
Minimum	40	2.5	2.5	830	300	75	70	1.1	17
Maximum	126	5.7	8	6,000	630	160	140	1.1	37.5
Ultimate	168	5.7	8	8,000	840	140	140	1.0	50

	Value	Units
Oxygen suppy		
Vapsa plant capacity	17	tpd
Minimum oxygen concentration	90	wt. %
Backup lox storage volume	13,000	gal
Ozone generation		
Number of generators	3	ea
Capacity (each) @12% wt. ozone concentration	2,000	lb/day
Capacity (each) @8% wt. ozone concentration	3,000	lb/day
Applied ozone dosage	5.7	mg/L
Number of future generators	1	ea

TABLE 2.4.6-1 ***(Continued)***

	Value	Units
Ozone contractors		
Number of contactors	4	ea
Capacity (each)	31.5	mgd
Number of cells	10	ea
Detention time	10	min
Water depth	22	ft
Number of diffusers in first cell	70	ea
Number of diffusers in second cell	70	ea
Typical gas flow per diffuser	1	scfm
Number of future ozone contactors	2	ea
Ozone destruction		
Number of ozone destruct units	3	ea
Capacity (each)	300	scfm
Vacuum on contactors	3-5	in H_2O
Number of future ozone destruct units	1	ea
Cooling water		
Flow per ozone generator	420	gpm
Flow per power supply unit	60	gpm
Flow to Vapsa system	35	gpm
Maximum inlet temperature	85	°F
Total cooling water flow (design)	1,355	gpm
Total cooling water flow (future)	1,795	gpm

Chemical		Dosage (mg/L)		
	Point of Application	Minimum	Average	Maximum
Ferric chloride (Pri)	Primary flash mix	20	30	65
Cationic polymer (Alt)	Primary flash mix	0.5	2	4
Lime (Alt)	Primary flash mix	3	8	20
Sodium hypochlorite (Alt)	Primary flash mix	4	5	10
Ferric chloride (Alt)	Secondary flash mix	20	30	65
Cationic polymer (Pri)	Secondary flash mix	0.5	2	4
Lime (Pri)	Secondary flash mix	3	8	20
Anionic Polymer (Pri)	Flocculation (2nd stage)	0.1	0.25	0.5
Anionic Polymer (Alt)	Flocculation (3rd stage)	0.1	0.25	0.5
Sodium Bisulfite (Pri)	Ozone contactor effluent	0.11	0.44	2.2
Nonionic polymer (Pri)	Filter influent	0.01	0.02	0.05
Sodium hypochlorite (Pri)	Clearwell influent	4	5	10
Sodium hydroxide (Pri)	Clearwell influent	15	30	45
Hydrofluosilicic acid (Pri)	Clearwell influent	0.5	1	1
Aqueous ammonia (Pri)	Clearwell influent	1	1.25	2.5
Sodium hypochlorite (Alt)	Clearwell effluent	4	5	10
Aqueous ammonia (Alt)	Clearwell effluent	1	1.25	2.5
Zinc orthophosphate (Pri)	High service pump discharges	0.5	2	4
Zinc orthophosphate (Pri)	Main service pump discharge	0.5	2	4
Nonionic polymer (Alt)	Washwater supply	0.1	0.25	0.5
Sodium hypochlorite (Alt)	Washwater supply	4	5	10

With the exception of tropical regions, the ambient and water temperatures change significantly with the seasons. Thus the plant design must account for thermal stresses and the expansion and contraction of the structures. The local weather conditions determine the temperature range, and this in turn strictly dictates the spacing of the expansion joints for both concrete and metal structures.

The expansion joints for reinforced concrete structures are usually spaced every 33 to 65 ft (10 to 20 m), depending on factors such as if the structure is exposed to sunlight, filled with water, or located underground. These joints should provide a space of 0.5 to 1 in. (12 to 25 mm) to allow movement of the structures. Each expansion joint must have a water stop, joint filler material and sealant, and special measures that allow the reinforcing concrete bars to move at the joint (e.g., a reinforcing bar in a pipe sleeve filled with grease). Although currently not considered, engineers should be aware that investigation of recent earthquake damage has identified a standard expansion joint design using a plastic sheet across the joint as a water stop barrier, as a weak point in water-holding structures during strong earthquakes. The Northridge, California earthquake of 1994 (magnitude 7.3) and the Kobe, Japan earthquake of 1995 (magnitude 7.9) damaged over 50% of the plastic water stop barriers used in tanks and filters, thereby resulting in flooding of the filter pipe galleries.

With respect to concrete and reinforced concrete structures, the water/cement ratio is a critical factor in both the strength of the concrete, corrosion of concrete surface, and the degree of shrinkage. This ratio is generally limited to less than 0.5. This figure also maintains the compression strength of the walls and slabs at a minimum of 4000 psi (280 kg/cm^2). The allowable shrinkage rate of water treatment concrete tanks is usually limited to 0.04 to 0.05%. When designing treatment plants for regions experiencing earthquakes, engineers must use the proper seismic acceleration coefficients (those recommended by the local building codes and geotechnical consultants) in all structural design computations. Should a plant site have high levels of sulfate in the soil and/or underground water, type 2 or type 5 cement may be considered. General-purpose type 1 cement is usually not capable of withstanding the corrosive conditions.

2.4.10 Plant Waste Handling and Disposal

Water treatment plants yield two basic types of waste: recoverable waste and nonrecoverable waste. The first type include filter wash wastes, wastewater from the filter-to-waste process, the supernatant of the sludge drying beds, and plant overflow. The latter include sludge from both the clarifiers and the filter wash-waste holding tanks, sanitary and chemical wastes, and wastes produced by other unit processes, such as the sludge press or ion exchanger. The acceptable modes of waste handling and processing are determined by the Water Pollution Control Act Amendment of 1972 and the National Primary Drinking Water Regulation (1986 Amendments). Thus the preliminary engineering study must be conducted with these regulations in mind.

In the past, the recoverable wastes were collected in a holding tank, then recycled to the head of the plant without further treatment, except in cases where natural plain sedimentation was employed. However, the new EPA drinking water treatment guidelines now consider this practice to be unacceptable. Since the wastes usually contain high levels of hazardous materials, such as microorganisms, SOCs, DBPs, and heavy metals, the wastes must now be treated (flocculation, sedimentation, and disinfection) prior to recycling. The bottom line is that the quality of the recycled water must be better than or as good as the raw water entering the plant. However, the Water Pollution Control Act also allows the treated recoverable wastes to be discharged to a nearby water course as long as a permit is acquired from the regional administrator.

Handling of nonrecoverable wastes is a more complicated issue. Several alternatives exist. The simplest is discharge into the sewer system. However, this method is often impractical because of the expensive fines levied by wastewater departments, particularly when they are not located in the same city as the treatment plant. The fundamental methods of processing and handling nonrecoverable wastes are concentration of the wastes, treatment and disposal of separated liquids, disposal of concentrated materials, and the recovery of by-products from the wastes. By-products that may be recovered are aluminum sulfate and calcium oxide; the wastes also yield inert compounds that can be used to produce bricks and aggregates for road construction. The most common handling and processing practices are gravity thickening, physical and chemical separation, heat treatment, and other types of chemical engineering processes.

Generally, only a limited number of alternatives exist for the disposal of the final residual wastes. Thus the engineer must carefully research the availability of appropriate disposal sites, requirements for disposal, methods of transporting the final residue, and the costs associated with final disposal. The preliminary engineering study must evaluate all feasible alternatives for the treatment, handling, and ultimate disposal of all waste produced by the water treatment plant. Furthermore, the study must present a realistic recommendation based on cost considerations and the effectiveness of the proposed waste processing and handling scheme.

2.4.11 Instrumentation and Control Systems

The general motives for installing a sophisticated instrumentation and control system for a treatment plant and water supply system are (1) the continuous production and supply of safe drinking water, (2) the automatic execution of corrective measures and automatic response to potentially disastrous situations, (3) minimizing the potential for human error, (4) the capability to quickly solve analytic problems, and (5) the ability to diagnose problems in remotely located equipment before a malfunction occurs.

Because of cost constraints, managers of all utilities, including those in the water supply business, are seeking means to cut both the operation and maintenance costs; the reduction of personnel is usually their target. Moreover, managers of water production and supply facilities are troubled by the new stringent water quality standards set by the EPA, including the requirement for frequent or possibly continuous monitoring of water quality for certain parameters.

Modern water treatment and supply control systems are commonly referred to as Supervisory Control and Data Acquisition (SCADA) systems. Although the SCADA systems have powerful features, it is important to be certain that these features are truly beneficial to the operation and maintenance of the facilities. SCADA systems that are poorly designed or those with no previous applications, such as a newly designed system, can potentially cause numerous problems and the expenditure of unnecessary amounts of money. Engineers should not be swayed by the promotional sales pitch to acquire state-of-the-art equipment because these systems may become obsolete within a period of 10 years, thus making replacement parts hard to find or even unavailable.

Another consideration is the complexity of the water treatment and supply systems. In contrast to most industrial facilities, the water system is relatively simple and the qualifications to become a plant operator are less stringent. In light of this fact, the real need for a high-level SCADA system is often questionable; the only appropriate situation would be if the facility is very large and has an exorbitant amount of data that must be processed quickly. Although the new computer-controlled SCADA technology is exciting, project engineers must avoid the potential pitfalls.

Four basic types of plant instrumentation and control systems are currently available: manual, semiautomatic, automatic, and supervisory. Their respective definitions are as follows: manual initiation of a function by the operator; manual initiation of an automatic function; the use of sensors, limit switches, timers, analytical instruments, controller, and control logic devices such as relays or programmable controllers; and remote control over plant operations while on the plant site or from locations away from the plant site.

The preliminary engineering study should evaluate all four levels of the instrumentation and control system based on the local conditions, size and complexity of the treatment and water supply systems, the management philosophy of the owner, and the anticipated financial constraints. The final recommendation should be somewhat based on owner preference. However, final selection must also be based on sound engineering judgment.

2.4.12 Preliminary Cost Estimates

The best system among the various water treatment alternatives may be selected on the basis of cost-effective construction and the costs associated with plant maintenance and operation. The preliminary cost estimate of the treatment plant is also necessary to assess the capital that must be raised to construct the plant.

The construction cost estimate is comprised of all costs associated with the construction contract, the overhead and profit of the contractor, but excluding the engineering and legal expenses. This estimate should also assume competitive bidding.

A common method in formulating preliminary cost estimates is to use the cost estimation curves developed by the EPA (*Estimating Water Treatment Costs,* EPA/600/2-79-162b, August 1979). Other methods employ construction cost estimation data, such as Figure 2.4.12-1, supplied by reputable consulting engineering

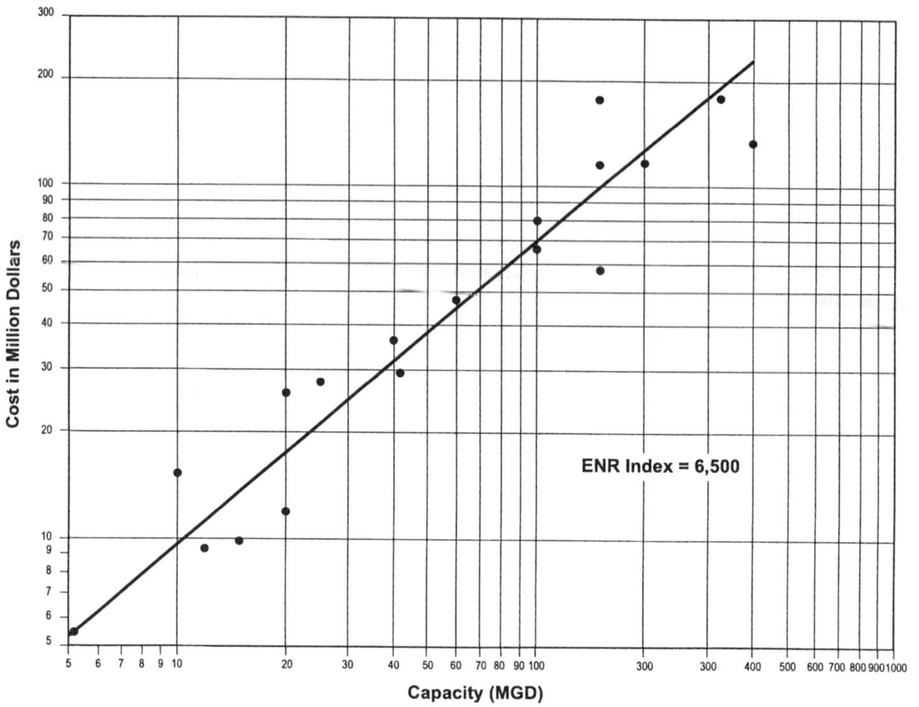

Figure 2.4.12-1 Construction cost for conventional water treatment plant.

firms. The cost figure obtained from the curves may be adjusted to a geographical area in the United States and to current standards through the application of a special cost index. The two best recognized cost indexes are the *Engineering News-Record* Construction Index and the Handy–Whitman Index of Water Utility Construction Costs. Figure 2.4.12-1 is based on the ENR Index of 6500. This index was created in 1931 and developed for 20 different cities in the United States and two Canadian cities, each representing a local geographic area. This index is designated as 100 and appears in the *Engineering News-Record,* a magazine published two to four times a year by McGraw-Hill Book Company.

It is important that engineers realize that preliminary estimates cannot account for all factors unique to the plant site. The preliminary estimates should therefore be considered as a budget estimate with an expected accuracy of approximately +30% to −15%. Cost estimates prepared for the purpose of comparing two or three alternative systems must take into account the useful life, salvage value, and annual revenue of the system, in addition to the initial construction costs. This can be achieved either by converting all costs to equivalent, uniform annual costs and income or by converting all costs and revenues to present worth at time zero. *Present worth* is defined as the yield attained at a later date of money invested at time zero × the required costs and revenues at a specified interest rate. If the cost-effectiveness of a proposed treatment system for a community is to be evaluated, the engineer must also perform a value analysis and appraise the treatment plant and related facilities (e.g., intake, low service pumping facility, and high service pumping stations). Furthermore, the location of the water distribution reservoirs and the water distribution pipe system must be analyzed.

The operation and maintenance costs of a water treatment plant typically include costs associated with (1) labor, supervision, and administration; (2) chemicals; (3) power; (4) maintenance and repair; and (5) miscellaneous supplies and services. Additional factors that affect actual operation and maintenance costs are the policy of the owner, the complexity of the system, and the local environment and weather. These costs rise on an annual basis due to continuing inflationary trends in labor, power, and equipment. Figure 2.4.12-2 is a cost estimation curve for expenses associated with operations and maintenance. The curves are established on data obtained from a basic, conventional water treatment plant processing raw water of fairly good quality with no significant buildup of scale in the pumping facilities. It is important to remember that these cost estimates are highly site specific and are only guidelines.

2.4.13 Recommended Treatment Processes

After the first 11 steps of the preliminary engineering study have been completed and thoroughly evaluated, the project engineer is ready to make a recommendation

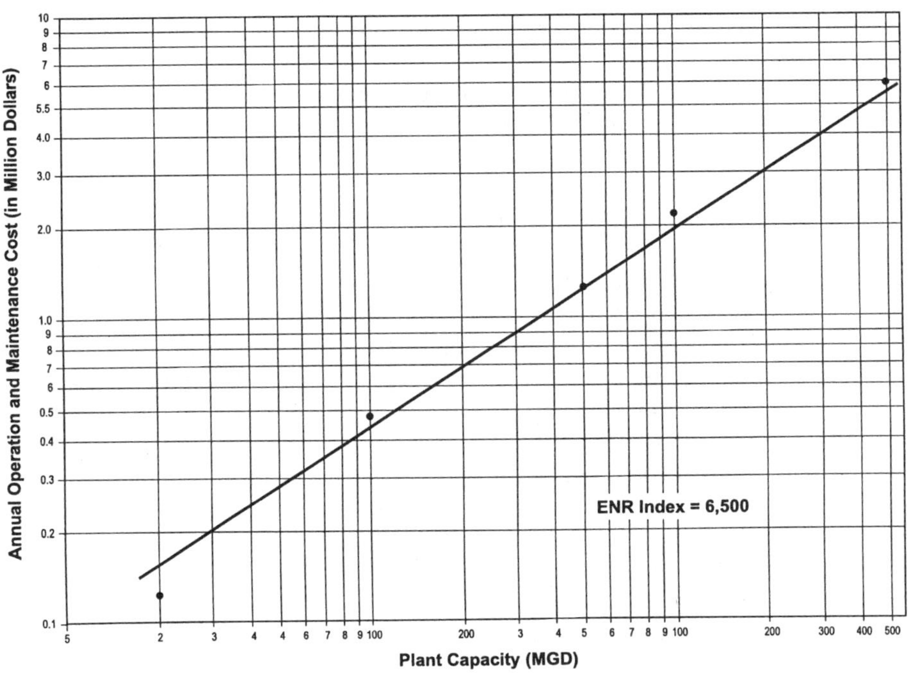

Figure 2.4.12-2 Plant operation and maintenance cost of conventional plant.

to the owner in regard to the most appropriate type of water treatment system. The steps discussed up to this point may be grouped into four basic procedures: data collection and problem formulation; synthesis and analysis of various subsystems and system alternatives; value analysis; and appraisal of the alternatives chosen. The recommended treatment system must be effective, reliable, simple in both design and operation, cost-effective, and a proven type of system with minimal operation and maintenance costs.

2.4.14 Plant Layout and Architectural Design

The establishment of the plant layout is one of the final steps of the preliminary design study. Once the alternative treatment processes are narrowed down to one or two, efforts to design the plant layout for the proposed site or sites should begin. The engineer must attempt to obtain input from qualified architects for aesthetic reasons, from superintendents of local waterworks for their operational preferences, and from the owner for his or her particular preferences on this issue. The use of a computer-aided design and drafting system (CADD) is the most effective method for conducting this task.

Although the layout is greatly influenced by the general site topography and the extent of future plant expansions, there are three basic plant layouts: cluster, satellite or college campus, and an intermediary form of the cluster and satellite plant layouts. Figures of the different plant layouts are shown in Section 3.2.2. Regardless of the layout pattern, the water should flow throughout the entire process train by means of gravity unless pumping between the unit processes is absolutely necessary.

From an engineering standpoint, there are eight important characteristics that must be considered during work on the layout. The first is the minimization of costs associated with the civil works. However, the design engineer may not delete the careful study of the site conditions, natural water course, cut-and-fill requirements, access roads to the site, the slope of these roads, and the drainage and protection of the proposed site against flooding.

The second item to be considered is the ease of construction. This item addresses issues such as providing a simple arrangement of water treatment units, providing only the minimum number of required units, allowing easy access to and between the units for construction equipment, avoiding a yard piping design that is too deep or that cross over each other, and selection of simple and uncomplicated individual structures.

Automatic, equal hydraulic loading to each tank or filter is the third consideration. This goal can only be attained through proper layout of the units. Over 50% of all existing water treatment plants have some degree of imbalance in their hydraulic loadings from one treatment unit to another, primarily as the result of improper design of the layout.

The fourth item is the centralization of control and operation. Veteran plant operators unanimously prefer this type of layout because it allows for easier and more effective plant supervision and control. Experience has shown that it is more effective to arrange subsystems requiring close supervision, such as filters, chemical feeders, and the flocculation process, in close proximity to the control/administration building rather than in a widely scattered pattern.

The fifth consideration is the physical separation of the majority unit process structures. Massive concrete structures of different depths and weight must be physically separated to prevent the formation of serious cracks due to uneven subsidence and to facilitate easy access by construction equipment for the installation and repair of the plant equipment. Physical separation is very important for plant sites that have soft ground or frequent earthquake activity. The majority of plant sites have significant differential ground subsidence between existing and newly constructed structures; this fact must be given careful consideration.

The development of a master plan for the ultimate plant layout and yard piping scheme is the sixth consideration. Engineers must allocate areas for future process units in order to minimize any interference and inconvenience to the existing treatment system and daily operational activities. From a safety standpoint, two independent means of access and two alternative roads within the site should be provided to the plant. This scheme is beneficial for both future plant expansion and the transportation of chemicals in large trailer trucks during normal plant operations.

The seventh significant consideration is the climatic conditions that exist at the plant site. For example, treatment plants located in very cold regions must house all unit processes to protect them from freezing. Consequently, the layout must be a compact scheme that avoids large separations between the major process units to minimize the size of the housing structures. If the unit processes must be separated, underground passage tunnels should be considered; it therefore follows that the process units must be arranged in a way that minimizes tunnel construction costs.

The eighth and last, but not least, consideration is architectural design. The aesthetic features of the plant layout and building and the landscaping of the plant site are superficial but very important issues. Most water treatment plants in the United States are publicly owned. They are frequently visited by public groups, schoolchildren, and engineers from other municipalities, states, or countries. Furthermore, the plant sites are often near residential areas or areas of natural preservation. Thus the proposed treatment plant may meet resistance from groups of local citizens or other organizations. The plant must therefore be functional and safe and must also blend into the surrounding environment, as well as be aesthetically pleasing when viewed from the surrounding neighborhoods. It is very important to consider spending at least 2 to 3% of the total design cost on improving the aesthetics of the plant.

2.4.15 Management Planning of the Design Phase

For the design of a water treatment system to be executed effectively and successfully, a balanced team of highly qualified specialists must be assembled. Preliminary selection of experts and management planning of the project must be accomplished at the end of the preliminary engineering study phase. One tool widely used in project management is the bar chart. However, the Program Evaluation and Review Technique (PERT), the Integrated System of Project Management (ISPM), and the

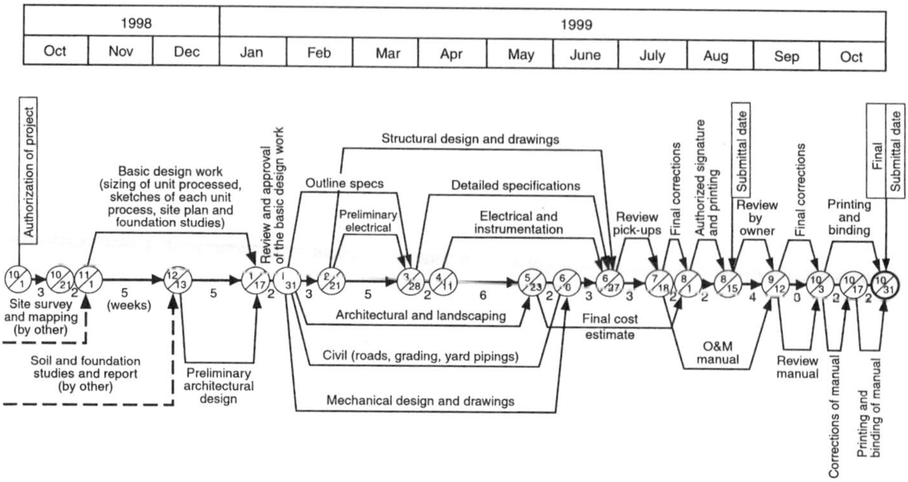

Figure 2.4.15-1 Example of project control schedule.

Critical Path Method (CPM) have also been used in the management of complex projects; these methods may also be used in project scheduling. PERT emphasizes the control phase of project management; CPM defines the duration of a project and thereby minimizes the cost, and ISPM stresses the aspect of project cost control. Figure 2.4.15-1 is an example of a project control schedule.

2.4.16 Environmental Impact Statement

In accordance with the National Environmental Protection Act of 1969, all water treatment and supply projects are required to file an environmental impact statement (EIS) prior to implementation. The EIS must include detailed studies and an analysis of the environmental impact of the facility. A nonbuild alternative is also required. Preparation of an EIS requires a team of many specialists (a biologist, a hydrologist, archeologists, and economists) to both develop and complete the studies. Projects that improve the existing environment or those with a negative impact statement, indicating no environmental impact by the proposed project, can proceed to the design stage rather quickly. However, under certain circumstances, an EIS may require a significant amount of time and money. The findings of the EIS may also have a tremendous impact on the design, construction schedule, and total cost of the project.

2.5 WATERSHED MANAGEMENT

The Surface Treatment Rule not only demands the enhancement of drinking water quality through improvements in water treatment processes and monitoring of treated water quality, but also emphasizes the protection of water sources. Municipalities supplying unfiltered water with disinfection need only demonstrate a fail-safe watershed management program in order to omit filtration. The Delaware and Catskill water supply systems of New York City are prime examples of this scenario. For water purveyors that face the prospect of adding new or upgrading existing treatment processes, a watershed protection program can play an important role in maintaining or improving raw water quality. Additionally, a robust watershed management is a very cost-effective option, as it may circumvent the need to construct expensive advanced water treatment processes.

Although surface water is primarily considered to be the main source of water, underground water, including the infiltration gallery, should be evaluated as an alternative source. The geographic conditions affecting the underground water, and possible contamination by human waste as well as livestock-farming, agricultural wastes, and industry wastes, are important issues. Additional issues that must be evaluated are potential ground subsidence and seawater intrusion into underground aquifers along the coastal areas. Both phenomena are due to the overdrafting of groundwater.

It is unfortunate that most water sources in the world have become polluted at an accelerated rate due to population explosion and the rapid development of various industries without regard for adequate waste control. In the United States and Canada, a few lakes and reservoirs (sources) have been well protected. However, urban development and other land-use activities now threaten these watersheds.

The most effective way to ensure the long-term protection of water sources is land ownership by water purveyors and cooperative public jurisdictions. However, the cost of purchasing valuable forests and highly productive agricultural lands surrounding the watershed is often too expensive for this to occur. This is especially true for smaller systems suffering financial problems due to rising operational costs and the need to comply with new regulations. Furthermore, the political pressures to develop the land may be too strong to overcome. Consequently, only a few water purveyors own or control most or all of the land within their water supply watersheds. The most practical means of meeting EPA and state requirements for high-quality drinking water is a combination of watershed control, reservoir management, and optimization of water treatment processes.

In general, there are four basic measures in achieving watershed control: general control measures, agricultural land control, forest management, and control of urban development. *General controls* are applicable to most watersheds. These measures control land use by restricting the use of reservoirs and river and reservoir buffers; in addition to land acquisition, trespass control, watershed inspection programs, plan review, public education and involvement programs, and written agreements with landowners.

Agricultural land control is the second basic measure in watershed control. Soil conservation practices encompass various practices, ranging from soil conservation of cultivated cropland, to conservation structures such as farm ponds and gully control, to animal waste control and management.

The third basic measure, *forest management,* can be difficult to manage since logging activities are generally dispersed within the watershed. Nonetheless, the management plan should include buffer strips, skid trails, and landings. Most attention during the planning, construction, and monitoring stages should be directed to the roads and skid trails since they cause the most significant ground erosion of all logging activities.

The last basic measure of watershed control is *urban development.* Strict enforcement of state public health codes and local ordinances in regard to location, design, installation, and maintenance of septic systems in urbanized watersheds is required. Local regulations should establish zoning ordinances that restrict the type, location, and intensity of development. In addition, pollution control structures such as detention basins, infiltration systems, and stormwater diversions designed to divert or reduce the flow of contaminants to the watershed should be provided.

Watershed protection plans are site specific. The success of the plans rely heavily on the support from elected officials, cooperating agencies, special interest groups, and public awareness.

BIBLIOGRAPHY

AWWA, *Water Treatment Plant Design,* American Water Works Association, New York, 1969.

AWWA and ASCE, *Water Treatment Plant Design,* 3rd ed., McGraw-Hill, New York, 1997.

Carney, M., "European Drinking Water Standards," *J. AWWA,* 83:6:48, (June 1991).

Cook, G. D., and Carlson, R. E., *Reservoir Management for Water Quality and THM Precursor Control, AWWA Research Foundation* Report, American Water Works Association, Denver, CO, 1989.

Daniel, F. B., et al., "Toxicological Studies on MX, a Disinfection By-product," *J. AWWA,* 86:3:103 (March 1994).

EPA, "National Primary Drinking Water Regulations: Interim Enhanced Surface Water Treatment: Final Rule," *Federal Register,* 40 CFR Parts 9,141 and 9,142 (December 16, 1998).

Flynn, K. C., and Williams, T., "Watershed Management," *EE&T,* pp. 36–40 (July 1994).

George, C. E ., "Effective Project Scheduling," *Water Sewage Works,* p. 386 (December 1971).

Gollntz, W. D., "Source Protection and the Small Utility," *J. AWWA,* 80:6:52 (June 1988).

Grant, E. L., *Principles of Engineering Economy,* 3rd ed., Ronald Press, New York, 1950.

Kawamura, S., *Integrated Design of Water Treatment Facilities,* Wiley, New York, 1991.

Kawamura, S., "Pilot Studies of Mixing and Settling," *AWWA Seminar Proceedings: Design of Pilot Plant Studies,* American Water Works Association, Denver, CO, 1982, p. 25.

Merritt, F. S., ed., *Standard Handbook for Civil Engineers,* 3rd ed., McGraw-Hill, New York, 1983.

Montgomery, J. M., Consulting Engineers, *Water Treatment: Principles and Design,* Wiley, New York, 1985.

Olson, R. V., "Critical Path Method of Work Scheduling," *J. AWWA,* 61:9:447 (September 1969).

Pontius, F. W., "New Horizons in Federal Regulation," *J. AWWA,* 90:3:38 (March 1998).

Robbins, R. W., et al., "Effective Watershed Management for Surface Water Supplies," *J. AWWA,* 83:12:34 (December 1991).

CHAPTER 3

Design of Basic Treatment Process Units

3.1 COMMON BASIC WATER TREATMENT PROCESSES

3.1.1 Design Procedure

Efficient and successful design work can only be accomplished through strict adherence to the standard procedures of the project as established by the organization. The key activities are as follows:

1. Assign job numbers.
2. Establish budget and project schedules (see Table 1.4.2 and Figure 2.4.15-1).
3. List all required drawings.
4. Organize a design team.
5. Allocate a budget for each discipline of the project team.
6. Firmly establish the process and process and instrumentation (P&I) diagrams during the early stage of the project.
7. Perform all necessary calculations and establish a reasonably detailed basic sketch for each process unit. These should be circulated to each of the disciplines (refer to Figure 3.1.1-1).
8. Establish the details of all subordinate plant facilities, including the chemical feed system and the waste and sludge handling system.
9. Discuss and finalize items 6 through 8 with the head of each discipline before releasing them to the design teams.
10. Begin design work based on the project schedule.
11. Establish tight coordination among the various project disciplines.
12. Maintain strict control of the overall project.
13. Edit all specifications written by other members of the team. Use the standardized specifications of the firm whenever possible.
14. Carefully review the design calculations, drawings, and specifications with the help of experts. Do not sign documents until each item has been checked and corrected.

15. Estimate the construction cost and arrange to have the construction documents produced.

The project engineer will complete the design project successfully if the following control methods are understood and followed.

1. Have a clear understanding of the project objectives and the project goal.
2. Establish a good and balanced design team.
3. Have available an adequate budget, reasonable design period, and sufficient personnel to complete the project.
4. Coordinate the various design disciplines.
5. Keep to a minimum changes to and deviations from the originally established design concept and sketches.
6. Follow the project control procedure strictly. Section 1.4 presents a detailed example of this item.
7. Both the project engineer and project manager must be knowledgeable, experienced, willing to devote their time to the project, and bring a positive attitude to the design team.

Tables 3.1.1-1 through 3.3.1-3 and Figure 3.1.1-1 may aid inexperienced engineers to grasp some of the basic design requirements and may ultimately help create realistic project control schemes. Table 3.1.1-1 provides an estimate of the total number of drawings required for certain-sized plants using a conventional complete

TABLE 3.1.1-1 Anticipated Number of Drawings for a Water Treatment Plant Design

Discipline	Percentage of Total	Number of Drawings if Total Is 200 Drawings
General	5	10
Civil	10	20
Architectural	8	16
Structural	30	60
Mechanical	20	40
Electrical and instrumentation	25	50
Landscaping	2	4

Notes: (1) The total number of drawings may vary depending on the conditions of the plant site, local weather conditions, the complexity of the design, and the plant size.
(2) Small conventional plants, those processing less than 10 mgd, may have 70 to 120 total drawings depending on the degree to which the proprietary treatment units are specified.
(3) Medium sized conventional plants (10 to 50 mgd) require anywhere from 120 to 200 drawings, but generally over 150.
(4) Large conventional plants (over 50 mgd) require 200 to 350 sheets of drawings.
(5) If an intake pumping station, high-service pumping station, or ozonation system is included in the plant design, the total number of drawings will be increased approximately by 7%, 7%, and 9%, respectively.

TABLE 3.1.1-2 Normal Design Time per Drawing

Discipline	Total Hours/Drawing (Design and Drafting)	Breakdown of the Hours	
Project engineering	19	Specifications	6 h
		Cost estimate	3 h
		Engineering	10 h
Civil	85	Engineering	45 h
		Drafting	40 h
Architectural	80	Engineering	35 h
		Drafting	45 h
Structural	70	Engineering	35 h
		Drafting	35 h
Mechanical	75	Engineering	30 h
		Drafting	45 h
Electrical and instrumentation	60	Engineering	20 h
		Drafting	40 h
Landscaping	40	Engineering	15 h
		Drafting	25 h

Notes: (1) The engineering time for all disciplines includes the hours required to check the design calculations and drawings and the time required to coordinate efforts with other disciplines of the project.

(2) If a good design efficiency is to be maintained, the weighted average hours per drawing, for the engineering and drafting phase, are 80 h.

(3) The cost of one sheet of drawing ranges from $3000 to $7000 (1989 cost) depending on the nature of the drawing, the required design time, the complexity of the drawing, and the number of corrections that must be made.

treatment process. Table 3.1.1-2 lists the normal design time and work-hours required to complete a sheet of drawings for each of the various disciplines. Table 3.1.1-3 is a general breakdown of the construction costs. Figure 3.1.1-1 is an example of basic design sketches for a unit process; this provides a basis for detailed design by each of the disciplines of the design team.

3.2 BASIC APPROACH

A water treatment plant should be designed to produce a continuous supply of safe drinking water regardless of the raw water characteristics and environmental conditions. Thus, the application of space-age technology in the process design and subsystem selection is not always the best approach. The ultimate plant design has a system and subsystem that are proven to be simple, effective, reliable, durable, and cost-effective. This basic philosophy is applicable to all circumstances and cases. Design engineers should study the following basic rules prior to design work:

1. Evaluate local conditions carefully.
2. Create a reasonably conservative design that is cost-effective to construct.

TABLE 3.1.1-3 Breakdown of Water Treatment Plant Construction Costs (Approximate)

Item	Percent
Civil work (earthwork, grading, paving, fencing)	7.0%
Yard pipings	8.0%
Landscaping and irrigation	1.0%
Operations building (chemical feed system included)	10.0%
Flocculation and sedimentation basins	17.0%
Filters	20.0%
Clearwell	8.0%
Pumping facilities	7.0%
Meter vaults (L. S.)	2.0%
Filter washwaste holding and recycling	3.0%
Sludge drying beds	2.0%
Miscellaneous items	0.3%
Chemical storage facilities	1.0%
Electrical and instrumentation works	12.0%
Testing and disinfecting works	0.2%
Move on and move off (contractors)	1.5%

Notes: (1) The table does not include the overhead and profit of the contractor; these are generally 20% of the total cost shown above.
(2) The figures above are based on a high-rate conventional process.

3. Create a simple, reliable, effective, and proven system by applying the best knowledge and skill to the design.
4. Design a plant that is easy and safe to construct as well as simple and safe to operate.
5. Allow for maximum operational flexibility and minimal operations and maintenance costs.
6. Design a plant that is aesthetically pleasing and does not have a negative impact on the environment.
7. Acquire a plant site that adequately satisfies the basic criteria and is free of all potential disasters.
8. Do not be embarrassed to obtain help from qualified associates, consultants, or specialists; an engineer should only perform services that are in his or her area of competence.
9. Be certain that the project meets all pertinent legal requirements and engineering standards.
10. Respect the wishes of the owner.

A water treatment process is an assemblage of unit processes that effectively produces an abundant supply of safe drinking water. The treatment system generally consists of four or more interrelated and compatible components. Since the unit processes must be interrelated, the operation of each component affects the perfor-

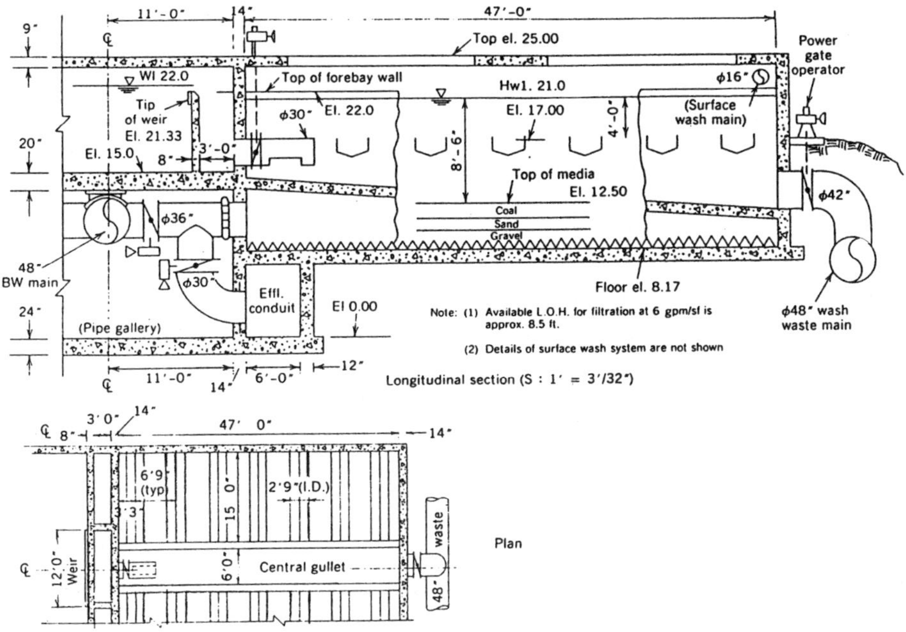

Figure 3.1.1-1 Dual-media bed alternative (16 filters) equal loading, constant-level control filters (surface wash by water jets).

mance of others, that is, the entire system. Most water treatment textbooks and design guide manuals presently on the market tend to view each component as an individual unit. Consequently, readers may erroneously believe that the most efficient overall plant performance is obtained by optimizing the design of the individual components. This type of suboptimization generally does not achieve optimum total plant design. Since the pivotal purification process is filtration, all pretreatment unit processes should be designed and operated to maximize the efficiency of the filtration process and the subsequent process of disinfection.

As mentioned previously, the design engineer should only work in his or her area of competence. The design team is usually composed of civil, structural, mechanical, electrical, and instrumentation engineers and engineers specializing in computers, soil foundation, and hydraulics, as well as an architect. The team may also include a chemical engineer if necessary. It is very important that both the project engineer and project manager be familiar with the outlines of work performed by each of the disciplines since they are responsible for coordinating the progress of the various sections on a daily basis (ideally).

Design engineers must remember to utilize commercially available components whenever feasible because this helps to keep down project costs. Components such as gates, valves, pumps, pipes, and heating and ventilation equipment, in addition to certain equipment such as flow meters, chemical feeders, mixers, sludge collection units, and instrumentation and control units, should be evaluated for substitution by commercially available units. The key to creating a good water treatment plant design is to pay attention to important details and avoid spending time on less significant items.

3.2.1 Introduction

The phrase *common water treatment processes* connotes a system that combines the processes of coagulation, flocculation, sedimentation, filtration, and disinfection with the necessary process control and instrumentation measures. Until the mid-twentieth century the design of water treatment plants was handled exclusively by civil engineers, and emphasis was placed on the civil, hydraulic, foundation, and structural aspects of design. It is therefore understandable why these early treatment plants tend to lack chemical engineering, microbiological, electrical, mechanical, and architectural considerations. Consequently, the performances of these plants do not always meet the expectations of the design and are not always aesthetically pleasing. The present trend is to have environmental engineers design the treatment processes and the overall plant design because they are familiar with a broader range of subjects.

The efficient and successful execution of a modern water treatment design is dependent on the concerted efforts of various specialists. The team members must therefore pool their knowledge and skills so that the performance and interaction of each component are designed in a complementary manner. This type of communication is essential because the design of each component affects the design of the entire treatment system.

The cost-effectiveness of the design phase may be improved by establishing a value engineering team, in addition to the design team. The project members may also opt to include both construction experts and experienced plant operators (of plants similar to the plant being designed). Figure 3.2.1-1 presents the members of a typical design team.

3.2.2 General Considerations

When designing basic treatment process units, the most important considerations are the fundamental design philosophy, the design procedures, and control over the design phase of the project; these issues are discussed in Section 3.1. However, certain

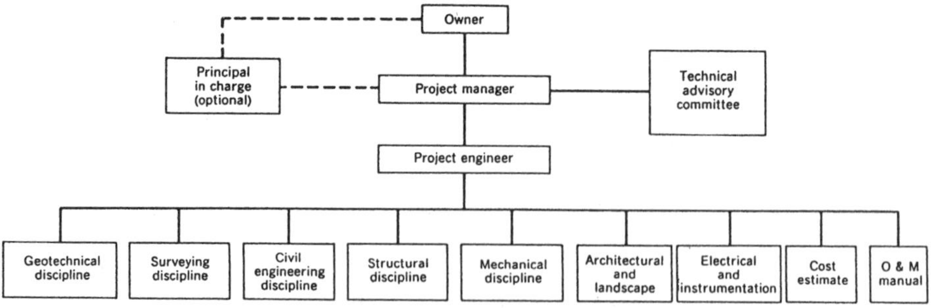

Figure 3.2.1-1 Typical water treatment system design team.

items must be established during the early stage of any design project: plant layout, process flow diagrams, hydraulic profile across the plant, process trains, and the issue of modifying and/or expanding the existing plant. The project engineer and manager are responsible for establishing these items based on some basic rules and considerations. A few of the more important general design concepts are listed below.

1. Create a master plan of the treatment plant site, including future process units and major yard piping.
2. Use the module expansion concept if the plant is to be expanded in stages. To avoid any operational confusion, each module should have the same size and elevation and ideally have identical processes.
3. Provide adequate access roads to each unit and module of the process. This will facilitate simple plant maintenance and allow easy installation of future modules.
4. Special attention must be given to the hydraulics of the plant. For example, provide equal hydraulic loading to each tank and every process unit and maintain a uniform backwash flow distribution to each filter.
5. Study the plant design to ensure that the safety of both plant operation and the environment have been addressed.
6. Research the flood level of the plant site, carefully plan the access roads to the plant site, and establish the various utility hookups.
7. Establish an adequate area for the plant site by calculating the area required by the sludge handling facilities and the area required for the future installation of new facilities, such as an ozonation system, GAC adsorption bed, and related facilities.

Plant Layout The plant layout is dictated primarily by the available area and topography of the plant site. Thus, poor site selection severely restricts the choice of alternative layouts. The design phase should only refine the preliminary layout that was established in the predesign stage. The design phase focuses on the anticipated traffic flow and operator movement, on establishing hydraulic balance among the process units and basins, on the aesthetics of the plant, and on the location of the main control building; the main control building should be situated as near as possible to the pretreatment process, filters, and chemical feed system.

Two important considerations must be addressed when working on the plant layout. The first is to provide a single chemical application point for all pretreatment process units. The second is to provide a single chemical feed point to the filter influent for all the filters and use this scheme for the combined filter water prior to the clearwell. Although special attention must be given to the yard piping and layout of the process units, the single-point application scheme is recommended because it simplifies the chemical feed system significantly and provides uniform water quality control. If the chemicals are divided to more than two points, the design will require additional feeders or flow-splitting devices such as rotameters, and the feed rate to each process unit must also be paced to the actual process flow rates.

There are two basic plant layouts, the campus type (Figure 3.2.2-1) and the cluster type (Figure 3.2.2-2). In general, the campus type layout has a few advantages over the cluster layout: (1) construction work is easier because each treatment process structure is readily accessible; (2) there is no structural damage between different unit processes of differing depth and weight due to uneven ground settlement; (3) new, advanced treatment process units are easily installed between existing processes if required; (4) civil work may potentially require less cut and fill if the site is not flat; (5) plant operators are provided with additional safety measures in case of chemical spills and/or fire; and (6) case histories show less structural damage due to earthquakes.

The campus layout also has disadvantages: (1) it has a larger plant site requirement; (2) it requires extra yard pipings, thereby yielding a higher loss of head across the entire process train; (3) plant operators must travel a greater distance between each process unit; and (4) construction costs may be slightly greater.

Regardless of the plant layout scheme, engineers are advised to provide two different access roads into the plant for safety and traffic considerations. The main access road should lead plant operators and visitors to the central control/administration building. The second access road is used primarily by chemical delivery trucks and sludge hauling and construction vehicles. Separation of the different activities is safer and aesthetically pleasing to the public.

In recent years the public has demanded safe and good drinking water, thus regulatory agencies are setting stringent standards for public water supply. In light of these changes, treatment plants need a flexible treatment process train that provides good reliable treatment. To meet these requirements, plants using water from reservoirs or lakes may consider the layout presented in Figure 3.2.2-3. This layout has two independent trains (reliable treatment) and operational flexibility. The plant may operate in four different treatment modes, depending on raw water quality: (1) in-line filtration with ozonation, (2) direct filtration with or without ozonation, (3) regular conventional treatment, and (4) regular conventional process with ozonation. The additional operational flexibilities of this layout are (1) different types of coagulant and coagulant dosages may be used for each process train, (2) each process train may have different hydraulic loading, and (3) the two process trains may have different modes of treatment. A suggested process unit layout is illustrated in Figure 3.2.2-4.

Process Flow Diagram The process flow diagram is the fundamental basis for the design work. This is especially true for the design of the instrumentation and control systems and the subsystems. A basic process flow diagram should be established during the preliminary design phase and refined at the beginning of the final design phase. The instrumentation and control specialist and the mechanical specialist may then begin their design work.

The process flow diagram should include the following items: (1) all unit processes in the correct sequence; (2) all major pipe connections with the flow directions; (3) all chemicals that are to be used and the application point(s) of each; (4) all major water sampling points that are necessary to maintain quality control; (5) the location and size of all major flow meters, valves, and connecting pipes in

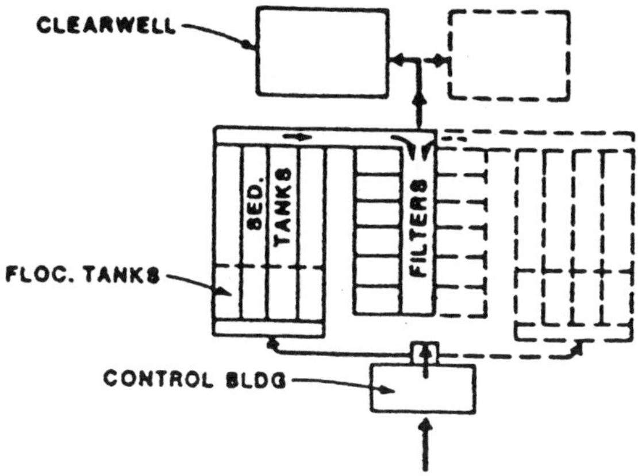

SCHEME 1

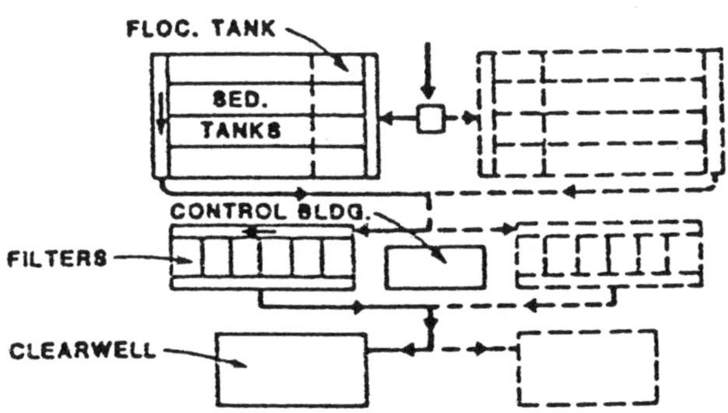

SCHEME 2

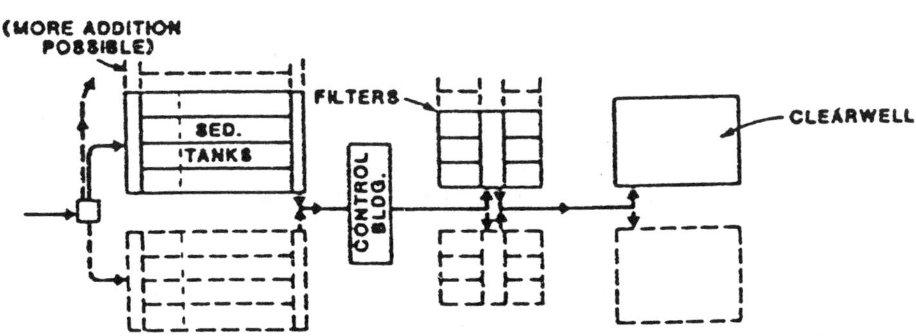

SCHEME 3

Figure 3.2.2.-1 Three basic campus plant layouts.

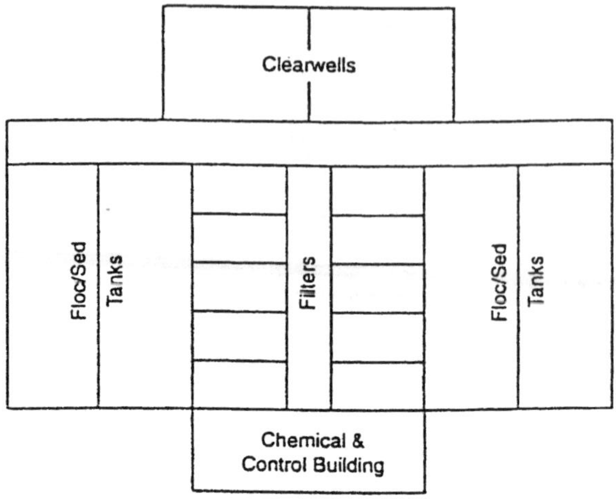

SCHEME 1

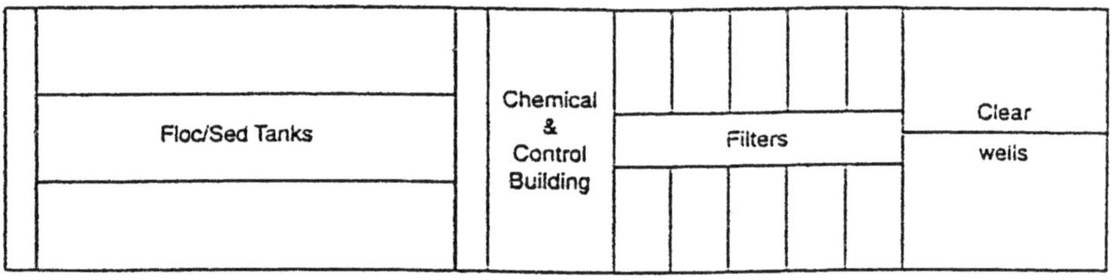

SCHEME 2

Figure 3.2.2-2 Two basic cluster plant layouts.

the process train; (6) the location of all major pumps, blowers, screens, and other such items of the process train; and (7) the control points for the pressure, water level, flow rate, and water quality of the process. All these items are essential for designing the process and instrumentation diagram (P&ID). Details of the P&ID are discussed in Section 3.2.11. Refer to Figure 3.2.2-5 for an example of a water treatment plant process diagram.

Preliminary Hydraulic Grade Line Across the Plant Once the process flow diagram is finalized, the design engineer should produce a diagram of the hydraulic grade line across the treatment process train. This diagram must show all unit processes

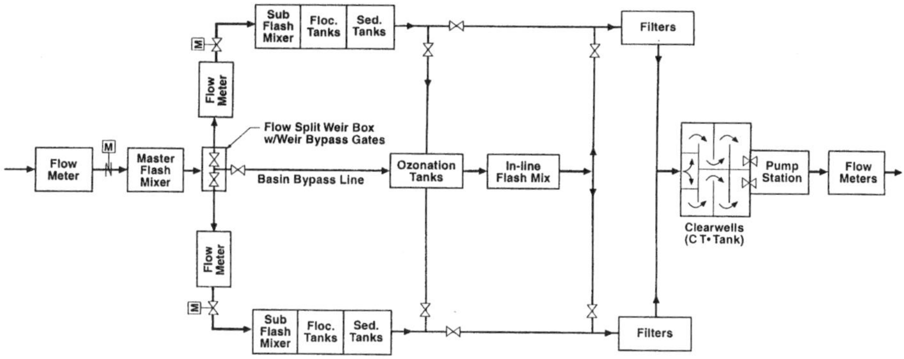

Figure 3.2.2-3 Process train scheme for maximum treatment flexibility.

in addition to the elevations of the walkway (the top of the structure), the water level, the bottom elevation, the invert, and possibly the crown levels of all connecting pipes and channels. Moreover, the finished ground level against each structure of the process unit must be shown to facilitate the work of the remaining disciplines of the design team, especially the civil, structural, and mechanical engineers. It must be emphasized that this diagram is a key component of the entire design project because it provides the basic information for the design team. Any significant modifications to this diagram during the middle or latter stages of the design period will have a negative impact on design efforts, especially with respect to the cost and deadline of the project.

Design of the Basic Processes The detailed design work of each process should begin as early in the design phase as possible. The basic design criteria of each process are based on the hydraulic loading, which should have been established during the preliminary design phase. Thus the size of each process unit and the type of major equipment are loosely established. However, at this stage all the details, including the dimensions and shape of each process, must be firmly set. Moreover, all major equipment must be selected to ensure that they will fit comfortably into each unit structure. The detailed design phase should therefore begin with the chief engineer of the mechanical discipline. This is particularly important with respect to process structures that house numerous pieces of equipment.

The process structures must be reviewed carefully because certain types of units require special foundations, a space around the unit, temperature control, dust control, complete isolation, a large water and power capacity, or an explosionproof room. All equipment, particularly the mechanical and electrical units, require servicing or may need to be replaced. Consequently, adequate spacing must be provided between the units. There have been a few cases where the valves in the filter pipe gallery were so inaccessible, due to a complicated piping scheme, that it was a major undertaking to try to service or replace them.

Design engineers should also design the unit process structures based on the largest and heaviest units available on the market because the exact manufacturer of the installed equipment will not be known until after the plant construction bid is awarded to

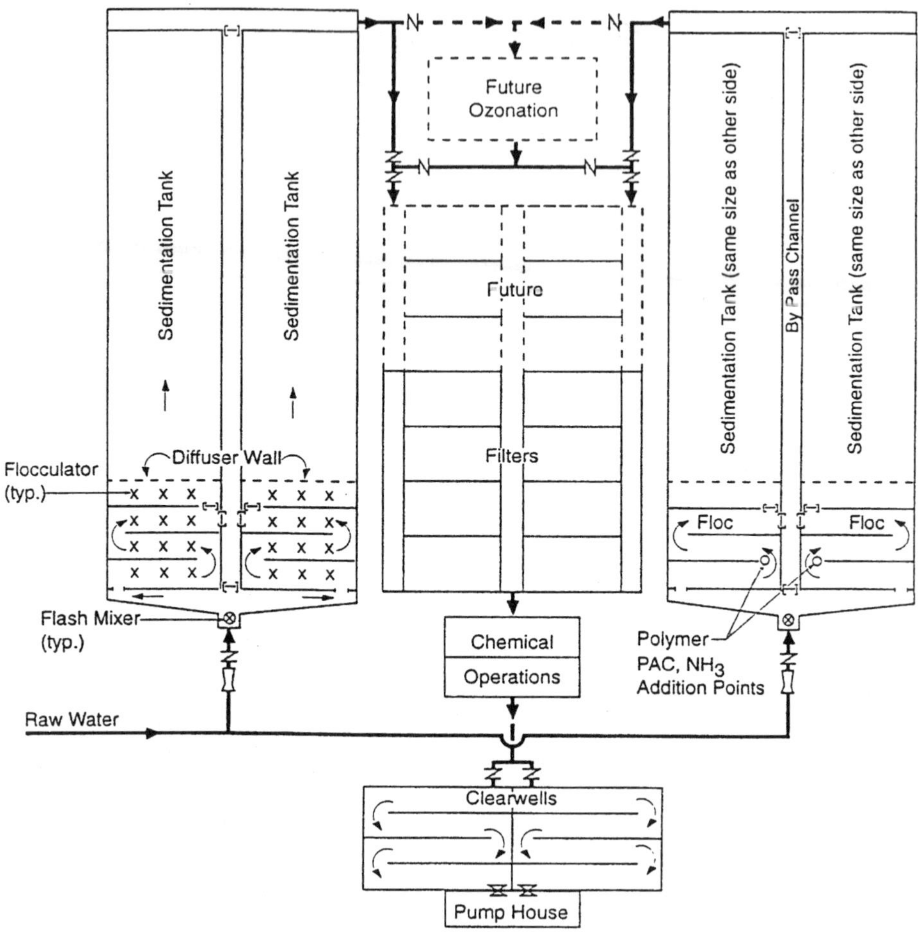

Figure 3.2.2-4 Suggested treatment process unit layout. The scheme provides (1) operational flexibility (conventional D, F, and in-line titration); (2) side-by-side comparison tests (coagulation, mixing, filtration rate, filtration rate, etc.); (3) compact arrangement and easy access to all process units; (4) future ozonation and filter addition (GAC adsorption related) are considered.

a contractor; contractors always select the most cost-effective unit that meets the design specifications. There have been several embarrassing cases where the equipment was delivered during construction but could not be installed in the structure because the door size was too small or where the replacement unit could not be installed due to the lack of adequate spacing between the equipment. These types of problems should be discovered and corrected during the checking procedure of the latter stages of the design phase. However, this requires the expenditure of more time and money. It is therefore extremely important to sort out these issues at the beginning of the design phase.

The project manager and the project engineer should also set up a good system of design documentation, including a format for the calculation sheets and filing system, and disseminate this program among the members of the design team. The

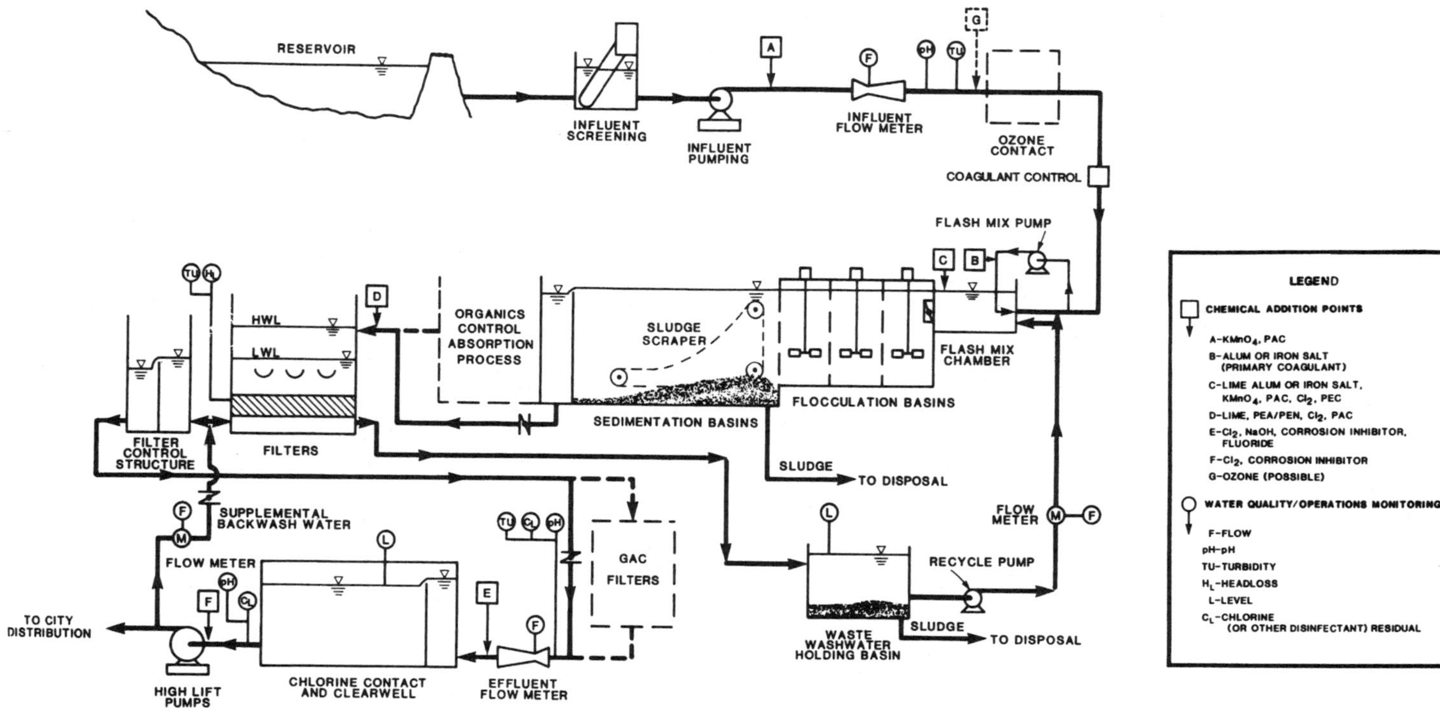

Figure 3.2.2-5 Example of a process flow diagram.

design calculations and the record of equipment selection are essential in case of an accident and/or lawsuit. The project engineer should be capable of handling all the basic design calculations and selection of the principal equipment. Yet he or she should only handle the area of his or her competence. Qualified experts should be consulted for the other areas; the project engineer must clearly convey the objectives and limitations of the system and subsystems. Table 3.2.2-1 contains some basic design criteria used to size certain unit processes and pipelines.

Modification of the Existing Plant An existing plant may be modified to expand plant capacity and/or be rehabilitated. The approach and considerations used in this situation are quite different from those used in designing a new treatment plant since the modification must be based on the original plant design. Difficult problems may arise if the original plant was designed by another firm many years ago, when governmental regulations were more lenient.

The modification of existing treatment plants is approached basically in three phases:

TABLE 3.2.2-1 Basic Design Criteria

Guide for Design Flow Rate for Waterworks	
Water treatment processes including intake facility	$Q_{\text{max-day}}$
Plant hydraulic capacity	$1.25\text{–}1.5 \times Q_{\text{max-day}}$
Clearwell capacity	Equal or larger than $0.15 \times Q_{\text{max-day}}$ or firefighting capacity (local code)
High-service pump station	$Q_{\text{max-day}}$ ($Q_{\text{max-hour}}$ in some cases)
Water distribution reservoir (in the city)	$Q_{\text{max-hour}}$

Notes: $Q_{\text{ave-day}}$ is an average annual daily flow rate.

$Q_{\text{max-day}} = 1.5 \times Q_{\text{ave-day}}$

$Q_{\text{max-hour}} = 1.5 \text{ to } 2.0 \times Q_{\text{max-day}}$

$Q_{\text{min-day}} = 0.25 \times Q_{\text{ave-day}}$

Guide for Pipe Size Selection for Waterworks	
Raw water main	6–7 ft/s
Flocculate water line	1.0–1.5 ft/s for conventional rapid sand filter with alum floc
	3.0 ft/s for direct filtration process or filters with reverse-graded media bed
Filter influent line	2.5–3.5 ft/s in general, 3.5–4.5 ft/s for polymer fed filter influent
Filter effluent line	5.0–6.0 ft/s
Filter wash-water main	8.0–9.0 ft/s
Wash-waste main	6.0 ft/s
Distribution main	6.0 ft/s
Pump suction line	4.0–6.0 ft/s
Pump discharge line	7.0–9.0 ft/s

first, evaluate the existing plant based on available documents and field trips; second, perform field testing of the raw water quality and the actual unit process efficiency, and study the local conditions; and third, evaluate and finalize the plant modification scheme based on the regulatory requirements of the local and national agencies.

The following steps must be performed when evaluating an existing plant:

1. Collect, analyze, and evaluate all available documents concerning the design, operation, and historical raw finished water quality of the plant.
2. Interview the chief operator of the plant regarding operational experiences and obtain a list of both positive and negative comments and changes that he or she would like to have implemented.
3. Obtain field measurements of the dimensions of all major process units and measure the elevations of the water surface level at selected points in the process train, at the designed flow rate of the plant, and at a different flow rate. Do not rely on the as-built drawings since the plant may have been modified beyond what the as-built drawings show.
4. Evaluate all existing major equipment, conduct an inventory of the major equipment, and evaluate their condition and utility.
5. Conduct an inventory of and evaluate the electrical, instrumentation, and control systems.
6. Inspect all major structures for structural integrity and evaluate their life expectancy.

Field testing is the next phase in modifying an existing plant. The following five steps must be performed:

1. Evaluate the treatability of the raw water by conducting bench-scale tests. Be sure to evaluate the type of coagulant, optimum coagulant dosage, optimum mixing conditions ($G \times t$), floc settling rate, and the filterability of the settled and flocculated water.
2. Evaluate the effectiveness of the existing filter washing system by obtaining core samples of the filter bed both before and after filter washing. Also analyze the actual backwash rate, effective sizes, uniformity coefficients, and specific gravities of the filter media. Furthermore, the mud deposition profile across the bed depth should be assessed.
3. Evaluate the actual flow-through time of the flocculation and sedimentation processes and the clearwell by conducting tracer tests.
4. Survey and investigate the geotechnical aspect of the locations where the additional process units will be constructed.
5. Verify the hydraulic bottlenecks and the performance of the unit processes at higher than designed flow rates by performing high-flow simulation tests on an isolated number of basins and filters.

The final phase in modifying an existing plant is evaluating the results obtained from the first two phases and establishing a plant modification scheme. Some of the more important considerations are as follows:

1. Evaluate and establish a balanced flow rate and process performance between the existing facilities and the new additions.
2. Try to use the existing filter washing system by specifying the same-size filters at the same elevation. This practice will also avoid any confusion for plant operators.
3. Check to see that the design criteria of all the unit processes meet the present requirements set by state, local, and federal agencies and those of the foreseeable future.
4. Verify that the owner's requests are incorporated in the modification scheme.
5. If any part of the old system is salvageable, try to integrate the old instrumentation and control system into the new scheme.
6. Establish a basic plant modification scheme so that the existing plant operations are minimally disrupted by the construction.

Lastly, for individuals who wish to evaluate their own plant conditions, the following steps are recommended.

1. Situation audit: addresses regulatory issues and analyzes the plant's compliance.
2. Plant capacity audit: evaluates the hydraulic loading of each unit process and analyze the actual hydraulic profile across the entire plant at the designed flow rate and also at higher plant flow rates to simulate conditions present during emergency plant operations.
3. Process facility audit: analyze the physical integrity of all process units and their conformance to all applicable regulations and codes.
4. Equipment audit: assess the adequacy and reliability of all major process equipment and auxiliary equipment under operating conditions and evaluate their compliance to all applicable codes, including safety and reliability in emergency situations.

BIBLIOGRAPHY

Kawamura, S., et al., "Improving Water Supply in Thailand," *J. AWWA,* 80:6:59 (June 1988).

Merritt, F. S., ed., *Standard Handbook for Civil Engineers,* 3rd ed., McGraw-Hill, New York, 1983.

3.2.3 Coagulation and Chemical Application Considerations

Coagulation is defined as the destabilization of charge on colloids and suspended solids, including bacteria and viruses, by a coagulant. Flash mixing is an integral part of coagulation. In this chapter we discuss the role of chemical application and flash mixing in effective water treatment.

Purpose The purpose of flash mixing is to disperse water-treating chemicals quickly and uniformly throughout the process water. Effective flash mixing is especially important when using metal coagulants such as alum and ferric chloride, since their hydrolysis occurs within a second, and subsequent adsorption to colloidal particles is almost immediate. In practical design, the dispersion of metal coagulants should therefore be completed in a fraction of a second. The time requirement for other chemicals, such as polymers (polyelectrolytes), chlorine, alkali chemicals, ozone, and potassium permanganate, is not as critical since they do not undergo hydrolytic reactions. Thus, from a practical point of view, the dispersion time of these chemicals may be completed within several seconds or less. Although flash mixing is an important unit process in water treatment, most conventional treatment plants are capable of producing quality water even with the use of a less than optimum flash mixer provided that coagulant is overfed in excess of 25 to 30%. It should be noted that flash mixing of coagulant is very different from the normal concept of liquid–liquid mixing used in chemical industries. Under normal conditions the ratio of liquid alum volume to raw water volume is 1:50,000 (in the water treatment field). Thus the instantaneous dispersal of such a small amount of alum into the raw water is a very unique situation.

A word of caution is necessary with regard to overemphasis of the coagulation theory and the ideal design. Many treatment plants with traditional complete treatment processes (which produces sweep floc), as well as reactor-clarifiers, which often lack a flash mixer, demonstrate good sedimentation and filter performance. There are also many case histories involving conventional treatment process plants that were poorly designed or constructed without flash mixing systems, that yield acceptable flocculation, sedimentation, and filter performance. Moreover, Kawamura, Amirtharajah and Mills, and Clark et al. (see the Bibliography) have found that flash mixing has little effect on settled water quality when operating under conditions of the sweep floc and enhanced coagulation mode.

Considerations The following items should be evaluated carefully when designing an effective flash mixing system: (1) type of coagulant to be employed, (2) number of chemicals to be fed and the characteristics of each, (3) local conditions, (4) raw water characteristics, (5) type of chemical diffusers, (6) available headloss for the flash mixer, (7) variations in the plant flow rate, (8) type of subsequent process, (9) cost, and (10) other miscellaneous items.

Type of Coagulant The selection of coagulant is extremely important in two respects: to set design criteria for the flash mixing system and for effective flocculation and clarification. The most commonly used coagulants are the metal salt coagulants: aluminum sulfate, ferric chloride, and ferric sulfate. Synthetic polymers such as polydiallyl dimethyl ammonium (PDADMA) and natural cationic polymers such as chitosan (made from crustacean shells) are also employed.

The difference between metal salt coagulants and cationic polymers is their hydrolytic reaction with water. Metal salts undergo hydrolysis when they are fed to the process water; yet all polymers do not. This hydrolytic reaction produces hydroxocomplexes such as $Al(H_2O)_6^{3+}$, $Fe(H_2O)_3^{3+}$, $Al(OH)^{2+}$, and $Fe(OH)^{2+}$ in a pH range found in the process trains of most water treatment plants. The formation of the hydrolytic products occurs in a very short period of time: less than 1 s. These products readily adsorb onto colloid particles and cause destabilization of their electrical charge. In turn, the hydrolytic products are quickly polymerized through hydrolytic reactions.

In summary, flash mixing or instantaneous mixing is important because (1) hydrolysis and polymerization are very fast reactions, (2) a uniform supply of coagulant and a uniform process water pH are essential for the production of hydrolytic products, and (3) adsorption of these species to colloidal particles occurs quickly. When cationic polymers are used as coagulants, instantaneous mixing is not critical because hydrolytic reactions do not occur; the rate of colloid adsorption is much slower because of their large physical size. Thus, from a practical point of view, the mixing time for polymers may be 2 to 5 s.

Number of Chemicals and Their Characteristics More than two or three chemicals are commonly fed into the flash mixer, which is located at the head of the treatment plant. The most frequently used chemicals are alum, cationic polymers, potassium permanganate, chlorine, powdered activated carbon (PAC), ammonia, lime or caustic soda, and anionic and nonionic polymers. Design engineers should evaluate the characteristics of each of these chemicals and classify them into two or three categories based on the following criteria: necessity of instantaneous mixing, potential chemical reactions that prevent the process chemicals from acting individually, and the chemical addition sequence (to the raw water) that yields the most effective results.

Proper chemical application sequence is very important because it produces remarkable results and also reduces total chemical cost. For example, when anionic polymers are fed to the process water during pinpoint floc formation, the size and weight of the alum floc are often improved dramatically. Generally, a 5- to 10-min lag time is essential for optimum floc formation. Table 3.2.3-1 illustrates this point.

Selection of an appropriate chemical application sequence is also vital in cases where the raw water does not have sufficient alkalinity. In this type of situation, alkali chemicals such as lime or caustic soda are fed to the water in conjunction with alum. The most reasonable chemical application sequence is first to lower the pH of the raw water through the addition of alum, thereby allowing divalent and trivalent aluminum hydrocomplexes to form; these complexes effectively reduce colloid charge. The next step is to adjust the pH of the water to the range of minimum aluminum solubility to allow aluminum hydroxide (floc) formation. Figure 3.2.3-1 presents the solubility equilibria of aluminum and iron hydroxides, while Figure 3.2.3-2 illustrates the steps involved in a proper coagulation and flocculation process.

Local Conditions The local conditions that should be investigated during planning of the process design are the availability of equipment and parts, the availability of services within the area, and the reliability of the power supply. Other conditions that influence the choice and design of the flash mixing system are the frequency of sandstorms and the winter temperatures. For instance, in remote areas and developing countries, the use of hydraulic jumps and in-line static mixers is the most appropriate choice. If the system is located aboveground, it should be protected by a building structure. However, if the plant is located in an area that experiences frequent sandstorms, severe winter weather, is in a quiet residential neighborhood or an area with a high rate of vandalism, and in some cases, the potential for military attack, the flash mixing system must be located in an underground vault.

TABLE 3.2.3-1 Importance of Time Delay Sequencing for Anionic Polymers as Flocculant Aids

(a). Jar Test of Surface Water for Turbidity Removal

Raw water: turbidity = 18 ntu, pH = 7.9, alkalinity = 84 mg/L

Time of Polymer Addition with Respect to Alum	Turbidity (ntu) for Alum Dosage (mg/L) of:			
	0	10	13	16
No polymer addition	18	10.5	6	4
Anionic polymer (8181) at 0.2 mg/L				
5 min before	—	—	7.5	4.5
Same time	—	—	6.5	3.5
5 min after	—	—	2.5	1.2
10 min after	—	—	1.2	0.7

(b). Jar Test of Iron Removal by Chlorine Oxidation for Well Water

Raw water: turbidity = 0.8 ntu, pH = 6.7, alkalinity = 198 mg/L
hardness = 188 mg/L, Fe = 4.4 mg/L
Chemical dosages: Cl_2 = 5 mg/L, NaOH = 15 mg/L

Time of Polymer Addition with Respect to Chlorine	Turbidity (ntu)	Iron (mg/L)	pH
No polymer addition	5	2.5	8.0
Anionic polymer (7763) at 0.2 mg/L			
30 s before	5	2.2	8.0
2.5 min after	3	1.6	8.0
5 min after	2	1.4	8.0

Raw Water Characteristics The design of the coagulation process is determined by the raw water characteristics. The most common problem associated with the process of coagulation is clogging of the chemical diffuser orifices and feed lines. This is related directly to the hardness of the process water (metal hydroxides) and, in part, to dissolved and suspended solids. If the water hardness exceeds 30 mg/L, the metal salt and alkali chemical feed lines and the diffuser orifices will become clogged within several months or less. The design of the system should therefore facilitate simple cleaning of the feed lines, injection nozzles, and orifices. Although low water temperatures have an adverse effect on chemical dispersion and reaction rates, under normal conditions this influence is minimal. However, raw water alkalinity and pH are critical factors in both coagulation and flocculation. Appropriate pH ranges for metal salt coagulants are illustrated in Figure 3.2.3-1.

Type of Chemical Diffuser The most commonly used water treatment chemicals may be categorized into three basic groups: gases, liquids, and solids. Most of these chemicals are converted into solution prior to being fed to the raw water, with the

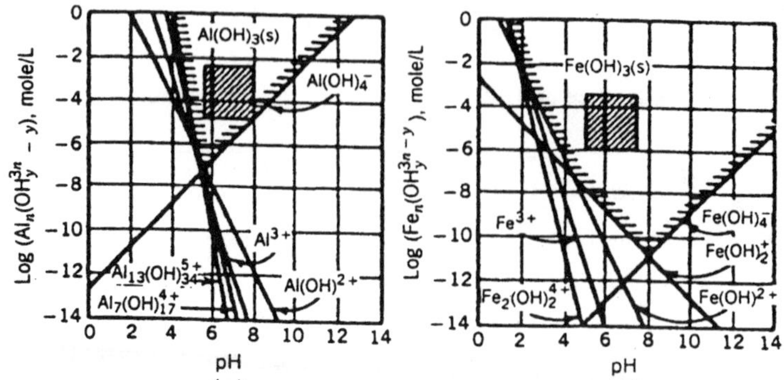

(a) (b)

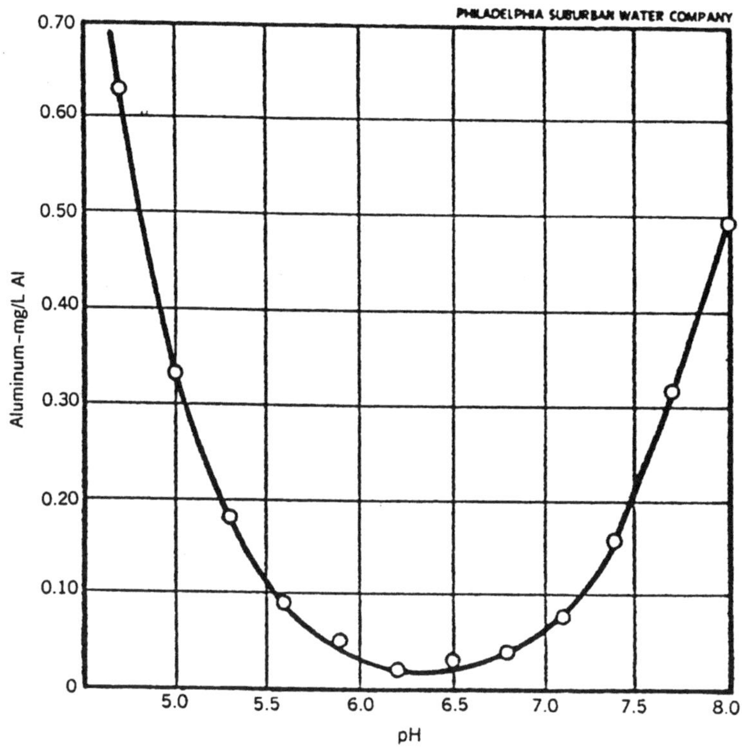

Effect of pH on solubility of alum floc
(constant alum dosage of 1.0 g.½gal(17.1 mg/L).
pH adjusted with acetic acid and sodium hydroxide
(c)

Figure 3.2.3-1 Solubility of common floc versus pH. (*a,b*) Equilibrium composition of solutions in contact with freshly precipitated $Al(OH)_3$ and $Fe(OH)_3$. Calculated using representative values for the equilibrium constants for solubility and hydrolysis equilibria. Shaded areas are approximate operating regions in water treatment practice; coagulation in these systems occurs under conditions of oversaturation with respect to the metal hydroxide. (From Stumm and O'Melia, 1968.) (*c*) Effect of pH on solubility of alum floc. [Constant alum dosage of 1.0 g/gal (17.1 mg/L). pH is adjusted with acetic acid and sodium hydroxide.]

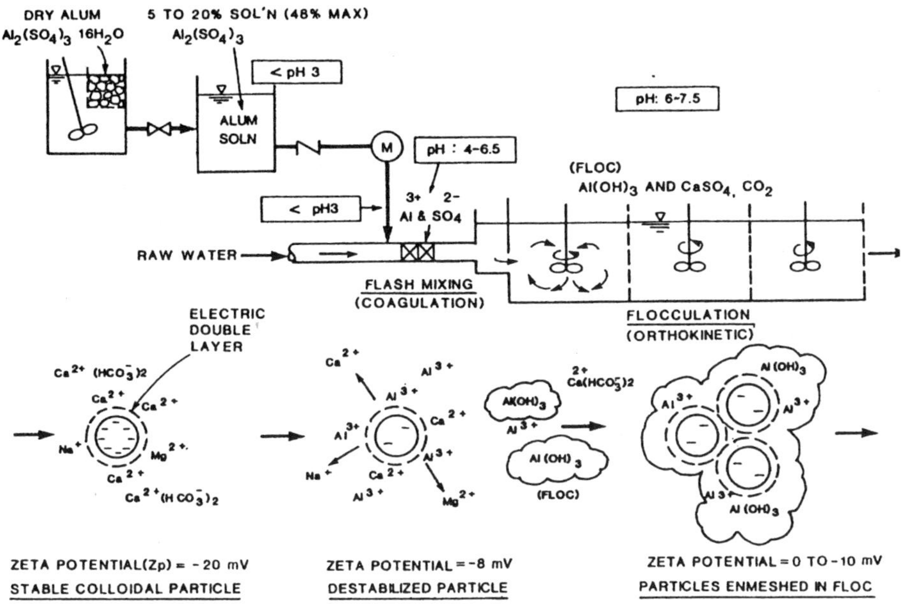

Figure 3.2.3-2 Alum coagulation and flocculation process.

exception of ozone, carbon dioxide, lime, and powdered activated carbon (PAC). Both ozone and carbon dioxide (with the exception of liquefied CO_2) are generally manufactured on site; these gases usually contain only a small percentage of the essential components and their solubility is very low. They are therefore fed to the process water through deep-channel contact tanks and are released from the tank bottom as fine bubbles to ensure efficient gas transfer. Further discussion of gas mixing is not covered in this chapter since flash mixing usually does not achieve this function effectively. Lime and PAC are generally fed to raw water in the form of a slurry. A few small plants may feed lime to the finished water as saturated calcium hydroxide water, provided that a lime saturator is used.

In summary, water treatment chemicals are fed to the flash mixing system in either of two forms, solution or slurry. For both pipe and channel flow application, the chemical diffuser (for solutions) is most often a pipe with multiple orifices. If the chemicals are in slurry form, two types of diffuser may be used: For pipe application the open end of the diffuser pipe acts as the diffuser, while channel flow application uses a trough with notches or holes at the bottom.

Special attention must be given to chemical solutions that produce scale since they can clog the diffuser orifices. Metal salt coagulants, such as alum or ferric chloride, and alkali chemicals (caustic soda, soda ash, lime-saturated solutions, and ammonia solutions) are scale-forming chemicals. When these chemicals are employed, a provision for periodic cleaning should be included in the design of the feed system; an additional diffuser must be specified so that plant operations are not interrupted when one diffuser is being cleaned. Figure 3.2.3-3 depicts the most com-

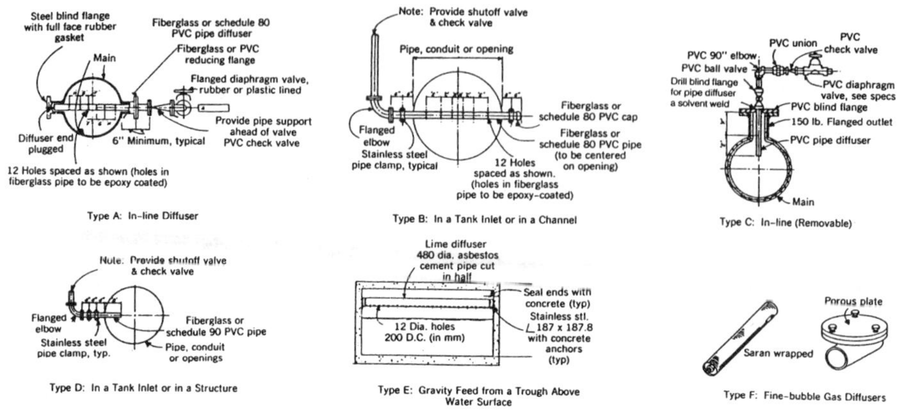

Figure 3.2.3-3 Chemical diffusers.

monly used chemical diffusers. Figure 3.2.3-4 illustrates the basic design criteria for multiorifice diffusers.

Available Headloss for the Flash Mixer Water treatment plants may occasionally have excess pressure in the raw water line. This pressure must be reduced by means of a special valve or electric power generator. When headloss can be afforded, a hydraulic mixing system such as an in-line static mixer, weir, or hydraulic jump should be considered. In any case, a headloss of approximately 2 to 3 feet (0.6 to 0.9 m) is generally required to ensure good mixing. However, if a plant has tight hydraulic conditions that cannot afford headloss, a special system such as pump diffusion should be employed. If instantaneous mixing is not absolutely necessary, a properly designed pipe diffuser system is the correct choice.

Since mechanical mixers do not produce significant headloss, they may be considered as a feasible alternative to the flash mixer. However, many studies indicate that most mechanical flash mixing systems designs are ineffective due to flow short-circuiting and frequent maintenance and repairs. In recent years, several articles have stressed the inability of mechanical mixing systems to achieve instantaneous mixing in tanks lacking the proper type of baffling.

Variations in Plant Flow Rate Variations in water demand cause fluctuations in the plant flow rate on a daily and seasonal basis. Since variations in plant flow rate can easily exceed the 1 : 2 ratio, the mixing energy of the hydraulic flash mixing system must also change in proportion to the plant flow rate. Recent studies in effective flash mixing have shown that the values for the product of mixing energy (G value expressed in seconds^{-1}) and mixing time (t in seconds) should range from approximately 300 to 1600. Hydraulic mixing can be used if $G \times t$ falls within this range. If the seasonal plant flow fluctuation is too great to satisfy the proper $G \times t$ range, the design must provide two parallel hydraulic flash mixing systems: The design criteria will be satisfied if only one of the two systems is used during a low-flow season.

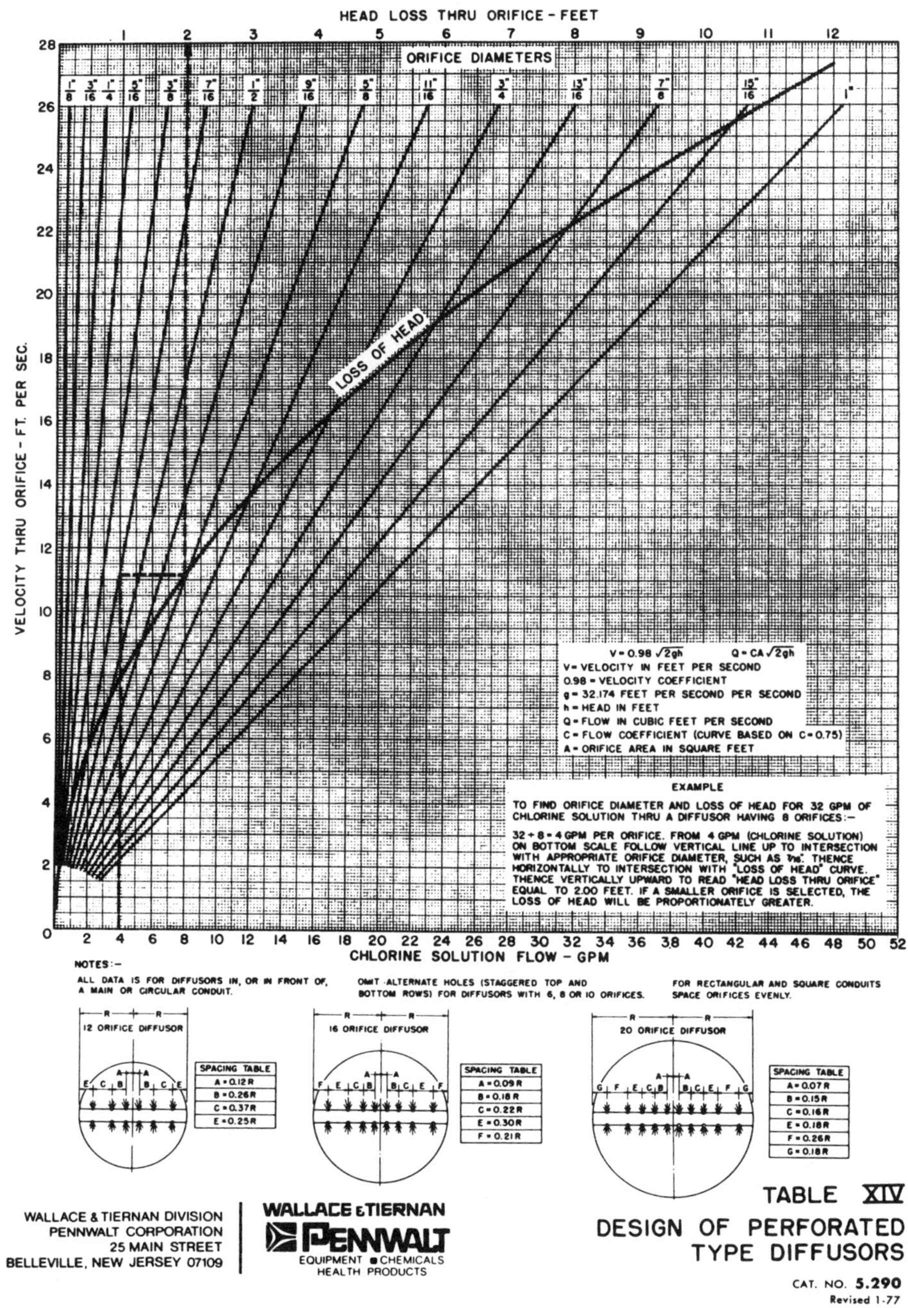

Figure 3.2.3-4 Diffuser pipe design guide.

Due to variation in mixing time, plant flow variation becomes an important design factor in other mixing systems, such as pump diffusion and mechanical mixers. The mixing energy must therefore be adjusted to satisfy the proper $G \times t$ range. The pump diffusion system uses either a flow-regulating valve or a pump with VFD (variable frequency drive) control to regulate the pump discharge flow, while mechanical flash mixers employ a variable mixing speed control unit.

Type of Subsequent Process The design engineer must carefully consider the type of treatment processes that will be subsequent to the flash mixing system since the coagulant dosage and the degree of pretreatment can be quite different. Aside from sludge conditioning and waste-wash-water reclamation, four basic process trains are used in modern water treatment: in-line filtration, direct filtration, two-stage filtration, and complete conventional treatment. Lime softening and high-level complete treatment processes, such as ozonation and granular activated carbon adsorption, are considered to be modified complete conventional treatments. Iron and manganese removal (through oxidation) may be achieved through either direct filtration or the conventional complete process train.

The in-line, direct, and two-stage filtration processes have two distinct characteristics; they have no sedimentation process and require very low levels of coagulant (2 to 6 mg/L). Inadequate flash mixing of coagulant results in higher filtered water turbidity because the small amount of coagulant fails to disperse uniformly. In contrast, the conventional complete process always includes sedimentation and thereby allows the formation of good settleable floc, provided that an adequate dosage of alum is applied (15 to 30 mg/L). Hence, if sufficient alum is fed to the process water, the role of flash mixing may not be obvious. Yet an effective flash mixing system requires 25 to 30% less alum. This reduction has been confirmed by several operational treatment plants after the installation of effective flash mixers.

If the conventional complete process is used in treating raw water with low turbidity, the benefits of the flash mixing system may not be apparent. However, if the raw water turbidity is high, use of an effective flash mixing system dramatically improves the turbidity of the settled water (Table 3.2.3-2).

Sludge conditioning and wash waste clarification must be treated differently from ordinary water treatment because of the high concentration of solids. The coagulant used in alum sludge conditioning is a polymer, not alum. Polymers do not require instant mixing. In fact, intense mixing may adversely disintegrate the existing floc structure within the sludge. Unfavorable conditions will not arise if the G value of sludge conditioning is 100 to 150 s^{-1} with a mixing time of 2 min or less.

Cost Cost-effectiveness is always an important consideration; it is particularly crucial when designing plants for developing countries. With this in mind, the flash mixing system should be reliable, effective, and easy to operate and maintain. The mechanism of the flash mixing system should be simple so that service and repair can easily be accomplished by the operational staff. Experience has shown that plant operators will not use the flash mixing system if it is overly complicated, requires frequent maintenance, or generates too much noise. Also, the operators will not use the

TABLE 3.2.3-2 Effect of Flash Mixing on Clarification

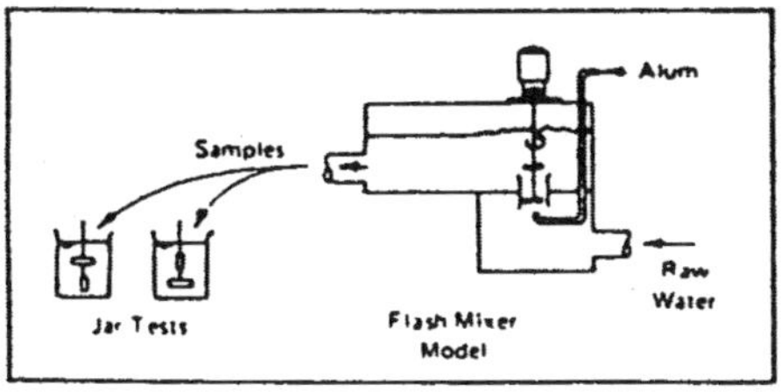

Value of the flash mixer was approximately 1,000 s^{-1}; mixing propeller rotation was set to give upward flow.

*Effect of Flash Mixing by a Bench-Type Test**

Factor	Test No.										Average
	1	2	3	4	5	6	7	8	9	10	
Quality of raw water											
Turbidity—*Jtu*	8	28	30	35	38	70	85	500	550	600	195
pH	6.7	7.0	7.1	6.9	6.9	6.8	6.8	6.8	6.6	6.5	6.8
Alkalinity—*mg/l* (as $CaCO_3$)	17	25	26	22	21	20	21	15	14	14	19.5
Alum dosage—*mg/l*	10	11	10	9	10	15	20	25	20	30	16
Time required to form visible floc—*min*											
Case 1	3.5	1.5	1.5	1.5	1.5	2.7	1.5	1.3	1	1.3	1.7
Case 2	3.5	1.5	2.5	1.5	1.5	3	2.5	2.3	3	2	2.3
Turbidity of supernatant—*Jtu*											
Case 1	1.0	1.5	2	9	7	3	1.8	12	8	6	5.1
Case 2	1.0	2	3	7	5	4	3.0	17	14	13	6.1

*Case 1 is with flash mixing and case 2 is without flash mixing.

Effect of Flash Mixing Tested on the Actual Plant—Kabutoyama Plant, Kobe, Japan*

Factor	Test 1	Test 2	Test 3
Quality of raw water			
Turbidity—*Jtu*	9.0	7.5	6.5
pH	6.6	6.65	6.75
Alkalinity—*mg/l* (as $CaCO_3$)	16.5	19	19
Alum dosage—*mg/l*	10	10	12
Turbidity of flocculating water—*Jtu*			
Flash mixer on	5.0	3.0	1.5
Flash mixer off	5.5	5.0	2.5
Turbidity of settling tank effluent—*Jtu*			
Flash mixer on	1.0	0.8	0.3
Flash mixer off	1.0	1.0	0.5

*1958—40 mgd capacity.

Source: Reprinted from *Journal American Water Works Association,* Vol. 65, No. 6 (June 1973), by permission.

flash mixer if they feel that it does not produce a discernible difference in the treatment process. For these reasons, the cost of the flash mixing system should be approximately 0.3% of the construction cost unless there are some justifiable reasons.

Miscellaneous Items Although some articles emphasize the adverse effect of backmixing during the flash mixing process, this theory is somewhat debatable. Backmixing implies the blending of chemicals to the mixture of raw water and recycled chemical-laden raw water. A typical example of backmixing is a tank with a mechanical mixer on center (for mixing alum), providing 1 to 2 min of detention time.

Operators of many plants using mechanical mixers as the flash mixing system observe no measurable benefits by operating the mixer. In fact, mechanical problems, such as noise and energy waste, and high maintenance cost are often cited as reasons not to run the mixer. Flash mixing by means of compressed air has been tried. However, this method has never become popular because there is some doubt that it can achieve instantaneous and uniform dispersion of coagulant and because of the noise generated by the compressor.

Type and Selection Guide In this section we present commonly used alternatives in the flash mixing process and guidelines for making the proper selections.

Available Alternatives Flash mixing may be achieved by means of a hydraulic, mechanical, or pump diffusion system. Figure 3.2.3-5 shows the most common types of flash mixer: mechanical mixer, hydraulic mixer, in-line mechanical mixer, in-line static mixer, diffusion mixing by pressured water jets, and miscellaneous—diffusion by pipe grid.

Selection Criteria Selection of the flash mixing system should be based on the considerations discussed earlier in this section.

Order of Preference When establishing the design of a flash mixing process, the order of preference is based on effectiveness, reliability, minimal maintenance, and cost. Thus the order of choice is as follows:

1. Diffusion mixing by pressured water jets
2. In-line static mixing
3. In-line mechanical mixing
4. Hydraulic mixing
5. Mechanical flash mixing
6. Diffusion by pipe grid

Discussion of Alternatives

Diffusion Mixing by Pressured Water Jets For various reasons, diffusion mixing by pressured water jets is the first choice among the alternative flash mixing processes. The advantage of this type of flash mixing system is that there is essentially no additional headloss by the mixer, it is very effective, it has a controllable degree of mixing, its power consumption is less than half that of a mechanical mixer system, it is cost-effective, and all parts of the system are off-the-shelf items.

The source of the pressured water may be either the plant utility water line, the discharge of the high service pumps, or pumped process water. The raw water should not be used with dilute and diffuse coagulant unless the turbidity of the raw water is continually low (less than 5 ntu) and virtually free of any suspended solids, including fish and freshwater clams. If the foregoing criteria are not met, serious nozzle clogging problems will occur unless a basket strainer is provided in the pump suction line. Ideally, settled or filtered water should be pumped. The pressure used to diffuse the chemicals must be a minimum of 10 psi (0.7 kg/cm^2).

One disadvantage of this alternative is the potential for coagulant and debris present in the pumped water to clog the nozzles. A second disadvantage is the difficulty in applying it to a system with extra large pipes or channels, for example, those having a diameter greater than 100 in. (2500 mm). In an extra large system, other means of dispersion, such as multiple nozzles or a pipe grid with injection nozzles, should be employed. To avoid clogging the injection nozzle with metal hydroxide scale, it is extremely important to ensure that the metal salt coagulants are not excessively diluted prior to injection. In general, alum solutions should not be diluted beyond 1% (5% for ferric chloride). The dilutions may be adjusted if the dilution water is very soft and if the high level of dilution does not shift the pH of the coagulant into the range of floc formation. These metal coagulants are best fed to the water jetting from the nozzle as neat solution by means of a separate feed line.

In-Line Static Mixers In-line static mixers are also known as motionless static mixers. Half a dozen different types of in-line mixers are currently available, and if selected and applied properly, they are quite effective in the coagulation process. The primary application of these units is in industry, because of their effectiveness. The advantages of these mixers are their lack of moving parts and the fact that they do not require external energy to be input into the system. However, two disadvantages exist: (1) the degree of mixing and the mixing time are a function of flow rate, and (2) the units are proprietary items and the design engineer must therefore rely on the performance claims of the manufacturer. From a practical point of view, a mixing time of 1 to 3 s and a maximum headloss of 2 to 3 ft (0.6 to 0.9 m) across the unit are normally quite acceptable for water treatment. The treatment plant design should have an intake screen upstream from the static mixer so that large debris will not obstruct the static mixer. It is also important to specify the mixing elements as removable to facilitate the necessary cleaning of debris and scale.

Hydraulic Mixing Parshall flumes, Venturi meters, and weirs are categorized as hydraulic mixing devices. When raw water flows into a plant through an open channel, the plant flow can be metered by either a Parshall or Palmer–Bowles flume since they provide hydraulic jumps and large-scale turbulence. However, if the raw water is piped to the plant, Venturi or orifice meters should be placed within the

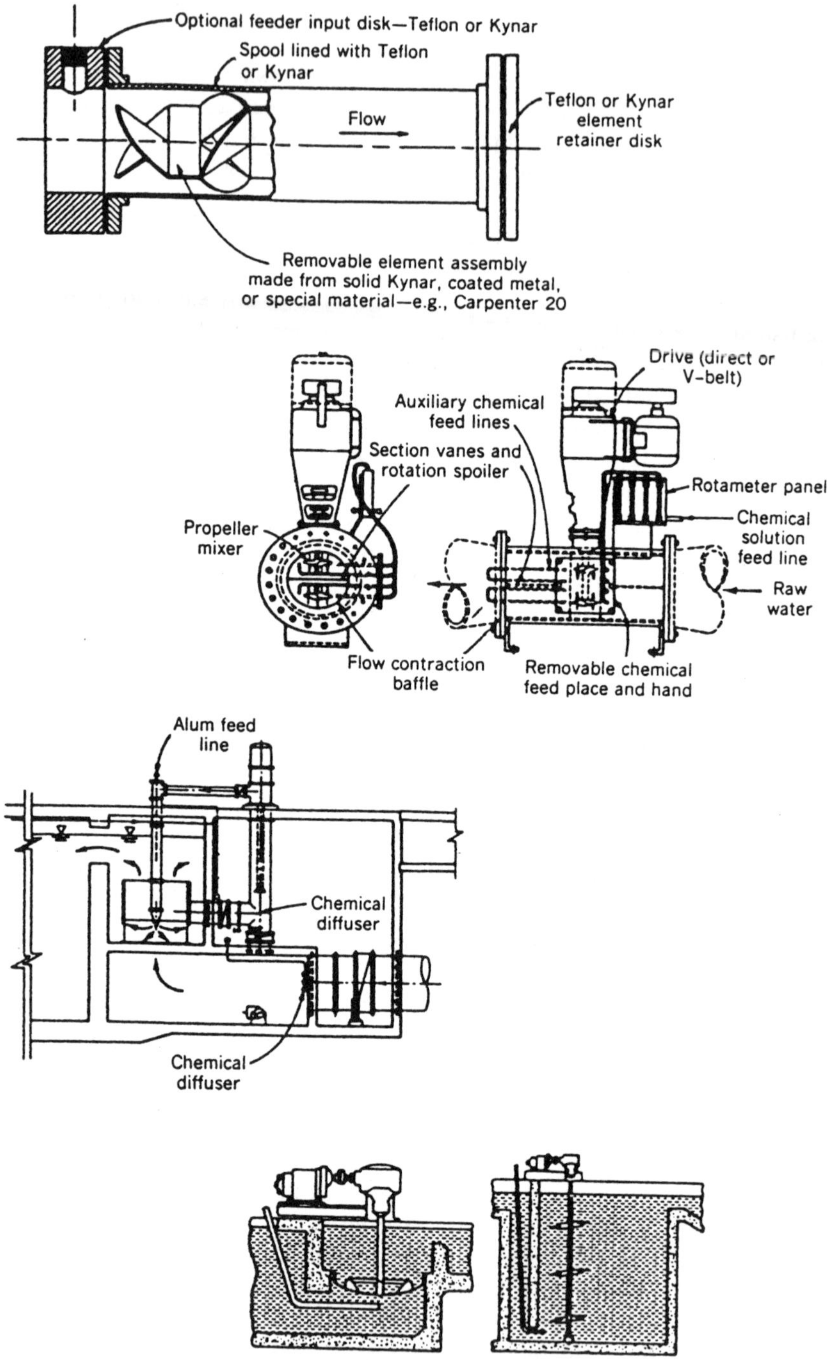

Figure 3.2.3-5 Various types of flash mixers.

influent pipeline; these meters produce differential head and result in a certain degree of mixing—more so with the orifice meter. Weirs are seldom used as flow-measuring devices in modern treatment plants; nevertheless, they also produce downstream turbulence.

These hydraulic flow-measuring devices may also be used for chemical flash mixing; they have been used in this capacity, particularly in developing countries. However, the drawback to this procedure is that the degree of turbulence is a function of plant flow rate, and thus there is no positive control over the degree of mixing.

Mechanical Flash Mixing The mechanical flash mixer is the most frequently used type of flash mixing unit in the water treatment industry. It consists of a tank or channel with one or more mechanical mixers. The common design parameters are as follows: $G = 300\ \mathrm{s}^{-1}$, a mixing time of 10 to 30 s, and a power requirement of 0.85 to 1.0 horsepower (hp) per million gallons per day. Recent studies have shown that this type of rapid mixing is not preferred for use in a continuous flow process because of (1) the lack of instantaneous mixing characteristics, (2) the production of significant flow short-circuiting, (3) a mixing period that is too long for metal salt coagulants, and (4) back-mixing, which may adversely affect coagulation. Furthermore, there is some difficulty in analyzing the various forces acting on the mixer shaft and impeller; this has resulted in shaft problems and gear drive failures in numerous installations, especially in large-scale plants. If marginal equipment is employed, it may create relatively high operation and maintenance costs and noise pollution. For the foregoing reasons, mechanical flash mixers rank low in system preference.

Diffusion by Pipe Grid This type of flash mixer depends on the wake turbulence created by the pipe grids; coagulant or other chemicals are added to the flow stream through injection orifices in the grid. Kaufman et al. (1972) have suggested that a minimum density of one orifice per square inch is required to yield satisfactory coagulation and flocculation results. However, the majority of actual installations employing this criterion experience problems with scale clogging the orifices after several months to a year of plant operation. Therefore, under normal conditions, it is not recommended that this parameter be included in plant design, except in pilot-scale installations.

Design Criteria The selection of an effective, yet trouble-free flash mixing system is not a simple task, especially when designing large-scale treatment plants. The appropriate design criteria for the flash mixer process must be selected carefully. Although the velocity gradient (G value) has been used as a design guide, it is somewhat ambiguous in determining the process because the effectiveness of coagulation is greatly affected by the following criteria: shape of the reactor tank, degree of flow short-circuiting, type of mixing element, degree of energy input, and effective mixing time. Appropriate design parameters should be based on (1) chemical diffusion rates, (2) the degree of turbulence associated with the inertial forces (characterized by the power number), and (3) the Reynolds number, which represents the viscous forces and flow characteristics. Unfortunately, most practical design criteria are not based on these parameters.

The widely accepted general design criterion for flash mixing is the velocity gradient G:

$$G = \left(\frac{P}{\mu V}\right)^{0.5}$$

where P is the power input, μ the absolute viscosity of water, and V the volume of the mixing zone. The most effective velocity gradient G (in seconds^{-1}) and mixing time t (in seconds), is

$$G \times t = 500 - 1600$$

If pump diffusion is used as the flash mixing system, the following design criteria should be used:

1. $G \times t = 400 - 1600$ (1000 average).
2. Mixing jet velocity of 20 to 25 ft/s (6 to 7.6 m/s) at the orifice.
3. If neat alum solution is diluted prior to application to the process water, the pH of the diluted alum should be lower than 3.0. If ferric salts are used as coagulant, the pH should be less than 2.0. However, dilution of metallic coagulant is not recommended unless absolutely necessary.

For in-line static mixers, the following design criteria should be used:

1. $G \times t = 350 - 1700$ (1000 average)
2. $t = 1$ to 5 s

The foregoing conditions may be satisfied with a single in-line static mixer if the plant flow rate variation is less than 50%.

Chemical Diffusers Chemical diffusers are a component of all flash mixing systems that most design engineers fail to note. Thus, many chemical diffusers become obstructed for various reasons. Metal salt coagulants, such as alum and ferric sulfate, and alkaline chemicals, such as lime, caustic soda, and ammonia, will clog small orifices within a short period of time unless the process water is extremely soft and has very low turbidity.

Selection of the proper chemical diffuser is a function of the coagulant and the type of process employed (refer to Figure 3.2.3-3 for the following discussion). For metal salt coagulants and alkaline chemicals, the correct type of chemical diffuser is generally a type D diffuser, since it minimizes the problem of clogging. Type E may also be used for aqua ammonia in cases where the area of chemical application is not easily accessible and if fumes are not allowed to escape into the environment through the installation of submerged curtain walls both upstream and downstream. Types A and B diffusers may be used if mixing is achieved solely by diffusion and not by means of positive flash mixing, since the diffusers may easily be disconnected. Types A and B diffusers may also be used with chlorine, polymers, potassium permanganate, and fluoride solutions because these chemicals usually do not produce scale at the orifices. Type E diffusers may be used with lime or PAC for channel flow application upstream of the flash mixer. Type F diffusers are used exclusively for gaseous chemicals such as ozone and carbon dioxide. Figure 3.2.3-4 illustrates the proper design of multiorifice diffusers.

Enhanced Coagulation Process Enhanced coagulation is loosely defined by the EPA as the addition of excess coagulant to improve removal of TOC. This process has been in practice for many years as a means of removing color from water containing a higher level of natural organic matter (NOM) such as humic acid and tannin. Removal of the hydrophilic NOM is attained by means of coagulation and flocculation using alum or ferric salts. It is well known that effective removal of color is achieved through the use of an optimum dosage of inorganic coagulant to make the zeta potential of the colored substance zero (the isoelectric point). The pH of isoelectric point for color removal is in the range 5.5 to 6.5 using alum as a coagulant.

As mentioned previously, the Surface Water Treatment Rule calls for enhanced coagulation to reduce the total organic carbon (TOC) of raw water under specific conditions. The isoelectric point varies depending on the nature of the colloid and the characteristics of water. Thus the pH of the isoelectric point is not necessarily 5.5, as suggested by the EPA. In general, the use of ferric chloride is more effective than alum in removing TOC, as shown in Figure 3.2.3-6. The pH of the isoelectric point is generally 6.0 or less for water treatment practice. The addition of acid to create a pH of about 6.0 (for enhanced coagulation process) often improves the efficiency of TOC removal and saves in coagulant dosage, as illustrated in Figure 3.2.3-7.

Negative aspects of enhanced coagulation include the production of very corrosive water for metals and concrete and the production of high levels of dissolved aluminum in settled water when alum is used as coagulant. Thus the pH of settled water must be corrected by alkali substance (lime or caustic soda) to prevent severe corrosion of submerged metals and concrete, and control high levels of dissolved aluminum. Compared to the conventional treatment mode, the production of two to three times more sludge is another drawback of enhanced coagulation.

Example Design Calculations

Example 1 Mechanical Flash Mixing System

Given

Plant flow rate	1.85 m^3/s (42 mgd) maximum per day
$G \times t$	1000
Mixing time	Approximately 1 s at 1.85 m^3/s
Average water temperature	20°C

Determine

(i) The size of the tank
(ii) The horsepower of the mixer and type of mixing blades
(iii) The application point of alum
(iv) The size of the inlet channel

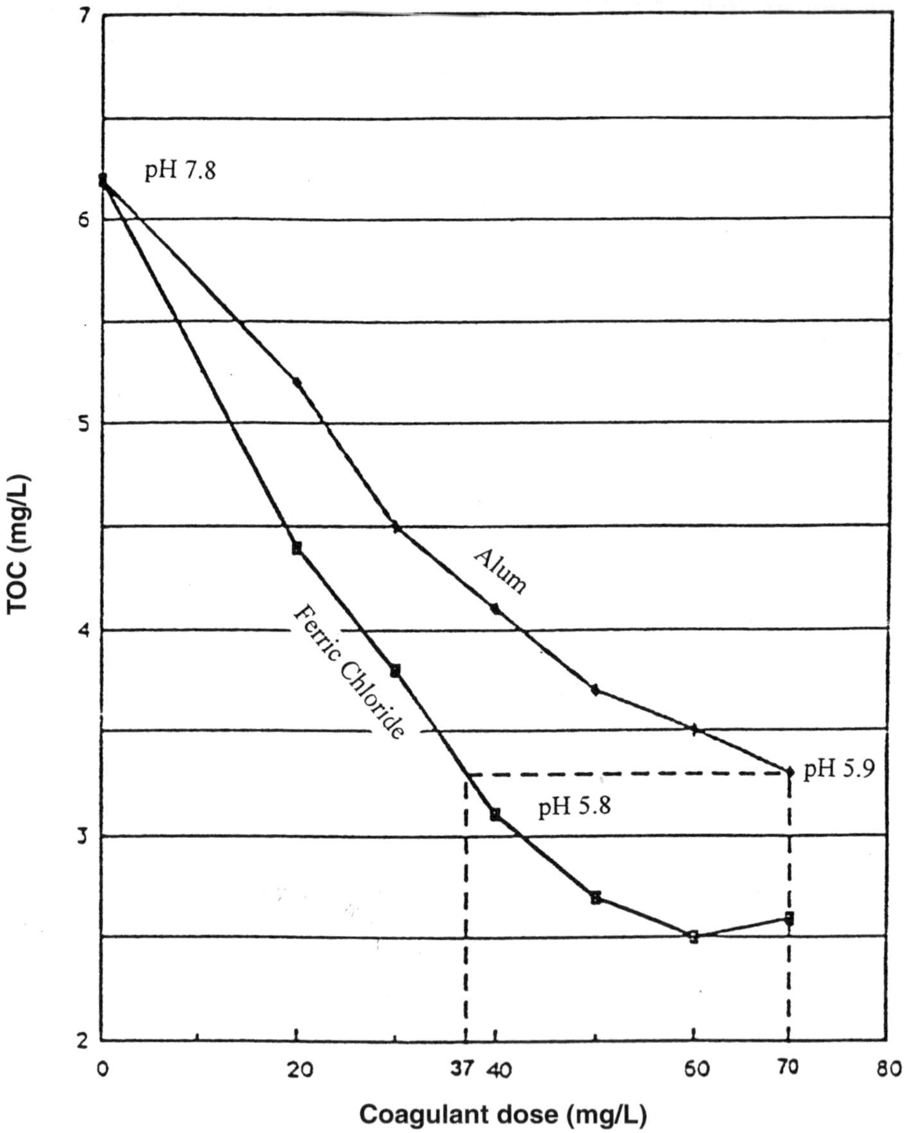

Figure 3.2.3-6 Example of enhanced coagulation jar test result.

Solution (i) Provide a flash mixing tank that is 10 ft (3 m) square and 13.5 ft (4.1 m) deep at the head of the plant. An intermediate slab containing a 4.5-ft (1.37-m)-diameter hole at midheight should be provided to create an effective flash mixing zone. Attach a 4-ft (1.22-m) length of pipe (Figure 3.2.3-8) at the opening to act as a mixing reactor tube. Flow velocity through the hole is

$$1.85 \div \frac{\pi \times (1.37)^2}{4} = 1.25 \text{ m/s}$$

and the mixing time is

$$1.22 \text{ m} \div 1.25 \text{ m/s} = 0.98 \text{ s}$$

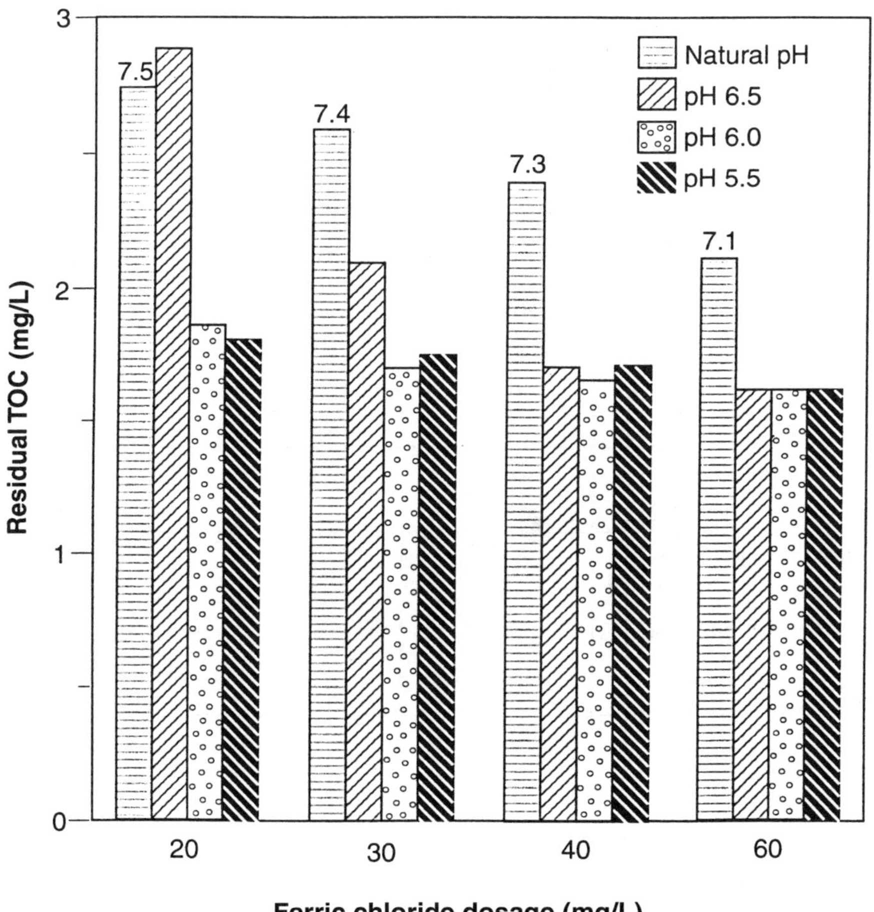

Figure 3.2.3-7 Effect of pH during coagulation process on TOC removal.

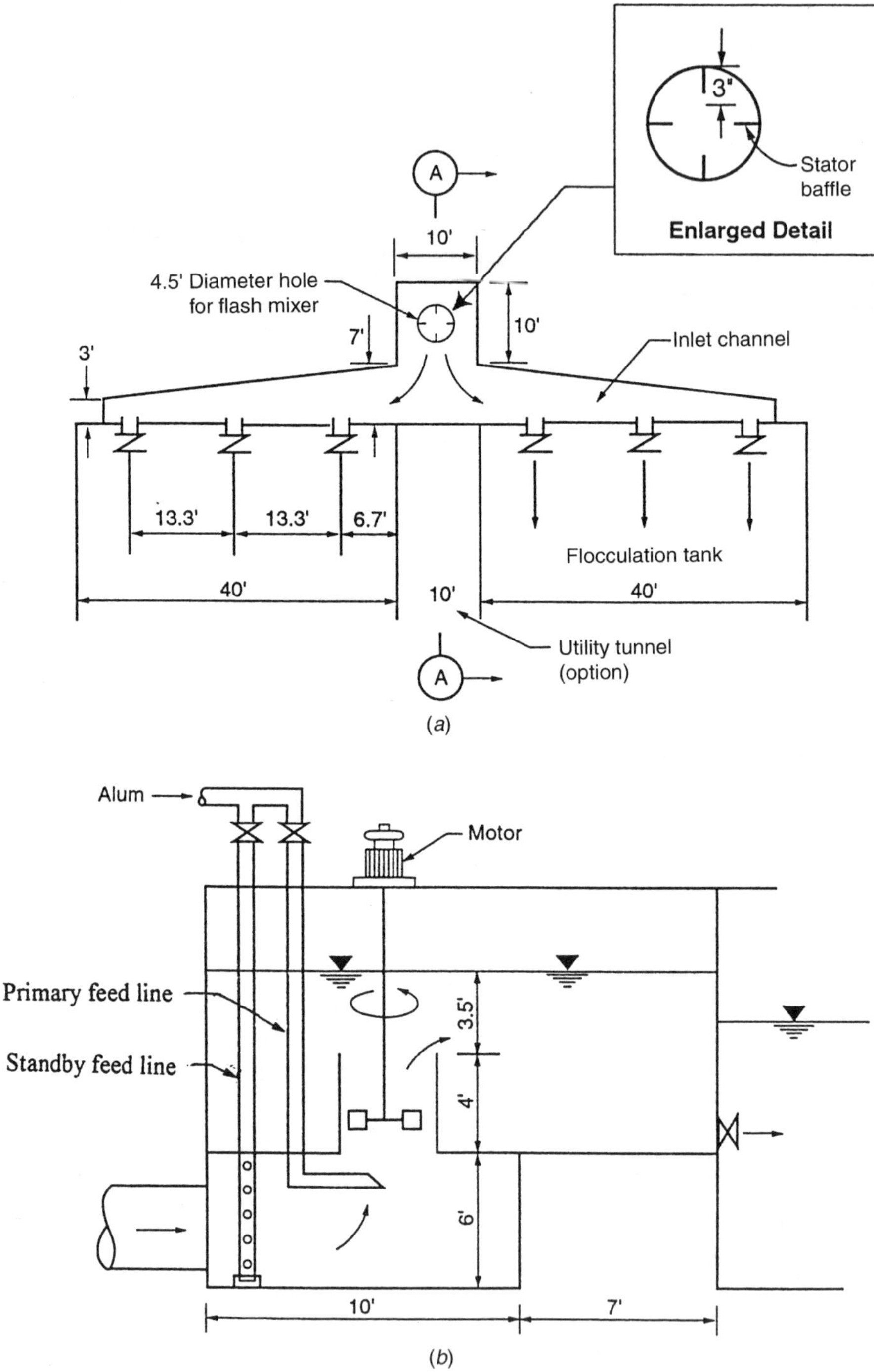

Figure 3.2.3-8 Flash mixing tank and inlet channel to the flocculation tank.

(ii) The power requirement for the mixer horsepower and mixing blades is $P = G^2V\mu$. When $G = 1000\ \text{s}^{-1}$,

$$V = \frac{\pi \times (1.37)^2}{4} \times 1.22 = 1.8\ \text{m}^3$$

Assume that the minimum water temperature is 5°C.

$$\mu = 0.0015 \cdot \text{Ns/m}^2 \text{ at } 5°\text{C}$$

$$P = (1000)^2 \times 1.8 \times (1.5 \times 10^{-3}) = 2700 \text{ joules (J)/s}$$

Since 1 kW is 1000 J/s,

$$P = 2.7 \text{ kW or } 3.6 \text{ hp}$$

Assuming an overall efficiency of 80%,

$$P = 3.6 \div 0.8 = 4.5, \qquad \text{say, 5 hp}$$

which is the minimum required mixer horsepower.

Four 45 pitched blade turbines (PBTs) with an overall impeller diameter of approximately 1.5 ft (0.76 m) should be selected for the mixing blades. The blades are located at the bottom of a 4.5-ft (1.4-m)-diameter flash mixing pipe. The interior of the mixing reactor tube must have four vertical stator baffles, each approximately 3 in. (75 mm) in width. The mixer shaft must rotate in such a manner that the water is lifted, minimizing the loss of head through the flash mixing system; shaft runout should be limited to $\frac{1}{8}$ in. (3 mm) and an underwater shaft bearing should not be provided at the end of the shaft. The details of the mixer design, including shaft rotation speed, should be referred to a reputable mixer manufacturer (based on the foregoing criteria).

(iii) The alum feed line is located below the mixing blade as shown in Figure 3.2.3-8. To prevent clogging of the feed line, employ a solution with a concentration greater than 5%. An additional alum feed line may be included at the entrance of the raw water line into the tank; this acts as an alternative application point should the main feed line become clogged or damaged. The end of the main chemical feed line has a simple 45° cut (Type D, Figure 3.2.3-3) to minimize clogging. The end of the auxiliary feed line is equipped with a diffuser pipe across the inlet pipe (Type B, Figure 3.2.3-3).

(iv) Assume that there are two flocculation tanks and that flash mixing occurs in the center of the two tanks. An influent channel is located between the flash mixer tank and the flocculation tanks as shown in Figure 3.2.3-8. To assess the size of the influent channel leading to the flocculation tanks, the rule of thumb is to maintain less than 0.3 in. (7.5 mm) of the velocity head existing within the influent channel when the system undergoes maximum flow rate. This practice ensures that the water surface elevation remains fairly constant across the entire channel. It is important to note that if one of the two flocculation tanks is isolated for maintenance or repair, a flow of 1.85 m^3/s may travel through one side of the channel.

Select a width of 7 ft (2.1 m) for the midpoint of the channel and 3 ft (0.9 m) for each end. The channels should not be designed with a width of less than 2.5 ft (0.75 m) since this causes difficulties in construction. The average water depth is assumed to be 7.5 ft (2.3 m).

	LOCATION WITHIN THE CHANNEL		
Center of Item	Entrance	Midpoint	Endpoint
Q	65 ft^3/s (1.85 m^3/s)	43 ft^3/s (1.23 m^3/s)	21.7 ft^3/s (0.62 m^3/s)
A	52.5 ft^2	37.5 ft^2	22.5 ft^2
υ	1.24 ft/s	1.15 ft/s	0.96 ft/s
$\upsilon^2/2g$	0.024 ft (7.3 mm)	0.02 ft (6.1 mm)	0.014 ft (4.3 mm)

The maximum velocity head of the channel is 0.29 in. (7.3 mm). Therefore, the rule-of-thumb criterion is met. When the two flocculation tanks are on-line, the flow velocity at the channel entrance is 0.62 ft/s and the velocity head is 0.07 in. (1.8 mm); the inlet condition to the flocculation tank is good.

Example 2 Pump Diffusion as the Flash Mixer

Given

Plant flow rate	1.5 m^3/s (34 mgd) maximum per day
$G \times t$	Approximately 1000 ($G = 750$ s^{-1})
Pipe diameter	1000 mm (40 in.)
Diffusion jet velocity	6 to 7.5 m/s (20 to 25 ft/s)
Minimum water temperature	10°C
Alum dosage	8 to 50 mg/L

Determine

(i) The mixing time
(ii) The water horsepower required
(iii) The pump capacity
(iv) The jet velocity and size of orifice required
(v) The type of nozzle
(vi) The horsepower of the pump motor
(vii) The location of the alum feed line

Solution (i) The assumed length (L) of the mixing zone is

$$L = 1.5D = 1.5 \times 1 = 1.5 \text{ m}$$

$$V = 0.785 \times D^2 \times L = 0.785 \times 1^2 \times 1.5 = 1.18 \text{ m}^3$$

The mixing time at the plant flow rate (t) is

$$t = 1.18 \text{ m}^3 \div 1.5 \text{ m}^3/\text{s} = 0.79 \text{ s}$$

(ii) The required water horsepower is found as follows:

$$P = G^2 2\mu V$$

$$\mu = 1.336 \times 10^{-3} \text{ N} \cdot \text{s/m}^2 \text{ at } 10°\text{C}$$

$$P = 750^2 \times (1.336 \times 10^{-3}) \times 1.18 = 887 \text{ J/s} = 0.89 \text{ kW} = 1.1 \text{ hp}$$

(iii) To achieve satisfactory diffusion of the primary coagulant in a short period of time, the rule of thumb is to design the pump flow rate to be 2 to 5% of the plant flow rate:

$$1.5 \text{ m}^3/\text{s} \times 0.03 = 0.045 \text{ m}^3/\text{s or } 710 \text{ gpm}$$

(iv) The velocity head of the flow issued from the blending nozzle is used in computing the mixing energy. The required mixing jet velocity is calculated by using the basic pump formula:

$$\text{kW} = \frac{9.81 \times Q \times H}{e}$$

where kW is the horsepower of the pump (in this case it is water horsepower), Q the pumping rate (m^3/s), H the total dynamic head (m), and e the efficiency—in this case 100% since the theoretical water horsepower is used to determine nozzle size. The mixing jet velocity is based on the water horsepower required to produce the given mixing energy of 0.89 kW and a pumping capacity of 45 L/s. Therefore,

$$0.89 = 9.81 \times 0.045 \times H$$

$$H = 2 \text{ m or } 6.6 \text{ ft}$$

Since the velocity head $H = \upsilon^2/2g$,

$$\upsilon = [2 \times (2 \times 9.81)]^{0.5} = 6.26 \text{ m/s or } 20 \text{ ft/s}$$

Multiple nozzles provide better efficiency of mixing. However, due to clogging problems caused by debris, use only one nozzle that sprays a full cone pattern of 90°. The diameter of the orifice is 3.75 in:

$$d = \left(\frac{Q}{0.785 \times \upsilon}\right)^{0.5} = \left(\frac{0.045}{0.785 \times 6.26}\right)^{0.5} = 0.095 \text{ m}$$

(v) Model 8HF700, a full jet nozzle, manufactured by Spraying System Company, is selected. This nozzle has 92° full cone spraying, type 316 stainless steel construction, a 3.6-in. (92-mm) nozzle diameter, a minimum free passage of 1.9 in. (48 mm) solids, 8-in. (200-mm) flange pipe connection, and a 700-gpm

(0.044-m^3/s) capacity at 7 psi (0.5 kg/cm^2) back pressure. Figure 3.2.3-9 illustrates this system and the nozzle.

We now check the energy input:

$$P(\text{hp}) = (0.97 C_d \, a v^3) \div 550$$

where P is the water horsepower, C_d the coefficient of discharge = 0.75, a the orifice area (ft^2), and v the jet velocity (ft/s).

$$P = [0.97 \times 0.75 \times 0.07 \times (22.3)^3] \div 550 = 1.02 \text{ hp}$$

Based on the G value, the required horsepower is computed to be 1.1 hp. The energy input is a little less but nonetheless acceptable.

(vi) We now determine the pump motor horsepower:

$$\text{TDH} = \text{required pressure at nozzle} + \text{exit loss of jet water} + \text{pipe friction loss}$$

NOTE: The static head for the pump is zero due to the siphon effect (see Figure 3.2.3-9), and the pipe friction loss is assumed to be 1 ft, but the actual value should be calculated. The exit loss of spray jet may be added but not included here.

$$\text{TDH} = (7 \times 2.31) + [1 \times (22.3^2 \div 64.4)] + 1 = 25 \text{ ft or } 7.6 \text{ m}$$

$$\text{BHP} = \frac{Q \times H}{3960 \times e} = \frac{700 \times 25}{3960 \times 0.85} = 5.2 \text{ hp} = 3.88 \text{ kW}$$

Therefore, use a 5-hp (3.7-kW) motor.

(vii) We now determine the location of the alum feed line.

FEED FLOW RATE OF ALUM SOLUTION

$$\text{Maximum} = 50 \text{ mg/L} \times 8.34 \times 34 \text{ mgd} = 14{,}178 \text{ lb/day}$$

$$\text{Average} = 15 \text{ mg/L} \times 8.34 \times (34 \div 1.5) = 2835 \text{ lb/day}$$

$$\text{Minimum} = 8 \text{ mg/L} \times 8.34 \times [34/(1.5 \times 4)] = 378 \text{ lb/day}$$

Assume that $Q_{max} = 1.5 \times Q_{ave}$ and $Q_{min} = 0.25 \times Q_{ave}$. Since commercial liquid alum contains 5.4 lb of dry alum per gallon, the actual feed flow rate is 2625, 525, and 70 gal/day for the maximum, average, and minimum flow rates, respectively.

Dosage	(1) Alum Feed Rate (gpm)	(2) Pumping Rate (gpm)	Col. (1) ÷ (2) × 100	Alum Solution Strength (%)
Maximum	1.82	700	0.26	0.13
Average	0.36	700	0.05	0.026
Minimum	0.05	700	0.007	0.0035

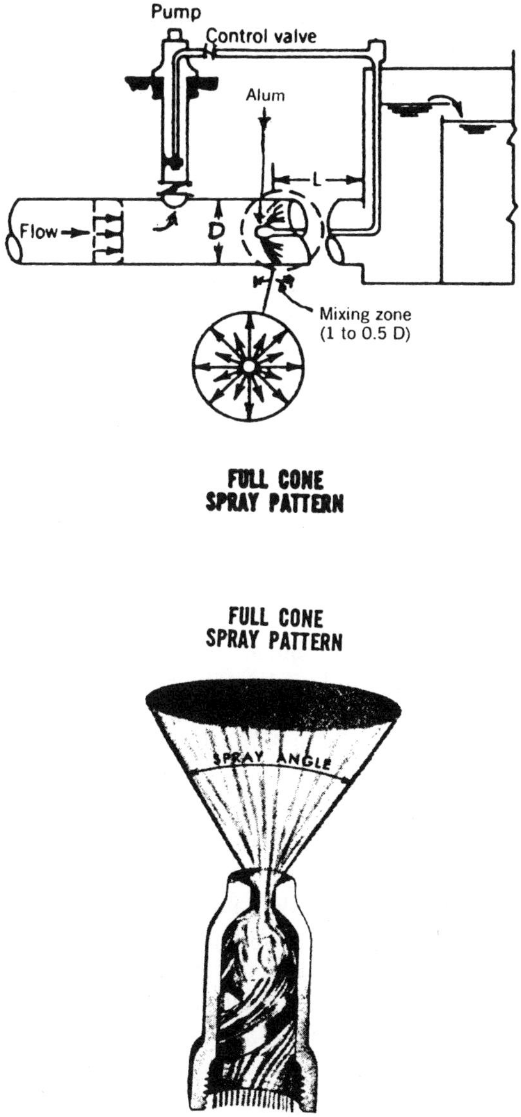

Figure 3.2.3-9 Pump diffusion flash mixing system and nozzle.

For most cases the strength of the alum solution should be greater than 1% upstream of the injection nozzle, and the pH of the solution should be less than 3.0. If these recommendations are followed, the nozzle and pipe may not become clogged by aluminum hydroxide scaling. Experience indicates that if these conditions are not met, heavy caking by alum hydroxides begins approximately 10 to 12 in. (0.25 to 0.3 m) downstream of where the alum feed line terminates into the pump discharge line. Therefore, the end of the alum feed line should be located at the flange of the injection nozzle. Alternatively, the feed line exit may be located just outside the nozzle. However, the dispersion of alum is not as effective as the first choice. Note that sludge caking in the discharge line clearly confirms that alum coagulation is completed within 0.1 seconds after application.

Example 3 Static Mixer as the Flash Mixer

Given

Plant flow rate	15 mgd (0.63 m^3/s) maximum per day
$G \times t$	Approximately 1500
Pipe diameter	24 in. (600 mm)
Maximum head loss	1.5 ft at 15 mgd
Minimum water temperature	10°C

Determine

(i) The type of in-line static mixer
(ii) The number of mixing elements
(iii) The energy input ($G \times t$)

Solution (i) Several types of in-line static mixers are available for water treatment application, but a simple unit with low-clogging characteristics should be selected. A provision for easy disassembly of the mixing baffles (for cleaning) must also be considered. Therefore, a motionless mixer, such as the one manufactured by Komax Systems, Inc., is preferred.

(ii) Under normal flow conditions the practical mixing time is limited to approximately 2 to 3 s for pipes with diameters larger than 5 ft (1.5 m). Thus, the total length of the mixing unit can be approximated by two times the flow velocity in the pipeline. The flow velocity for a plant influent line generally ranges from 6 to 8 ft/s (1.8 to 2.4 m/s). The total mixer length therefore ranges from 12 to 16 ft (3.6 to 4.9 m). A rule of thumb in estimating the length of one element is to designate the length as 1.5 to 2.5 times the pipe diameter. Based on this criterion, the length of one element is in the range 3 to 5 ft (0.9 to 1.5 m). Thus two or three elements may be used.

(iii) We now determine the energy input:

$$G = \left(\frac{P}{\mu V}\right)^{0.5} \quad \text{and} \quad P = Qwh$$

where P is the power input (lb · lb/s), Q the flow rate (ft^3/s), w the unit weight of water (62.4 lb/ft^3), and h the pressure drop (ft). If the total length of the mixing unit is assumed to be 8 ft (2.4 m), the volume of the unit is computed by the following formula: Assume a 24-in. (0.6-m) pipeline diameter:

$$[0.785 \times (2)^2] \times 8 = 25.12 \text{ ft}^3$$

Based on the manufacturer's information, a static mixer with two mixing elements would give the following results:

Flow Rate (mgd)	Head Loss (ft)	G at 10°C (s^{-1})	t (s)	$G \times t$
5	0.17	346	3.2	1107
10	1.34	1340	1.6	2144
15	3.0	2520	1.1	2770

However, the desired $G \times t$ value should be below 2000 at the maximum plant flow rate and approximately 1300 with an average flow rate. For this reason, the pressure drop characteristic of the element should be based on the desired $G \times t$ values; the manufacturer can reduce the pressure drop across each element by removing a portion of the baffle plates. The following table demonstrates how the desired pressure drop is achieved.

Flow Rate (mgd)	Approximate Headloss Across Mixer (ft)	G at 10°C (s^{-1})	t (s)	$G \times t$
5	0.028	140	3.2	450
10	0.46	806	1.6	1290
15	1.5	1782	1.1	1960

Example 4 Chemical Application

Given The following water treatment chemicals are selected to be fed at the head of the plant: alum, cationic polymers, anionic polymers, nonionic polymers, sodium hydroxide, chlorine, ammonia, PAC, and potassium permanganate.

Determine

(i) The chemical application sequence
(ii) The number of potential application points
(iii) The type of mixing system

Solution (i) Chemicals that potentially interact if fed together are alum and anionic polymers, alum and sodium hydroxide, cationic and anionic polymers, chlorine and PAC, ammonia with chlorine and alum, and chlorine with certain cationic polymers. Alum requires instantaneous dispersion. PAC and potassium permanganate must be fed ahead of the coagulant to enhance their effectiveness.

(ii) Based on the various considerations discussed earlier, the scheme shown in Figure 3.2.3-10 is established. Ideally, each application point should be separated

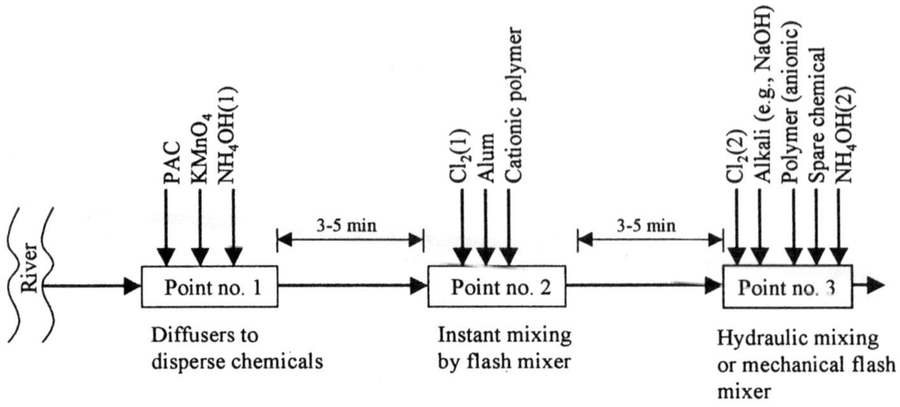

Figure 3.2.3-10 Scheme for Example 4.

by a time lag of at least 5 min, but this is often difficult to attain. The first application point should therefore be as far upstream from the plant as practical. The second point is located at the head of the plant. It is also difficult to place the third application point so that it meets the time lag criterion; thus, it should be situated as far as possible from the second point. Anionic and nonionic polymers are good aids in floc formation if added when pinpoint flocs are formed by alum. This process generally requires 5 min. For this reason, the application points for anionic and nonionic polymers are best located in the second stage of flocculation.

(iii) As illustrated in Figure 3.2.3-10, there are three application points:

- *Point 1:* dispersion of chemical by diffusers
- *Point 2:* instant mixing by static mixers or pump diffusion
- *Point 3:* mixing by mechanical or hydraulic mixer

Operation and Maintenance The most important operation and maintenance aspect of the initial flash mixing process is the selection of the proper chemicals and the application of the correct quantities to an ever-changing raw water quality and plant flow rate. Jar tests have been used effectively in optimizing coagulant dosage for conventional processes. However, they do not appear to supply the plant operator with the proper coagulant dosage for direct filtration and in-line filtration processes—hence the development of glass tube pilot filters. Although the glass tube pilot filter is widely used, approximately 50% of plants equipped with this coagulant control system do not use it because the system requires a great deal of operator judgment and attention.

Zeta-potential measurements have also been tried, with some success. In recent years, streaming current detector (SCD) units have improved on the zeta potential measurement and appear to be successful in coagulant control. SCD units can be used for any type of water treatment process using coagulant. Yet it still requires a significant amount of operator attention.

The second topic that must be stressed is the constant monitoring of the chemical feed and flash mixing systems. The most common water treatment problems arise from the clogging of chemical feed lines and diffusers. Malfunctioning of mechanical mixers can easily be detected by the plant operator. However, clogged static mixers, broken underwater chemical feed lines, and clogged diffuser orifices are not as easily detected unless the water is drained. Thus, scheduled plant shutdowns must be exercised during periods of low water demand.

The third consideration in proper operation and maintenance is adjustment of the mixing energy. Mechanical mixers and pump diffusion flash mixers must have their mixing energy altered with seasonal changes in raw water quality. Cold water is more viscous and theoretically requires a higher energy input to achieve the same mixing effect as obtained during summer months. Extra power is also required to disperse coagulant when processing raw water that is highly turbid or colored.

Treatment plants that are equipped with mechanical flash mixers should have the following items inspected on a regular basis: the motor and speed reducer, mixer shaft runout, vibrations, and noises. The lubrication oil should also be changed at certain time intervals, or by seasons, based on the recommendations of the manufacturer. It must be emphasized that coagulation, flocculation, and clarification processes are pretreatments to filtration. Plant operators generally tend to produce large flocs and attempt to make the settled water as clear as possible by applying rather large amounts of coagulant. Consequently, the effect of the flash mixer is often undetected and therefore turned off.

The role of the flash mixer may or may not be evident, depending on the type of process. The conventional water treatment process uses a large amount of coagulant that is more readily dispersed, forming large flocs and clear settled water. However, most modern high-rate filtration processes use 20 to 30% less alum than the conventional dosage; thus, vigorous flash mixing is essential for instant dispersion. Since direct filtration processes use such small amounts of alum (4 to 6 mg/L) compared to the conventional processes, they produce much smaller floc and high settled water turbidity. Most modern high-rate filters are capable of filtering settled water with turbidity greater than 1 to 2 ntu, and in many cases 3 to 6 ntu is considered to be optimum. The primary objective of flash mixing is to effectively coagulate colloidal matter present in raw water by applying the proper amount of coagulant through good flash mixing prior to the flocculation process. This is particularly important for direct and in-line filtration processes.

Coagulation Control Test The jar test is the most widely practiced and proven method of evaluating, as well as controlling, the coagulation, flocculation, and clarification processes. Although widely used, jar tests provide the operator with little information as to the mode of filtration (direct or in-line filtration) or whether the tests are being performed improperly. When designed and performed correctly, jar tests yield very meaningful data that can assist engineers in designing new treatment plants or in modifying existing plants. Plant operators also benefit from jar tests since they will be able to optimize the efficiency of the treatment process. The jar test procedure recommended is presented in Appendix 10.

Jar tests have traditionally been performed under fixed test conditions to determine the relation between coagulant dosage and settled water turbidity. However, it is very important to evaluate the filterability of the settled water by filter paper filtration (simulated filter) since water with the lowest turbidity is not necessarily the best filter influent with respect to filter efficiency.

There are five basic objectives for the jar test: (1) optimization of coagulant, (2) optimization of the chemical application sequence, (3) optimization of the mixing energy and time ($G \times t$), (4) evaluation of clarifier and filtration performance, and (5) evaluation of corrosive characteristics of settled water.

OPTIMIZATION OF COAGULANT To optimize the coagulation process, operators must choose the most appropriate type of coagulant (alum, ferric chloride, cationic polymers) and evaluate the optimum dosage under standard mixing conditions (to be described later). Coagulant solution strengths should also be considered since they may affect the decision. For example, by using a combination of alum and cationic polymers, the amount of sludge production is often reduced. However, cationic polymers cost 6 to 10 times more than alum, and the polymer dosage is therefore usually limited to 2 mg/L as a matter of economics.

OPTIMIZATION OF THE CHEMICAL APPLICATION SEQUENCE Optimization of the chemical application sequence is a very important consideration, yet little attention has been given to this matter. For instance, the sequence of application is extremely important when using alkali chemicals and anionic or nonionic polymers (in conjunction with alum) as flocculation aids. As described earlier, these polymers help produce excellent floc only if they are fed a few minutes after alum is added to the process water.

OPTIMIZATION OF MIXING ENERGY, MIXING PATTERN, AND MIXING TIME The optimum mixing condition depends on the type of overall treatment process train, the type of filter bed of the plant, and the raw water characteristics. In general, a tapered mixing—the use of high-intensity mixing at the beginning, followed by a gradual reduction toward the end—helps promote better flocculation under any condition. Under actual plant conditions, it is common for the conventional treatment process to have a velocity gradient path of 50 s^{-1} at the beginning and 10 s^{-1} at the end of 30 to 40 min of mixing: $G \times t = 4 \times 10^4$ to 5×10^4. For the direct filtration process, the accepted range for the velocity gradient is from 65 to 15 s^{-1}; for 15 to 30 min of mixing, $G \times t = 4 \times 10^4$ to 6×10^4. These ranges should be kept in mind when conducting the jar test. Due to excessive shear force, flocculation times that are too long (over 45 min) generally produce floc types that are less than optimum for settling.

EVALUATION OF CLARIFIER AND FILTRATION PERFORMANCE The design criteria for new sedimentation tanks or the performance of existing clarifier units may be evaluated by measuring the floc settling velocity through use of the jar test. A settling velocity of 40 mm/min (1.5 in./min), which is the maximum loading rate of ordinary sedimentation tanks, for alum floc during the summer months. If the settling velocity of alum floc is much slower than the corresponding surface loading rate of the existing clarifier unit, the floc characteristics may be improved by either adding anionic polymer (0.15 to 0.3 mg/L) to the alum flocculation process or by substituting ferric chloride for alum; these must be evaluated by a series of jar tests. Refer to Camp (1946), Hudson (1981), and Singley (1981) for discussions of floc settling velocity measurement.

EVALUATION OF CORROSIVE CHARACTERISTICS OF SETTLED WATER It is important to evaluate the corrosivity of the settled water. Regular inorganic coagulant will reduce the pH of the settled water, thereby decreasing its alkalinity. Filter performance may be evaluated by passing the supernatant through Whatman No. 1 filter paper. The best filter pretreatment is one that results in filtrate turbidity at or below the turbidity goal (0.1 to 0.2 ntu) within the shortest filtration time.

As a rule of thumb, the following items should be recorded during the jar test:

1. Time at which pinpoint floc appears
2. Size of the floc at the end of the flocculation period
3. Turbidity, color (optional), pH, alkalinity, and temperature of the raw water and of selected settled water
4. Turbidity of filtered water and the time required to filter 50 mL of selected settled water

Figure A10-2 in Appendix 10 presents a scale for floc size identification. Table A10-1 is an example of jar test results.

BIBLIOGRAPHY

Amirtharajah, A., and Mills, K. M., "Rapid Mixing Design for Mechanism of Alum Coagulation," *J. AWWA,* 74:4:210 (April 1982).

ASCE, AWWA, and CSSE, *Water Treatment Plant Design,* American Water Works Association, New York, 1969, 1990, 1997.

AWWA Research Foundation Report, *Selection and Design of Mixing Process for Coagulation,* American Water Works Association, Denver, CO, 1994.

Camp, T. R., "Sedimentation and the Design of Settling Tanks," *ASCE Trans.,* 3:895 (1946).

Chao, J. L., and Stone, B. G., "Initial Mixing by Jet Injection Blending," *J. AWWA,* 71:10:570 (October 1979).

Clark, M. M., et al., *Selection and Design of Mixing Process for Coagulation,* AWWA Research Foundation, American Water Works Association, Denver, CO, 1994.

Culp, W., and Culp, R. L., *Handbook of Public Water Systems,* Van Nostrand Reinhold, New York, 1986.

Fair, G. M., et al., *Elements of Water Supply and Disposal,* Wiley, New York, 1971.

Guven, O., and Benefield, L., "The Design of In-Line Jet Injection Blenders," *J. AWWA,* 75:7:357 (July 1983).

Hudson, H. E. Jr., *Water Clarification Process,* Van Nostrand Reinhold, New York, 1981.

Kaufman, W. J., et al., *Initial Mixing and Coagulation Process,* Report 72-2, University of California, Berkeley, CA, February 1972.

Kawamura, S., "Coagulation Considerations," *J. AWWA,* 65:6:417 (June 1973).

Kawamura, S., "Considerations for Improving Flocculation," *J. AWWA,* 68:6:328 (June 1976).

Kawamura, S., "Optimization of Basic Water Treatment Processes—Design and Operation: Coagulation and Flocculation," *J. Water SRT-Aqua,* 45:1:35–47 (1996).

Montgomery, J. M., Consulting Engineers, *Water Treatment: Principles and Design,* Wiley, New York, 1985.

Sanks, R. L., *Water Treatment Plant Design for Practicing Engineers,* Ann Arbor Science, Ann Arbor, MI, 1979.

Singley, J. E., "Coagulation Control Using Jar Tests," *AWWA Proceedings,* No. 20155, American Water Works Association, Denver, CO, 1981.

Stumm, W., and O'Melia, C. R., "Stoichiometry of Coagulation," *J. AWWA,* 60:5:514 (May 1968).

Weber, W. J., *Physicochemical Processes,* Wiley-Interscience, New York, 1972.

3.2.4 Flocculation Process

Purpose Flocculation is the gentle mixing phase that follows the rapid dispersion of coagulant by the flash mixing unit. Its purpose is to accelerate the rate of particle collisions, causing the agglomeration of electrolytically destabilized colloidal particles into settleable and filterable sizes.

The terms *coagulation* and *flocculation* are sometimes used interchangeably in technical literature. However, the aggregation of particulate material is actually a two-step process. The initial step involves the addition of coagulant, to reduce or eliminate the interparticulate forces responsible for the stability of the particulates. This process is called *coagulation.* The subsequent particulate collisions and enmeshment of particulates into flocs occur as the result of molecular motion and the physical mixing of the liquid. This is *flocculation.*

Considerations The following factors must be taken into consideration when designing a proper flocculation process: (1) raw water quality and flocculation characteristics, (2) treatment process and finished water quality goals, (3) available hydraulic headloss and plant flow variations, (4) choice of flocculation tank inlets, (5) local conditions, (6) cost, (7) relation to existing treatment facilities, and (8) miscellaneous items. These items are discussed in detail.

Raw Water Quality and Flocculation Characteristics These are the first considerations when designing a flocculation facility. The design engineer should ideally have access to data on the year-round quality of the raw water, spanning a period of 5 years or even longer. The seven important water quality aspects are turbidity, total organic compounds, pH, alkalinity, color, algae counts, and temperature. The nature of the colloids, particularly the colloidal organic compounds, and the particle size distribution of the turbidity are characteristics that are preferably known since these factors greatly affect the flocculation characteristics. The magnitude of the raw water turbidity alone is not likely to be a surrogate to coagulant dosage requirements (see Figures 3.2.4-1 and 3.2.4-2).

If historic water quality data are not available, an immediate sampling should be established prior to or as close as possible to the beginning of the predesign study period. Data on the seven basic water quality parameters should then be collected for a period of 1 year. However, if this schedule cannot be implemented, an attempt should be made to at least obtain the water quality data in both dry and wet seasons.

Flocculation characteristics may be evaluated by the jar test, a process commonly practiced in water treatment plants. The jar test must be conducted in a systematic fashion so that it will yield the necessary basic data for the design of the flocculation process, as discussed in Section 3.2.3. The duration of the test should be as long as can be afforded. Yet, as a minimum, repeatability of the results must be demonstrated.

Treatment Process and Finished Water Quality Goals The overall treatment process and finished water goals are the second most important considerations because both affect flocculation. For instance, in a direct filtration process, the flocculation tanks should not produce large settleable flocs since a sedimentation process is not involved. Instead, the flocculation basins should produce an appropriate type of floc so that in-depth filtration can be achieved by a reverse-graded filter bed. In contrast, the conventional complete and lime softening processes require thorough flocculation to produce good settleable floc with a filter influent turbidity of less than 3 ntu, regardless of the raw water quality. This is particularly true for single sand rapid filters that are characterized by surface filtration.

The finished water quality goals also influence the degree of flocculation. Excessive amounts of the following substances in the raw water may effectively be removed by improving the flocculation and sedimentation steps of the process train: color as a precursor to DBPs, algae as a source of taste and odor, asbestos fibers (limitations), and certain toxic metals and compounds.

Available Hydraulic Headloss and Plant Flow Variations Other considerations in the design of the flocculation process are the allowable headloss across the flocculation process and the magnitude of plant flow variation. If the allowable headloss is limited, hydraulic flocculation methods are ruled out. Thus, the design engineer is forced to use mechanical methods. Similarly, plant flow rate fluctuation is an important consideration. If the fluctuation is relatively minor year-round, that is, 50% variation from the daily average flow rate, hydraulic flocculation is applicable and will perform effectively.

Choice of Flocculation Tank Inlets Almost all proprietary reactor-clarifiers are designed with a single pipe inlet to the flocculation zone. Yet most municipal wa-

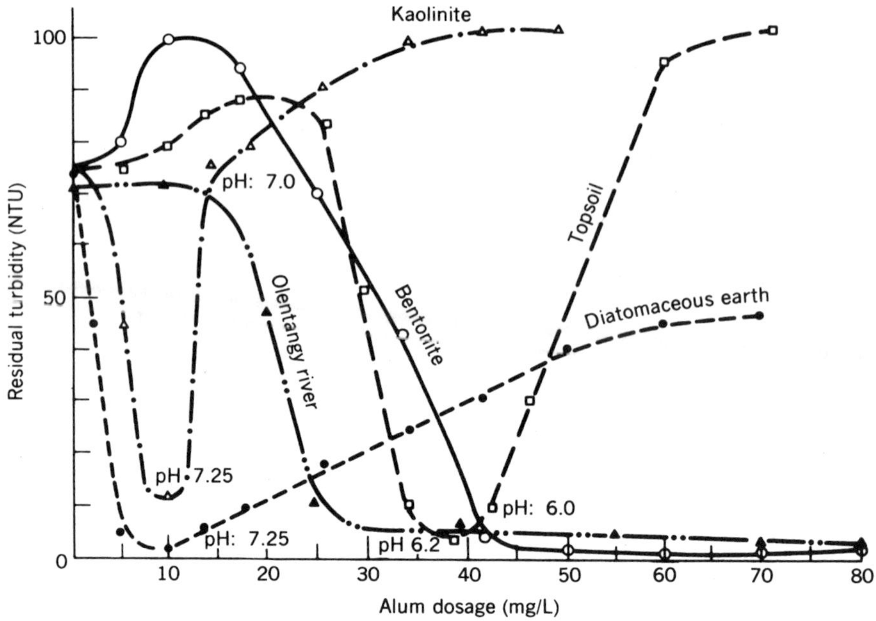

	Turbidity NTU	pH	Alkalinity meq/L
Bentonite	75	8.3	0.4
Diatomaceous	75	9.0	0.4
Kaolinite	75	8.7	0.4
Topsoil	75	6.7	0.4
Olentangy R.W.	65	8.02	3.26

(*a*)

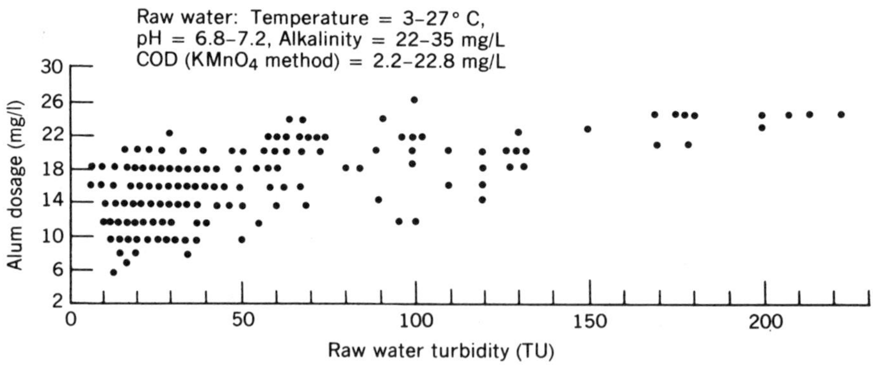

(*b*)

Figure 3.2.4-1 (*a*) Nature of turbidity and alum dosages on flocculation. [Adapted from S. Kawamura, *J. AWWA,* 68(6):328 (June 1976).] (*b*) Relation between raw water turbidity and optimum alum dosage for Yodo River, Japan. [Adapted from S. Kawamura, *J. Jpn. WWA,* 239:17 (September 1954).]

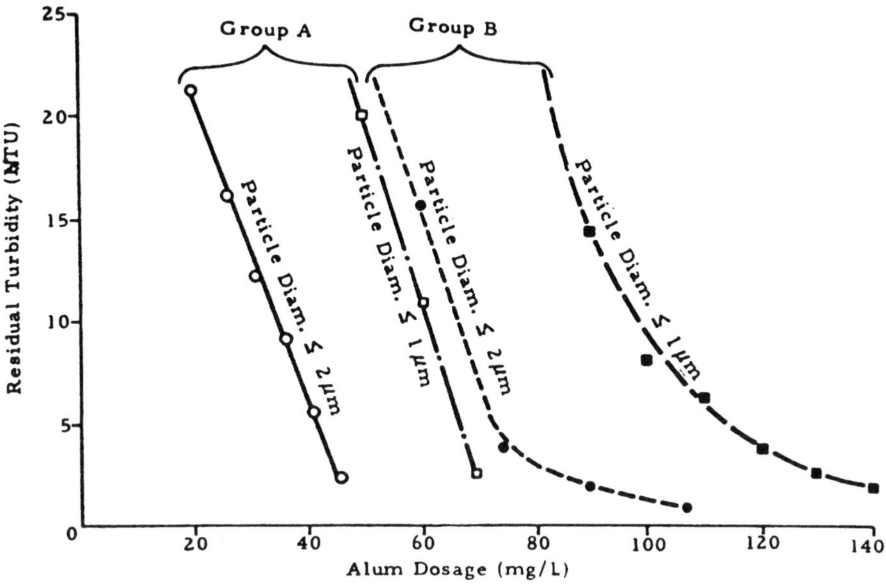

Raw Water

	Group A	Group B
Turbidity	30 NTU	165 NTU
Alkalinity	0.64 meq/L	1.0 meq/L
pH	6.5	9.4

Figure 3.2.4-2 Effect of particle size on flocculation. (From S. Kawamura, Thesis, Ohio State University, 1961.)

ter treatment plants have a traditional rectangular flocculation tank located upstream of the sedimentation tanks or flotation tanks (DAF system). In this case the most simple and cost-effective design is a single inlet to each flocculation tank. This scheme can be used if the flow pattern in the flocculation tank is a serpentine type (see Figure 3.2.4-8). However, many design engineers provide more than two inlets to each flocculation tank from the influent channel or pipe. Regardless of the number of inlets to each tank, the most important design issue is equal hydraulic loading to each tank.

Engineers have two basic choices when designing the tank inlet: a series of identical individual weirs set at the same elevation or a series of submerged orifices each fitted with an isolation valve or gate. The weir inlet is attractive from a theoretical viewpoint, but there are four basic problems associated with this choice: (1) the inlet channel must be large enough to maintain nearly equal water surface elevation for the entire length of the channel; (2) heavy suspended solids accumulate in the channel due to the large size of the channel and the low flow velocity within the channel; (3) each tank is often isolated by means of stop logs which are not leak-tight and create extra work for plant operators unless down-opening slide gates are used (down-opening slides do not reliably reset at exactly the same elevation after they are raised for isolation); and (4) slight changes in water depth (in the channel) result in significant changes in flow rate over the weir whenever all the weirs are set at the same elevation; some plants set the weirs at different elevations along the inlet channel, but this is not recommended, as the set elevation is valid for only a certain plant flow rate.

The second choice in tank inlet design, submerged orifices fitted with isolation valves/valve operators or gates, are furnished near the bottom elevation of the influent channel. Compared to the weir inlet design, this type has two drawbacks, that of a slightly higher capital cost and slightly more loss of head. However, this design has four advantages over weir inlets: (1) the flow rate is not significantly affected by slight changes in water level from one end of the influent channel to the other, (2) the size of the influent channel may be much smaller and a flow velocity of 2 to 3 ft/s (0.6 to 0.9 m/s) may be allowed, (3) there is minimum deposition of suspended solids in the influent channel due to higher flow velocity and because the flow exits near the bottom of the channel, and (4) the isolation valves or gates provide virtually driptight closure.

Sensitivity of flow rate variation due to reduction or drop in water elevation in the influent channel are presented in Table 3.2.4-1, a comparison of the results for both weir inlet and submerged orifice inlet. As illustrated in the table (weir inlet), a drop of 1 in. (2.54 cm) in water level at the last inlet to the basin correlates to a flow rate over the weir of 83.7% and becomes only 66.3% with a 2-in. (5.1-cm) drop. In contrast, an 18-in. (45-cm) orifice yields a flow rate of 96.3% with a 1-in. drop and 92.5% with a 2-in. drop in water level. In summary, the most practical (best) inlet from the influent channel to the flocculation tank is the orifice inlet.

Local Conditions Local conditions should also be analyzed when selecting an appropriate type of flocculation process. The five major factors that should be evaluated are site topography, climatic conditions, availability of services, capability of the operational personnel, and level of the local technology.

Cost Cost is always an important consideration. Both the initial cost (capital) and the operation and maintenance expenses should be evaluated.

Relation to Existing Treatment Facilities In the case of plant expansion, the relation of the new flocculation process to the existing process is a realistic problem. The basic rule is to make all the flocculation units identical so that uniform performance and operation and maintenance procedures can be maintained.

Miscellaneous Items Miscellaneous considerations, such as the hydraulic characteristics of the flocculation tank and scum and sludge removal, must be incorporated during the design phase. The prevention of flow short-circuiting, the creation of effective eddies and turbulence within the tank, and minimization of high shear

TABLE 3.2.4-1 Flow Variations Due to Available Heads (Water Depths)

	Water Level (Available Head) in the Influent Channel				
	Full Head	0.5 in. Lower	1 in. Lower	2 in. Lower	3 in. Lower
Rectangular weir with 6-ft length	10.66 at H = 8 in. (100%)	9.88 at H = 7.5 in. (92.7%)	8.92 at H = 7 in. (83.7%)	7.064 at H = 6 in. (66.3%)	5.38 at H = 5 in. (50.5%)
Three sizes of orifices					
Case 1: 24-in. size	10.66 at H = 4.38 in. (100%)	10.03 at H = 3.88 in. (94.1%)	9.37 at H = 3.38 in. (87.9%)	7.85 at H = 2.38 in. (73.6%)	5.98 at H = 1.38 in. (56.1%)
Case 2: 18-in. size	10.66 at H = 13.86 in. (100%)	10.47 at H = 13.36 in. (98.2%)	10.27 at H = 12.86 in. (96.3%)	9.86 at H = 11.86 in. (92.5%)	9.44 at H = 10.86 in. (88.6%)
Case 3: 16-in. size	10.67 at H = 22.2 in. (100%)	10.55 at H = 21.7 in. (98.9%)	10.43 at H = 21.2 in. (97.8%)	10.173 at H = 20.2 in. (95.4%)	9.92 at H = 19.2 in. (93%)

Notes: (1) Weir formula: $Q = 3.33\ LH^{1.5}$

where Q is the flow rate (ft^3/s), L the length of weir (ft), and H the head (water height) over the crest (ft).

(2) Orifice formula: $Q = CA\sqrt{2gH}$

where Q is the flow rate (ft^3/s), A the orifice area (ft^2), C the coefficient (0.7 for the orifice and islolation valve full open), and H the differential (driving) head (ft).

(3) Numerical figures in this table are the flow rates (ft^3/s) under certain driving head's (ft).

(4) A 20-in orifice is the proper size selection in this case because of a reasonable head loss (9.1″) and the flow rate yields 94.3% with a 1″ drop, and 88% with a 2-in. drop of water level.

forces created by the mixing blades are all important hydraulic problems. Important operational problems include the ease of scum removal from the water surface and of sludge and silt from the bottom of the basin.

Some chemists also emphasize the effect of ion concentrations, found within the raw water, on flocculation as the salt levels of the water influence the coagulant dosage requirements. Since the drinking water quality standards for the TDS of raw water already limits the presence of excessive amounts of ions, this issue is not of great concern for most design engineers. In recent years, the process of ozonation has become popular due to the stringent limitation on DBPs (THMs and HAA5) in drinking water. Ozone is capable of both oxidizing substances and flocculating certain organic and inorganic substances. Therefore, the flocculation process is generally more effective if it is preceded by ozonation.

Type and Selection Guide In this section we describe the commonly available alternatives to the flocculation process, as well as guidelines for making the proper selections.

Available Alternatives Flocculation mixing may be provided by either mechanical mixers or baffles. These units reduce flow short-circuiting and induce the collision and agglomeration of particles. Figure 3.2.4-3 presents the most common types of mechanical mixer and baffle used in the field of water treatment. The typical categories of mixing systems are listed below.

1. Mechanical mixing
   (a) Vertical shaft with turbine or propeller blades
   (b) Paddle type with either horizontal or vertical shafts
   (c) Proprietary units such as Walking Beam, Flocsilator, and NU-Treat
2. Baffled channel basins
   (a) Horizontally baffled channels
   (b) Vertically baffled channel
3. Reactor clarifier proprietary systems
4. Contact flocculation (gravel-packed filter)
5. Diffused air or water jet agitation

Selection Criteria Selection of the flocculation process should be based on the following criteria:

1. Type of treatment process: for example, conventional, dissolved air flotation, direct filtration, softening, or sludge conditioning
2. Raw water quality: for example, turbidity, color, and temperature
3. Flocculation characteristics in response to changes in mixing intensity and mixing times

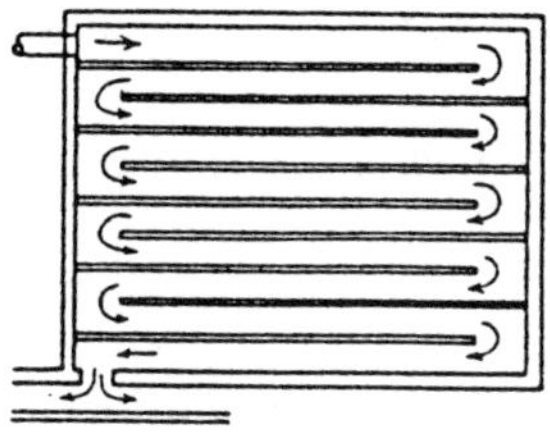

Round-the-end baffled channel type (plan)

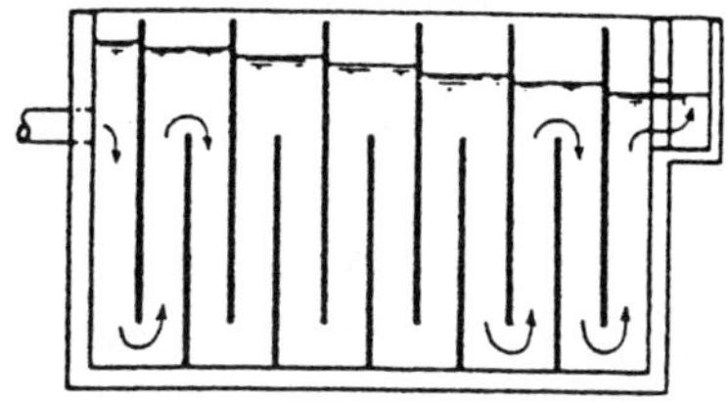

Over- and underbaffled channel type (section)

Baffled channel flocculation system

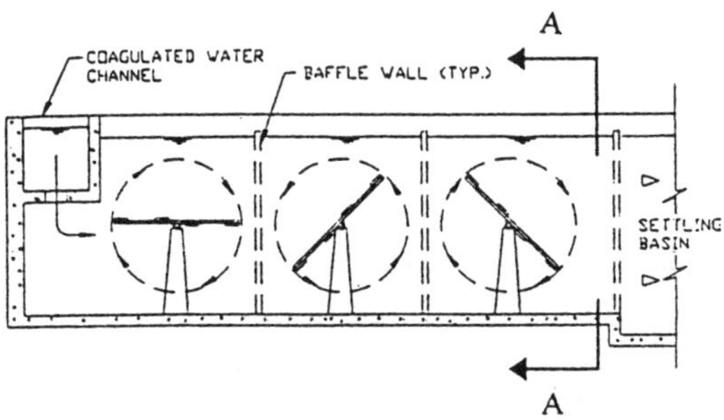

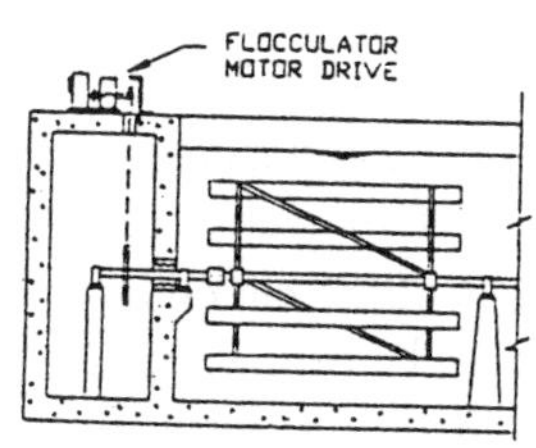

Section A- A

Horizontal shaft with paddle wheel flocculation system

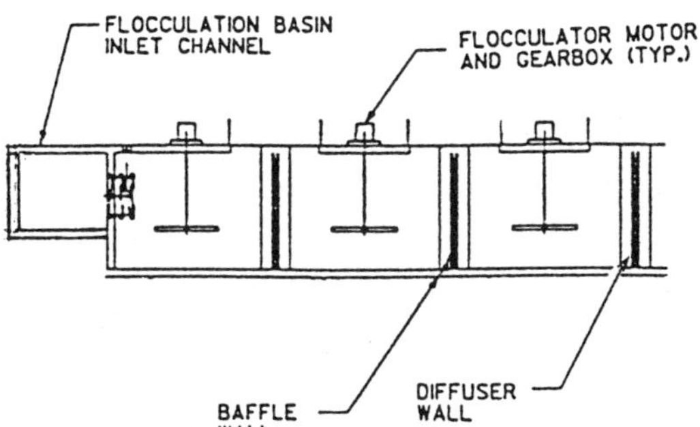

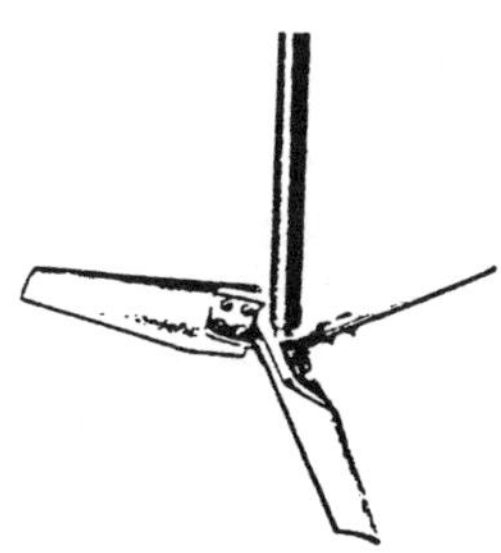

Mixing Blade

Vertical shaft with hydrofoil mixing blade flocculation system

Figure 3.2.4-3 Three basic types of flocculation systems.

The following criteria should be used when selecting the type of mixing:

1. Local conditions such as high wind, ice buildup, or increased viscosity of the gear oil during extremely cold periods
2. Available headloss across the plant
3. Shape and depth of the basin
4. Capital and operation and maintenance (O&M) costs

Order of Preference Based on the design of the most recent installations, the order of preference for the selection of the flocculation process equipment is as follows:

1. Vertical-shaft flocculators (with hydrofoil mixing blades) in either direct flow through horizontal tanks with proper compartmentalization in each flocculation stage or a serpentine flow path, unless the total number of the units becomes excessive
2. Paddle flocculators in horizontal flow tanks with proper compartmentalization in each stage
3. A combination of baffled channels (helicoidal flow) in a serpentine flow path with flocculators in the last stage of the flocculation process
4. Baffled channel flocculation process (around-the-end type) provided that the mixing criteria of $G \times t$ exceeds 3×10^4 during the low-plant-flow period

Reactor-clarifiers and contact flocculation systems are in a category of their own; these are usually proprietary items of equipment manufacturers. Both systems may be considered under certain conditions based on plant size (usually small), local conditions, and the type of process (i.e., reactor-clarifier for lime softening and contact flocculation for direct filtration with multimedia).

Discussion of Alternatives

Mechanical Mixing Systems Selection of a proper flocculation unit depends on the overall unit process that is selected. For instance, if the raw water quality is good and direct filtration is feasible, the filter bed will probably be a reverse-graded high-rate filter bed such as a dual-media type. In this situation, the floc should be small, yet physically strong to resist the high shearing forces within the bed. Vertical-shaft high-energy flocculators produce a floc that meets these requirements. By contrast, when the raw water quality is poor (i.e., polluted or has medium to high turbidity), the suspended matter is removed from the raw water through the production of good floc and effective sedimentation; this is especially true for cold-water regions. In this case, paddle flocculators generally produce larger and heavier floc because of the larger number of blades, larger total blade perimeter, and larger blade surface area. This type of flocculator rotates at slow speeds and is capable of creating more localized eddies with lower shearing forces. However, a vertical-shaft flocculator with hydrofoil mixing blades is capable of producing large and heavy floc if an anionic polymer is used in conjunction with alum in a specific application sequence. These examples serve only to stress the importance of the relationship between the selection of process and the flocculation equipment.

When a mechanical mixer is selected as a flocculator, the mixer should have the following characteristics: (1) it must deliver the G values specified for each stage of flocculation; (2) it must provide sufficient eddies and turbulence so that the required velocity gradients are produced; (3) it must provide low shear forces at the edges of the mixing blades, especially during the last two stages of flocculation; and (4) it should have low maintenance and operational costs. Certain types of mixers, such as plate turbine and 45° pitched blades, produce a relatively uniform flow motion or a rotational motion because of their rapidly rotating blades. Yet the localized shear forces at the tips of the blades are high. Therefore, theses types of mixers are not suited for the actual flocculation process.

Regardless of the type of flocculator, tapered mixing across the flocculation tank is always an important consideration. The last stage of the flocculation process often requires a delicate and gentle mixing condition to promote the growth of floc size. For this reason, it is best to specify an infinitely variable speed control unit (frequency control) for the final-stage flocculators. Examination of records from various plants reveals that the first and second stages do not require frequent changes in speed. Therefore, a two-speed motor control mixer may be used.

Another aspect that should be considered is the reliability of the flocculation system. Vertical-shaft flocculation tanks contain a large number of mixer units. Failure of one or two of the flocculator drive units usually does not significantly affect the overall efficiency of the system. Conversely, the horizontal-shaft paddle flocculators generally have only one drive unit for several paddles; failure of one drive will terminate the movement of several flocculator paddles. It is therefore important to specify an extremely reliable drive unit when using horizontal-shaft flocculators.

In summary, all other factors being equal, vertical-shaft floculators are usually the first choice for the following reasons: (1) minimal maintenance, (2) operational flexibility, (3) very little headloss across the tank, (4) easy control of mixing intensity, (5) effectiveness, and (6) minimal impact to the overall performance if one unit malfunctions. The drive units should have hydrofoil mixing blades, and all shaft bearings should be located above the water surface of the basin. Independent of the type of mechanical mixer are the requirements for sufficient walkway space [4 ft (1.2 m)] around each mixing unit for the control panels, for power connections, and for the installation, removal, and regular maintenance of equipment. One drawback of this system is that large plants may require many mixing units. If a plant requires over 50 vertical-shaft flocculators, it would generally be considered to be excessive both in quantity and cost compared with other types of proven flocculation systems.

Baffled Channel Basins The main factors in determining the applicability of the baffled channel flocculation process are the local conditions and the presence of a moderate amount of allowable headloss across the tank. For example, many developing countries commonly have problems with proper maintenance of the equipment. Since the baffled channel flocculation process does not have any mechanical parts, it should be considered seriously in this type of situation.

When designed properly, both the horizontally baffled (around-the-end flow) and the vertically baffled (over- and underflow) channels perform well and exhibit good plug flow characteristics. However, traditional baffled channels have two disadvantages: There is a significant headloss of 2 to 3.3 ft (0.6 to 1 m) across the tank, and mixing intensity is a function of plant flow rate. Helicoidal or tangential flow baffled channels create less than a 50% loss of head than traditional designs without sacrificing performance. Many engineers who evaluate baffled channel flocculation systems generally believe that the system is inadequate when they see a low G value across the system under low-flow conditions. However, this impression is often incorrect because they have not calculated $G \times t$. In general, good flocculation occurs when $G \times t$ ranges from 3×10^4 to 2×10^5 (Figure 3.2.4-4). Thus an average G of the system may be as low as 12 s^{-1}, but if mixing time t is no less than 45 min (2700 s), $G \times t$ will be greater than 3×10^4; therefore, the system should perform well.

This discussion is not meant to imply that use of the baffled channel flocculation process is restricted to developing countries. If the conditions of flow rate variation and available headloss can be met, this process should also be considered for use in developed countries. Some of the newer plants in both the United States and Japan use baffled channel flocculation tanks.

Baffled Channels and Flocculators Combination System This concept combines the best features of both systems. It offers a large reduction in mechanical flocculator units so there is a savings in capital and maintenance costs, and increased reliability. The flocculator units should be located in the final stage of the flocculation process. This allows floc to remain in suspension and supplies additional mixing energy during low process flow. Design engineers should provide at least 1.5 ft (0.45 m) of loss of head across the flocculation system at a flow rate designed for helicoidal flow baffled channels. Figure 3.2.4-5 depicts this flocculation system.

There are currently two large plants in the United States using the combination baffled channel and flocculator system; the Houston East plant [150 mgd (6.5 m^3/s)] began operating in 1993. These plants have a settled water turbidity as low as 0.3 ntu. The New Mohawk plant [100 mgd (4.4 m^3/s)] of Tulsa, Oklahoma does not have problems with the water in the channel freezing even though the plants are exposed to ambient temperatures below 6°F (−15°C) during the winter. Plant operational data from these two plants demonstrate that process efficiency does not change during the winter when plant flow rate is 30% of design flow rate. However, operators often isolate a certain number of basins during periods of low water demand to maintain a flow rate that is 50% of designed flow rate to prevent floc deposition within the channels.

Contact Flocculation Contact flocculation (gravel-packed filter) is another design that requires minimal maintenance and is similar to the in-line filtration process.

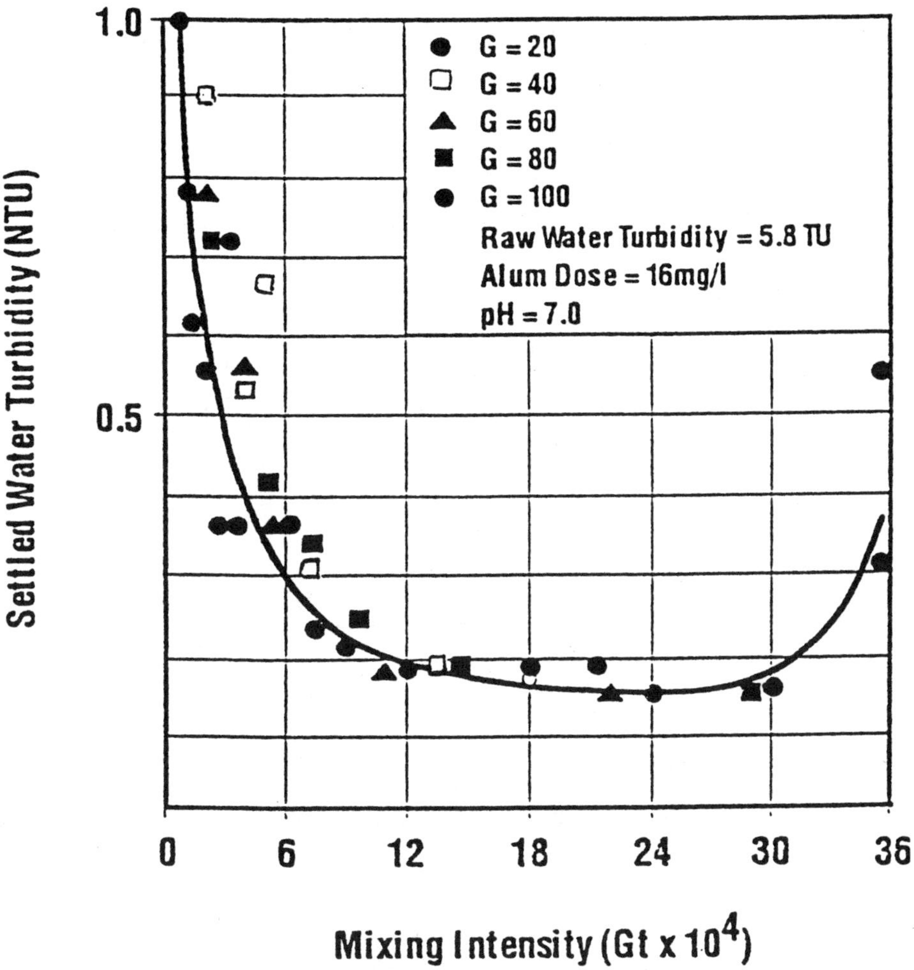

Figure 3.2.4-4 Evaluation of flocculation conditions based on $G \times t$ as a parameter.

Some equipment manufacturers, such as Neptune Microfloc and Culligan, have promoted this system in their packaged water treatment units. The advantage of contact flocculation is its compact size and its lack of moving parts. Hydraulic loadings of 10 to 15 gpm/ft^2 (25 to 37.5 m/h) are possible and the system requires an empty bed contact time of only 3 to 5 min to produce a floc with the proper characteristics. This type of flocculation system depends on plant flow rate and requires minimal variations in flow rate and water temperature for adequate mixing intensity. Contact flocculation also requires frequent backwashing with air scour.

Diffused Air or Water Jet Agitation Of all the processes, the least preferred are diffused air and water jet flocculation. The diffused air system is characterized by a high rate of energy consumption and is also an inefficient flocculation process.

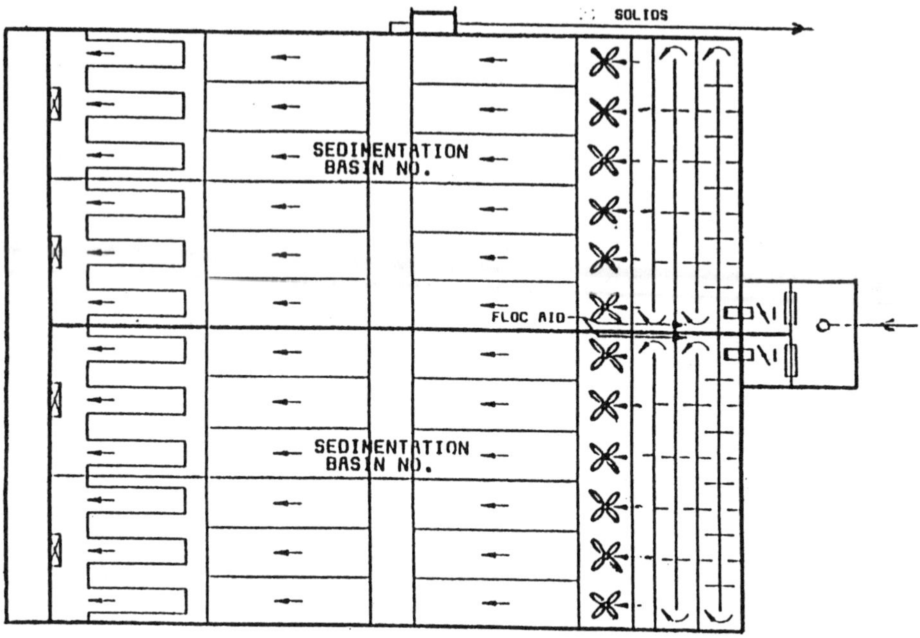

Figure 3.2.4-5 Flocculation system which has a combination of baffled channels and mechanical flocculators.

Therefore, this system should serve as an auxiliary or temporary measure. Use of a water jet mixing system is not recommended for flocculation because the high shearing force on the jet path restricts the growth of floc size.

Design Criteria When designing a flocculation process, selection of the mode of mixing and determination of the physical relations and characteristics of the flocculation tanks and clarifiers (sedimentation tanks) are among the first decisions to be made. As described previously, either hydraulic mixing or mechanical mixing may be chosen. In evaluating the physical characteristics, a decision must be made as to whether the flocculation tank will be built as part of the clarifier or as an independent structure. After establishing these two issues, the design may be refined; establish the number of tanks; consider the total number of tanks when the plant capacity reaches the ultimate capacity; define the number of mixing stages and the mixing energy level in each stage; and determine the type of baffle that will be used to minimize flow short-circuiting.

One of the most efficient and economical flocculation tank designs incorporates the basin into the clarifier unit. The majority of equipment manufacturers design their units exclusively in this manner. For custom design, the flocculation basin may be combined with the clarifier by placing it at the influent end of a rectangular sedimentation tank. This arrangement offers many advantages, such as minimal land requirements, minimal floc breakage between the two processes, simplicity of design, and above all, simple and economical construction.

The design of the flocculation tank is commonly based on only two criteria: detention time and mixing energy level. The energy level is the G value or velocity gradient and is defined by Camp et al. (1943) as

$$G = \left(\frac{P}{\mu V}\right)^{0.5}$$

where P is the power input to the fluid, V the volume of the tank, and μ the absolute viscosity of the fluid.

The flocculation process must be designed to provide maximum interparticle contact. These contacts are a function of the velocity gradients within the fluid; the velocity gradients are produced by hydraulic or mechanical mixing (orthokinetic flocculation). The number of particle contacts is expressed as

$$N = n_1 n_2 \frac{G}{6} (d_1 + d_2)^3$$

where N is the number of contacts between n_1 particles of diameter d_1 and n_2 particles of diameter d_2 in a unit of time. Therefore, the rate of flocculation increases with the number and size of the particles and with the power input, but decreases with the viscosity of the fluid.

The mean velocity gradient (G value) of certain types of flocculation systems may be computed as follows. For the baffled channel,

$$G = \left(\frac{gh}{\nu t}\right)^{0.5}$$

where ν is the kinematic viscosity of fluid, t the mean detention time, g the gravity acceleration constant, and h the headloss across the tank.

For the mechanical mixers with paddles:

$$G = \left(\frac{C_D A v^3}{2\nu V}\right)^{0.5}$$

where C_D is a drag coefficient that depends on the paddle shape and flow conditions (use 1.8), A the cross-sectional area of the paddles, v the relative velocity of the paddle with respect to the fluid, ν the kinematic velocity of the fluid, and V the volume of the flocculation tank. It is estimated that v is from 0.5 to 0.75 of the peripheral velocity of the paddle.

Proper application of the parameters presented above is described in the following section.

General The general design criteria for a basic rectangular flocculation tank are as follows:

Energy input	$G \times t = 3 \times 10^4$ to 2×10^5 t is in seconds (5×10^4 average) or $G = 10$ to $70\ s^{-1}$ ($30\ s^{-1}$ average)
Detention times	20 to 30 min at maximum daily flow rate
Water depth	10 to 15 ft (3 to 4.5 m)
Flocculation stages	Two to six stages (commonly three or four)

The following factors will determine the number of flocculation stages to be included in the design.

1. Type of subsequent treatment unit (sedimentation, filter, or sludge drying bed) and the type of overall process (direct filtration or complete process)
2. Quality and treatability of the raw water: turbidity levels, the nature of the turbidity, and water temperature during the cold months
3. Degree of flow short-circuiting across the flocculation basin (with or without baffles) and the type of baffles
4. Local conditions: regulatory requirements and the operation and maintenance of the flocculation process

Vertical-Shaft Flocculators In a mechanical mixer the diameter of the mixing blade (D), in relation to the tank diameter (equivalent diameter T), is important and D/T should be greater than 0.35. Another item to watch is the maximum flow velocity induced by the mixing blade: It should be less than 8 ft/s (2.5 m/s) in the first stage and less than 2 ft/s (0.6 m/s) in the last stage of the flocculation process if good and settleable floc formation is to be expected. Mixing equipment manufacturers have established the design criteria for mechanical mixers based on their research and experience. However, their main objectives are for liquid–liquid, liquid–gas, and solid–liquid mixing as a batch operation. Thus, manufacturers often do not have a good understanding of flocculator design for a continuous-flow mixing process and the production of fragile flocs. When produced properly, the flocs should have the characteristics and appearance of snowflakes.

There are two basic philosophies in specifying mechanical flocculators. The first is simply to specify the function and a few physical characteristics, such as G value, tip speed (excluding hydrofoil blades), and the D/T ratio. The other is to write detailed specifications for the equipment, installation procedures, and exact power requirements, including the dimensions and elevation of the mixing blades. It appears that the first approach is better for the majority of engineers since the proper design of vertical-shaft flocculators is somewhat state-of-the-art. When using hydrofoil mixing blades, tip speed is irrelevant since the blades lift or push down on the water, depending on the setting. Figure 3.2.4-6 shows the energy transmittal efficiency for some mixing blades when used as flocculator units.

Horizontal-Shaft Flocculators The advantage of the horizontal-shaft unit is that one shaft can operate a number of agitators and thereby reduce the number of drive

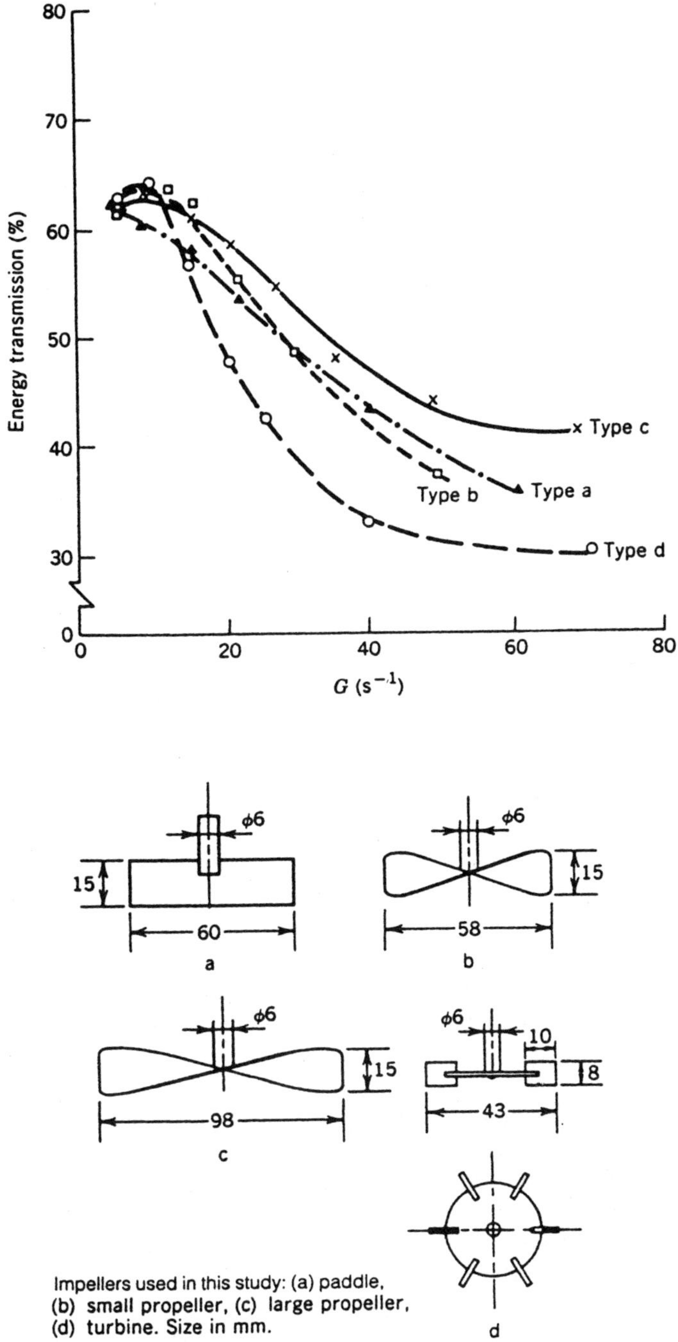

Figure 3.2.4-6 Mixing efficiency as a function of *G* value and type of impeller. (After Leentraar, 1980.)

units. However, its reliability is low because failure of one shaft results in a 25 to 30% loss in mixing capability. The horizontal-shaft units can be equipped with either paddles or turbine blades. The basic design criteria for horizontal-shaft flocculators are the same as those for the vertical type, except for the type of paddle units.

1. The total paddle area should be 10 to 25% of the tank cross-sectional area. If the values are over 30%, the paddles will produce a distinct rotating motion in the water, causing a reduction in local eddies and turbulence along the edge of the paddles.
2. Each arm should have a minimum of three paddles, so that dead space, especially near the shaft, will be minimized.
3. The peripheral speed of the paddles should be between 0.5 and 3.3 ft/s (0.15 to 1 m/s).
4. The designed G value is lower than that of vertical-shaft flocculators; it should initially start at about 55 s^{-1} (highest), then be reduced to about 10 s^{-1} in the last stage of the flocculation tank. Higher than 60 s^{-1} will result in a high shaft rotation speed; often, this leads to severe wear and tear of the shaft bearings within a short period of time. The workmanship of shaft installation is of utmost importance for paddle flocculators.
5. For a direct filtration process, which does not have clarifiers, the flocculation tanks may be designed with around-the-end channels. This type of design minimizes flow short-circuiting for horizontal-shaft flocculators. If the flocculation tank is part of a rectangular clarifier tank, the last stage of the flocculation tank should be designed to approximate the flow direction found in the clarifier; this will prevent tangential inflow into the clarifier (see Figure 3.2.5-4).
6. Most drive unit failures are caused when the unit is started at the top rotational speed. Like all equipment, it should be started at a low speed to avoid very high torque force and a high power requirement.

Tables 3.2.4-2 and 3.2.4-3 present the general design criteria and compare the major characteristics among baffled channel, horizontal-shaft, and vertical-shaft flocculators.

BAFFLED WALLS When mechanical mixers are selected for the flocculation process, the use of baffles, for compartmentalization, improves flocculation. The shape and location of the baffles are determined by the design engineer because there are no practical design guides. Design engineers are strongly advised to compartmentalize each stage of the mechanical mixer flocculation process with a proper baffled wall. Each baffle should have orifices that are 4 to 6 in. (0.1 to 0.15 m) in diameter uniformly distributed across its vertical surface and a velocity of 1.2 to 1.8 ft/s (0.35 to 0.55 m/s) should be produced through the orifice holes of each baffle at maximum flow rate; the higher velocity of 1.8 ft/s should be applied to the first baffle and the lower velocity of 1.2 ft/s to the last baffle.

TABLE 3.2.4-2 General Design Criteria

	Baffled Channel	Mechanical Flocculators: Horizontal Shaft with Paddles	Mechanical Flocculators: Vertical Shaft with Blades
G (s^{-1})	50–10 tapered	50–10 tapered	70–10 tapered
T (min)	30–45	30–40	20–40
Flocculation stages (number of channels or number of horizontal shafts)	6–10	3–6	2–4
Mix energy control	Flow passage variation	Variable mixing speed	Variable mixing speed
Maximum flow velocity or mixer tip speed (fps)	3	3	6–9
Blade area/tank area (%)	—	5–20	0.1–0.2
Blade: D/T	—	0.5–0.75	0.2–0.4
Shaft rpm	—	1–5	8–25
Major application	Conventional complete treatment	Conventional complete treatment	Direct filtration and conventional complete treatment

In general, each baffle is placed across the tank perpendicular to the flow path. The top of the baffled wall is slightly submerged [0.5 in. (1.3 cm)] to allow the scum to flow over, and the bottom of the wall should have a space of 1.0 in. (2.5 cm) for easy drainage and sludge removal. To prevent flow short-circuiting, engineers should not design a large passageway; the passage may be large enough to accommodate a person if it is blocked off during normal operation or if a portion of the diffuser wall is capable of swinging open (by hinges) only during the cleaning operation. If timber is used to construct the baffle, it will tend to warp if the span is too long [over 10 ft (3 m)] or if is too thin [less than 2 in. (5 cm) net].

If the flocculation tank is designed as part of the sedimentation tank, the baffled wall located between the tanks is called the diffuser wall. In this case, the baffled wall should produce a flow rate velocity of no more than 0.8 ft/s (0.25 m/s) to avoid excessive breakage of the developed floc. Another advantage of this type of baffled wall is the production of additional mixing energy. Baffles that are based on the criteria above provide a G value of 5 to 25 s^{-1}, depending on the flow velocity through the baffles. Figure 3.2.4-7 illustrates the results of a scale-model hydraulic study on the determination of an effective baffled wall system.

STATOR BAFFLES The efficiency of vertical-shaft mixers can be improved through the installation of stator baffles. The purpose of stator baffles is to increase the effective turbulence as well as to prevent rotational fluid motion within the tank. A proper stator baffle design specifies the installation of two or four pairs of baffles on opposite sides of the compartment. The width of each individual stator baffle should be $\frac{1}{8}$ to $\frac{1}{12}$ of the distance across the individual mixer compartments.

TABLE 3.2.4-3 Comparison of Three Basic Flocculation Processes

		Mechanical Flocculators	
	Baffled Channel	Horizontal Shaft with Paddles	Vertical Shaft with Blades
Flocculation	Good to excellent	Good to excellent	Fair to Good
Reliability	Good	Fair to good	Good
Operation flexibility	Moderate to poor	Good	Good
Capital cost	Low	Moderate to high	Moderate to high
Construction	Easy	Moderate	Easy to moderate
Maintenance	Low	Moderate	Low to moderate
Flow condition	Near plug flow	Short circuiting	Short circuiting
Major advantages	(1) Simple and effective (2) Low O&M costs (3) No moving parts	(1) Good floc formation (2) Effective mixing system with good turbulence (3) No headloss	(1) High-energy input (2) Low maintenance (3) No headloss
Major disadvantages	(1) Mixing energy is a function of flow rate (2) Approx. 2 to 3 ft headloss	(1) Precise installation required (2) Limited high-energy input (3) Regular maintenance required	(1) Many units required (2) Turbine and 45° PBT has high shear at blades (3) Inadequate amount of turbulence and eddy

Baffled Channel (Hydraulic) Flocculators The around-the-end baffled channel and the over-and-under baffled channel are the two basic types of baffled channel flocculators. The around-the-end baffled channel is considered to be more practical since the latter has more problems with the accumulation of scum and silt/grit behind the baffle.

When baffled channels are designed to provide uniform flow velocity through each stage of channels (plug flow type), the headloss per 180 bend is approximately 3.2 to 3.5 times the velocity head in the adjacent channels. However, if the baffled channels are designed to produce the required flow velocity at only the narrowed paths, such as slits or submerged ports, the headloss will be approximately 1.5 times the velocity head of the restricted area; this type of baffled channel is called helicoidal or tangential flow because the stream of water enters tangentially into each chamber, producing a helicoidal flow pattern toward the outlet downstream of each turn. To facilitate easy construction and cleaning, the distance between the baffles should be a minimum of 2.5 ft (0.75 m). The minimum water depth should be 3.3 ft (1 m).

In general, a minimum of six channels should be provided and the headloss across the flocculation tank must be approximately 1 to 2 ft (0.3 to 0.6 m) so that an average G value of 30 to 40 s^{-1} will be established. The process should have a tapered mixing pattern and a minimum residence time of 20 min at maximum plant flow. The G value of the baffled channel flocculation process may be estimated quickly using the following formula (using English units):

$$G = 178 \left(\frac{h}{t}\right)^{0.5} \qquad \text{at } 4^\circ\text{C}$$

where t is the residence time (min), h the headloss (ft), and G the velocity gradient (s^{-1}). Tracer tests have indicated that there is little flow short-circuiting in baffled channels. Case 8 in Figure 3.2.4-7 illustrates this condition.

Example Design Calculations

Example 1 Mechanical Mixer (Vertical-Shaft Flocculators)

Given

Plant flow rate	20 mgd (0.88 m^3/s) daily average
Detention time	20 min at maximum daily flow rate
Number of tanks	Two rectangular tanks
Stage of flocculation	Three stages
Energy input	60 to 20 s^{-1} for both stages 1 and 2, G = 30 to 10 s^{-1} for the last stage
Type of flocculator blade	Vertical shaft with either 45° pitched turbine (PTB) or hydrofoil propeller blades; variable-speed drive control
Minimum water temperature	50°F (10°C)

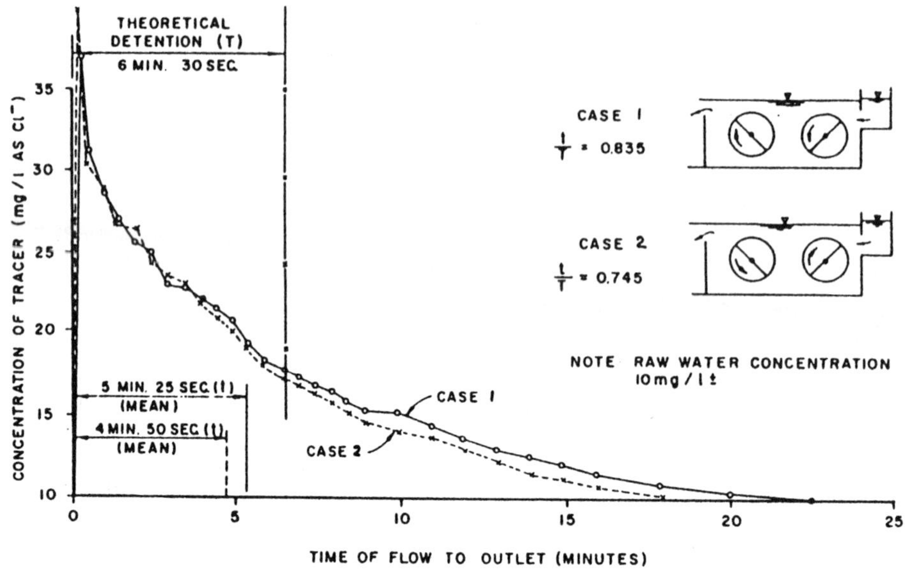

(a)

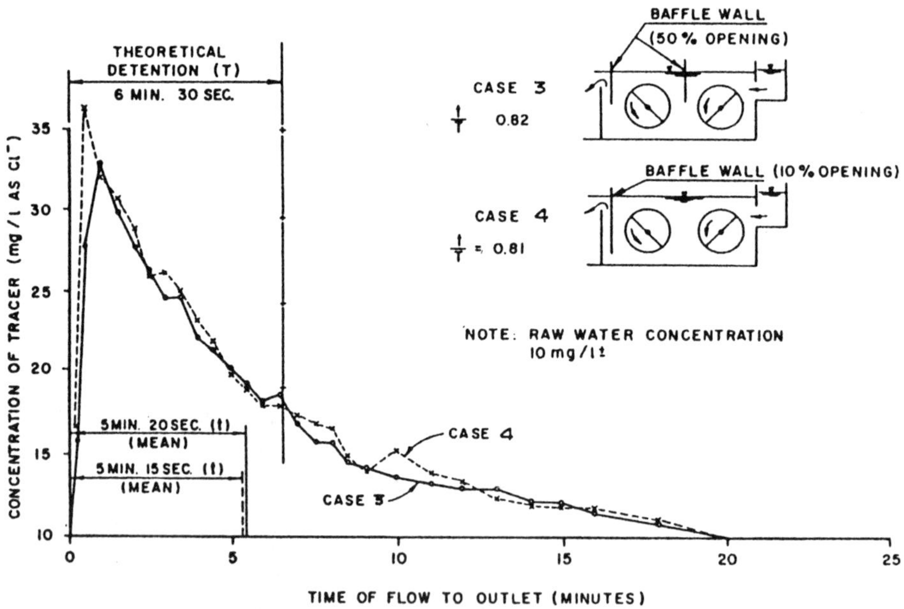

(b)

Figure 3.2.4-7 (*a,c*) Flocculator arrangements versus detention time; (*b,d*) flocculators with baffles versus detention time.

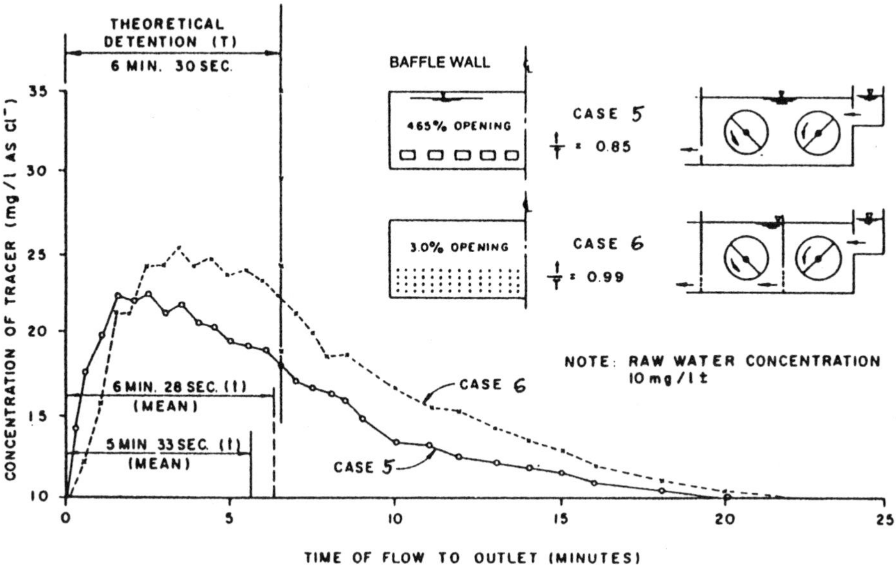

(*c*)

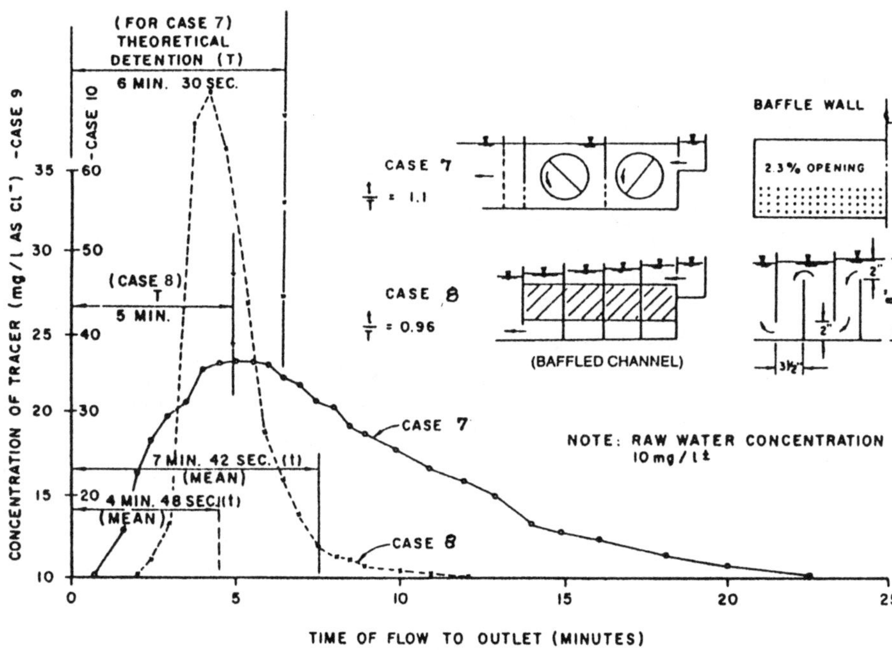

(*d*)

Figure 3.2.4-7 *(Continued)*

Determine

(i) The size of each flocculation basin, including water depth
(ii) The total number of flocculators
(iii) The motor horsepower of each flocculator
(iv) The details of the diffuser walls

Solution (i) Assume that the maximum daily flow rate is 150% of the daily average flow rate. The volume required for each tank is

$$\frac{1.5 \times 0.88}{2} \times 60 \times 20 = 792 \text{ m}^3 \text{ (27,970 ft}^3\text{)}$$

Make the average water depth 14 ft (4.25 m). The total tank area required is

$$792 \div 4.25 = 186.4 \text{ m}^2 \text{ (2005 ft}^2\text{)}$$

The size (shape) of each basin is as follows (see Figure 3.2.4-8):

- *Alternative 1:* 8 m × 24 m ($W \times L$) or 26 ft × 78 ft
- *Alternative 2:* 11 m × 16.5 m ($W \times L$) or 36 ft × 54 ft
- *Alternative 3:* 11 m × 18.5 m ($W \times L$) or 36 ft × 61 ft

Alternative 1 is the most economical design, but it is not the most effective tank. Alternative 2 is not as cost-effective to build as alternative 1. However, the larger number of flocculators gives it a greater safety factor and more effective flocculation. Therefore, adopt alternative 2. Alternative 3 is composed of a series of complete mixing reactors that minimize the problem of flow short-circuiting. Even though this is the best type of flocculation tank, this scheme requires a flow distribution channel to be constructed, leading to the sedimentation tank. The flow direction in this channel is not ideal because it is perpendicular to the flow direction in the clarifier. However, the use of a clarifier inlet diffuser wall, which is described in Chapter 4, can produce a reasonably good inlet flow condition.

(ii) The total number of flocculators is 12 since alternative 2 has been selected because of its cost-effectiveness.

(iii) Motor horsepower of the flocculators. For all flocculators in stages 1 and 2,

$$P = G^2 \mu V$$

where $\mu = 0.0013 \text{ N} \cdot \text{s/m}^2 \text{ (}2.73 \times 10^{-5} \text{ lb} \cdot \text{s/ft}^2\text{) (N, newton} = \text{kg} \cdot \text{m/s}^2\text{)}$

$$P = (60)^2 \times (1.3 \times 10^{-3}) \times [(5.5)^2 \times 4.25]$$
$$= 602 \text{ J/s per flocculator}$$

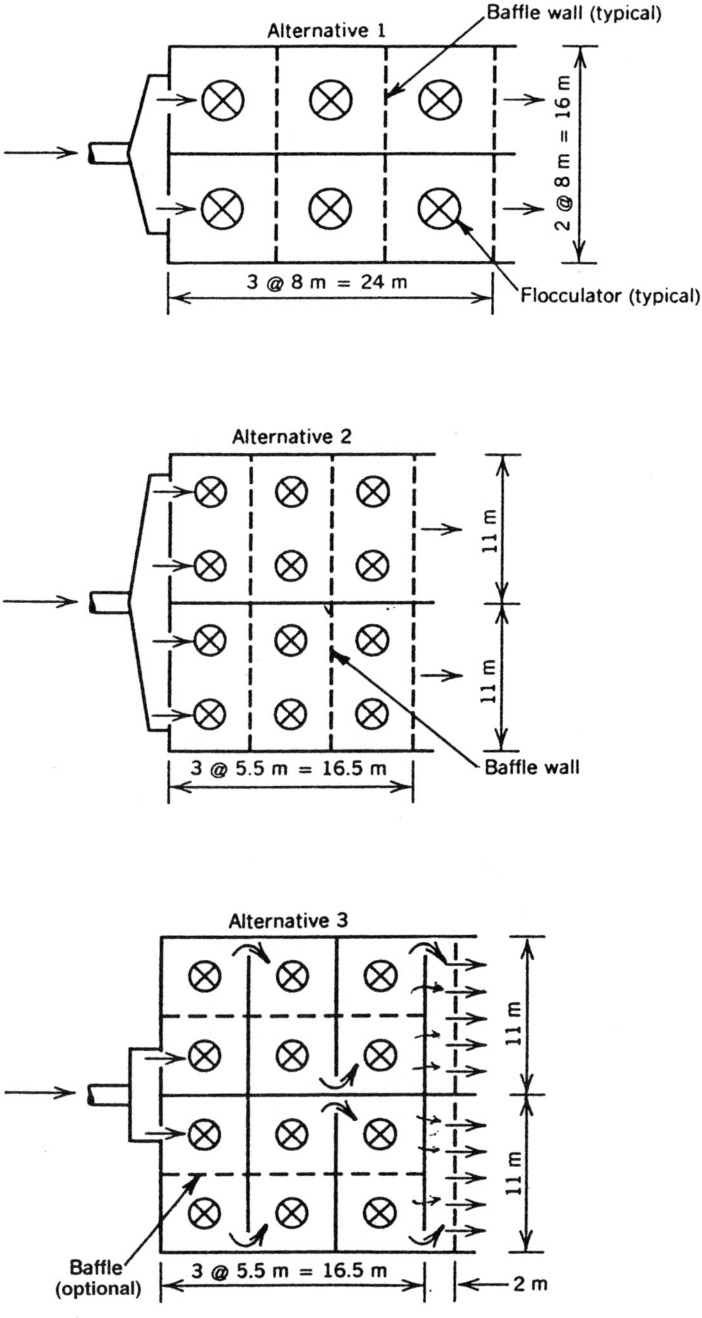

Figure 3.2.4-8 Alternative basins for Example 1.

Since 1 kW = 1000 J/s, P = 0.6 kW or 0.8 hp (1 hp = 0.746 kW). Assuming that the overall efficiency of a mechanical mixer is 75%, the minimum required motor horsepower for each flocculator in stages 1 and 2 is

$$0.6 \div 0.75 = 0.8 \text{ kW (1 hp)}$$

For the flocculators in stage 3, the motor horsepower required can be computed in the same way.

$$P = (30)^2 \times (1.3 \times 10^{-3}) \times [(5.5)^2 \times 4.25] = 150 \text{ J/s}$$

Therefore, P = 0.15 kW ÷ 0.75 = 0.2 kW ($\frac{1}{4}$ hp), minimum for each flocculator in stage 3.

(iv) There will be three baffles: call them No. 1, No. 2, and No. 3 baffles, counting from the inlet side. The flow velocity through each baffle will be set as shown and the total orifice area required can be computed.

	Baffled Wall 1	Baffled Wall 2	Baffled Wall 3
Maximum flow velocity (m/s)	0.55	0.45	0.35
Total orifice area for both tanks (m^2)	2.4	2.93	3.77

The size of a single orifice should be 0.1 to 0.15 m in diameter to prevent clogging by algae or suspended matter. A reasonable distance must also be kept between the orifices so that the structural integrity of the wall is maintained. Thus select 0.15 m as the diameter for each orifice. The following table lists the total number of orifices per tank, the total orifice area, and the headloss per baffle for both the maximum and average daily plant flow rates.

	Baffled Wall 1	Baffled Wall 2	Baffled Wall 3
Total number of orifices	140	168	216
Total orifice area (m^2)	2.47	2.97	3.82
Velocity at Q_{max} (m/s)	0.53	0.44	0.35
Velocity at Q_{ave} (m/s)	0.36	0.30	0.23
Headloss at Q_{max} (m)	0.023	0.016	0.010
Headloss at Q_{ave} (m)	0.010	0.007	0.004

Figure 3.2.4-9 shows the details of three baffled walls in each tank. The ratio of the orifice area to the cross-sectional area of the tank is 2.5%, 3.2%, and 4.1% for walls No. 1, No. 2, and No. 3, respectively. The headloss through the orifice is computed from the formula $Q = CA(2gh)^{0.5}$, where $C = 0.8$.

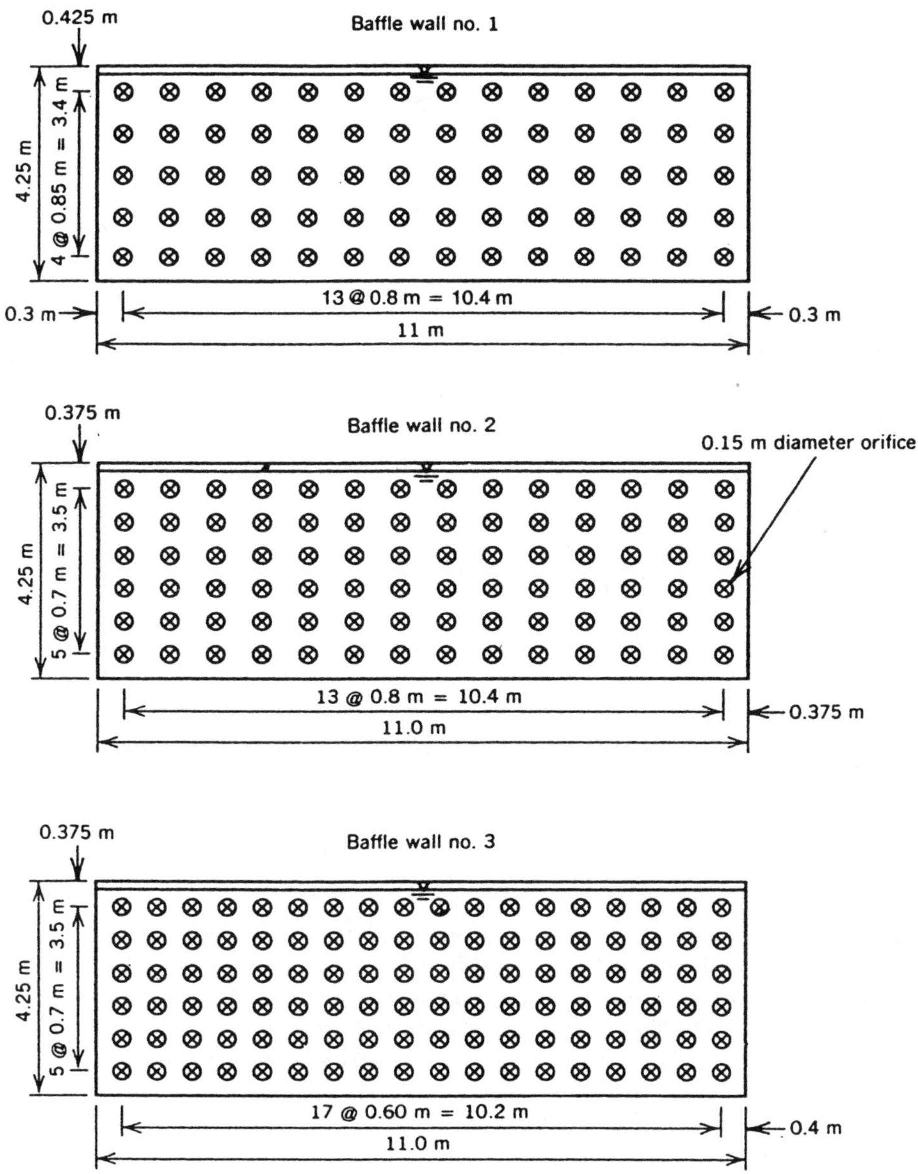

Figure 3.2.4-9 Details of the three baffled walls of each tank.

Example 2 Horizontal Shaft with Paddle Flocculator

Given

Plant flow rate	1.85 m^3/s (42 mgd) maximum per day
Number of tanks	Three rectangular tanks
Stage of flocculation	Two stages due to site limitations
Detention time	32 min at 1.85 m^3/s
Shape of each tank	5 m (W) $\times$ 24 m (L) $\times$ 5.0 m ($W \cdot D$)
Type of flocculator	Horizontal shaft with paddles with variable-speed control
Energy input	(1) $G = 50\ s^{-1}$ maximum (2) Displacement factor (Bean, 1953) = 30 maximum
Minimum water temperature	10°C (50°F)

Determine

(i) The number of paddles per shaft
(ii) The details of the paddle
(iii) The maximum number of shaft revolutions per minute
(iv) The required horsepower of each drive unit

Solution (i) Considering the number of shaft supports and the integrity of the paddle, a reasonable figure would be eight paddles per shaft, or a total of 48 paddles for this flocculation process.

(ii) After a few trials the shape and dimensions of the paddle shown in Figure 3.2.4-10 are adopted.

NOTES:

1. Since there are only two stages of flocculation, each stage is compartmentalized with baffled walls similar to baffled walls 1 and 2 of Example 1 (see Figure 3.2.4-9).
2. The number of paddles used is as many as is practical because the drag force and eddies produced by the paddles are proportional to the total length of the paddle edges.
3. The paddles are arranged diagonally to minimize flow short-circuiting; the travel speed of the paddle is minimal near the shaft.
4. Each paddle wheel contains 3.97 m^2 of the paddle area. This is just under 25% of the cross-sectional area of the tank.

(iii) Since each paddle wheel has more than one blade, G is computed by

$$G = C_D\left(\sum \frac{Av^3}{2\nu V}\right)^{0.5}$$

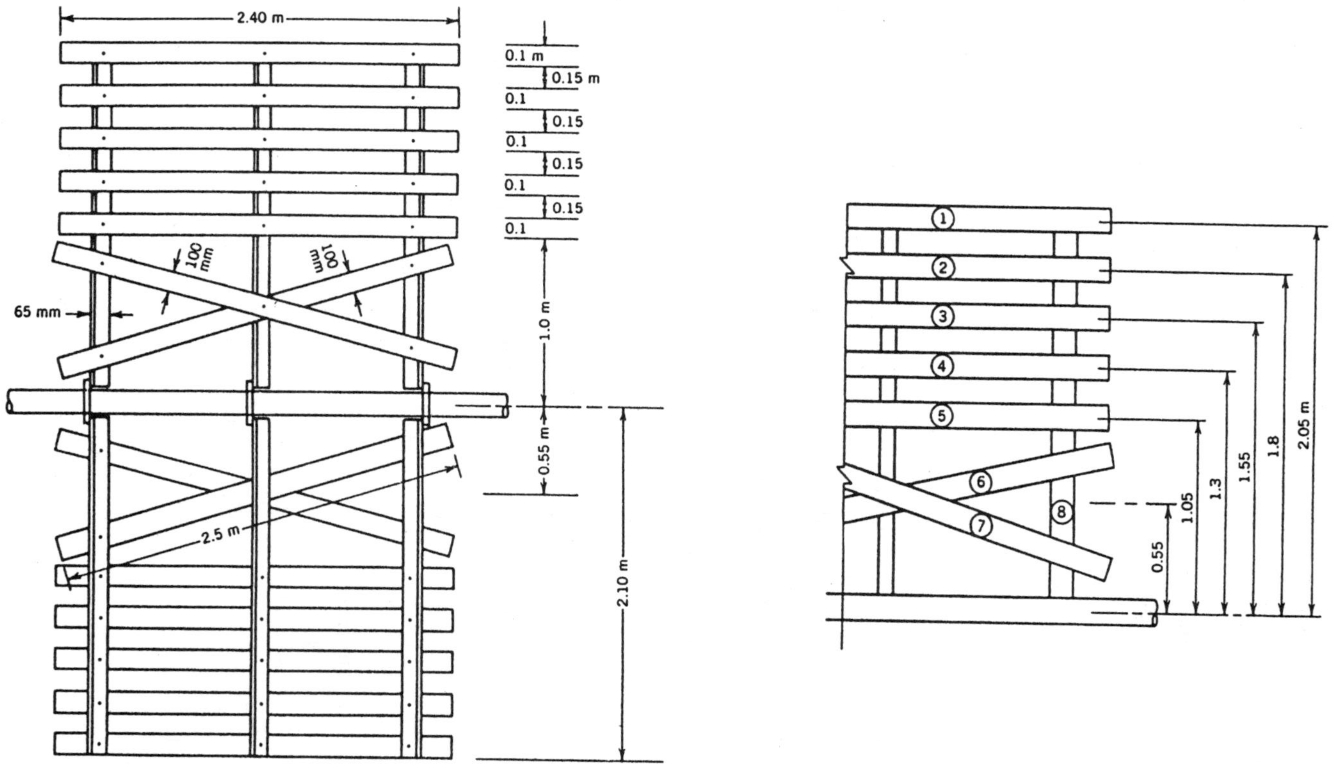

Figure 3.2.4-10 Detail of a flocculator paddle.

where A is the area of each paddle perpendicular to flow, v the relative velocity of each paddle to water, ν the kinematic velocity of the fluid, and V the volume of the tank.

$$v = (0.75 \times 2\pi rN) \div 60 = 0.0785rN$$

where N is the rotational speed (rpm) and r is the radius. Try 0.5 m/s as the peripheral speed of each paddle wheel. Incidentally, the rpm of the paddle at a velocity of 0.5 m/s is $(0.5 \times 60) \div (2\pi \times 2.1) = 2.27$. The first paddle's

$$\begin{aligned} G &= \sqrt{[1.8 \times (0.24 \times 2 \times 48) \times 0.372^3] \div [2 \times (1.31 \times 10^{-6}) \times 3600]} \\ &= 15.04\ \text{s}^{-1} \end{aligned}$$

In like manner:

The second paddle's	$G_2 = 12.5\ \text{s}^{-1}$
The third paddle's	$G_3 = 10.0\ \text{s}^{-1}$
The fourth paddle's	$G_4 = 7.7\ \text{s}^{-1}$
The fifth paddle's	$G_5 = 5.6\ \text{s}^{-1}$
Two diagonal paddles	$G_6 = 4.3\ \text{s}^{-1}$

Thus, the G value for the total paddle wheels in the tank is calculated as the sum of the above:

$$G = \sum_{i=1}^{i=6} Gi = 55.3\ \text{s}^{-1} = 55\ \text{s}^{-1}$$

Make the maximum rotational speed of the shaft 2.33 rpm to meet the required G value of 55 s^{-1}.

$$G \times t = 55.3 \times (32 \times 60) = 1.06 \times 10^5$$

The power requirement is

$$\begin{aligned} P &= G^2 \mu V \\ &= 55.3^2\ (1.31 \times 10^{-3}) \times 3600 \\ &= 14.437\ \text{J/s} \\ &= 14.4\ \text{kW} \end{aligned}$$

(iv) Assume that overall mechanical efficiency is 75%. Thus $P = 14.4 \div 0.75 = 19.2$ kW (25 hp). Based on Bean's design criteria of the displacement factor,

$$\frac{D}{F} \geq 30$$

where

$$D = \sum d$$
$$d = a \times s$$
$$S = 2\pi r \times \text{rpm}$$

and

d = displacement (pumping) capacity of each paddle (ft^3/min)

a = area of each paddle (ft^2)

s = traveling distance of the paddle area A (ft/min)

r = distance (ft) between the center of gravity of the paddle area (a) and the center of the shaft

F = flow rate (ft^3/min)

Since a paddle speed of 2.33 rpm satisfies the G value, use this speed to check the displacement factor.

$$d_1 = 0.24 \times 2\pi \cdot 2.33 \times 2.05 = 7.20 \text{ m}^3/\text{min}$$
$$d_2 = 0.24 \times 2\pi \cdot 2.33 \times 1.8 = 6.37$$
$$d_3 = 0.24 \times 2\pi \cdot 2.33 \times 1.55 = 5.44$$
$$d_4 = 0.24 \times 2\pi \cdot 2.33 \times 1.3 = 4.56$$
$$d_5 = 0.24 \times 2\pi \cdot 2.33 \times 1.05 = 3.69$$
$$d_6 = 0.25 \times 2\pi \cdot 2.33 \times 0.55 = 2.01$$
$$d_7 = 0.25 \times 2\pi \cdot 2.33 \times 0.55 = 2.01$$
$$d_8 = 3 \times [(0.065 \times 2.1) \times (2\pi \cdot 2.33 \times 1.05)] = 6.29$$
$$2 \times \sum_{i=1}^{i=8} d_i = 2 \times 37.52 = 75.04 \text{ m}^3/\text{min for one paddle}$$

Since there are 48 paddles in this flocculation process,

$$D = 75.04 \times 48 = 3602 \text{ m}^3/\text{min}$$
$$F = 1.85 \times 60 = 111 \text{ m}^3/\text{min}$$

Thus $D/F = 3602 \div 111 = 32.1 > 30$. Assuming that the overall equipment efficiency is 70%, we find that there are eight paddles mounted per shaft. A large initial torque is created when the motor is switched on. For this reason, a 22-kW (30-hp) motor is selected for each of the six drive units.

NOTE: Example 2 is based on an actual plant design. The flocculation system was constructed as outlined above and has been producing excellent floc since 1957.

Example 3 Baffled Channel Flocculation Process

Given

Plant flow rate	1 m^3/s (23 mgd) maximum daily
Detention time	25 min at 1 m^3/s
Number of tanks	Two tanks
Stage of flocculation	Three stages
Energy input (at 1 m^3/s)	$G = 70\ s^{-1}$ for stage 1
	$G = 34\ s^{-1}$ for stage 2
	$G = 20\ s^{-1}$ for stage 3
Type of channel	Around-the-end type
Minimum water temperature	10°C or 50°F

Determine

(i) The size of the tank and the number of channels
(ii) The size of the channels and the arrangement of baffles
(iii) The headloss and the openings at each baffle

Solution (i) The total volume required is $(1 \times 60) \times 25 = 1500\ m^3$. Since there are two tanks, the volume of one tank is 750 m^3. Establish the average water depth as 2 m and provide six channels (2.2 m wide) per tank. The length of each channel should be

$$750 \div [(2.2 \times 6) \times 2] = 28.4\ m$$

Make the length 28.5 m.

(ii) Figure 3.2.4-11 illustrates all the information necessary for the design: size of the channels and arrangement of baffles.

(iii) Stage 1 flocculation channels require that G be 70 s^{-1} at 0.5 m^3/s per tank. Therefore, the headloss across the stage 1 channels is

$$h_1 = \frac{G^2 \nu V}{gQ} = [70^2 \times (1.3 \times 10^{-6}) \times 250] \div (9.81 \times 0.5)$$
$$= 0.325\ m$$

Stage 2 channels require that G be 35 s^{-1} at 0.5 m^3/s. Therefore,

$$h_2 = [35^2\ (1.3 \times 10^{-6}) \times 250] \div (9.81 \times 0.5) = 0.081\ m$$

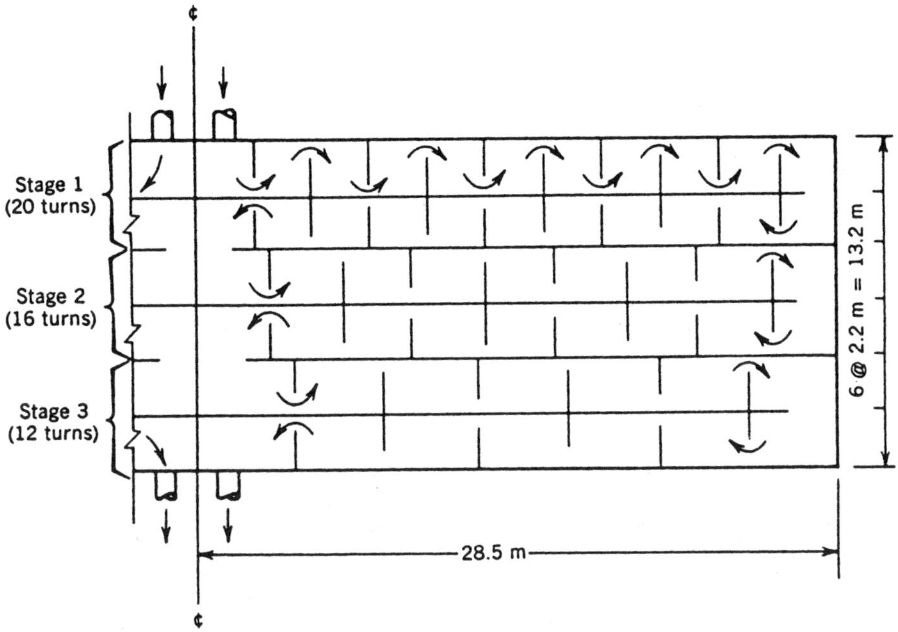

Figure 3.2.4-11 Baffled channel flocculation system.

Stage 3 channels require that G be 20 s^{-1}. Thus

$$h_3 = [20^2 \times (1.3 \times 10^{-6}) \times 250] \div (9.81 \times 0.5) = 0.027 \text{ m}$$

The total headloss across the tank is

$$\sum_{i=1}^{i=3} h_i = 0.433 \text{ m}$$

The opening dimension at each baffle of the stage 1 channels may be computed from the required headloss in each turn. This particular example contains a total of 20 turns, thereby producing a headloss of 0.325 m (a headloss of 0.0163 m per turn). The headloss per turn is computed as

$$h = \frac{Kv^2}{2g}$$

where $K = 1.5$ because the mean flow velocity of each channel is only 0.114 m/s and the flow enters tangentially into the next chamber; the flow enters in such a manner that a helicoidal flow pattern is developed and a clear 180° plug flow bend does not develop. The velocity required at each slit in the stage 1 channels is

$$v_1 = \left(\frac{2gh_1}{1.5}\right)^{0.5} = \left(19.62 \times \frac{0.00163}{1.5}\right)^{0.5}$$

$$= 0.462 \text{ m/s}$$

The required width for each slit in the stage 1 channels is calculated to be

$$w_1 = \frac{Q}{vH} = \frac{0.5}{0.462} \times 2 = 0.54 \text{ m}$$

In stage 2 flocculation there is a total of 16 turns and the total headloss across the stage is 0.081 m. Thus

$$v_2 = 0.258 \text{ m/s}$$

and the width of each slit is $w_2 = 0.97$ m. Stage 3 flocculation has 12 baffles and must produce a total headloss of 0.027 m; 0.00223 m at each baffle. Therefore, $v_3 = 0.17$ m/s and the width of the slit should be 1.47 m. The following table summarizes the characteristics of the baffled channel hydraulic flocculation process.

	Maximum Day, 1 m³/s			Average Day, 0.667 m³/s		
	Stage 1	Stage 2	Stage 3	Stage 1	Stage 2	Stage 3
Slit size (m)	0.54	0.97	1.47	0.54	0.97	1.47
Flow velocity at slit (m/s)	0.47	0.26	0.17	0.31	0.17	0.11
Headloss (m) per slit	0.0163	0.0051	0.0022	0.0073	0.0023	0.001
Number of slits/stage	20	16	12	20	16	12
G-value/stage (s^{-1})	70	35	20	38	19	11
Headloss (m) across process		0.43			0.19	

NOTE: The design presented in Example 3 is based on the assumption of 100% efficiency. The actual design should allow for 80% efficiency. Therefore, stages 2 and 3 should have 10 turns per channel, not 8 and 6 as shown. The total number of turns for the entire tank becomes 60, not 48 as shown (Figure 3.2.4-11).

Operation and Maintenance The three basic operational procedures common to all flocculation systems are checking floc size by way of visual observation, removal of scum from the water surface, and control of algae growth on the tank walls and baffles. If mechanical mixing is provided, the mixer speed control should also be inspected.

Floc observation during the night shift requires a specific type of lighting. The regular lamps located at the walkway are inadequate because the light is reflected;

therefore, proper floc observation requires a strong spotlight to be aimed toward the tank floor. This light should be located slightly above the waterline. If these requirements are met, the floc can be visualized clearly due to the Tyndall effect. However, the preferred setup is to have an underwater lamp placed 12 to 18 in. below the water level.

Recently, all major treatment plants in Japan have installed underwater television cameras in the flocculation tanks to monitor floc size with high magnification on a continuous basis. Computers automatically compare actual floc size, as observed by the cameras, to a preprogrammed standard size floc and warn plant operators when floc size is less than adequate in addition to automatic correction of coagulation feed rate. This type of system provides reliable plant operations and should be considered for new treatment plants.

The majority of water treatment plants experience problems with scum accumulation unless the flocculation tank has been designed to avoid this. Although the scum does not have a significant effect on the water treatment process, it is clearly visible and is unsightly to most visitors. Another problem common to many plants is the growth of algae on the tank walls and baffles. Algae growth is unsightly and can give the water an objectionable odor. Algae growth may be prevented through the use of prechlorination: A constant level of 0.3 mg/L or more of residual chlorine should be present at all times. However, for many treatment plants in the United States the use of prechlorination is limited due to restrictions placed on THM formation. The growth of algae is most severe in the settling basin. One solution to this problem is the periodic application of a high dose of chlorine during the night for a short duration (shock treatment).

In regard to maintenance, the baffled channel process requires only an annual dewatering of the tank (every 12 months), during which any necessary repairs can be performed and sludge removal can be accomplished. Sludge removal is easier with the horizontal turning (around-the-end) baffled channels than with the vertical turning (over-and-under) channels. The mechanical mixer flocculation units are significantly more maintenance intensive than the baffled channel type. These mixers typically require the lubrication oil of the speed-reducing units to be changed on a seasonal basis, repair of any oil leaks, and the repair and/or adjustment of the underwater shaft bearings when necessary.

A survey of the operational staff for 30 plants revealed that most operators did not frequently change the flocculation speed. At the majority of the plants surveyed, the mixing speed was not even adjusted with changes in the season, despite an obvious change in water temperature. The main reason given by the operators was that the efficiency of the process did not appear to improve even when attempts were made to optimize the speed. Consequently, certain mechanical speed-reducing units were often "frozen" at one speed and were no longer capable of changing speeds.

The infinite range speed-reducing units used for the flocculators require the most maintenance. Consequently, a significant number of plants have replaced them with two-speed motor control units. However, some recent infinite range speed reducers are more reliable as well as cost-effective. A variable-frequency control system for the motor speed adjustment is an example of this type of unit.

BIBLIOGRAPHY

Amirtharajah, A., "Design of Flocculation Systems," in *Water Treatment Plant Design for the Practicing Engineer,* R. L., Sacks, ed., Ann Arbor Science, Ann Arbor, MI, 1979, pp. 195–229.

ASCE, *Water Treatment Plant Design,* ASCE Manual of Engineering Practice 19, American Society of Civil Engineers, New York, 1952.

Bean, E. L., "Study of Physical Factors Affecting Flocculation," *Water Wastewater Eng.,* p. 33 (January 1953).

Brown, G. G., "Agitation," Chapter 34 in *Unit Operations,* Wiley, New York, 1953.

Camp, T. R., "Flocculation and Flocculation Basins," *Trans. ASCE,* 20:1 (1955).

Camp, T. R., et al., "Velocity Gradients and Internal Work in Fluid Motion," *J. Boston Soc. Civil Eng.,* 30:219 (1943).

Cheng, R. C., et al., "Enhanced Coagulation: A Preliminary Evaluation," *J. AWWA,* 87:2:91 (February 1995).

Crozes, G., et al., "Enhanced Coagulation: Its Effect on NOM Removal and Chemical Costs," *J. AWWA,* 87:1:78 (January 1995).

Davis, C. V., "Water Treatment," Chapter 38 in *Handbook of Applied Hydraulics,* 3rd ed., McGraw-Hill, New York, 1973.

Dempsey, B. A., et al., "Polyaluminum Chloride and Alum Coagulation of Clay–Fulvic Acid Suspensions," *J. AWWA,* 77:3:74 (March 1985).

Dempsey, B. A., et al., "The Coagulation of Humic Substances by Means of Aluminum Salts," *J. AWWA,* 76:4:141 (April 1984).

Dentel, S. K., et al., "Using Streaming Current Detectors in Water Treatment," *J. AWWA,* 81:3:85 (March 1989).

Edwards, M., "Predicting DOC Removal During Enhanced Coagulation," *J. AWWA,* 89:5:78 (May 1997).

Fair, G. M., and Geyer, I. C., *Water Supply and Wastewater Disposal,* Wiley, New York, 1954.

Fair, G. M., et al., *Water and Wastewater Engineering,* Vol. 2, Wiley, New York, 1968.

Haberer, K., and Schmidt, S. N., "The Haberer Process: Combining Contact Flocculation, Filtration and PAC Adsorption," *J. AWWA,* 83:9:82 (September 1991).

Hanson, A. T., and Cleasby, J. L., "Effect of Temperature on Turbulent Flocculation: Fluid Dynamics and Chemistry," *J. AWWA,* 82:11:56 (November 1990).

Hudson, H. E., Jr., *Water Clarification: Processes, Practical Design, and Evaluation,* Van Nostrand Reinhold, New York, 1981, pp. 40–64, 54–74, 75–100, 111–122.

Hundt, T. R., and O'Melia, C. R., "Aluminum–Fulvic Acid Interaction: Mechanics and Applications," *J. AWWA,* 80:4:176 (April 1988).

James, C. R., et al., "Considering Sludge Production in Selection of Coagulants," *J. AWWA,* 74:3:148 (March 1982).

Kawamura, S., "Application of Colloid Titration Technique to Flocculation Control," *J. AWWA,* 59:8:1003 (August 1967).

Kawamura, S., "Coagulation Considerations," *J. AWWA,* 65:6:417 (June 1973).

Kawamura, S., "Optimization of Basic Water Treatment Processes—Design and Operation: Coagulation and Flocculation," *J. Water SRI-Aqua,* 45:1:35–49 (1996).

Kawamura, S. "Removal of Color by Alum Coagulation," Parts 1 and 2, *Water Sewage Works,* p. 282 (August 1967), p. 324 (September 1967).

Kawamura, S., "The Fundamentals of Alum Flocculation as Applied to Water Purification," thesis, Ohio State University, Columbus, OH, 1961.

Leentraar, J., et al., "Some Dimensionless Parameters of Impeller Power in Coagulation–Flocculation Process," *Water Res.,* 14:140 (1980).

Montgomery, J. M., Consulting Engineers, *Water Treatment: Principles and Design,* Wiley, New York, 1985, pp. 116–134, 514–523.

Najm, I., et al., "Optimization EC with PAC Addition: A Case Study," *J. AWWA,* 90:10:88 (October 1998).

Oldshue, J. Y., "Flocculator Impellers: A Comparison," *Chem. Eng. Prog.,* p. 72 (1983).

Oldshue, J. Y., *Fluid Mixing Technology,* McGraw-Hill, New York, 1983.

Schulz, C. R., et al., "Evaluating Buoyant Coarse Media Flocculation," *J. AWWA,* 86:8:51 (August 1994).

Tekippe, R. J., and Ham, R. K., "Velocity Gradient Paths in Coagulation," *J. AWWA,* 63:7:439 (July 1971).

Thompson, J. D., et al., "Enhanced Softening: Factors Influencing DBP Precursor Removal," *J. AWWA,* 89:6:94 (June 1997).

Von Essen, J. A., "Energy Saving with Exotic Vessel and Impeller Design," *Proceedings of the 9th Engineering Foundation Conference on Mixing,* New England College.

Vrijenhoek, E. M., et al., "Removing Particles and THM Precursors by Enhanced Coagulation," *J. AWWA,* 90:4:139 (April 1998).

Walker, J. D., "High Energy Flocculation Unit," *J. AWWA,* 60:11:1271 (November 1968).

White, M. C., et al., "Evaluating Criteria for Enhanced Coagulation Compliance," *J. AWWA,* 89:5:78 (May 1997).

3.2.5 Sedimentation (Clarification) Process

Purpose One of the most common water and wastewater treatment unit processes is *sedimentation,* also known as *clarification.* Sedimentation is broadly defined as the separation of a suspension into a clarified fluid and a more concentrated suspension. The sedimentation process is designed to remove a majority of the settleable solids by gravitational settling, thereby maximizing the downstream unit processes such as filtration. Floatation also separates liquids from solids within a suspension. However, buoyancy is the key to this process.

The sedimentation process is divided into two classifications: grit chamber (plain sedimentation) and sedimentation tanks (clarifiers). The criteria for these classifications are the size, quantity, and specific gravity of the suspended solids to be separated. In this section we discuss the design of a sedimentation process based on both theory and the experience of the author, in addition to field operational data. It must be emphasized that the efficiency of the sedimentation process is greatly influenced by the level and adequacy of raw water conditioning upstream from the sedimentation process. The key to effective clarification is proper coagulation and flocculation of suspended matter within the raw water.

There are three main configurations for sedimentation tanks: horizontal rectangular basins, upflow sedimentation tanks, and upflow reactor clarifiers with sludge blanket. A rectangular tank with a horizontal flow is generally the favored tank configuration due to its hydraulic stability (if properly designed) and its tolerance of shock loadings. This type of tank also has a predictable performance and is capable of withstanding a flow rate that is twice the ordinary design rate recommended by the regulatory agencies, without significant deterioration in the quality of the settled water. Furthermore, the horizontal rectangular basins are simple to operate and adapt easily to the installation of high-rate settler modules.

Considerations There are many important considerations that directly affect the design of the sedimentation process: (1) overall treatment process, (2) nature of the suspended matter within the raw water, (3) settling velocity of the suspended particles to be removed, (4) local climatic conditions, (5) raw water characteristics, (6) geological characteristics of the plant site, (7) variations in the plant flow rate, (8) occurrence of flow short-circuiting within the tank, (9) type and overall configuration of the sedimentation tank, (10) design of the tank inlet and outlet, (11) type and selection of high-rate settling modules, (12) method of sludge removal, and (13) cost and shape of the tank. These subjects, along with other miscellaneous items, are discussed in detail.

Overall Treatment Process When designing an overall treatment process, the source of the water must be evaluated carefully. If the source is a river with episodes of flash floods, the treatment process train should include a grit chamber so that sand and silt will settle out near the intake. A grit chamber is a plain sedimentation process that removes discrete particles larger than 15 μm in diameter. If the filter bed is designed as an ordinary rapid sand bed, the degree of clarification is important because the turbidity of the filtered water is directly proportional to the settled water turbidity (Figure 3.2.5-1). Thus a conservatively designed sedimentation process should be used to obtain a settled water turbidity of less than 2 ntu.

As discussed in Section 3.2.4, flocs that are both small in size and physically strong, obtained by adding polymer as a filtration aid, are needed for multimedia or coarse, deep filter beds. With these characteristics in mind, the criteria used in the design of the sedimentation basin must be different from those used in the design of classic rapid sand filters, so that the overall process efficiency is maximized.

Data obtained from many existing plants demonstrate that dual-media filters remain very efficient despite filter influent turbidities ranging from 5 to 7 ntu, provided that solids in the raw water are properly conditioned by coagulant (Figure 3.2.5-2). The direct filtration process also supports this characteristic. The sedimentation process can therefore tolerate much higher hydraulic loading rates when using multimedia or coarse, deep media filter beds. In general, when dual-media filters are used, the hydraulic loading rate of the sedimentation process can be twice or even greater than the rate used in rapid sand filter system.

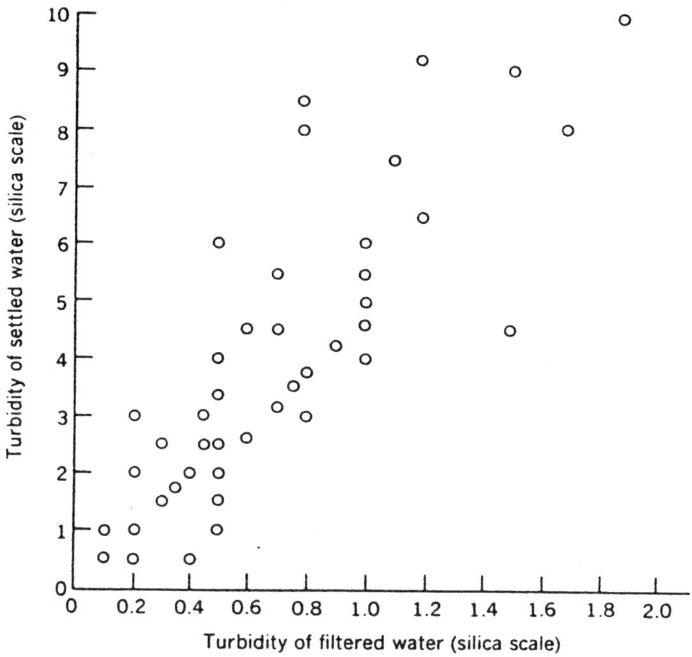

Figure 3.2.5-1 Relationship between turbidity of settled and filtered water on conventional rapid sand filter. *Notes:* (1) Filter rate ranged from 4.2 to 6.7 m/h (1.7 to 2.7 gpm/ft^2). (2) Alum flocculation and sedimentation before filtration. No filter aid was applied. [Reprinted from *Journal American Water Works Association,* Vol. 67, No. 10 (October 1975), by permission. Copyright © 1975, American Water Works Association.]

Nature of Suspended Matter Raw water contains two basic types of suspended matter: particles that are discrete in nature (nonflocculable) and colloidal suspensions. Examples of *nonflocculable* particles are sand and silt. *Colloidal* suspensions include clay, microorganisms, and substances that cause color. Colloidal suspensions must be flocculated with coagulant so that their particles will be removed during sedimentation and filtration. Discrete particles may be removed by a cyclone separator, plain sedimentation, or through the use of a grit chamber. However, colloidal suspensions require a different type of settling basin (clarifier), one that has a much more conservative hydraulic loading rate. This is explained in the section covering design criteria.

Settling Velocity of the Particles The sedimentation process is based on gravitational settling of the particles. For this reason, the design engineer should know the settling velocities of each particle that is to be removed in the anticipated water temperature range. The settling velocity of each particle in still water may be measured under controlled conditions within the laboratory. Bench-scale flocculation tests (jar tests) and the measurement of floc settling velocities are therefore strongly recom-

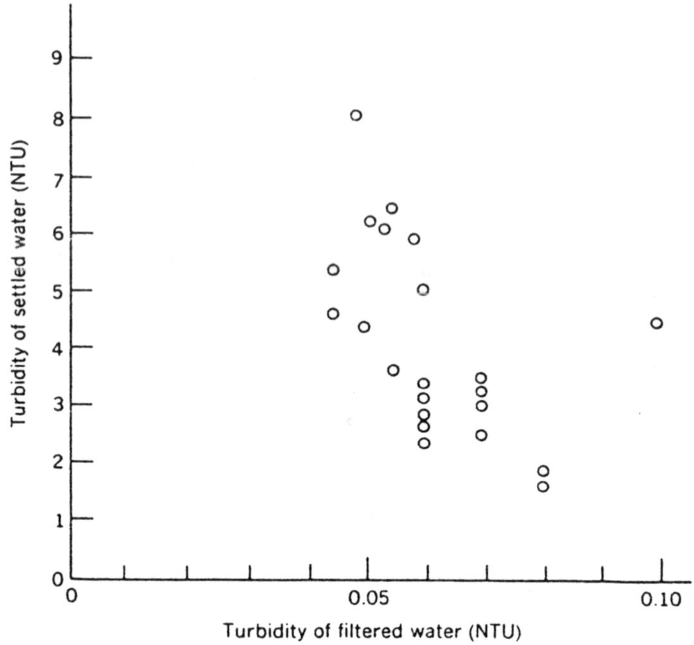

Figure 3.2.5-2 Relationship between turbidity of settled and filtered water on typical dual-media filter. *Notes:* (1) Filter rate ranged from 10 to 24 m/h (4.1 to 9.4 gpm/ft^2). (2) Polyelectrolyte (as filter aid) dosage ranged from 7 to 23 ppb. [Reprinted from *Journal American Water Works Association,* Vol. 67, No. 10 (October 1975), by permission. Copyright © 1975, American Water Works Association.]

mended. As previously mentioned, an appropriate safety factor should be applied since floc settling velocities can vary widely depending on several parameters: type of coagulant employed, mixing conditions during the flocculation process, and the nature of the colloidal matter contained in the raw water. If these tests cannot be performed, for whatever reason, the numbers listed in Table 3.2.5-1 may be used.

According to Hazen (1904), the efficiency of an ideal horizontal flow sedimentation tank is a function of the settling velocity (v_0) of the particle to be removed, the surface area of the tank (A), and the rate of flow (Q) through the basin. This relation can be expressed as

$$v_0 = \frac{Q}{A}$$

Q/A is commonly known as the surface loading or overflow rate and is expressed in m/h or gpm/ft^2. Once the settling rate of the suspended matter (being removed) is determined and the rate of flow through the tank is known, a physical basin size (area of the tank) can be calculated. A minimum safety factor of 1.5 should be included in the calculation of the area.

TABLE 3.2.5-1 Setting Velocities of Particles (at 10°C) and Compatible Hydraulic Surface Loading Rates of Sedimentation Tanks

						Compatible Tank Surface	
		Particle Size		Settling Rate		Loading	Approximate
Type of Particle	Specific Gravity	Mesh	mm	mm/s	fpm	m/h	gpm/ft^2
Sand	2.65	18	1.0	100	19.7	360	144
Sand	2.65	20	0.85	73	14.3	263	105
Sand	2.65	30	0.6	62	12.2	223	89
Sand	2.65	40	0.4	42	8.2	151	60
Silt	2.65	70	0.2	21	4.1	76	30
Silt	2.65	100	0.15	15	3.0	54	22
Silt	2.65	140	0.10	8	1.6	29	12
Silt and clay	2.65	200	0.03	6	1.2	22	9
Silt and clay	2.65	230	0.06	3.8	0.75	14	5.6
Silt and clay	2.65	400	0.04	2.1	0.41	7.5	3
Clay	2.65	—	0.02	0.62	0.12	2.3	0.9
Clay	2.65	—	0.01	0.154	0.03	0.54	0.2
Alum floc	1.001	—	1–4	0.2–0.9	0.04–0.18	0.71–3.3	0.3–1.3
Lime floc	1.002	—	1–3	0.4–1.2	0.8–0.23	1.5–4.3	0.6–1.7

According to Hazen, the efficiency of the tank is independent of its depth and detention time. However, a shallow depth theoretically favors settling of the particles. Detention time also affects tank efficiency since flocculated particles become larger and heavier with time, due to the mixing effect in the tank, and therefore settle much faster. Consequently, the detention time of the basin is significant with regard to flocculant particles.

Local Climatic Conditions Several local climatic conditions must be considered when designing a sedimentation process: daily and seasonal temperature fluctuation, water and air temperature range, degree of rainfall, and intensity and direction of the winds. If the daily temperatures fluctuate widely, within a day or a week, the basin will most likely experience flow short-circuiting, due to a density current. Moreover, regions with severe winter conditions should generally have the tank either covered or housed in a heated building.

Water treatment plants located in regions of warm or hot weather will have problems with algal bloom and have a heavy growth of vegetation in the watershed. If the region is both warm and sunny, the treatment plant will always have problems with algal growth on the basin walls and launders. This can be prevented by either covering the basin or providing sufficient chlorine residual to the influent water. However, chlorine should be added with extreme caution because of the likelihood that THM formation will exceed the tolerable limits.

Treatment plants located in regions of high rainfall tend to have large amounts of suspended matter in the raw water. The design should therefore include a continuous sludge removal system that will remove the accumulated sludge before it reduces the effective tank volume and before it becomes septic. In cases where the raw water contains a high concentration of grit and silt, a presedimentation basin must be included in the design.

In regions other than the tropics, a majority of the local vegetation sheds leaves into the water source, as well as the basin, during the winter, causing organic color to be imparted to the water, creating low pH and clogging the filter bed. Floc formed from highly colored water through the use of high concentrations of alum or ferric salts (enhanced coagulation) is very fragile and difficult to settle. Therefore, both submerged and surface metals and concrete surfaces will corrode unless special design alterations are made.

Regions that experience strong winds continually have design problems of their own. Strong winds can greatly disturb the water flow within the clarifiers and may induce flow short-circuiting, depending on the direction of the wind and orientation of the basins. Furthermore, if trees are located near the settling basins, the winds may also blow leaves and branches into them.

Raw Water Characteristics One of the most dramatic effects of raw water characteristics with regard to both horizontal flow and upflow clarifiers is the sudden change in water temperature. A change of 1°F (0.5°C) can cause a severe density flow (Figure 3.2.5-3). If the influent water is colder, it will travel along the bottom

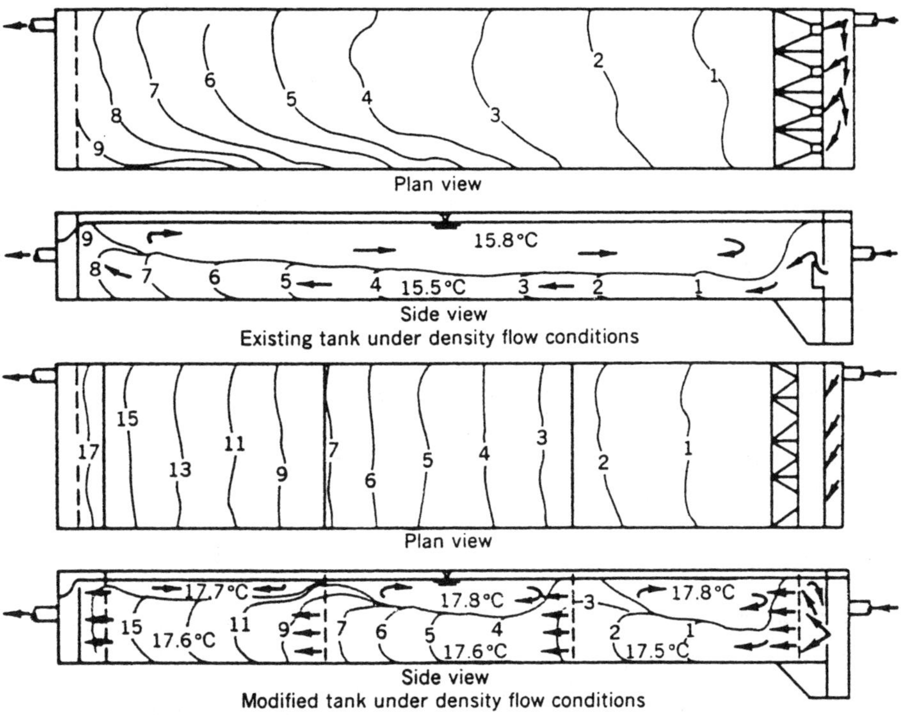

Figure 3.2.5-3 Original and modified tanks under density flow conditions. Observations are based on a hydraulic scale model study. *Notes:* (1) Computed detention time is 49 min or 4 h 15 min for the actual tank. (2) Observed flow-through time is 9 min for the original tank model and 17 min for the modified model. [Reprinted from *Journal American Water Works Association,* Vol. 73, No. 7 (July 1981), by permission. Copyright © 1981, American Water Works Association.]

and rise upward toward the end of the tank. If the water is warmer, flow short-circuiting will occur at the surface of the tank. This condition is particularly serious when a treatment plant has more than one source and these sources are often switched, or if the blending ratio of two or three sources is frequently changed.

Turbidity spikes also cause density currents that are identical to those created by cold influent water. Highly turbid water produces a large volume of floc after flocculation. Since these flocs have a higher specific gravity than the water in the tank, the influent dives to the bottom, creating a distinct density flow. A raw water turbidity of over 30 ntu usually produces this type of density flow.

Attention should also be given to the aggressiveness of the raw water to submerged parts (both concrete and metal). The design engineer should either choose an appropriate type of cement and provide protection by applying a special coating or use noncorrosive material such as type 18-8 stainless steel. Submerged steel parts must be properly painted and preferably have cathodic protection. Although type 18-8 stainless steel is significantly more expensive, its use generally will eliminate the need for costly operation and maintenance procedures, such as painting and cathodic protection.

Soil Characteristics Consideration must also be give to the soil characteristics of the plant site. Of special interest are the bearing capacity of the native soil, the presence of geological faults, the level of the underground water table, and the salt content of the soil. Inadequate geotechnical studies have much too often resulted in the subsidence of the clarifiers and the development of large cracks in the basin walls and floor slabs.

A high salt content within the soil quickly corrodes metals, including reinforcing steel bars embedded in concrete. This problem may be minimized by providing a concrete cover that is greater than 2 in. A soil resistivity below 1500 $\mu\Omega$/cm is generally considered to be highly corrosive, and the use of steel shell tanks should be avoided unless special measures for corrosion protection are implemented. Under these conditions, special cement (type II or V) and a low-shrinkage concrete mixture must be specified for the concrete.

Variation in Plant Flow Rate Sudden changes in plant flow rate, also known as *hydraulic shock loads,* may affect the flow condition and disturb the settling efficiency of the tank. This is especially true with sharp increases in flow rate during short periods of time. Clarifiers that are most susceptible to hydraulic shock loading are the upflow clarifier and the sludge blanket reactor clarifier. The rectangular horizontal flow clarifiers generally tolerate hydraulic shock loadings much better than the types mentioned previously.

Flow Short-Circuiting Three basic types of flow short-circuiting may occur within the clarifiers. The first is caused by improper design. The second, although not serious, is a chronic flow short-circuiting that becomes evident when a significant amount of floc is carried over to the filters. The third type is the creation of a density flow within an ordinary sedimentation tank. This density flow is caused by a difference in density between the influent water and the water within the tank; the influent water may be colder or warmer, contain a higher concentration of suspended solids, or create a salinity pulse, that is, an influx of NaCl due to the intrusion of seawater.

The first type of flow short-circuiting is typified by a tank design that has both a poor inlet design and a short distance between the inlet and the outlet. Other factors that exacerbate this problem are a large tank width in conjunction with a short tank length and a deep water depth. A tank designed in this manner will always have a high Reynolds number and very low flow stability.

The second type of flow short-circuiting is observed in many sedimentation tanks whose plants are based on the prevailing tank design criteria set by the regulatory agencies. The ordinary plant visitor may not notice the flow short-circuiting, but plant operators observe that (during the day) the influent water tends to dive down at the inlet of the tank and rise at the effluent, carrying much floc with it. The operators call this phenomenon *noonday turnover.* In actuality, the most serious flow short-circuiting tends to occur mainly from midnight to dawn. In this case, the influent water short-circuits at the surface due to its slightly higher water temperature. As long as the top of the tank is exposed to the weather, both the first and second types of flow short-circuiting cannot be controlled effectively by the placement of diffuser walls at both the inlet and outlet of the tank.

The third type of flow short-circuiting is a severe case of the second. This density flow problem is commonly caused by switching from one source of water supply to another, suddenly changing the mixing ratio of two and sometimes three different sources, abruptly shifting the reservoir intake elevation, and having a high raw water turbidity (over 50 ntu). This particular type of flow short-circuiting may also be induced by strong winds blowing along the longitudinal axis of the tank or by a salinity pulse. Nothing can stop this type of density flow once it has started.

This third type of flow short-circuiting can be minimized by installing intermediate diffuser walls perpendicular to the flow direction, in the middle or at two-thirds of the tank length. The minimization is achieved by increasing the detention time and mixing effect in front of each intermediate diffuser wall (see Figure 3.2.5-3). The magnitude of the density current may be evaluated by *Harleman's formula:*

$$v = \left(8g\frac{\Delta\rho}{\rho}\frac{hs}{f(1+\alpha)}\right)^{0.5}$$

where

v = velocity of the density flow (ft/s)

g = acceleration of gravity (32.2 ft/s^2)

$\Delta\rho$ = density difference between the two liquids (lb/ft^3)

ρ = density of the influent (lb/ft^3)

h = depth of the density current flow (ft)

s = slope of the channel bottom

f = Darcy–Weisbach friction factor

α = correction factor for kinetic energy ranging from 0 to 1 (0.43 for turbulent flow)

A similar equation is recommended by the Tokyo Metropolitan Government Report (1963) for computing the velocity of the density flow:

$$v = \left(2g\,\frac{\Delta\rho}{\rho}\,\frac{\Delta h\alpha}{k}\right)^{0.5}$$

where

v = velocity of the density flow (m/s)

$\Delta\rho$ = difference in density (g/cm^3)

ρ = density of the influent (approximately 1.0 g/cm^3)

g = acceleration of gravity (9.81 m/s^2)

Δh = depth of the density flow (m)

α = correction factor for kinetic energy (approximately 0.5)

k = dilution factor (normally 1.5 to 2)

According to the field measurements conducted by this author, the difference in water temperature (near the surface of the tank and the influent water) is usually 0.2 to 0.5°C during the day and the flow velocity of the density flow is 2.6 to 6 ft/min (0.8 to 1.8 m/min) despite a design flow velocity of 1.3 ft/min (0.4 m/min).

Type of Sedimentation Tank Upflow clarifiers and reactor clarifiers perform very well if both the raw water characteristics and the hydraulic loading rates are constant. The reactor clarifier is best suited for lime or lime with soda ash softening of the water because of the seeding effect of the unit. Consequently, lime softening of underground water is the perfect application for the reactor clarifier. However, these two types of sedimentation tanks are very susceptible to hydraulic and solids shock loading. Wind and uneven exposure to sun can also disturb the effectiveness of the clarification process.

The upflow clarifier and the reactor clarifier are usually compact and predesigned by the equipment manufacturer. The overall cost may therefore be cheaper than using custom-designed rectangular tanks. Furthermore, the sludge can easily be removed from these tanks, due to the continuously fluidized sludge blanket. On the other hand, a significant amount of metal is used on the inside of these tanks and the problems of corrosion and maintenance must be addressed. It is necessary to emphasize that upflow and reactor clarifiers are not capable of functioning at rates well over the designed rates because floc removal is based on the balance between particle settling velocity and vertical flow velocity.

Unlike the upflow and reactor clarifiers, the horizontal rectangular basin can tolerate both hydraulic and solid shock loadings. In fact, rectangular basins can handle a flow rate that is 50 to 100% over the design rate without significantly degrading the quality of the settled water. The flow condition of the basin, under higher flow rates, is more stable because of an increase in the Froude number. Most large municipal water treatment plants use horizontal rectangular clarifiers primarily because of their flexible performance, predictable settling efficiency, and minimum maintenance cost. A comparison between the upflow and rectangular clarifier is presented in Table 3.2.5-2.

Overall Configuration The preferred configuration of the sedimentation basins is a battery of rectangular tanks sharing common sidewalls. The upstream portion of each basin contains a flocculation process, and each rectangular tank is connected by a common influent and effluent channel; both channels span all the basins. This configuration offers several advantages: a reduction in construction cost, minimum site requirement due to the lack of space between the basins, easy distribution of the plant influent water to each basin, minimum floc breakage, minimum floc settlement between the flocculation and sedimentation basins, simple refilling of the basins with clean settled water from the effluent end via a valve, easy single-point application of water treatment chemicals to the combined settled water (prior to the filters), and minimum yard piping.

Another configuration that is often adopted is a circular array of reactor clarifier basins arranged in a grid point fashion or that follow the topography of the plant site. The only disadvantage with this type of configuration is that each tank requires individual inlet and outlet yard pipings, and adequate spacing must therefore be provided between the basins. The reactor clarifier basins have a flocculation zone in the center and sedimentation zone in the outer portion; they are proprietary units designed by the equipment manufacturers. The reactor clarifier basin can be either circular or square in shape. The configuration of the basin is strongly influenced by the shape, topography (slope), and geographic condition of the plant site. Thus the design engineer may not have much of a choice in determining the basin configuration.

Inlet and Outlet of the Basin Regardless of the shape and the type of clarifier, hydraulic control of the basin influent is one of the most important design factors. As expounded in many technical articles, any flow imbalance at the basin inlet will lead to flow short-circuiting, jetting, turbulences, and an overall lack of hydraulic stability within the settling zone. Although various baffling methods have been tested for distributing the water to the settling basin at the basin inlet, the most simple and effective method to date is the perforated baffle. The most effective perforated baffle, for a specific basin shape, can only be determined through hydraulic scale model studies. However, certain general rules may be used in designing the baffle walls for most basins. The author conducted an extensive hydraulic-scale model study in the late 1950s and determined the general criteria for the perforated baffle. Based on this study the major design requirements are as follows:

1. The ports should be uniformly distributed across the baffle wall, which in turn covers the entire cross section of the basin.
2. A maximum number of ports should be provided so that the length of the jets is minimized and the dead zone between ports is reduced.
3. The headloss through the ports should be approximately 0.12 to 0.35 in. (0.3 to 0.9 mm), to equalize the flow distribution across the entire cross section of the tank inlet with minimum floc breakage.

TABLE 3.2.5-2 Selection Guide for Some Basic Types of Clarifiers

Type of Clarifier	Some Design Criteria	Advantages and Disadvantages	Proper Application
Rectangular basin (horizontal flow)	Surface loading: 0.34–1 gpm/ft^2 (0.83–2.5 m/h) Water depth: 3–5 m Detention time: 1.5–3 h Width/length: >1/5 Weir loading: <15 gpm/ft (11 m/m· h)	1. More tolerance to shock loads 2. Predictable performance under most conditions 3. Easy operation and low maintenance costs 4. Easy adaptation to high-rate settler modules a. Subject to density flow creation in the basin b. Requires careful design of the inlet and oulet structures c. Usually requires separate flocculation facilities	Most municipal and industrial water works Particularly suited to larger capacity plants
Upflow type (radial-upflow type)	Circular or square in shape Surface loading: 0.5–0.75 gpm/ft^2 (1.3–1.9 m/h) Water depth: 3–5 m Settling time: 1–3 h Weir loading: 10 gpm/ft (7 m/m · h)	1. Economical compact geometry 2. Easy sludge removal 3. High clarification efficiency a. Problems of flow short-circuiting b. Less tolerance to shock loads c. A need for more careful operation d. Limitation on the practical size of the unit e. May require separate flocculation facilities	Small to mid-sized municipal and industrial treatment plants Best suited where the rate of flow and raw water quality are constant

Reactor clarifiers	Flocculation time: approx 20 min Settling time: 1–2 h Surface loading: 0.8–1.2 gpm/ft^2 (2–3 m/h) Weir loading: 10–20 gpm/ft (7.3–15 m/m · h) Upflow velocity: <50 mm/min	1. Incorporates flocculation and clarification in one unit 2. Good flocculation and clarification efficiency due to a seeding effect 3. Some ability to take shock loads a. Requires greater operator skill b. Less reliability than conventional due to a dependency on one mixing motor c. Subject to upsets due to thermal effects	Water softening (1.5–2 gpm/ft) A plant that treats a steady quality of raw water
Sludge blanket clarifiers	Flocculation time: approx 20 min Settling time: 1–2 h Surface loading: 0.8–1.2 gpm/ft^2 (2–3 m/h) Weir loading: 10–20 gpm/ft (7.3 to 15 m/m · h) Upflow velocity: <10 mm/min Slurry circulation rate: up to 3–5 times the raw water inflow rate	1. Good softening and turbidity removal 2. Compact and economical design 3. Tolerates limited changes in raw water quality and flow rate a. Very sensitive to shock loads b. Sensitive to temperature change c. Several days required to build up the necessary sludge blanket d. Plant operation depends on a single mixing flocculation unit motor e. Higher maintenance costs and a need for greater operator skill	Water softening Flocculation/sedimentation treatment of raw water with a constant quality and rate of flow Plant treating a raw water with a low content of solids

Notes: (1) The reactor clarifiers and the sludge blanket type clarifiers are often considered to be in the same category.
(2) Surface loading: m/h = ($m^3/m^2 \cdot d$) ÷ 24; therefore, 60 $m^3/m^2 \cdot d$ = 1 gpm/ft^2.
(3) Weir loading: m/h · m = ($m^3/d \cdot m$) ÷ 24; therefore, 175 $m^3/d \cdot m$ = 10 fpm/ft.

4. The headloss through the ports should be less than 0.4 in. (1 cm) to prevent floc breakage.
5. The size of the ports should be uniform in diameter, 3 to 8 in. (0.075 to 0.2 m), to avoid clogging by algae and other debris.
6. The ports should be spaced approximately 10 to 20 in. (0.25 to 0.5 m) on center to provide structural strength for the diffuser wall.
7. Port configuration should be arranged in such a manner that the parallel jets will direct the flow toward the basin outlet.
8. If the basins are fed from a common channel or if there is an inlet feed channel zone between the flocculation and sedimentation basins, the inlet diffuser wall should be located approximately 6.5 to 8 ft (2 to 2.5 m) downstream of the inlet. The inlet zone significantly increases floc size by allowing the proper gently mixing effect to occur (Figure 3.2.5-4).

The water exiting the sedimentation basin should be uniformly collected across an area that is perpendicular to the proper flow direction. Beginning in the late 1950s a number of long launders (skimming trough) were installed in the sedimentation tanks in an attempt to achieve uniform collection of the settled water. This practice is quite varied for the upflow clarifiers but generally not for the rectangular settling tanks. The installation of long launders in horizontal flow rectangular tanks is virtually ineffective when either a bottom-flowing or surface-running density current exists in the tank.

As discussed previously, most sedimentation basins exhibit a bottom-flowing density current; the basin end wall deflects this current, which then rises to the surface, carrying floc, and flows into the end portion of each launder. This phenomenon has been observed in both operational basins and in hydraulic-scale model studies. In terms of design, this phenomenon suggests that transverse launders placed away from the end wall of the basin might be more effective than finger (long) launders. A few operational plants have supported this proposal. Yet long launders have three major advantages when used in conjunction with rectangular basins: the water level of the tank remains virtually constant regardless of changes in the plant flow rate; wave action created by winds is minimized; and high-rate settler modules, such as tube settlers, are easily hung from the launder system.

V-notched weir plates are generally attached to the launders. The weirs create a free discharge flow that enters the launder system, along with any floating matter. Consequently, there is a tendency to break the fragile alum floc. Unlike the weirs, submerged orifices do not allow floating scum to flow into the launder and they also tolerate less accurate installation. Therefore, in some cases, submerged orifices are drilled into the sidewalls of the launders. The floc breakage characteristics of submerged orifices may be slightly better than those of weirs, but the flow velocity through the orifice must range from 1.6 to 2.5 ft/s (0.5 to 0.75 m/s) in order to pro-

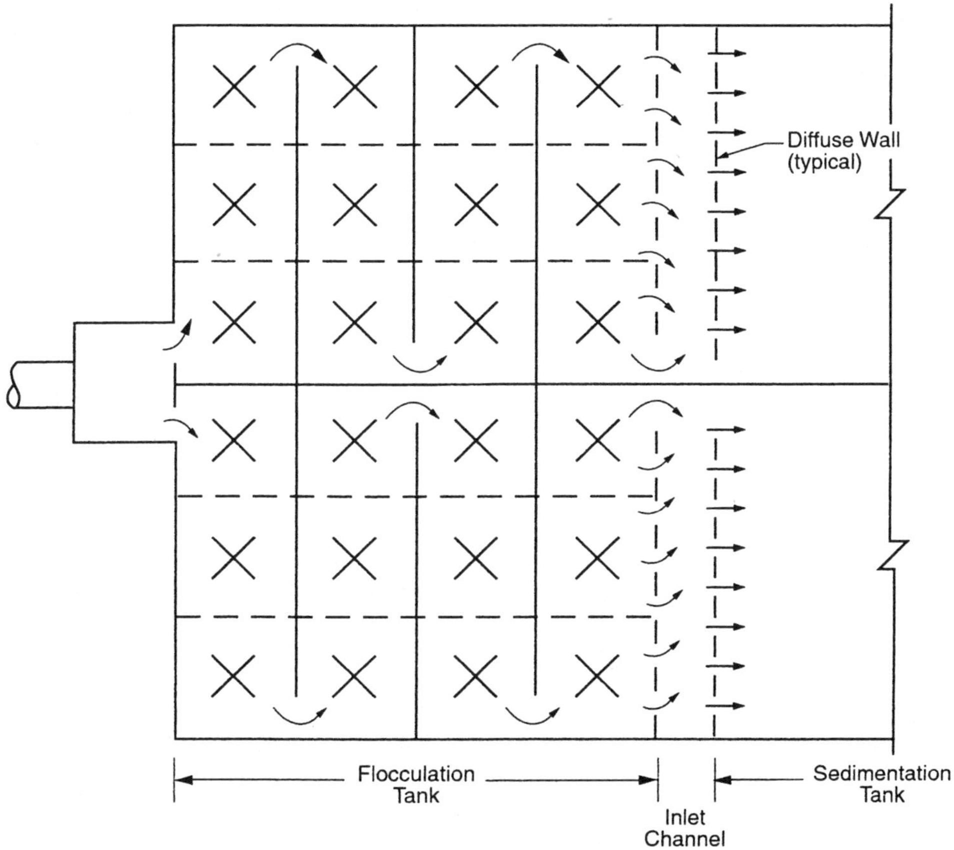

Note: The perforated baffle walls shown in between the flocculators are optional. When hydrofoil type mixing blades are used, there is no need for those baffles.

Figure 3.2.5-4 Inlet channel between flocculation and sedimentation tanks.

vide a controlled headloss of approximately 1.5 in. (38 mm). Some floc breakage is therefore unavoidable.

The previously held idea that the installation of perforated baffles at the tank outlet is equivalent to the installation of inlet baffles (to collect settled water uniformly) still persists. On the surface this concept appears to be reasonable and sound. However, observations of hydraulic-scale model studies and operational plants demonstrate that the effluent baffle is unexpectedly ineffective. The primary problem is that the effluent baffle is only capable of providing an insignificant correction factor for a density current. It is therefore common to see numerous flocs downstream of the effluent baffle, although the water upstream may be very clear. For this reason, the use of the effluent baffle at the end part of the basin is not recommended. In fact, the effluent baffle is now seldom employed.

High-Rate Settling Modules The term *high-rate settler* refers to any small inclined tubes or tilted parallel plates which permit effective gravitational settling of suspended particles within the modules. The detention time in the settling modules ranges from 5 to 20 min, depending on the type of module and the surface loading rate. Surface loading is conveniently calculated from the area of the basin that is covered by the settling module and is generally 2 to 3.5 gpm/ft^2 (5 to 8.8 m/h). The settling efficiency of modules with such high hydraulic loading values is equivalent to a conventional sedimentation basin with a designed detention time of 3 to 4 h and a surface loading rate of 0.5 gpm/ft^2 (1.3 m/h).

The concept of an ideal settling tank (and the advantages of shallow tanks) was first advocated by Hazen (1904) at the beginning of the twentieth century. Camp (1953) later created the "double-try" settling tank while applying this concept. Around 1955 the first high-rate settling module was commercialized in Japan under the name Uno Separator; the module consisted of sloping parallel plates spaced 25 to 50 mm apart. Similar modules were later marketed in Sweden under the name Lamella Separator and a tube settler module was also marketed by Micro Floc, Inc. (in 1969) and other firms (United States).

Shortly after release of the first high-rate settler, a modified parallel-plate module was released. This version had a series of deflector baffles attached to one side of the plates, which were designed to promote flocculation within the module (Figure 3.2.5-5) to permit greater efficiency in clarification. However, data from side-by-side pilot studies suggest that the effectiveness of this module appears to be only marginally better than modules without baffles.

Although the high-rate settlers are designed to deliver a certain level of performance, they do not always function according to their specifications. The primary reasons for this aberrant behavior are (1) poor flocculation, (2) uneven flow distribution to the inlet of the settler module, (3) heavy algal growth in the modules, and (4) heavy scale ($CaCO_3$) deposits on the module. The heavy scaling clogs the flow paths and, more important, eventually leads to corruption of the system.

Depending on the type of parallel-plate settling module, water enters either horizontally and parallel to the plates or upward along the 60° tilted plates. In the tube settlers the direction of flow is always upward at 60° to the horizontal plane, although some earlier prototypes have their plates at a 15° angle. Under normal circumstances, the settled floc slides down the inclined plates or tubes to the tank bottom for subsequent mechanical sludge removal. The design of the high-rate settler modules must address the following issues: settling velocity and characteristics for the suspended matter, flow velocity within the settler module, surface loading, Reynolds number and Froude number for the flow within the module, selection of the appropriate sludge collection unit (to be installed beneath the settler module), spacing of the launder when installed above the settling module, and the supporting system.

Experience has shown that data from small-scale pilot plants (area of 1 m or less) cannot be used as meaningful design criteria because pilot studies often give very optimistic loading rates; the studies are usually conducted under carefully controlled and ideal conditions. The actual design criteria should be more conservative because

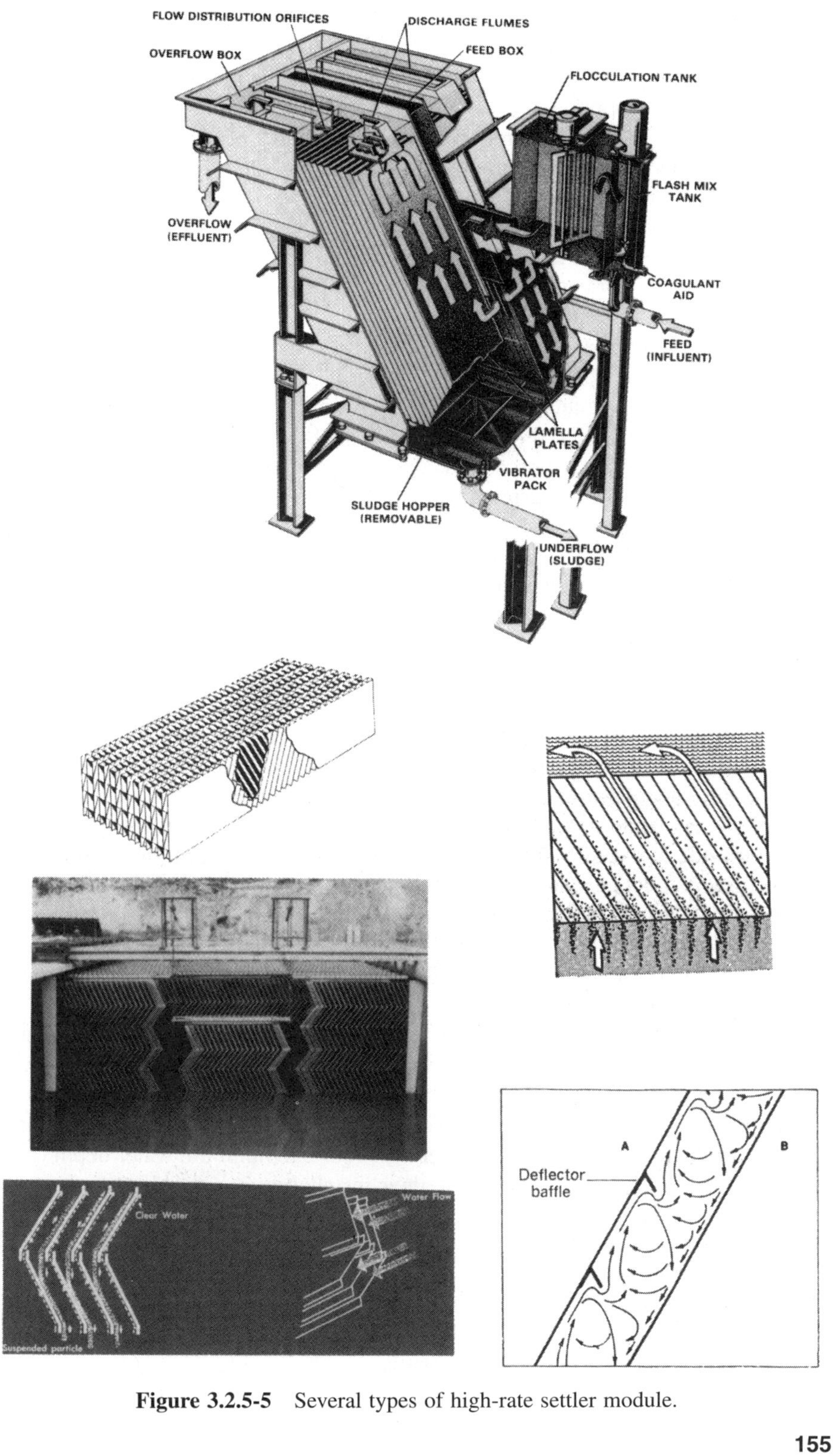

Figure 3.2.5-5 Several types of high-rate settler module.

of the potential for nonuniform flow conditions at the inlet to the settler modules as the result of density flow, improper tank design, or unfavorable coagulant dosage or flocculation efficiency. Depending on the surface area of the tank that is covered by the settler modules, the maximum surface loading value should be limited to 2 to 2.5 gpm/ft^2 (5 to 6.3 m/h) for cold regions and 3 to 3.5 gpm/ft^2 (7.5 to 8.8 m/h) for warm-to-hot regions.

Selection of the proper material for the fabrication of settler modules is very important. A module fabricated with paper-thin PVC plates often deteriorates within 10 to 15 years due to weathering and wave actions in the basins. Some settler modules are made with thick plastic plates and have a module height over 6 ft (2 m). Although more durable, the cost is very high and often $250 to 300 for each 1 ft^2 (0.3 m^2) of tank surface area to cover the module (late 1990 cost on the U.S. west coast). The preferred material is thin stainless steel plates (type 304) because of the structural strength and noncorrosive nature, with a smooth surface to minimize the adherence of floc and algae growth.

The design of the inlet (to the settler module) is very important because it is essential to have a uniform flow into the entire module. In the case of a rectangular tank, the first quarter of the basin length is generally devoid of settler modules. This allows for the installation and maintenance of sludge collection systems such as chain-and-flight units, and for heavy floc to settle naturally and also provides an inlet buffer zone that improves the inlet flow condition to the settlers. In tube settler modules, an average flow velocity of approximately 0.5 to 0.65 ft/min (0.15 to 0.2 m/min) that is normally used in settling most alum floc. Moreover, an approaching flow velocity of approximately 2 ft/min (0.6 m/min) should be used in the tank upstream of the module installation.

The detention time of the settler module varies as a function of the type of module. Tube settlers generally have a detention time of 3.5 to 5 min, while tilted plate settlers provide a longer period of time; some units even require 15 to 20 min. The Reynolds and Froude numbers are not considered to be part of the design criteria, yet they act as good guides in designing the high-rate settlers. The Reynolds number for the settlers should be kept below 200, preferably around 50, and the Froude number should be greater than 10^{-5}.

When selecting a continuous sludge removal system, the number of available alternatives is rather limited. The most applicable types of sludge collector, for the sedimentation process, are the chain-and-flight (fiberglass-reinforced plastic) and the cable-operated underwater bogies with squeegees. It should be noted that these types of sludge collecting systems generally require a minimum water depth (in the tank) of 12 ft (3.7 m) to accommodate their installation.

Use of the skimming trough system is required because it uniformly collects clarified water from the area of the tank that is covered by the settler modules. However, the settler modules may be hung from the launders, provided that the launders are designed to tolerate this strain. The launders are usually spaced 10 to 13 ft (3 to 4 m) on center, and hydraulic loading is usually less than 17 gpm/ft (12.5 m^3/m · h). Figure 3.2.5-5 depicts a few high-rate settlers.

Sludge Removal Method The settling characteristics of alum ferric floc are typified by three stages: the hindered settling, transition, and compacting stages (Figure 3.2.5-6). During the hindered settling stage the floc slides down the smooth surface of the plates (or tubes), which are inclined at a minimum of 60° from the horizontal plane. In the transition and compaction stages the floc no longer slides down as easily, and the angle of repose (of the sludge) is close to 90°. In most cases only a few hours are necessary for the settled floc to reach the compaction stage.

A design flaw that is commonly made by inexperienced engineers is to provide a series of shallow-angle sludge hoppers on the bottom of the basin. In this scheme, each hopper is fitted with a pipe containing a shutoff valve as the means of sludge removal. According to this design, the valve is opened two to three times a day and the sludge is washed out with the water. However, even when the sidewalls of the hoppers are built 45 to 60° from the horizontal, this particular design does not work. The reason for the design failure is that a water passage, the same size as the pipe diameter, is created in the settled sludge when the valve is opened and the majority of the sludge therefore remains settled and undisturbed. To salvage this design, the sludge must be kept in a fluid condition at all times. Consequently, the bottom of the sedimentation tank should be designed as a flat surface with or without a slight slope toward the location of the sludge hopper, depending on the method of mechanical sludge removal.

In the United States the three continuous sludge removal systems that have been proven and widely used in recent years are center-pivoted rotating rakes, the chain-and-flight collector, and the traveling bridge. The center-pivoted rakes have a good history of service and low maintenance cost. Yet there are four major drawbacks to this particular system: (1) a large amount of metal is used for the raking assembly, creating corrosion and associated maintenance problems; (2) the center-pivot portion of the tanks must be very deep (over 20 ft) with a conically shaped bottom, which significantly increases tank construction cost; (3) even if the expandable arms are used to cover the corners of the square area, there is inadequate sludge removal in the corners; and (4) the capital cost for the system can potentially be high if the rectangular tank has a length/width ratio of 3 : 1 or larger. Furthermore, plant

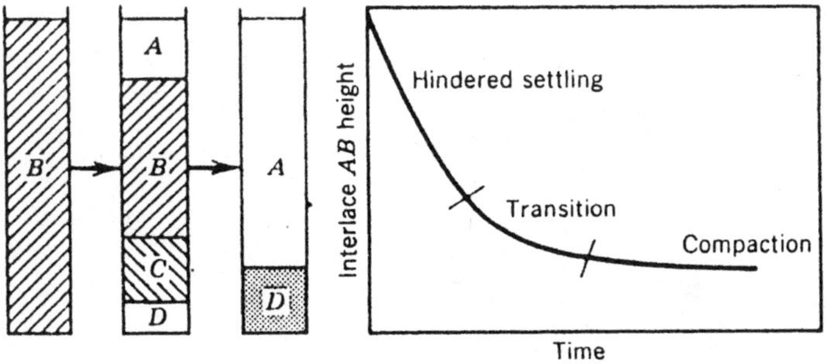

Figure 3.2.5-6 Settling characteristics of alum floc. *A,* Supernatant zone; *B,* discrete settling zone; *C,* hindered settling zone; *D,* compression zone.

operators of treatment plants that have only the front half of the tank covered with the rotating rakes are not satisfied with its performance. The chain-and-flight collector has a maximum length of approximately 230 ft (70 m) per unit. The traveling bridge collector can be installed to any length of tank. It must be emphasized that a tank equipped with high-rate settler modules requires a compatible continuous sludge removal unit, due to the problem of rapid sludge accumulation. This high rate of sludge accumulation may completely obstruct modules such as tube settlers unless the sludge is removed continuously. Figure 3.2.5-7 depicts the sludge removal units that are available from reputable equipment manufacturers.

Cost When the geological and civil works are normal for a plant site, an ordinary rectangular sedimentation basin will cost from \$80 to \$90 per square foot of the basin surface area (late 1990 cost on the U.S. west coast). As a rule of thumb, the cost of a flocculation/sedimentation basin process is approximately 20% of the total plant construction cost. The reactor clarifiers are rather compact units and manufacturers claim that they are cost-effective, rendering a 10 to 20% savings in cost. Studies of water treatment plant sites show that the actual costs of these two types

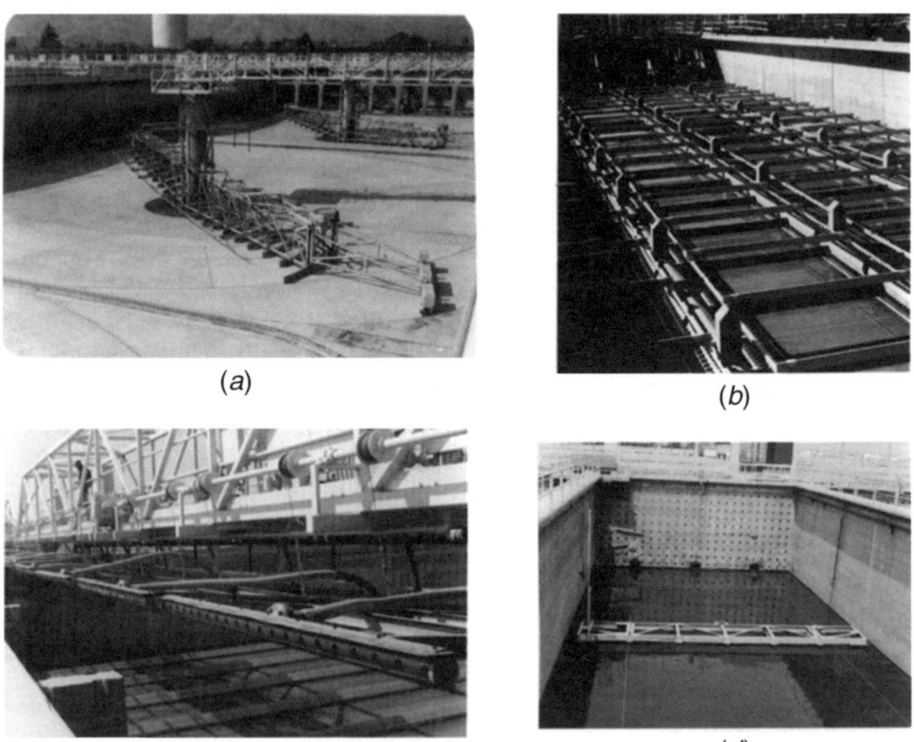

(*a*) (*b*) (*c*) (*d*)

Figure 3.2.5-7 (*a*) Center-pivoted rotating rake collector; (*b*) chain-and-flight collector; (*c*) traveling bridge collector (suction type); (*d*) cable-operated underwater bogie collector.

of clarifiers are generally the same. Yet the reactor clarifiers are preengineered by the manufacturers, up to 150 ft (45 m) in diameter, and the design cost of these units is therefore virtually eliminated, excluding the foundations and pipings.

High-rate settlers can be effective if they are both properly designed and installed. There is a vast difference in cost between the tube settler module system ($100/ft^2$) and some parallel-plate settler systems ($350/ft^2$; late 1990 cost on the U.S. west coast). Thus, selection should be based on a sound engineering evaluation.

Selection of the sludge removal units is similarly affected by cost. If the length of the rectangular sedimentation tank exceeds 230 ft (70 m), the traveling bridge is more cost effective: two chain-and-flight units (in series) are required to service the same distance. A traveling bridge unit can easily span 100 ft (30 m), and the capital cost for the equipment is significantly reduced by having one bridge span two or three tanks. This type of arrangement can be achieved because the width of each tank is usually less than 40 ft (12 m). Conversely, the chain-and-flight unit has a maximum width of approximately 20 ft (6 m). The tank width should therefore be the aliquot of 16 to 20 ft (5 to 6 m) in order to economize on equipment cost. Generally, the chain-and-flight unit is more economical than the traveling bridge only if the length of the tank is less than 230 ft (70 m); even if an additional cross collector is placed at one end of the tank, there is still a savings of approximately 30 to 50%.

Intermediate Diffuser Baffle Walls A rectangular tank with one or two intermediate diffuser walls was first introduced in Japan by the author in 1952, and a number of functional large basins were designed soon after. The intermediate diffuser walls help to minimize the formation of density currents because the gentle vertical mixing effect upstream of the diffuser wall reduced the temperature difference between the influent water and the water in the tank. This gentle mixing also significantly promotes the growth of floc size and weight. Consequently, the floc settles very well, both in the first compartment and upstream of each diffuser wall. Furthermore, stable flow patterns occur downstream of each diffuser wall and the wind effect on the flow in the tank is reduced. It must be emphasized that overuse of intermediate diffuser walls created difficulties during installation of the continuous sludge removal equipment. The aforementioned type of basin was installed in the cities of Tokyo, Kobe, and Kanagawa (Japan) under bogie-type sludge collectors. The collectors were placed between the diffuser walls (across the flow direction), in contrast to the longitudinal direction of most ordinary tanks.

Okuno et al. (1980) (Tokyo Metropolitan Public Works) compared the operational efficiency of four types of full-scale wastewater sedimentation tanks; a central feed and peripheral collecting circular tank, a peripheral circular tank, a rectangular tank with a tray (double-decked), and a rectangular tank with two intermediate diffuser walls. The results of that study showed that tanks with two intermediate diffuser walls were far superior and that the double-decked tanks did not perform satisfactorily (Figure 3.2.5-8), due primarily to the unstable flow characteristics of the tank.

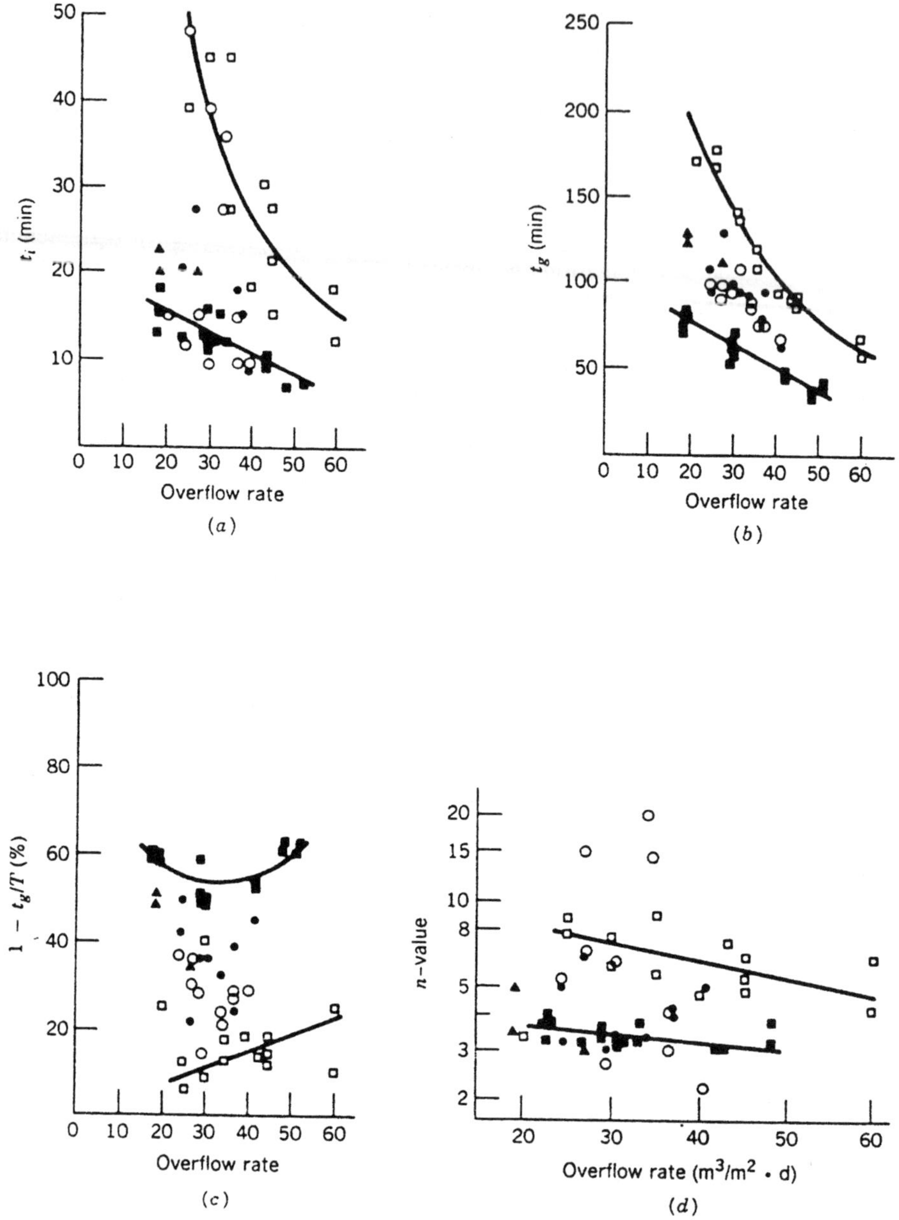

Figure 3.2.5-8 Hydraulic characteristics of five types of clarifier: ○ CFPC, central feed peripheral collect; ● PFPC, peripheral feed peripheral collect; ▲ RDD-1, rectangular double deck-1; ■ RDD-2, rectangular double deck-2; □ RTMD, rectangular tank with multidiffuser wall. (*a*) Short circuit; (*b*) mean retention time; (*c*) dead space ratio; (*d*) Fair's *n* (stabilization index). (Adapted from N. Okuno et al., *Analysis of Existing Clarifiers and Design Considerations for Better Performance.* Tokyo Metropolitan Government Report, 1980.)

Shape of the Tank When first designing a rectangular tank, it appears that there is much freedom in selecting the shape of the basin. However, experience has shown that basins that are both wide and deep tend to exhibit flow instability and a distinct density flow pattern. On the other hand, a basin that is narrow, shallow, and long will have flow stability and a minimal amount of flow short-circuiting. Rectangular tanks with a 180° turn at the midlength and crescent-shaped tanks are both ineffective because of flow short-circuiting; these designs can be salvaged only if a very long detention time is provided. The actual basin design must also take the following items into consideration: the installation of sludge removal equipment, the potential installation of high-rate settler modules, the presence of turbulence caused by winds, the scouring of bottom sediments, and the cost of the tank.

The following guidelines can help to determine the physical shape of the sedimentation tanks: (1) the water depth should be 10 to 16 ft (3 to 5 m); (2) the ratio between the length (L) and the width (W) should be $L/W = 6:1$, with a minimum of $4:1$; (3) the ratio between width (W) and the water depth (H) should be $W/H = 3:1$, with a maximum of $6:1$. From an economical as well as a practical viewpoint, the preferred height of the tank freeboard is approximately 2 ft (0.6 m). The freeboard acts as a wind barrier and helps to prevent wind-induced waves from splashing on the walkway.

The flow characteristics of the regular sedimentation basin can be estimated by the Reynolds (Re) and Froude numbers (Fr):

$$\mathrm{Re} = \frac{vR}{\nu} < 20{,}000$$

$$\mathrm{Fr} = \frac{v^2}{gR} > 10^{-5}$$

where

v = displacement flow velocity (m/s)

R = hydraulic radius, or $R = a/P$ (m)

a = area of flow passage (m^2)

P = wetted perimeter (m)

ν = kinematic viscosity (m^2/s)

g = gravity constant (9.81 m/s^2)

It is rather surprising to discover that an ordinary basin has a Reynolds number that is over 25,000 and a Froude number that is less than 10^{-6}. Consequently, the flow characteristics of an ordinary basin are far inferior to that most engineers would expect. The flow characteristics of an ordinary basin can be improved significantly if longitudinal baffles are installed parallel to the sidewalls of the basin. This arrangement divides the basin into a series of narrow channels, thereby decreasing the flow passage area (a). The proper installation of longitudinal baffles in this manner will reduce the Reynolds number to 50% of that of ordinary basins and will increase the Froude number to the desired range of 10^{-5}.

The minimum width between the longitudinal baffles should be 10 ft (3 m), to allow for the installation of sludge removal equipment, easy maintenance, and economic considerations. The baffles may be composed of thin concrete walls or even wooden planks. One of the least desired shapes is a basin that has a 180° turn at its midlength. A significant number of operational basins have been designed in this manner, primarily for space-saving reasons. These tanks have either a horizontal or vertical turn at the midlength and always exhibit poor efficiency. This inefficiency is due to the presence of turbulence near the area where the water must make the turn and the existence of large dead spaces within the basin.

Upflow and Reactor Clarifiers The upflow and reactor clarifiers are proprietary units that have their basic size and blueprints preestablished by the equipment manufacturers (based on the flow rates). These units can be grouped into three categories: simple upflow clarifiers, reactor clarifiers, and sludge blanket reactor clarifiers. The upflow and reactor clarifiers offer several advantages over the rectangular basins: (1) compact design and economical use, since no design effort is required by the engineer; (2) simpler sludge removal; (3) efficient clarification due to the seeding effect—this seeding effect is achieved through recirculation of the floc and adsorption of the floc by the sludge blanket, as long as the flow rate and raw water quality are fairly constant; (4) better clarification efficiency due to improved flocculation and adsorption effect by the sludge blanket; and (5) capability of delaying degradation of effluent water quality, in case of improper coagulant dosage or even a short interruption of coagulant feed, due to the buffering effect of the sludge blanket zone.

However, there are also several disadvantages: (1) compared to the horizontal tanks they require more stringent operational control; (2) there is a rapid loss of efficiency during hydraulic and solid overloading and hydraulic shock loading; (3) frequent temperature fluctuations induce distinct flow short-circuiting; (4) if the unit is not equipped with a positive raking mechanism, it is difficult to remove the sludge during periods of high turbidity; and (5) more time is required to produce the sludge blanket whenever the sludge blanket clarifier is drained. Moreover, a significant amount of metal is used on the interior of these tanks and the problem of corrosion and maintenance must be addressed.

These types of clarifiers are usually employed in industrial and municipal applications, such as lime softening of groundwater and clarification of turbid water. It must be emphasized that the flow rates and the raw water quality in these situations are nearly constant at all times. Table 3.2.5-3 is a selection guide for these units. Refer to Figure 3.2.5-9 for illustrations of each type of unit.

High-Speed Microsand Settling Process Although the principle is not new, in recent years the process has also been called the ballasted sand process. The principle was applied to water treatment in the 1970s by a French firm and marketed as the Cyclofloc process. This was then improved by Kruger, Inc. and patented in the United States (Actiflo). US Filter has commercialized a similar process.

Examination of regular plant treatment records indicate that better flocculation is achieved, without increasing alum dosage in proportion to the level of turbidity,

TABLE 3.2.5-3 Selection Guide of Proprietary Clarifiers

Type of Unit	Advantages	Disadvantages
Detritus tank	Potential low capital cost	Flow short-circuiting, potential clogging of the grit removal pipe, corrosion of the raking system
Center feed and peripheral collection	Low capital cost, easy sludge removal	Flow short-circuiting, limitations on the practical size of the unit, less tolerant of shock loading
Peripheral feed and peripheral collection	Good performance for raw water with a high suspension of solids	Potential for flow short-circuiting, performance relies on the manufacturer's design, limitations on the practical size of the unit
Reactor clarifier with high recirculation and mechanical sludge plow	Compact and economical, tolerant of shock loading, good clarification due to a seeding effect	Corrosion of underwater metal parts, less reliable due to its dependency on one drive unit, requires extensive yard piping to connect each unit, practical size limitations
Reactor clarifier with sludge blanket zone and mechanical sludge plow	Good lime softening and turbidity removal, compact and economical design, several manufacturers to choose from	Very sensitive to shock loading, requires 2–4 days to build up the necessary sludge blanket, requires greater operator skills, corrosion of underwater metal parts
Reactor clarifier with hydraulic flocculation and sludge removal	Good performance under the proper conditions, no maintenance of mechanical equipment	Sludge will choke the unit if heavy solids are present in the raw water, no operator control over the flocculation process, process functions will terminate if the water dousing system fails, produced by only one manufacturer (patented)

whenever there is a higher quantity of suspended particles and a higher proportion of larger suspended solids in the raw water due to rain. Figure 3.2.4-1*b* illustrates this phenomenon; the flocculation theory also proves this point. The process exploits this phenomenon by feeding microsand (20 to 200 μm) together with inorganic coagulant and cationioc polymer, thereby allowing the floc to attach to the surface of the microsand particles. Since the settling velocity of the microsand is 20 to 60 times higher than that of alum floc, the hydraulic surface loading of the clarifier may be 20 to 60 times higher than that of regular clarifiers.

The higher-speed microsand settling process includes both flocculation and clarification stages. The flocculation stage has 9 to 10 minutes of mixing time with a G value of 400 to 150 s^{-1} and tapered mixing by vertical-shaft flocculators. The clarification stage has 10 minutes of settling time with lamella plate settler modules with hydraulic surface loading of 14 to 25 gpm/ft^2 (35 to 62 m/h). In general, the microsand feed rate is 0.15 to 0.4% of the processing flow rate and the amount of

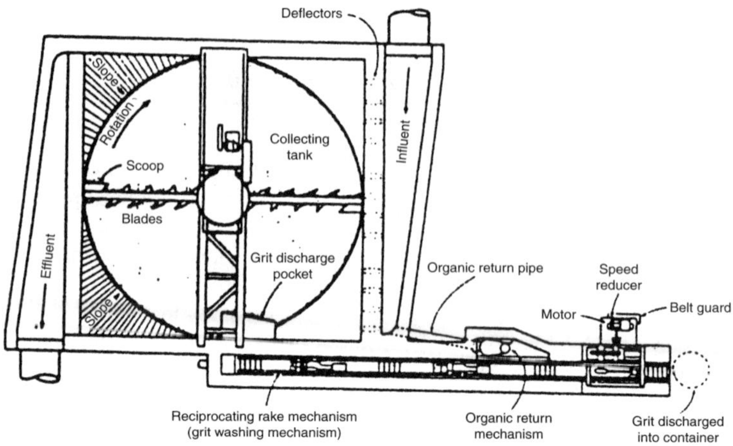

(*a*)

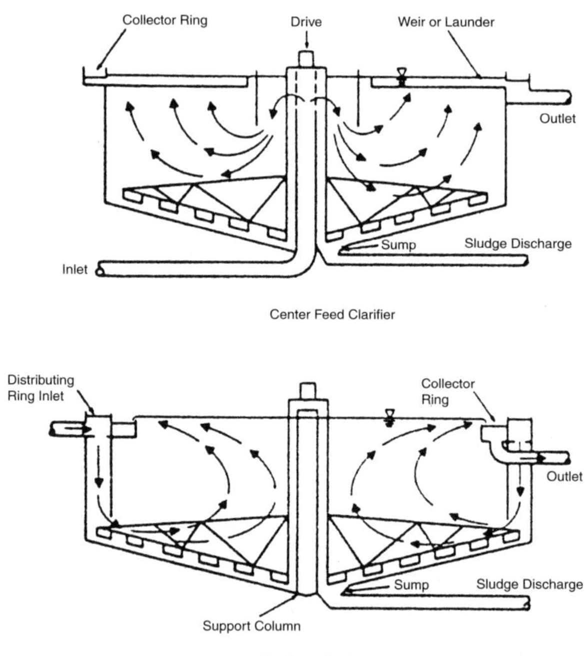

(*b*)

Figure 3.2.5-9 Various types of clarifier: (*a*) detritus tank with grit washer (Dorr–Oliver); (*b*) concept of center feed and peripheral feed clarifiers; (*c*) reactor-clarifier with high recirculation (Infilco); (*d*) sludge blanket reactor-clarifier (Graver Water Company); (*e*) reactor-clarifier with hydraulic flocculation and sludge removal (Degremont).

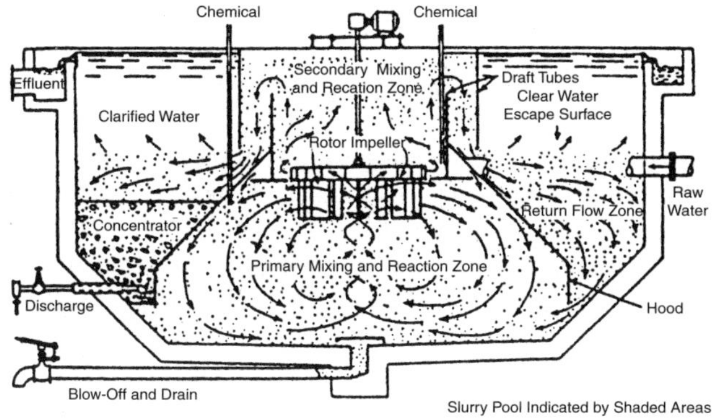

(c)

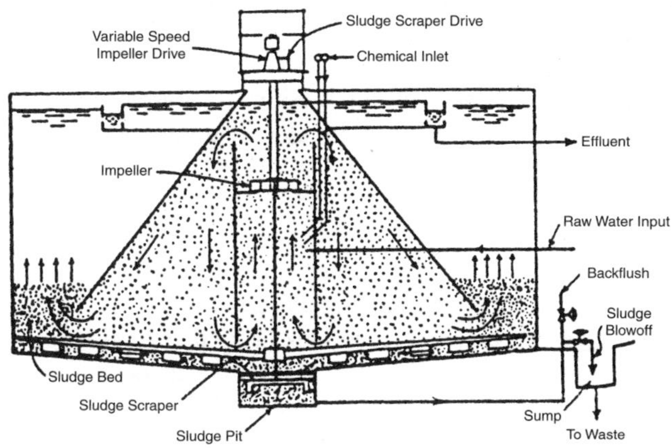

(d)

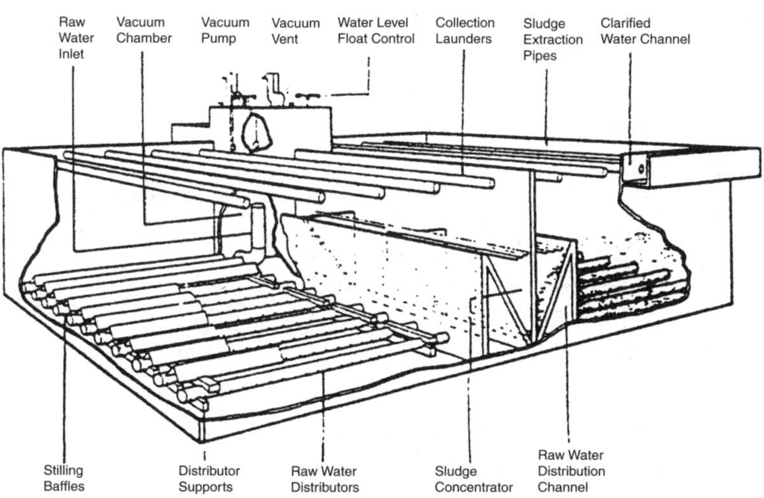

(e)

Figure 3.2.5-9 *(Continued)*

sand in the sludge zone is 10 to 12%. Figure 3.2.5-10 illustrates the principle of this process.

The advantages of the high-speed microsand settling process include: (1) requires a very small footprint; (2) has good performance; (3) has a very quick process startup, about 15 min; and (4) enjoys a potential savings in capital costs. However, the disadvantages are: (1) heavy dependence on mechanical equipment and a short processing time; (2) the entire process must be shut down when there is a power outage lasting more than 10 minutes; (3) requires a higher coagulant dosage than conventional processes with a high proportion of polymers; and (4) use is restricted by patents.

The potential applications for this process would be in situations where there is a very limited site area for the clarification process, if the filter wash waste is to be clarified prior to recycling, and for the clarification of combined sewer overflow (CSO).

Type and Selection Guide In this section we present a brief summary of the typical sedimentation tanks used in the water treatment process and a few guidelines for making the proper selection.

Available Alternatives

Horizontal flow	Long rectangular tanks
Horizontal flow	Center-feed tanks (circular or square)
Horizontal flow	Peripheral-feed circular tanks
Upflow clarifiers	Proprietary units
Reactor clarifiers	Units: proprietary, recirculation, sludge blanket
High-rate settler modules	Proprietary items, tube settlers, parallel-plate modules
High-speed microsand	Proprietary patented items, very high hydraulic loading settling process, and a short startup

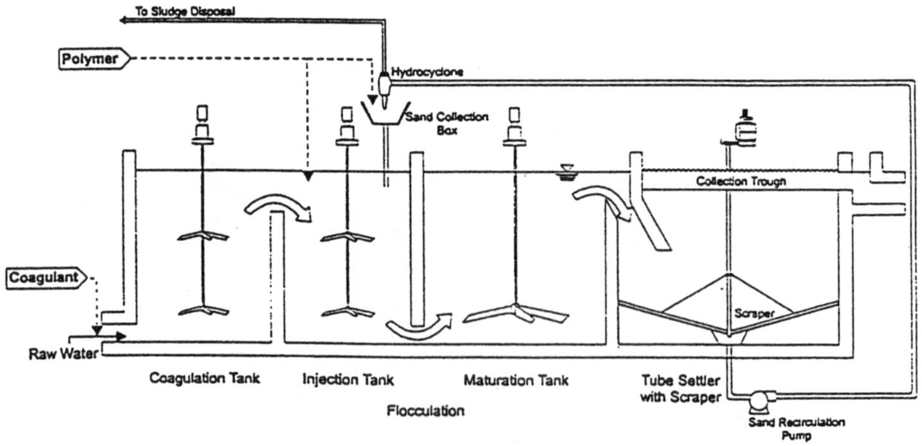

Figure 3.2.5-10 Concept of high-speed microsand settling process.

Selection Criteria Selection of the sedimentation tanks should be based on the following criteria:

1. The type of suspended matter that is to be removed: for example, grit, silt (plain sedimentation), chemically flocculated suspended matter, or biological floc
2. The overall process train and the role of sedimentation: for example, pretreatment for rapid sand filters, pretreatment for multimedia high-rate filters, clarification of lime-softened floc prior to filtration, or removal of sand and silt upstream of the intake pumps
3. The topography and ground conditions of the plant site
4. The potential for hydraulic shock loading and the degree of fluctuation in the quality of the influent water
5. The requirements for the quality of the settled water with respect to subsequent processes
6. The nature and amount of sludge that will be produced
7. The local climatic and geological conditions
8. The utilization of site dimensions and considerations with respect to future plant expansion
9. The capital required for the sedimentation tanks and the operation and maintenance costs
10. The time period that is allowed for the design and building of the treatment plant

Order of Preference In selecting the sedimentation process, the order of preference is established by both the performance and the operation and maintenance costs of each type. Based on past experience, the order of preference among the various types of plain sedimentation tank is (1) a long rectangular tank, (2) a rectangular tank containing high-rate settler modules, and (3) proprietary units. Similarly, the order of preference among the clarifiers is (1) a long rectangular tank, (2) a rectangular tank containing high-rate settler modules, and (3) a reactor clarifier or high-speed microsand clarifier.

A rectangular tank with a horizontal flow is generally favored because of its hydraulic stability (if properly designed) and its tolerance of shock loadings. This type of tank also has a predictable performance and is capable of withstanding a flow rate that is twice the designed rate without significant deterioration in the quality of the settled water. Additionally, this type of tank is easily operated, has low maintenance, adapts easily to the installation of high-rate settler modules, and uses space efficiently, especially with the common-wall design.

The use of reactor clarifiers should be considered and evaluated if (1) the plant flow rate and raw water quality are relatively constant; (2) the raw water quality is relatively constant, particularly with no sudden changes in water temperature, and the raw water turbidity is always between 5 and 300 ntu; (3) using lime softening or other processes that benefit from the seeding effect; (4) the units are installed in a housing; (5) there are time constraints on both the design and construction, as well as budget constraints; and (6) the owner prefers their use.

Three types of clarifiers are not recommended for removing alum floc because of their hydraulic instability: horizontal flow with center feed, peripheral feed, and the simple upflow clarifiers. The use of rectangular tanks with trays (including the double-deck type) is generally not recommended unless the available site area is very limited since they do not meet the expected level of settling efficiency, have high construction costs, and present difficulties when installing mechanical sludge removal equipment on each tray (unless the clearance is over 6 ft).

Discussion of Alternatives

Horizontal Flow and Long Rectangular Basins As discussed previously, in the section "Overall Configuration" under "Considerations," the most cost-effective and space-saving configurations are rectangular basins arranged longitudinally and side by side. The important design requirement is a uniform distribution of feedwater at the entrance of the basin. Effluent control by means of a series of long skimming troughs and a baffle wall is not as effective as some textbooks and design guides published by regulatory agencies suggest. These subjects are described in the section "Inlet and Outlet of the Basin" under "Considerations."

Two major pieces of basin equipment commonly used are the sludge collection system and high-rate settler modules (see the earlier discussion). Proper installation of the tube settler modules requires a minimum tank depth of 12 ft (3.6 m), depending on the type of sludge collection system under the modules. A typical installation profile is illustrated in Figure 3.2.5-11.

Sludge Collection Systems The manufacturers of rectangular and circular basins offer several different types of continuous mechanical sludge collectors. The main types are (1) chain-and-flight, (2) traveling bridge with sludge scraping squeegees and a cross collector at one end of the tank, (3) traveling bridge with sludge suc-

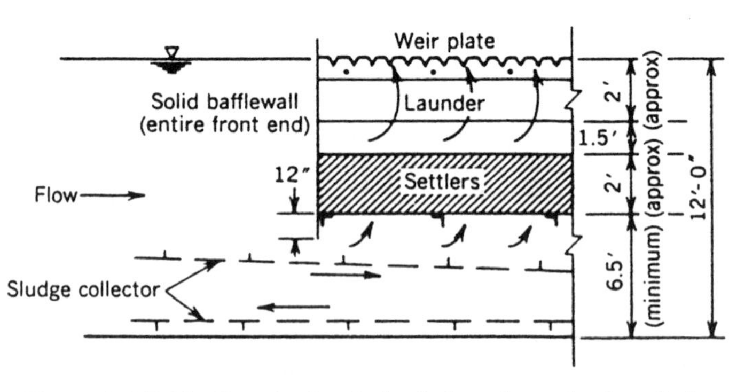

Figure 3.2.5-11 Required tank depth for tube settler installation.

tion headers and pump or siphon mechanism pulled by wires, (4) float-supported suction headers, (5) underwater bogies with squeegee (pulled by wires), and (6) center-pivoted rake.

Selection of the proper sludge collection system depends on the local weather conditions, the nature and quantity of suspended solids in the raw water, the type and amount of coagulant used, the shape of the tank, the installation (or lack) of high-rate settlers, aesthetics, and cost. In regions of cold weather the formation of ice on the surface of the water prohibits the use of moving parts. Therefore, the traveling bridge and float-supported sludge headers should not be used in these regions unless the tanks are either covered or housed in a building. There are even a few case histories where the traveling bridge has been derailed by gale force winds.

Other important factors that should be considered are the nature of the suspended solids and the quantity of solids expected to settle in the tank. For example, (1) water containing large amounts of freshwater clams will have shells settle in the grit chamber and settling tanks; (2) suction sludge collector systems require the orifices of their headers to be cleaned periodically because the size of the orifices is usually less than $\frac{3}{4}$ in. (20 mm), and therefore the orifices become frequently clogged; (3) lime softening of hard water tends to produce large amounts of heavy sludge, which may become troublesome for suction units; and (4) the chain-and-flight units require frequent maintenance whenever large amounts of abrasive grit are handled on a continuous basis (frequent maintenance may also be indicative of improper installation and/or use of inferior parts).

The shape of the tank also limits the type of sludge collection unit that may be installed. A unit with a center-pivoted rotating rake is best suited for circular or square basins. Since most sedimentation tanks are long and rectangular in shape, traveling bridge or chain-and-flight units are commonly employed. The choice between the two types is determined by cost, aesthetics, and the presence of settler modules in the tank. Most chain-and-flight units can service a tank length of approximately 200 ft (60 m) if it is run on a continuous basis to remove a heavy accumulation of sludge. However, one manufacturer claims that it can provide a unit that can cover 300 ft (90 m), but it must be operated in an intermittent fashion, such as in the removal of a light accumulation of sludge. If the basin length is longer than these limits, two sets of collectors are required. Basins containing high-rate settlers will have space limitations and therefore require either the chain-and-flight or underwater bogies to be installed.

Unlike chain-and-flight modules, traveling bridge collectors can service any length of tank. However, it is more cost-effective if the basin length exceeds 260 to 300 ft (80 to 90 m). Yet residents living near these types of plant may complain that traveling bridge collectors are unsightly. The bridge often travels on rails with a cog drive mechanism but can be on solid rubber tire wheels without rails.

A suction system with a traveling bridge is generally capable of withdrawing more concentrated sludge than the scraper. This type of unit also brings the sludge to a channel that is above the water level, thereby allowing the plant operator to visually inspect the consistency of the sludge and smell any extraordinary odor. Telescoping valves achieve the same effect when used in conjunction with a sludge scraper unit (Figure 3.2.5-12). If the sludge is sucked up by means of pump or siphon, the discharge rate of the sludge should be designed to be approximately 10 gpm per linear foot (120 L/m · min) of the tank width.

In summary, a traveling bridge with suction headers is the best choice for large tanks that do not have high-rate settlers but have normal levels of alum or ferric sludge deposits. In regions with moderate to warm weather, plant sites far from residential areas are prime candidates for a traveling bridge suction collector. Chain-and-flight units may be used under any conditions. However, their application to very long and wide tanks is not cost-effective because the maximum length of one chain-and-flight unit is usually limited to 200 ft (60 m) and width of the flights is limited to 20 ft (6 m). Additionally, water containing large amounts of abrasive grit would necessitate a high maintenance rate. If cost control is a major issue, float-supported suction headers and underwater bogie units may be considered as alternatives.

The traveling speed of a continuous sludge removal unit is usually 1 ft/min (0.3 m/min). This speed ensures that sludge resuspension does not occur. Units that can lift the sludge collection header onto a return path may travel at speeds that are 5 to 10 times higher than on the sludge collection path. For developing countries or small plants in isolated areas, the bottom of the sedimentation tank should have a slope of at least 1 : 300 toward the tank inlet to ensure easy manual sludge removal through the use of water jets. If mechanical sludge scraper equipment is used, the

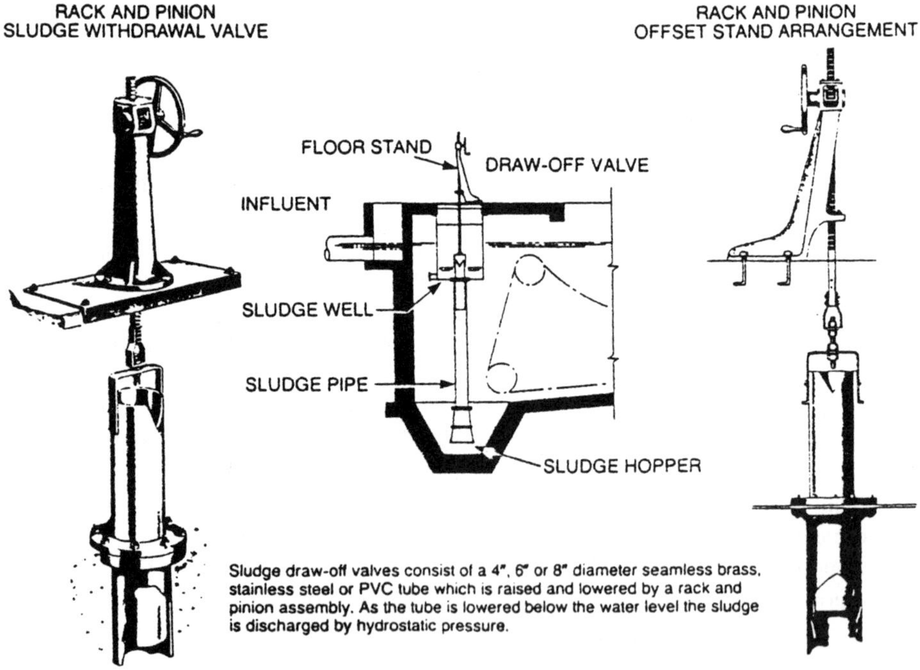

Figure 3.2.5-12 Telescopic values for sludge withdrawal.

slope may range from 1 : 300 to 1 : 600. When a sludge suction mechanism is adopted, the entire bottom of the tank may be level. Operators of plants using this type of system seldom express adverse comments regarding the design of the basin. The underflow rate (associated with sludge removal) of horizontal flow and long rectangular basins is usually 0.1 to 0.2% of the plant flow rate. The concentration of solids in the sludge is 0.5 to 2%, depending on the type of mechanical unit employed and operational adjustments.

Design Criteria This section presents the basic design criteria for three types of clarifier: grit chamber, rectangular sedimentation tank, and sedimentation tank with high-rate settlers.

Grit Chamber (Removal of sand and silt)

Type	Rectangular tank with horizontal flow
Minimum size of grit to be removed	0.1 mm
Minimum number of tanks	Two (One tank with a bypass channel may be used if the anticipated amount of grit is small.)
Water depth	10 to 16 ft (3 to 5 m)
Mean flow velocity	10 to 15 ft/min (3 to 4.5 m/min)
Detention time	6 to 15 min
Surface loading	4 to 10 gpm/ft^2 (10 to 25 m/h)
Length/width ratio (L/W)	4 : 1 to 8 : 1
Water depth/length ratio	Minimum of 1 : 8

Rectangular Sedimentation Tank (Removal of flocculated suspended matter)

Minimum number of tanks	Two
Water depth	10 to 15 ft (3 to 4.5 m)
Mean flow velocity	1 to 3.5 ft/min (0.3 to 1.1 m/min)
Detention time	1.5 to 4 h
Surface loading	0.5 to 1.0 gpm/ft^2 (1.25 to 2.5 m/h)
Length/width ratio (L/W)	Minimum of 4 : 1
Water depth/length ratio	Minimum of 1 : 15
Launder weir loading	12 to 18 gpm/ft^2 (9 to 13 m^3/m · h) (could possibly be 20 times higher)
Sludge collector speed	
For the collection path	1 to 3 ft/min (0.3 to 0.9 m/min)
For the return path	5 to 10 ft/min (1.5 to 3 m/min)

Sedimentation Tank with High-Rate Settler (Tube settler)

Minimum number of tanks	Two
Water depth of the tank	12 to 15 ft (3.6 to 4.5 m)

Surface loading of the tank area covered by the settler	1.5 to 3 gpm/ft^2 (3.8 to 7.5 m/h)
Mean velocity in the tube settler module	Maximum of 0.5 ft/min (0.15 m/min)
Detention time in the settler module	Minimum of 4 min
Launder weir loading	5 to 10 gpm/ft^2 (3.7 to 7.5 m^3/m · h)

NOTE: The design criteria of surface loading, as well as the inlet and outlet design, are predetermined by the manufacturer. These proprietary products generally have surface loading of 3 to 5 gpm/ft^2 (7.5 to 12.5 m/h) for the tank surface area covered by the settler modules. In some cases the space between the plates and the length of the plates may be customized.

Example Design Calculations

Example 1 Estimated Density Flow Velocity

Given

Water depth of the sedimentation tank	13.1 ft (4 m)
Water temperature	20°C (68°F)
Water temperature of the influent	19°C (66°F)

Determine The estimated velocity of the density flow in the sedimentation tank

Solution Use the density flow formula recommended by the Japan Society of Civil Engineering (1963):

$$v = \left(2g\,\frac{\Delta\rho}{\rho}\,\frac{\Delta h \propto}{k}\right)^{0.5}$$

Assume that

$$\begin{aligned} \alpha &= 0.5 \\ k &= 2 \\ \Delta h &= 2\text{ m} \\ \Delta\rho &= 0.99843 - 0.99823 = 0.0002\text{ g/cm}^3 \end{aligned}$$

Therefore,

$$v = \left(2 \times 9.81 \times \frac{0.0002}{0.99823} \times \frac{2 \times 05}{2}\right)^{0.5} = 0.044\text{ m/s}$$

$$= 2.66\text{ m/min}$$

Example 2 Grit Chamber Design

Given

Plant flow rate	100 mgd (4.4 m^3/s)
Minimum size of the sand to be removed	0.1 mm
Water temperature	10°C (50°F)

Determine

(i) The number and shape of the tanks
(ii) The size and water depth of each tank
(iii) The configuration of the inlet valves and diffuser wall

Solution (i) Provide two rectangular tanks for the most cost-effective design.

(ii) The size of each tank is calculated as follows. The settling velocity of 0.1 mm of sand is v_0 = 1.57 ft/min (8 mm/s) (refer to Table 3.2.5-1). Make the water depth of each tank 12 ft (3.7 m) and the width of each tank 40 ft (12 m). The mean flow velocity is

$$Q = (100 \times 1.55) \div 2 = 4650 \text{ ft}^3/\text{s}$$

$$v = \frac{Q}{a} = 4650 \div (12 \times 40) = 9.7 \text{ ft/min}$$

$$L = K\frac{h}{v_0 v} \qquad \text{where } K = 1.5$$

$$L = (1.5)\left(\frac{12}{1.57}\right) \times 9.7 = 111.2 \text{ ft}$$

Say 110 ft (33.5 m) for the tank length. The detention time is

$$\begin{aligned} V/Q &= (12 \times 40 \times 110) \div 4650 \\ &= \text{ft}^3 \div (\text{ft}^3/\text{min}) = 11.3 \text{ min} \end{aligned}$$

and the surface loading is

$$\begin{aligned} Q/A &= 34{,}750 \div (40 \times 110) \\ &= \text{gpm} \div \text{ft}^2 \\ &= 7.9 \text{ gpm/ft}^2 < 10 \text{ gpm/ft}^2 \qquad \text{OK} \end{aligned}$$

Check calculations via the Reynolds (Re) and Froude (Fr) numbers. For the Reynolds number:

$$\text{Re} = \frac{vR}{\nu}$$

where

$$R = \frac{a}{P}\, 12 \times \frac{40}{40 + (2 \times 12)} = 7.5 \text{ ft}$$

$$\nu = 1.41 \times 10^{-5} \text{ ft}^2/\text{s}$$

$$v = 9.7 \div 60 = 0.162 \text{ ft/s}$$

Therefore,

$$\text{Re} = \frac{0.162 \times 7.5}{1.41 \times 10^{-5}} = 86{,}200 \qquad \text{OK for grit chamber}$$

For the Froude number:

$$\begin{aligned} \text{Fr} &= \frac{v^2}{gR} \\ &= \frac{0.162^2}{32.2 \times 7.5} \\ &= 1.09 \times 10^{-4} > 10^{-5} \qquad \text{Good} \end{aligned}$$

(iii) For the inlet valves and diffuser walls, use two 36-in. rectangular butterfly valves at each tank inlet to facilitate easy removal of grit in the inlet channel. Avoid using sluice gates because they tend to "freeze up" with infrequent operation. The velocity through the valve is

$$v = \frac{Q}{A} = (50 \times 1.55) \div (2 \times 3^2) = 4.3 \text{ ft/s}$$

$$\begin{aligned} \text{headloss through the valve} &= \frac{1}{2g}\left(\frac{38.75}{0.7 \times 9}\right)^2 \\ &= 0.59 \text{ ft } (0.18 \text{ m}) \end{aligned}$$

Provide a straight-line diffuser wall that is parallel to the end wall of the inlet with a space of 6 ft (1.8 m). A total of 420 orifices [6 in. (0.15 m) in. diameter] are required for the diffuser wall to allow a flow velocity of 2 ft/s (0.6 m/s) through each orifice at a flow rate of 100 mgd. The orifices should be spaced 15 in. (0.4 m) apart (center to center) in vertical rows and 21 in. (53 cm) (center to center) for the horizontal direction. The ratio of the orifice area to the tank cross section is approximately 10%. The headloss through the orifice is approximately 1.5 in. (3.8 cm) at 100 mgd. Figure 3.2.5-13 illustrates the configuration of the grit chamber.

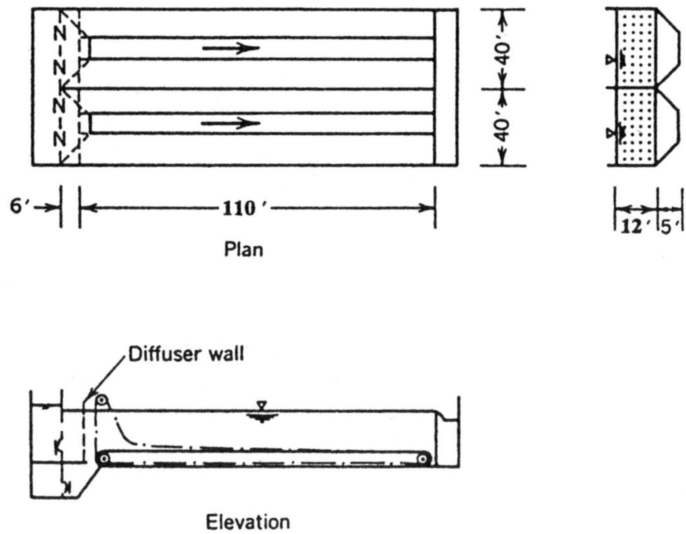

Figure 3.2.5-13 Grit chamber.

Example 3 Rectangular Sedimentation Tank Without a Tube Settler

Given

Plant flow rate	2 m^3/s (46 mgd) maximum daily rate
Suspended solids to be removed	Alum floc
Measured floc settling velocity	30 mm/min at 10°C (50°F)
Expected settled water turbidity	2 ntu or less for rapid sand filters
Minimum ambient temperature	−3°C (27°F)
Potential future plant expansion	Up to 4 m^3/s
Location of plant site	Far from residential areas

Determine

(i) The number of tanks
(ii) The size of each tank
(iii) The configuration of the tank inlet and diffuser wall
(iv) The type of sludge removal system
(v) The configuration of the effluent structure

Solution (i) Either two or three tanks may be selected. Although three tanks provide more optional flexibility (one tank may be off-line due to repairs), two tanks should be chosen for this example because (1) future plant expansion will provide the additional tanks, (2) the daily maximum flow rate is 2 m^3/s and it is safe to assume that the daily winter flow rate will be approximately 1.0 to 0.7 m^3/s (more suited to two tanks), and (3) the two-tank design is more cost-effective.

(ii) Considering that the standard maximum width of the chain-and-flight sludge collector is 6 m (20 ft), an aliquot of 6 m should be evaluated for the width of the tank: 12, 18, and 24 m (40, 60, and 80 ft, respectively).

For a traveling bridge sludge collector, a tank width in the range 12 to 30 m (40 to 100 ft) is considered to be cost-effective. The tank should be 18m (60 ft) wide to maintain a good tank length/width ratio. Based on floc settling velocity measurements (bench-scale studies conducted during the winter season), the hydraulic surface loading rate of the tank can be set at 1.8 m/h (0.72 gpm/ft^2) with a floc settling velocity of 30 mm/min. Although one might think that this value is too conservative for the winter season (a flow that is less than 1 m^3/s), the plant may use only one tank to treat the plant flow, thereby allowing the tank loading rate to remain normal. Consequently, 1.9 m/h (0.75 gpm/ft^2) is used as the design criterion of the tank. This number includes a comfortable safety margin for all seasons.

The total tank surface area required is

$$(2 \times 3600) \div 1.9 = (\mathrm{m^3/h}) \div (\mathrm{m/h}) = 3790 \ \mathrm{m^2}$$

Since each tank has a width of 18 m (60 ft), the length required would be

$$3790 \div (2 \times 18) = (\mathrm{m^2}) \div (\mathrm{m}) = 105 \ \mathrm{m} \quad (343 \ \mathrm{ft})$$

The water depth of each tank should be 4 m (13 ft) to facilitate the possible installation of tube settler modules in the future.

Check the values:

$$\text{Detention time at } Q_{\max} = \frac{(18 \times 105 \times 4) \times 2}{2 \times 60}$$

$$= \frac{\mathrm{m^3}}{\mathrm{m^3/min}}$$

$$= 126 \ \text{min} \ (2 \ \text{h} \ 6 \ \text{min})$$

$$\text{Detention time at } Q_{\text{ave}} = 126 \ \text{min} \ 1.5^*$$

$$= 189 \ \text{min} \ (3 \ \text{h} \ 9 \ \text{min})$$

*The maximum daily flow rate is usually 1.5 times the average daily flow rate.

$$\text{Tank length/width ratio} = 105 \ \text{m to } 18 \ \text{m}$$

$$= 5.8{:}1 > 4{:}1 \qquad \text{OK}$$

$$\text{Tank length/water depth ratio} = 105 \text{ m to } 4 \text{ m}$$
$$= 26.3{:}1 > 15{:}1 \qquad \text{OK}$$

$$\text{Mean flow velocity} = \frac{Q}{A} = (2 \times 60) \div 2 \times (18 \times 4)$$
$$= (\text{m}^3/\text{min}) \div \text{m}^2$$
$$= 0.83 \text{ m/min} < 1.1 \text{ m/min} \qquad \text{OK}$$

$$\text{Reynolds number:} \quad \text{Re} = \frac{vR}{\nu}$$

where

$$v = 0.83 \text{ m/min} = 1.39 \times 10^{-2} \text{ m/s}$$
$$R = \frac{4 \times 18}{(2 \times 4) + 18} = 2.77 \text{ m}$$
$$\nu = 1.31 \times 10^{-6} \text{ m}^2/\text{s at } 10°\text{C}$$

Therefore,

$$\text{Re} = \frac{1.39 \times 10^{-2} \times 2.77}{1.31 \times 10^{-6}}$$
$$= 29{,}620 \qquad \text{No}$$

Re is too high and the flow in the tank is highly turbulent. The preferred Reynolds number is less than 18,000.

$$\text{Froude number:} \quad \text{Fr} = \frac{v^2}{gR}$$

where

$$v = 1.39 \times 10^{-2} \text{ m/s}$$
$$R = 2.77 \text{ m}$$
$$g = 9.81 \text{ kg/s}^2$$

Therefore,

$$\text{Fr} = \frac{(1.39 \times 10^{-2})^2}{9.81 \times 2.77}$$
$$= 7.1 \times 10^{-6} < 10^{-5} \qquad \text{No}$$

The number is too small and the flow is therefore unstable.

How much will the flow characteristics in the tank improve if longitudinal baffles (partition walls) are installed in each tank?

Trial 1: One Longitudinal Wall in the Middle

$$R = \frac{(4 \times 9) \times 2}{(9 + 2 \times 4) \times 2} = \frac{\text{m}^2}{\text{m}} = 2.12 \text{ m}$$

Since v is practically unchanged,

$$\text{Re} = \frac{1.39 \times 10^{-2} \times 2.12}{1.31 \times 10^{-6}} = 22{,}670 > 20{,}000$$

This Reynolds number is better than before but still too high.

$$\text{Fr} = \frac{(1.39 \times 10^{-2})^2}{9.81 \times 2.12} = 9.2 \times 10^{-6} < 10^{-5}$$

The Froude number is still a little too low.

Trial 2: Two Longitudinal Walls 6 m Apart (on Center) Forming Three Channels

$R = 1.71$ m	v is practically unchanged
$\text{Re} = 18{,}820 < 20{,}000$	Tolerable
$\text{Fr} = 1.2 \times 10^{-5} > 10^{-5}$	OK

Trial 3: Six Longitudinal Walls 3 m Apart (on Center) Forming Seven Channels

$R = 1.1$ m	Flow velocity (v) is a little faster but practically unchanged
$\text{Re} = 11{,}850 < 20{,}000$	Good
$\text{Fr} = 1.8 \times 10^{-5}$	Good

As demonstrated, a tank with a long and narrow channel is hydraulically superior. It is therefore understandable why a tube settler or a parallel-plate settler performs well. Thus, from a practical perspective with cost as a factor, the choice would be to use two longitudinal baffles to create three channels that are 20 ft (6 m) in width. The revised tank length/width ratio (L/W) is 105 m to 6 m or 17.5 : 1.

In summary, the features of each tank are as follows:

Average water depth	4 m (13 ft)
Tank width	18 m (60 ft) with two longitudinal baffles
Tank length	105 m (343 ft)
Detention time	2 h 6 min
Surface loading	1.9 m/h (0.75 gpm/ft^2)
Mean flow velocity	0.83 m/min (2.7 ft/min)

(iii) When the tank is fed by a common inlet channel, two basic types of settling tank inlet may be used: flow splitting weir and submerged ports with valves. If the flocculation tank is the head end of the sedimentation tank, a diffuser wall may be used to separate the two tanks. Where the tanks are fed directly from a feed channel, the design of the channel should be large enough to avoid producing significant headloss from one end of the channel to the other. This condition is particularly important for the flow splitting weir.

The weir inlet has problems not only with leakage every time a tank is isolated with stop logs but also with sludge accumulation in the feed channel. From a practical aspect, the best decision is to place submerged ports at the bottom elevation of the inlet channel by means of butterfly valves. The inlet ports should provide at least 4 in. (0.10 m) of controlled headloss to ensure even flow distribution to each tank. The design of the inlet channel is delineated in "Choice of Flocculation Tank Inlet", Section 3.2.4.

Provide three submerged holes with butterfly valves as the tank inlet. Each hole should be 0.6 m (24 in.) in diameter so that the flow velocity through each orifice is

$$v = \frac{Q}{\Sigma a} = \frac{2}{6 \times (0.785 \times 0.6^2)} = 0.708 \text{ m/s (2.3 ft/s)}$$

$$\begin{aligned}\Delta h &= \frac{1}{2g}\left(\frac{q}{ca}\right)^2 = \frac{1}{19.62}\left(\frac{2 \div 6}{0.7 \times 0.283}\right)^2 \\ &= 0.144 \text{ m (5.7 in.)} \qquad \text{OK}\end{aligned}$$

The diffuser wall should provide approximately 0.23 m/s (0.75 ft/s) of flow velocity through each port at the maximum daily flow rate.

$$\text{Required total port area} = \frac{Q}{v} = \frac{2 \text{ m}^3\text{/s}}{0.23 \text{ m/s}} = 8.7 \text{ m}^2$$

The total number of 0.12-m (diameter) ports that are required is

$$\frac{8.7 \text{ m}^2}{0.785 \times (0.12 \text{ m})2} = 770$$

Provide 10 rows of ports in the vertical direction with a spacing of 0.4 m and 39 rows in the horizontal direction spaced 0.45 m apart (13 rows in each 6-m segment). In doing this, the actual total number or ports is 390 for each tank, or a total of 780 (total area of 8.82 m^2). See Figure 3.2.5-14 for the design.

Check the values:

$$\begin{aligned}\text{Flow velocity in each port at } Q_{\max} &= 2 \text{ m}^3\text{/s} \div 8.82 \text{ m}^2 = 0.227 \text{ m/s} \\ &< 0.25 \text{ m/s} \qquad \text{OK}\end{aligned}$$

$$\begin{aligned}\text{Headloss through diffuser wall} &= \frac{1}{2g}\left(\frac{2}{0.7 \times 8.82}\right)^2 \\ &= 0.0053 \text{ m (0.2 in.)} \simeq 0.0055 \text{ m} \qquad \text{OK}\end{aligned}$$

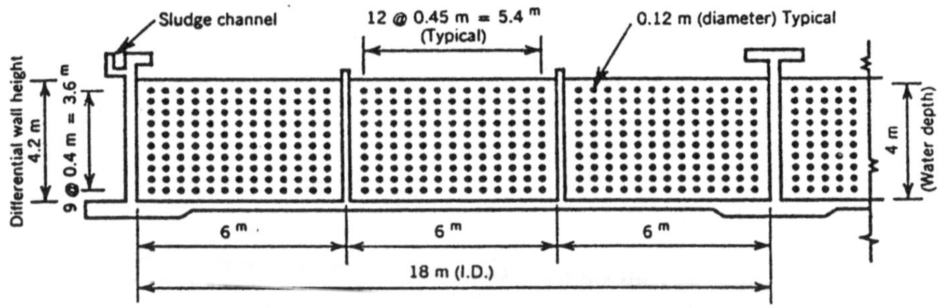

Figure 3.2.5-14 Inlet diffuser wall (elevation).

(iv) The traveling bridge with pump suction is the most preferred type of sludge removal system because (1) the tank length is over 100 m and therefore the traveling bridge is the most cost-effective; (2) one unit can span two 18-m (width) tanks, and this is very cost-effective; (3) heavy ice formation (which prohibits movement of the bridges) is not an issue since the lowest anticipated temperature is only $-3°C$; (4) the tank bottom does not have to be sloped with this system, thus simplifying the design and minimizing construction cost; and (5) the aesthetics of the bridge (the industrial equipment look) is not an important issue because the plant is far from residential areas.

(v) Some regulatory agencies require an extensive launder system to be included in the plant design as part of the configuration of the effluent structure. An acceptable hydraulic weir loading rate is approximately 10.8 $m^3/m \cdot h$ (15 gpm/ft) for the launders. The required weir length (based on the criterion) is

$$(2 \times 3600) \div 10.8 = 666 \text{ m for two tanks}$$

A few engineers also believe that the installation of a baffle wall, one similar to the influent diffuser wall at the effluent end of the tank, minimizes the density flow. However, experience indicates that neither the long launders nor the baffle wall (located at the effluent end) significantly improves the settling efficiency of long rectangular settling tanks. Consequently, the end walls of the tanks may be built as an overflow of the effluent wall.

It is also necessary to provide three 0.2-m (diameter) holes equipped with butterfly valves in the end wall near the bottom elevation of the tank. These valves are used in backfilling the tank from the common effluent channel after it is cleaned and backflashing during the cleaning process. Backfilling of the tank is especially important for flocculation–sedimentation combination tanks that have vertical-shaft flocculators; the flocculators will not function until the water level in the tank reaches half the depth of the tank. If turbid water is introduced from the inlet side during backfilling of the tank, the unflocculated water will flow to the end of the tank until the water level reaches half the depth of the tank. The untreated water will therefore flow directly into the filters after the tank is filled.

Example 4 Rectangular Tank with Tube Settlers

Given The same criteria apply as in Example 3, with the addition of the following:

Dual-media filters	After the sedimentation process
Minimum water temperature	5°C
Minimum ambient temperature	−10°C (14°F)
Location of plant site	Near residential areas; site area is small

Determine

(i) The number of tanks
(ii) The tank area that will be covered by the settler modules
(iii) The size of each tank
(iv) The configuration of the tank inlet and diffuser wall
(v) The configuration of the launders and sludge collection system

Solution (i) The number of tanks is the same as in Example 3—two tanks.

(ii) The tank area that is to be covered by the tube settler module may be calculated. The following relation can be established from Figure 3.2.5-15:

$$S_0 = \frac{Q}{A} \frac{W}{h \cos \alpha + W \cos^2 \alpha}$$

Since tube settler modules have $\alpha = 60°$,

$$S_0 = \frac{Q}{A} \frac{W}{0.5h + 0.25W}$$

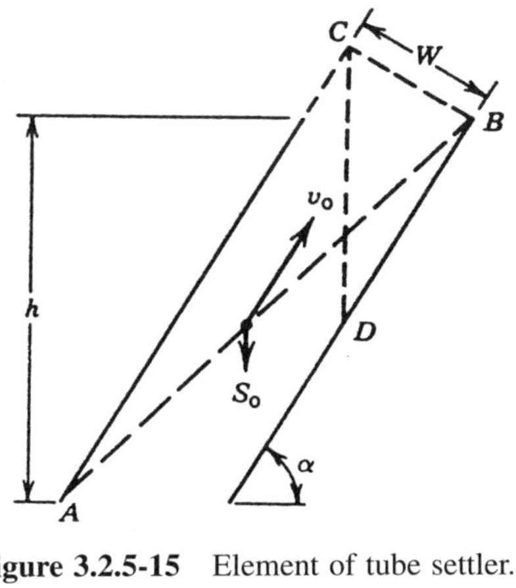

Figure 3.2.5-15 Element of tube settler.

where A is the surface area of the settling tank that is covered by the settler module (m^2) and Q is the flow rate of the tank (m^3/s).

A standard tube settler module has the following dimensions: $h = 0.55$ m (1 ft 9 in.) and $W = 0.05$ m (2 in.). Therefore,

$$S_0 = \frac{Q}{A} \frac{0.05}{(0.5 \times 0.55) + (0.25 \times 0.05)} = \frac{Q}{A}(0.174)$$

The average floc settling velocity, as measured in the laboratory, is 0.0005 m/s (refer to Table 3.2.5-1). However, actual tank conditions are not as well controlled, and therefore a value of $S_0 = 0.00025$ m/s is used, providing an approximate safety factor of 2.0:

$$S_0 = 0.00025 = \frac{0.174 \times 1}{A}$$

$$A = 696 \text{ m}^2 \text{ or } 7490 \text{ ft}^2 \text{ for each tank}$$

Since the width of the tank is assumed to be 18 m, the length of the tank that is to be covered by the tube settler is 696 m^2 ÷ 18 m = 38.7 m. Make it 38 m. The actual tank area for the tube settler is therefore 18 m × 38 m = 684 m^2 per tank.

Check the values. The surface loading for the tank area that is covered by the settler module is

$$\frac{Q}{A} = \frac{1 \times (60)^2}{684} = 5.3 \text{ m/h } (2.1 \text{ gpm/ft}^2) \qquad \text{OK}$$

Flow velocity in the settler module is

$$v_0 = \frac{Q}{A \sin \alpha} = \frac{1 \times 60}{684 \times 0.866} = 0.1 \text{ m/min } (0.00167 \text{ m/s})$$

The hydraulic radius of the tube settler module is

$$R = \frac{a}{P} = \frac{(0.05)^2}{4 \times 0.05} = 0.0125 \text{ m}$$

$$\text{Re} = \frac{(1.67 \times 10^{-3}) \times (1.25 \times 10^{-2})}{1.519 \times 10^{-6}} = 14 < 50 \qquad \text{Very good}$$

$$\text{Fr} = \frac{(1.67 \times 10^{-3})^2}{9.81 \times (1.25 \times 10^{-2})} = 2.3 \times 10^{-5} > 10^{-5} \qquad \text{Good}$$

Since a standard-size module has tubes that are 0.61 m (2 ft) long, the detention time is 0.61 m ÷ 0.1 m/min or 6.1 min.

(iii) Two identical tanks that are 18 m wide will be used as discussed in Example 3. It is common practice to have 75% of the tank area covered by the settler modules and the remaining 25% left as open space, to settle heavy floc, to improve the flow inlet condition to the settler, and provide easy access to the mechanical sludge removal system. The total tank length should therefore be

$$L = 38 \text{ m} \div 0.75 = 50.7 \text{ m (166 ft)}$$

Say 50 m (164 ft). The water depth in the tank should be 4 m (13 ft), as illustrated in Figure 3.2.5-11. With respect to the longitudinal baffle walls, the flow characteristics of the tank are no longer critical because the flow characteristics in the tube settler are significant. Thus, only one longitudinal baffle wall is required at the centerline of the tank.

In summary, the features of each tank are as follows:

Tank width	18 m (59 ft)
Tank length	50 m (164 ft)
Water depth	4 m (13 ft)
One longitudinal baffle wall (L/W)	5.5 : 1

Tube settler modules are installed over 75% of the tank area (38 m from the end wall).

(iv) The tank inlet and diffuser wall are the same as in Example 3.

(v) Since an extensive skimming launder system is required with the tube settler modules, provide six launder troughs for each basin. Each launder must cover the entire length of the settler module; thus, the length of each launder should be 38 m. The flow rate in each launder trough is 1 ÷ 6 = 0.167 m^3/s. For a trough with a rectangular section and a level invert,

$$Q = 1.38\ bh_0^{1.5}$$

where b is the width (m) and h_0 is the water depth at the far upstream of the trough.

Select 0.5 m as the width for the launder trough:

$$h_0^{1.5} = \frac{0.167}{1.38 \times 0.5} = 0.242$$

$$h_0 = 0.39 \text{ m}$$

Make the interior height of the trough 0.5 m (0.11 m freeboard). Each trough should be supported by pedestal columns spaced 3.8 m (12.5 ft) on center.

The chain-and-flight sludge collection system must be employed in this design because of the use of tube settler modules, the formation of thick ice during severe winter conditions, and for aesthetic reasons. Since one tank is partitioned into two channels (9 m wide), two sets of chain-and-flight units are installed in each channel. The end wall of each tank should have two 0.25-m (diameter) holes with butterfly valves for backfilling of the basin. Ideally, the entire basin should be housed to protect the system from extremely cold temperatures. An illustration of the sedimentation tank is provided in Figure 3.2.5-16.

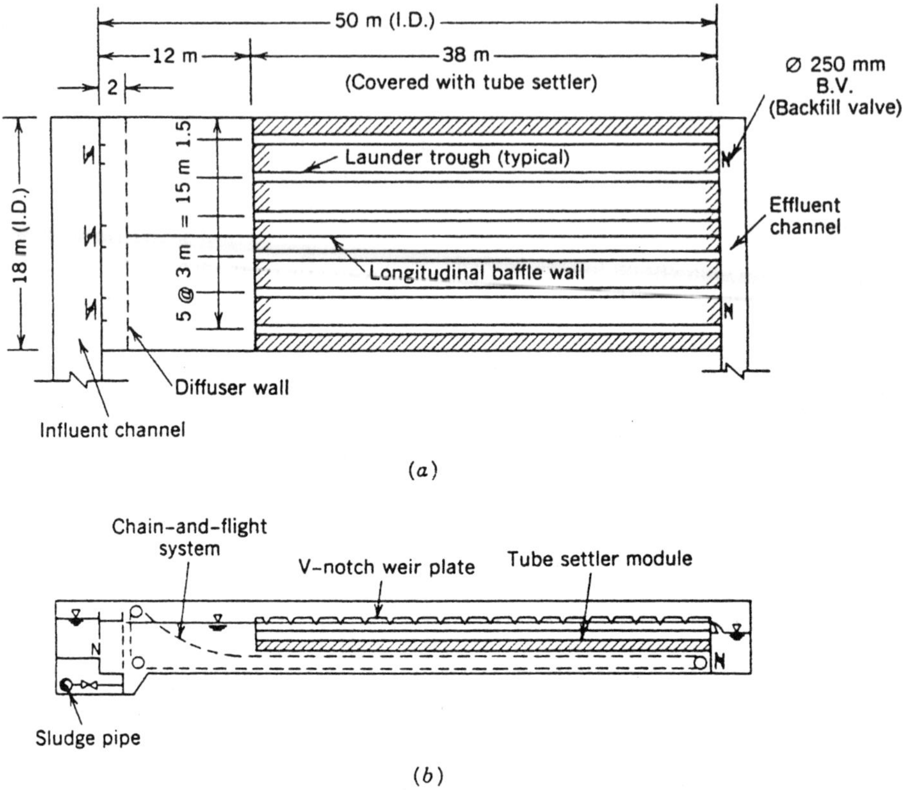

Figure 3.2.5-16 Sedimentation tank with tube settler modules: (*a*) plan; (*b*) elevation.

Operation and Maintenance The operational staff of water treatment plants should be aware of the following items:

1. Floc settling conditions in the tank
2. Abnormal phenomena
3. Equal hydraulic loading to each tank
4. Schedule for sludge removal
5. Special design features of the tank

If the suspended solids are well flocculated, a majority of the floc should settle in the first half of the tank. Consequently, visual observation of the floc settling condition (from the walkways) is important and the coagulant dosage or flocculator mixing intensity should be adjusted accordingly; inadequate flocculation will allow small floc to leave the flocculation tanks. However, turbulence in the basin increases floc size in the area from the middle to the end of the tank. As a result, there is a heavy carryover of floc to the filters.

For many plants it is common for the water in the middle of the tank to be clear, yet have a lot of floc rise at the end of the tank. This type of pattern indicates the presence of a distinct density flow. Plant operators have only three possible choices to counteract this situation. The first is to block off the last 10 to 20 ft (3 to 6 m) of V-notches or submerged orifices, on the launders. This will work only if launders are used to collect both settled and clear water from the midsection of the tank. This simple modification often improves the situation significantly. The second choice is to feed 0.2 to 0.5 mg/L of anionic polymer to the second stage of the flocculation process. The polymer acts as a flocculation aid that helps to produce readily settleable floc. The last choice pertains to plants that have more than one source of water: Minimize switching water sources or minimize changes in the blending ratio of the two sources. By doing so, the flow short-circuiting through the basin (the product of a density current) can be reduced.

It is very important that plant operators detect and evaluate any abnormal phenomena occurring within the tank. This includes the sudden appearance of floating sludge (sludge bulking), scum, midge fly larvae, and algae growth on the tank walls and baffles. Sludge bulking occurs when gas accumulates in the settled sludge as a result of anaerobic decomposition by bacteria or supersaturation of the raw water (by air) so that it floats rather than sinks. Sludge bulking is usually controlled by prechlorination: 2 to 3 mg/L of chlorine. Midge fly larvae bloom occurs during certain seasons if the sludge is not removed frequently. Algae will grow on the tank walls, launders, and baffles during warm months, especially in nutrient-rich water. Both the algae and larvae problems can be controlled effectively by prechlorination (1 to 2 mg/L). However, recent limitation on the use of THM restricts the use of prechlorination for many plants. A feasible alternative is to provide occasional shock treatments with 2 to 3 mg/L of chlorine.

The accumulation of heavy scum can be attributed to many causes; changes in raw water quality, changes in coagulant (from alum to ferric chloride), and supersaturation of raw water by air or gas. Supersaturation is the most difficult problem to solve. The only practical solution is to prevent the entrainment of air in the raw water.

Another phenomenon that should be noted on a regular basis is corrosion. Concrete surfaces and submerged metal parts will severely corrode with repeated exposure to a combination of aggressive water, chlorine, and alum. In this situation, a proper coating must be applied to these surfaces or, in the case of the metals, cathodic protection should be introduced. A severe pitting type of corrosion may occasionally be observed in submerged stainless steel parts (type 316). This occurs only under anoxic conditions and can be minimized by specifying the use of type 304 stainless steel or by making the conditions aerobic.

The design of the tank is based on equal hydraulic loading to all tanks. Yet this may not occur due to erroneous design or improper adjustment of the weir and valve openings. Plant operators should therefore check the flow rate through each tank by visual observation or by measuring the water height over the effluent weirs. An uneven flow rate to each tank can be roughly corrected by adjusting either the flow-splitting weirs at the tank inlet or the inlet valve openings.

Certain factors can create undesirable flow patterns. For example, uneven leveling of each launder trough or effluent end weir wall can cause a skewing flow to be created. Uneven tank subsidence may tilt the tank and thereby create horizontal flow short-circuiting. Improper design of the inlet channel and the tank inlet will also cause a great imbalance in the flow rate among several tanks that are in parallel. In certain plants a difference of as much as 50% has been observed, even though the elevations of the launder troughs at each end of the tank were virtually the same.

Perhaps one of the most important issues for the sedimentation tank is optimization of the sludge withdrawal process. In general, tanks that lack high-rate settler modules will have 60 to 80% of the sludge deposited in the front half of the tank, provided that the flocculation process is functioning well. Sludge sweeping should therefore be performed more frequently for the front half of the tank; a traveling bridge can be programmed to sweep the front half two or three times before running the entire length (in order to sweep the last half). In contrast, other types of sludge scraping mechanisms, such as a chain-and-flight unit, cannot perform partial sweeps.

The proper sludge sweep cycle and the duration may be determined from several trial runs, by noting the concentration of the solids and the texture and smell of the sludge (for freshness). The most appropriate types of unit for this test are the sludge suction sweep units and the telescoping valves (for sludge discharge) with chain-and-flight collectors. These units allow the plant operator to determine the proper sludge sweeping cycle by allowing easy observation and testing of sludge consistency. Other methods of testing include the installation of magnetic or ultrasonic sludge density and flow meters in the sludge discharge line or the installation of a piece of transparent tubing in the sludge line to allow visual observation. However, the second method (transparent pipe) is simply not practical because the pipe becomes coated with sludge in a short period of time.

Table 3.2.5-4 is an example of a plant operator's operational guide for a traveling bridge fitted with sludge suction pumps. In areas where the water table is high, the tank cannot be drained at will unless the tank is designed to resist uplift. For this reason, some plant designs include an underdrainage system for each tank. Each underdrain line has an observation point where the operator can visually inspect the flow of water to see if the water is running or standing. This type of system allows the operators to check any tank for leaks and allows them to note unusually high underground water tables.

Finally, the operator should be familiar with certain design features of the tank. For example, the diffuser wall should be designed to have a restricted water passage (facilitated by uniformly distributed ports) to produce uniform distribution. In a few cases, the plant operators had a passageway cut in the diffuser wall [3 by 6 ft (0.9 by 1.8 m)] to allow easy access for maintenance crews. In another case, the bottom 1 to 2 ft (0.3 to 0.6 m) of the baffle wall were removed completely in an attempt to make the task of sludge removal and tank drainage more convenient. All these modifications caused the diffuser wall to become nonfunctional.

Since the majority of plant operators are not fully aware of the actual flow characteristics of conventional sedimentation basins, Appendix 9 (tracer test) has been

TABLE 3.2.5-4 Guide for Traveling Bridge Operation

(a) A Half-length (front half) sweeping (47 metres)
(Table shows sweeping cycle and corresponding sweeping speed)

		Sweeping Speed (m/min)				
		0.75	1.0	1.5	2.25	3.0
Total sweeping Time (1) (minutes)		62	47	31.3	21	15.7
Volume of Sludge Removed (2) (1) x 4.8 m^3/min		297.6	225.6	150.2	101	75.4
Required cycle of 1/2 length sweeping per day.	1% solid sludge (1310 m^3/d)	4.4	5.8	8.7	-	-
	1.5% solid sludge (873 m^3/d)	2.9	3.8	5.8	8.6	-
	2% solid sludge (655 m^3/d)	2.2	2.9	4.3	6.5	8.7
	2.5% solid sludge (524 m^3/d)	1.8	2.3	3.5	5.2	7.0

Recommended Range

(b) A Full length sweeping followed by 1/2 length sweeps
Total pump capacity = 4.8 m^3/min (Total of two bridges)
(Table shows an appropriate sweeping speeds under certain conditions)

		Sludge Conentration (% solids)			
		1%	1.5%	2%	2.5%
Total volume of sludge to be removed (m^3)	Case 1*	218.3	145.5	109.2	87.3
	Case 2+	327.5	218.3	163.8	131
Req'd pumping time to remove the sludge by this path (min.)	Case 1*	45	30.3	22.8	18.2
	Case 2+	68.2	45.5	34.1	27.3
Req'd sweeping speed (m/min) for each case (approx)	Case 1*	2.1	3.1	4.1x	4.8x
	Case 2+	1.4	2.1	2.8	3.4x

Note: *Case 1: Five consecutive 1/2-length sweepings before the full length sweeping.
+Case 2: Three consecutive 1/2-length sweepings before the full length sweeping
xBeyond the capability of travelling bridge

added to this book. As shown in the figures in Appendix 9, the actual flow-through time (t) is only about 30% of the mean detention time (T); the performance of most sedimentation tanks is represented by these graphs. Although plant operators cannot improve this situation, they should be aware of this fact and should try to improve the settling efficiency of the tank by improving flocculation.

BIBLIOGRAPHY

ASCE, AWWA, and CSSE, *Water Treatment Plant Design,* American Water Works Association, Denver, CO, 1969, 1990, and 1997.

Bernardo, L. D., et al., "Use of Perforated Baffle at the Inlet of Rectangular Settling Basins," *J. AWWA,* 72:9:528 (September 1980).

Camp, T. R., "Sedimentation and Design of Settling Tanks," *ASCE Trans.,* 111:895–958 (1946).

Camp, T. R., "Studies of Sedimentation Basin Design," *Sewage Ind. Wastes,* 25:1–12 (1953).

Culp, K. Y., et al., "Tube Clarification Process: Operating Experiences," *ASCE Proc.,* p. 1179 (January 1955).

Degremont, *Water Treatment Handbook,* 6th ed., Lavoisier Publishing, Paris, 1991.

Edzwald, J. K., et al., *Removal of Humic Substances and Algae by Dissolved Air Flotation,* EPA/600/2-89-032, U.S. Environmental Protection Agency, Washington, DC, 1989.

Hazen, A. M., "On Sedimentation," *ASCE Trans.,* 63:45 (1904).

Hong, S. N., "Treatment of Combined Sewer Overflow with High Speed Microsand Settling," *WEFTEC '96,* Dallas, TX, 1996.

Hong, S. N., et al., "ACTIFLO Process for Drinking Water and Storm Water Treatment," *WEAUTECH '96,* St. George, UT, June 1996.

Hudson, H. E. Jr., *Water Clarification Processes: Practical Design and Evaluation,* Van Nostrand Reinhold, New York, 1981, pp. 123–138.

Huisman, L., *Sedimentation and Floatation,* Delft University of Technology, Delft, The Netherlands, 1973.

Ingersol, A. C., et al., "Fundamental Concepts of Rectangular Settling Tanks," *ASCE Proc.,* 1179 (January 1955).

Kawamura, S., "Hydraulic Scale Model Simulation of the Sedimentation Process," *J. AWWA,* 73:7:372 (July 1981).

Kawamura, S., et al., "Evaluation of Launders in Rectangular Sedimentation Tanks," *J. WPCF,* 58:12:1124 (December 1986).

LePonder, N., and Binot, P., "Treatment of Combined Sewer Overflow (CSO) with High Speed Microsand Settling," *WEF Special Conference,* Montreal, Ontario, Canada, July 1995.

Monk, R. D. G., et al., "Designing Water Treatment Facilities," *J. AWWA,* 79:2:45 (February 1987).

Montgomery, J. M., Consulting Engineers, *Water Treatment: Principles and Design,* Wiley, New York, 1985, pp. 135–151, 523–532.

Okuno, N., et al., *Analysis of Existing Clarifiers and Design Considerations for Better Performance,* Tokyo Metropolitan Government Report, 1980.

O'Melia, C. R., "Particles, Pretreatment, and Performance in Water Filtration," *J. Environ. Eng. Div. ASCE,* 111:6 (December 1985).

Reed, G. D., and Mery, P. C., "Influence of Floc Size Distribution on Clarification," *J. AWWA,* 78:8:75 (August 1986).

U.S. Patent 5,730,864, "Installation for Treating an Untreated Flow by Simple Sedimentation After Ballasting with Fine Sand," March 24, 1998.

Walker, J. D., "Sedimentation," in *Water Treatment Plant Design for the Practicing Engineer,* R. L. Sanks, ed., Ann Arbor Science, Ann Arbor, MI, 1979, pp. 149–182.

Weber, W. J., *Physiochemical Process for Water Quality Control,* Wiley-Interscience, New York, 1972, pp. 111–138.

Yao, K. M., "Design of High Rate Settlers," *ASCE Sanit. Eng. Div.,* 99:621–637 (October 1973).

Yee, L. Y., et al., "Inlet Design for Rectangular Settling Tank by Physical Modeling," *J. WPCF,* 57:12:1168 (December 1985).

Zabel, T., "The Advantages of Dissolved Air Flotation for Water Treatment," *J. AWWA,* 77:5:46 (May 1985).

3.2.6 Dissolved Air Flotation Process

Purpose A significant number of water treatment plants obtain raw water from lakes and reservoirs. In general, these sources contain low-density suspended solids such as algae and organic compounds, and the water is often highly colored. Although turbidity of lake/reservoir water is normally low, the hypolimnion zone of the lake/reservoir is nutrient rich; overturning of the stored water during the spring and fall of each year often results in algae blooms in addition to dissolved iron, manganese, and taste and odor problems. The regular conventional flocculation and sedimentation process is not very effective because the flocs are light and settle slowly unless a very low surface loading rate is specified and costly polymers are fed to the process train.

Dissolved air flotation (DAF) is an effective alternative to sedimentation because it uses minute air bubbles to float light flocs; the floated solids are skimmed off, leaving clear water near the bottom of the tank. The time required for flocculation is shorter than for the conventional settling process, and the hydraulic surface loading rate of the flotation tank is generally 10 times or more that for conventional sedimentation tanks. In addition, the solids concentration of the produced sludge is generally much higher (2 to 3%) than that of the sludge produced by sedimentation (0.5%). The DAF process has been widely used to thicken waste/activated sludge in biological wastewater treatment as well as in the pulp/paper and mining industries in the United States. Application of the DAF process to drinking water treatment began in the United States in 1993, however, the process has been used successfully in European countries since the 1960s. Figure 3.2.6-1 shows a general idea of the DAF system.

Considerations Most of the considerations described previously under regular sedimentation are also applicable to the DAF process. However, a few specific issues are important to this process. The issues include (1) raw water quality, (2) floc-

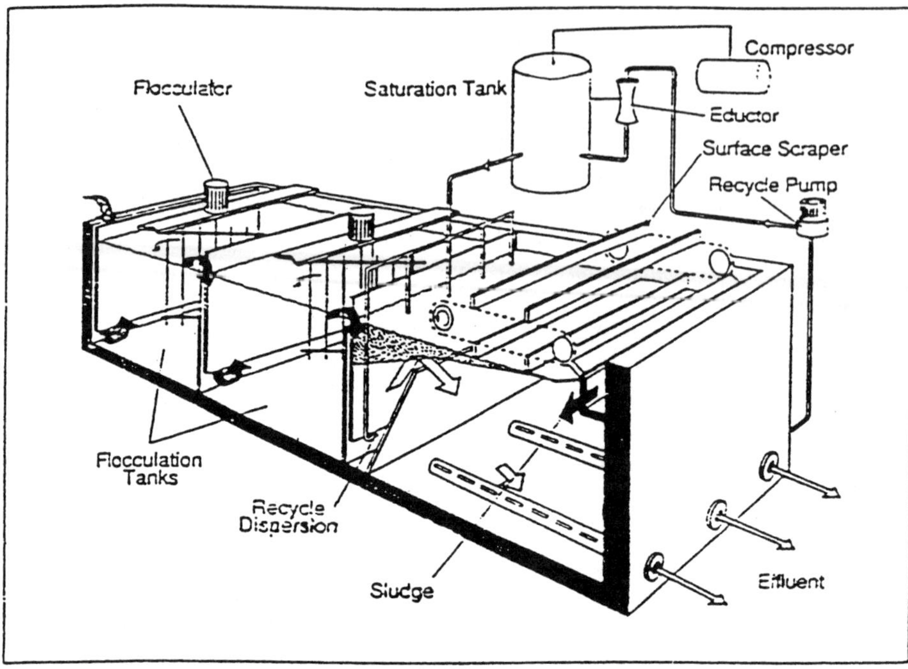

Figure 3.2.6-1 Dissolved air flotation system. (Courtesy of PURAC Engineering, Inc.)

culation process design, (3) flotation tank inlet design, (4) recycling need and the system design, (5) housing needs, air saturation system selection, (6) floating solid (float) removal, (7) flotation tank effluent design, (8) limitations, and (9) cost. The major limitations of DAF are the inability to treat highly turbid water and the necessity to house the flotation tanks.

Raw Water Quality The DAF process is most suited for treatment of algae-laden, highly colored water which also has relatively low turbidity. In general, turbidity of up to 50 ntu is comfortable and the upper limit is as high as 100 ntu with a short duration. The TOC level of raw water is important because the DAF process does not operate effectively in the enhanced coagulation mode because many flocs become too heavy and will not float due to a high dosage of inorganic coagulant and polymer. A TOC value of 8 to 10 mg/L should be the maximum level for the DAF process. However, removal of *Cyrptosporidium* oocysts and *Giardia* cysts can be attained, and according to recent studies, a 2 to 4 log removal may be achieved.

Flocculation Process There are a few significant differences in the design of the flocculation process for a DAF versus regular sedimentation process. Unlike sedimentation, the DAF process must produce small and light flocs which attach to a large number of air bubbles that float to the surface. Three important criteria must be met to achieve this objective: (1) high-energy mixing, (2) shorter mixing period, and (3) installation of an inclined baffle at the effluent of the tank so that the bubble–floc agglomerate are directed toward the surface of the flotation tank.

A mixing energy level of 30 to 120 s^{-1} of *G*, which is double that of a conventional flocculation process, should be used in the design of the DAF flocculation process. The mixing time is generally 15 to 20 min, and this short mixing time requires a good baffle wall system to compartmentalize each stage of flocculation. In general, only two-stage flocculation is provided. The effluent design is quite different from that of a conventional sedimentation process design. A critical design parameter is the placement of an inclined baffle at the effluent of the tank which is 60 to 70° to the horizontal.

Flotation Tank The width of the flotation tank is limited by the floated floc (sludge) removal equipment, and the length of the tank is dictated by the distance that the bubble–floc agglomeration remains afloat. Therefore, in general, the tank should not be longer than 40 ft (12 m). The maximum depth of the tank is normally 8.5 ft (2.6 m); however, a tank that is too shallow may carry air bubbles into the effluent due to influent momentum.

The general practice is a detention time of 10 to 15 min with a surface loading of 4 to 6 gpm/ft^2 (10 to 15 m/h). Recent pilot studies in North America have shown that a surface loading of 6 to 8 gpm/ft^2 (15 to 20 m/h) may be used when polymer is added in conjunction with regular inorganic coagulant. Therefore, the footprint of the flotation tanks is about 10% that for regular sedimentation basins.

The inlet to the flotation tank plays a critical role in the efficiency of the flotation process. The inclined baffle wall inlet should introduce flocculated water uniformly along the width of the tank into the recycle dispersion zone near the bottom of the tank. The top of this baffle should be approximately 1 ft (0.3 m) below the water surface in the tank. Recycle flow is introduced continuously through a distribution system of proprietary nozzles, orifices, or other means. When the recycle flow pressure of 60 to 90 psi (415 to 620 kPa) is suddenly reduced to near atmospheric pressure, the saturated air in the recycle flow is released in the form of microbubbles with an average size of 40 to 50 μm. Water containing microbubbles has a milky appearance. These microbubbles are adsorbed on the flocs and these bubble–floc agglomerates float to the surface of the water in the flotation tanks.

It is important to remember that the flotation tanks must be covered by a housing, as both rain and wind will break up the floated flocs. Moreover, the flotation tank must have a drain valve for removal of water as well as of accumulated sludge.

Air Saturation System Air saturation of water may be attained by various means: injectors, a pressure vessel saturator, or a pressured vessel with packed materials. Of the three, the packed saturator system is considered to be the most cost-effective. The recycling rate is generally 7 to 8% of the processing flow rate, which corresponds to approximately 8 to 10 g of air per cubic meter of processing water. The air/solids ratio is approximately 380 mL of air to gram solids for surface water; this is about 15 times higher than that for activated sludge thickening.

Floated Solids Removal Floating solids may be removed intermittently by mechanical skimming devices or through flooding. A mechanical skimmer is expensive but creates sludge with a higher content of solids (1 to 3%). The most common types are chain and flight, oscillating, reciprocating skimmers, and rotating skimmers. The use of flooding eliminates the need for mechanical equipment and has minimal effect on treated water quality. However, this method creates a very low sludge solids concentration (0.2 to 0.5%) and produces a high percentage of wastewater (up to 2%).

It is important to note that raw water quality can dictate the selection of the appropriate sludge removal system. Sludge produced from soft and highly colored water is easily breakable and must be skimmed off within 30 minutes of formation; a full-length reciprocating skimmer should therefore be used frequently to skim off the sludge at a speed of 1 to 2.5 ft/min (0.3 to 0.76 m/min). Sludge produced from turbid water or algae-laden water is stable, with sludge solids concentrations as high as 2 to 3%. Therefore, the sludge may be removed on an intermittent basis.

Effluent Design The clarified water in the flotation tank must be collected from near the tank bottom under a curtain wall or by means of pipe laterals with orifices. It is important to keep the water level in the flotation tank nearly constant to facilitate effective float skimming, and an effluent water-level control weir must be provided downstream of the clarified water collection system. Moreover, the effluent collection level must be at least 1 ft (0.3 m) above the floor of the flotation tank to minimize the chance of scouring any sludge settled on the bottom.

Cost The DAF system has a much smaller footprint than that of the conventional flocculation and sedimentation process. This aspect makes the capital cost of DAF less than that of the conventional flocculation and sedimentation process, however, the addition of mechanical equipment and housing may make the cost equivalent to that of a conventional process. The operational costs of a DAF process is also higher than that for conventional sedimentation systems, due to the higher power use of its air saturators, recycling pump, and sludge skimmers; the DAF power requirement is approximately 3.5 kWh/mgd (2.5 kWh per 10,000 m^3/d), in contrast to 1.0 kWh/mgd (0.7 kWh per 10,000 m^3/d) for regular flocculation and sedimentation processes. Coagulant dosage is also somewhat higher—the process requires 13 to 15 mg/L of alum—even when the raw water turbidity is less than 3 ntu. Overall, the cost of DAF is similar to or higher than that of a conventional sedimentation process.

Type and Selection Guide There is a special application of the DAF process, together with the filtration process, which results in the accumulation of floc on the surface of the water above the filter bed, allowing for the floc to be skimmed off and the clarified water to be filtered. This combined DAF and filtration process has four basic drawbacks: (1) design and operation of the filtration system is complicated; (2) DAF loading is controlled by the filtration rate, so the DAF process cannot be optimized; (3) plant operators are not able to visually inspect the quality of the clarified water (filter influent), due to floating solids; and (4) each flocculation tank must be coupled with a filter. Therefore, it is generally better to design the DAF process as an independent process unit.

A packed saturation tank system with a reciprocating or a full-length chain-and-flight float skimmer is often selected. However, the former system is preferable. Pumps required for sludge removal require special consideration, due to the large amounts of air contained in the sludge; a WEMCO or screw impeller pump is recommended.

Design Criteria

Flocculation Process

Minimum number of tanks	Two
Number of stages per tank	Less than three (two normal)
Water depth	12 to 15 ft (3.6 to 4.5 m)
Width of each tank	Up to 30 ft (10 m): match to flotation tank width
Detention time	15 to 20 min
Energy input range (G)	30 to 120 s^{-1}
Type of flocculator	Vertical-shaft flocculator with hydrofoil blades

Flotation Tank

Minimum number of tanks	Two
Tank width	Up to 30 ft (10 m)
Tank length	In general, 40 ft (12 m) maximum
Water depth	8.5 ft (2.6 m) normal
Surface loading	4 to 8 gpm/ft^2 (10 to 20 m/h)
Detention time	10 to 15 min

Recycling System

Recycling rate	6 to 10% of clarified water
Maximum air loading	10 g/m^3
Saturator loading rate	25 gpm/ft^2 (62.5 m^3/m^{2}h)
Operating pressure	60 to 90 psi (415 to 620 kPa)

BIBLIOGRAPHY

Braghetta, A., et al., "DAF Pretreatment: Its Effect on MF," *J. AWWA,* 89:5:78 (May 1997).

Edzwald, J. K., et al., "Flocculation and Air Requirements for Dissolved Air Flotation," *J. AWWA,* 84:3:92 (March 1992).

Johnson, B. A., et al., "Pilot Plant Testing of DAF for Treating Boston's Low Turbidity Surface Water Supply," *IAWQ–IWSA–AWWA Joint Specialized Conference,* Orlando, FL, April 1994.

Krofta, M., and Wang, L. K., "Flotation and Filtration," *J. AAWA,* 74:6:305 (June 1982).

Plummer, J. D., et al., "Removing *Cryptosporidium* by Dissolved Air Flotation," *J. AWWA,* 87:9:85 (September 1995).

Zabel, T., "The Advantages of Dissolved Air Flotation for Water Treatment," *J. AWWA,* 77:5:42 (May 1985).

3.2.7 Granular Medium Filtration Process

Purpose In a water treatment process train, the fundamental system that removes particulate matter is filtration. The most common filtration process employs a granular medium of a certain size and depth. The pretreated water passes through the filter bed, where a majority of the particulates are removed in the top portion, as well as throughout the entire depth of the bed.

The primary objective of the design engineer should be to deliver a filter design that provides a steady production of high-quality water with minimal capital and minimal operation and maintenance costs. When designing filters, most engineers rely on either their own experience, the experience of others, or textbook design criteria. However, engineers should expand certain design variables, such as the size and thickness of the medium, filtration rate, filter underdrains, filter washing conditions, and available headloss for filtration. These variables must be developed based on the raw water quality, type of filter bed, and desired filtered water quality.

It should be noted that there is a basic common denominator among the coagulation, flocculation, and filtration processes. Under certain raw water conditions, adequate treatment of the raw water can be carried out in the filter alone, and the need for ordinary flocculation and sedimentation processes may be eliminated.

In this section we focus on the common granular media filters. Other types of filters are discussed in Sections 3.2.8, 7.7, and 7.8.

Considerations Several factors must be taken into consideration when designing a proper granular medium filtration process. They are (1) local conditions, (2) design guidelines set by regulatory agencies such as the state department of health, (3) site topography, (4) plant size, (5) raw water quality, (6) type of pretreatment process, (7) new and proven types of filter, (8) provisions for future modification or addition of filters, (9) type of filter wash system, (10) control of the filtration rate, (11) type of filter bed, (12) chemical application points, and other miscellaneous items. These subjects are discussed in detail.

Local Conditions Perhaps the single most important issue in filter design are the local conditions. Factors such as the availability of qualified technicians and raw materials, the local climate, and the soil and geotechnical characteristics of the plant site all affect the design of the granular medium filtration process. For example, a basic type of filter bed should be chosen for a treatment plant located on a small resort island or in a town of a developing country. Under these conditions, the best type of filter is one that has a simple control system. It is tragic to see that many design engineers fail to take this into consideration by designing treatment plants with expensive filters and sophisticated control systems for regions lacking both qualified plant operators and support personnel. Not only are these plants producing substandard quality water, they are rapidly deteriorating, due to the lack of qualified maintenance personnel. The filter bed design should also be determined by the availability of filter materials at a reasonable cost.

Climate is the second issue that must be studied prior to filter design. In regions of cold weather, the filters must be either covered or housed in a building to prevent the water from freezing. Although the control cabinets are supposed to be hermetically sealed, plants located in regions experiencing severe dust storms should have a housing over the filter control system. The filter bed should also be covered to prevent debris and sand from entering the filter, thereby preventing both operational and maintenance problems.

Finally, the soil and geotechnical characteristics of the plant site should be analyzed. Of primary interest are the level of the underground water table and the condition of the soil because the filters are relatively deep and heavy structures. There are a few documented cases where the filters became tilted or subsided as a direct result of a high groundwater table and a poor foundation.

Large cities that are surrounded by a wealth of technology and qualified technicians are prime candidates for treatment plants equipped with both a first-class filter system and a sophisticated backup system. Some of the filters used in the treatment plants of these large metropolitan areas are capable of operating at a rate of 8 to 13 gpm/ft^2 (20 to 33 m/h), which is well above the average filtration rate in the United States of 5 to 6 gpm/ft^2 or 13 to 15 m/h. However, both pilot studies and actual plant operational data show a significant increase of number of particles (2 to 10 μm) in filtered water at filtration rates above 8 to 10 gpm/ft^2 (20 to 25 m/h). Optimization of filtration has attained grave importance in recent years, due to the need to remove *Cryptosporidium* oocysts and *Giardia* cysts. Most water purveyors are now limiting their filtration rate to 6 gpm/ft^2 (15 m/h), especially during cold weather months.

Guidelines Set by the Regulatory Agencies Guidelines for designing treatment plants can be found on both the state and county levels. They are especially stringent with respect to filter design. A few of the requirements are worth noting: maximum filtration rate, continuous monitoring of the filter effluent turbidity, rewash (filter-to-waste) requirements, and the types of filter that can be employed. For example, the state of California discourages the use of pressure filters in public water treatment plants because the condition of the filter bed cannot readily be observed. Therefore, the bed could potentially be in poor condition without the operator's knowledge. California also discourages the use of declining-rate filtration control systems because of the likelihood for high-turbidity breakthrough at the beginning of the filter cycle because of a filtration rate that is twice or more the nominal rate at the beginning of the process.

A few states also dictate the physical features of filters, such as media size, depth of the filter bed, spacing and height of the wash trough, and double-walled construction of the upper and lower gullet with a drainage pipe to assist operators in detecting leakage of unfiltered water from the upper gullet. The *Ten States Standards* and the rules set by the state of Utah are examples of this type of design control. On the national level, the EPA is concerned primarily with the removal of harmful microbes, such as *Giardia lamblia, Cryptosporidium,* and *Legionella,* and viruses from the water supply. This can be achieved through coagulation followed by filtration and disinfection.

Site Topography The topography of the plant site must be evaluated carefully prior to design since it determines the location, type, and arrangement of each unit in the filtration process. For example, the headloss across a water treatment plant typically ranges from 15 to 25 ft. Therefore, if the plant site is flat, the filter structure and clearwell must be located deep within the ground unless a pump is placed in the middle of the process train. If a gravity flow is to be maintained across the plant, the only other alternative would be to elevate the pretreatment unit processes above the ground.

Topography also dictates the type of filter backwash facility that may be used. If a suitable hill exists on the plant site, the wash tank can be built on it. However, a direct pump wash or self-backwash system should be seriously evaluated if the plant site is flat. This minimizes problems with freezing and avoids the unsightly appearance of the elevated wash tank. Additionally, the energy effectiveness of the self-backwash filters is gained. The ideal plant site has a constant slope of 2 to 3%, which allows the filter wash tank, filters, and wash-waste holding tank to be situated easily and built economically since there is no need for excessive excavation. The slope also helps to satisfy the water level required for each process. A steeply sloped topography not only severely restricts the arrangement of the unit processes but greatly increases the probability of landslide.

Plant Size The size of the water treatment plant is instrumental in determining both the physical and performance requirements of filters. Plant size usually dictates the total number of filters, the size of each filter, the degree of instrumentalization of the filter control system, and the method by which the filtration rate is controlled. Some existing large plants, those processing over 200 mgd (9 m^3/s), have a total of more than 30 filters. However, the total number of filters may be limited if either the size of each filter is increased [up to a practical limit of 2000 ft^2 (100 m^2)] or if a high filtration rate such as 8 gpm/ft^2 (20 m/h) is adopted, or if both methods are implemented.

The capacity of the treatment plant also aids the design engineer in selecting the most appropriate type of filter. For example, plants operating on a small scale [less than 3 mgd (130 L/s)] can generally use proprietary filters supplied by equipment manufacturers, also known as *package filters.* These types of filters are generally cost-effective, easy to operate and maintain, and perform well for plants of this size as long as the package filters are selected properly, that is, based on a rigorous engineering evaluation.

Raw Water Quality and the Type of Pretreatment The raw water characteristics should also be scrutinized when designing the filter bed since these characteristics determine the type of pretreatment, which in turn determines the type of filter bed. For instance, if the underground water contains high levels of iron or manganese, the best method for removing these compounds is by oxidation with chlorine or potassium permanganate, followed by pressure filters; the safe yield from a well is generally less than 3 mgd (0.13 m^3/s). This method is the primary choice for well-water application because it preserves the pressure of the pumped water and allows for backwashing of the filters using the effluent from other pressure filters on the line; additional backwash pumps are therefore not required. Iron and manganese removal may also be achieved through direct filtration with preoxidation by chlorine, ozone, or permanganate. Green sand has been used as a filter medium in this type of situation, although sand may be used as an alternative.

To optimize the performance of the filtration process, the filter bed design should complement the pretreatment process. For instance, under conditions where direct filtration or in-line filtration is applicable, the filter bed should be designed with a larger storage capacity for suspended solids than that for conventional beds. Therefore, a reverse-graded or coarse and deep bed should be selected. The presence or lack of a high concentration of suspended solids in the raw water requires special consideration when selecting the type of filter bed. Filtration of algae-laden water is very difficult even after flocculation and sedimentation, due to rapid clogging of the filters by algae. Use of a coarse deep bed or a reverse-graded filter bed should be evaluated under these circumstances. Conversely, a common monosand bed (often referred to as a *rapid sand bed*) may be considered when the filter influent is very good, the result of a conservatively designed pretreatment process. Since the bed depth of these types of filter is relatively shallow [less than 24 in. (0.6 m)], surface filtration, not in-depth filtration, plays a major role in the monosand beds. Two-stage filtration should also be evaluated if the raw water quality is good (i.e., has low turbidity), thereby allowing both the flocculation and sedimentation processes to be eliminated.

Water temperature has a great impact on any filtration process. For example, during the winter when water temperatures are below 40°F (4.5°C) the filter run length is almost half that of the summer months (over 70°F or 21°C) at the same filtration rate. Filtered water turbidity is also higher in the winter. Moreover, in the winter, filter bed expansion is almost twice that of the summer under the same backwash rate. Therefore, a significant amount of media loss may occur during the winter unless the wash rate is adjusted.

Due to the limited amount of water for domestic use in large urban regions, tertiary treatment of sewage for water reuse has become an important issue. The filter for biologically treated sewage behaves somewhat differently from that used in the water supply field, although the physical characteristics are the same. The graphs in Appendix 18 illustrate the typical performance of tertiary filters.

New and Proven Types of Filters Like any other type of equipment, new forms of filtration systems, filter media, and filter control systems become available on the market at any given time. Yet the design engineer must be extremely careful with regard to using the new products. Although these new products are appealing, there is likelihood that problems will emerge after a year or two of continuous operation. For example, one company marketed small aluminum chips as a filter medium. According to the manufacturer, chlorine (added to the filter influent) slowly dissolved the aluminum and allowed coagulation and flocculation to occur within the filter bed, eliminating the need for coagulant. The concept was very appealing and the filters performed well during the first 3 to 6 months of operation. However, after 9 to 12 months the filter beds became a large chunk of fused aluminum chips. Similar unforeseen problems have also occurred with new types of filter underdrain systems.

Despite convincing laboratory test results, engineers should remember that real filters are subjected to many different conditions that cannot be duplicated in either a laboratory or a piloting facility. Since most critical problems occur within a year or two of operation, it is prudent to wait until these problems are isolated and resolved. Design engineers should not hesitate to research new products, but they are strongly advised to select items that are proven to be reliable, with low maintenance and ease of repair.

Allowing for Future Filter Modifications or Additions The design engineer should anticipate an increasing demand on the plant output by including a provision for future modifications of the filters or addition of more filters. The process should therefore be designed so that future construction will have minimal impact on plant operations. In many cases, water treatment plants increase their operating capacity by adding basins and filters in stages. If the original treatment unit processes are not designed as a module, the new filters must be added onto the original filter bank. In situations such as this, the filter influent channel, effluent channel, and backwash system must be designed to accommodate the new flow rate, but based on the original plant specifications.

Construction during plant expansion should have minimal impact on the existing facilities. For this reason, the module expansion scheme must have a carefully conceived master plan for the yard pipings. The plan should maintain equal distribution of water flow to each module and should not have any physical interference between the major pipelines. The need to provide for future filter modifications has become more of a reality due to the EPA's proposal to make future water treatment regulations more stringent. It is also probable that the old monosand bed will be replaced with dual-media reverse-graded beds, such as anthracite and sand, to increase the output of the filters.

Presently, the most important consideration is the possibility that granular activated carbon (GAC) beds will be required to adsorb objectionable organic compounds—an EPA rule that is anticipated to be implemented in the near future. One method would be to replace the existing filter medium with GAC medium. However, the GAC bed requires an empty bed contact time (EBCT) of at least 10 min, thus necessitating a bed depth of 6 to 8 ft as long as the filtration rate is at a level rate of 4 to 6 gpm/ft^2 (10 to 15 m/h). Under no circumstances can such a depth be provided by ordinary filters unless the filter cells are designed initially with the option to convert to GAC adsorption filters.

Deep bed filters also demand special design features, such as air and water backwash (also referred to as *air-scouring wash*); the air-scouring backwash system requires a specific type of filter underdrain system. Moreover, recent prototype studies indicate that deep bed filters should be designed with both an air-scour wash and surface wash system whenever polymers are used as part of the coagulant to prevent mudball formation.

Filter Washing System There are four basic schemes in filter washing: elevated wash tank, direct pump, self-backwash (Greenleaf type), and continuous backwash (Hardinge or Dyna Sand filters). The first two systems are the traditional and proven filter washing schemes. The self-backwash filter may be either a proprietary item or custom designed by the engineer. Continuous backwash filters are proprietary items. Filter selection is based on plant capacity, site topography, local conditions, and other factors, such as energy efficiency, the wash-waste handling situation, and owner preference.

The elevated wash tank and direct pump wash systems have ample flexibility in adjusting the backwash rate. The elevated tank scheme is particularly suited for a plant site with high ground on which to locate the tank. The advantages of the elevated tank scheme are its effective use of energy, a small-capacity pump that fills the tank within an hour or two, and emergency reserves of pressurized water in case of power outage. However, the elevated tank system requires a large tank to satisfy the two filter wash requirements: a large and long backwash pipe and a minimum elevation of 33 ft (10 m) above the filter wash troughs. Conversely, the direct pump wash system eliminates the need for a large elevated wash tank, and control of the wash rate is relatively easy. However, both the size of the pump and the horsepower required become quite large for large filters. The direct pump system is therefore best suited for small filters or air-scouring wash systems since these must change the backwash rate two or three times during a washing cycle.

With respect to the washing system, the basic alternatives are backwash alone, a combination of surface wash and backwash, ordinary air-scour wash and simultaneous air and water wash. Backwash alone cannot adequately maintain clean filter bed conditions unless coagulation and flocculation are achieved with a minute amount of alum (without polymers). Rapid sand filters, ordinary dual-media beds, and tertiary filters are all cleaned effectively by backwash supplemented by surface wash. In contrast, coarse deep bed and tertiary filters are more suited to ordinary air-scour wash or simultaneous air and water wash. The latter requires special buffer plates around each wash trough (see Figure 3.2.7-6).

Some authors claim that the air-scouring wash (see Figure 3.2.7-7) is a far superior system than surface wash. But this claim loses credibility after reviewing the performance of several operational plants: the data show that a surface wash system that is properly designed and operated can match or even exceed the performance of an air-scouring system. For example, a review of four plants in California revealed that tertiary filters fitted with dual-arm surface agitators were devoid of mud balls and had clean bed conditions despite 15 years of operation. In contrast, filters in both the United States and Europe that employed air-scour wash systems occasionally had a significant mass of mud balls whenever polymers were used as coagulant or filtration aid.

Design engineers should not be misled by the impressive boiling action at the surface of the filter bed during air-scouring wash because this violent boiling action actually occurs in only the top 6 to 8 in. (0.15 to 0.2 m) of the filter bed. Therefore, mud balls that have descended below this boiling zone remain unbroken. It is worthwhile to mention that several water treatment plants in the United States, originally designed with ordinary air-scouring filters, have added the surface wash system in response to the problem of mud ball formation.

Filtration Rate Control There are basically two modes of filtration: constant-rate filtration and declining-rate filtration. Despite claims to the contrary, both constant- and declining-rate filters are capable of producing filtrates with less than 0.1 ntu turbidity. However, the colloidal particles must be conditioned properly in the pretreatment process, and the filter bed must be both designed and maintained properly.

Only four basic types of rate-control systems have supplied consistent and reliable performance: (1) constant-rate filtration with a flow meter and a flow modulation valve; (2) constant-level filtration with equal-flow splitting inlet weirs, a water-level sensor, and a flow modulator valve; (3) declining-rate filtration with a submerged inlet and a flow restriction in each effluent pipe that may or may not be fitted with a weir to control the effluent level; (4) constant-rate filtration with equal-flow splitting inlet weirs and a weir to control the common effluent level. These four basic types of systems are represented in Figure 3.2.7-1.

Each of these control systems has both good and bad features. The last two systems are simple and require very low maintenance. However, only a plant with more than six filters can fully maximize these positive features. The first two systems have been the standard flow control measures for the past 50 years and are still quite popular among design engineers. Therefore, a large percentage of operational filters are equipped with these types of systems, despite their high capital cost of approximately $25,000 (average) per filter.

Of the four rate control systems, the first (historically) has the highest maintenance cost and requires frequent operator attention. Another negative aspect is that the filtration mode changes from the designed constant-rate filtration to declining-rate filtration when the control system fails. Many old constant-rate filters have been operating in the declining-rate mode because their flow control systems are beyond repair. This is particularly true for treatment plants in developing countries.

The declining-rate filtration system is very simple to design and build and generally produces good-quality water. However, to meet the present high standards for filtered water quality, this mode of filtration requires vigilant plant operators because the water level in the filter cells tends to fluctuate widely and creates the potential for partial drainage, overflow, and potential initial turbidity breakthrough at beginning of the filtration cycles, unless the design has been made very carefully. It is precisely because of these problems that some regulatory agencies, such as in

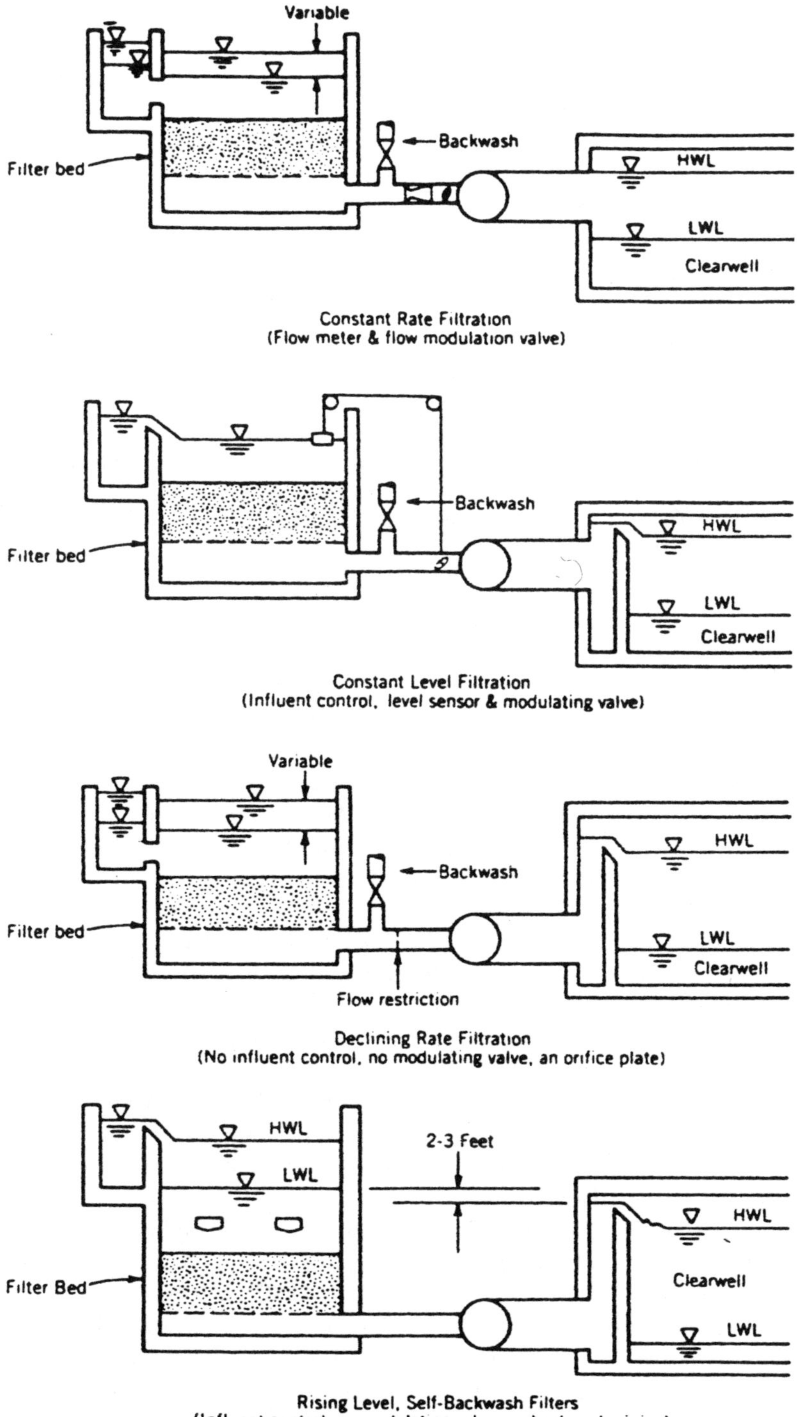

Figure 3.2.7-1 Basic filter control systems. (Adapted from *Water Treatment: Principles and Design* by J. M. Montgomery, Consulting Engineers, Wiley, New York, 1985.)

the state of California, discourage the use of declining-rate filtration in new filter designs.

One operational problem associated with declining-rate filters is that there are no clear parameters for filter washing: the headloss for all the filters on-line is nearly equal, there is no flow metering device to indicate the filtration rate of each filter, and the turbidity of the filtered water does not increase with time. Thus operators usually wash the filters based on the length of the filter run. It is an undeniable fact that very few filters in industrialized countries are designed as declining-rate filtration system, because there are no positive means of controlling the filtering conditions. Yet the declining-rate system still satisfies the needs of many filtration plants, especially in developing countries, because of its simplicity and low maintenance requirements and the less stringent governmental requirements.

The fourth type of system, constant rate with equal loading to each filter with a rising water level, is capable of backwashing a filter by using the effluent from the rest of the filters on the line. These are commonly referred to as *self-backwash filters*. A few equipment manufacturers supply these types of filter with four filters per module. The major drawbacks of the self-backwash filters are the limited flexibility in adjusting the backwash rate, lack of filter to waste capacity, and the deeper filter cells; they are at least 4 ft deeper than the normal filter design. However, there are positive features to this type of system: the lack of a flow modulation valve; the simple opening and closing of the inlet and waste-wash valves to initiate filter washing; the ability to function without the system of backwash pumps, tanks, and wash lines; and the easy determination of headloss through the filter (by measuring the water level in a filter cell). Since the water depth increases with the duration of the filter run, the problem of air binding is minimized. The self-backwash filter is therefore particularly suited for situations where the raw water is saturated with air that is entrained in the water. Moreover, this type should always be evaluated because of its limited use of mechanical equipment, including backwash pumps and valves. Also, the filter-to-waste feature can be incorporated fairly easily. Presently, there are three large plants, with capacities exceeding 400 mgd (19.5 m/s), using this type of filter: two in California and another in Manila, Philippines.

Type of Filter Bed The selection of a filter bed is first contingent on the availability of bed material. Other factors, such as the primary purpose of filtration, the type of filtration process, anticipated filter run time, and the filtration rate of the system, also influence the decision.

The first step in selecting a filter bed is to study the local conditions. In the United States, almost any type of common material is available at a reasonable cost: sand, crushed anthracite coal, granular activated carbon (GAC), and even garnet sand. This is not the case for most developing countries and a few industrialized nations. For instance, a large part of Africa and Southeast Asia use sand and gravel as the primary filter materials because other types are simply too expensive. Although only three types of granular media are generally used, filter media can be composed of other materials, such as pumice or synthetic material, provided that there is no measurable leakage of unwanted substances from the media, no significant attrition loss, and the cost is reasonable. For instance, pom-pom-like synthetic fiber balls called *Marimo* are used in a patented filter system manufactured in Japan.

The second step in selecting an appropriate filter medium is to delineate the primary purpose of the filtration process. Sand and the combination of sand and anthracite coal are the two basic and proven alternatives with respect to ordinary water purification. Alternatively, the more expensive garnet grains may be used in conjunction with sand or anthracite coal. If the main purpose is the filtration of sewage, as in the tertiary filtration of sewage, monosand or mono-anthracite coal is the most appropriate filter bed medium. A dual-media filter bed may also be considered because of its proven performance, but the disadvantage is that it tends to have problems with sludge accumulation in the sand–coal interface. However, this problem may be eliminated by employing air-scouring backwash or dual-arm surface agitators to wash the sludge out of the interface; but the air-scouring wash system requires the filter bed to be restratified with a high backwash rate because the sand and coal are intermixed during the wash, so the advantages of smaller backwash pipe lines, valves, and waste-wash tank are lost. Conversely, a monomedium bed does not require restratification (i.e., may not require a wash rate step-up control system), and the advantages of the air-scouring wash can be fully utilized.

The filtration process also affects the selection of the filter bed because of the special requirements of each type of process. The direct and in-line filtration processes must have filter beds with a large floc holding capacity. A reverse-graded filter bed, such as a dual-media or coarse deep bed, satisfies this requirement. In two-stage filtration the filter bed of the first stage acts as a roughing filter and carries out the flocculation process. Data obtained from pilot filter tests and actual installations using the two-stage filtration process indicate that the first-stage filter bed may be designed in the same fashion as an ordinary filter.

The fourth step in selection of the filter bed is based on the filtration rate. For slow sand filtration the most appropriate choice is a fine-grained monosand bed since the filtration rate for slow sand filters is less than 0.2 gpm/ft^2 (0.5 m/h). Rapid sand filtration, with filtration rates ranging from 2 to 3 gpm/ft^2 (5 to 7.5 m/h), usually have medium-sized sand (0.5 mm E.S.). Filters displaying filtration rates of 5 to 10 gpm/ft^2 (12.5 to 25 m/h), commonly known as *high-rate filters,* always consist of a reverse-graded filter bed or a deep, large-grain, monomedium bed. Details of these filters are discussed elsewhere in this chapter.

Chemical Application Points Disinfectant, filter aids, adsorbent (for taste and odor control), and alkali chemicals (for pH control) are the most common chemicals added to the filter influent to aid the process of filtration. Three important design considerations must be addressed with regard to this subject: Minimize the number of chemical application points, recognize that certain chemicals have the potential to "break through" into the filtered water, and ensure that the chemicals are adequately distributed in the water. Finally, the design engineer should be cognizant that the locations of the application points determine the type of chemicals that may be applied.

The first item is to minimize the number of chemical application points. High-rate filtration often employs a polymer as a filtration aid. Polymer and chlorine are fed to the filter influent to promote filter efficiency and keep the filter bed clean. If the plant design provides only one water passage between the pretreatment process units and the filter bank, these two chemicals may be fed to the water at this single location via a mixing device. In many treatment plants there are several passages between the settled water channel and the filters. This requires the chemicals to be distributed to these several application points and necessitates the use of additional chemical feeders and flow-splitting equipment.

The second important issue is the potential breakthrough of the added chemicals into the filtered water; this phenomenon degrades the quality of the filtered water. Some treatment plants routinely add alkali chemicals such as lime or sodium hydroxide to the filter influent to raise the pH to 8 or above, following flocculation with alum and sedimentation. This practice always produces a high aluminum concentration in the finished water due to the pH and the basic characteristic of the aluminum hydroxide (see Figure 3.2.3-1). Another common practice is to add powdered activated carbon (PAC) to the filter influent as a means for taste and odor control. However, a significant amount of carbon breaks through the filter and appears in the finished water. Particle counters may be used to monitor the filter effluent, but nepherometric turbidimeters are virtually incapable of detecting carbon leakage.

The third issue, adequate mixing of the chemicals at the application point, can be resolved by including either an appropriate type of chemical diffuser pipe in conjunction with a baffled channel or an in-line static mixer (for the pipe).

The last issue, with respect to the application of chemicals to the process water, is the selection of the feeding points. The filter influent, the filter effluent, and the backwash line are the three potential points. Chemicals that may be fed to the influent are chlorine, polymers, and to some extent PAC. Chlorine, alkali chemicals, corrosion inhibitors such as sodium hexametaphosphate, and fluoride compounds (for fluoridation) should ideally be applied after filtration, because of their function and effectiveness. The two chemicals that may be fed to the backwash water to disinfect and condition the filter beds are chlorine and polymers. It should be noted that the recent EPA standards may force many treatment plants with filters lacking a filter-to-waste function to feed polymers to the backwash line to minimize the initial turbidity breakthrough of the filters.

Miscellaneous Items Design engineers must also consider several additional issues when designing a proper granular medium filtration process: the use of wash troughs; the amount of allowable headloss for filtration; and the type of filter underdrain, type of filter, and waste-wash-water handling facility. In the United States, wash troughs are an essential part of the filter design, due to the predominant use of a high backwash rate. Filter backwashing, with or without surface jets as the auxiliary scouring system, necessitates the use of wash troughs because prompt removal of the dirty wash waste is essential in minimizing both the washing time and the formation of mud balls in the filter bed. Conversely, an ordinary air-scouring wash system uses a low backwash rate, and the washing time is therefore twice as long as for high-rate backwashing. Most air-scouring systems in Europe utilize sequential wash and do not have wash troughs since the filters are effectively cleaned, provided that the width of the filters is less than 12 ft (4 m), a monosand filter bed is used, and no polymer is applied to the filter influent. This type of system has operated in Europe since the early twentieth century (see Figure 3.2.7-7). The choice between the two types of washing systems is contingent on filter type (especially the size and depth of the filter bed) and designer preference.

When designing the filtration process, engineers should realize that the various filters provide different amounts of allowable headloss. Gravity filters generally have an available headloss of about 8 to 10 ft (2.5 to 3 m). Other types of filters, such as proprietary, valveless, or monovalve packaged filters, do not provide more than 6 ft (1.8 m) of available headloss. Automatic backwash filters (Hardinge type) typically provide a maximum of 1.0 ft (0.30 m) of available headloss.

Filter underdrain systems differ primarily in their filter washing design and the types of filter required by each system. Most municipalities in the United States use large filters and operate with a wide variety of underdrain systems, ranging from clay to reinforced fiberglass plastic blocks, precast concrete laterals, false bottoms with strainer, to porous plates. To realize the full potential of the treatment plant processes, design engineers should select an underdrain system that is most compatible with the size of the filter, the physical characteristics of the filter media, and the type of filter washing system that is implemented. If air scouring is chosen as the washing system, the most common and accepted type of underdrain is a false bottom fitted with long leg strainers. However, there are several disadvantages in using this type of system: the slits of the strainer heads may become clogged by scale, filter media or debris left in the pipes, and the false bottom during construction. There have also been a few instances when false-bottom floors became loose and lifted up after repeated filter washings. To avoid clogging the strainer heads, it is essential to clean the underdrains and pipes thoroughly after construction. Moreover, a slit size of a minimum of $\frac{5}{36}$ in. (0.35 mm) or an orifice size (underdrain blocks) larger than $\frac{5}{18}$ in. (7 mm) in diameter should be selected whenever possible.

The four essential factors in selecting a filter underdrain system are uniform flow distribution of the backwash, durability, reliability, and cost-effectiveness. The flow distribution of the backwash water in the filter cell is kept uniform by (1) making the orifice or slits of the filter underdrain system small enough to introduce a controlled headloss, and (2) decreasing the flow velocity in the pressurized conduit upstream of the underdrain system so that the hydraulic grade and energy lines of flow entering the underdrain system are fairly uniform. The false-bottom underdrain system usually provides the best performance and has minimal maintenance costs.

The overall capacity of a water treatment plant determines the type of filter that may be included in the filtration process. For instance, automatic backwash filters, self-backwash filters, and pressure filters are frequently used for small plants, those processing less than 15 mgd (0.65 m^3/s). These types of filter are proprietary items marketed by a number of manufacturers. They offer advantages such as low capital cost, minimal headloss across the filters (except in pressure filters), good filtered water quality, and simple operation and maintenance. Medium to large plants, those processing over 10 mgd, generally use gravity downflow filters with dual-media filters constructed of reinforced-concrete structures.

Upflow filters, where the water passes upward through the filter bed, have also been used on a limited basis. One obvious advantage of upflow filters is that they can achieve coarse to fine filtration with a single medium such as sand, and thereby eliminate the need for more costly multimedia beds. However, one serious disadvantage of an upflow filter is that the filter bed expands as it becomes clogged and allows previously removed solids to escape into the effluent. An extra deep bed (minimum of 6 ft) containing coarse sand (1.5 to 2 mm) or a system of restraining grids, placed at the surface of the filter bed, may be used to overcome this problem if normal-sized medium is used. This particular type of design is seldom considered for municipal treatment plant applications because of the extreme care that must be exercised during operation. Therefore, its use is restricted by regulatory agencies.

If a treatment plant undergoes expansion, the design engineer should try to match both the size and elevation of each new filter to existing filters so that the old backwash facility may still be used. This not only reduces the cost of expansion but also maintains continuity in operating procedures. For ordinary filters, the size of the waste-wash-water handling facility is directly proportional to the plant capacity and the size of each filter cell. The automatic backwash filters sequentially clean small portions of the filter bed and thereby constitute a continuous filtration process; this type of filtration process produces a small amount of wash waste on a continuous basis. For this reason, the waste can easily be handled by a very small facility. In contrast, the large filters used at large plants require a sizable handling facility because these filters produce a significant amount of instantaneous waste flow during a washing period. Regardless of the type of filter, 2 to 3% of the plant flow rate is waste flow and the designer must therefore evaluate alternative types of filters and the size of each filter whenever the plant site does not have enough areas to accommodate a large waste handling facility.

Type and Selection Guide In this section we briefly summarize the common filter units and provide guidelines for making the proper selection.

Available Alternatives There are three basic types of granular medium filters: (1) slow sand filters, (2) rapid sand filters, and (3) high-rate filters with either a coarse deep monomedium or reverse-graded multimedia bed. Both the rapid sand and high-rate filters require chemical coagulation and, in most cases, clarification prior to filtration so as to optimize filter performance. Coagulation and clarification pretreatment remove excessive amounts of suspended solids, color, and some mineral components. The pretreatment also removes a large portion of microorganisms such as algae and *Giardia*.

A commonly encountered design flaw is that the type and degree of pretreatment do not match the filter requirements. For instance, if the source is a large lake with consistently good water quality, a slow sand filtration or a direct in-line filtration system, as well as a two-stage filtration system with high-rate filters, may be adopted. However, if the quality of the raw water is generally good with occasional high turbidity spikes, algal bloom, or taste and odor problems, a complete treatment process must be considered; the treatment process train should include a DAF process or a settling process. In this type of situation the sedimentation tanks may be either bypassed or used as a large channel between the flocculation and filtration process, but only during periods of good raw water quality, by using the direct filtration mode and by employing small amounts of coagulant. From these examples it is evident that design engineers must review both the raw water characteristics and the seasonal variations in the water quality and should seriously consider running a bench-scale or pilot study to evaluate the treatability of the raw water. Only then can a proper pretreatment scheme be established.

The filters can be classified into four descriptive categories based on the filtration rate, the type of filter washing required, the type of filtration rate control, and the type of filter bed selected. As discussed previously, slow sand, rapid sand, and high-rate filters are the three commonly recognized classifications. Yet from the viewpoint of the design engineer, a more detailed division is required.

The first group comprises gravity filters: slow sand, rapid sand, and high-rate filters containing a multimedia or a coarse, deep medium bed. The second division consists of pressure filters, which encompass proprietary items such as the horizontal single cell, horizontal multicell, vertical single cell, and vertical biflow filters. The third group is that of automatic backwash filters: these are exclusively composed of proprietary items. Finally, custom-designed filters include standard rapid sand filters, multimedia high-rate filters, coarse deep bed high-rate filters, self-backwash gravity filters, constant-rate gravity filters, declining-rate gravity filters, GAC adsorption filters, and ion exchange resin filters.

Selection Criteria Before designing a filtration system, engineers must establish the following items: type of filter, size and number of filters, filtration rate and terminal headloss, control of the filter flow rate, characteristics of the filter bed, type of filter wash system and wash trough, filter underdrains, auxiliary scouring, and filter appurtenance systems. Specification of these design elements should be based on the following conditions:

1. Local conditions, such as weather, the level of available technical service, the capability of operators, and the topography of the plant site
2. Regulatory constraints
3. Plant capacity [Does the plant process more than or less than 10 mgd (0.5 m^3/s)?]
4. The quality of the water source: turbidity, color, temperature, amount of dissolved air, algae count, taste and odor, and fluctuations in the water quality
5. The final quality of the process water, that is, the Drinking Water Quality Standards
6. The type of overall treatment process train: conventional complete treatment, direct filtration, iron and manganese removal, organic compound removal, or demineralization

7. The condition of the plant site and the effective use of space
8. The amount of available hydraulic headloss across the filters
9. The initial capital cost and the cost of operation and maintenance
10. Owner preference
11. The potential for future plant expansion and modification

TYPE OF FILTER Small to medium-sized plants processing less than 15 mgd (0.65 m^3/s) may use proprietary filters; these should be evaluated along with custom-designed filters. The proprietary units offer several advantages: lower capital cost, easy operation and maintenance, a filtered water quality ranging from acceptable to good, and shorter plant design and construction periods due to the predesigned feature of these units. The main disadvantage is that they generally have a limited life span of 20 to 30 years. There is also a possibility that future use of certain proprietary units may be restricted because they may not be able to meet the more stringent guidelines of the regulatory agencies. Other disadvantages include their questionable aesthetics and limited flexibility with respect to filter design and operation. Automatic backwash filters, self-backwash filters with a module design, and steel-shelled gravity or pressure filters are all proprietary types of filter.

The most common types of filter that are included in the design of large treatment plants, those processing a minimum of 15 mgd, are the gravity-downflow and dual-media (sand and coal) filters. The dual-media filter with a reinforced-concrete cell structure is the standard design that is used by engineers because its life cycle spans 50 years or more.

SIZE AND NUMBER OF EACH FILTER The size of each filter and the total number of filters are interrelated subjects. The maximum filter size is dictated by the difficulties encountered in providing uniform flow distribution of the backwash water over the entire filter bed, by the economically feasible size of the filter backwash tank and pumps, and the cost of the waste-wash handling facilities. In general, the practical maximum size of an individual filter bed is approximately 1500 ft^2 (150 m^2). A minimum of four filters should be provided for medium-sized to large plants, those processing over 15 mgd. In contrast, very small plants processing a maximum of 2 to 3 mgd (0.09 to 0.13 m^3/s) may use as few as two filters if there are financial constraints. The fewer the total number of filters, the more cost-effective the plant is to construct. However, this means that the larger plants require very large individual filters. Refer to the design criteria section of this chapter for the equation used to determine the required number of filters in relation to the plant capacity. If a self-backwash type of filter is selected, the minimum number of filters per module must be four, regardless of plant size, and preferably six. Otherwise, the system may not be capable of producing the required volume of backwash water (to wash one filter) during periods of low plant flow. However, these numbers may be disregarded if either an auxiliary backwash pump or a tank is provided to supplement the inadequate production of wash water.

It is important to remember that if a plant is designed with a small number of filters, the filtration rate in the remaining filters increases substantially whenever one or two of the filters are placed off-line due to washing or repair. The design engineer must therefore select an adequate number of filters and the proper size for each filter, so that the filtration rate of the remaining units will not be excessive when at least one is out of service. It is recommended that no more than 33%, and preferably a maximum of 15 to 20%, of hydraulic surcharge be allowed to flow to the remaining filters whenever one unit is off-line. This practice minimizes any turbidity breakthrough that may occur during shock loading.

FILTRATION RATE AND TERMINAL HEADLOSS Contrary to popular belief, the filtration rate of high-rate filters does not significantly affect the quality of the filtered water for rates up to 10 gpm/ft^2 (25 m/h). This is true only if the suspended colloids (in the raw water) are properly coagulated by a combination of coagulant and a small amount of polymer and if the excess suspended solids are removed during pretreatment. However, the rate at which the headloss is developed during higher filtration rates is quite dramatic, so much so that filtration rates over 10 gpm/ft^2 are considered to be impractical for normal dual-media filters. Figures 3.2.7-2 and 3.2.7-3 illustrate the performance characteristics of high-rate filters. When processing physically weak floc such as alum floc, the quality of the filter effluent tends to degrade

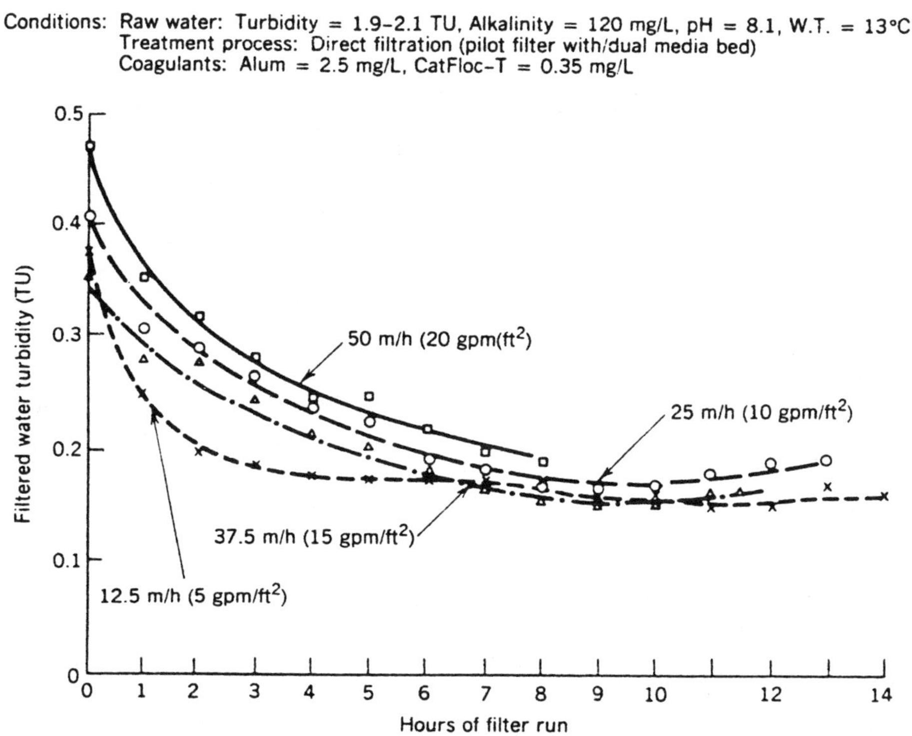

Figure 3.2.7-2 Filtered water turbidity versus hours of filter run. (From MWD Southern California, Internal Report, 1978.)

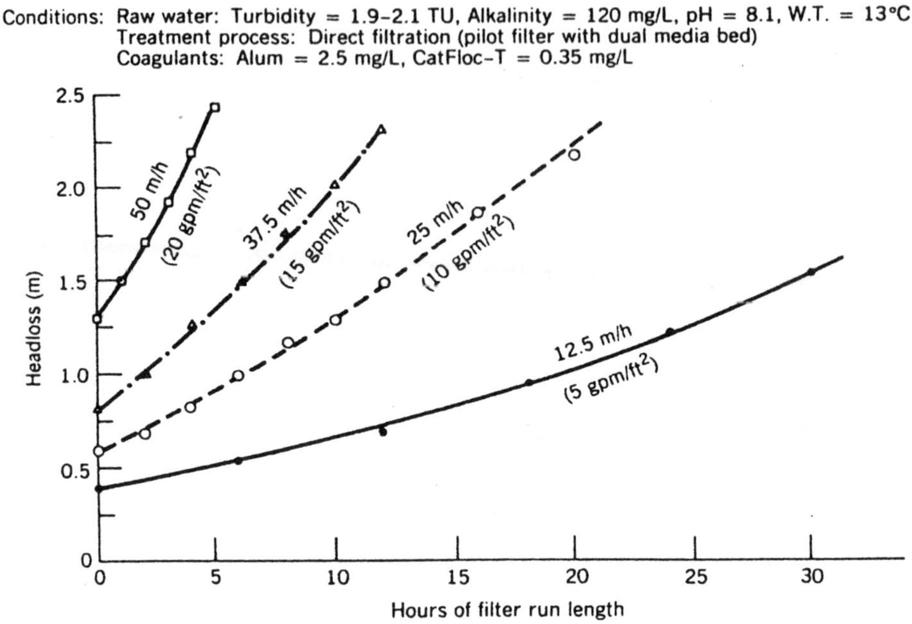

Figure 3.2.7-3 Headloss versus filter run at four different filtration rates. (From MWD Southern California, Internal Report, 1978.)

as the filtration rate rises above 4 gpm/ft^2 (10 m/h). This phenomenon is more frequent during periods of cold weather and is more apparent for rapid sand filters. Consequently, a designed filtration rate of 6 gpm/ft^2 (15 m/h) is generally recommended for ordinary high-rate filters and 3 gpm/ft^2 (7.5 m/h) for rapid sand filters. A filtration rate of 10 gpm/ft^2 (25 m/h) is reasonable for coarse deep bed high-rate filters if a small amount of polymer is applied as a filter aid. However, the design engineer should discuss the filtration rate chosen with the local regulatory agency prior to design.

Ordinary rapid sand filter beds generally have a filtration rate of 2 gpm/ft^2 (5 m/h), which translates to a Reynolds number of approximately 1.0; this number is well within the limits of a laminar type of flow through a granular bed. The typical high-rate filter with a standard dual-media filter bed has a filtration rate of 6 gpm/ft^2 (15 m/h) and a Reynolds number of approximately 10, which is still within the upper limits of a laminar flow characteristic. Both coarse deep bed and first-stage roughing filters of a two-stage filtration process have filtration rates of 15 gpm/ft^2 (37.5 m/h). The Reynolds numbers for these two types of unit are 30 to 40; these figures border on the laminar flow regime (far end) and the turbulent type of flow (beginning). This transitional type of flow may explain why high-rate filter beds have remarkable flocculation characteristics, in addition to the ability to retain a majority of the flocculated suspended matter when a small amount of coagulant or flocculant (polymer) is added to the filter influent. These characteristics are generally attributed to semilaminar types of flow.

The available headloss for filtration is generally designed to be 8 to 10 ft (2.5 to 3 m) for gravity filters but only 1 ft (0.3 m) for automatic backwash filters; the latter are predesigned by equipment manufacturers. Pressure filters usually provide a terminal headloss of over 30 ft (9.3 m), but their application is not practical for large treatment plants because of the large number of units required.

The filtration rate and terminal headloss for a particular type of filter and filter medium design should be selected after thoroughly analyzing the economic tradeoffs: the total required area of the filter bed, the available hydraulic loss across the filter process, the anticipated terminal headloss prior to turbidity breakthrough in the filter bed, and the length of the filter run. All these factors must fall within the limits dictated by the quality of the effluent water. It is important to mention that the effluent turbidity of gravity filters often begins to increase when the net headloss is over 6 ft (1.8 m). Consequently, the filters must be backwashed, despite an available headloss of 8 ft (2.4 m) or more, if the turbidity of the filtrate is to be maintained at less than 0.25 ntu. Regardless of the conditions, filter washing should be initiated automatically or manually by any of the following: turbidity breakthrough, filter run time, or terminal headloss.

CONTROL OF THE FILTER FLOW RATE The two basic schemes for controlling filter flow rate are constant- and declining-rate filtration. In recent years, the most popular scheme is the constant-rate filtration system since it provides better operational control over the filters, has proven performance, and is preferred overwhelmingly by plant operators. Declining-rate filtration has some merits but lacks positive filtration control measures. Thus, owners of modern filtration plants seldom choose this type of system.

A typical scheme for the constant-flow type of control, developed in recent years, is equal-loading self-backwash filtration (see Figure 3.2.7-1). This design scheme should be evaluated for any situation because of its simplicity, efficient use of energy, uncomplicated instrumentation, simple filter control system, and simple operation and maintenance. This system also eliminates the need for backwash headers, valves, pumps, and an elevated wash-water storage tank.

DEPTH, SIZE, AND COMPOSITION OF THE FILTER BED Filter efficiency is determined by such factors as certain physical characteristics of the filter bed, the conditions of the bed, and the effectiveness of coagulation pretreatment. The design engineer only has control over the physical parameters of the filter bed: media size and shape, bed porosity, total surface area of the filter media grains, and bed depth/media grain size ratio. Once the size of the media and the depth of the bed are selected, both the bed porosity and the total surface area of the media grains becomes fixed. Thus, selection of the appropriate size and depth of media are of crucial importance to the design engineer.

Three basic methods are used to select the proper depth and size of media: on the basis of existing data obtained from filters treating the same or a similar type of raw water, on the basis of data obtained from pilot-plant studies, and on the accepted ratio between the bed depth and media size. The first option is often used because it is a foolproof method. However, it is both crude and unscientific. The second approach is tempting but requires a minimum of 6 months to obtain reliable data based on the different raw water qualities exhibited during both warm and cold weather. Consequently, a significant sum of money (at least \$100,000) must be budgeted to complete the pilot study. Moreover, a pilot study may not always be feasible for various reasons: a limited schedule for completing the project, unavailability of fresh raw water because of the remoteness of the project site, and incomplete water sources due to factors such as the drilling of wells or the construction of a new reservoir (dam).

The third method is based on the relationship between the established size of the media grain and the filter bed. This relationship has been studied since the mid-twentieth century by two or three researchers using pilot filter columns in conjunction with actual filter performances. This method is considered to be the proven technique in selecting the appropriate depth and size of the media, and a detailed discussion is presented in the section covering design criteria. The relationship between the size and the depth of the filter media implies that the available surface area of the filter bed media grains is a significant factor in selecting the type of filter bed. Ordinary high-rate filters have a porosity ratio of 0.45, a sphericity of 0.8, and a bed depth/effective media size ratio of 1000. Thus the total surface area of the filter media grains per unit area is approximately 3000 ft^2 (3000 m^2) in area per square foot (square meter of bed area). When a dual- or multimedia filter bed is specified, the size of the medium grains and the specific gravity of each layer become extremely important with respect to filter backwashing and the potential for media loss. If an inappropriate combination of media is specified, only a fraction of the filter bed will be cleaned adequately. The remaining part of the bed will be either compacted and dirty or washed away at a particular backwash rate. To ensure that the two filter media particles with sizes d_1 and d_2 and densities ρ_1 and ρ_2 have the same settling velocity, the following relationship should exist:

$$\frac{d_1}{d_2} = \left(\frac{\rho_2 - \rho}{\rho_1 - \rho}\right)^{0.667}$$

where ρ is the density of water.

Figure 3.2.7-4 will aid engineers in selecting the proper combination of common filter materials at a particular backwash rate. Note that Figure 3.2.7-4 does not apply to a sequential system using air-scouring backwash. The figure is applicable to dual- or trimedia beds with air-scouring wash because the bed must be fluidized in order for it to become restratified. Figure 3.2.7-5 compares the performances of five different types of filter bed.

FILTER WASH SYSTEMS AND WASH TROUGHS There are three basic types of filter wash systems: (1) fluidized-bed backwash with or without surface wash, (2) sequential air-scouring wash, and (3) concurrent air-scouring wash. Most European filters use sequential air-scouring wash; Americans use concurrent wash. The effectiveness of both types of air-scouring system is about the same, but the application of sequential wash is generally limited to monosand beds that are less than 3.5 ft (1 m) deep.

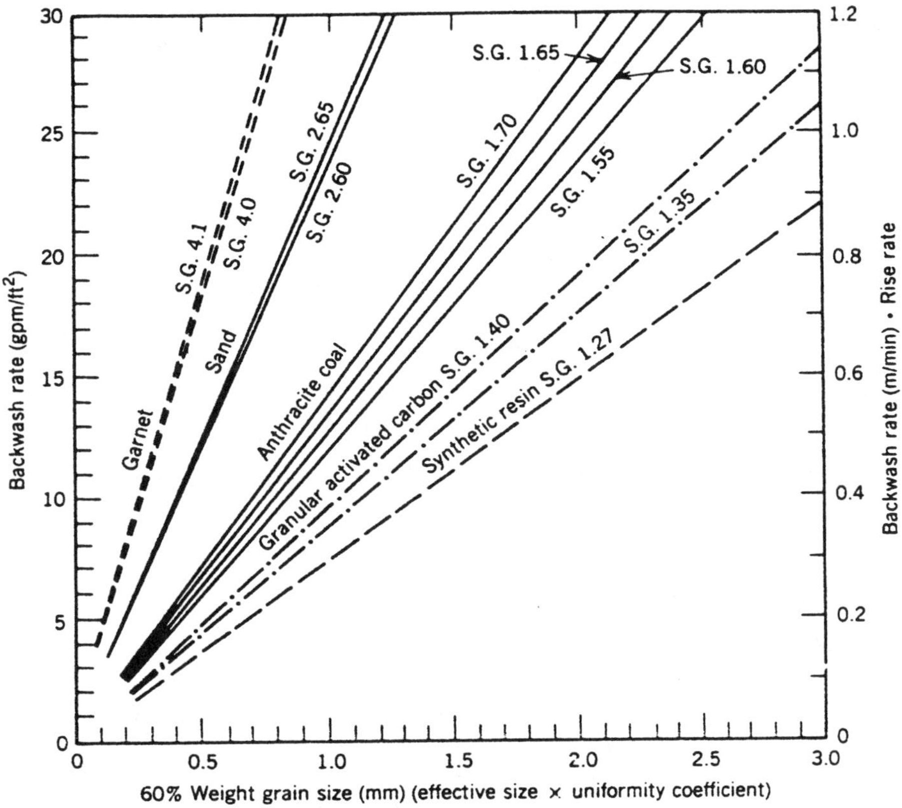

Figure 3.2.7-4 Appropriate filter backwash rate at water temperature of 20°C (68°F).

The sequential air-scouring wash begins with 3 cfm/ft^2 (0.9 m/min) of air scouring without backwash for 4 to 5 min, followed by backwash (rinse wash) for 10 to 15 min at a rate of 8 to 12 gpm/ft^2 (20 to 30 m/h), depending on the media size. The concurrent air-scouring system starts with 6 to 8 gpm/ft^2 (15 to 20 m/h) of slow backwash together with approximately 3 cfm/ft^2 (0.9 m/h) of compressed air for 4 to 6 min, followed by 18 to 22 gpm/ft^2 (45 to 55 m/h) backwash to purge air bubbles trapped in the filter bed and to also restratify the dual- or trimedia filter bed. This type of system has been used extensively in the United States.

The first type, fluidized-bed backwash with or without surface wash, has been used extensively in the United States, South America, and Japan. The backwash troughs are arranged 6 to 8 ft apart (1.8 to 2.4 m) and 2.5 to 4 ft (0.8 to 1.2 m) above the filter bed. The second type is practiced primarily in Europe and typically does not provide wash troughs above the filter bed. Rather, a single overflow side wall is provided for the discharge of the backwash waste into a channel, and the top of the sidewall is usually about 20 in. (0.5 m) above the filter bed. The third type provides backwash troughs similar to the first but often with one or two fewer troughs. The system of continuous concurrent water and air washing (air scouring) has very limited applica-

Filter Media Specifications

Name	Material	Effective Size	Depth (in.)	Uniformity Coefficient	*L/D*
A—Small dual media	Anthracite	1.0	20	1.45	1016
	Sand	0.5	10	1.3	
B—Intermediate dual media	Anthracite	1.48	30	1.5	1023
	Sand	0.75	15	1.2	
C—Large dual media	Anthracite	2.0	40	1.5	1016
	Sand	1.0	20	1.25	
D—Mixed media	Anthracite	1.0	18	1.45	1306
	Sand	0.42	9	1.5	
	Garnet	0.25	3	1.25	
E—Monomedium	Anthracite	1.0	40	1.4	1016

Note: *L/D* corresponds to the ratio of the depth of the medium (*L*) to the effective size of the medium (*D*).

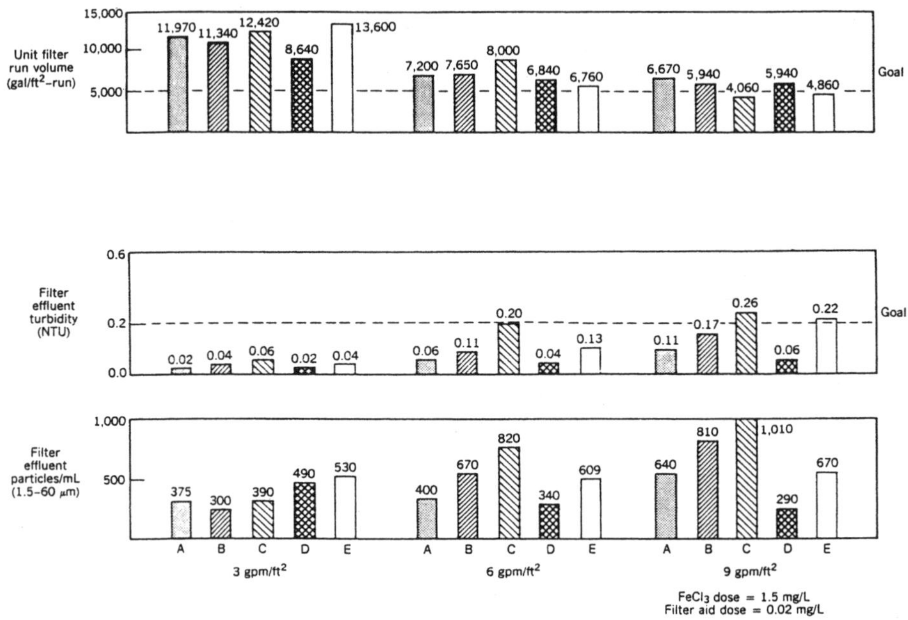

Figure 3.2.7-5 Results of filtration rate and medium comparison tests. (Adapted from Elekutria WTP pilot study report by J.M. Montgomery, Consulting Engineers.)

tion. This method applies air and water during the entire wash period, the backwash rate is high enough to fluidize the bed, and special baffled plates are provided on either side of each trough to minimize the loss of media during air scouring (Figure 3.2.7-6). This air-scouring wash system is suited for washing tertiary filters of sewage treatment plants that filter strong biological floc. If a fluidized-bed backwash system is selected, the spacing and height of the troughs must be determined carefully to ensure that dislodged floc is efficiently washed away and so that the process of backwash and surface wash does not flush filter media over the sides of the troughs.

According to the *Ten States Standards* (1968), the distance from the top of the trough to the surface of a rapid sand filter bed should be equivalent to the rate of backwash expressed in inches per minute. The rules also specify that the troughs

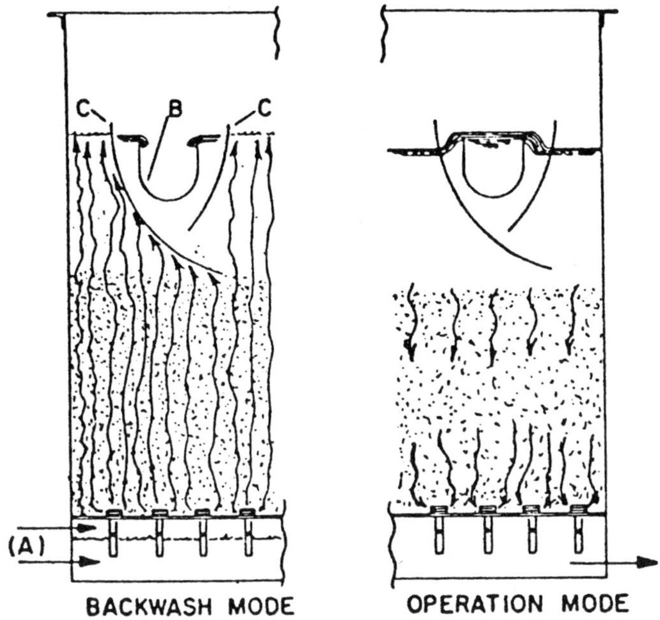

Figure 3.2.7-6 Special baffles for concurrent air and water-wash troughs. Schematic diagram shows the operation of the Multiwash filter process. During backwash (left) water and air are simultaneously passed through the media (A). The dirty backwash water is collected by the backwash trough (B) after air is separated from the water stream by the exclusive Multiwash baffles (C). After backwash the filter is placed back into normal service (right). (Courtesy of General Filter Company.)

should be spaced in such a manner that the floc does not travel more than 3 ft (horizontally) before reaching a trough. These criteria are sufficient when designing rapid sand filter beds. If the troughs are designed according to these standards, there will be excessive loss of the anthracite coal layers. Therefore, for dual-media filters, the top of the troughs should generally be located 3.5 to 4 ft (1.1 to 1.2 m) above the surface of the filter bed and should be spaced 6 to 10 ft (1.8 to 3 m) apart.

There are basically two types of wash troughs: those with a shallow and wide cross section and those whose cross section is both narrow and deep with a U-shaped or slightly V-shaped bottom. Wash troughs with a wide cross section tend to have higher upflow velocities whenever the backwash flow exceeds the elevation of the trough bottom; thus the suspended solids are quickly removed. Troughs that are both narrow and deep may be constructed with thinner walls, due to a higher moment of inertia and greater structural integrity. However, the bottom of the wash troughs should not be flat (for both types) because froth and sludge tend to accumulate beneath the trough. This accumulated mud then falls back onto the filter bed and forms mud balls.

Engineers must design wash troughs so that they will accommodate the maximum expected wash rate for both backwash and surface wash, in addition to a free-fall into the main collection gullet at the lower end of 4 to 6 in. (10 to 25 cm). The bottom of the trough (interior) may be either horizontal or sloping. A horizontal bottom is much simpler to construct and install. The width of the filter cell should be less than 20 ft (6 m) so that off-the-shelf types such as plastic troughs reinforced with fiberglass may be used. The required height for a trough with level inverts and a rectangular cross section may be computed by using the following formula:

$$h_0 = \left(\frac{Q}{2.49B}\right)^{0.667}$$

where Q is the total rate of discharge per trough (ft^3/s), B the interior width of the trough (ft), and h_0 the depth (ft) of the process water at the upstream portion of the trough. A freeboard of at least 3 in. (7.5 cm) should be added to h_0. If metric units are used, the formula above becomes

$$h_0 = \left(\frac{Q}{1.38B}\right)^{0.667}$$

where Q is in m^3/s and h_0 and B are in meters.

The distance between the lip of the trough and the top of the bed (conventional filter bed) should be

$$(0.5L + D) < H_0 < (L + D)$$

Spacing of the troughs from center to center is generally determined by

$$1.5H_0 < S < 2.5H_0$$

where L is the depth of the filter bed media, D the depth of the trough, H_0 the distance between the lip of the trough and the top of the bed, and S the spacing distance of the troughs from center to center.

Unlike the fluidized-bed backwash system, the European standard air-scouring wash does not use a series of wash troughs but has a single V-shaped trough along the sidewall opposite the wash-waste overflow wall, which provides the surface sweeping function (Figure 3.2.7-7). If the wash waste is to be removed effectively without the use of wash troughs, there is a limitation to the width of each filter cell: 13 to 15 ft (4 to 4.5 m) is considered to be the maximum width. The design engineer should also be aware that filter media migration has frequently been reported in cases where the filters lack the regular wash troughs; in this case, media migration is caused by the momentum of the filter influent.

Since the 1960s, many old rapid sand filters have been modified to become high-rate filters by converting the sand bed to a dual-media bed. A common problem caused by this change is excessive loss of anthracite coal medium under regular filter washing practices. A pilot study guided by the author found that a simple baffle configuration on both sides of each wash trough can reduce media loss by as much as 70% (Kawamura et al., 1997).

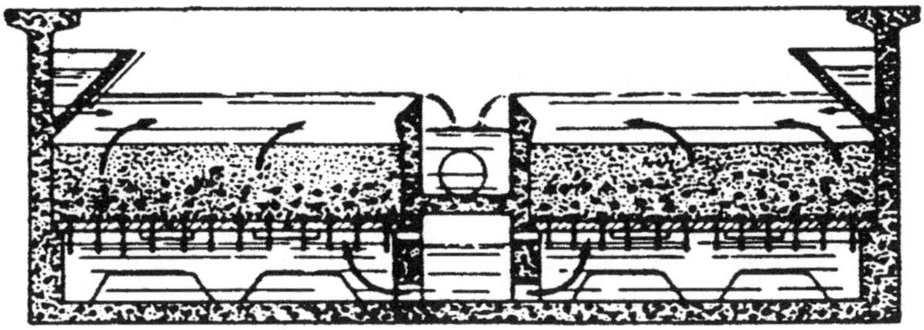

Figure 3.2.7-7 Air-scouring wash filter. (From Degremont, *Water Treatment Handbook,* Halsted Press, New York, 1979.)

FILTER UNDERDRAINS The selection of the appropriate type of filter underdrain must be based on whether or not air-scouring wash is chosen as the filter washing system. The underdrain system used in air-scouring wash is designed specifically for this purpose. It is therefore prudent to use one of the proprietary types, one that has a long history of success, instead of a custom-designed underdrain; this will ensure that the requirements of the system are met without failure. Figures 3.2.7-8 and 3.2.7-9 depict a few commercially available types of underdrain: false bottoms with long leg strainers and fiberglass-reinforced plastic blocks with channels (interior).

The strainer system is the predominant type of air-scouring wash filter underdrain because of its long history of effectiveness and dependability. It is extremely important to select strainers that are physically strong. Furthermore, they preferably have a mechanism for adjusting the stem height, even if this feature requires the expenditure of more money. Strainers that are physically weak or brittle tend to break during construction or filter backwashing, resulting in costly and lengthy repairs. Even when the bottom slab of the filter underdrain is not level, due to imprecise construction or uneven settlement of filter structures, certain types of strainer allow the height of the strainer legs to be adjusted by approximately $\frac{3}{4}$ in. (20 mm). Another important feature of the strainer type of underdrain is that it has a false floor on which the strainer nozzles are mounted.

Engineers are advised to design the air-scouring underdrain as a monolithic reinforced concrete floor since this will reduce the potential for underdrain upset. As described previously, a significant number of filters have had problems with large amounts of air emanating from the structural joints during air scouring, and their floors have consequently become damaged. This was especially true in cases where precast concrete blocks were secured with anchor bolts and the area between the blocks was cement grouted. The plenum should have a minimum height of 2 ft (0.6 m) to facilitate inspection of the nozzle legs and the condition of the false bottom. This height also allows the plenum to be constructed easily.

Finally, the design engineer must give special consideration to the size of the strainer head slits. If these slits are less than 0.1 in. (0.25 mm) wide, they will easily become clogged by fine media grains and other debris that may be left in the backwash line or plenum during construction. The engineer is therefore advised to

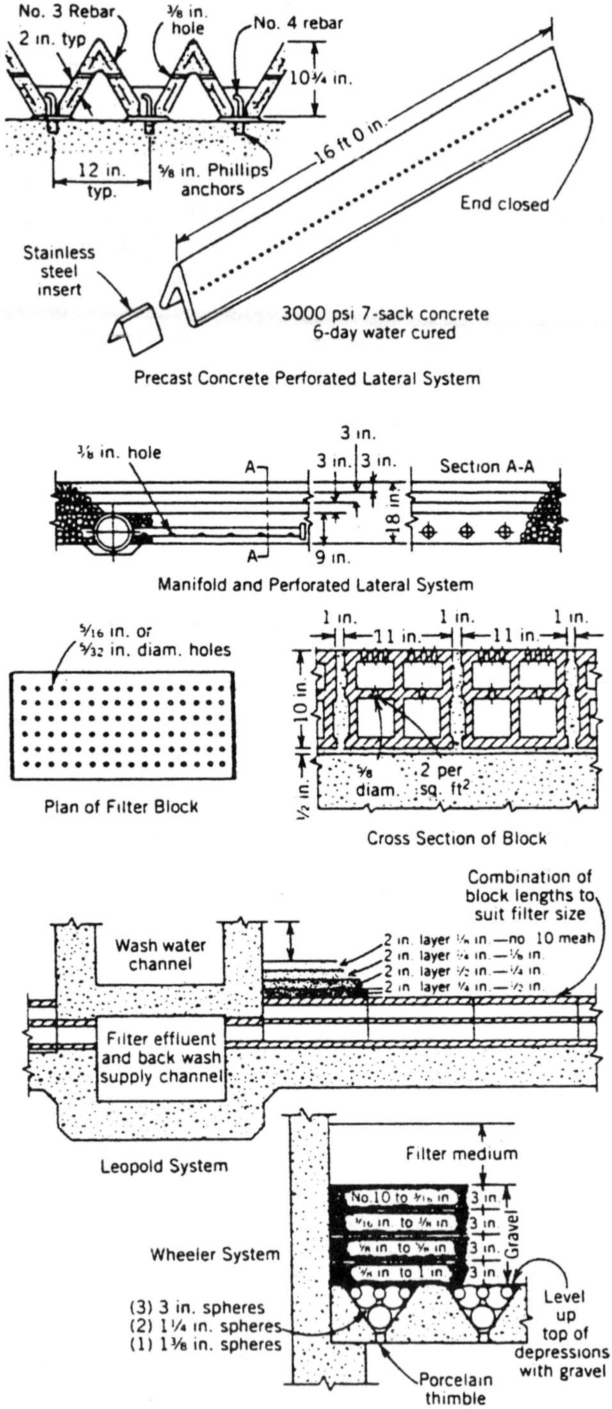

Figure 3.2.7-8 Filter underdrains with non-air scouring. (Adapted from *Water Treatment: Principles and Design* by J.M. Montgomery, Consulting Engineers, Wiley, New York, 1985.)

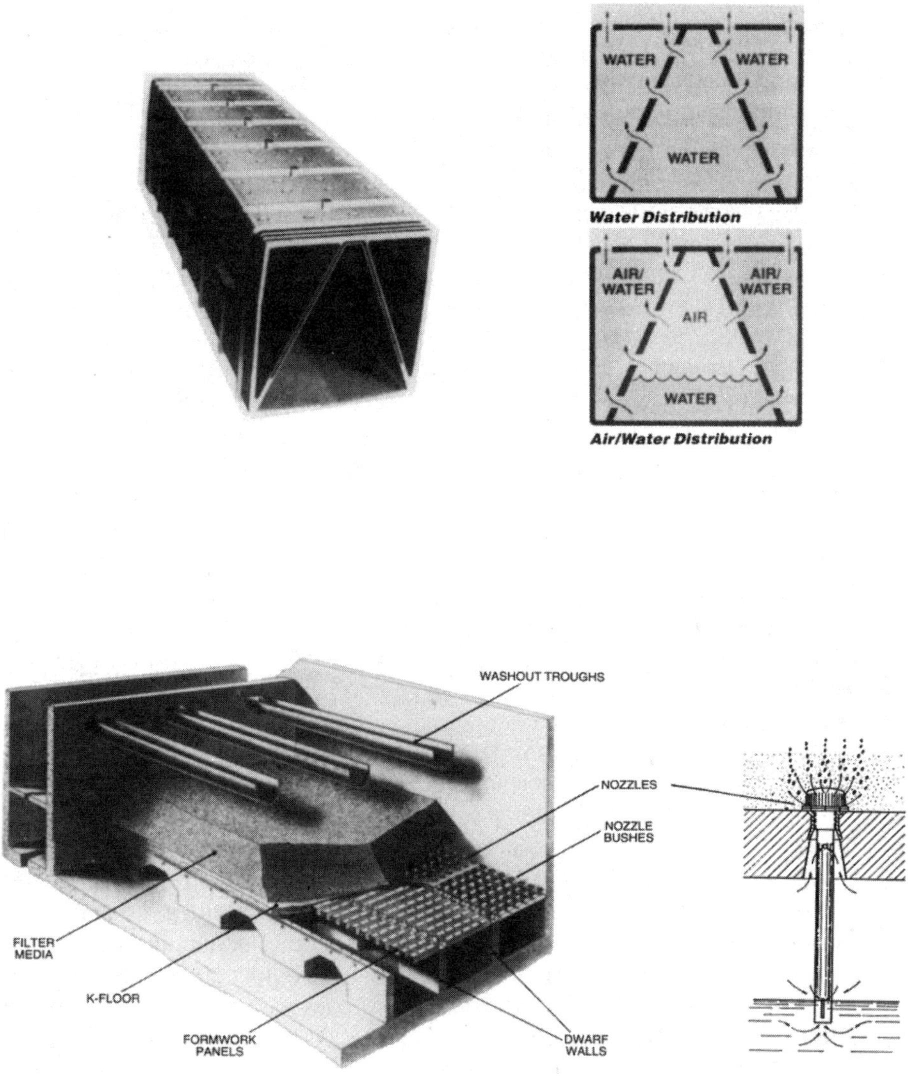

Figure 3.2.7-9 Filter underdrains for air scouring: (*a*) Leopold's universal underdrain block (courtesy of F.B. Leopold Company, Bulletin ASV-100); (*b*) PCI's K-Floor with nozzles (courtesy of Paterson Candy International).

select the largest possible width. The appropriate slit size is half the effective size of the filter medium covering the strainer heads. Thus, if the effective size of the sand is 1 mm, the size of the slits should be 0.5 mm. Manufacturers often recommend covering the strainers with 6 in. (150 mm) of pea-sized gravel because this practice minimizes clogging of the strainers as well as deformation of the slits by the scrubbing action of the filter media, and distributes the backwash flow more efficiently, but only for certain types of strainer nozzle.

The plastic block type of underdrain of the air-scouring wash system has been used successfully since 1980. This system requires a specifically graded gravel bed, with an approximate depth of 12 to 15 in. (30 to 38 cm), to be located directly above the underdrain because the size of the orifice is too large to retain the filter media. However, plant operators must strictly follow the manufacturer's instructions for both the wash rates and the sequence of the air and water. Otherwise, the gravel bed will be upset and the expensive process of reconstructing the filter gravel bed will be required.

Recently, a few manufacturers have introduced a porous plate in lieu of a specifically graded gravel bed. The porous plates attach directly to the top of the plastic underdrain blocks. This type not only eliminates mounding and disturbance of the gravel bed but also gains 12 to 18 in. (0.3 to 0.45 m) of extra filter bed depth due to the elimination of the gravel bed. However, potential clogging of the porous plate by scale, especially by calcium carbonate scale, is a concern.

A number of alternative types of underdrain systems are available for the regular backwash system: Precast concrete laterals, pipe laterals, and Leopold's tile block system (Figure 3.2.7-8) are proven types that are used extensively. Unlike the air-scouring wash, these types of underdrain are not limited to the proprietary types, provided that they are properly designed. Moreover, any reputable contractor can fabricate most of these underdrains on the job site. The design of the filter underdrain system is based exclusively on the hydraulics of the filter backwash. The details are discussed elsewhere.

AUXILIARY SCOURING OF THE FILTER WASH Auxiliary scouring is absolutely essential during filter backwash whenever coagulants are used in the pretreatment process, especially if polymer is used as a filter aid. Backwash alone cannot maintain the filter beds in a reasonably clean condition for longer than several months, regardless of adjustments in the backwash rate and control of the bed expansion rate.

Regular surface wash systems are capable of adequately cleaning ordinary rapid sand and dual-media filters. The advantages of this type of system are its simplicity, effectiveness, and easy operation and maintenance. Unlike air-scouring wash, it does not require a delicate auxiliary scouring and backwash sequence, nor does it require the backwash rates to be adjusted during a wash cycle. The air-scouring wash system should be employed for any deep bed filters with coarse media and may be the preferred system for ordinary filter beds processing biologically pretreated sewage.

The two basic types of surface wash systems are the fixed grid and rotating arms. Both types have been used since the mid-twentieth century and have been proven effective when properly designed, fabricated, installed, and operated. The fixed grid type of surface wash may be designed by the engineer and requires much less maintenance because of the lack of moving parts. A rotating arm system should neither be designed by engineers nor fabricated in the machine shop because of its sensitive design and specifications.

The three types of rotating agitators are surface agitators, subsurface agitators, and dual-arm agitators. The first is used primarily in conjunction with rapid sand filters.

The second and third types are designed specifically for dual- or multimedia filter beds. In the United States, only one manufacturer has had over 20 years of successful operation for the three rotating arm systems. The mechanical rake system has been tried, but its bulky mechanism, the limited freedom in design, and its frequent maintenance requirements have made this system unpopular. Figure 3.2.7-10 illustrates the effectiveness of surface wash.

AIR-SCOURING FILTER WASH In recent years, some publications and equipment manufacturers have given readers the impression that the air-scouring filter wash system is the only effective means of maintaining clean filter beds for any filter type. This is, however, not always true. The air-scouring wash vigorously agitates only the top 6 to 10 in. (0.15 to 0.25 m) of the filter bed. This boiling phenomenon misleads plant operators and visitors to believe that the violent agitation occurs throughout the filter bed. However, some mud balls that have fallen below the boiling zone remain unbroken. When utilizing strainer nozzle underdrains it is important to know that compressed air emanating from the bottom strainer nozzles rise directly above each nozzle and does not spread sideways. Bed media, located in between the uprising streams, is pulled down (mass balance law). This behavior is more distinct when a slow backwash is applied with compressed air (Figure 3.2.7-11). Both pilot filter studies and core sampling of actual filters using air-scouring wash confirm this phenomenon, and there is a slight increase in sludge deposition at the bottom of the bed.

In a worst-case scenario, oocysts of *Cryptosporidium* deposited on top of the filter bed are dragged downward, breaking through the filter bed. Therefore, it is worthwhile to consider the use of a surface wash system in conjunction with an air-scouring system that has strainer nozzles for the underdrain. This scheme will break up the top several inches (20 cm) of compacted media before air scouring when sur-

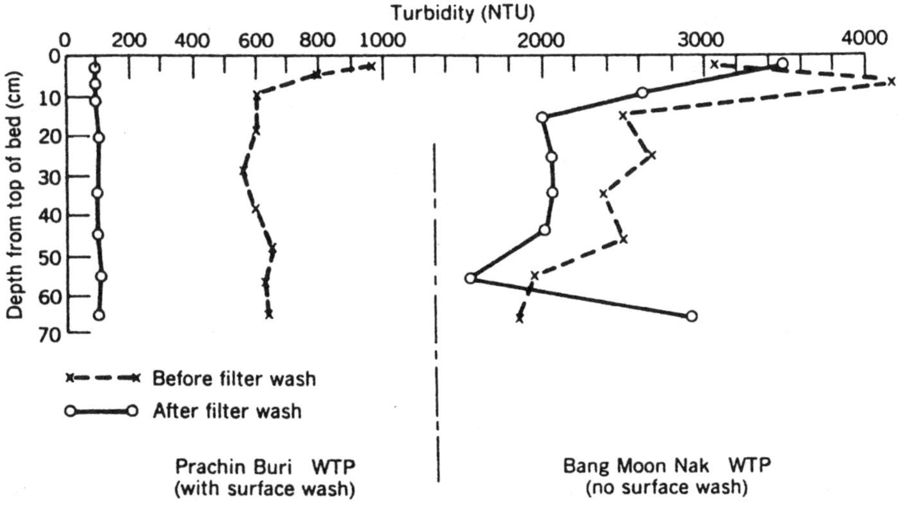

Figure 3.2.7-10 Sludge profile in filter beds. Amount of sludge is shown as turbidity (ntu).

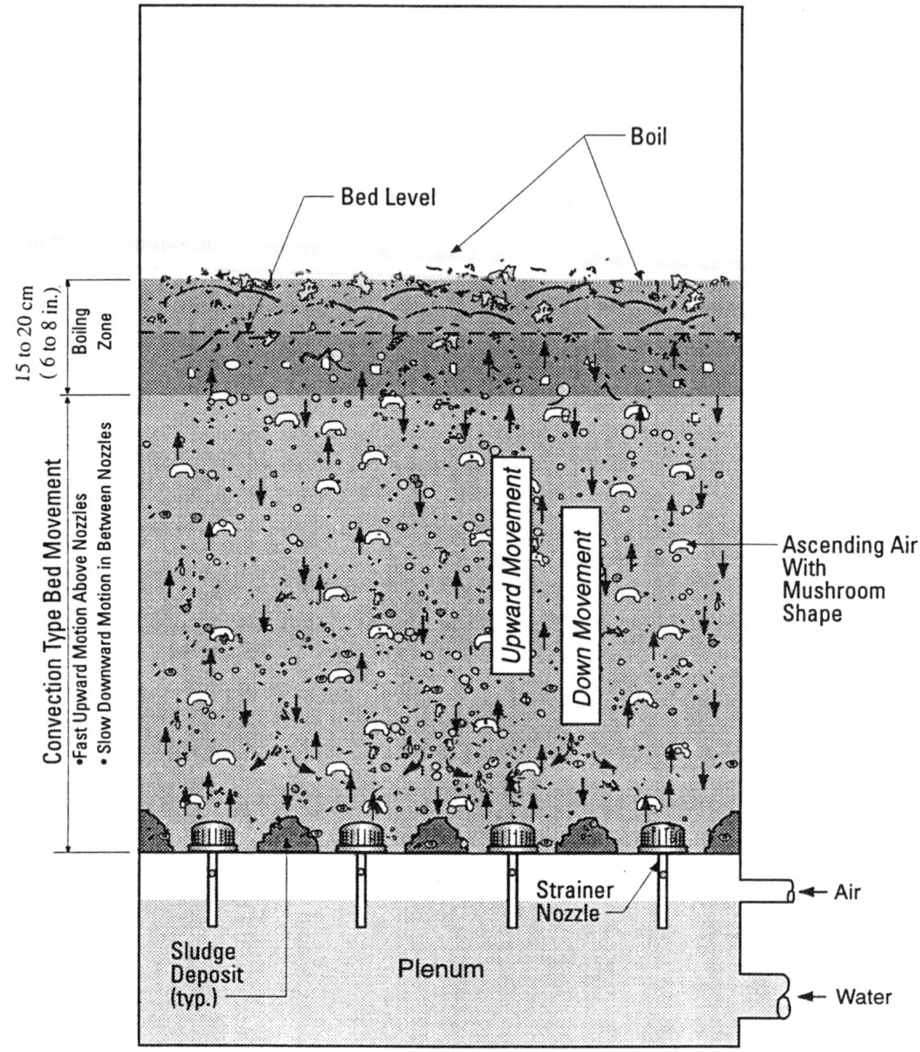

Figure 3.2.7-11 Filter bed movements under air-scouring wash.

face wash is applied first, and slow backwash can keep suspended solids above the bed at all times.

Choosing an air-scouring system complicates not only the design and construction of the filters, but also filter operation: air blowers, air piping, and air and backwash sequence control. Filter washing time is also affected, twice to three times as long as the surface wash system. Furthermore, there have been several catastrophic underdrain upsets due to the compressive and explosive nature of compressed air. It is therefore strongly recommended that a standpipe no smaller than 12 in. (0.3 m) in diameter be placed in the backwash pipe main as a pressure relief measure for any air-scouring wash system. The required height of the overflow pipe is normally 6 ft (1.8 m) above the top slab of filter structure; 2 to 3 ft above the hydraulic grade line during backwash.

It is the author's opinion that any filter bed with a depth less than 30 in. (0.75 m) does not require an air-scouring wash system because surface wash (fixed nozzle type) can provide adequate cleaning. A surface wash system designed by John Baylis and installed at South District Water Treatment Plant [500 mgd (22 m^3/s)] in the 1950s has kept all of the filter bed very clean after almost 50 years of operation. However, when the depth of filter bed exceeds 3 ft (1 m), installation of an air-scouring wash system is a proper choice.

The important issues for selecting an air-scouring wash systems design are (1) a long and proven history of satisfactory performance, (2) an adequate anchor system, (3) manufactured by a company with good quality control, and (4) strictly supervised construction of the underdrain system by an experienced construction supervisor throughout the duration of the job. If a plenum type of underdrain system is selected, it should have a monolithically cast reinforced concrete false floor. Avoid the use of precast elements that must be bolted onto pedestals, as this will present the problem of massive leakage of air around the blocks due to the failure of the cement grout after over 10 years of use.

APPURTENANT SYSTEMS The major filter appurtenant systems are the filter backwash tank, pumps, auxiliary scouring systems, wash flow control system, and wash-waste handling systems. Since the last item is a subject in itself, it is discussed separately in Section 4.2. The filter backwash tank must have a large enough capacity to hold a sufficient amount of water for two filter washes. The tank must also provide an adequate depth of water above the tank exit so that a vortex will not be formed toward the exit pipe. Formation of a vortex leads to the entrainment of air in the wash water, which in turn reduces the flow rate and hinders the functioning of the flow rate controller. Therefore, it is always a good practice to provide an antivortex assembly in the wash tank at the entrance of the wash main.

The lowest water level in the backwash tank should generally be about 40 ft (12 m) above the top of the wash-water troughs located in the filter bank. Under normal conditions, the fixed grid type of surface wash can obtain water from the backwash tank. In contrast, the rotating arm surface wash system requires a minimum of 80 psi (5.5 kg/cm^2 or 540 kPa) and therefore usually requires a booster pump system.

The source of the filter backwash may be one of the following: an elevated wash tank, the pressurized main water distribution line, the water from the direct pump of the clearwell, or the effluent from the rest of the filters on the line (as in the self-backwash filters). The choice among these alternatives depends on factors such as site topography, cost considerations, and aesthetics. The elevated tank is the most common source of backwash. Depending on the topography of the site, the tank is either on stilts or situated on a hill and may also be installed over the control building. The tank may also be filled by a small-capacity pump between backwashes.

An alternative to the elevated tank system is the direct pump backwash. In general, this system is rather expensive to operate due to its high electric consumption.

In fact, some plants cannot operate the filter backwash pumps during periods of peak power demand. Therefore, the direct pump wash system must be accompanied by a standby power generator. It is also advisable to place a pressure relief standpipe in the pump discharge line to avoid filter underdrain failure, which may result from excessive pressure. The power demand of the direct pump wash system may be reduced by designing the filter to backwash only half a filter cell at a time. This allows the plant to use a wash pump that is only half the usual size.

Air-scouring backwash filters are often designed with a pump backwash system for two reasons: the backwash rate is half that of an ordinary fluidized backwash system, and the wash rate is a two-step system. Two-step backwash may be achieved by using one pump for the first wash and a second pump during the second step rather than an elaborate flow rate control system. Air-scouring wash is attained by employing air blowers. The most common type of blower is the positive-displacement blower (Roots type), despite its high level of operational noise.

The backwash flow control system is necessary for setting the proper washing rate regardless of the type of auxiliary scouring employed during backwash. The two most common systems are the mechanical flow rate controller and the master flow control valve located in the backwash main. The mechanical flow rate control system is a combination of a flow meter and a flow-modulating butterfly valve. The second system is generally a plug or ball valve which is opened in a slow and controlled manner to purge air out of the filter bed at the beginning of a wash and to increase the wash rate slowly to a maximum (preset) rate; the preset rate is controlled by a stopper located at the valve operator. Use of the master flow control valve is recommended for most cases. However, the mechanical flow rate controller should be employed if the water level in the tank varies over 10 ft (3 m) during a filter wash or if the backwash pipeline is both large in size and short in length.

Otherwise, an insufficiently controlled headloss and a large variation in the backwash rate (during the washing cycle) will result. It is important to note that the backwash valve should be opened slowly to avoid disturbing the gravel bed (filter) and the underdrain system. If a self-backwash filter design is adopted, the filter washing system may be simplified dramatically since it does not require a pump system, a backwash flow control system, a backwash line, or valves.

FILTER-TO-WASTE The purpose of the filter-to-waste system is to divert inferior-quality filtered water to the waste line. Inferior filtered water may be generated under both normal and abnormal conditions. Under normal conditions, turbidity breakthrough is often observed at operational treatment plants whenever there is inadequate pretreatment and/or when the filter beds are in unripened condition.

Inferior filter water may be produced when the following abnormal plant conditions exist: (1) immediately after the installation of new filter media due to the existence of an unripened bed (i.e., the bed is not properly conditioned to effectively capture suspended solids by means of a small amount of polymer fed as filtration aid or the deposition of appropriately charged suspended solids and the correct quantity of floc on the filter media), (2) after disturbance of the filter bed due to repair or modification of the gravel bed or underdrain system, (3) accidental initiation of the surface wash or air scouring during the filtration cycle, (4) liquefaction of the filter beds resulting from strong earthquakes, and (5) administration of incorrect coagulant dosage. Although virtually all existing filters are equipped with a drain pipe, this pipe is too small to divert normal filtered water flow to the waste line.

In recent years regulatory agencies, including the U.S. Environmental Protection Agency (EPA), are deeply concerned about the presence of cysts and oocysts of protozoans, especially *Cryptosporidium*. Consequently, the goal of filtered water is now set at 0.1 ntu. The Department of Health of some states, including California, set more stringent guidelines by advocating the wasting of filtered water with turbidity greater than 0.25 ntu. The California Department of Health Services (DOH) also endorses slow starting of filters after filter washing in order to circumvent initial turbidity breakthrough.

It is important to note that EPA and DOH of a handful of states allow the addition of coagulant or polymer to the backwash water as an alternative to filter-to-waste. This practice conditions the filter bed prior to starting the filtration cycle. Unfortunately, it is not a robust method and therefore not always successful, a fact supported by both experienced pilot filter operators and engineers. A few treatment plants also report problems with floc formation and sludge deposition in the clearwells when using this alternative. Yet despite the drawbacks, both design engineers and plant managers often choose to add polymer as coagulant because it is a cheaper method of meeting regulatory requirements.

It is possible to add polymer at the influent of each filter at the beginning of the filtration cycle rather than the backwash water. This will help to effectively minimize initial turbidity breakthrough with a simple setup. The disadvantage is the high number of application points. If a filter requires an influent turbidity of less than 1 ntu in order for it to produce good filtered water consistently, it is important to investigate whether the filter is suboptimally functioning due to insufficient coagulation of suspended solids in the raw water and/or inadequately conditioned filter beds.

It is the author's opinion that each filter should be fitted with a filter-to-waste piping system to manage initial turbidity breakthrough under both normal and abnormal plant operations. Although there are several methods to implement filter-to-waste, the system depicted in Figure 3.2.7-12 should be considered.

Basic Hydraulics One of the most important items of filter design as well as operation is the hydraulics through the filter structure. Design engineers must know the headloss across the filter bed, filter bed expansion rate during backwash, and a common flow velocity for filter pipings and channels. This section will aid the engineer to determine these items.

Initial Loss of Head Loss of head at the beginning of the filtration cycle is called *initial headloss:* the resistance created by granular media to the passage of water. The initial headloss through a filter bed at various filtration rates under different water temperatures can be estimated by *Darcy's equation:*

$$\Delta h = \frac{LQ}{kA} \qquad \text{(Darcy)}$$

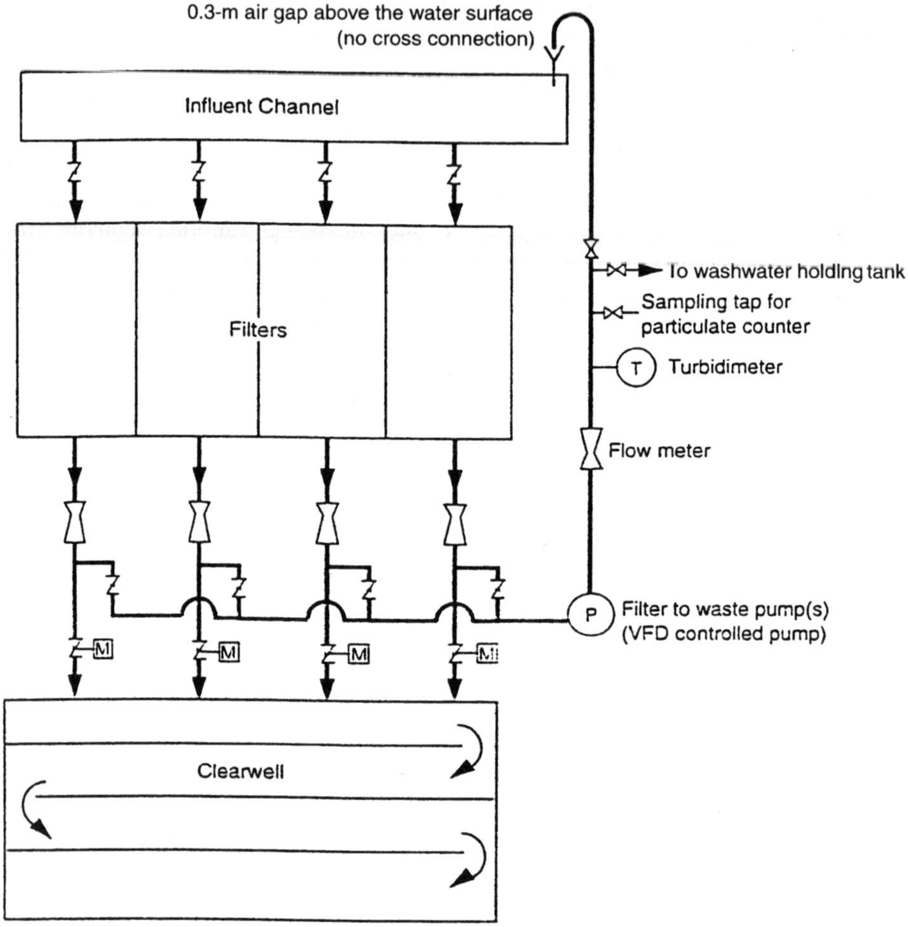

Figure 3.2.7-12 Filter-to-waste schematic. Special features: (1) No additional capacity for wash-water holding tank(s); (2) slow filter starting can be achieved (3 $\rightarrow$ 4 $\rightarrow$ 5 gpm/ft^2 or 7.5 $\rightarrow$ 10 $\rightarrow$ 12.5 m/h); (3) filter-to-waste rate can be matched to actual seasonal filtration rate; (4) instrumentation/control can be simplified; (5) turbidimeter in the recycle line can determine the FTW time; (6) discharge the FTW-to-waste line when its quality is questionable.

and

$$k = c(0.7 + 0.03t)de^2 \qquad \text{(Hazen)}$$

where

$$h = \text{headloss (cm)}$$

$$L = \text{depth of the filter bed (cm)}$$

$$Q = \text{flow rate (cm}^3\text{/s)}$$
$$A = \text{area of the bed (cm}^2\text{)}$$
$$k = \text{permeability coefficient (cm/s)}$$
$$c = \text{coefficient}$$
$$t = \text{water temperature (°C)}$$
$$de = \text{effective size of filter medium}$$

NOTE: For most filter media sizes, c is 124 and varies only from 123 to 125 for effective size (de) of 0.3 to 2.0 mm.

Example Compute the headloss for a 30-in. (76-cm)-deep filter bed with 0.75 mm (0.075 cm) effective size medium under a 6 gpm/ft^2 (15 m/h) filtration rate at 68°F (20°C).

Solution Compute the headloss through a unit surface area (1 cm^2) of the bed.

$$k = 124(0.7 + 0.03 \times 20)0.075^2$$
$$= 0.906$$

$$\text{Filtration rate} = (15 \times 100) \div 60^2 = 0.417 \text{ cm}^3\text{/s}$$

Thus

$$\Delta h = (76 \times 0.417) \div (0.906 \times 1)$$
$$= 34.5 \text{ cm (1.13 ft)}$$

As a quick reference, the approximate initial headloss across the common filter beds at 60°F (15°C) is presented below.

Filtration Rate		Headloss		
gpm/ft^2	m/h	ft	m	Type of Filter Bed
2	5	1.0	0.3	Regular rapid sand
3	7.5	1.45	0.45	Regular rapid sand
4	10	1.0	0.3	Standard dual media
6	12.5	1.45	0.45	Standard dual media
8	20	1.9	0.58	Standard dual media
10	25	2.45	0.75	Standard dual media

Filter Bed Expansion Rate and Loss of Head by Backwash The bed expansion rate under optimum backwash rate can be estimated by the following equation, where f is the porosity ratio of the bed:

$$\text{bed expansion rate} = \frac{(0.1)^{0.22} - f}{1 - (0.1)^{0.22}} = \frac{0.6 - f}{0.4}$$

In general, the porosity ratio of the sand bed is 0.43 to 0.45 and 0.48 to 0.50 for anthracite coal.

Some treatment plants employ an alternative method to determine the appropriate backwash rate: 50% filter bed expansion rates during backwash. During the late 1950s Baylis and others emphasized that 50% sand expansion should be achieved for rapid sand filters. This requirement is justified when the sand bed is composed of rather small (E.S. = 0.40 to 0.45 mm) grains, yet it is important to note that a majority of filters were not fitted with auxiliary scouring (surface wash) systems during the late 1950s. However, even with 50% bed expansion during backwash, most of the rapid sand filter beds constructed in the 1950s and 1960s are extremely dirty. Without proper surface wash or air scouring, the filter beds accumulate mud balls after 2 to 3 years of operation. The targeted bed expansion rate for modern high-rate filters during backwash with surface wash is approximately 37% for a typical sand bed with a porosity ratio of 0.45; the expansion rate for an anthracite coal bed with a porosity ratio of 0.5 is about 25%. These figures have been verified by well-operated actual filtration plants.

The depth of the expanded bed during backwash may be estimated using the equation

$$\frac{l_e}{l} = \frac{1 - f}{1 - f_e}$$

where l is the bed depth at rest, l_e the depth of the expanded bed, f the porosity ratio of the bed at rest, and f_e the porosity ratio of the expanded bed. When the sand bed is expanded 37% during the backwash, f_e becomes approximately 0.6. Also, when an anthracite coal bed is expanded 25%, f_e of the bed is about 0.6, under appropriate backwash rate for both cases (refer to Figure 3.2.7-4 for the appropriate backwash rate).

Loss of head for the expanded bed can be estimated as

$$\frac{h}{l_e} = (\rho_m - 1)(1 - f_e)$$

where h is the headloss and ρ_m is the specific gravity of the filter medium.

Example A dual-media filter bed comprised of 12 in. (0.3 m) of sand with an effective size of 0.5 mm, porosity of 0.45, and specific gravity of 2.65 and 24 in. (61 cm) of anthracite coal with an effective size of 1.0 mm, porosity of 0.5, and specific gravity of 1.65 is specified with a surface washing system. What is an estimated loss of head under an appropriate backwashing conditions? Assume that the sand bed expands 37% and a 25% expansion for the anthracite coal bed.

Solution The headloss by the sand bed is

$$h = (12 \times 1.37)\,(2.65 - 1)\,(1 - 0.6)$$
$$= 10.8 \text{ in. } (27.4 \text{ cm})$$

and the headloss by the anthracite bed is

$$h = (24 \times 1.25)\,(1.65 - 1)\,(1 - 0.6)$$
$$= 7.8 \text{ in. } (19.8 \text{ cm})$$

Thus the total headloss is approximately 18.6 in. (47 cm).

Flow Velocity The maximum flow velocities (approximate) sustained by the filter piping or channels and valves are as follows:

	Ordinary Filters		Self-Backwash Filters	
	ft/s	m/s	ft/s	m/s
Influent channel/pipe	3	0.9	1.5	0.45
Inlet valve	5	1.5	5	1.5
Forebay channel	0.5	0.15	0.5	0.15
Effluent valve	5	1.5	2	0.6
Effluent channel	5	1.5	2	0.6
Backwash main	8	2.4	3	0.9
Backwash valve	8	2.4	6	1.8
Surface wash line	8	2.4	8	2.4
Wash-waste main	8	2.4	8	2.4
Wash-waste valve	6	2.4	6	1.8
Filter underdrain lateral	4.5	1.4	4.5	1.4

NOTE: The flow velocity in the influent pipe or channel may be 4 ft/s (1.2 m/s) if flow rate controllers are located in each filter effluent line or if the filtration system is a declining-rate system. If the filter inlet is a weir for equal loading design (see Figure 3.2.7-1), the flow velocity in the influent channel should be less than 1.5 ft/s (0.45 m/s).

Design Criteria This section covers the basic design criteria for granular medium filters: the number of filters, the size of each type of filter, the characteristics of the filter bed, the filtration rate, the headloss across the filters, the type of filter backwashing system, the specifications for the gravel support bed, the basic hydraulics, and a basic filter design scheme.

Number of Filters The minimum number of filters for very small plants processing less than 2 mgd (90 L/s) is two. If the capacity of the plant exceeds 2 mgd, the minimum number of filters becomes four. Note that the required number of filters is calculated from the equation

$$N = 1.2Q^{0.5}$$

where N is the total number of filters and Q is the maximum plant flow rate in mgd (1 mgd = 3785 m^3/d).

Size of Each Filter The dimensions of four types of filter are discussed below.

ORDINARY GRAVITY FILTERS

Width of a filter cell	10 to 20 ft (16 ft or 4.9 m average) (3 to 6 m)
Length/width ratio	2 : 1 to 4 : 1 (3 : 1 average)
Area of a filter cell	250 to 1000 ft^2 (600 ft^2 or 56 m^2 average) (25 to 100 m^2)
Depth of the filter	15 to 25 ft (17 ft or 5.2 m average) (4.5 to 7.6 m)

NOTES:

1. The filters for medium-sized to large plants, those processing a minimum of 20 mgd (0.9 m^3/s), are generally composed of two cells per filter and have a central gullet down the center.
2. The water depth above the filter bed should be equal or deeper than the headloss provided for filtration in order to minimize air binding. Since 6 to 8 ft (1.8 to 2.4 m) is a common headloss for filtration, a water depth of 8 ft is preferred.

SELF-BACKWASH FILTERS (GRAVITY FILTER)

Width of a filter cell	10 to 20 ft (16 ft average)
Length/width ratio	2 : 1 to 4 : 1
Area of a filter cell	250 to 800 ft^2 (25 to 80 m^2)
Depth of the filter	18 to 25 ft (22 ft average) (5.5 to 7.5 m)

NOTE: The minimum number of filters is four, but six to eight are preferred. Each filter normally consists of only one cell due to limitation in the backwash water supply.

AUTOMATIC BACKWASH FILTERS (HARDINGE TYPE)

Width of a filter	Standard 16 ft (5 m)
Length of a filter	Up to approximately 120 ft (37 m)
Depth of a filter cell	7 to 11 ft (2.1 to 3.3 m)

Width of each cell	8 to 25 in. (0.2 to 0.64 m)
Depth of filter bed	11 to 48 in. (16 in. or 0.4 m standard) (0.25 to 1.2 m)
Traveling speed of the washing carriage	21 in./min (0.53 m/min) standard and indexing available for deeper than 16 in. bed depth

PRESSURE FILTERS

Diameter of a filter cell	4 to 20 ft (10 ft or 3 m average) (1.2 to 6 m)
Length of a filter cell	8 to 50 ft (20 ft or 6 m average) (2.4 to 15 m)
Depth of a filter bed	2 to 3 ft (0.6 to 0.9 m)

NOTE: Both horizontal and vertical cells are available. However, vertical cells are more common for larger units because they are cleaned to a greater extent during filter washing. The multicell (four-cell) horizontal units have the following advantages: a self-backwash system, space-saving features, and cost-effectiveness.

Filter Bed: Types of Medium and Depth Silica sand and anthracite coal are the most commonly used types of filter medium. The American Water Works Association has set the quality standards for both these materials (refer to AWWA Standards B100). Garnet, ilmenite, pumice, and synthetic materials may also be used as filter media but to a lesser extent due to their limited availability and relative expense. General silica sand and anthracite coal are the most commonly used types of filter medium. The depth of the filter bed is a function of medium size and generally follows the relationship shown below:

$$
l/d_e \geq \begin{cases} 1000 & \text{for ordinary monosand and dual-media beds} \\ 1250 & \text{for regular trimedia (coal, sand, and garnet) beds} \\ 1250 \text{ to } 1500 & \text{for most coarse deep monomedium beds } (d_e \text{ is 1.2 to 1.4 mm}) \\ 1500 \text{ to } 2000 & \text{for very coarse monomedium beds } (d_e \text{ is 1.5 to 2.0 mm}) \end{cases}
$$

where l is the depth of the filter bed (mm) and d_e is the effective size of the filter medium (mm).

NOTES:

1. When the size of the medium exceeds 1.5 mm, the space between the filter grains becomes very large compared to those in regular filter beds. The void triples when the diameter of the grain is doubled. Therefore, the l/d_e ratio should be used as an estimate only when the medium is greater than 1.5 mm, and actual figures should be obtained through pilot study. Also, it is strongly recommended to place a thin [12 in. (0.3 m)] sand layer at

the bottom of the coarse deep monomedium bed to act as a barrier to prevent small particles such as oocysts of protozoans from breaking through.

2. If the turbidity of the filtered water must be less than 0.1 ntu, without the use of polymer as a filtration aid, it is recommended that the l/d_e ratio be increased by 15%. Table 3.2.7-1 illustrates the general design criteria for common filter beds.
3. Design engineers should anticipate stringent drinking water standards for future plant operation and are therefore advised to include a provision for replacing the regular filter beds with GAC medium. Due to the longer contact time required by the GAC bed, a minimum EBCT of 10 min, the depth of the regular filter bed must be made deeper to accommodate the GAC medium. A depth of 5.5 ft (1.7 m) is needed to provide 10 min of contact time at a loading rate of 4 gpm/ft^2 (10 m/h).

Filtration Rate As mentioned previously, regulatory agencies restrict the maximum filtration rate for a filter design. However, the agencies will consider a higher filtration rate if a well-managed long-term pilot study can prove that the higher rate satisfies all the requirements. The proven and accepted filtration rates for common filter beds are listed in Table 3.2.7-1.

Engineers should realize that an acceptable filtration rate is a function of factors such as the size of the medium, the degree of pretreatment, and whether polymer is used as a filtration aid. Design engineers who must comply with very conservative design criteria should include a provision for the possible increase in filtration rate. Modern dual-media filter beds are capable of accommodating rates that are 6 to 8 gpm/ft^2 (15 to 20 m/h), whereas coarse deep beds may accommodate filtration rates of 8 to 13 gpm/ft^2 (20 to 30 m/h) with or without preozonation. Hence the size of the influent valve, effluent valves, pipes, and channels must be selected properly when implementing the higher flow rate.

Headloss Across the Filter

TOTAL HEADLOSS ACROSS EACH FILTER

Ordinary gravity filters	9 to 15 ft (12 ft or 3.7 m average) (2.7 to 4.5 m)
Pressure filters	50 to 100 ft (75 ft or 23 m average) (15 to 30) m)
Automatic backwash filters	2 to 3 ft (0.6 to 0.9 m)

NET HEADLOSS AVAILABLE FOR FILTRATION

Ordinary gravity filters	6 to 12 ft (9 ft or 2.8 m average) (1.8 to 3.6 m)
Pressure filters	25 to 50 ft (7.5 to 15 m)
Automatic backwash filters	0.5 to 1.0 ft (1.0 to 0.3 m)

Filter Washing

BACKWASH WITH SURFACE WASH The appropriate backwash rate should be determined by the specific gravity of the medium, the size of the media grains, and the water temperature. Figure 3.2.7-4 may be used in determining the backwash rate. The effect of water temperature on the backwash rate is significant, as shown in Table 3.2.7-2.

BACKWASH RATE

Ordinary rapid sand bed	15 to 18 gpm/ft^2 (37.5 to 45 m/h)
Ordinary dual-media bed (trimedia included)	18 to 22 gpm/ft^2 (45 to 55 m/h)
Ordinary GAC bed	12 to 16 gpm/ft^2 (30 to 40 m/h)

SURFACE WASH RATE

Fixed nozzle type	
Flow rate	3 to 4 gpm/ft^2 (7.5 to 10 m/h)
Pressure	8 to 12 psi (55 to 83 kPa)
Rotating arm type	
Single Arm	
Flow rate	0.5 to 0.7 gpm/ft^2 (1.25 to 1.75 m/h)
Pressure	70 to 100 psi (480 to 690 kPa)
Dual arms	
Flow rate	1.3 to 1.5 gpm/ft^2 (3.25 to 3.75 m/h)
Pressure	80 to 100 psi (550 to 690 kPa)

NOTES:

1. The pressures shown are at the water jet discharge points.
2. Place the nozzles 2 to 3.5 in. (5 to 9 cm) above the top of the bed [rotating arm type is 2 in. (5 cm)].

AIR-SCOURING BACKWASH

BACKWASH RATE

Air-scouring stage for ordinary beds	2 to 4 gpm/ft^2 (5 to 10 m/h)
Air-scouring stage for coarse deep beds	8 to 10 gpm/ft^2 (20 to 25 m/h)
Rinse stage for ordinary beds	10 to 12 gpm/ft^2 (25 to 30 m/h)
Rinse stage for coarse deep beds	16 to 24 gpm/ft^2 (40 to 60 m/h)

AIR-SCOURING RATE

For ordinary filter beds	2.5 to 3 ft^3/min/ft^2 (46 to 55 m^3/hr)
For coarse deep beds	3 to 4 ft^3/min/ft^2 (55 to 73 m^3/hr)

Underdrain System Selection of the underdrain system varies with the type of filter system that is chosen. Normal backwash and self-backwash filters may use the same type of underdrain system but with one important modification—the latter has a larger orifice size. Air-scouring backwash filters are most compatible with plastic blocks or a strainer type of underdrain system. Although many other types of filter underdrain systems are available, those listed here are considered to be the proven and most effective types.

TABLE 3.2.7-1 Types of Medium and Applications

Filter Medium	Type of Filter	Medium Design Criteria	Advantages/Disadvantages
Fine sand	Slow sand filter 0.05–0.17 gpm/ft^2 (0.13–0.42 m/h) filtration rate	Effective size: 0.25–0.35 U.C.: 2–3 Depth: 3.3–4 ft (1.0–1.2 m) S.G.[a] ≥ 2.63	1. Simple design and construction 2. Good effluent quality without pretreatment a. Requires a large filter bed area b. Applicable only for good quality raw water c. Requires frequent scraping off of surface layer (every 20–30 days)
Medium sand	Rapid sand filters 2–3 gpm/ft^2 (5–7.5 m/h) filtration rate	Effective size: 0.45–0.65 U.C.: 1.4–1.7 Depth: 2–2.5 ft (0.6–0.75 m) S.G. ≥ 2.63	1. A proven and widely accepted filtration process 2. A wide application range if pretreatment is provided a. Rather short filter runs due to surface filtration b. Always a need for coagulation pretreatment and an auxiliary washing system
Coarse sand	High-rate filters 5–12 gpm/ft^2 (12.5–30 m/h) filtration rate Direct filtration	Effective size: 0.8–2.0 U.C.: 1.4–1.7 Depth: 2.6–7 ft (0.8–2 m) S.G. ≥ 2.63	1. An effective high-rate filtration process with very long filter runs 2. A wide application range with polymer pretreatment a. Auxiliary wash system is limited to air-scour type b. Requires deep filter cells and a special underdrain
Multimedia coal-sand dual or coal-sand-garnet trimedia	High-rate filters 4–10 gpm/ft^2 (10–25 m/h) filtration rate Direct or in-line filtration	*Sand* Effective size: 0.45–0.65 U.C.: 1.4–1.5 Depth: 1 ft (0.3 m) *Anthracite coal* Effective size: 0.9–1.4	1. An effective high-rate filtration process with long filter runs 2. A proven and widely accepted filtration process a. Either surface wash or air-scour wash is required as an auxiliary washing system and a polymer is required as a filter aid

		U.C.: 1.4–1.5 Depth: 1.5 ft (0.45 m) S.G. ≥ 1.5 to 1.6 *Garnet* Effective size: 0.25–0.3 U.C.: 1.2–1.5 Depth: 0.25 ft (0.0075 m) S.G. ≥ 4.0 to 4.1	b. Proper selection of each medium is important c. Requires a high backwash rate for restratification
Granular activated carbon (GAC)	Removal of organic contaminants 3–6 gpm/ft^2 (7.5–15 m/h) filtration rate Contact time: 10–15 min	Effective size: 0.5–1.0 U.C.: 1.5–2.5 Depth: 6–12 ft (1.8–3.6 m) S.G. ≥ 1.35 to 1.37	1. A proven and accepted process for specific removal of organic contaminants (i.e., taste and odors, THMs, and pesticides) 2. Can also operate effectively as a conventional filter a. Must be regenerated or replaced when adsorption capacity is depleted b. High initial and maintenance costs
Proprietary type media	Variety of types, including green sand and synthetic media	Depends on the purpose	1. Design and efficiency guaranteed by the manufacturer a. Limited number of suppliers b. Mostly patented items

[a] S.G., specific gravity.

TABLE 3.2.7-2 Backwash Rate Adjustment for Various Water Temperatures

Wash-Water Temperature (°C)	Water Viscosity (cP)[a]	12 gpm/ft² (30 m/h) at 22°C	15 gpm/ft² (37.5 m/h) at 22°C	18 gpm/ft² (45 m/h) at 22°C	21 gpm/ft² (52.5 m/h) at 22°C
1	1.728	9.8	12.3	14.8	17.3
2	1.671	9.9	12.4	14.9	17.5
3	1.618	10.0	12.6	15.1	17.7
4	1.567	10.1	12.7	15.2	17.9
5	1.519	10.2	12.8	15.4	18.1
6	1.472	10.3	13.0	15.6	18.3
7	1.428	10.5	13.1	15.7	18.4
8	1.386	10.6	13.2	15.9	18.6
9	1.346	10.7	13.4	16.0	18.8
10	1.307	10.8	13.5	16.2	19.0
11	1.271	10.9	13.6	16.3	19.2
12	1.235	11.0	13.8	16.5	19.4
13	1.202	11.1	13.9	16.6	19.5
14	1.169	11.2	14.0	16.8	19.7
15	1.139	11.3	14.1	16.9	19.9
16	1.109	11.4	14.3	17.1	20.1
17	1.081	11.5	14.4	17.2	20.2
18	1.053	11.6	14.5	17.4	20.4
19	1.027	11.7	14.6	17.5	20.6
20	1.002	11.8	14.8	17.7	20.8
21	0.9779	11.9	14.9	17.8	20.9
22	0.9548	12.0	15.0	18.0	21.1
23	0.9325	12.0	15.1	18.1	21.3
24	0.9111	12.1	15.2	18.3	21.4
25	0.8904	12.2	15.4	18.4	21.6
26	0.8705	12.3	15.5	18.5	21.8
27	0.8513	12.4	15.6	18.7	21.9
28	0.8327	12.5	15.7	18.8	22.1
29	0.8148	12.6	15.8	19.0	22.2
30	0.7975	12.7	15.9	19.1	22.4

[a]cP, centipoise.

NORMAL BACKWASH FILTERS: WITH OR WITHOUT SURFACE WASHING

Type	Headloss at Ordinary Backwash Rates	Orifice Size (Diameter)	Remarks
Precast concrete laterals (teepees)	3 to 5 ft (1 to 1.5 m)	$\frac{5}{16}$ to $\frac{3}{8}$ in. (8 to 10 mm)	12-in. lateral spacing 3-in. orifice spacing on either side of the lateral; maximum

			lateral length of 16 ft (4.9 m)
Dual-parallel lateral blocks (tile or RFP blocks)	2 to 6 ft (0.6 to 1.8 m)	$\frac{5}{32}$ to $\frac{1}{4}$ in. (4 to 6 mm)	12-in. lateral spacing 48 or 18 orifices/ft^2 area; maximum lateral length of 50 ft (15.3 m)
Pipe laterals	3 to 5 ft (0.9 to 1.5 m)	$\frac{1}{4}$ to $\frac{3}{8}$ in. (6 to 10 mm)	12-in. lateral spacing; orifices are spaced 3 to 4 in. apart and 45° down-angle from the horizontal on both sides of the lateral; maximum lateral length of 20 ft (6 m)
Strainer nozzle	4 to 7 ft (1 to 2 m)	0.25 to 0.75 mm	Plenum or lateral bottom strainers spaced 6 to 10 in. apart; a space less than 10 in. for the lateral

Self-Backwash Filters The only difference in design criteria for these types of underdrain, as compared to normal backwash filters, is the required headloss across each filter. This in turn affects the orifice size of the underdrain system. The required headloss, at the designed backwash rate, should be 0.5 to 1 ft (0.15 to 0.3 m) while maintaining no more than a ±5% of water flow uniformity across a filter bed area. In the case of precast concrete laterals, this requirement is satisfied by using orifices that are $\frac{1}{2}$ in. (12 mm) in diameter and spaced 3 in. or 7.5 cm (center to center) on each side of the laterals. All other types of underdrain must be designed specifically to meet the required headloss.

Air-Scouring Backwash Filters The backwash filters listed below are the proven and most effective types available.

Type	Headloss During Washing	Orifice Size	Remarks
Dual-lateral blocks (plastic blocks)	25 and 40 in. under simultaneous air and water backwash at 3 ft^3/min per square foot constant air rate and 15 to 20	$\frac{1}{4}$ in. (6 mm)	12-in. lateral spacing; 23 dispersion orifices/ft^2 of bed area; maximum jointed lateral length is 50 ft (15.3 m); gravel layers

Type	Headloss During Washing	Orifice Size	Remarks
	gpm/ft^2 backwash rate		required on top of the blocks
Strainer	15 to 24 in. under 20 gpm/ft^2 wash rate	0.25 to 0.77 mm (slit size)	The preferred height of the plenum bottom is 3 ft; durable and proven types of strainer should be selected; a gravel bed is usually not required above the strainers

NOTES:

1. The porous plate type of underdrain system has been used successfully in conjunction with the automatic backwash filter, a proprietary unit. However, this type of underdrain will eventually become clogged if moderate to hard water (over 150 mg/L of CaCO3) is filtered continuously. Periodic cleaning of the underdrain with acid, 5% solution of HCl, is therefore mandatory.
2. The lateral type of underdrain should provide air release holes or slits at the top, end, and midpoint of each lateral. Moreover, each end of the lateral should be anchored to prevent the laterals from moving during air scouring.
3. The most important design consideration for the underdrain system is the hydraulics during backwash. The flow velocity at the entrance to each lateral should not exceed 3.5 ft/s (1 m/s), and the velocity at the entrance of the manifold is preferably less than 5 ft/s (1.5 m/s), so that a nearly uniform flow distribution is maintained in the filter bed.
4. A plenum beneath the filter underdrain system ensures better flow distribution. Yet the underdrain slab should be monolithic in nature, to minimize the problems of the underdrain system.

Gravel Support Bed For most underdrain systems the following values are commonly used.

Layer Number	Passing Screen Size	Retaining Screen Size	Depth of Layer	Note
1	$1\frac{1}{2}$ in. (40 mm)	$\frac{3}{4}$ in. (20 mm)	4 to 6 in. (100 to 150 mm)	Bottom layer
2	$\frac{3}{4}$ in. (20 mm)	$\frac{1}{2}$ in.(12 mm)	3 in. (75 mm)	
3	$\frac{1}{2}$ in. (12 mm)	$\frac{1}{4}$ in. (6 mm)	3 in. (75 mm)	
4	$\frac{1}{4}$ in. (6 mm)	No. 6 sieve (3 mm)	3 in. (75 mm)	

5	No. 6 sieve (3 mm)	No. 12 sieve (1.7 mm)	3 in. (75 mm) Total depth: 16 to 18 in. (400 to 450 mm)	Top layer

NOTE: A few departments of health in the midwestern region of the United States demand 2 to 3 in. of 0.8- to 2.0-mm pea gravel (torpedo sand) as the top layer of the gravel bed, as shown above.

For the dual-parallel lateral block type of underdrain (Leopold, Inc.), the following values are the most appropriate:

Layer Number	Passing Screen Size	Retaining Screen Size	Depth of Layer	Note
1	$\frac{3}{4}$ in.	$\frac{1}{2}$ in.	3 in.	Bottom layer
2	$\frac{1}{2}$ in.	$\frac{1}{4}$ in.	3 in.	
3	$\frac{1}{4}$ in.	No. 6 sieve	3 in.	
4	No. 6 sieve	No. 12 sieve	3 in. Total depth: 12 in. (300 mm)	Top layer

For the dual-lateral block type of underdrain with air-scouring wash (Leopold, Inc.), the following values are recommended:

Layer Number	Passing Screen Size	Retaining Screen Size	Depth of Layer	Note
1	$\frac{3}{4}$ in.	$\frac{1}{2}$ in.	2 in.	Bottom layer
2	$\frac{1}{4}$ in.	$\frac{1}{8}$ in.	2 in.	
3	$\frac{1}{8}$ in.	No. 10 sieve (2 mm)	2 in.	
4	$\frac{1}{4}$ in.	$\frac{1}{8}$ in.	2 in.	
5	$\frac{1}{2}$ in.	$\frac{1}{4}$ in.	2 in.	
6	$\frac{3}{4}$ in.	$\frac{1}{2}$ in.	2 in. Total depth: 12 in. (300 mm)	Top layer

NOTES:

1. Some manufacturers and textbooks recommend using a gravel bed that is shallower than 16 in. for ordinary underdrains, including the Wheeler bottom. However, numerous case histories have proved that shallow gravel beds tend to become disrupted.
2. Larger gravel, such as the bottom two layers, must be at least 3 to 4 in. (7.5 to 10 cm) in depth to compensate for any inaccuracies that may occur during construction.

3. The size and depth of the first layer should be adjusted according to the type of underdrain that is selected. Precast teepees and pipe laterals [6 in. (15 cm) in diameter] must have a depth of 6 in. for the first layer to ensure that the laterals are covered.
4. Most strainer nozzle types of underdrains can eliminate the need for a gravel bed if the nozzles are spaced closely together. However, if the slit size is greater than $\frac{2}{100}$ in. (0.25 mm), it is recommended that the nozzle domes be covered by a layer of pea-sized gravel not only to improve good distribution of backwash flow but also to avoid erosion and deformation of the slits by the sandblasting effects caused by fast-moving filter media during backwash.
5. If air in the underdrain system and main backwash line is not purged prior to backwash, the composition of the gravel bed may be disrupted during backwashing, by the explosive movement of large volumes of compressed air. This is the main cause of gravel bed upset for regular backwash filters.
6. A reverse-graded gravel bed of the type illustrated for the dual-lateral block type of underdrain with air-scouring wash, developed in Chicago by John Baylis, is called *hourglass gradation*. It effectively prevents gravel bed upset by using controlled amounts of airflow with backwash. However, if this type of gravel bed is used in conjunction with ordinary backwash, the fine gravel (middle) layer is likely to accumulate sludge.
7. The quality of the gravel must meet the standards set by the AWWA (refer to AWWA Standards B100). The use of limestone must be avoided, especially when the filter bed is acidified in preparation for cleaning.

Miscellaneous Items

1. The water depth above the filter bed must be a minimum of 6 ft (1.8 m) to minimize potential air binding of the filter bed.
2. The lowest water level in the backwash tanks should be approximately 35 ft (10.7 m) above the lip of the wash-water troughs. The capacity of the backwash tank must be large enough to hold water for two filter backwashings.
3. The elevation of the weir, which controls the filter effluent water level, should be set at a height that produces the appropriate hydraulic gradient at the top of the filter bed at the designed filtration rate of the filters.
4. The maximum water level in the wash-waste holding tank should be properly set so that the backup of water in the forebay of the filter is prevented when the most remote filter is undergoing backwashing. The end of each wash trough must have free flow at all times to maintain even flow conditions through each trough.

Copying the European design (Figure 3.2.7-7), wash troughs need not be used if air-scouring wash is adopted. However, a coarse deep bed filter with air scouring will require the use of wash troughs; the troughs may be spaced beyond what is allowed by the formula since it applies to regular backwash systems. The wash troughs improve distribution of the filter influent, and consequently, media migration does not occur. This is possible because filtration rates above 8 or 10 gpm/ft^2 (20 or 25 m/h) induce a high inflow rate, thus changing the hydraulic characteristics of the influent. The filter-to-waste system is important for filter design. This system may not be used all the time under normal operational conditions but is needed under abnormal situations. Abnormal conditions include bed reconstruction, disturbance of the filter bed (including gravel bed), turbidity breakthrough, and repairs to the underdrain system.

Basic Filter Design Scheme In this section we provide a few general tips and describe briefly some basic filter designs.

GENERAL The possibility of contaminating the treated water, due to leakages, cross connections, or operator error, is minimized by providing a clear physical separation between the unfiltered water line and the filtered water line. Engineers are therefore advised not to locate these pipelines or conduits in the same filter pipe gallery. When designing a filtration system, the engineer must consider the accessibility of the various mechanical items. The pipe gallery should have a clear and wide passage so that operators can maneuver vehicles, which may be necessary to provide adequate maintenance and repair work for large-scale plants. All major valves and valve operators should be situated for ease of visual observation and service by the plant operator. Additionally, an access hole should be provided in the filter underdrain to facilitate inspection. The backwash main in the filter pipe gallery should be located below the water level within the filter cells to avoid the accumulation of air in the backwash main. Also, an air release valve should be furnished at the end of the wash-water line, and air relief valves should be supplied for the filtered water line and the main wash-waste lines. If air is allowed to accumulate within these pipe lines, the flow may be reduced by an air blockage and the explosive escape of the air could damage the filters.

The top of the forebay walls should be higher than the highest water level in the filter cell. This design feature minimizes filter media migration, which is initiated by the inertia of the influent flow. However, most European air-scouring wash filters do not follow this criterion and the majority of these filters therefore have problems of varying degrees with filter media migration. Engineers should try to reevaluate the advantage of the flow rate controller in each filter effluent line because of the recent requirement of practicing filter-to-waste. The flow rate controller is almost an essential item when switching from filter-to-waste to filtration mode in order to avoid hydraulic shock loading (refer to Figure 3.7.2-12).

Problems associated with air binding are minimized when a minimum water depth of 6 ft (2 m) is provided above the filter bed. This depth also induces a higher available headloss for filtration. The problem of air binding may also be prevented by installing a control weir for the effluent level of the filter; this scheme will provide positive back pressure in the entire bed at all times. Additionally, this control weir prevents the disturbance of the gravel bed by entrained air during backwash by continually flooding the filter bed. A more detailed discussion of air binding is presented later in this chapter.

Based on the considerations described above, a typical filter arrangement will be similar to that illustrated in Figures 3.2.7-13 and 3.2.7-14.

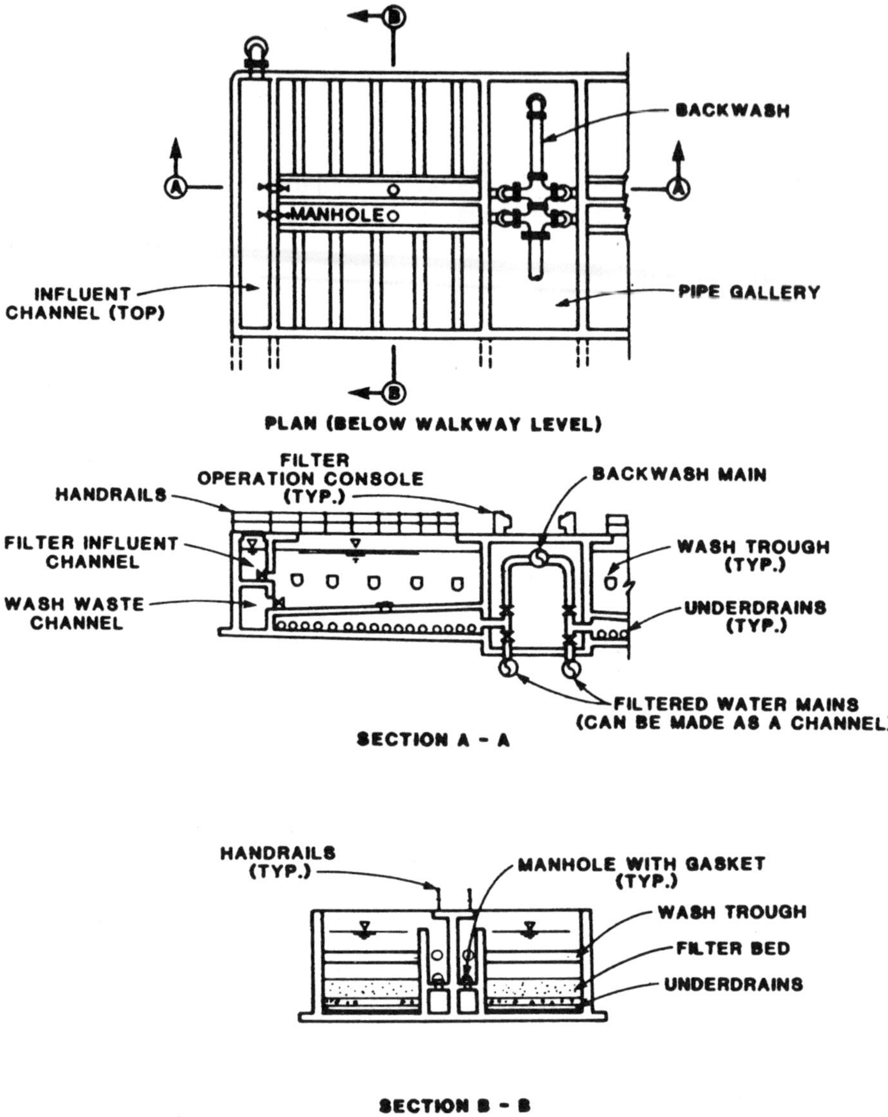

Figure 3.2.7-13 Basic filter design scheme.

BASIC FILTER STRUCTURES When the total number of filters exceed six, they are generally arranged on both sides of the filter gallery, primarily for cost-saving reasons. Figure 3.2.7-14 depicts five basic design schemes. Alternatives 1 and 2 pertain to the self-backwash filters and differ only in the design of the filtered water collection main. Alternative 3 is a conventional type of filter with an equal-loading, constant-level control system. Each filter in this scheme has a flow-modulating valve in its effluent line but lacks a flow metering device. Alternatives 4 and 5

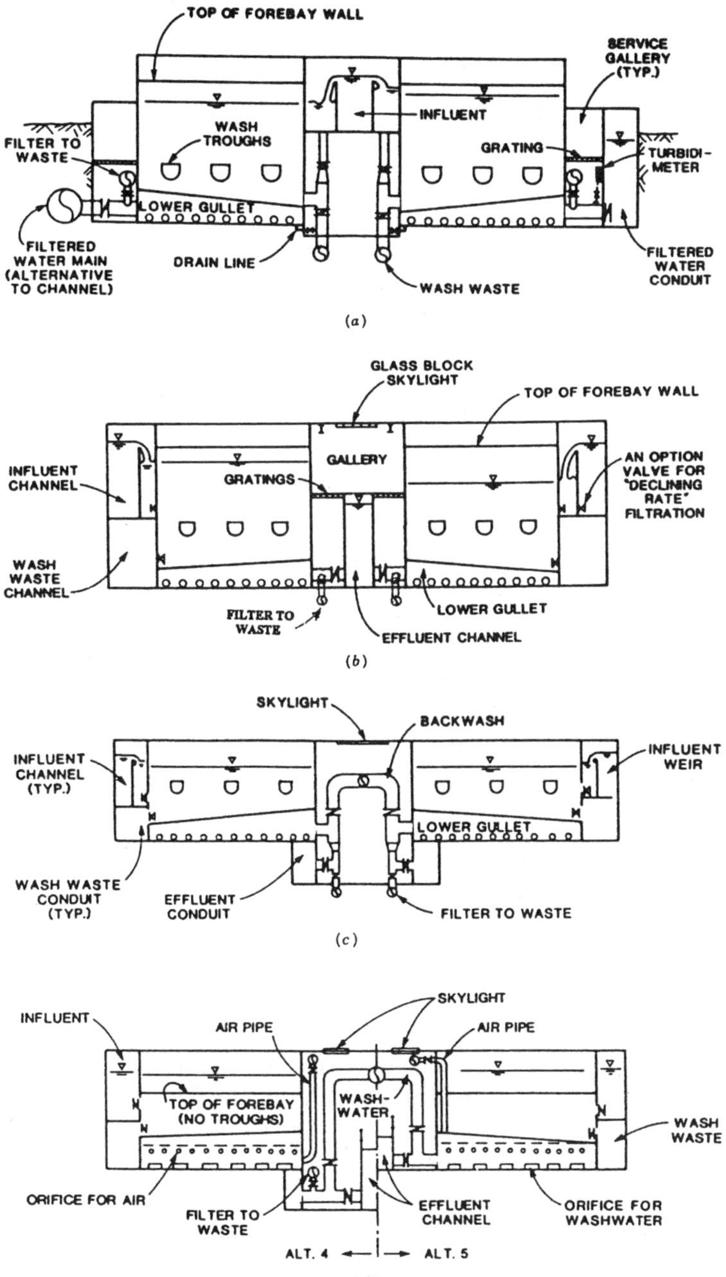

Figure 3.2.7-14 (*a*) Alternative 1: self-backwash filter with surface wash. A clear separation of filtered and unfiltered pipings; the structure is approximately 25 ft deep. (*b*) Alternative 2: self-backwash filter with surface wash. A central filtered water channel arrangement; the structure is approximately 25 ft deep. (*c*) Alternative 3: conventional filter with surface wash. Approximately 17 ft for filter cell and 25 ft height for pipe gallery. (*d*) Alternatives 4 and 5: air-scouring wash filters. Approximately 17 ft for filter cell and pipe gallery, except alternative 4, which has 25 ft for height of gallery.

relate to air-scouring wash filters. Although the filter-to-waste line is not shown in alternative 5, it can easily be provided at the intersection of the backwash and filter effluent valves. One important consideration is elevation of the air supply headers. The headers should be located at least 2 ft (0.6 m) above the highest water level in the filter cells, thereby reducing the chance of water back-siphoning into the air line.

See Appendix 24 for an example of filter design.

Example Design Calculations Three examples are discussed in this section. They concern selection of the most appropriate filtration process and filter design and present the best method for determining the appropriate combination of filter media for multimedia beds.

Example 1 Filtration Process

Select the most appropriate type of filtration process and overall process train based on the source of the water and the following local conditions.

Case 1. A treatment plant in the Midwest has an ultimate capacity of 90 mgd (4 m^3/s). The source is a large river that exhibits high-turbidity spikes, up to 500 ntu, during the summer months. The source also has an average coliform count of 780 MPN (most probable number), an average total organic carbon (TOC) value of 3.5 mg/L, a seasonal algae bloom, and associated taste and odor problems that are primarily a function of diatoms.

Case 2. A 150-mgd (6.5-m^3/s) plant located in the Northwest obtains water from a large lake that consistently has good water quality and an average turbidity of 1.3 ntu. However, turnover of the lake water occurs twice a year. During these periods the turbidity level is expected to rise as high as 10 ntu and the iron and manganese levels may become as high as 1.2 and 0.2 mg/L, respectively. Additionally, the algae count can be expected to be as high as 1000 areal standard units (ASU)/mL.

Case 3. A small town in a developing Asian country requires a 3-mgd (0.13-m^3/s) plant. The raw water may be obtained from a local lake. The quality of this water is good except during the monsoon season. Although accurate records are not available, the turbidity is expected to be approximately 50 ntu during the monsoon season.

Case 4. A town proposes to treat 5 mgd (0.22 m^3/s) of water obtained from three wells. The well water contains an average of 2.2 mg/L iron and 0.4 mg/L manganese and has a hardness of 550 mg/L. Other elements, including turbidity, are always under the maximum contaminant level (MCL) of the drinking water standards. The plant is located in Alaska.

Case 5. A large camp site, located in a national park in the Central Mountain region, requires a maximum of 3 mgd (0.13 m^3/s) of drinking water during the tourist season. The water source is a local impoundment fed by streams. The water quality is reasonably good but has occasional turbidity spikes of up to 50 ntu during thunderstorms. The water contains moderate coliform and algae counts and may contain *Giardia*. The plant is shut down from November to April.

Solution *Case 1*. A coarse deep monomedium filter bed with air-scouring wash is the most appropriate choice and a provision for future conversion of these filter beds to GAC beds should be considered seriously. The bed should have 6 to 8 ft (2 to 2.4 m) of sand or anthracite coal monomedium, with an effective size of 1.2 to 1.5 mm, depending on the depth of the bed. The monomedium should have a uniformity coefficient below 1.4 and a filtration rate of 6 to 8 gpm/ft^2 (15 to 20 m/h); these figures are reasonable choices. A filter bed that is 6 ft (1.8 m) deep provides an empty bed contact time (EBCT) of 15 min at a filtration rate of 3 gpm/ft^2 (7.5 m/h). However, if the rate is 6 gpm/ft^2 (15 m/h), the depth of the bed must be 8 ft (2.4 m) to maintain 10 min. contact time. Placement of a 12-in. (0.3-m) sand barrier (0.7 mm effective size) at the bottom of the filter bed should also be considered.

The overall process train should be preozonation, coagulation, flocculation, sedimentation, filtration, and chloramination. The provision for converting the filter bed to GAC adsorption should include the addition of filter cells to compensate for the lower allowable filtration rate of the GAC filters. Since the GAC filter effluent contains very high counts of bacteria, additional postozonation may be implemented prior to postchloramination. A provision for feeding hydrogen peroxide during preozonation and the addition of a surface wash system may also be included in the design to enhance the efficiency of ozonation and to eliminate the problem of mud ball formation within the deep filter bed.

Case 2. In this case the recommended system is a standard dual-media filter bed with surface wash as the auxiliary scouring during backwash. The self-backwash type of filter is selected because of its simplicity, energy efficiency, and proven performance. These characteristics minimize operation and maintenance efforts. The treatment process train includes prechlorination, coagulation, and roughing filters prior to the dual-media filters and chlorination process; this type of process train is called two-stage filtration. A bypass channel should be provided around the roughing filters because the plant will be operated as an in-line filtration process during most of the year. However, the roughing filter must be employed during seasons with high algae counts and periods of high turbidity.

A provision for future addition of the preozonation and chloramination processes, as well as granular GAC capping of the filter beds, should be included to control taste and odor problems. Direct filtration is not the appropriate choice for this situation since this process has no defense against very short filter runs due to algae blooms. The roughing filter of the two-stage filtration performs as a flocculation process and also removes 50 to 75% of suspended matter. The process of two-stage filtration is discussed in detail in Section 3.2.8.

Case 3. Considering the local conditions and the raw water quality, slow sand filtration with a presedimentation reservoir should be selected as the treatment scheme. This system is simple to operate and maintain and does not require costly maintenance and repairs. The presedimentation reservoir should provide a storage period of 1 to 2 days because this will facilitate the removal of large suspended particles during periods of turbidity spikes (in the raw water). An alternative choice is a membrane filtration system.

Case 4. The appropriate choice for this situation is a standard dual-media bed or a medium-sized monosand bed, with surface wash scouring during filter wash. The first choice for the overall treatment process is lime softening (lime flocculation and clarification) by means of a sludge blanket type of reactor clarifier, followed by acid pH control or carbon dioxide (for recarbonation), then clarifier, all prior to the filters. Lime softening reduces the water hardness to a range of 80 mg/L and effectively removes iron and manganese at the levels dictated in the problem.

The second choice for the overall treatment process begins with the oxidation of iron and manganese with either chlorine or potassium permanganate, or both. The second step is a regular pretreatment to the dual-media bed filtration, followed by sidestream softening via an ion exchange process. Finally, the softened and filtered water are blended to produce a water with medium hardness. Flocculation and sedimentation are preceded by oxidation of the reduced metals.

The pH of the treatment process water should be kept in the range 8 to 8.5 by using ferric iron coagulant, cationic polymer, or both. Water softening may be achieved through a cation exchange resin process or a membrane technique, with proper pH adjustment and the addition of a scale inhibitor, due to the rather small output of the treatment plant. All processes must be housed since the ambient temperatures are extremely cold during the winter months.

Case 5. If enough land is available, a slow sand filtration process should be considered because of its simplicity, reliability, and low operation and maintenance costs. However, it is very unlikely that the area required by the slow sand filtration will be available in the mountainous location. Due to the size of the plant, the seasonal operation, and the limited maintenance services, a proven type of semiautomatic plant, including a membrane filtration process, is recommended.

The filter bed should be a simple monosand bed (rapid sand bed) because of the low initial cost and simple operation. The primary choice for the process is membrane microfiltration. The second choice is a package type of flocculation/sedimentation pretreatment with a self-backwash type of pressure filter. The final choice is a roughing filter followed by proprietary automatic backwash filters. The last two processes require skilled and knowledgeable operators. Selection of the treatment process is strongly influenced by the requirements set by the local regulatory agency.

Example 2 Filter Design

Given A 50-mgd (2.2-m^3/s) filtration plant with an ultimate capacity of 100 mgd (4.4 m^3/s) is located in a quiet and good residential area in the Northeast. The plant site has an average 2% slope (one way). The treatment process train is a conventional complete process and pumps lake water that has a reasonably good quality throughout the year.

Determine

(i) The total number of filters
(ii) The type of filter bed, the filtration rate, and the filter washing scheme
(iii) The size of each filter
(iv) The filter arrangement
(v) The size and type of the major filter valves
(vi) The filter influent channel, pipings, and central gallery
(vii) The type of filter media, gravel bed, and underdrain
(viii) The backwash rate, the type of wash troughs, and auxiliary scour wash
(ix) The plan and sections of the filter
(x) The elevation of the filter effluent control weir and the configuration of the effluent channel
(xi) The device used to measure the backwash flow rate
(xii) The filter piping scheme that allows for the filter-to-waste feature and con tinuous sampling of the filter effluent

Solution (i) The total number of filters in the final stage of this plant is $N = 1.2(100)^{0.5} = 12$. Therefore, six of these filters are constructed in the initial stage.

(ii) Due to the relatively good quality of the raw water, a standard dual-media filter bed is preferred. This type of filter bed has a history of reliable performance when used in conjunction with surface wash and does not employ noisy air blowers for the air-scour process, thereby preserving the peace of the neighborhood. The designed filtration rate, with all the filters on line, is 6 gpm/ft^2 (15 m/h). Under the given conditions, this rate is widely accepted by most regulatory agencies. Note that the size of the major pipes and channels should be 50% larger than the designed rate to accommodate a potentially higher loading rate to each filter during future operation.

Filter washing is achieved through normal backwash: a self-backwash filter design with surface wash as the auxiliary scouring during backwash. The self-backwash scheme is selected because it resolves important factors, such as aesthetic and noise considerations that are dictated by the neighborhood, the possibility of the water freezing should an elevated tank be selected, and the slope of the site. The average 2% slope reduces the need for deep excavation since the filters must be located at a lower elevation. The most energy-efficient wash tank system is the elevated wash tank. However, it should be considered only if the plant site has a suitable hill on which it can be situated. If the wash tank is an elevated tank on legs or a high cylindrical tank, this results in a negative visual impact on the neighborhood.

(iii) The total required filter bed area at 100 mgd is 11,583 ft^2 (1080 m^2) or (100 $\times$ 695) gpm $\div$ 6 gpm/ft^2. Since the total number of filters is 12, each filter bed area is 965 ft^2 (90 m^2). Each filter will have two filter cells and a central gullet; this configuration ensures favorable hydraulic characteristics during backwash. The filter area of each filter cell is $965 \div 2 = 482.5$ ft^2 (45 m^2). Therefore, make each filter cell 15 ft (4.5 m) wide and 32 ft (10 m) long, so that the area of each cell is 480 ft^2 (45 m^2). If the regulatory agency requires a filtration rate of 6 gpm/ft^2 when one filter is off line, each filter cell should have a bed area of 520 ft^2 (48.5 m^2). We check the actual filtration rate by calculating:

$$69{,}500 \text{ gpm} \div (480 \times 2 \times 12) \text{ ft}^2 = 6.03 \text{ gpm/ft}^2 = 15 \text{ m/h} \qquad \text{OK}$$

(iv) The filters are arranged as illustrated in alternative 1 of Figure 3.2.7-14. A double row of filters is constructed along a central pipe gallery where the piping for the filter influent and wash waste are located. The filtered water channels are situated opposite from the central gallery so that there is no possibility of contaminating the filtered water by cross connection. The filtered water may be collected into a channel instead of a large, expensive pipeline and fittings because of the high flow rate of this system and to allow the valves to be accessible for easy maintenance and repair.

(v) The size and types of major filter valves are listed below.

$$\text{Influent flow rate} = 6.03 \text{ gpm/ft}^2 \times 960 \text{ ft}^2 = 5790 \text{ gpm} = 12.9 \text{ ft}^3\text{/s}$$

$$\text{Backwash flow rate} = 22 \text{ gpm/ft}^2 \times 960 \text{ ft}^2 = 21{,}120 \text{ gpm} = 47.1 \text{ ft}^3\text{/s}$$

$$\text{Surface wash flow rate} = 3 \text{ gpm/ft}^2 \times 960 \text{ ft}^2 = 2880 \text{ gpm} = 6.4 \text{ ft}^3\text{/s}$$

Based on these data, the following selections are made:

Valve	Size and Type	Velocity Through the Valve (ft/s)
Filter influent	24-in. butterfly valve	4.1 OK
Filter effluent	36-in. butterfly valve	1.83 OK
Backwash (maximum)	Filter effluent value is also the backwash value for the self-backwash filters	6.7 OK
Surface wash	12-in. butterfly valve	8.2 OK
Filter-to-waste	20-in. butterfly valve	5.9 OK
Wash-waste drain	36-in. butterfly valve	7.6 for back and surface wash OK

NOTE: The sluice gate and the gate valve are bulky and expensive items that may require frequent maintenance and repair, in contrast to the butterfly valves specified by the AWWA standards.

(vi) To establish an approximately equal water level in each inlet channel, the flow velocity at the beginning of the influent channel must not exceed 1.5 ft/s (0.45 m/s). Since each filter is equipped with identically sized rectangular weir inlets at the same elevation, this ensures that all filters on the line receive almost the same flow rate (i.e., equal loading).

The influent channel is located directly above the pipe gallery and should be approximately 22 ft (6.7 m) in width to allow for a low flow velocity. The gallery therefore has ample space for inspection and maintenance of the valves and other equipment. For this type of filter, the total depth of the filter structure is approximately 25 ft (7.6 m), which allows sufficient water depth to be provided above the bed. As illustrated in Figure 3.2.7-15, the channel width, the height of the gallery, and the location of the inlet and wash-waste pipelines are all dictated by the dimensions of the pipe fittings and the valves, in addition to the thickness of the concrete walls.

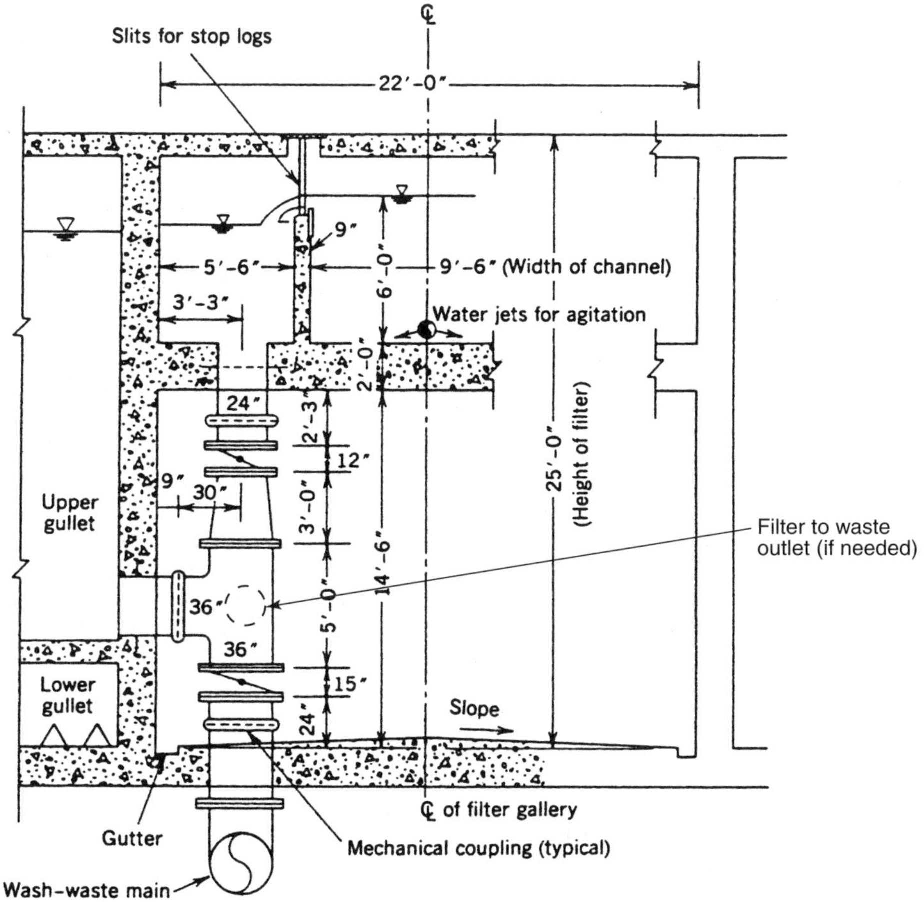

Figure 3.2.7-15 Filter piping, influent, and pipe gallery design (only the left half, from the centerline, is shown).

Constructability, the ease with which the valves can be removed during major repairs, the wall pipes for leakage water stops, and the slope of the floor toward the drainage gutters are all important design considerations. Unfortunately, many inexperienced design engineers disregard topics such as providing a minimum space requirement of 9 in. (22.5 cm) between the wall and the adjacent pipe flanges, for easy installation of the nuts and bolts, for supplying mechanical couplings to facilitate minor adjustments in pipe alignment and easy removal of the valves, and finally, for adequate drainage of water from the gallery floor.

The influent channel has a cross-sectional area that is $6 \times 9.5 = 57$ ft^2 (5.3 m^2). Since the settled water is divided among the six first-stage filters and the six final-stage filters, the 100-mgd flow is split into two directions. The maximum flow velocity in the influent channel is therefore

$$(50 \times 1.55) \text{ ft}^3/\text{s} \div 57 \text{ ft}^2 = 1.36 \text{ ft/s or } 0.4 \text{ m/s} \qquad \text{OK}$$

(vii) The specifications for the filter media area as follows:

	Anthracite Coal (Top Layer)	Filter Sand (Bottom Layer)	Remarks
Effective size (mm)	1.0 ± 0.05	0.55 ± 0.02	For purchase specifications
Uniformity coefficient	≤ 1.4	≤ 1.5	For purchase specifications
Specific gravity	1.65 ± 0.05	≥ 2.63	After 24 h of soaking
Depth (in.)	20	10	After scraping off the top layer of fine particles

NOTES:

1. When each medium is first placed in the filters, an extra 1.5 to 2 in. should be added to the depth to allow for removal of the fine particles (scraping) after backwashing, as specified below. The process of scraping is vital for proper functioning of the filter bed. If the fine particles are not removed from the surface of the bed, it will result in very short filter runs.
2. Following the initial placement of the sand, the filter should be backwashed twice at a rate of 22 gpm/ft^2 (55 m/h). After the first backwash, the top $\frac{1}{2}$ in. of medium should be removed. Repeat the backwash process and scrape an additional $\frac{1}{2}$ in. off the top. Thus the final depth of the sand layer will be 10 in. (25 cm).
3. Place the coal layer on top of the sand layer and backwash the filter bed three times at 22 gpm/ft^2 (55 m/h), scraping off the top layer after each wash. Remove the top 1 in. after the first wash, $\frac{1}{2}$ in. after the second wash, and $\frac{1}{4}$ in. after the third wash.
4. The AWWA filter material specifications (AWWA standards B100) should be used for the other necessary properties of the media. The gravel bed is composed of five layers of graded gravel and the total depth is 18 in. (45 cm). See item B(a) for the details.

For this example the filter underdrain system of choice is the precast concrete laterals with a triangular cross section (Figure 3.2.7-8). Dual lateral blocks or pipe laterals may be employed if the owner prefers their use. Regardless of the type of underdrain, the headloss should not exceed 12 in. (0.3 m) at the designed backwash rate. In addition, a uniform flow distribution must be maintained over the entire filter bed; the difference should be less than 5%. The higher the headloss, the deeper the filter structure. Precast concrete laterals that are 14.5 ft (4.4 m) long are placed at 12-in. (0.3-m) intervals, and both sides of the sloped walls should have orifices that are $\frac{1}{2}$ in. in diameter and 3 in. on center.

Each filter cell is 32 ft (9.6 m) in length and there are 64 laterals in each filter. The lower gullets should be designed with a reasonably large cross section for three reasons: (1) to provide adequate space for construction workers to work, (2) to allow easy inspection and cleaning of the inside of each underdrain lateral, and (3) to provide a low backwash flow velocity in the gullet to ensure uniform flow distribution to each lateral. As a rule of thumb, the backwash flow velocity in the gullet should be less than 2 ft/s. Also, the velocity should be less than half of the inlet flow velocity to each underdrain lateral. Thus, make the lower gullet size 5 ft in width and 5 ft in height (for the wash-water connection side) and 3 ft in height at the other end.

The maximum backwash flow rate for each lateral is

$$21{,}120 \text{ gpm} \div 64 = 330 \text{ gpm or } 0.735 \text{ ft}^3/\text{s}$$

The cross-sectional area of the lateral (a triangle cross-section) is

$$(0.615 \times 0.6) \div 2 = 0.185 \text{ ft}^2$$

The flow velocity at the inlet of the lateral is

$$0.735 \div 0.185 = 4.0 \text{ ft/s} < 2 \times 1.9 \text{ ft/s} = 3.8 \text{ ft/s} \qquad \text{OK}$$

Make the size of the lower gullet 5 ft in width and 5 ft in height at the effluent channel side (the entrance of the backwash water). The maximum flow velocity in the lower gullet during backwash is

$$47.1 \text{ ft}^3/\text{s} \div (5 \text{ ft} \times 5 \text{ ft}) = 1.9 \text{ ft/s}$$

The total number of orifices ($\frac{1}{2}$ in. diameter) per lateral is

$$2 \times 58 = 116 \text{ or } 7424 \text{ orifices per filter}$$

The flow rate from each orifice is

$$0.735 \div 116 = 0.0063 \text{ ft}^3/\text{s}$$

The cross-sectional area of each orifice is

$$(0.5/12)^2 \times 0.785 = 0.00136 \text{ ft}^2$$

The backwash flow velocity through each orifice is

$$0.0063 \div 0.00136 = 4.63 \text{ ft/s}$$

The headloss through each orifice hole is

$$\Delta h = K \frac{v^2}{2g} = 2.4 \frac{4.63^2}{64.4} = 0.80 \text{ ft} < 1 \text{ ft} \qquad \text{OK}$$

It is imperative that each lateral be anchored to the floor even if the weight of the lateral exceeds the normal uplift force that is generated during an average backwash. The orifices should be located in the upper portion of the inner triangular section and directed horizontally so that air will not be trapped in the lateral and so that the water jets are not directed toward the surface of the filter bed. Both ends of each orifice should be reamed to provide rounded edges, minimizing the headloss. Computer analysis of this type of underdrain reveals an approximate maldistribution of $\pm 5\%$.

(viii) The designed backwash rate is the maximum wash rate and should only be used during the summer months when the water temperature is high. Based on Figure 3.2.7-4 and the specified filter media, the maximum backwash rate is 22 gpm/ft^2 (55 m/h). The maximum backwash flow rate should be at least 30 mgd if the maximum backwash rate is used. Due to the design of the self-backwash filter, 27 mgd must be filtered to produce a backwash flow rate of 19 gpm/ft^2. Whenever the plant is operating at less than the minimum plant flow rate required to backwash one filter, the required rate may be met simply by increasing the plant flow rate to the rate required to wash one filter.

Each filter cell is provided with four wash troughs composed of fiberglass-reinforced plastic. The lightweight and durable troughs facilitate easy installation and simple adjustment of their level during construction and provide trouble-free maintenance. The maximum flow rate for each trough is

$$(21{,}120 + 2{,}880) \text{ gpm} \div 8 = 3{,}000 \text{ gpm}$$

By using the trough sizing diagram (Figure 3.2.7-16), a trough with a width of 21 in. and an overall height of 24 in., with a half-circle-shaped bottom, is selected.

The elevation of the trough should be high enough so that media grains kicked up by the surface wash jets are not easily washed out during the simultaneous surface wash and backwash period. The height (H_0) of the troughs above the filter bed is (refer to the discussion of filter wash troughs)

$$(0.5 \times 2.5) + 2 < H_0 < (2.5 + 2)$$

$$3.25 \text{ ft} < H_0 < 4.5 \text{ ft}$$

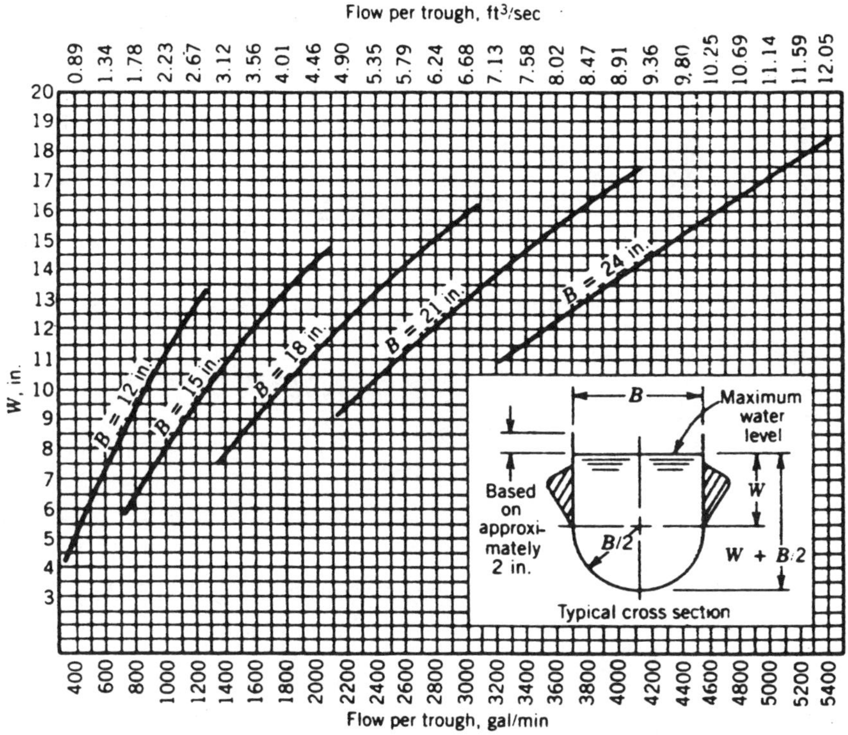

Figure 3.2.7-16 Wash trough sizing diagram. (Courtesy of F.B. Leopold Company.)

Make it 4 ft.

The spacing (S) of the trough is

$$(1.5 \times 4) < S < (2.5 \times 4)$$

$$6 \text{ ft} < S < 10 \text{ ft}$$

Use a spacing of 8 ft on center.

Auxiliary scouring will be provided by a surface wash system. The first choice is either a fixed nozzle or rotating arm type. The second choice is either a surface wash or a subsurface wash unit. Despite its effectiveness, the Baylis type of auxiliary scouring (Figure 3.2.7-17), consisting of a series of vertical pipes with nozzle caps at each end, is not adopted because an intense jet of water will hit the gravel bed directly and upset the stratification whenever the caps are lost; several case histories exist for this problem.

In this application a fixed nozzle type of auxiliary scouring is used simply because of fewer maintenance requirements. Three-inch-diameter schedule 80 PVC or stainless steel lateral pipes should be horizontally arranged at 4-ft intervals and 3 in. above the surface of the bed. The orifices must be $\frac{1}{4}$ in. (6 mm) in diameter at

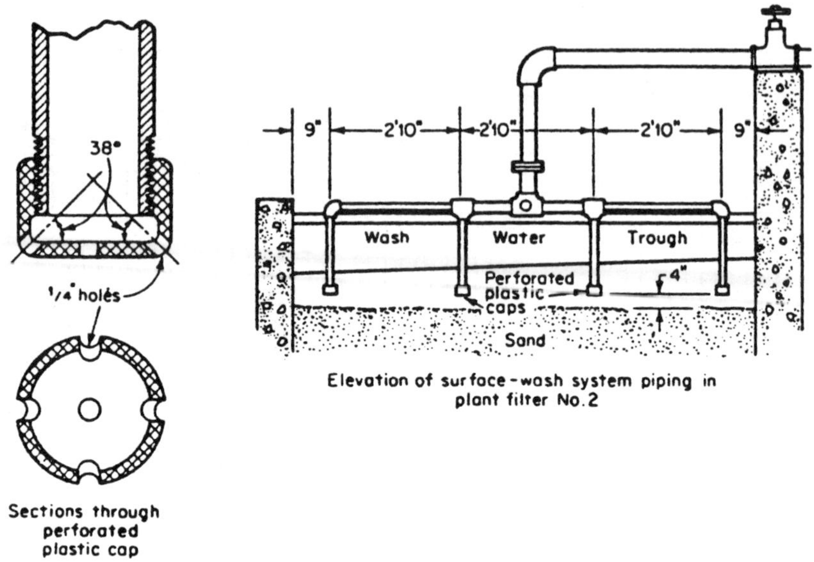

Figure 3.2.7-17 Example of the Baylis surface washing system.

approximately 8-in. intervals. The jets should be directed 25 to 35° below the horizontal plane so that the coal and sand interface will be scoured.

The subsurface wash systems have a slightly better performance when used in conjunction with dual-media beds. Yet experience over the last 20 years has shown that the difference is marginal. In fact, the subsurface wash system potentially has problems with the nozzles becoming clogged whenever the rubber caps are lost. The flow rate for each filter during surface wash is 2880 gpm or 6.4 ft^3/s (based on 3 gpm/ft^2). Since there are 8 laterals in each cell and 16 laterals in each filter, the flow rate to each lateral is

$$2880 \div 16 = 180 \text{ gpm } (0.40 \text{ ft}^3/\text{s})$$

Therefore, the flow velocity in the 3-in. lateral is 8.2 ft/s, which is a reasonable figure.

Each lateral in the surface wash system should have $\frac{1}{4}$-in. orifices drilled in a staggered fashion at 8-in. intervals on either surface of its 14.5-ft length. The entire width of the filter cell, 15 ft, may not be covered by the orifices due to the maneuvering space that is required to install fittings such as elbows and end caps, and possible repair of the piping system.

The total number of orifices per lateral is

$$[(14.5 \times 12) \div 8] \times 2 = 44$$

The flow rate through each orifice is

$$0.40 \div 44 = 0.0091 \text{ ft}^3/\text{s}$$

The area of each orifice is

$$(0.25 \div 12)^2 \times 0.785 = 0.00034 \text{ ft}^2$$

while the flow velocity at each orifice is

$$0.0091 \div 0.00034 = 26.8 \text{ ft/s}$$

To satisfy the design criteria, the jet velocity for surface scouring should be between 20 and 30 ft/s (6 and 9 m/s).

The loss of hydraulic head through the orifice is

$$\Delta h = 2.4\left(\frac{26.8^2}{64.4}\right) = 27 \text{ ft or } 11.7 \text{ psi}$$

The source water may be either settled or filtered water. The latter is more common and must be supplied by a pump; the use of filtered water also requires a special design in order to avoid the possible backflow of wash waste into the filtered water tank. The total dynamic head of the pump is a function of the size and length of the supply pipe and the elevation of the water source. The pump generally has a minimum total dynamic head of 65 ft (20 m).

(ix) The plan, sections, and details of the designed surface wash system are shown in Figure 3.2.7-18.

(x) The elevation of the effluent level control weir is an important factor in setting the filter backwash rate control weir. The hydraulics of the filter backwash is evaluated on a wash rate of 22 gpm/ft^2 (48.2 ft/s). The 36-in. butterfly valve (BV) entrance loss, BV loss, and exit loss are given by

$$h_1 = (0.5 + 0.25 + 1.0)\,\frac{v^2}{2g} = 1.75\,\frac{(6.7)^2}{2g} = 1.22 \text{ ft}$$

Headloss in the lower gullet is

$$v = 47.1 \div 5^2 = 1.9 \text{ ft/s maximum} \quad \text{(negligible loss)}$$

Entrance to the underdrain lateral is

$$h_2 = 0.5\,\frac{v^2}{2g} = 0.5\,\frac{(4.0)^2}{2g} = 0.12 \text{ ft}$$

Headloss in the underdrain lateral is negligible. Headloss at 0.5-in. orifices of the underdrain is given by

$$h_3 = 2.4\,\frac{(4.63)^2}{2g} = 0.80 \text{ ft}$$

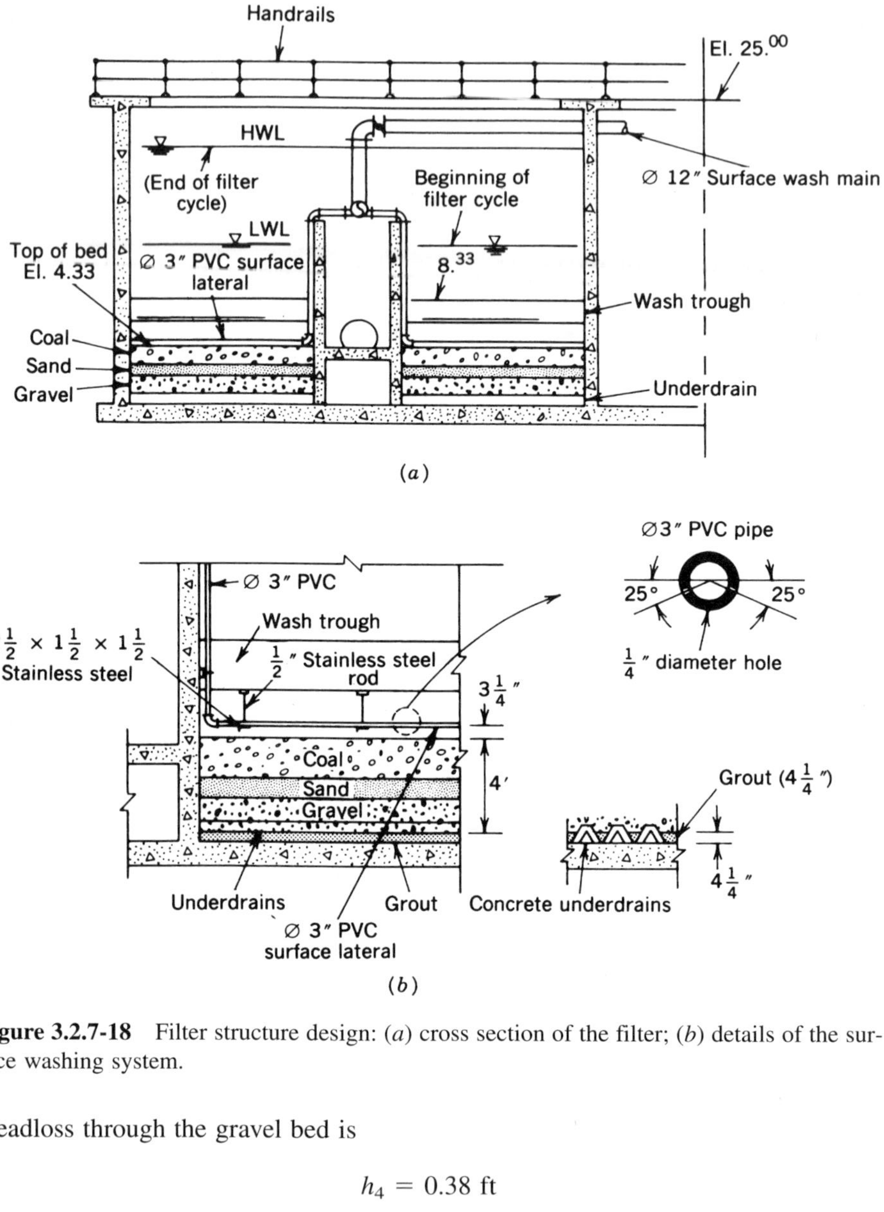

Figure 3.2.7-18 Filter structure design: (*a*) cross section of the filter; (*b*) details of the surface washing system.

Headloss through the gravel bed is

$$h_4 = 0.38 \text{ ft}$$

obtained from a pilot study conducted at 16°C. Headloss through the filter bed is

$$h_5 = 1.42 \text{ ft}$$

obtained from a pilot study conducted at 16°C. Headloss over the wash trough (troughs as weirs) is

$$h_6 = 0.17 \text{ ft}$$

The height of water flows into the wash troughs is 2 in. (0.167 ft). Thus

$$h_7 = 0.17 \text{ ft}$$

Total headloss is

$$\sum_{i=1}^{i=7} i = 4.2 \text{ ft}$$

Since a headloss of approximately 4.2 ft (1.28 m) will be provided during one backwashing at a rate of 22 gpm/ft^2 (55 m/h), the effluent level control weir should be set so that the water level in the effluent channel is 4.2 ft (1.28 m) higher than the top of the wash-water troughs. To adjust the overflow elevation to a 4-ft span, two weirs 10 ft long (3 m), and two down-opening slide gates are installed in the wall of the filter effluent channel. As illustrated in Figure 3.2.7-19, section A–A, the top of the notched concrete wall should be set at an elevation of 10.25, that is, 1 ft 11 in. above the top of the wash troughs, which are located in the filters. The operator may manually raise the down-opening slide gates from an elevation of 10.25 to 13.25 whenever the backwash rate must be adjusted. Under these circumstances the gate acts as a weir. The required hydraulic head for a plant flow rate of 50 mgd, using two weirs that are 10 ft (3 m) in length, is

$$H^{1.5} = \frac{(50 \times 1.55) \div 2}{3.33 \times 10} = 1.16$$

Thus H is 1.1 ft.

The top edge of the slide gates should roughly be set at an elevation of 11.43 (say, 11.45) if the plant is to produce a wash rate of 22 gpm/ft^2 (55 m/h) when the plant flow rate is 50 mgd. The effluent channel is 6 ft (1.8 m) wide with an average water depth of 12 ft (3.6 m). At a plant flow rate of 50 mgd, the maximum flow velocity in the channel is

$$(50 \times 1.55) \text{ ft}^3/\text{s } (6 \times 12) \text{ ft} = 1.08 \text{ ft/s} < 2 \text{ ft/s}$$

NOTE: As described earlier, a new filter bank, identical to this filter bank, will be added at a future date. The settled water will flow into the filters between the two filter banks. In the future, this second filter bank will have an identical filter effluent level control system and the filter effluent from the two banks will be combined. The flow rate will be measured using a Venturi meter.

(xi) The rate of backwash flow is measured by a venturi meter that is placed in the 54-in. filter effluent main. When one filter is undergoing backwash, the meter will indicate a decrease in flow rate. This decreased flow rate is the actual wash rate for the filter. The backwash rate may be adjusted by either raising or lowering the down-opening slide gates of the filter effluent channel.

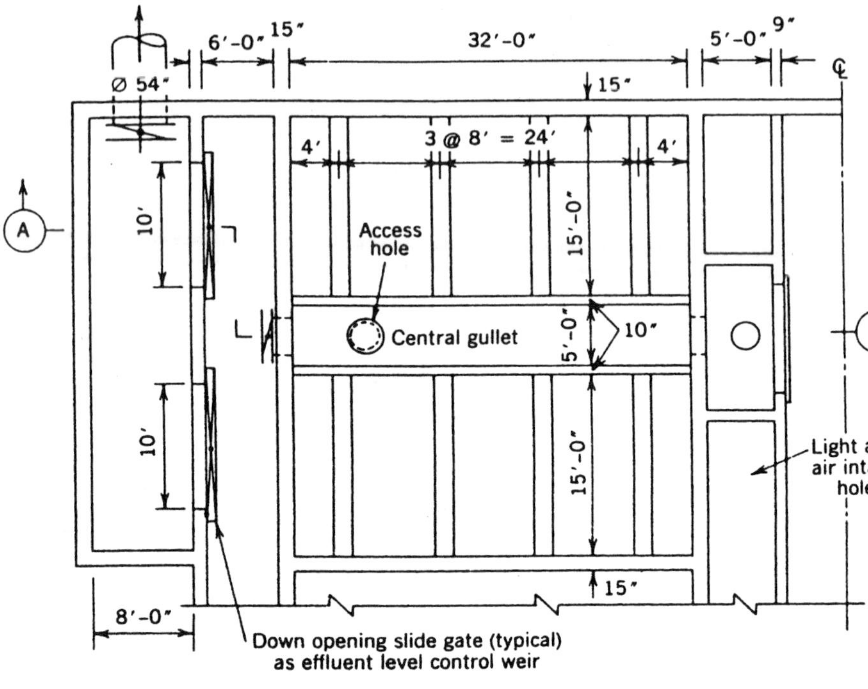

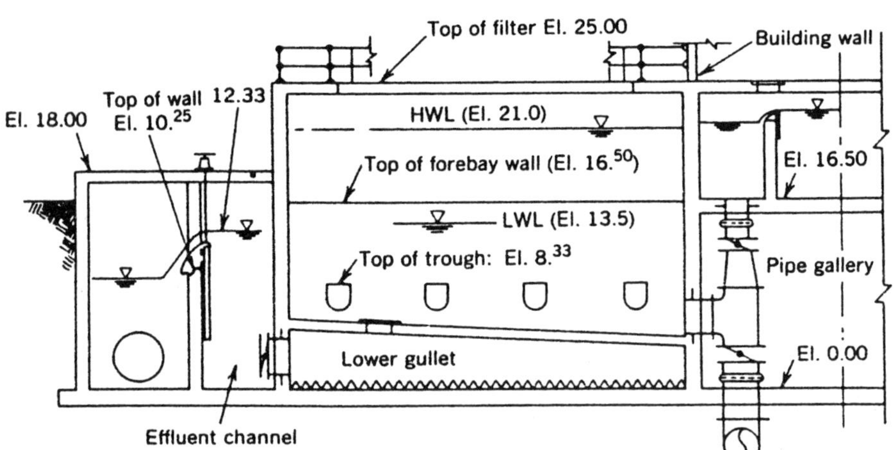

Figure 3.2.7-19 Filter structure design: (*a*) plan at elevation 17.00; (*b*) section A–A.

(xii) Alternative 1 of Figure 3.2.7-14 should be adopted if filter-to-waste is practiced at the beginning of each filter cycle or if there is continuous sampling of filtered water from each filter. The design in Figure 3.2.7-19 should be modified to provide a service gallery. The volume of the wastewater can be quite large; however, the quality of this water is quite good. Therefore, it is better to recycle it back to the filter influent rather than directing it to the wash-waste holding tank and thereby reducing the effective holding capacity of the tank.

Example 3 Selecting the Proper Combination of Filter Media for Multimedia Beds

Given A dual-media filter bed is to be designed. However, the local regulatory agency has limited the effective size of the sand to 0.45 mm and the uniformity coefficient to less than 1.7. The available filter medium, anthracite coal, has a specific gravity of 1.63, determined by the AWWA's standard test method.

Determine The effective size and uniformity coefficient for the coal media

Solution First, the design engineer should select the lower uniformity coefficient of sand; this figure may be obtained from the supplier. Once this is obtained, the optimum filter performance can be anticipated and the sand and coal mixing zone can be reduced. Any supplier of filter media should be able to provide sand and coal with a uniformity coefficient of 1.4 without increasing the unit cost. However, a uniformity coefficient of 1.3 is now available with a slight increase in unit cost. Engineers should set the sand specifications at an effective size of 0.45 mm, a uniformity coefficient of 1.5, and a specific gravity of 2.63. The matching coal size may be determined by either computation or by using the nomograph:

$$\frac{d_1}{d_2} = \left(\frac{\rho_2 - \rho}{\rho_1 - \rho}\right)^{0.667}$$

where $d_1 = 0.45$ mm; d_2 is the size that must be determined; $\rho_1 = 2.63$, the density of particles of size d_1; $\rho_2 = 1.63$, the density of particles of size d_2; and $\rho = 1.0$, the density of the fluid.

$$\frac{0.45}{d_2} = \left(\frac{1.63 - 1}{2.63 - 1}\right)^{0.667}$$

or $d_2 = 0.85$ mm, the effective size of coal.

Alternatively, Figure 3.2.7-4 may be used. First, locate the size of 0.675 on the horizontal scale. Then move vertically to a specific gravity of 2.63, which may be established between 2.60 and 2.65 for sand. Move horizontally to the right to a specific gravity of 1.63, which is assumed to be between 1.6 and 1.65 of anthracite coal. Finally, move down vertically to the horizontal scale and find 1.27. The effective size is therefore $1.27 \div 1.5 = 0.85$ mm.

Operation and Maintenance In this section we discuss the various aspects of (1) evaluating regular filter performance, (2) optimizing the pretreatment process, (3) performing jar test procedures, and (4) evaluating and adjusting the filter washing process. Other important operational parameters and common filter problems are also covered.

Regular Filter Performance To optimize the filtration process, plant operators should routinely evaluate filter performance. The three indicators used in evaluating plant performance are the filtered water turbidity, the length of the filter run, and the ratio of the volume of backwash water used to the volume of filtered water. Under normal conditions, the filtered water turbidity should always be less than 0.1 ntu, with 0.1 ntu as the target turbidity level. The second item, filter run length, is a function of the size of filter medium and the influent water quality, which in turn is a function of the solids content, water temperature, filtration rate, and the condition of the filter bed. A treatment plant that is both well designed and operated should have a filter run length longer than 24 h whenever the filtration rate is less than 6 gpm/ft^2 (15 m/h). If a reverse-graded filter bed is selected, the filters should be run for an average of 2 days at a rate of 4 to 6 gpm/ft^2 (10 to 15 m/h), as long as the treatment process is not a direct or in-line type of filtration.

Shorter filter run lengths can be attributed to eight major factors: the accumulation of fine media particles on the surface of the filter bed, an effective size of filter medium that is too small for the filtration rate, too much floc and suspended matter (in the filter influent), a filter bed that is dirty and full of mud balls, and overabundance of filter-clogging algae in the water source, air binding, a higher than optimum dosage of polymer (as either a flocculant aid or a filter aid), and recycling substandard clarified wash waste to the filter influent. With respect to direct and in-line filtration, extremely short filter runs can easily be induced by unusually high levels of suspended matter in the raw water and/or overfeeding of coagulant.

One of the best indicators of filter performance is the ratio of the water used for filter washing to the amount of water that is filtered prior to filter washing. Under normal conditions, this ratio is less than 3%; a ratio that is less than 2% is considered to be very good. If the ratio exceeds 5%, the filter performance is judged poor. Operators should realize that filter performance is much better during months with warm ambient temperatures, due to faster floc formation and quicker floc settling in the settling tanks. A well-maintained filter may have a ratio of 2.5 to 3% during the winter but only 1.5 to 2% in the summer.

Another indicator of filter performance is the volume of water that is filtered per unit area of filter bed during a filter run. This is also referred to as the *unit filter run volume* (UFRV), which is the product of the filtration rate (gpm/ft^2) and the filter run length (minutes) and is expressed as gallons per square foot of filter bed area. In general, a UFRV that is less than 5000 gal (200 m^3/m^2) is unacceptable even with direct filtration because of the extremely short filter run length. A UFRV that is over 10,000 gal (410 m^3/m^2) is indicative of better performance, and over 15,000 gal (610 m^3/m^2) is good filter performance for conventional complete treatment processes.

Pretreatment Optimization Plant operators have two methods by which they can optimize filter performance. The first involves the coagulant, while the second concerns the mixing energy level and mixing time. In both cases, jar tests are used to evaluate the various factors, especially for the conventional treatment process. To optimize the coagulant, the operator must evaluate both the type of coagulant and the dosage. Alum and polymerized aluminum chloride generally do not produce floc that is physically strong enough to endure high-rate filters. Therefore, small amounts (15 to 30 μg/L) of nonionic polymer are usually added to strengthen the floc and to condition the filter beds. Ferric iron salts, such as ferric chloride or ferric sulfate, should always be evaluated as a primary coagulant whenever the raw water has a pH that is over 7.5 (high) and an alkalinity that is over 50 mg/L. The smaller dose of ferric coagulant produces floc that is both physically stronger and heavier than those produced with alum. Moreover, ferric floc is virtually insoluble in water over a wide range of pH: a range of 5 to 11 for ferric floc, compared to 5.5 to 7.2 for alum floc.

Whenever the required dosage of metallic coagulant is low (less than 6 mg/L), the solubility of the floc becomes an important factor in the filtered water quality. Many treatment plants using a low dosage of alum in treating water with a high pH exhibit high concentrations of residual aluminum in the filtered water. Treatment plants that practice direct filtration or in-line filtration should therefore give special consideration to this issue. Plant operators should remember that a high dosage of alum during periods of high turbidity also creates a high concentration of residual alum because of suppressed pH (a pH that is less than 5.5). Consequently, proper pH control by alkali chemicals is required during the flocculation process.

Finally, evaluation of cationic polymers, approved by the EPA for the purpose of water treatment, is essential in selecting the type of coagulant to be employed. The positive aspects of the cationic polymers are that they are not affected by the pH of the regular process water; they produce a tougher floc; and the sludge is easier to handle than those produced by alum. Since the effective dosage is small, rarely exceeding 2 mg/L, the volume of sludge that must be handled and disposed will be reduced.

In many cases, a combination of alum and cationic polymer is used because of better clarification and cost-effectiveness. Most cationic polymers cost approximately eight times more per unit weight of the product than alum. Consequently, the polymer dosage is usually limited to 0.5 mg/L and is supplemented by 2 to 4 mg/L of alum. This combination provides both good performance and cost-effectiveness when applied to a direct filtration process. As indicated earlier, final selection of the appropriate coagulant is a function of the effectiveness of the chemical(s) as a coagulant, the cost of treating the water, problems associated with sludge disposal, and the ease with which the coagulant is handled by plant operators.

The second important factor in optimizing the coagulant is determination of the optimum dosage. Generally, the optimum coagulant dosage for a water treatment plant is the dosage that results in a filter influent turbidity that is less than 1 ntu, or the dosage that produces the lowest supernatant turbidity after jar testing. Due to cost considerations, plant operators usually select the minimum dosage required to produce the second lowest supernatant turbidity (based on jar test results). This practice is justifiable for rapid sand filtration because a straight relationship exists between the filtered water turbidity and the settled water turbidity. However, modern filter beds, such as the reverse-graded beds and coarse deep beds, require a very different way of thinking. The modern high-rate filters require polymer to be applied to the filter influent with mixing to maximize their efficiency; the second flocculation (incidental flocculation) actually occurs in the filter inlet channels, pipes, and filter bed (primarily in the filter bed). The polymer may be used with or without metal salt coagulants, such as alum, and the dosage that produces a supernatant turbidity (after jar testing) of less than 1 ntu is generally not the best for the overall filter performance. Data obtained from operational high-rate filters indicate that the optimum coagulant dosage, that which produces properly coagulated water (does not produce large floc), is only one-half to one-third the optimum dosage required by rapid sand filtration. The proper coagulant dosage for high-rate filters generally yields a supernatant turbidity of 2 to 4 ntu by the jar test. Figure 3.2.7-20 summarizes the relation between proper coagulant dosage and types of filter (treatment process).

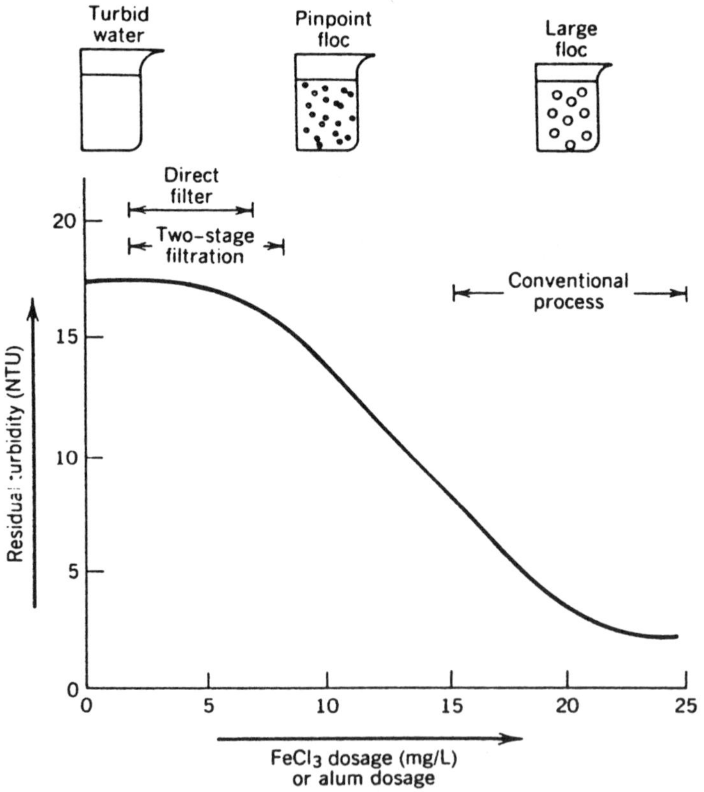

Figure 3.2.7-20 Proper coagulant dosage range for the three treatment processes.

The most difficult situation in selecting the optimum coagulant dosage is encountered with the in-line and direct filtration process. The dosage at which pinpoint floc is produced is already an overdose that will induce a short filter run. Several equipment manufacturers and researchers have suggested that for these types of filtration process, plant operators use pilot filter column tests to determine the proper coagulant dosage. However, experience has shown that very few operators of plants equipped with the pilot filter columns use them regularly enough to warrant the cost. Plant operators respond that the pilot columns are simply not practical enough.

Another option is to use a small glass tube pilot filter. This type of test filter is composed of a half dozen 1-in.-diameter glass tubes. The glass tubes are filled with small-grained sand to a depth of 1 ft (0.3 m). The filterability of the supernatant, obtained from jar tests, is determined by passing it through the pilot filter. Yet plant operators rarely employ this apparatus on a regular basis because it is both unreliable and time consuming. The small glass pilot filter yields unreliable data because the sand in the glass columns is not ripened, as in actual filter beds. An alternative to the sand has been employed by the author and others. In this case, filter paper (Whatman No. 1) is used as the filter medium. This option is much quicker and the results correlate reasonably well with actual filter bed performance. It must be emphasized that the ordinary jar test procedure is a good control tool for the plant process, but only for conventional processes. It does not work well for direct or indirect filtration, unless the flocculated water is first filtered through filter paper, such as Whatman No. 1 or a similar type of paper with a pore size of approximately 10 μm. Incidentally, the pore size of regular rapid sand filter beds is considered to be approximately 15 μm.

The second method of optimizing filter performance involves the mixing energy level and the mixing time ($G \times t$). Although these parameters are supplied to the operator by the design engineer, there is still some room for adjustment.

Recommended Jar Test Procedures Although some plant operators believe that jar testing does not yield reliable results, it is a useful test for controlling coagulant dosage. Factors such as mixing conditions, alum solution strength, method of chemical feeding, and the flash mixing system all affect the correlation between the jar test results and actual plant performance.

If the mixing conditions ($G \times t$) of the jar tests are not similar to those of the actual process, the results may be quite different. Most modern treatment plants use a 50% solution of liquid alum. However, a 1% solution of liquid alum is often used in the jar tests, potentially distorting the correlation between the jar test and the actual plant process. Experience has shown that a 0.1% a solution of alum is better suited for the jar tests as long as it is freshly made (Kawamura, 1973). The feeding condition of the alum (i.e., continuous feed or intermittent feed), the actual type of flash mixing system, and the presence of flow short-circuiting in the actual tank also greatly affect the correlation between the jar test results and plant performance. Refer to Appendix 10 for a recommended jar test procedure.

Evaluation and Adjustment of Filter Washing Procedure To optimize the filtration process, the condition of the filter bed must be evaluated routinely. Operators may maintain the proper bed conditions by adjusting the filter washing procedure. There are three basic methods by which to evaluate the effectiveness of the filter washing procedure: (1) visually inspecting the filter bed before and after filter washing, (2) measuring the turbidity of the backwash waste at 1-min intervals after initiating backwash, and (3) core sampling of the filter bed both before and after filter washing.

A filter bed that is properly maintained is clean (i.e., practically devoid of mud balls) but is also in ripened condition; the media grains of a ripened filter bed are coated with a proper amount of floc or polymer. During visual inspection of the filter bed, plant operators must look for mud balls, cracks, and any other unusual conditions. The bed is in poor condition, with mud balls and mud accumulation, if the operator finds lumps and cracks on the surface of the filter bed prior to filter washing and lumps even after washing. The most obvious indicators of a dirty filter bed are the existence of worms and debris on its surface. If a filter cell exhibits any mounding, concaved areas, or obvious shifts in the media, there is either a scouring action by filter influent, problems with the surface washing system, or a maldistribution of the backwash water and boiling action in the filter bed. The operator should immediately investigate the causes of these unfavorable phenomena.

The second method, measuring the turbidity of the backwash waste, is used to evaluate the effectiveness of the filter washing procedure in removing floc from the filter bed. A valuable tool in assessing the washing efficiency and the optimum duration of the filter washing procedure is the turbidity profile of the backwash waste plotted against wash time. A low profile with a low peak curve is indicative of ineffective washing. A high profile curve with a high peak curve is characteristic of effective washing. Figure 3.2.7-21 illustrates these two types of curves.

Many plants wash their filters for an excessively long period of time, until the operator can clearly see the surface of the bed. This method of washing is actual detrimental to the filtration process because the filter often exhibits a distinct initial turbidity breakthrough, for a period of 15 to 30 min or longer, until the bed regains its ripened condition. Studies based on both pilot-scale and actual filters indicate that this initial turbidity breakthrough may be minimized or even eliminated by terminating filter washing when the turbidity of the wash waste ranges from 10 to 15 ntu; this usually occurs after 5 to 6 min of regular washing (not air-scouring washing). Adverse effects, such as a shorter filter run length and the accumulation of mud within the filter bed, will not occur if the filter wash rate is properly selected and a surface wash is implemented.

The energy input generated by the backwash process, and by either surface wash or air scour, may be estimated by using the following equations. Energy input due to the backwashing process is

$$G = \frac{\rho g v_b \, \Delta h}{\mu L}$$

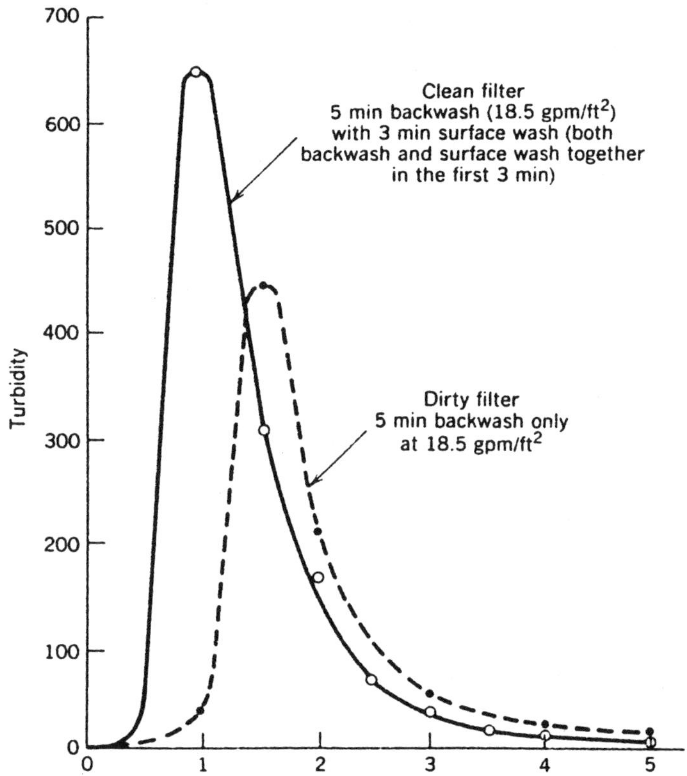

Figure 3.2.7-21 Backwashing waste characteristics (conventional sand filters). (*Note:* The water level in the filter cell was lowered approximately 6 in. above the top of the filter bed prior to filter washing.)

where G ranges from 300 to 400 s^{-1}. Energy input due to the water jets of surface wash is

$$G = \frac{\rho g v_s \, \Delta\rho}{\mu \alpha L}$$

where G ranges from 1000 to 1300 s^{-1}.

When backwash is applied to a filter bed, the headloss begins to increase with an increase in the wash rate. Yet when a specific backwash rate is reached and the wash rate continues to increase, the hydraulic headloss will begin to decrease due to fluidization of the filter bed. This characteristic is illustrated in Figure 3.2.7-22. Data obtained from both pilot and actual filter studies show that the optimum backwash rate is approximately twice the wash rate at which maximum headloss is attained. For instance, in Figure 3.2.7-22 the proper wash rates for sand II and sand I are 20 and 15 gpm/ft^2 (50 and 37.5 m/h), respectively. Beyond these optimum rates there are no appreciable benefits, primarily because the filter bed is highly fluidized;

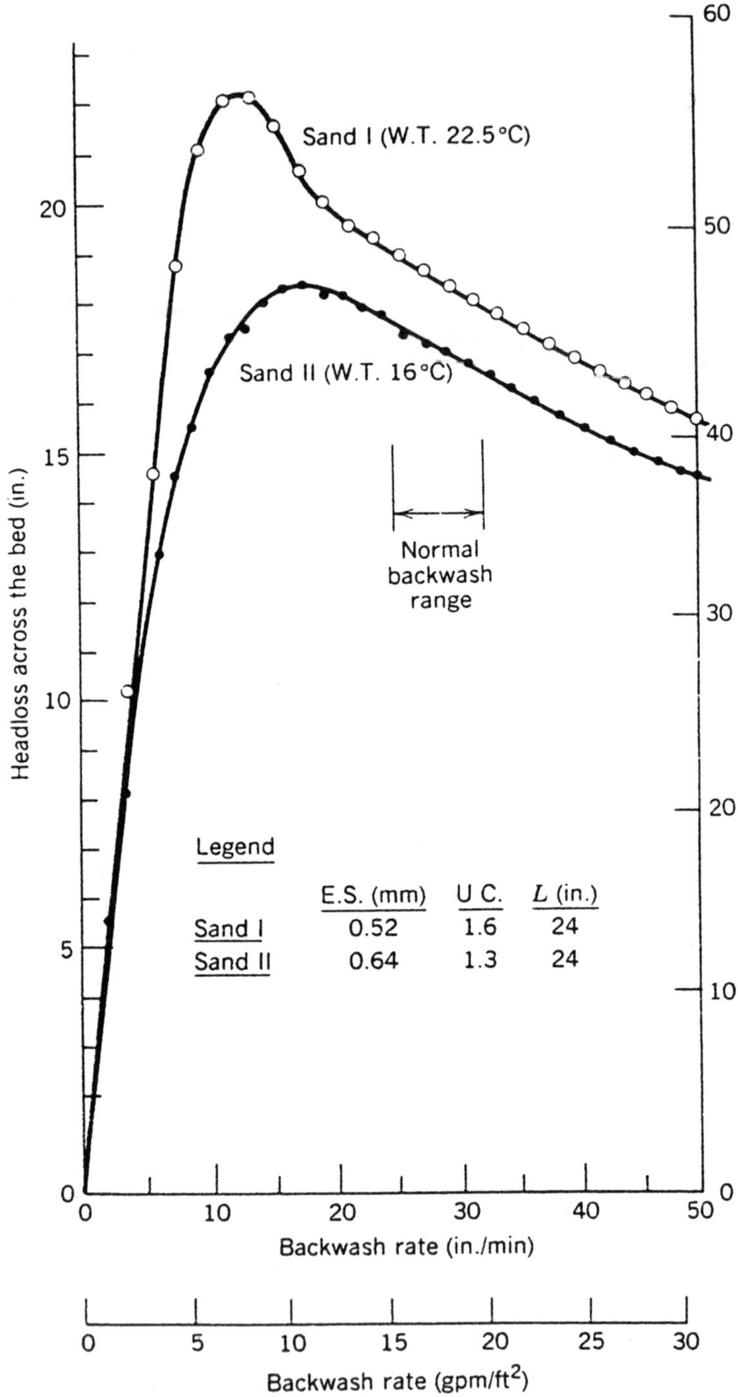

Figure 3.2.7-22 Relation between backwash rate and headloss. (Adapted from P. A. Norman, "Backwash Instability in Rapid Sand Filters," M.S. thesis. University of New South Wales, Australia, 1970.)

the media grains are no longer in contact with each other and the scrubbing action of the filter bed is therefore minimized. This same logic can be applied during the rinse period after air-scouring wash. However, operational experiences indicate that the backwash rate at which the maximum headloss is produced is the most effective concurrent washwater rate because the majority of the sludge, attached to the media grains, is stripped during air scouring and transported into the water. The recommended filter washing procedures, with auxiliary scouring, are illustrated in Figure 3.2.7-23. The mud deposition profile of Figure 3.2.7-10, both before and af-

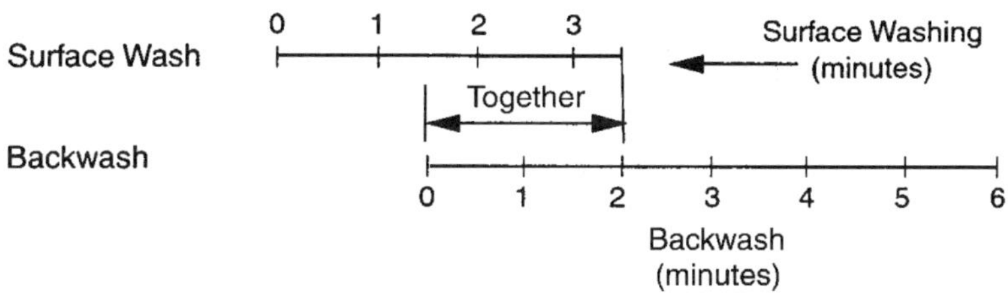

Notes: (1) Draw down water level to approx. 4 inches (10 cm) above the bed before surface wash starts.

(2) Longer concurrent wash will result in higher media loss.

(*a*)

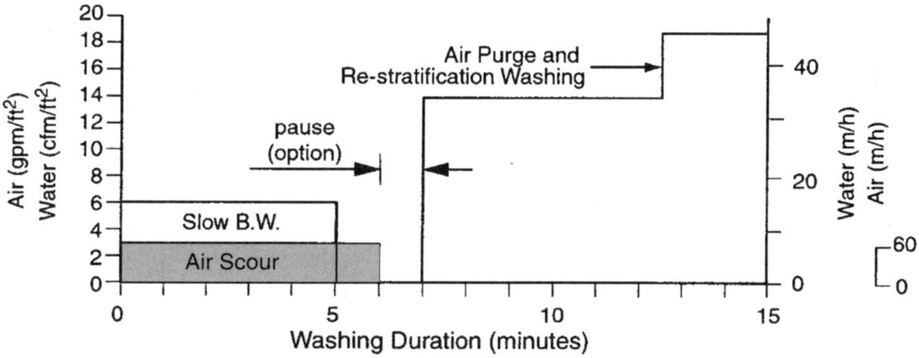

Notes: (1) Draw down water level to approx. 4 inches (10 cm) above the bed.

(2) The first slow backwashing must be stopped before the water level gets up to six cinches below the top waste weir.

(3) Backwash in rinsing stage would require approx. 18 gpm/sf (45 m/h) to purge air in coarse media bed as well as re-stratification of dual media bed.

(*b*)

Figure 3.2.7-23 Recommended filter washing sequence: (*a*) recommended surface and backwash sequence; (*b*) normal concurrent air and water air-scouring wash for common-size media bed.

ter filter washing, clearly demonstrates the effectiveness of the washing procedure. Based on the discussion above, plant operators should be able to evaluate their present washing practices and make the necessary adjustments. In most cases the filter bed expansion rate during backwash ranges from 20 to 35%. However, bed expansion is less important if a surface wash system is installed; when air-scouring wash is utilized, the importance of the bed expansion is almost diminished.

The third method used to evaluate the effectiveness of washing conditions is core sampling, which allows plant operators to establish a mud profile for the entire depth of the filter bed. Core samples should be obtained from selected filter beds at regular intervals according to a rigorous schedule. This procedure is very important because it allows the operator to check the actual depth of the medium, to obtain a medium-sized distribution profile across the entire depth of the filter bed, to evaluate filter washing efficiency, to check filter bed conditions, and to evaluate the movement of filter gravel. These data can only be obtained by taking core samples from multiple locations within the filter cell. Moreover, it is crucial to obtain samples at various depths from the same location. If one core sample is taken across the entire depth of the filter bed, from one location, it may not be representative of the filter bed since it does not show the accumulation of fine medium at the surface of the bed. The recommended sampling techniques and a simple quantitative analysis of the mud that accumulates on the surface of the filter bed are described in Appendix 11.

Mud Deposition Profile Both the effectiveness of the filter backwash process and the condition of the filter bed may be clearly evaluated from the mud profile graphs. For example, a filter without a surface washing system cannot adequately wash out floc that is retained during a filter cycle, thus allowing the sludge to accumulate slowly. Conversely, a filter with a surface wash system is effectively cleaned throughout the entire depth of the bed, as indicated in Figure 3.2.7-10. Many case histories have shown that both regular rapid sand filters and standard dual-media filters can be kept clean with installation of the proper surface wash system and application of the correct backwash rate. Improperly designed dual-media filters with air-scouring wash systems have problems with mud accumulation in the coal and sand mixed zone. For example, when the size of the two filter media is not hydraulically matched and the backwash rate after air scouring is inadequate to fluidize the bed (such as 12 gpm/ft^2 or 30 m/h), a large portion of the coal and sand mixed zone will develop heavy mud accumulation. Figure 3.2.7-24 is an example of this problem at a lime softening plant in California.

A filter bed that is properly conditioned is also in a ripened stage; each grain of media is coated with a thin film of coagulant hydroxide or polymer. A bed that is too clean usually exhibits a distinct turbidity breakthrough at the beginning of each filter cycle, lasting anywhere from 30 to 60 min. Based on the author's experience, a turbidity of 30 to 60 ntu, determined after filter washing by the sludge retention analysis test method, is indicative of a clean and ripened bed if polymer is continually fed to the filter influent. A filter bed that is slightly dirty (less than ideal but no need for concern) will have a turbidity between 60 and 120 ntu. However, a filter bed exhibiting turbidity over 120 ntu is dirty, and both the filter washing system and filter

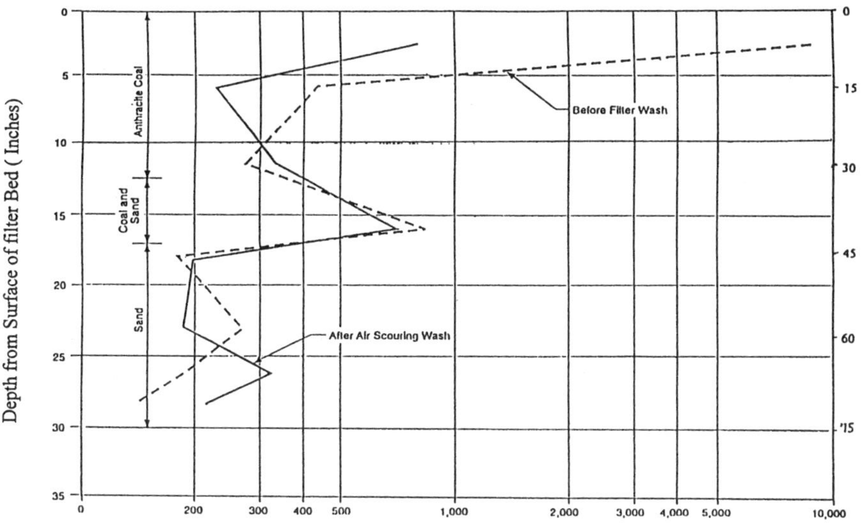

Figure 3.2.7-24 Floc retention profile for a filter bed with a large coal and sand mixed zone.

washing conditions must be reevaluated. Turbidity above 300 ntu is indicative of a mud ball problem. Figure 3.2.7-25 can be used to judge the cleanliness of the filter bed. The mud deposition analysis procedure is described in Appendix 11.

When mud balls are found in the filter bed, measure the percent volume of mud balls in the bed. If the percentage is less than 0.1%, the filter bed is clean. If the mud balls are 0.1 to 0.5% by volume, the filter bed is in good condition; 0.5 to 1.0% is indicative of a fairly clean bed; 1 to 5% indicates that the filter bed is in bad con-

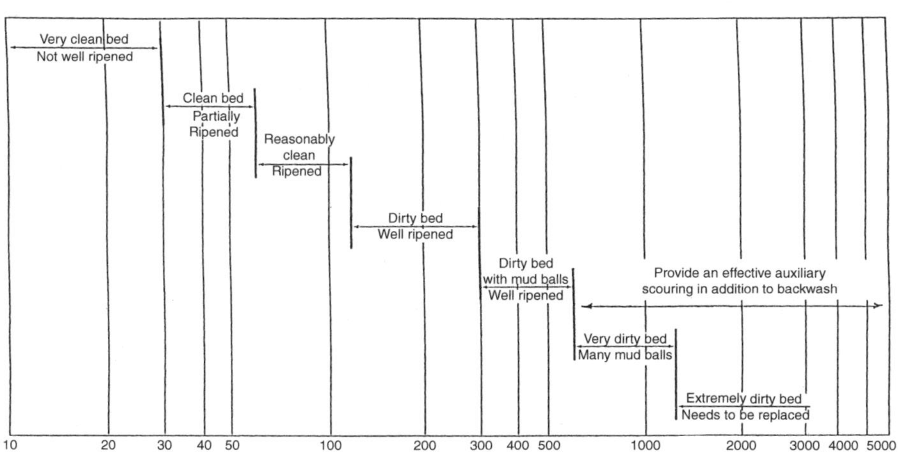

Figure 3.2.7-25 Filter bed condition range: cleanliness evaluation.

dition. If the volume of mud balls is over 5%, the bed must be replaced with new media.

Restoration of the Filter Bed Filter bed medium is commonly lost after repeated filter washings. Rapid sand filters generally do not lose a significant amount of sand; a normal loss of 1 to 2% of the total bed depth (original) occurs annually. Yet the loss of anthracite coal is much more significant; an annual loss of 5 to 7% is not unusual for ordinary dual- or trimedia beds.

Excessive media loss is often caused for the following reasons: air binding in the filter bed; an unnecessarily long and overlapping surface wash and backwash (over 3 min); an excessive backwash rate, especially during months of cold water temperatures; a mismatch in the size of the coal and sand; a wash trough elevation that is too low; improper sequential control of both the air-scouring and rinsing backwash processes; and leakage of the filter media through the underdrain system. If the media loss is excessive, the plant operator should first locate and correct the cause. The original depth of the filter bed should be restored by adding the appropriate media to the surface of the filter bed and backwashing the filters a minimum of two runs; this will fluidize the bed and ensure a proper gradation profile.

Many engineers and plant operators ask what the tolerable amount of media loss should be. Losing up to 20% of the original depth, or a loss of 6 in. (15 cm) in depth, whichever is attained first, is generally a tolerable loss if total bed depth is approximately 30 in. (76 cm). Filters with surface wash systems begin losing the advantages of surface washing when the distance between the top of the bed and the elevation of the jet nozzles exceeds 6 in. Filter performance also begins to decrease when media loss is greater than 20% of the original depth.

The purchase of replacement sand for rapid sand filters is quite simple; specify the same size medium grain as the original but with a lower uniformity coefficient (1.5), to avoid the short filter run lengths characteristic of fine medium grains. However, dual- and multimedia beds must have the replacement media properly matched in physical characteristics so that all the layer will fluidize to the same degree during backwash. The specific gravity and size of both anthracite coal and garnet grains are very important considerations. Based on past experience, the specific gravity of sand is always near 2.63, but this is not the case for commercially available anthracite coal and garnet grains. Over the past 30 years the specific gravity of anthracite coal, when used as filtering material, has ranged from 1.50 to 1.72. Therefore, the size of the available anthracite coal must be adjusted to match that of the sand. Figure 3.2.7-4 will help determine the complementary media sizes. First, locate the representative sand size ($0.55 \times 1.5 = 0.83$ mm). Then move vertically up to the specific gravity of the sand (2.63). Next, move horizontally to the right, to a specific gravity of 1.65 (assuming that this is the available type of anthracite), and vertically down to 1.53, the representative particle size of the coal. Assuming that the uniformity coefficient is specified as 1.5, we find that the effective size of the coal is

$$1.53 \div 1.5 = 1.02 \text{ mm}, \qquad \text{say, } 1 \text{ mm}$$

The following formula may also be used to compute the matched size of two different types of filter bed medium:

$$\frac{0.55}{d_2} = \left(\frac{1.65 - 1}{2.63 - 1}\right)^{0.667}$$

or $d_2 = 1.015$: say, 1.0 mm.

If other kinds of filter material, such as garnet or GAC, are to be used in conjunction with sand, the appropriate media sizes may be found in the same manner. As was demonstrated in Example 1, it is imperative that the media specifications not allow for a wide range of media sizes and specific gravities.

Gravel Bed Upset in the Filter Major disturbances in the gravel bed are usually caused by errors in the design phase, construction phase, or during operation. Plant operators have no control over the first two. Improper design is the responsibility of the design engineer. The obvious design flaws are improper specification for the gradation and thickness of each gravel layer. Based on case histories, disturbances in the gravel bed and leakage of filter media through the supporting gravel bed are generally caused by improper construction of the gravel bed, including the omission of one or even two layers. Gravel bed upsets are also caused by improper operation of the backwash process.

The most common operational error is to quickly initiate full-scale backwashing for a filter containing no water. This type of error usually occurs when the filter is being filled with water for the first time by someone who does not have adequate knowledge of the filter, by the contractor, or by an inexperienced operator. The composition of the gravel bed may also be disturbed by backfilling a filter that has been drained at full backwashing rate without slowly operating the backwash valve. In either case, the combination of the quickly escaping compressed air (in the filter underdrain system), the short-circuiting of the water, and the escape of air can cause major disturbances in the arrangement of the gravel bed. It may even damage or break the underdrain system. Once the gravel bed is disturbed, it is irreversible until reconstructed. It is therefore essential to backfill the filters slowly until the water level covers the surface of the filter bed. The initial backfilling rate should not exceed 5 gal/ft^2 (12.5 m/h). The backwash valves must also be opened very slowly to avoid lifting the filter media bed off the gravel layer. Otherwise, an open space will be formed between the media bed and the gravel bed. If this should occur, the filter bed will break at several points, causing a boiling action, and the upper portion of the supporting gravel bed will be disturbed (Figure 3.2.7-26). If the plant operator notices a distinct boiling action during filter backwashing, the filter must be drained and the condition of both the gravel bed and the inside of the underdrain system must be inspected. The filter bed should be dug up by using a 4 ft × 4 ft (1.2 m × 1.2 m) bottomless box as a shoring frame. Failure of the gravel bed is also clearly evidenced when filter media accumulation is found both inside and downstream of the underdrain system.

In the mid-twentieth century, John Baylis proposed the use of a reverse-graded gravel bed. Although this type of bed is effective in preventing the movement of

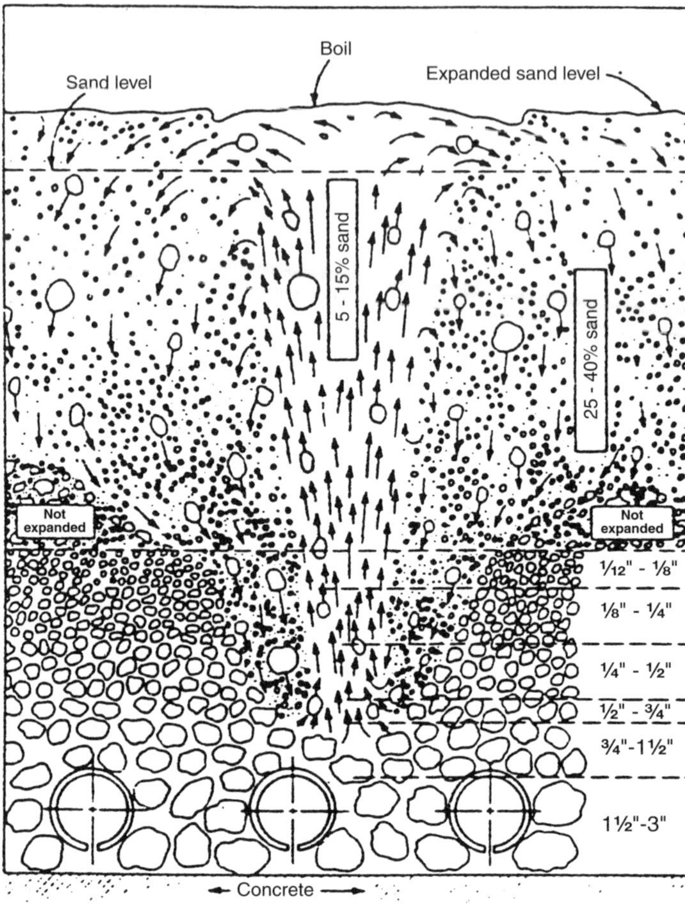

Figure 3.2.7-26 Filter gravel bed upset by operational error. (From AWWA, *Water Quality and Treatment,* 3rd ed., McGraw-Hill, New York, 1971.)

gravel, it has problems with mud accumulation in the fine gravel layers of the gravel bed (middle layer) for ordinary backwashing filters.

Air Binding Air binding is a phenomenon in which large amounts of air bubbles accumulate in the filter bed. Air dissolves in the water near or even over the saturation point. The pressure in the upper portion of the filter bed may be reduced to below atmospheric pressure due to the creation of negative pressure during the latter stages of filtration. When this occurs, the air in the water no longer remains dissolved but accumulates within the filter bed, rapidly increasing the headloss throughout the bed. Figure 3.2.7-27 illustrates the development of negative pressure within the filter bed. If a high water level is maintained in the filter, air binding will be delayed. Therefore, a shallow filter with a water depth that is 4 to 5 ft (1.2 to 1.5 m) above the filter bed and a filter that lacks an effluent water control weir tend to pro-

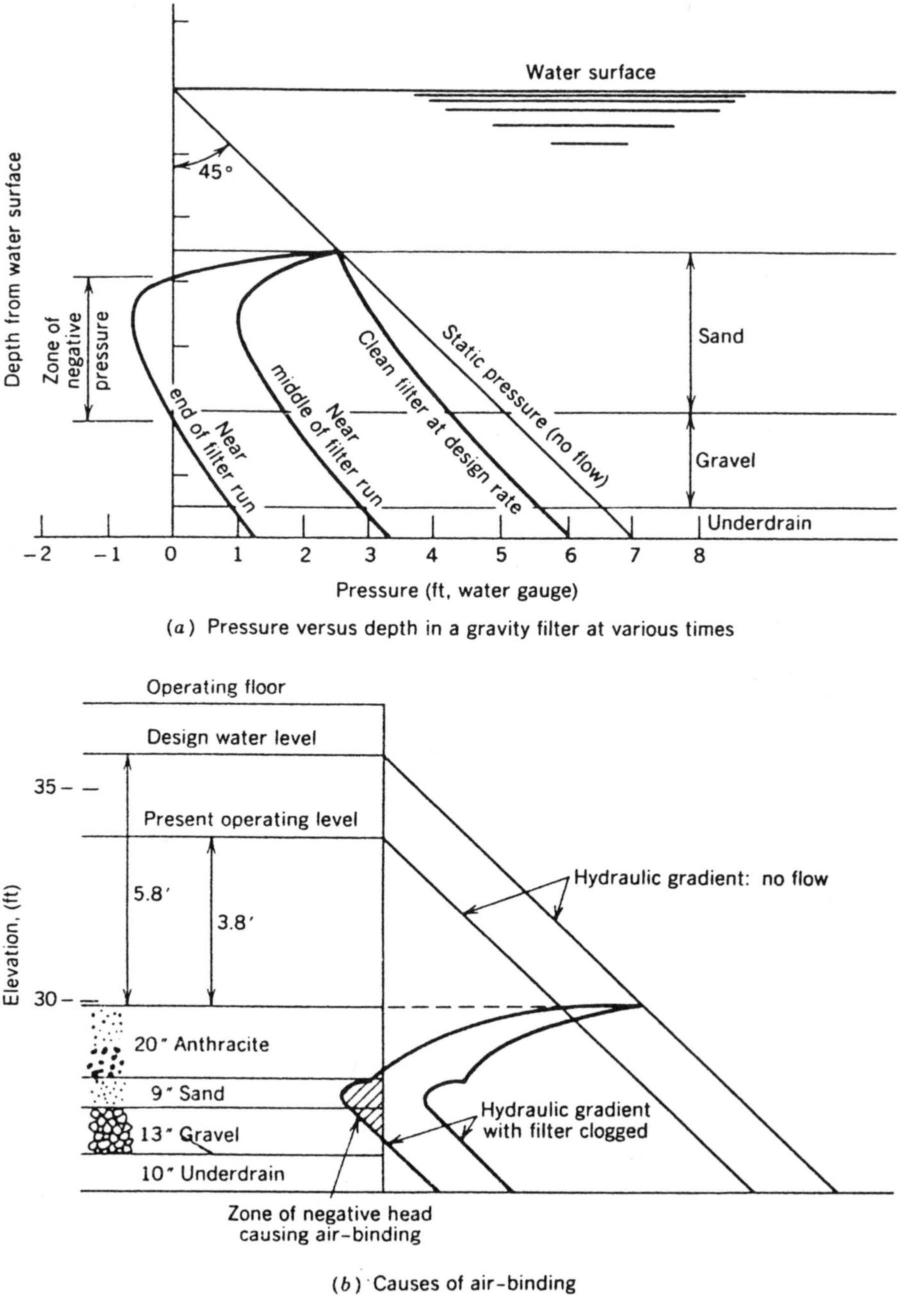

Figure 3.2.7-27 Air binding in the filter bed. [(*b*) From H. E. Hudson, *Water Clarification Process,* Van Nostrand Reinhold, New York, 1981.]

mote air binding. This type of filter is shown in Figure 3.2.7-1 and represents a constant-rate filtration scheme.

When air binding occurs in the filter bed, gaseous air violently escapes when the filter effluent valve is closed. This phenomenon causes media loss. Therefore, the water in the filter should be lowered to a level below the top of the wash troughs, and the filter waste valve should not be opened until the majority of the air has escaped from the filter bed. Moreover, filter washing should be delayed until most of the air has escaped. The degree of air binding may be reduced or even eliminated if filter washing is frequently initiated whenever the headloss reaches 4 to 5 ft (1.2 to 1.5 m). This practice prevents the creation of negative pressure in any part of the filter bed. In many cases, air binding tends to occur during certain seasons, such as spring, when there is a high degree of dissolved air. Air binding may also be a function of the degree of aeration due to hydraulic power plant activity upstream of the plant, suction of air from air vacuum protection valves in the raw water main, or the degree of cascading at the intake.

Air binding will not occur if the filters are designed properly. For example, the declining-rate filters and the rising level self-backwash filters shown in Figure 3.2.7-1 can eliminate air binding due to a high water depth (above the bed) and/or because the effluent water level control weir of the clearwell is situated high enough so that negative pressure is not created during a normal filtering cycle.

Filter Performance Surveillance The four basic tools used in monitoring filter performance are the turbidimeter, pH meter, particle sizing and counting equipment, and filter headloss profile monitor. Each type is discussed in detail. Turbidimeters are used not only to monitor the filter effluent turbidity of individual filters but also to obtain samples from the middle (depth) of the filter bed and the filter wash waste. Monitoring the turbidity of the water at the coal–sand interface, located approximately halfway down the filter bed, is recommended because the turbidity of this water can give the operator advance warning of a turbidity breakthrough across the filter bed. The filter turbidity of the wash waste should be measured because it is a good parameter for adjusting the duration of filter washing.

Nephelometric measurement of turbidity is required because the standard turbidity measurement in the United States is expressed in ntu. The nephelometric technique measures the degree by which the incident light beam is scattered by the suspended particles; the light beam is at a right angle to the particles. Several companies manufacture turbidimeters, but the most popular is produced by the Hach Company (United States). Extreme care must be taken not to introduce any air bubbles to the turbidimeter because this will cause inaccurate readings. Each individual filter effluent line should be monitored by a reliable turbidimeter that is also capable of measuring very low levels of turbidity; the new EPA drinking water standards demand a turbidity of less than 0.3 ntu 95% of the time. If this primary standard is violated, a harsh penalty will be enforced by the EPA.

The pH meter is indirectly related to filter performance because it detects the potential for corrosion and gives early warning of other problems in filter performance. If the pH of the filtered water is above 8, a significant amount of dissolved aluminum may break through the filter bed. The same situation will occur if the pH is below 5. A low pH also gives rise to corrosion problems, and special attention should be directed to the problem of lead corrosion. A high filtered water pH will also reduce the effectiveness of chlorine disinfection. It is standard practice to measure the pH of the combined filtered water for the reasons given above.

A unique tool in evaluating filter performance, in conjunction with the turbidimeter, is particle sizing and counting equipment. Such an analyzer is especially valuable in evaluating the filtered water characteristics that cannot be detected by the turbidimeter. For example, the turbidity readings may be the same for the filtered water of plant A and plant B. Yet the particle size and the number of particles in each of the waters may be quite different. Since the design of the turbidimeter is based on nephelometry—that is, it measures the degree to which the incident light is scattered by the suspended particles—the size, shape, and color of the suspended particles greatly affect the reading. The turbidimeter therefore is incapable of measuring the true content of dark suspended particles such as PAC. However, a good correlation generally exists between the turbidity and the particle count (Figure 3.2.7-29). Based on the Hiac/Royco Model 8000, MET ONE Model 210, HACH 1900, or equivalent units, particle counts of less than 50 mL^{-1} are generally considered excellent. Particle counts (larger than 2 μm) of 50 to 150 mL^{-1} are good, but counts over 200 are unfavorable.

The newest particle sizing and counting equipment sold by Met One, for example, is able to measure in the range 0.5 to 15 μm. This is important because most protozoa (*Giardia* and *Cryptosporidium*) are in this size range, but most filters with granular beds are considered to be somewhat ineffective in removing the particulates in this range. These units are rather expensive, over $20,000 per unit; they are actually best suited for study and research and may be too expensive to be used in each filter as a surveillance tool for daily filter performance. However, it is advantageous to have one unit in the plant process control laboratory or on one selected filter.

The filter headloss profile monitor supplies the plant operator with a headloss profile across the filter bed. The monitor should be arranged so that there are several pressure taps at various depths of the filter bed. These pressure taps should be connected to glass tubes, creating a piezometer board. If high-tech equipment is included in the plant design, the headloss profile may be shown on a cathode ray tube (CRT) in the control room, thus allowing for continuous observation and giving early warning of air-binding conditions. Although this unit is a good performance monitoring tool, it is not necessary for all the filters to be supplied with one. One select filter should be designed to receive this unit if the plant owner expresses interest in its use.

Common Filter Behavior and Control Measures In contrast to all the other processes in an ordinary water treatment process train, the filtration process is generally a batch process. The filtering cycle usually lasts from 1 to 3 days, followed by filter washing. The washing cycle lasts a total of 15 to 30 min, including the drawdown of water level prior to washing and refilling with water. After the washing cycle is completed, the filter cycle is repeated. The special performance behavior of the filter may be attributed to the batch operation process. As illustrated in Figures 3.2.7-28 and 3.2.7-29, the filtered water often exhibits a sharp turbidity breakthrough at the beginning of the filter cycle. It is usually short, 10 to 20 min. After this period the turbidity of the filtered water drops below 0.3 ntu and eventually settles in the vicinity of 0.15 ntu; this level is both the practical and the upper turbidity limit defined by most regulatory agencies. However, the turbidity of the filtered water gradually increases with time and exceeds 0.3 ntu after 24 h. The turbidity breakthrough at both the beginning and late stages of a filtering cycle is characteristic of most constant-rate filters employing alum coagulation. Design engineers and plant operators must be cognizant of this particular filter behavior and should analyze the methods used to control the turbidity breakthrough.

The initial turbidity breakthrough may be controlled by any of four methods: wasting the filtered water, slowly opening the filter effluent valve, adjusting the filter washing practice, or adding polymer to the filter backwash water. The first two methods are related to the design of the filter system. The last two are operational manipulations, except proper design of the polymer feed system to the backwash water. The first method isolates the filtered water from the clearwell for a period of 10 to 20 min, until the turbidity of the water falls below 0.25 ntu. The second slowly opens the filter effluent valve, taking 10 to 20 min to arrive from a fully closed valve position to the appropriate open position, a position that permits the designed filtration rate to occur. The third method, adjusting the filter washing practice, terminates the filter wash cycle when the turbidity of the wash waste is approximately 10 to 15 ntu, thus maintaining the ripened condition of the filter bed. The last method applies approximately 0.3 mg/L of nonionic polymer during the last 3 min of backwash. This last method is more of a trial-and-error type of technique and is not always successful because the polymers are very selective with respect to the characteristics of the suspended matter. In some cases, the addition of 0.2 to 0.3 mg/L of cationic polymer to the filter influent for 30 min, from the beginning of the filtration cycle, can control the breakthrough.

Turbidity breakthrough in the late stages of the filter cycle may be controlled by strengthening the floc and increasing the adsorption capability of the filter bed. This may be achieved by feeding cationic polymer as a coagulant, with or without alum, or by adding minute amounts of nonionic polymer to the filter influent (as a filtration aid); alum must be the sole coagulant. Plant operators must be careful to limit the dosage of the nonionic polymer (filtration aid) to 0.015 to 0.025 mg/L, since overfeeding will result in extremely short filter run lengths. According to Figure 3.2.7-30, the late-stage turbidity breakthrough of the previous filter is eliminated completely when the coagulant is switched from alum to cationic polymer.

Operational Procedures Following Plant Shutdown If the treatment plant is shut down for more than 3 days during any season (for any reason) or if the plant flow rate is only 10 to 20% of the designed rate for a period of 1 month or more, the majority of the filters may be drained or all the filter effluent valves may be closed, allowing the water to stand over the filter bed. However, if the plant is shut

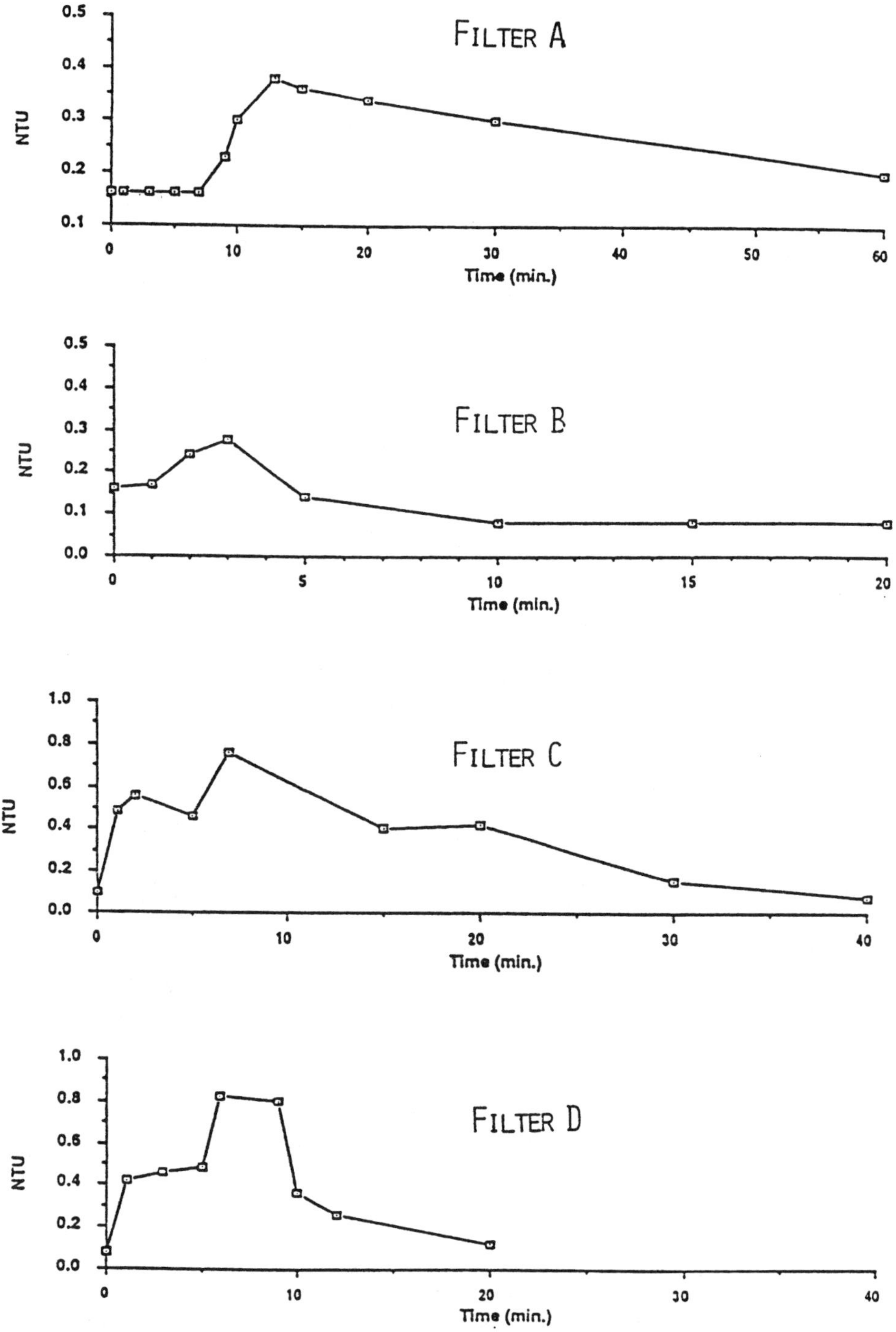

Figure 3.2.7-28 Turbidity breakthrough at beginning of filter run.

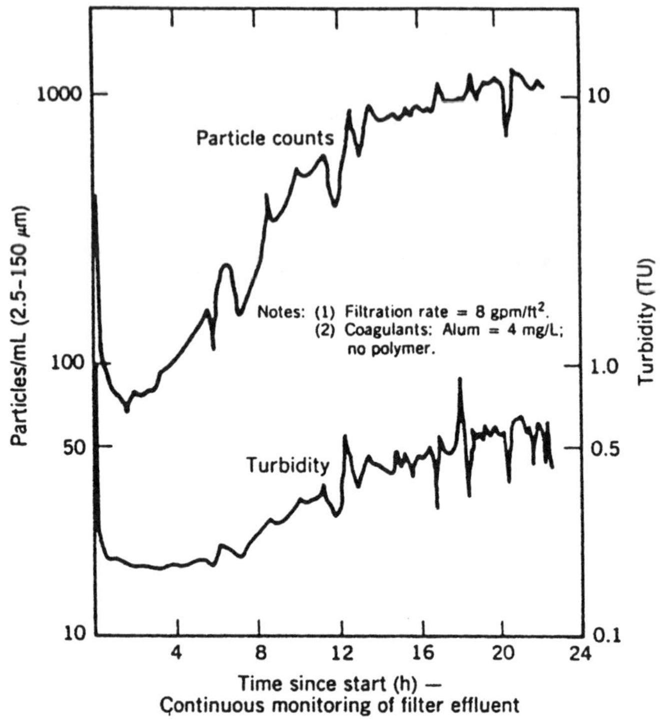

Figure 3.2.7-29 Filter performance evaluation by turbidity and particle counts with continuous monitoring of filter effluent (Adapted from MWD Southern California, Internal Report, 1978.)

down for only 2 or 3 weeks, the filters should not be drained because it will take a considerable period of time to regain the proper (mature) filter bed conditions if the bed has been allowed to dry up. When a plant is treating an extremely low flow rate for long periods of time, a few operators choose to use only one or two filters and the rest are left on standby; these standby filters contain standing water. However, it is better to use all the filters, allowing water to move through all the filters, since this will prevent an explosive growth of microorganisms in the filter bed and will prevent anaerobic conditions from occurring in the filter bed. Anaerobic conditions often lead to objectionable taste and odor problems as well as the reduction of oxidized iron and manganese in the filter bed. In any case, if the filters are left standing for over 3 days without being drained (during warm months), they should be backwashed prior to reintroduction into the filtration cycle. If the filters cannot be backwashed, due to an insufficient amount of water, it is essential to drain the standing water completely and to backwash the filters for approximately 3 min or more after backfilling with backwash water. Since a rate of 2 gpm/ft^2 (5 m/h) is equivalent to 3.2 in./min (8 cm/min), it would take approximately 30 min to drain a filter completely, assuming that the water depth in the filter is 8 ft (2.4 m) and the average rate of the drain is 16 ft/hr (5 m/h).

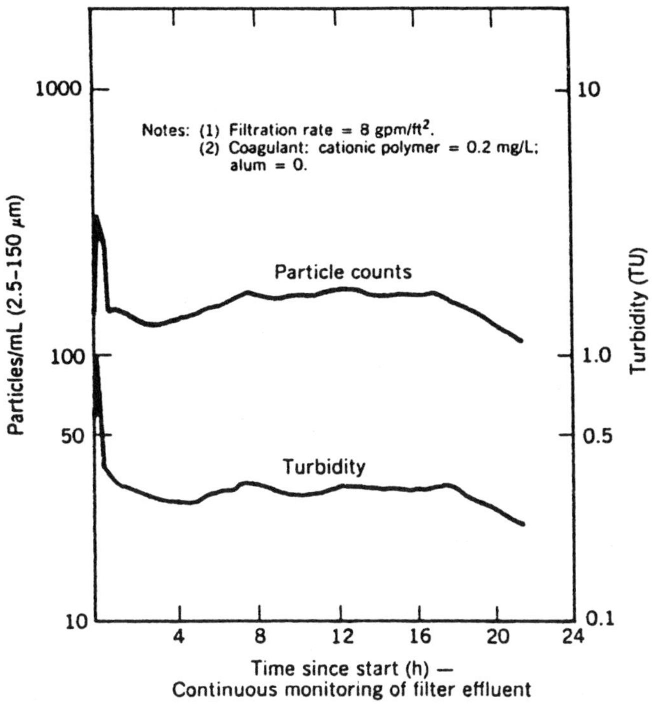

Figure 3.2.7-30 No turbidity breakthrough except at beginning. There is continuous monitoring of filter effluent. (Adapted from MWD Southern California, Internal Report, 1978.)

If the filter underdrain system or the gravel beds are reconstructed or disturbed by workers, the filter-to-waste line should also be employed. For example, if a portion of the filter underdrain is repaired, a filtered water turbidity of almost 30 ntu or more may result from the disturbance of the gravel bed, which is usually quite filthy after several years of service. The design of the filter system should therefore include a filter-to-waste line, even though it may not be used for normal daily operations.

Liquefaction of Filter Bed Due to Strong Earthquakes An interesting and important behavior of granular medium filters is the potential for high-turbidity breakthrough during strong earthquakes (magnitudes above 5) due to scrubbing and liquefaction phenomena. Plant operators should be aware of this fact, and whenever these phenomena are observed, they should (1) stop plant operations, if possible, or (2) route the filter effluent through the filter-to-waste line. The importance of these emergency procedures is illustrated by the case of two modern water treatment plants in southern California, located near the epicenter of a 5.5 magnitude earthquake, which recorded an effluent turbidity as high as 5.0 ntu immediately after the quake.

Before a system is put back into operation, each filter should be lightly backwashed for about 3 min.

BIBLIOGRAPHY

Ahmad, R., et al., "Effect of Backwashing on Biological Filters," *J. AWWA,* 90:12:62 (December 1998).

AWWA, *Water Quality and Treatment,* 3rd ed., McGraw-Hill, New York, 1971.

Barrett, E., et al., "Comparing Plant Scale Dual and Mixed Media Filters, *J. AWWA,* 84:6:76 (June 1992).

Baylis, J. R., "Nature and Effect of Filter Backwashing," *J. AWWA,* 51:11:1433 (November 1959).

Brigano, F. A., et al., "Dual-Stage Filtration Proves Cost Effective," *J. AWWA,* 86:5:57 (May 1994).

Clark, S. L., et al., "Contact Filtration: Particle Size and Ripening," *J. AWWA,* 84:12:61 (December 1992).

Cleasby, J. L., "Declining-Rate Filtration," *J. AWWA,* 73:9:484 (September 1981).

Cleasby, J. L., et al., "Effectiveness of Backwashing of Wastewater Filters," *J. Environ. Eng. Div. ASCE,* 104:4:79 (August 1978).

Degremont, *Water Treatment Handbook* 5th ed,. Halsted Press, New York, 1979.

Forgel, D., et al., "Removing *Giardia* and *Cryptosporidium* by Slow Sand Filtration," *J. AWWA,* 85:11:77 (November 1993).

Fujita, K., "Hydraulics of Filtration Theory," *J. Jpn. WWA,* 445:2 (August 1973).

Ginn, T. M., Jr., et al., "Effects of Particle Detachment in Granular Media Filtration," *J. AWWA,* 84:2:66 (February 1992).

Hancock, C. M., et al., "Assessing Plant Performance Using Microscopic Particulate Analysis," *J. AWWA,* 88:12:24 (December 1996).

Hargesheimer, E. E., et al., "Tracking Filter Performance with Particle Counting," *J. AWWA,* 90:12:32 (December 1998).

Hozalski, R. M., et al., "TOC Removal in Biological Filtraters," *J. AWWA,* 87:12:40 (December 1995).

Hudson, H. E., *Water Clarification Processes,* Van Nostrand Reinhold, New York, 1981.

Ives, K. J., *The Scientific Basis of Filtration,* Noordhoff-Leyden, Groningen, The Netherlands, 1975.

Kawamura, S., "A Study on Filter Washing Efficiency," *Proceedings of the 6th Annual Japan WWA Conference,* May 1955.

Kawamura, S., "A Pilot Study on Filter Bed and Underdrain System," *Proceedings of the 14th Annual Japan WWA Conference,* May 1963.

Kawamura, S., "Design and Operation of High-Rate Filters," *J. AWWA,* 91:12:77 (1999).

Kawamura, S., "Design and Operation of High-Rate Filters," *J. AWWA,* 67:10:535 (October 1975).

Kawamura, S., "Optimization of Basic Water Treatment Processes—Design and Operation: Sedimentation and Filtration," *J. Water SRT-Aqua,* 45:3:130–142 (1996).

Kawamura, S., "Two-Stage Filtration," *J. AWWA,* 77:12:42 (December 1985).

Kawamura, S., et al., "Modifying a Backwash Trough to Cut Media Loss During Filter Washing," *J. AWWA,* 89:12:60 (December 1997).

Lang, J. S., et al., "Investigating Filter Performance as a Function of the Ratio Filter Size to Media Size," *J. AWWA,* 85:10:123 (October 1993).

Maran, D. C., et al., "Particle Behavior in Deep Bed Filtration," Parts 1 and 2, *J. AWWA,* 85:12:69 and 85:12:82 (December 1993).

Mayhugh, J. R., et al., "Filter Media Rehabilitation at a Lime-Softening Plant," *J. AWWA,* 88:8:64 (August 1996).

Monk, R. D. G., "Design Options for Water Filtration," *J. AWWA,* 79:9:93 (September 1987).

Monk, R. D. G., "Improved Methods of Designing Filter Boxes," *J. AWWA,* 76:8:54 (August 1984).

Nieminski, E. C., and Ongerth, J. E., "Removing *Giardia* and *Cryptosporidium* by Conventional Treatment and Direct Filtration," *J. AWWA,* 87:9:96 (September 1995).

Niquette, P., et al., "Shutdown of BAC Filters: Effects on Water Quality," *J. AWWA,* 90:12:32 (December 1998).

Norman, P. A., "Backwash Instability in Rapid Sand Filters," M.S. thesis, University of New South Wales, Australia, 1970.

Pontius, F., "Regulating Filter Backwash Water," *J. AWWA,* 89:8:14 (August 1997).

Suthaker, S., et al., "Optimization of Filter Ripening Sequence," *J. Water SRI-Aqua,* 47:3:107–118 (1998).

Tobiason, J. E., et al., "Pilot Study of the Effect of Ozone and Peroxone on In-Line Direct Filtration," *J. AWWA,* 84:12:72 (December 1992).

Trussell, R. R., and Chang, M., "Review of Flow through Porous Media as Applied to Head Loss in Water Filters," *J. Env. Engrg.,* ASCE, 125:11:998 (Nov. 1999).

Valade, M. T., et al., "Particle Removal by Floatation and Filtration: Pretreatment Effects," *J. AWWA,* 88:12:36 (December 1996).

Visscher, J. T., "Slow Sand Filtration: Design, Operation and Maintenance," *J. AWWA,* 82:6:67 (June 1990).

Wang, J. Z., et al., "Biofiltration Performance," Parts 1 and 2, *J. AWWA,* 87:12:55 and 64 (December 1995).

Weber, W. J., *Physicochemical Processes,* Wiley-Interscience, New York, 1972.

3.2.8 Specific Types of Filters

This section discusses the slow sand filters, proprietary filters (in general), the two-stage filtration process, and a miscellaneous type of proprietary filter—the automatic backwash filter.

Slow Sand Filters Slow sand filters have been used in water treatment since the early nineteenth century. They have been proven to be effective under various conditions as long as the filters are properly designed and applied to appropriate situations. The system is simple, cost-effective, reliable, and easy to operate. Thus, their use should be considered seriously for remote areas or small to midsized water supply systems in developing countries. The filtration rate of slow sand filters is 50 to 100 times slower than that of ordinary rapid sand and high-rate filters. Consequently, a considerably larger area (filter bed) is necessary to produce the same amount of water.

A unique feature of slow sand filters, as described in much of the literature, is the presence of a thin layer of medium on the surface of the filter bed, known as the *schmutzdecke,* loosely translated as "dirty mat." This special layer contains a large variety of microorganisms and enables these filters to remove bacteria by a factor of 10^3 to 10^4 and *E. coli* by a factor of 10^2 to 10^3. The schmutzdecke also removes organic matter and reduces the turbidity of the raw water.

Slow sand filters offer several advantages: (1) they are simple to build and operate and are cost-effective, (2) the filtration system is reliable and produces good-quality drinking water, (3) the system does not require highly trained operators and has minimal power requirements, and (4) the filters can tolerate hydraulic and solid shock loadings provided that they are not excessive. Yet the disadvantages tend to restrict their use: (1) they require a large amount of land, (2) the filters are easily clogged by excessive amounts of algae, and (3) intermittent operation of the filters may degrade the quality of the filter effluent by promoting anaerobic conditions within the filter bed.

The slow sand filter system is a batch process and the filters must therefore be cleaned periodically. Since the majority of the suspended solids (in the raw water) are removed in the top 0.8 to 1.2 in. (2 to 3 cm) of the bed, the filter must be cleaned by scraping a thin layer of sand off its surface. Backwashing is not employed since this will restratify the bed hydraulically. The filter bed may require scraping every 2 to 6 months, depending on the turbidity and the concentration of algae in the raw water. The filter cleaning operation is usually performed on a manual basis and completed in 1 to 2 days. Since manual cleaning of the filters does not require highly trained personnel, this particular feature is best suited for regions with an abundance of inexpensive labor. Industrialized nations should use mechanized systems, such as tractors with special skimming buckets or traveling bridges fitted with scraper mechanisms, to make the cleaning operations more economical.

After the bed is cleaned, a period of up to 1 week is required to ripen the filter bed (build the schmutzdecke) so that an acceptable quality of water is produced. Treatment plants that use slow sand filters and process raw water with seasonally high turbidity spikes and algal blooms must use pretreatment with chemical flocculation and settling, in conjunction with prechlorination.

Acceptable Raw Water Quality for Slow Sand Filtration The acceptable raw water quality for slow sand filters is generally based on the same criteria as those used for the direct filtration process of high-rate filters. The criteria are listed in Table 2.4.5-1. Slow sand filters can tolerate raw water with turbidities greater than 15 ntu as long as the turbidity spikes are less than 50 ntu and last no more than 2 to 3 days. However, if the spikes exceed these criteria or if frequent high algal counts are to be expected, an appropriate pretreatment process train must be implemented.

General Design Criteria The general design criteria for the slow sand filter system are listed below.

Required filter bed area	0.5 to 1.6 ft^2 per capita per day (0.5 to 0.15 m^2 per capita per day)
Number of filters	$n = 0.25Q^{0.5}$, where n is the number of filters and is greater than or equal to 2, and Q is the plant flow rate (m^3/h)
Area of each filter bed	Less than 32,300 ft^2 with an average of 1075 to 2150 ft^2 or less than 3000 m^2 with an average of 100 to 200 m^2
Filtration rate	0.04 to 0.08 gpm/ft^2 (0.1 to 0.2 m/h)

Filter bed depth	3 to 5 ft (1 to 1.5 m)
Support gravel depth	1 to 1.5 ft (four layers) 0.3 to 0.45 m
Water depth above the bed	3.3 to 5 ft (1 to 1.5 m)
Height of the freeboard	≥8 in. or 0.2 m
Height of the filter box	8.2 to 13 ft, average of 10.5 ft; (2.5 to 4 m, average of 3.2 m)
Filter sand	
Effective size	0.15 to 0.35 mm
Uniformity coefficient	Less than 3, preferably near 2
Filter gravel	
Top layer	0.4 to 0.6 mm with a depth of 10 cm (4 in.)
Second layer	1.5 to 2.0 mm with a depth of 10 cm
Third layer	5 to 8 mm with a depth of 10 cm
Bottom layer	15 to 25 mm with a depth of 10 cm
Underdrains	The slow sand filtration system generally has a main manifold and many laterals; the laterals may be porous drain tiles or glazed pipes laid with open joints, but PVC pipes are more often used; 4 m (13 ft) maximum lateral spacing
Filter effluent-level control weir	The effluent weir should be at least 0.2 m above the top of the initial height of the filter bed

Figure 3.2.8-1 is a longitudinal section of a slow sand filter with a filter effluent level control weir and major valves. The filter effluent level control weir may be a down-opening slide gate, for more flexible operation, or a telescoping tube type of valve.

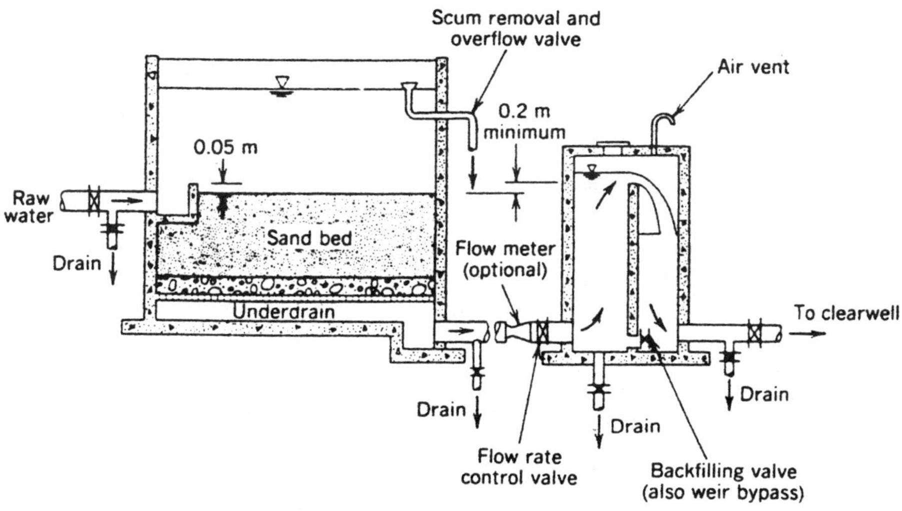

Figure 3.2.8-1 Section view of slow sand filter.

Operation and Maintenance of Slow Sand Filters In this section we discuss the (1) startup, (2) filtration, (3) filter cleaning, and (4) resanding processes.

STARTUP When construction of the filter is complete, the sand in the bed is both clean and dry. In this condition, the filter is incapable of water purification because it must first undergo a maturation process to build up the schmutzdecke. The maturation process has two distinct phases: backfilling and gradual initiation of filtration.

The first step, slow backfilling, drives the air out of the sand bed. When the raw water reaches a level of approximately 3 ft or (1 m) above the surface of the bed, backfilling is terminated and the filter inlet valve is opened, introducing raw water from the inlet instead of the outlet. The backfilling process is extremely important because it prevents the filter bed from being scoured or disturbed by the cascading effect of the raw water onto the filter bed. When the water reaches normal operational levels, the second step is initiated.

The second step in the maturation process is gradual initiation of filtration. This procedure is achieved by partially opening the filter effluent valve; the suggested filtration rate is 10 to 30% of the designed filtration rate. During this step the filtered water must be continuously removed via a drain valve until the quality of the water meets the drinking water quality standards. This step usually takes several weeks in regions of hot weather and considerably longer in colder regions, due to the suboptimal levels of bacterial nutrients and low temperature.

The maturation or ripening period may be shortened by seeding the new filter sand with schmutzdecke from another filter. Filter bed maturation may be detected as a slight, yet distinct, increase in the headloss across the filter bed. Once the bed becomes ripe, the effluent valve should be opened to produce the designed flow rate, and the quality of the effluent water should be rechecked. When the effluent water meets the drinking water quality standards, the drain valve is closed and all the filtered water is directed to the water supply system. The quality of the effluent water must always be checked whenever the filtration process is interrupted for more than 1 or 2 weeks.

FILTRATION PERIOD The main task of the operator is to control the flow rate through each filter by first adjusting the opening of the filter effluent valve and then by making the necessary adjustment at the filter influent valve. At the beginning of each filtration cycle, the filter effluent valve should be at a partially closed position since there is very little headloss across the filter bed. The headloss increases with the accumulation of suspended matter on the surface of the bed—with the creation of the schmutzdecke. It is therefore necessary to provide some type of flow indicator near the filter effluent valve, which allows the operator to gauge the flow rate and adjust the valve opening so that the proper flow rate can be met. The simplest method is to install a gauge staff upstream of the filter effluent-level control weir. It may be a gauge glass or a float in a floatwell, but it must be located outside the box housing the control weir since the operator must be able to check the reading easily while making the flow adjustments. Alternatively, a flow meter may be located immediately upstream of the filter effluent valve. This is an ideal configuration, and all filters should implement this scheme to provide more accurate flow control, with a few exceptions. The exceptions, due to cost considerations, are small plants and developing countries. The filter influent valve should also be adjusted periodically to keep the influent and effluent flow rates in balance. If the operator fails to do so, an exorbitant amount of water will flow through the overflow pipe or the water above the filter bed will be very shallow.

Plant operators must maintain the filters in good working order by routinely monitoring the filter system. For instance, algal blooms in the raw water should be controlled at the source by adding 0.15 mg/L of copper sulfate, or through other means, before the filters become clogged. Operators should routinely clean the filters and their surroundings, in addition to removing any floating scum, leaves, and debris. The operator should also routinely analyze samples of both the raw water and the filtered water at prescheduled intervals.

FILTER CLEANING As mentioned previously, slow sand filters may require cleaning every 2 to 6 months, depending on the algal concentration and turbidity of the raw water. The cleaning process is initiated at the end of the last filter run and is composed of three phases: draining the filter bed, removing the sand, and reripening the bed. The end of a filter run is characterized by a rapid increase in headloss across the bed that is simultaneous with a reduction in the filter effluent flow rate, despite a fully opened valve. There will also be a slight degradation in the quality of the filter effluent water. The filter influent valve is closed at the end of a filter run and the effluent valves are opened to drain the water level; the water level above the bed will drop to the height of the outlet weir overnight. The filter effluent valve should then be closed and both drain valves, just downstream of the filter influent valve and upstream of the filter effluent valve, should be opened to speed the drainage process. The water level must be 6 to 8 in. (15 to 20 cm) below the surface of the bed before the cleaning operation can be initiated. As soon as the top of the bed is sufficiently dry, the second step in the cleaning operation may be started. The scraping process should not be delayed because the guano of scavenging birds contaminates the filter bed to a greater depth.

The scraping operation removes the top 0.8 to 1.2 in. (2 to 3 cm) of the bed by means of manual labor or mechanical equipment. In either case, the schmutzdecke and the surface sand adhering to it should be quickly stripped and carefully stacked into ridges. If the bed is cleaned manually, the waste material should be removed with square blade shovels and the debris transported by wheelbarrow or conveyor belt. If the filter beds are large and the cost of personnel is high, the scraping may be performed by a tractor fitted with a specifically designed scraping blade. Both mechanical equipment and wheelbarrows must be run on protective planks so that the top portion of the filter bed is not damaged and the microorganisms (in the bed) are not killed. Upon completion of the scraping operation, the surface of the filter bed should be leveled.

If the schmutzdecke is composed largely of filamentous algae, the cleaning operation will be easy. However, if the predominant species are nonfilamentous organisms, such as diatoms, the cleaning process will be more difficult. Workers must take precautions to avoid being exposed to harmful microorganisms contained in the schmutzdecke. Similarly, the workers must be careful not to contaminate the filter bed with bodily excretions or spit.

The third step in the cleaning process is reripening of the filter bed. This process is the same as described in the startup process, although the maturation period is greatly accelerated.

RESANDING After approximately 20 to 30 scrapings, the depth of the filter bed is reduced to 20 to 28 in. (0.5 to 0.7 m). This is the minimum depth that should remain above the supporting gravel bed. When the bed is reduced to this level, the filter should be drained and new sand must be added to restore the original medium depth. The resanding operation does not simply add new sand on top of the original bed. First, 12 to 20 in. (0.3 to 0.5 m) of the old bed is moved to one side of the filter. Next, new sand is added on top of the old sand; the depth of this old sand layer should be approximately 8 in. (0.2 m). Finally, the new sand layer is covered by the old sand that was initially set aside. This practice retains most of the valuable microorganisms, thus shortening the reripening period.

In regions where sand is difficult to obtain or is expensive, it is common to recycle the sand that was removed during the cleaning process. This contaminated layer is washed by water jets or by a hydraulic educator system and sorted for future resanding operations. The scrapings must be washed immediately after removal or it will become putrid and produce an objectionable odor; it is quite difficult to remove the coating of debris if the scrapings are allowed to dry. Sludge from the scrapings contains numerous bacteria and organic substances, thus is very odoriferous and attracts many flies. It is therefore prudent to schedule the recycling and resanding operations during the winter season.

Proprietary Filters Many water treatment equipment manufacturers supply various types of filter, mainly for industrial purposes, but a few market filters have been especially designed for water treatment plants. In the United States, the proprietary filters are used primarily in small plants (those processing less than 15 mgd). However, this is not necessarily true for some overseas countries.

The two basic types of proprietary unit used in water treatment are the gravity filter and pressure filter. Gravity filters may be composed of a variety of materials, but pressure filters are cylindrical pressure vessels that are fabricated exclusively from steel plates. The use of pressure filters may be considered if the high water pressure of the plant inlet is to be preserved, this avoids having to repump to obtain the water pressure necessary to distribute the filtered water. Since pressure filters use an equivalent of approximately 15 psi (1 kg/cm^2) of line pressure, provided that the inlet pressure is over 35 psi (2.5 kg/cm^2) of water head, the filter effluent line still has enough high pressure to backwash the filters without the use of wash pumps. Additionally, there is enough line pressure remaining in the system to transmit filtered water to a remote distribution reservoir. If the gravity filters are used under these same conditions, the filter effluent line virtually has no line pressure left. Engineers should therefore evaluate the possible use of pressure filters when designing small to midsized plants under the aforementioned conditions, despite the fact that state regulatory agencies generally discourage the use of pressure filters because of the difficulty in inspecting filter bed conditions.

In the water supply business, one of the best applications for pressure filters is the filtration of pumped, deep-well water containing iron and manganese. This type of filter will remove these compounds through oxidation by chlorine or potassium permanganate, and the well pumps usually supply a high level of extra pressure to the pump discharge line. The proprietary filters that are generally used in public water treatment plants are open-top gravity filters, since these allow easy observation and maintenance of the filter bed. Yet each manufacturer markets a somewhat unique filter: self-backwash, continuous automatic washing, air-scouring wash, bumping bed, coarse media bed for oil removal, ultra-high-rate, and so on. The size and details of the proprietary filter units are generally preengineered by the manufacturer based on a wide range of flow rates; these units are not cataloged based on filter size and appurtenances, but on flow rates. Since the manufacturer has predesigned the blueprints, the completion time of the project is much shorter compared to using custom-designed filters. Additionally, many of these filters are self-contained units that require rather simple power and pipe connections to make them functional. Another advantage of these units is that the manufacturers always give the customers a limited guarantee on both the performance of the units and the equipment. However, the customer usually cannot request significant design modifications without forfeiting the guarantee. This can be a potential problem when the job site is in a remote area; repair and replacement of certain parts may be difficult if they are patented items. Moreover, many proprietary filters are fabricated with metal plates, and corrosion protection is therefore a serious problem. Although proprietary filters may be custom designed, this negates the advantage of using these units: the savings in time and money.

The decision to use proprietary filters versus custom-designed units is the responsibility of either the engineer or the owner. If the engineer is authorized to make the decision, he or she must conduct a thorough evaluation of the units that are on the market. Special attention should be given to the number of years the unit has been in operation, feedback from plant operators using these types of unit, quality control, follow-up servicing, cost of each type of unit, and the financial stability of the manufacturer. The engineer may choose to conduct prequalification of certain manufacturers prior to bidding in order to avoid purchasing an inferior unit, although the unit may meet the specifications. Figure 3.2.8-2 depicts an example of a proprietary pressure filter.

Two-Stage Filtration The two-stage filtration process has historically been applied to the slow sand filtration process to improve the overall efficiency of filtration. Not until the late 1970s did this process gain attention as a substitute for the high-rate filters used in both direct filtration and in-line filtration: these two types of filter are susceptible to turbidity spikes and algal blooms in the raw water. The two-stage slow filtration process has also been used successfully in Europe in the

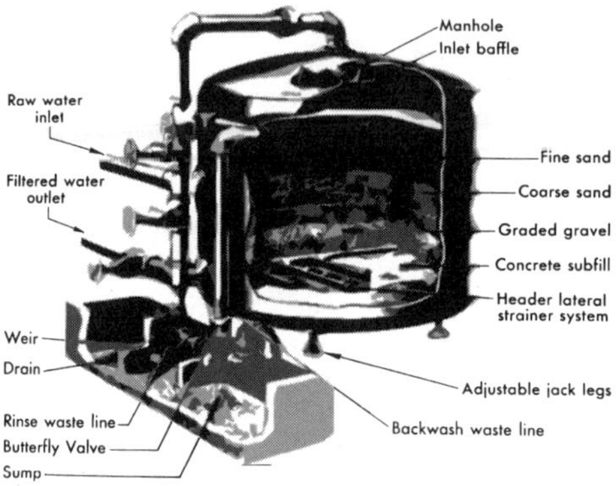

Permutit Type E Vertical Filter
(Permutit Co)

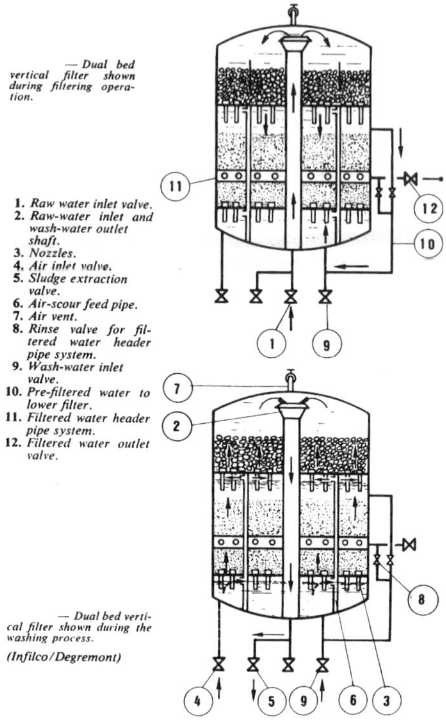

Figure 3.2.8-2 Proprietary pressure filters. (From bulletins of Permutit Company and Infilco-Degremont International.)

treatment of poor-quality raw water, as well as raw water with seasonal degradations in quality.

The two-stage filtration process for high-rate filtration is composed of two distinct filtration steps: first-stage filters and final-stage filters. These stages consist of a bank of coarse bed roughing filters and a bank of ordinary high-rate filters, respectively. The ability of the coarse media bed to act as a flocculation system has been recognized and demonstrated in a number of studies since the early 1970s. The function of the roughing filters is flocculation: to produce floc with the proper characteristics for the subsequent high-rate filters, in addition to removing approximately 50 to 80% of suspended matter. The filters therefore depend on coagulant dosage and the nature of the raw water.

The major advantage of the two-stage process are the substantial savings in coagulant—only 4 to 6 mg/L of alum is required—and the small amount of sludge production. In contrast to the direct and in-line filtration processes, the two-stage process is also capable of withstanding turbidity spikes of short duration and algal blooms. A few equipment manufacturers have recognized these advantages and have commercialized this system with proprietary designs and patents. The one disadvantage of these filters is that wash waste is produced by two sets of filters. Refer to Table 2.4.5-1 for a general idea of the applicable raw water quality for the two-stage process. Figure 3.2.7-20 contains the proper range of coagulant dosage for the two-stage fil-

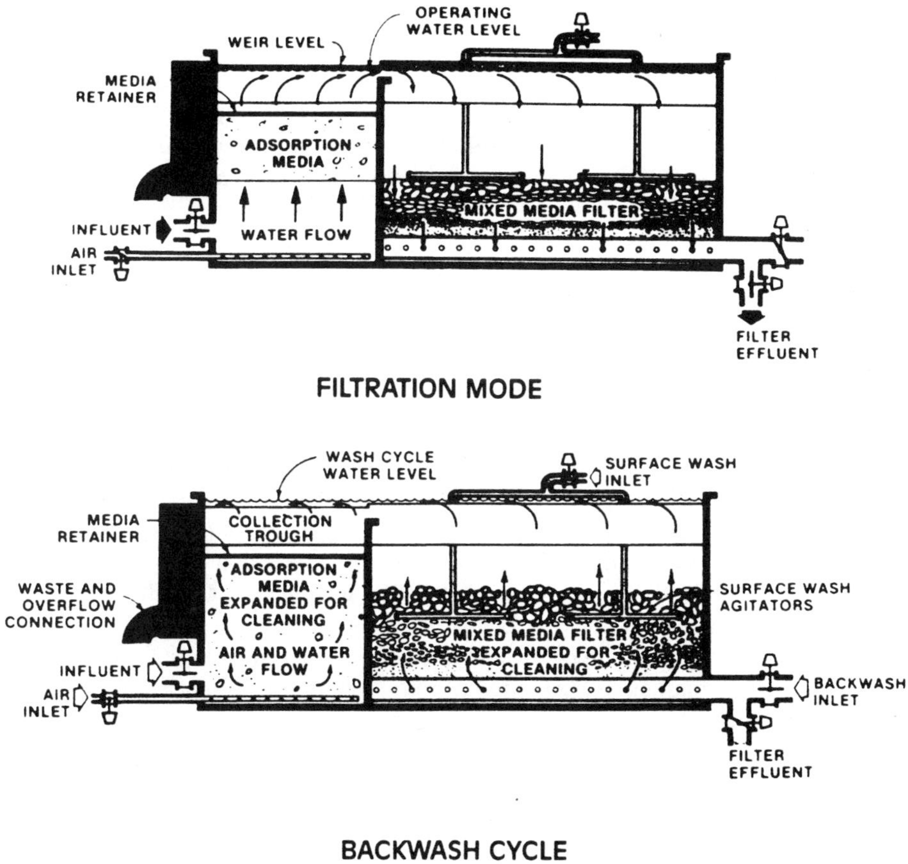

Figure 3.2.8-3 Commercialized two-stage filtration system. (Courtesy of Neptune Microfloc, Inc.)

tration process in relation to direct filtration and a conventional complete treatment process. Figure 3.2.8-3 is an example of the two-stage filtration system.

General Design Criteria

ROUGHING FILTER (FIRST-STAGE FILTERS)

Filtration rate	12 to 15 gpm/ft^2 (30 to 37.5 m/h)
Filter medium	Pea gravel, crushed anthracite, or plastic chips
Size of medium, d	3 to 6 mm, effective size
Uniformity coefficient	$\leq$1.5
Bed depth, l	$500 < l/d > 250$
Filter washing	Air-scouring backwash
Backwash rate	15 to 20 gpm/ft^2 (37.5 to 50 m/h)
Air-scouring rate	3 to 5 ft^3/min per square foot (55 to 90 m/h)

Maximum available headloss for filtration	3.3 ft (1 m)
Underdrains	Air-scouring type
Wash-water troughs	Use of the troughs is preferred
Direction of flow	Downflow is normal; however, upflow and horizontal flow have been applied

FINAL FILTER (SECOND-STAGE FILTERS)

The design criteria for the final filters are exactly the same as those of the high-rate filters (refer to Section 3.2.7).

Miscellaneous Filters Diatomite, biflow, and automatic backwash filters such as the Hardinge filter, ABF (manufactured by Infilco-Degremont and other companies), Dyna Sand (by Parkson Corp. and other firms), and diatomaceous earth precoat filters are the other types of filter that are used with limited applications in the processing of the public drinking water supply. Of these units, the ABF and diatomaceous earth precoat filter are the only types that have considerable acceptance for industrial applications, as well as potable water plants for small public water supplies, when the use of package plants is appropriate.

When the patent on the ABF design expired in the 1960s, a total of four manufacturers began marketing very similar units. This competition helped to generate a much wider acceptance of this type of filter, especially for use in tertiary wash-water filters. Yet certain states discourage the use of ABF for potable water because of its declining-rate mode of filtration and the lack of a provision for filter-to-waste mechanism. However, a recent unit is equipped with filter-to-waste coupled with a turbidimeter and has the option of adding polymer to each cell after backwashing.

The original ABF design has a depth of only 11 in. (0.28 m) of 0.6 mm (effective size) sand. However, a depth of 16 to 24 in. (0.4 to 0.6 m) of dual media is now available. The ABF filter bed is supported by a porous plate underdrain and is divided into numerous filter cells that are approximately 9 in. wide. Each cell is washed individually by a traveling bridge equipped with a backwash pump and a wash-waste pump, thus allowing the rest of the filter cells to remain in service. If the filter bed depth exceeds 16 in., the traveling bridge stops at each filter cell (indexing) to complete filter backwash. The standard width of one automatic backwash filter is 16 ft (5 m), and the recommended maximum length of each filter is 150 ft (45 m).

The basic principle on which the automatic backwash filters operate is a surface filtration that facilitates shallow penetration of floc into the filter bed. The filter therefore operates at a low headloss of 6 to 12 in. (0.15 to 0.3 m) with a filter rate of 2 to 3 gpm/ft^2 (5 to 7.5 m/h). Filter backwashing is initiated by either a headloss or a timer. A typical wash cycle is every 2 to 6 h and the surface of the bed never becomes compacted because of this frequent wash cycle. Consequently, surface wash is not required and air binding does not occur.

The advantages of this filter are that it does not need a backwashing facility, in the regular sense, and the wash waste is very easy to handle due to the low flow rate: only 200 gpm (0.23 m^3/h) for a period of 30 to 90 min. One drawback with respect to potable water production is the need for a housing over the filter bank. The housing is necessary because of the potential for contamination, since the filtered water flows in an open-top channel.

The maintenance requirements of the ABF are very simple: periodic inspection and lubrication of the drive mechanism and pumps, and occasional touch-up painting. Yet if the process water has a medium to high hardness, the porous plate underdrains will become clogged by calcium carbonate deposits over a period of 15 to 20 years, depending on the conditions. Experience has shown that an acidification treatment—application of 5 to 10% hydrochloric acid—restores the capacity of the filters. Figure 3.2.8-4 presents a schematic view of an ABF. Discussion of the other types of miscellaneous filters will not be attempted since the application of these filters to the public water supply is very limited due to their serious disadvantages.

One of the newly emerging filtration techniques for potable water treatment is the membrane technique: microfiltration, ultrafiltration and nanofiltration. The pore size of these membranes is small enough to separate solids from the raw water and capable of filtering out bacteria, viruses, *Giardia,* and *Cyrptosporidum*. It is important to note that humic substances and THMs can be controlled by nanofiltration. These new water treatment systems have been evaluated by pilot-scale plants during the late 1980s, and the results are promising. Membrane filtration is discussed further in Section 7.8.2

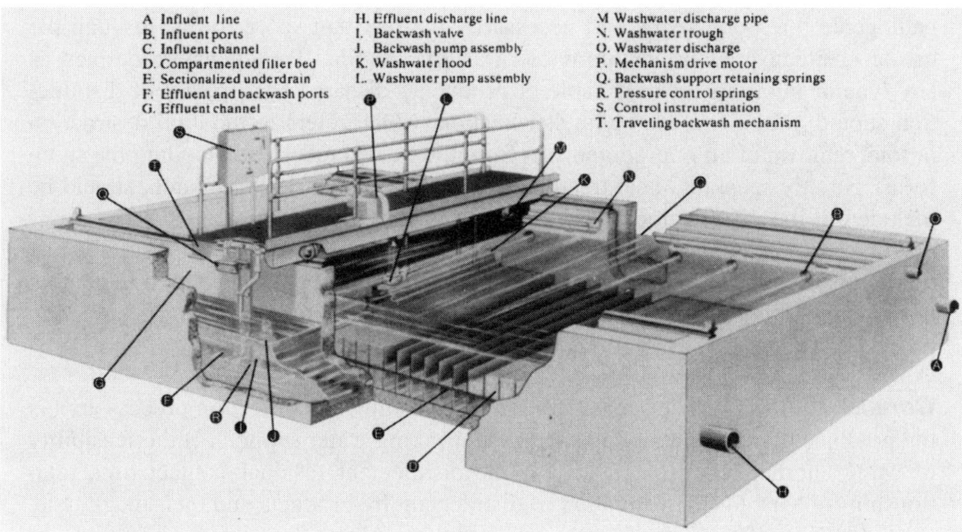

Figure 3.2.8-4 Automatic backwash filter. (Courtesy of Infilco-Degremont.)

BIBLIOGRAPHY

Bersillon, J. L., et al., "Ultrafiltration Applied to Drinking Water Treatment: Case of Small System," *AWWA Annual Conference,* Los Angeles, June 1989.

Burns, D. F., et al., "Particulate Removal on Coated Filter Media," *J. AWWA,* 63:2:121 (February 1970).

Cleasby, J. L., "Slow Sand and Direct In-Line Filtration of a Surface Water," *J. AWWA,* 76:12:44 (December 1986).

Degremont, *Water Treatment Handbook,* 5th ed., Halstead Press, New York, 1979.

Dillingham, J. L., et al., "Optimum Design and Operation of Diatomite Filtration Plants," *J. AWWA,* 58:6:657 (June 1966).

Graham, N. J. D., ed., *Slow Sand Filtration,* Ellis Horwood, New York, 1988.

Huisman, L., et al., *Slow Sand Filtration,* World Health Organization, Geneva, 1974.

Jacangelo, J. G., et al., "Assessing Hollow-Fiber Ultrafiltration for Particulate Removal," *J. AWWA,* 81:11:68 (November 1989).

Kawamura, S., "Two-Stage Filtration," *J. AWWA,* 77:12:42 (December 1985).

Sanks, L. R., ed., *Water Treatment Plant Design for the Practicing Engineer,* Ann Arbor Science, Ann Arbor, MI, 1978.

Schulz, C. R., and Okun, D. A., *Surface Water Treatment for Communities in Developing Countries,* Wiley, New York, 1984.

Vaillant, C. J., "Simple Water Treatment Processes," *IWSA Southern Asia Regional Conference,* Bombay, October 1981.

3.2.9 Disinfection Process

Purpose The purpose of the disinfection process is to kill the growing form of pathogenic microorganisms, not necessarily the resistant spore forms, through the use of chemicals, ozone, or ultraviolet (UV) irradiation. Pathogens are defined as any type of microorganism capable of producing disease. The process of disinfection should not be confused with sterilization, which refers to the total destruction or total removal of all microorganisms from the treated medium, including the spore forms. Strictly speaking, heat treatment, and gamma or x-ray irradiation, should be included in the disinfection process. However, these processes are generally practiced by various industries and are not used in treating the public water supply, primarily because they do not have the residual disinfecting power required to combat postcontamination within the distribution systems. Thus, these types of disinfecting processes, except UV irradiation, are not discussed in this book.

Considerations Major considerations in selecting a disinfection process are (1) the presence of surrogate organisms in the drinking water supply, (2) the feasibility of using alternative disinfectants, (3) the disinfectant residual–contact time relationship ($C \times t$), (4) the formation of disinfectant by-products and their magnitude, (5) the quality of the process water, (6) safety problems associated with the disinfectants, and (7) the cost of each disinfection alternative. Each of these considerations is discussed in turn.

Surrogate Organisms Coliform organisms, particularly *Escherichia coli* (*E. coli*), are continuously present in the human intestine in large numbers. Billions of these organisms are excreted by an average human being or animal each day. Thus their presence in the process water is indicative of fecal pollution. The coliform group of bacteria is defined as all aerobic and facultative anaerobic, gram-negative, non-spore-forming bacilli that produce acid and gas through the fermentation of lactose.

The most common method of assessing the safety of the drinking water supply is the coliform test. However, microbiologists generally agree that coliforms may be more rapidly inactivated at much lower disinfectant dosages than some enteric viruses and protozoan cysts. Consequently, the EPA has proposed that the standard plate count be instituted as a routine test to screen for bacteria regrowth. The presence of a large population of organisms, as indicated by the plate count (heterotrophic bacteria), would certainly be indicative of an ineffective disinfection process. The heterotrophic plate count (HPC) for water in distribution systems should be less than 500 colonies/mL. The EPA is also emphasizing the inactivation of protozoan cysts such as *Entamoeba histolytica* and *Giardia lamblia,* as well as Cryptosporidium, Legionella, *and enteric viruses.*

Alternative Disinfectants Chlorine, chloramines, chlorine dioxide, and ozone are practical alternative disinfectants. The selection process is based on the effectiveness of the alternatives as a disinfectant, formation of harmful by-products, residual disinfecting capabilities, and cost-effectiveness.

CHLORINE Chlorination of potable water has been practiced in the United States since 1903; therefore, much knowledge is available concerning the application technology and its effectiveness. Despite growing concerns regarding THM, a suspected carcinogen that forms during the chlorination process, chlorination will undoubtedly continue to be the primary water disinfectant for years to come, especially in developing countries. The basis for this reasoning is that all potable water should provide disinfectant residual in the distribution system, and chlorine can easily satisfy this requirement. Additionally, chlorine has been proven to be effective as a disinfectant at reasonable cost for almost a century.

Chlorine is generally applied to the process water as an aqueous chlorine solution: Chlorine gas is hydrolyzed to form hypochlorous acid (HOCl) and hypochlorite ion (OCl^{-}) in ammonia-free water. Hypochlorous acid in water is commonly known as *free residual chlorine*. The relative concentration of these two species is a function of the pH of the water and, to some degree, the temperature (Figure 3.2.9-1). The source of the chlorine is usually liquefied chlorine gas. However, regions that prohibit the use of chorine (for safety reasons) and small plants commonly use calcium hypochlorite, sodium hypochlorite, or chlorinated lime as sources of chlorine. Free residual chlorine reacts rapidly and indiscriminately with many substances in the water, including microorganisms. Thus, effective disinfection and minimum THM and haloacetic acid (HAA) formation may be achieved if chlorine is added to clarified water with low TOC such as filtered water.

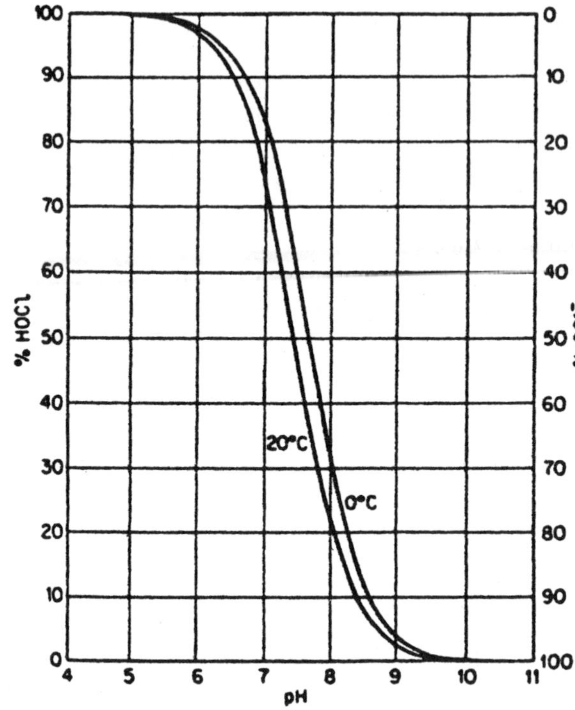

Figure 3.2.9-1 Effect of pH on the distribution of hypochlorous acid and hypochlorite ion in water.

Many factors affect the disinfection efficiency: oxidizable substances in the process water, particulate concentration, pH, temperature, contact time, and the level of the residual chlorine. The ideal disinfecting condition has a low pH (6 to 7), a relatively high water temperature (20 to 25°C), a longer contact time (over 30 min), and higher levels of residual chlorine (over 0.5 mg/L). According to Morris (1975), hypochlorous acid is 100 times more effective than hypochlorite ion as a germicide for enteric bacteria and amoebic cysts. Yet hypochlorous acid is less effective in combating viruses, as well as amoebic cysts and bacterial spores, compared with other types of disinfectant, such as ozone or chlorine dioxide. Although hypochlorite ion is much less effective than hypochlorous acid compared with monochloramine, its germicidal power is greater by a factor of 2.

Chlorination is a proven, reliable, and mature technology. However, THM formation and public safety are major concerns for both the engineer and the general public. Studies have shown that THM formation is reduced substantially when the application point of chlorine is changed from the head of the plant to postfiltration. The potential leakage of chlorine gas has been minimized with improvements in the chlorine feeders and chlorine storage. Improvements in both technology and operating practices will provide a more reliable and safe disinfecting process.

CHLORAMINES Chloramination of drinking water has been used successfully in many water utilities and has proven to be an effective disinfecting process. However, chloramines are less effective than chlorination as bactericides and viricides. Moreover, neither the enteric viruses nor protozoan cysts are effectively inactivated in short periods of time (less than 30 min) at low dosages (1 to 2 mg/L). Yet chloramination does have the advantage of producing very small amounts of THM—levels that are insignificant to human health. During the late 1970s, the EPA proposed that chloramination not be used as the primary disinfectant, due to its suboptimal performance as a bactericidal and viricidal agent, but this proposal was later modified, leaving implementation to the discretion of each state. Thus, the chloramination process may be used as a predisinfectant and/or a postdisinfectant prior to distribution.

One of the most sensible uses of chloramination is in combination with other disinfectants, particularly in the case of postdisinfection, since the chloramine residuals last much longer than free chlorine and do not form significant amounts of THM. Chloramination is normally practiced at a ratio of approximately 1 part of ammonia to 4 parts of chlorine (using a mg/L base) to ensure monochloramine formation. Thus, if a postchlorinated water has 1.0 mg/L of free residual chlorine, 0.25 mg/L of ammonia is to be added. However, if 0.3 mg/L of ammonia is added prior to postchlorination, a minimum of 1.2 mg/L of chlorine must be added.

CHLORINE DIOXIDE Chlorine dioxide (ClO_2) has been used as a primary disinfectant by several large systems in Europe. However, it has had limited application as a water disinfectant in the United States; it has been used primarily as a means of taste and odor control, oxidation of iron and manganese, and control of TTHMs and HAAs. The basic reaction for the production of chlorine dioxide is

$$\underset{\text{sodium chlorite}}{2NaClO_2} + \underset{\text{chlorine}}{Cl_2} \rightarrow \underset{\text{chlorine dioxide}}{2ClO_2} + 2NaCl$$

Therefore, 1.34 lb of sodium chlorite reacts with 0.5 lb of chlorine to produce 1.0 lb of chlorine dioxide. However, excess chlorine oxidizes chlorine dioxide to chlorate ion:

$$HOCl + ClO_2 \rightarrow ClO_3^- + HCl$$

Chlorine dioxide in water produces chlorate ion and chlorite ion with a slow reaction:

$$2ClO_2 + H_2O \rightarrow \underset{\text{chlorate}}{ClO_3^-} + \underset{\text{chlorite}}{ClO_2^-} + 2H^+$$

Chlorine dioxide has a high oxidation power. It is an effective bactericidal agent and its killing power of microorganisms is equal to or better than hypochlorous acid in water. For example, compared to free chlorine, chlorine dioxide is more effective as a bactericide and viricide at a higher pH range (8.5 to 9) than at a pH of 7. One definite advantage of chlorine dioxide is that it requires the pH of the filtered water to be above 8, thereby minimizing lead corrosion and the corrosion of other metals, in accordance with the recent EPA guideline. Also in this category is the disinfection of softened water by the lime softening process. Another advantage of chlorine dioxide is its ability to maintain a residual in the distribution system for an extended period of time.

The disadvantages of chlorine dioxide are its higher operational costs ($2.2 per pound for sodium chlorite) relative to chlorine ($0.13 per pound for chlorine)—late 1990s cost. Additionally, excess chlorine added to the water will form THMs and undesirable by-products whenever chlorine is used in excessive amounts to ensure complete reaction with sodium chlorite. This practice was commonly employed in the United States to produce chlorine dioxide.

Since the EPA has set the criteria for chlorine dioxide residual to be less than 0.8 mg/L and a MCL for chlorite at 1.0 mg/L, whenever chlorine dioxide is used as a disinfectant or in taste and odor control, the dosage of chlorine dioxide and optimization of the generator performance are extremely important. However, ClO_2 and ClO_2^- can be controlled by ferrous iron additon. Chlorine dioxide has potentially explosive characteristics with a slight change in environment. Therefore, it should not be stored in pressurized containers but should be produced on site. Preengineered and packaged chlorine dioxide–generating equipment is offered by a few equipment manufacturers.

OZONE Among the commonly available oxidants used in water treatment, ozone is the most powerful oxidant and disinfectant. Ozone has been used as a disinfectant worldwide in over 1100 plants, including very small installations, primarily in Europe. In practice, ozone is often used in conjunction with a secondary disinfectant such as chlorine or chloramine, since it is not capable of providing a residual disinfectant to combat the proliferation of microorganisms within the distribution system. Although ozone is used as the sole disinfectant in a few European countries and other parts of the world, the use of ozone alone is not allowed as the final disinfection before distribution in the United States, precisely because of the lack of protection against postcontamination.

The effectiveness of ozone disinfection is a function of the pH and temperature of the process water and the method of ozone application. The required residual—contact time relation ($C \times t$) of ozone disinfection is approximately 1.5% of free chlorine residual. Yet the type of mixing, the degree of mixing, and the rate of mass transfer greatly influence the effectiveness of ozonation because ozone has a low solubility in water.

The advantages of using ozone as a disinfectant are its strong disinfecting power, potentially better flocculation and filter performance, a significant reduction in the formation of THM precursors prior to final chlorination, a limited number of by-products (aldehydes are a by-product that may have adverse effects on human health), and the low dosages (1 to 2 mg/L) required to complete disinfection. However, ozone also has the following disadvantages: the process does not provide residual protection after ozonation due to the rapid decomposition of ozone; the ozone must be produced on-site because it cannot be stored; and the ozonation system has a high capital cost ($3000 per pound of ozone production per day for most medium-sized plants) and frequent high levels of bacteria regrowth in the distribution line if adequate residual chlorine is not used. If cost is not a factor, the effectiveness of ozone disinfection may be enhanced by the use of a catalyst such as UV light or hydrogen peroxide.

ULTRAVIOLET LIGHT UV light is in the region of the electromagnetic spectrum between visible light, which has wavelengths greater than 400 nm, and x-rays (less than 100 nm). UV has three spectral bands: (1) 100 to 280 nm, which is short-wave UV; (2) 280 to 315 nm, middle-wave UV; and (3) 315 to 400 nm, long-wave UV. Disinfection by UV light is performed by short-wave UV, which is capable of rupturing the deoxyribonucleic acid (DNA) of cells. This disruption of DNA prevents cell reproduction, thereby inactivating pathogenic microorganisms.

Regular low-pressure UV lamp systems should use 40- to 60-UV lamps per million gallons per day (3790 m^3/d). The power requirement for each lamp ranges from 65 to 80 W. The intensity of the UV light emitted by a lamp is expressed as milliwatts per square centimeter (mW/cm^2). UV dose is calculated by intensity times the exposure time in seconds and is expressed as mW · s/cm^2. The dose required to disinfect water varies with the quality of the water, but generally ranges from 24 to 45 mW · s/cm^2. The lamps operate at temperatures ranging from 40 to 50°C and produce nearly monochromatic radiation at 254 nm.

Medium-pressure UV lamp systems have been employed in Europe since 1980 and is gaining popularity in the United States because this system requires fewer lamps than the low-pressure UV lamp system. These medium-pressure lamps operate in a temperature range of 600 to 800°C and produce a high-intensity broad-spectrum light. Suppliers offer a choice of three transformer power level settings in the range 60 to 100%, thereby allowing plant operators to operate the lamps at a lower setting (save energy) during periods of low plant flow.

The pulsed UV system is an emerging UV disinfection system that delivers high-intensity microsecond bursts of broadband emission with a power flux of 10 million mW/cm^2. Pulsed UV flash lamps contain xenon, which is intermittently energized to form plasma. The temperature in the lamp rises to approximately 10,000 K, which can be compared to sunlight. At this temperature the plasma emits an intense, broadband spectrum (UV, visible, and infrared wavelengths). However, the outer surface of the lamp is reduced to near-ambient temperature by an internal cooling system, to minimize problems associated with scaling. A recent study conducted by the University of South Florida and the Electric Power Research Institute demonstrated the effectiveness of the pulsed UV disinfection system:

	Log Inactivation		
Total Dose (mW · s/cm^2)	Bacteria	Virus	*Cryptosporidium*
30	6	2	4.3
50	>9	3	4.8
100	>9	5.6	5.6

Some manufacturers of low-pressure UV systems are Atlantic UV Corp. and Calgon Carbon Corp. Low-pressure, high-intensity lamps are manufactured by

UltraGuard in Canada and WEDECO in England. Trojan Technologies of Canada and Aquionics of Erlanger, Kentucky produce medium-pressure UV systems. Pulsed UV systems are supplied by Pure-Pulse of San Diego, California and Inovatech of El Cajon, California.

Disinfectant Residual–Contact Time Relationship Two factors are of primary importance in disinfection: the concentration of the disinfectant residual and the contact time. The relationship may be expressed as

$$C \times t = \text{constant}$$

where C is the disinfectant residual (mg/L) and t is the contact time (min). The important issue is that a low concentration of disinfectant, with a long contact time, accomplishes the same goal as using a high residual concentration with a short contact time. However, other factors, such as the pH and temperature of the process water, also affect the efficiency of the disinfection process.

According to the Surface Water Treatment Rule (final guidance manual issued by the EPA in 1989) all water treatment systems must comply with the $C \times t$ requirements. Yet compliance is a function of the type of treatment used. The EPA recommends selecting the disinfection by inactivation process based on the type of filtration system that is used by the filtration plant. Table 3.2.9-1 presents removal/inactivation as a function of the type of filtration. Table 3.2.9-2 illustrates 1 log (90%) and 0.5 log (68%) inactivation of *Giardia* at various temperatures and pH levels. Table 3.2.9-3 presents 2 to 4 log virus inactivation.

Let us assume that a treatment plant has a conventional process, the winter water temperature is 0.5°C (33°F), and the pH of the process water is 8.0. Based on Table 3.2.9-1, 2.5 log for *Giardia* and 2.0 log for viruses are credited for the treatment processes; therefore, inactivation (disinfection) will require 0.5 log for *Giardia* and 2.0 log for viruses. From Table 3.2.9-2*b* the required minimum $C \times t$ (mg/L · min) for *Giardia* by free chlorine disinfection is 51 and 0.48 for ozone. According to Table 3.2.9-3, the $C \times t$ requirement for 2 log inactivation of viruses by free chlorine is 6 and 0.9 by ozonation. However, if chloramines are used for the final disinfection, the $C \times t$ must be 690 for *Giardia* and 1243 for viruses. The maximum level of residual allowed by the EPA is 4 mg/L for both free chlorine and chloramines.

Disinfection By-Products All disinfectants react with any element that is oxidizable in water. Thus the water supply industry is now faced with the problem of controlling disinfection by-products. THMs and other halogenated organics are produced by chlorination. Aldehydes are formed by ozonation; other compounds, such as haloacids, haloacetonitriles, haloketones, and bromate, may also be formed, depending on the type of elements present in the process water. Although the magnitude of the ozonation by-products is generally below the level of health concerns, it may be only a matter of time before science discovers which ozonation by-products significantly affect human health. The EPA has currently set the maximum allowable level of THM as 0.08 mg/L. However, this level will probably be reduced to 0.04 mg/L in the near future.

TABLE 3.2.9-1 Recommended Minimum Log Removal/Inactivation by Disinfection and Filtration Schemes

	Assumed Log Removal by Filtration		Recommended Minimum Log Inactivation by Disinfection	
Filtration Type	*Giardia*	Viruses	*Giardia*	Viruses
Conventional	2.5	2.0	0.5	2.0
Direct	2.0	1.0	1.0	3.0
Slow sand	2.0	2.0	1.0	2.0
Diatomaceous earth	2.0	1.0	1.0	3.0

Source: Adapted from 54 FR 27511 (June 29, 1989).

Quality of the Water Being Processed If the water being processed contains turbidity, color, ferrous iron, manganese, and organic substances, all disinfectants described in this section will react with them. Consequently, the effectiveness of disinfection becomes very low and a significant amount of disinfection by-products will probably be formed. Turbidity and the presence of other particulate matter in

TABLE 3.2.9-2 $C \times t$ Values to Achieve *Giardia lamblia* Inactivation

		Temperature (°C)					
	pH	0.5	5	10	15	20	25
		(a) 1 log (90%) Inactivation					
Free chlorine[a]	6	49	35	26	18	13	9
	7	70	50	37	35	19	12
	8	101	72	54	36	27	18
	9	146	104	78	52	39	26
Ozone	6–9	0.97	0.63	0.48	0.32	0.24	0.16
Chlorine dioxide	6–9	21	8.7	7.7	6.3	5	3.7
Chloramines (preformed)	6–9	1270	735	615	500	370	250
		(b) 0.5 log (68%) Inactivation					
Free chlorine[a]	6	25	18	13	9	7	4
	7	35	25	19	13	9	6
	8	51	36	27	18	14	9
	9	73	52	39	26	20	13
Ozone	6–9	0.48	0.32	0.23	0.16	0.12	0.08
Chlorine dioxide	6–9	10	7	5	3.3	3	1.7
Chloramines (preformed)	6–9	690	363	337	250	181	130

Source: Adapted from EPA's *Guidance Manual* (March 31, 1989).

[a] $C \times t$ values will vary depending on the free chlorine concentration. Values indicated are for 1.0 mg/L free chlorine. (For other free chlorine concentrations, see EPS's *Guidance Manual*.)

TABLE 3.2.9-3 $C \cdot t$ Values to Achieve Virus Inactivation

	Log Inactivation at pH 6–9	Temperature (°C)					
		0.5	5	10	15	20	25
Free chlorine	2.0	6	4	3	2	1	1
	3.0	9	6	4	3	3	1
	4.0	12	8	6	4	3	2
Ozone	2.0	0.9	0.6	0.5	0.3	0.25	0.15
	3.0	1.4	0.9	0.8	0.5	0.4	0.25
	4.0	1.8	1.2	1.0	0.6	0.5	0.3
Chlorine dioxide	2.0	8.4	5.6	4.2	2.8	2.1	1.4
	3.0	25.6	17.1	12.8	8.6	6.4	4.3
	4.0	50.3	33.5	25.1	16.8	12.6	8.4
Chloramines (preformed)	2.0	1243	837	643	428	321	214
	3.0	2063	1423	1067	712	534	356
	4.0	2883	1988	1491	994	746	497

Source: Adapted from EPA's *Guidance Manual* (March 31, 1989).

the process water are very significant with respect to effective disinfection because the microorganisms living within these materials may not be killed. Thus, water entering the disinfection process must have both high quality and low turbidity. The water should therefore be filtered prior to final disinfection so that the disinfecting agents can work effectively without producing significant amounts of by-products.

Safety of Operation All disinfectants are harmful to human health and animal life. The design and handling of the disinfectant system should therefore adhere strictly to the local, state, and federal (OSHA) safety rules and codes, in addition to those listed in the chemical handling manuals provided by the chemical industries. If a chlorination system is located in a residential area, a neutralization system (for chlorine gas) should seriously be considered, to combat chlorine gas leakage. The decision to integrate a neutralization system into the overall plant design should be weighed for the sake of public safety, even though the governmental codes do not specify their use. Certain municipalities also prohibit the transport of chlorine gas to specific areas of the city because of safety considerations. Under these circumstances, the choice of disinfectant is limited to hypochlorites of calcium or sodium. However, hypochlorites may be generated electrolytically from brine or salt solutions using specially designed cells and an ordinary power source. The locations of the ozone manufacturing facilities and the ozone contact basins are also dictated by safety considerations. These are generally isolated from the operations building because of the potential for fire and health hazards. No matter the type of disinfectant, plant operators should plan and frequently practice safety maneuver drills.

Cost Cost is a prime consideration in selecting the type of disinfection process. The approximate construction costs for the chlorination, chloramination, chlorine dioxide, and ozone disinfecting systems, as established by the EPA and slightly modified, are listed in Table 3.2.9-4. The cost does not include the construction of supporting systems, nor does it include the construction of the contact tank. As Table 3.2.9-4 clearly indicates, construction of the ozonation system is by far the most expensive of the four alternatives. Refer to Table 3.2.9-5 for the total cost of disinfection, including the operation and maintenance of the feed system for each of the four alternatives. These costs were developed by the EPA.

Design Criteria The design criteria for the chlorination, chloramination, and chlorine dioxide systems are discussed below. The fourth alternative disinfection system, ozonation, is discussed in Section 4.8. Therefore, no attempt is made to describe it here.

Chlorination System (*Note:* 1 kg = 2.2 lb.)

Dosage	1 to 5 mg/L (2.5 mg/L average).
Number of chlorine feeders	Minimum of two, one standby is required. Standard-sized feeders are 10, 20, 30, 50, 75, 100, 150, 250, 500, 1000, 1500, 2000, 3000, 4000, 5000, 6000, 8000, and 10,000 lb/day and in the 20 : 1 feed range.
Evaporators	These are normally required if the daily dosage exceeds 1200 lb/day. A safe yield of chlorine gas from one 1-ton cylinder is 400 lb/day without an evaporator under normal temperature.
Gas cylinder weighing scale	Minimum of one, as required by regulatory agencies.
Chlorine expansion tank	Required for liquid chlorine line.
Chlorine gas strainer	Required.
Chlorine gas pressure regulation valve	Required.
Chlorine gas pressure gauges	Required.
Chlorine eductors	Required.

TABLE 3.2.9-4 Cost of Constructing Disinfection Systems ($)

Plant Size (mgd)	Chlorine (2 mg/L)	Chloramine (2 mg/L)	Ozone (1 mg/L)	Chlorine Dioxide (1 mg/L)
1	15,000	20,000	55,000	25,000
10	32,000	40,000	150,000	50,000
100	130,000	145,000	800,000	170,000
150	150,000	170,000	1,200,000	200,000

Source: Data from EPA Publication 600/2-79-162b, August 1979. The costs are based on ENR Index 2850.

TABLE 3.2.9-5 Costs of Disinfection (¢/1000 gal)

Plant Size (mgd)	Chlorine (2 mg/L)	Chloramines (2 mg/L)	Ozone (1 mg/L)		Chlorine Dioxide (1 mg/L)
			Air Feed	Oxygen Feed	
1	3.8	7	6	7.8	8.5
10	1.7	1.9	2.2	2.3	2.2
100	0.48	0.9	1.4	1.36	1.7
150	0.45	0.8	1.2	1.15	1.5

Source: Data taken from EPA Publication 600/2-81-56, 1981. The costs do not include the contact tanks.

Water supply to the eductor	Dual supply of pressurized water or a standby booster pump is required. Additionally, a device to prevent backflow should be provided.
Chlorine solution rotometers	Optional—as needed.
Chlorine gas leak detectors	Required.
Chlorine residual analyzers	Required.
Chlorine containers	One-ton cylinders are preferred whenever the plant uses over 200 lb/day of chlorine. A 1-ton cylinder weighs 1.9 tons, including the weight of the cylinder. A 20-ton stationary tank or a 17-ton tank trailer or a 90-ton tank car system is used for large plants.
Housing	Chlorine storage and feeder rooms must not be located underground. The cylinders and feeders should be located in separate rooms. The inventory should include enough chlorine cylinders for a minimum of 15 days of operation.
Ventilation	Each room must have an air ventilation system that provides one complete air change per minute. The air outlets should be located near the floor and exhausted in a safe direction. The air inlet should be through gravity louvers located near the ceiling.
Inspection window	The door and wall of both the chlorine storage and feeder rooms should have clear glass windows to permit viewing without having to enter the rooms.
Access door	All access doors must be outside the building. If access is required from within the main building, the door must have an airtight seal.
Heating	The chlorination building must have a heating system that will maintain an ambient tempera-

	ture of 15°C (60°F). This feature is vital if a system does not have a chlorine evaporator.
Safety equipment	Eye wash, shower, and gas masks of the canister type must be provided in compliance with OSHA regulations.
Chlorine neutralization system	Optional (Ceilcote Co., Inc., of Ohio, Ametec, Inc., of Ontario, Canada, and EST Corp., of Quakertown, Pennsylvania manufacture the scrubber). See Appendix 13.
Chlorine cylinders	All cylinders must be anchored by chains for safety reasons. This measure must be implemented in plants located in earthquake zones.
Application points	Chlorine application points must be provided at the head of the plant, the filter influent, and the filtered water, even though chlorine may not be applied on a continuous basis.

Figure 3.2.9-2 is a schematic of a chlorine feed system.

The measured chlorine gas released from the feed system is generally dissolved in water by means of a hydraulic eductor, then transported to the application point as chlorine solution (approximately 3000 mg/L). However, in cases where the distance between the feeder and feed point is over 150 ft (45 m), the eductor is placed at the injection point. Under these circumstances, the chlorine gas is pulled through the PVC feed pipe under a slightly negative pressure for reasons of safety (in case the pipe is cracked). This scheme becomes problematic in regions experiencing cold

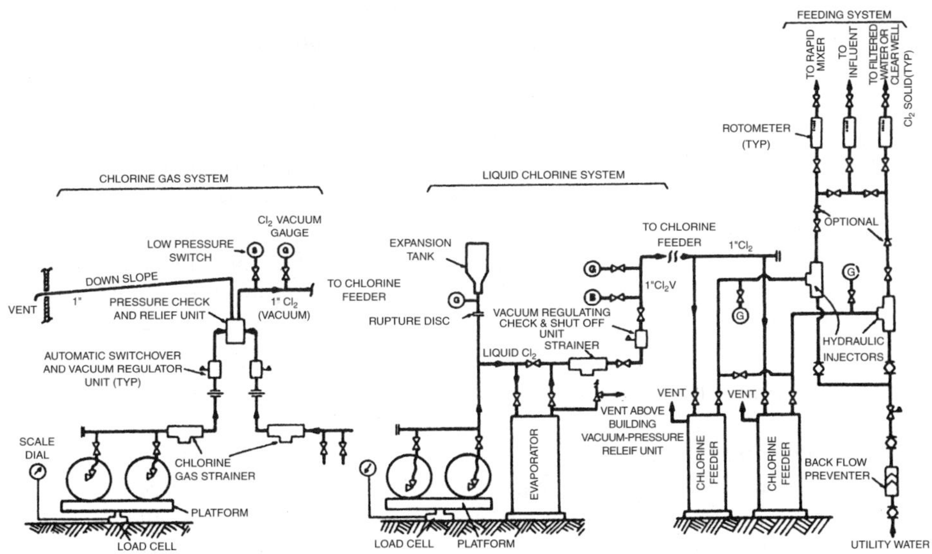

Figure 3.2.9-2 Chlorine feed system.

winter weather because the cold temperature around the feed pipe causes the warm gas to condense, forming a very strong chlorine solution. Consequently, the gas flow is interrupted. Moreover, the potential for a severe chlorine leak is created due to the dissolving of the PVC joint solvent. For these reasons, design engineers must specify the feed pipes to be well insulated, and the feed line must be encased by an additional pipe.

Chloramination System In the chloramination system, ammonia is added prior to or after chlorination. The source of ammonia may be anhydrous ammonia, aqua ammonia, or ammonium salts such as ammonium sulfate. However, the use of ammonium salts is limited to small installations because of their hygroscopic nature and high cost; the cost does not outweigh the fact that ammonium salts are safe to handle. Generally, aqua ammonia is safer to handle than anhydrous ammonia and is also generally more cost-effective, due to its wide availability. In regions where the ambient temperature exceeds 90°F (32°C), it is recommended that 25% or even 20% ammonia solution be stored in a tank located outside to avoid problems associated with ammonia vaporization.

The main design criteria for an aqua ammonia feed system are listed below.

Dosage	0.2 to 1.5 mg/L (25 to 30% of chlorine residual).
Number of ammonia feeders	Minimum of two; one standby is required.
Type of feeder	Solution feeder (metering pump) for 25 or 30% strength aqua ammonia.
Storage tank(s)	A minimum of 15 days' storage: a minimum shipment of 4000 gal (15 m^3) if in bulk shipment. The tank(s) must be composed of steel or fiberglass. The tanks must be painted white and have a sun shield to prevent the contents from heating up. The tank will be slightly pressured (25 psi or 1.75 kg/cm^2) to prevent the escape of ammonia vapor.
Safety features	Eyewash, shower, and a dike around the storage tank for containment purposes.
Feeding ammonia	Ammonia will react with the hardness of water and form scale. Thus the ammonia solution is fed from above the water level, if possible, to avoid clogging the feed pipe.

Figure 3.2.9-3 illustrates the aqua ammonia feed system.

Chlorine Dioxide System Chlorine dioxide is prepared by reacting sodium chlorite with chlorine gas in a glass column packed with ceramic Rashig rings. A self-contained chlorine dioxide feed system is commercially available. Bottled chlorine dioxide gas is not commercially available, due to its explosive nature. Theoretically, a sodium chlorite solution containing 1.34 kg of sodium chlorite will react with 0.5 kg of chlorine to produce 1 kg of chlorine dioxide. If the reaction is to be com-

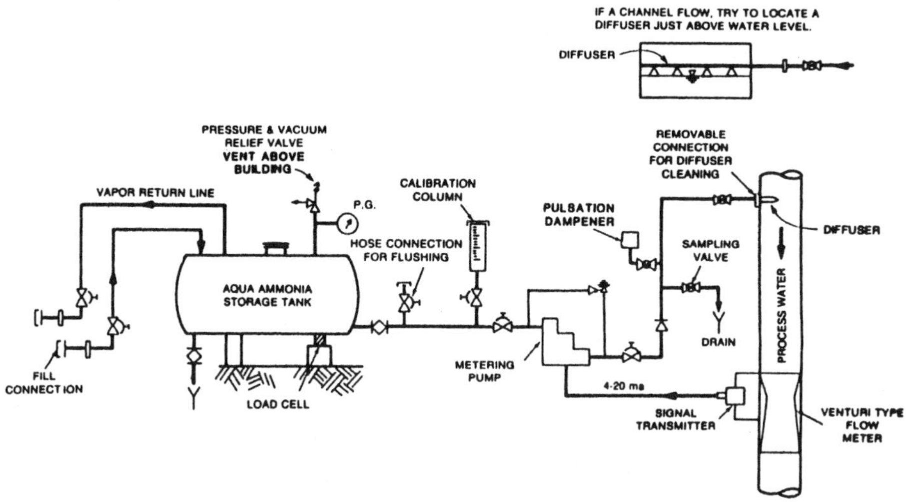

Figure 3.2.9-3 Aqua ammonia feed system.

pleted in a short period of time, say, less than 1 min, the pH of the mixture should range from 2 to 4. Acid feeding is therefore usually required.

Dosage	1.5 to 3 mg/L, but depends on the chlorine dioxide demand of water (EPA limits the maximum residual to 0.8 mg/L and 1.0 mg/L for chlorite).
Number of feeders	Minimum of two; one standby is required.
Sodium chlorite solution tank	A fiberglass tank that is large enough to produce a 1-day supply of 17% solution at maximum dosage.
Number of chlorite tanks	Minimum of two.
Chlorite feeders	A solution metering pump with a minimum of two feeders; one standby is required.
Safety features	Eyewash, shower, and gas masks of the canister type as dictated by OSHA regulations. Additionally, some caution must be exercised when handling sodium chlorite because it becomes combustible on contact with organic matter.

Refer to Figure 3.2.9-4 for a schematic representation of the chlorine dioxide feed system.

Hypochorination Hypochlorination has been practiced in Europe since the mid-nineteenth century. Yet hypochlorination only began in the United States starting in the late 1920s, when a dry bleaching chemical called calcium hypochlorite (containing about 70% available chlorine) became obtainable under the name of HTH.

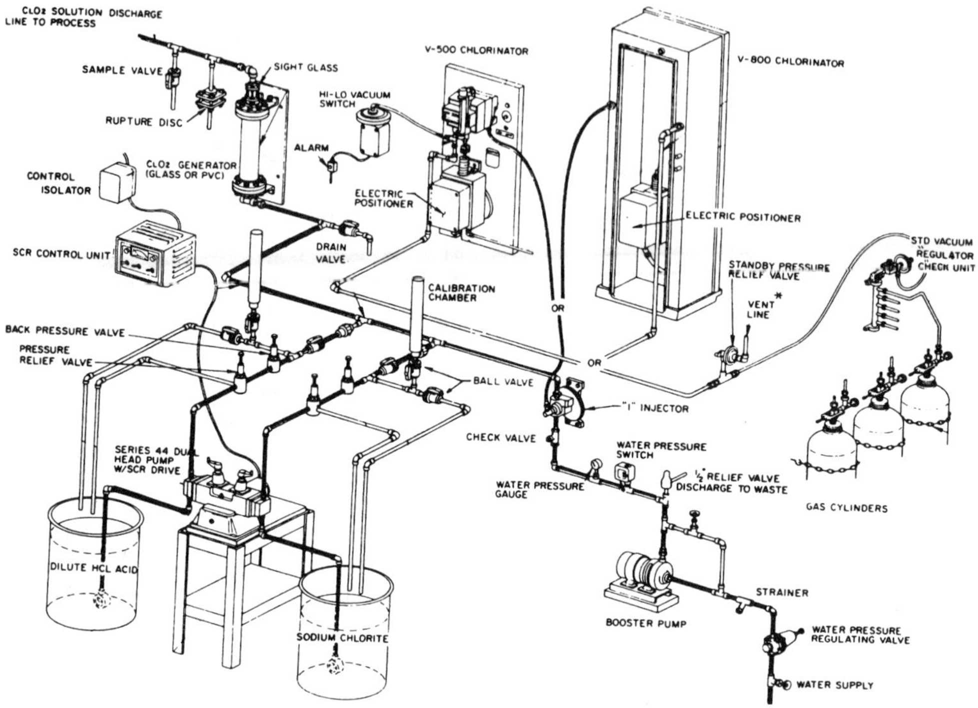

Figure 3.2.9-4 Schematic illustration of chlorine dioxide feed system produced by Wallace & Tiernan Company.

Liquid bleach (sodium hypochlorite) also became available about the same time and is widely used today (Clorox) for home laundry and general disinfection purposes. Although sodium hypochlorite is approximately three times more expensive than chlorine gas and deteriorates rapidly in a warm to hot environment, many water treatment and wastewater treatment plants have converted from chlorination to hypochlorination in recent years. The two main reasons are (1) safety concerns with regard to chlorine by regulatory agencies, backed by the public; and (2) increasing difficulty in the use of chlorine due to the risk management rule promulgated by the EPA.

$$\text{Sodium hypochlorite:}\quad NaOCl + H_2O \rightarrow HOCl + NaOH$$

$$\text{Calcium hypochlorite:}\quad Ca(OCl)_2 + H_2O \rightarrow 2HOCl + Ca(OH)_2$$

However, HOCl will convert to hypochlorite ion in water with a high pH (refer to Figure 3.2.9-1).

Almost all water purveyors utilize sodium hypochlorite because it is easier to handle and poses fewer problems when spilled. Its cost ranges from \$0.27 to \$0.35 per pound of available chlorine, depending on the conditions. Sodium hypochlorite is produced by a reaction between caustic soda and chlorine gas:

$$2NaOH + Cl_2 \rightarrow NaOCl + NaCl + H_2O$$

Since additional caustic soda is added as a stabilizer, the commercial product has a very high pH (greater than 11.2). The common strength of sodium hypochlorite used in hypochlorination is either 12.5% or 15% solution. Solution strength is usually expressed as percent by volume (trade percent). However, the weight percent of available chlorine is important for the disinfection process. Weight percent is defined as the trade percent divided by the specific gravity of the solution. For instance, a 15% solution (trade) has a specific gravity of 1.168 at 20°C (68°F); thus the weight percent of this solution is 12.84% and it contains approximately 1.25 lb of available chlorine per gallon. The most popular strength of sodium hypochlorite for disinfection purposes is 12.5 % since it contains approximately 1 lb of available chlorine per gallon and the stability of the solution is better than a 15% solution.

A major problem with the sodium hypochlorite solution is its significant loss of available chlorine within 1 week unless the solution strength is less than 5% and the temperature is below 10°C (50°F). The deterioration rate increases with higher solution strength and higher temperatures. The 12.5% solution will lose a fourth (25%) of its available chlorine content within 20 days during summer months. The stability of the solution is also greatly affected by light, pH, and heavy metal cations such as iron, copper, and nickel. Therefore, the dilution water should not have more than 0.3 mg/L of iron and no more than 1 mg/L of copper. The freezing temperature of sodium hypochlorite are −12°C or 10°F for a 10% (trade concentration) solution, −17°C or 2°F for a 12.5% solution, and −22°C or −8°F for a 15% solution.

The use of granular (dry) calcium hypochlorite for hypochlorination is limited primarily to developing countries, where the supply of sodium hypochlorite is scarce or, in the United States, to the disinfection of newly constructed filters, clearwells, and distribution lines. The cost of 100 lb of calcium hypochlorite (70% solution) ranges from $110 to $130. Under normal conditions the solution will lose only 3 to 5% available chlorine per year. The decomposition releases oxygen and produces high heat (177°C or 350°F). It should be noted that hypochlorination can make the pH of the water higher than its original pH when the buffer capacity of the water is low (low alkalinity), thereby affecting the optimum pH of alum flocculation under prechlorination as well as the $C \times t$ value of final disinfection.

Dosage	1 to 5 mg/L (3 mg/L) average.
Number of feeders	Minimum of two; one standby is required.
Type of feeders	Diaphragm pump, peristatic pump, metering pump with 100 : 1 motor speed control, magnetic flow meter with needle valve flow control with pump recycling (padded or pressurized storage tank is not recommended, based on experience).
Type of valves	Diaphragm valve. If a ball valve is used, placement of a vent between the valves is necessary to release the pressure caused by oxygen accumulation.
Pipe lines	Schedule 80 PVC is most commonly used. Provide air (oxygen) vents at strategic points to release gas pressure. Hydraulic eductors identical to the chlorination system to dilute the solution. Transporting to the application points

	can cause heavy scale formation in the pipeline if water hardness exceeds 100 mg/L; softened water is required.
Diffuser	In general, provide a perforated pipe across the pipe or channel (see Figure 3.2.3-3, type A or B). However, high hardness of the treating water will form scale at each orifice hole. Provide a special means of cleaning scale or two diffusers so that one diffuser may be removed for cleaning when it is clogged.
Storage tank	Minimum of two tanks for maintenance. Capacity of tank: *Alternative A:* Enclosed room with air conditioning: 10°C or 50°F maximum. 6000 gal (23 m^3) per tank for tank truck delivery or 11,000 gal (42 m^3) per tank for large plants. *Note:* The tank truck has either a 4000-gal (15-m^3) or 5000-gal (19-m^3) tank. *Alternative B:* Tanks are exposed to ambient temperatures. One tank capacity should be limited to 10 days' usage during summer months.

NOTES:

1. Both sodium and calcium hypochlorite produce oxygen under certain conditions; therefore, the piping system becomes pressurized, giving rise to potential leakage and breakage problems.
2. Conversion of regular chlorination systems to hypochlorination systems requires certain considerations and evaluation of the existing system, including storage, the type of valves, type of metering pump, size of the feed line, vents, and the quality of the dilution water.

Disinfectant Contact Tank The design of the disinfectant contact tank is contingent on the type of primary disinfectant selected. The $C \times t$ values provided by the EPA clearly indicate that the size of the contact tank drastically varies among chloramines, ozone, and chlorine. Nonetheless, the fundamental design rule is to minimize flow short-circuiting within the contact tank.

A recent EPA draft guideline manual suggests that the actual contact time of a contact tank be verified by running tracer tests. When these tests are performed, they establish that most existing clearwells, which serve as contact tanks, have a severe flow short-circuiting problem and that the actual flow-through time is often only 10% of the calculated detention time during conditions with lowest water depth and highest water demand. The contact tank must therefore be designed to have as near a plug flow condition as possible. One of the best tank designs is identical to a baffled channel type of flocculation tank (refer to Section 3.2.4). Refer to Appendix 9 for a tracer test.

Number of tanks	An absolute minimum of one; two tanks are preferred.
Type of flow path	A serpentine flow path with or without baffles.
Water depth	12 to 20 ft (3.5 to 6 m).

End of the tank	An overflow wall must be provided to maintain a constant water depth.
Detention time	This depends on the type of disinfectant and the $C \times t$ requirement.

Dechlorination Recent regulations do not allow water with residual chlorine to be discharged into public water courses, to protect fish inhabiting these waters. Dechlorination is required when residual chlorine is too high; MCL is 4 mg/L for both free residual and chloramine residuals. The most practical method of dechlorination is by sulfur dioxide (SO_2) or sulfite compounds such as sodium bisulfite ($NaHSO_3$) and sodium metasulfite ($Na_2S_2O_5$). The sulfur dioxide dechlorination process is shown below. The first reaction is the formation of sulfurous acid in the sulfonator injector:

$$SO_2 + H_2O \rightarrow H_2SO_3$$

Sulfurous acid then reacts with various chlorine residual species as illustrated below.

For the free chlorine residual: $HOCl + H_2SO_3 \rightarrow HCl + H_2SO_4$

For monochloramine: $NH_2Cl + H_2SO_3 + H_2O \rightarrow NH_4Cl + H_2SO_4$

As shown above, acids are produced, thereby lowering the pH of water, and 2.8 mg/L of alkalinity is lost for each part of chlorine removed.

The stoichiometric relationship requires 0.9 part of SO_2 to remove 1 part of chlorine residual, but a 1 : 1 ratio is used for the actual declorination practice. The reaction between sodium bisulfate and chlorine residual is as follows:

$$NaHSO_3 + Cl_2 + H_2O \rightarrow NaHSO_4 + 2HCl$$

Each part of the chlorine residual reacts with 1.46 parts of sodium bisulfate.

Activated carbon can adsorb chlorine so that dechlorination can be achieved. However, this is not the most practical method, due to capital cost and other limitations.

Example Design Calculations

Example 1

Given A new 50-mgd or 2.2-m^3/s plant (daily average flow rate) located in Denver, Colorado, requires a new chlorine disinfection system. Assume that the minimum, average, and maximum dosages are 1.5, 2.5, and 5 mg/L, respectively.

Determine

(i) The considerations for the system design
(ii) The number and capacity of the chlorine feeders and the appurtenant units
(iii) The size and number of chlorine containers (storage)

Solution (i) There are various design considerations for the new chlorine disinfection system: the potential interruption of the chlorine supply due to severe winter weather conditions; the extreme toxicity of chlorine; the need to adhere strictly to the safety codes and EPA regulations; specifying a large curvature for the access road, in addition to providing adequate maneuvering space in front of the chlorine cylinders; providing adequate height (14 ft) below the monorail to facilitate the unloading of chlorine cylinders from delivery trucks; and providing a heating system in the chlorine storage and feeder rooms.

(ii) The number and capacity of the feeders are as follows:

$$\text{maximum plant flow rate} = 1.5 \times 50 = 75 \text{ mgd}$$

$$\text{maximum dosage} = (5 \times 8.34) \times 75 = 3128 \text{ lb/day } (1422 \text{ kg/d})$$

$$\text{average dosage} = (2.5 \times 8.34) \times 50 = 1043 \text{ lb/day } (474 \text{ kg/d})$$

$$\text{minimum dosage} = (1.5 \times 8.34) \times (0.25 \times 50) = 156 \text{ lb/day } (71 \text{ kg/d})$$

$$\text{ratio of maximum to minimum dosages} = 3128 : 156 = 20 : 1$$

NOTE: Commercially available chlorine feeders have a 20 : 1 feed range for manual control and a 10 : 1 feed range for automatic control.

Based on these figures, select three identical 1500-lb/day (700-kg/day) feeders. This choice allows for maximum and minimum dosages, and the third feeder serves as an automatic control standby unit. Additionally, provide two identical 4000-lb/day (1800-kg/day) chlorine evaporators. The maximum chlorine dosage cannot be handled without the use of evaporators; a safe yield of chlorine gas from a 1-ton cylinder is 400 lb/day (180 kg/d) without heating.

(iii) The optimum storage of chlorine is based on the maximum dosage of the disinfectant at the average plant flow rate. Since the delivery of chlorine may be interrupted during winter months due to snowstorms, a 2-week stockpile must be provided under the conditions described above; the plant flow rate during the winter months is less than half the average annual rate. Thus actual storage (under normal dosage) will be approximately 60 days.

$$[(5 \times 8.34) \times 50]\ 15 = 15.6 \text{ tons}, \qquad \text{say, 16 tons}$$

The storage area should also accommodate the empty 1-ton cylinders. Since each truck carries a minimum of eight cylinders, provide eight additional spaces for the empty cylinders. Consequently, a total of 24 storage racks (for the 1-ton cylinders) should be provided in the storage room. Figure 3.2.9-5 is an illustration of both a typical chlorine storage room and a feeder room.

Operation and Maintenance The operation of the disinfection process is contingent on the type of disinfectant that is employed, the plant size, regional characteristics, and the raw water characteristics. Since chlorine is the primary disinfectant

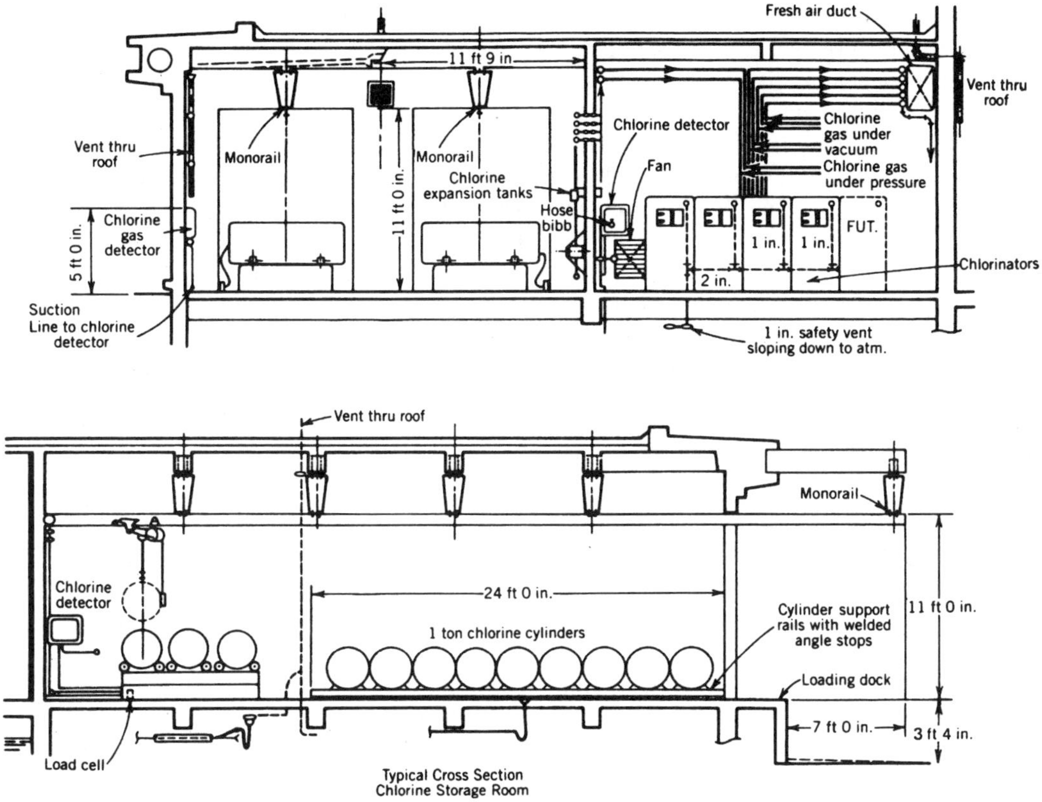

Figure 3.2.9-5 Typical cross sections of chlorine storage room (bottom) and chlorinator room (top).

in water treatment, this discussion is centered on the process of chlorination. If prechlorination can be practiced without producing significant levels of THM and HAA, a residual chlorine concentration of approximately 0.3 mg/L should be maintained in the settled water at all times. This practice will prevent the growth of algae on the basin walls and will prevent the growth of microorganisms in the filter bed.

The level of the residual chlorine in the clearwell should be at least 0.5 mg/L so that the chlorine residual at the end of the distribution system will be 0.2 to 0.3 mg/L at all times. Plant operators must record the level of residual chlorine in the clearwell and distribution system at regular intervals, since a log of these figures must be submitted to the local regulatory agency on a monthly basis. The chlorine feeders and appurtenant systems must be operated and maintained in strict accordance with the operation manual furnished by the equipment manufacturers and the manual provided by the Chlorine Institute; this includes the handling of chlorine cylinders. If the chlorine feed system should fail, the unchlorinated water must not be allowed to enter the distribution system because the Safe Drinking Water Regulations demand a minimum residual chlorine of 0.2 mg/L to be present at the end of the distribution system.

Maintenance of the chlorination system requires a thorough understanding of the literature provided by the manufacturer, in addition to the operation and maintenance manual prepared by the design engineers. The daily, weekly, and monthly operation and maintenance procedures are generally described in these manuals. It is imperative that operators not attempt to perform any unfamiliar maintenance work or work that is beyond their qualifications. Many serious accidents have occurred because someone has ignored this very basic rule.

In the case of chlorine leakage, corrective actions must only be taken by trained operators wearing the appropriate safety equipment. It is essential that all plant operators be trained to repair chlorine leaks. Operators should always work in pairs when searching for and repairing chlorine leaks. All other personnel must be evacuated until the conditions are safe. If the chlorine leak is large, the fire department and police must be notified immediately; all inhabitants in the adjacent area must be warned and evacuated.

Alkali chemicals such as caustic soda and lime may be used to neutralize very small chlorine gas leaks. For example, stoichiometrically, 1.13 lb of caustic soda is required to react 1 lb of chlorine. Therefore, a 1-ton chlorine leak would theoretically require 2260 lb of caustic soda. However, the scrubber manufacturers recommend 3600 gallons of 7% caustic soda. In the case of hydrated lime, 125 lb is required to neutralize a 100-lb cylinder of chlorine, and 2500 lb is required to neutralize 1 ton of chlorine under ideal mixing conditions.

It is extremely important that treatment plants establish a formal safety program. All personnel should perform "hands-on" training periodically using the safety equipment. The American Water Works Association provides training materials, including videos. The Chemical Transportation Emergency Center (CHEMTREC) provides immediate advice for those at the scene of an emergency and quickly alerts experts for more detailed assistance and appropriate follow-up whenever the situation becomes uncontrollable. CHEMTREC has a toll-free number (1-800-424-9300) that all plant operators should keep in a readily accessible place. An example of a chlorine gas scrubbing system is shown in Figure A13-2.

As described in Section 4.1, if any plant uses, handles, or stores more than 2500 lb (1135 kg) of chlorine and 5000 lb (2270 kg) of sulfur dioxide at the plant site, the plant operators must be familiar with and comply by the Risk Management Planning (RMP) rule established by the EPA; this rule has been in effect since June 1999.

BIBLIOGRAPHY

Aieta, E. M., et al., "A Review of Chlorine Dioxide in Drinking Water Treatment," *J. AWWA,* 78:6:62 (June 1986).

AWWA Committee Report, "Disinfection," *J. AWWA,* 74:7:376 (July 1982).

Barnes, D., "Alternatives to Chlorination for Water Disinfection," *Water J. (Australia),* p. 12 (September 1983).

Chiu, K. P., et al., "Effect of UV System Modifications on Disinfection Performance," *J. Environ. Eng. Div. ASCE,* 125:5:459–469 (May 1999).

Clancey, J. L., et al., "UV Light Inactivation of *Cryptosporidium* Oocysts," *J. AWWA,* 90:9:92 (September 1998).

Dyksen, J. E., et al., "Advanced UV Light for Inactivating *Cryptosporidium* Oocysts," J. AWWA, 90:9:103 (September 1998).

EPA, *Municipal Wastewater Disinfection: Design Manual,* EPA/625/1-86-021, U.S. Environmental Protection Agency, Washington, DC, October 1986.

Glaze, W. H., et al., "The Chemistry of Water Treatment Processes Involving Ozone, Hydrogen Peroxide, and Ultraviolet Radiation," *J. Ozone Sci. Eng.,* 9:335 (1987).

Hoff, J. C., *The Relationship of Turbidity to Disinfection of Potable Water,*" EPA/370/9-78-00c, U.S. Environmental Protection Agency, Washington, DC, 1977.

Hunter, G. L., et al., "Medium-Pressure Ultraviolet Lamps and Other Systems Are Considered for Wastewater Application," *J. Water Engineering and Technology,* pp. 41–44 (June 1998).

Longley, K. E., et al., "The Role of Mixing in Wastewater Disinfection," *Proceedings of the ASC Meeting,* Anaheim, CA, 1978.

Marshall, T., "Deadly Pulses," *J. Water Engineering and Technology,* pp. 36–41 (April 1999).

Metcalf & Eddy, *Wastewater Engineering: Treatment, Disposal, Reuse,* 2nd ed., McGraw-Hill, New York, 1979.

Morris, J. C., *Aspects of the Quantitative Assessment of Germicidal Efficiency Disinfection: Water and Wastewater,* Ann Arbor Science, Ann Arbor, MI, 1975.

Parrotta, M. C., and Bekdash, "UV Disinfection of Small Ground Water Supplies," *J. AWWA,* 90:2:71 (February 1998).

Shull, K. F., "Experience with Chloramines as Primary Disinfectants," *J. AWWA,* 73:2:101 (February 1981).

White, G. C., *Handbook of Chlorination,* Van Nostrand Reinhold, New York, 1972.

3.2.10 Filtered Water Reservoir (Clearwell)

Purpose In the United States the finished water reservoir in the water treatment plant is customarily called the *clearwell.* There are four basic reasons for requiring clearwells: (1) to meet water peak demands, (2) to provide a sufficient volume of water for plant operations including filter washing, (3) to ensure adequate chlorine contact time ($C \times t$), and (4) to store enough water for firefighting. Water treatment processes are designed based on the maximum daily water demand rate, not the maximum hour rate; therefore, an adequate amount of water must be stored in the clearwell to meet the peak demand (maximum hour rate). Usually, the maximum hour water demand is 1.5 to 2.0 times the maximum day demand, and its duration and demand vary due to the local conditions. Treatment plants must use a certain amount of water to service the operational needs, and filter washing requires the largest use of water; filter wash can be several filters in succession. All water treatment plants must meet the $C \times t$ requirement for the disinfection process by the Surface Water Treatment Rule established by the EPA. In general, a clearwell should provide the required chlorine contact time and have proper baffling inside the clearwell to minimize flow short-circuiting and dead water zones. The storage of sufficient water for firefighting is a very important factor for small communities that do not have adequate local water distribution tanks. Adequate water storage in the clearwell improves filtered water quality and filter operation since the filters do not require frequent changes in filtration rates to meet the peak demands.

Considerations Clearwell storage should be sized in conjunction with the distribution system storage in the city. The mass diagram or Ripple method for the determination of storage requirements for an impounding reservoir may be applied to size a clearwell. One of the important considerations is the location of the clearwell. The clearwell should be located above the possible flood level and the bottom slab should be placed above the groundwater table. At least 50% of the water depth should be above grade, if possible. Sewers, sludge lagoons, a wash-waste holding tank, or a similar source of possible contamination should be kept away from the clearwell [a minimum of 50 ft (15 m)] to avoid any contamination of finished water. Recent EPA regulation does not allow the use of an open-top reservoir to store finished water, the clearwell must be covered, and the top of the clearwell should be at least 2 ft (0.6 m) above the finished grade of the plant or any possible flood level. The clearwell cover should be a suitable watertight roof that prevents the entry of birds, animals, insects, and excessive dust.

The clearwell also requires drains, overflow protection, and access with hand railings. The vents should prevent the entrance of surface water and rainwater (use of an inverted-U pipe is recommended), and the opening level should be about at least 2 ft (0.6 m) above the top of roof; the opening should be covered with a fine SS-316 insect screen (No. 18 or 1-mm openings).

Pipe connections to the clearwell should have a reliable flexible coupling or an articulated pipe arrangement to avoid breakage of the pipe connection to the sidewall of the clearwell. Many pipe connections to the reservoir have broken in earthquake-prone regions such as California and Japan, due to inadequate flexibility in the pipe connection to the sidewall of the clearwell. The top of the clearwell may be utilized as a public service area, for example, a tennis court, baseball court, and even a parking lot. However, safety against vandalism and potential pollution of finished water is a concern. The local laws and regulations of the site area should be examined during the planning stage.

Treatment plants that incorporate ozonation as a part of the treatment process do not need to meet the regular $C \times t$ requirement for the required final disinfection by chlorination because of the highly effective disinfection by ozone upstream of the final chlorine disinfection process. Therefore, a few treatment plants provide a smaller clearwell capacity than the regular chlorination practice. However, this practice is not a robust design; ozonation systems can fail due to natural disasters such as earthquakes, tornadoes, hurricanes, and even sabotage. For instance, a 600-mgd (26-m^3/s) water treatment plant in Los Angeles could not operate its ozonation system after the 1995 Northridge earthquake (magnitude 7.3) for about 3 weeks due to damage of the oxygen production facilities and the power supply system to the ozone generators. Regardless of whether or not an ozonation system is included in the treatment process, the size of the clearwell should be large enough to perform chlorine disinfection in case the ozonation system fails; this design provides good reliability. If the clearwell is an aboveground tank fabricated from steel plate, the design should follow the current AWWA Standards. Use of steel tanks requires that they have a protective coating both inside and outside, and engineers should also consider to provide cathodic protective system to prevent corrosion of steel plates. The painting system should be in accordance with NSF Standard 61 and be accepted by the local state authority.

All new clearwells must be disinfected in accordance with current AWWA Standard C652. However, the heavily chlorinated water should be carefully disposed of since it can kill the fish if disposed of into the water course without dechlorination. Also, the water may contain a high level of DBPs. One of the purposes of a clearwell is to fulfill the final disinfection process with the necessary $C \times t$ requirement. However, traditional clearwell design has often disregarded the flow characteristics in the tank so that severe flow short-circuiting as well as stagnant zones are found. Knowing this condition, SWTR allows only 10% of hydraulic detention time as the reaction time of t unless a tracer test to come up with actual flow-through time of existing clearwells. SWTR also shows the baffling classifications table as a general guideline of actual disinfectant time efficiency factor in the clearwell, which is expressed as T_{10}/T, as shown below, where T_{10} is the actual detention time (flow-through time) and T is the hydraulic detention time, which is the volume of tank divided by the flow rate.

BAFFLING CLASSIFICATIONS

Baffling Condition	T_{10}/T	Baffling Description
Unbaffled (mixed flow)	0.1	None, agitated basin, very low length/width ratio, high inlet and outlet flow velocities
Poor	0.3	Single or multiple unbaffled inlets and outlets, no intrabasin baffles
Average	0.5	Baffled inlet or outlet with some intrabasin baffles
Superior	0.7	Perforated inlet baffle, serpentine or perforated intrabasin baffles, outlet weir, or perforated launders
Perfect (plug flow)	1.0	Very high length/width ratio (pipeline flow), perforated inlet, outlet, and intrabasin baffles)

The rule of thumb for designing a good clearwell (to provide a good flow characteristic for disinfection) is to furnish 2 to 4 longitudinal baffles dividing the tank into 3 to 5 channels, assuming that the shape of the clearwell is either square or rectangular. This arrangement gives each channel a length/width ratio of 5 : 1 to 15 : 1, so that the total ratio of the tank is between 25 and 50, with a minimum water depth of 10 ft (3 m). The chlorine contact tank portion of the clearwell should provide a minimum of 30 min of net contact time with 1 mg/L free chlorine residual for

Giardia inactivation, and the water depth is maintained by the overflow wall at the end of the contact tank. However, if chloramination is the disinfection process, a net contact time of 210 min should be provided at 0.5°C (33°F) with 3 mg/L of residual for *Giardia* inactivation. Therefore, to determine the correct detention time, the water depth must be monitored constantly to calculate the volume of water in the basin (entire clearwell volume) necessary to meet the required $C \times t$. Effluent overflow cannot be used because the clearwell is a flow balancing basin; thus the water level must be fluctuated. When modifying existing clearwells to meet $C \times t$ requirements, engineers should provide as many baffles as practical and also consider installing a perforated diffuser wall at the beginning of each channel wherever the flow turns 180°. The minimum number of clearwells is two, thereby satisfying reliability and maintenance needs.

Additional design considerations for the clearwell are the amalgamation (combination) of clearwell structure and the pumping station, which provides high service pumps and utility pumps, including filter wash-water supply pumps. This scheme not only reduces construction costs but also makes a compact site layout. The preferable design is to locate the vertical shaft turbine pumps on top of a concrete slab of the end portion of the clearwell, with or without housing. Since the turbine pumps must be pulled out vertically for service and repair, engineers should furnish a removable hatch on the roof of the housing directly above the pump to prevent creating a building with a very high ceiling structure. In most cases, plants in the warm weather regions do not provide any housing over the pumps. Even in regions with cold winter weather, the housing is deleted if the ambient temperature is not as severe and snow and ice are not great problems. This type of pumping station design provides easy access to the pumps and avoids the need for a dry well, pump priming system, and floor drainage pumps; thus the construction and maintenance costs are generally minimized. However, a dry pump room utilizing the endwall of the clearwell as a part of the room with the installation of horizontal centrifugal pumps is also a good proven design. Centrifugal pumps usually last longer than turbine pumps. However, this type of design is generally more costly than the design scheme described above and there is the potential problem of flooding in the dry room.

Type and Selection Guide There are three basic types of clearwell: (1) ground level up to approximately 30 ft (9 m) in height and a steel or reinforced concrete tank—these are mostly cylindrical in shape; (2) ground level, a rectangular or square deep basin [25 ft (7.6 m)] with a reinforced concrete structure having vertical sidewalls or a trapezoidal cross section with sidewalls of rather thin concrete and peripheral walls composed of reinforced concrete that are 8 ft (2.4 m) high to support the roofing system or an anchored floating membrane cover; and (3) a rectangular, rather shallow [10 ft (3 m)] reinforced-concrete tank, located directly underneath the filter structure.

The first type of clearwell is usually built in cases where the groundwater level is near the ground surface. A cylindrical tank is chosen because of less capital cost.

However, in most cases, transfer pumps are required to transfer filtered water into the tanks. The second type is suitable in cases where the groundwater level is reasonably low and a rather large volume of water must be stored. In general, filtered water flows into the clearwell by gravity. Unless designed carefully, there is a potential danger that the finished water could be contaminated by floodwater or contaminated groundwater, and there is the risk of damage to the structure or floating due to high groundwater levels during rainy seasons. The third basic type, clearwell underneath the filter module, is an old design concept that has two serious drawbacks: (1) it is very difficult or impossible to calculate reliable disinfection time due to filter effluent flows into the clearwell at many locations, including adjacent to the effluent side of the clearwell, and (2) clearwell capacity is often inadequate for disinfection for *Giardia* and especially *Cryptosporidium* oocysts.

A few clearwell designs furnish only one pipe connection to the clearwell; therefore, the water level in the tank is floating and is dependent on the amount of inflow and outflow. Consequently, the flow characteristics in the tank is the worst for disinfection. Under certain conditions, such as the peak-hour period, chlorinated water does not flow into the tank and no disinfection time is provided by the clearwell. This type of pipe connection is not recommended. A clearwell should have both inlet and outlet connections at strategic locations, and the tank should have a proper baffle wall system in order to make the T_{10}/T ratio at least 0.70. Figure 3.2.10-1 shows an example of a ground-level clearwell with a reinforced concrete structure design.

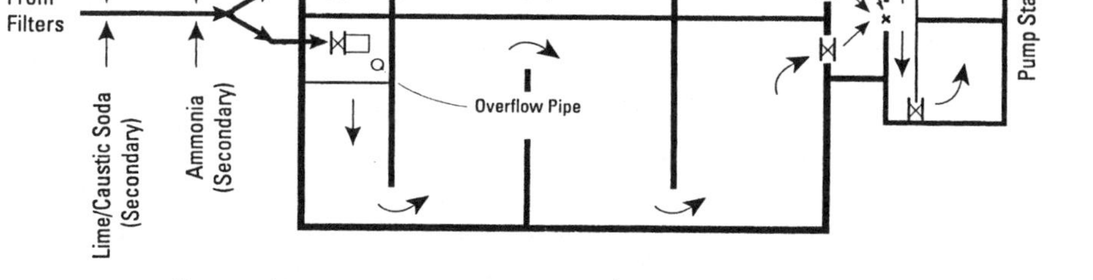

Notes: 1.) Submerged weir controls water level in the filter beds.
2.) Advantages of this scheme are:
- Smaller C x T due to free chlorine residual in the clearwell.
- Smaller C x T because of the lower pH in the clearwell.
- Single chemical feed location in both inlet and outlet of clearwell.

3.) Provide drain valve at a proper location of each tank.

Figure 3.2.10-1 Suggested clearwell design and chemical feed locations.

Design Criteria

Number of clearwells	Minimum of two, or one large clearwell with a divider wall to make it two cells
Capacity	Approximately 15% of the plant design flow rate per day
Baffling	Minimum $T_{10}/T = 0.7$ with proper baffling

NOTES:

1. Capacity can be estimated as the sum of the volume for 2 h of plant operational needs, including two filter washes, and the volume of water necessary to meet the $C \times t$ requirement by free chlorine residual and the unusable bottom portion of water [usually, the bottom 4 to 5 ft (1.2 to 1.5 m) water depth].
2. Miscellaneous required items are described in the "Considerations" section.

BIBLIOGRAPHY

Bishop, M. M., et al., "Improving the Distinction Detention Time of a Water Plant," *J. AWWA,* 85:3:68 (March 1993).

Boulos, P., et al., "Hydraulic Mixing and Free Chlorine Residual in Reservoirs," *J. AWWA,* 88:7:48 (July 1996).

Kennedy, M. S., et al., "Assessing the Effects of Storage Tank Design on Water Quality," *J. AWWA,* 85:7:78 (July 1993).

Sykes, R. G., and Coate, A. R., "PCBs in Sealants in Water Distribution Reservoirs," *J. AWWA,* 87:12:10 (December 1995).

3.2.11 Instrumentation and Process Control

Purpose Due to increased public awareness on the subject of safe drinking water, regulatory agencies have made the drinking water standards more stringent. Consequently, water treatment plants must have more sophisticated processes and equipment, and plant operators are forced to operate and maintain systems that are vastly more complicated than in the past. As the costs of energy, chemicals, and personnel continue to increase, managers of water purveyors must be concerned with the quality of the water supply as well as improving the efficiency of production and supply. One solution to these issues is to provide the treatment plants and distribution systems with modern supervisory control and data acquisition systems.

The advantage in computerizing treatment plants is that plant operators are supplied with timely and accurate information on which to base their decisions, thus allowing them to achieve effective process control. Process variables such as the raw water quality, plant flow rate, chemical feed rate, and finished water quality may be monitored by on-line instruments, which relay the information to the operator on a continuous basis. This type of instrumentation may also help to reduce equipment damage because the status of each system is monitored continuously, and should the situation go wrong, the plant is automatically shut down.

Computerization of the treatment plant frees plant operators from monitoring the various gauges or charts and from updating the operational logs continuously. The system required by water treatment plants is generally simple in nature, unlike the complex systems used in the oil, chemical, and rocket industries. The safety and security of plant personnel and the plant itself may be improved by installing a closed-circuit television monitoring system. The installation of modern instrumentation in a treatment plant permits operators to concentrate on the overall control of the treatment plant and on effective plant operation.

Symbols and Manufacturers The symbols used in designing the instrumentation system are based on those provided by the Instrument Society of America (ISA), the Institute of Electrical and Electronic Engineers (IEEE), and the National Electrical Manufacturers Association (NEMA). Refer to Figure 3.2.11-1 for an example. The field of instrumentation is a rapidly evolving technology in which new components and equipment are increasingly being geared toward the field of water treatment. Thus, it is very rare that a single manufacturer will be able to furnish the bulk of the necessary equipment. Yet it is very important to place the overall responsibility of installation and system startup with a single manufacturer. Firms that have been dominant providers of service in the United States are Fisher and Poter Company, Bristol Babcock, Inc., Westinghouse Company, and Robert Shaw Controls Company. Major manufacturers of instrumentation control equipment such as Foxboro, Leeds, and Northrop may or may not be involved in water treatment plant projects, due to the level of competition and the required bidding procedures for services of this nature. However, this is beginning to change.

Considerations The major considerations in designing the instrumentation and control system are (1) the local conditions, (2) the benefits of the system, (3) the selection of the system, (4) the needs of the system, (5) the function of the instrumentation and control system, (6) the availability of the system, and (7) the understanding and commitment of the operators and management to the system. Each issue is discussed in detail.

Local Conditions The instrumentation system is a highly sophisticated item. For this reason, the following issues must be evaluated carefully: availability of the equipment, availability of qualified operators, follow-up services, weather conditions, and the living standard of the region, because this dictates the drinking water quality standards. Due to the high cost of systems acquisition, as well as the cost of repair and replacement parts, the system should be operated only by highly skilled and knowledgeable operation and maintenance personnel. Moreover, the environment of the control room must be strictly controlled to optimize the performance of the computers (microchips); the room must be maintained at a specified temperature and humidity and have a dust-free environment. Extensive instrumentation and

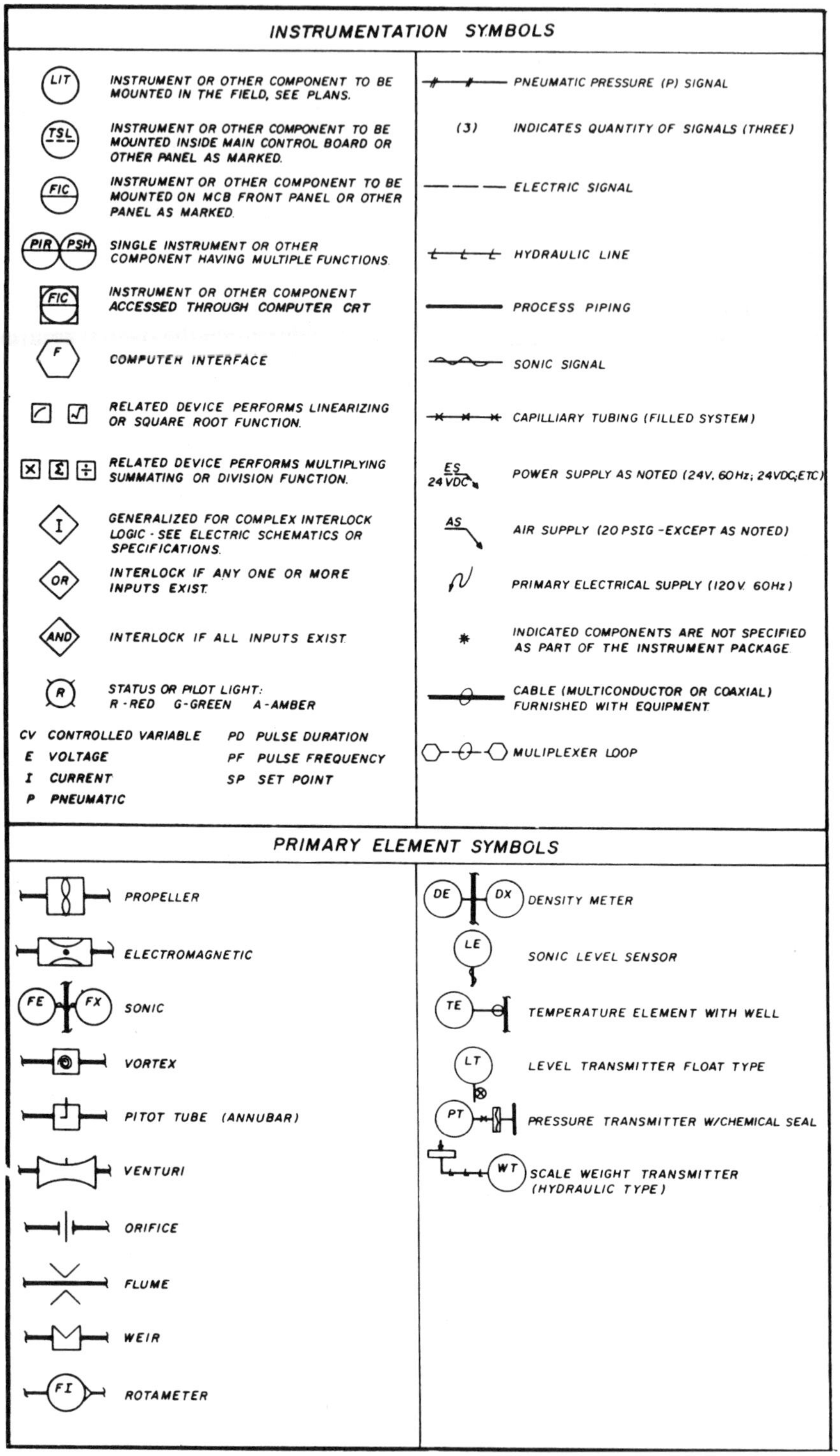

Figure 3.2.11-1 Instrumentation symbols.

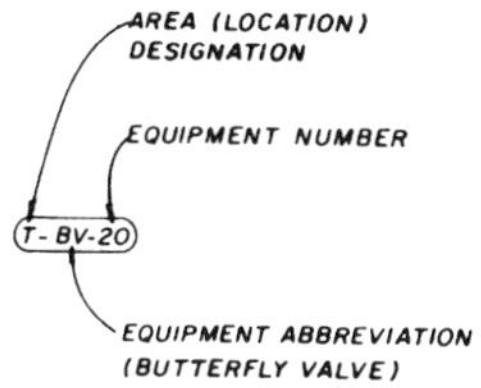

MECHANICAL EQUIPMENT CALLOUT

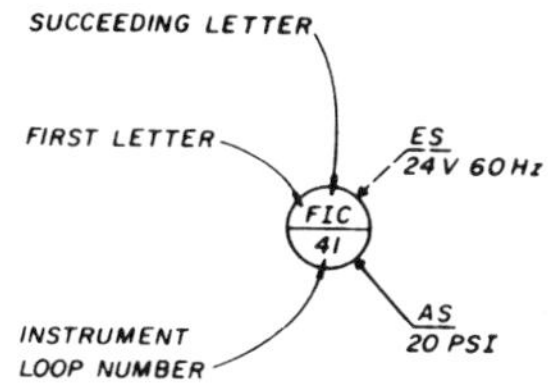

INSTRUMENT TAG NUMBER

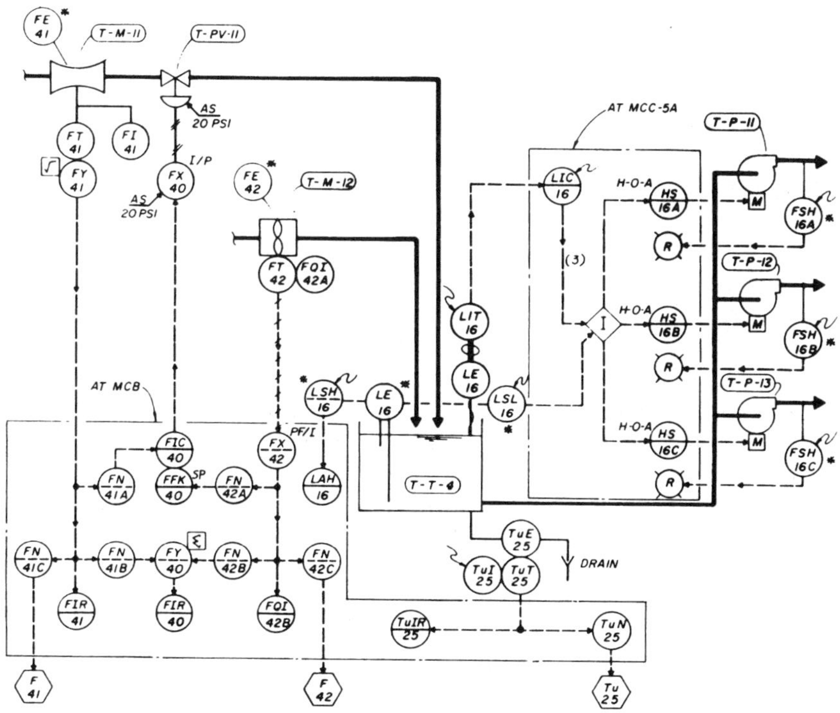

SAMPLE LOOP OR P & I DIAGRAM

(ISA STANDARD S5.1)

SAMPLE INSTRUMENT TAG NUMBERS	
TAG NUMBER	FUNCTION
FE - 41	FLOW SENSING ELEMENT (LOOP 41)
FI - 41	FLOW INDICATOR
FIC - 41	FLOW INDICATING CONTROLLER
FIR - 41	FLOW INDICATING RECORDER
FFK - 41	FLOW RATIO STATION
FN - 41	FLOW SIGNAL ISOLATOR
FQI - 41	FLOW INTEGRATING TOTALIZER
FSH - 41	FLOW SWITCH HIGH
FSLL - 41	FLOW SWITCH LOW LOW (EXTREME)
FT - 41	FLOW TRANSMITTER
FX - 41	FLOW SIGNAL CONVERTER
FY - 41	FLOW SIGNAL MATH FUNCTION

Figure 3.2.11-1 *(Continued)*

control systems are not recommended for developing countries or small plants located in remote areas since they generally cannot meet the criteria listed above.

Benefits of the System Although all plant operators would like to have a modern instrumentation and control system, in each case the benefits of the system must be weighed against the cost factor. For instance, a simple water treatment process such as slow sand filtration cannot justify the installation of high-level instrumentation. Yet an ozonation system, reverse osmosis process, pumping plant, filter process, and water distribution system all greatly benefit from a highly sophisticated instrumentation and control system. Manual OM-6, *Process Instrumentation and Control Systems,* by the Water Pollution Control Federation (1978), lists a number of benefits derived from instrumentation; refer to Table 3.2.11-1.

Selection of the System Although there are various ways to improve plant operations and to achieve the benefits listed in Table 3.2.11-1, modification of an existing, outdated system is usually very difficult, due primarily to outdated sensors, transmitters, and control signalers. In fact, complete replacement of an old system is often more cost-effective than updating certain parts. However, the owner may wish to keep the existing instrumentation system, including the 7-day recording circular chart and the relay logic control system, as a backup to the new computerized system and/or a training tool for plant operators. In this type of situation, the design engineer should respect the wishes of the owner.

When an engineer is designing a new plant, he or she should evaluate the most appropriate degree of instrumentation and control for that particular plant. Although the state-of-the-art system may appeal to both the engineer and the owner, these exotic systems may become obsolete after several years due to rapid technological ad-

TABLE 3.2.11-1 Benefits of the Instrumentation System

Purpose	Benefits
Process	Improved process results
	Efficient use of energy
	Efficient use of chemicals
	Automation of some process adjustments
	Greater ability to control complex processes
	Process changes detected in a timely manner
Personnel	Timely and accurate process information
	Safer operation
	Increased security
	Efficient use of labor
	Allows for an overview of plant operations
	Decrease in manual paperwork
	More complete records
Equipment	Increase in running time
	Status is known at all times
	Automatic shutdown to prevent major damage

vances, and replacement parts may become unavailable. The basic rule in selecting a new instrumentation and control system is to anticipate the future needs of the treatment plant, including plant expansion, the potential addition of new treatment processes, and owner preference or corporate policy. If the new system must be compatible with the systems of neighboring districts, engineers should also consider the type of system employed by adjacent water districts and the system used by the city; for example, determine if the city has a main computer that controls the overall distribution of water.

It is important to remember that most water treatment unit processes are slow reacting in nature and have high system lag times. In fact, fast response times create unstable or undesirable operational conditions, such as "hunting" actions. Thus, design engineers should keep this fundamental nature of water treatment processes in mind when selecting the instrumentation and process control system. Also, the units and system to be selected should be "user friendly"; otherwise, the design system will never be used effectively.

Needs of the System Design engineers must collaborate with owners to identify the needs of the system. For instance, treatment plants consisting of a simple treatment process require instrumentation only for the plant flow rate, monitoring of raw and filtered water turbidity, and monitoring of the residual chlorine content of the finished water. Conversely, plants consisting of rather complex processes, such as ozonation or a large number of filters and pumps, require an extensive instrumentation and control system to simplify plant operation and to make the plant efficient and cost-effective.

The needs of a system may be classified into three categories: essential for plant operation, useful for plant operations, and luxury item. The essential items are plant flow rate control, chemical feed rate control, pump rate control, ozonation system control, filter control, analysis and recording of water quality, control and recording of residual chlorine, and detection of chlorine gas leakage (with an alarm). The useful items are those that reduce the mental and physical activity of the operators or those that can save operational costs. They include computerized data logging, programmable control of the filter backwashing process, closed-loop automatic control of residual chlorine and the pH of the water being processed, and devices that check the coagulant dosages such as streaming current detectors and automatic plant controllers with advanced data-handling capability (the latter requires extensive custom-designed software). All these items are not vital to normal plant operation. Finally, luxury items are defined as items that make plant operation easier without providing compensatory savings in cost. The design engineer should discuss the needs of the system with both plant operators and the owner prior to making any decisions.

Task of the Instrumentation and Control System After the preceding items have been considered, the type and purpose of the system must be defined. At a very basic level the system can be used to serve operational, maintenance, safety, and documentation purposes. The operational goals of the instrumentation and control system should be to provide the best control quickly and without human errors. The system can inform the operator of all current activities, where they are happening, and the duration. The system is also capable of making corrective or preventive actions. It can detect and warn the operator of changes in temperature, excessive vibrations, and abnormal levels of power consumption by individual operating equipment and thereby prevent serious damage, costly repairs, and downtime from occurring; these all relate to maintenance work. With respect to the safety aspect of plant operation, a surveillance system is useful in meeting the security requirements and the safety of the plant operator, who must operate the equipment alone. Fatal accidents may also be minimized by installing systems that can detect and warn the operator of chlorine gas, ozone, and oxygen leaks. Finally, documentation is very important for both the operator and plant manager in identifying the control modes and running time, in analyzing the history of repair work, and in taking inventory of available spare parts.

System Availability Three basic types of systems are currently available: digital, analog, and control instruments. Digital systems have primarily two positions: on–off, open–closed, high–low, or alarm–normal. This type of signal is used to indicate a change in status or to act as an alarm. The signals may be activated by a position, limit, float, or pressure switch. Analog data systems cover a range of values from liquid or gas flow rate, liquid levels, liquid or gas pressure, to the turbidity or pH of the water. The analog data may be sensed and transmitted in their analog form or may be converted into digital form. The control instruments range from a simple on–off switch to a programmable controller with a computer backup.

Many factors are involved in selecting the most appropriate type of system for each treatment plant. The characteristics and the need of the treatment process train may determine both the type of control and the complexity of the control system. Factors such as the frequency of information, technical limitation, and local conditions may also limit the type of system designed. Other important considerations are the capital cost of the system and the associated operation and maintenance costs. Finally, the ease and frequency of calibration, the difficulty in maintenance, and the availability of timely servicing must also be evaluated. A conventional monitoring and control instrumentation system uses 4- to 20-mA analog signals (standard) in transmitting measurement and control signals. Voltage or air pressure signals are rarely used today.

Understanding and Commitment For any instrumentation and control system to operate at its fullest potential, operators must have a complete understanding of the system. The facilities must therefore have well-organized training courses that not only familiarize operators with the system but also teach them the steps to be taken when the system fails: After all, the system is not infallible. Thus backup control and monitoring methods must be an integral part of the plant design, and operators must be familiar with these methods.

System maintenance must be conducted by either the operators or maintenance personnel. However, if the system is highly complex, it is in the best interest of the owner to sign a service contract with the equipment supplier or the service agency. Since the operation and maintenance of a sophisticated instrumentation and control system requires an adequate budget just to keep it in service, management must clearly understand the benefits of a properly functioning system. If management is committed to the system, it must also be committed to providing adequate resources for the system: financial resources and the personnel required to design, operate, and maintain it. Without both a clear understanding of the benefits and a thorough commitment to the system by the operators and management, the instrumentation and control system will not function at its full potential, and many future problems may be anticipated.

Common Measurements The level, pressure, flow rate, turbidity, pH, and chlorine residual of the process water are major process variables that are measured on a continuous basis in all water treatment plants. Each variable is discussed briefly.

Level measurements are commonly made by float-operated transmitters, pressure sensors, capacitance probes, sonic or ultrasonic units, or a bubble tube system. The pressure within a system may be measured as absolute pressure, where vacuum is zero pressure; as gauge pressure, where ambient atmospheric pressure is designated as zero; or as a differential pressure. The sensors may be U-tube manometers, bourdon tube gauges, bellow gauges, pneumatic or electric force-balance pressure transmitters, or strain gauge transmitters.

Flow measurement may be conducted by weirs and flumes, such as the Parshall flume, which is used to measure open-channel flow. However, the flow of liquid or gas in pipelines should be measured by differential flow meters, acoustic (ultrasonic) flow meters, magnetic flow meters, vortex shedding flow meters, or propeller or turbine flow meters. Yet Venturi flow meters are the most commonly used type of flow meter due to their high accuracy (0.5 to 2%), low headloss, and wide availability. Engineers should note that selection of the proper size or capacity of a flow meter is vital to obtaining accurate flow measurement. Most flow meters have a turndown (maximum to minimum flow rate) of 5 : 1 to 10 : 1. A common mistake of many inexperienced design engineers is to design the meter size based on future maximum flow rates; this practice does not allow for the measurement of low flows during the plant startup period and other special occasions.

In the United States, turbidity is measured by nephelometric turbidimeters. The principle of the turbidimeters is to detect and measure the scattered light beam or the portion of a light beam that has not been absorbed by suspended particles after an incident light has been passed through the process water. Thus the size, shape, refractive indexes, absorption capacity, and viewing angles of the suspended matter greatly influence the turbidity reading. The nephelometric turbidimeter may be used in situations where the turbidity of the water is approximately 30 ntu. It will not be very accurate in situations where high turbidities are expected (over 30 ntu) since the meter will go blind at these levels and give false readings.

Turbidity has a fairly linear direct response: zero signal at zero turbidity. True color does not register as turbidity but can cause negative errors. The surface scatter type of turbidimeter may be used for water with high levels of turbidity but must be installed in a vibration-free environment. This type of meter is not appropriate for measuring turbidities lower than 1 ntu because the response is nonlinear. Since any type of turbidimeter will give a false high reading whenever the water sample contains air bubbles, an air bubble trap should be installed in the sample line upstream of the meter.

pH meters are generally composed of three electrodes: a pH sensing electrode, a reference electrode, and an electrode that compensates for the temperature. These electrodes are usually mounted in a chamber and installed in the sample line or submerged in a tank or channel. The most recent models have a signal preamplifier within the electrode assembly to minimize the electrical noise of the high-impedance electrical circuits.

Finally, free and combined chlorine residuals may be measured by residual chlorine analyzers. These units are commercially available from several manufacturers. The principle used in measurement is amperometric titration through oxidation–reduction titration procedures.

Basic Controls When selecting the type and control mode for a specific application, the engineer must be familiar with the purpose and the hydraulic characteristics of the particular treatment process. There are three basic considerations that engineers should remember when selecting a control system: the nature of the required control, the degree of disturbance that the process can tolerate, and the cost. There are four types of control methods: manual, semiautomatic, automatic, and supervisory. Semiautomatic control requires manual initiation of the automatic function. A typical example would be manual startup of the automatic filter backwash sequence. The automatic control method uses sensors, limit switches, timers, analytical instruments, controllers, and logic control devices (i.e., programmable controllers) to control the process or equipment automatically. Supervisory control refers basically to a system that is remotely monitored and adjusted; either in-plant or remote from the plant.

The two basic schemes for the control loops are the open loop and the closed loop. Open-loop control does not possess a measurement device within the circuit, and the treatment process is not contained in the loop (Figure 3.2.11-2, scheme *c*). A prime example of this type of control is the flow-paced chlorine feed system because the chlorine feed rate is proportional to the flow rate of the processing water. Chemical feeders commonly use open-loop control because of the ease and effectiveness of the system. In contrast, the closed-loop control circuit includes a device that measures the effects of the process. This information is then relayed to the controller, which compares the result with a set point and performs the necessary adjustments. For example, a closed-loop chlorination system will be produced if a signal is relayed from the residual chlorine analyzer to the controller. Schemes *d, e,* and *f* in Figure 3.2.11-2 are examples of closed-loop circuits. As mentioned previously, most water treatment processes usually have a long lag time and are therefore not well suited for closed-loop control systems.

Several types of control modes may be used, depending on the purpose and hydraulic characteristics of the process. The three basic modes are proportional control and its variations, cascade control, and ratio control. Proportional control is defined

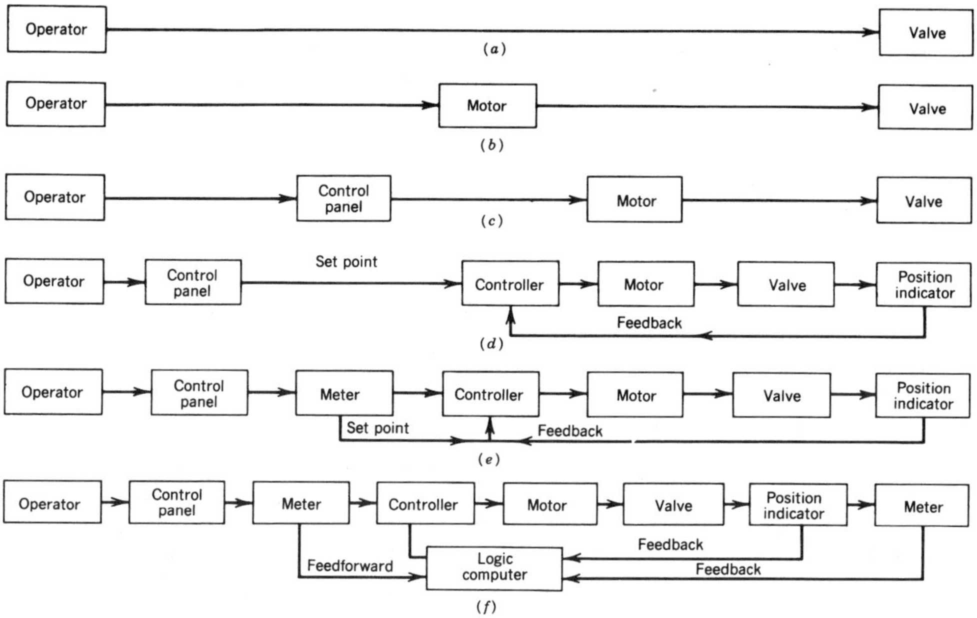

Figure 3.2.11-2 Control systems: (*a*) manual control; (*b*) manual control made easier; (*c*) remote manual control; (*d*) automatic feedback control; (*e*) automatic feedforward control; (*f*) automatic feedforward–feedback computer control. (Adapted from WPCF Manual OM-6, 1978.)

as a system in which the controller continuously adjusts a controlling device, such as a valve, to balance the process input with the process demand. An example of proportional control is control of the filtration rate of filters with respect to the water level in the clearwell. There are also several variations of proportional control: proportional control with reset control or proportional control with derivative control.

Cascade control is widely used in the water treatment field, especially in pumping control, where the output of one controller adjusts the set point of another controller. In contrast, the ratio control scheme maintains a constant ratio between two variables; the primary variable is normally uncontrolled. The ammonia feed rate of the chloramination process is controlled in proportion to the quantity of the residual chlorine.

Although engineers should select control systems that are simple and cost-effective, one type of control system should never be used for chemical feeding: on–off control. The impulse-driven control scheme is an on–off control that has been employed widely over many years, but it tends to feed chemicals only part of the time, and therefore a portion of the water remains untreated. This mode of chemical feeding is detrimental to coagulants such as alum or ferric chloride and to other chemicals, such as chlorine and chloramines.

Computer-Based Monitoring and Control Computer-based monitoring and control equipment uses digital signals. A large number of signals may be transmitted by installing a remote terminal unit (RTU) at each major unit process. The RTU receives analog or digital signals from nearby sensors, then sequences transmission of these signals so that a single coaxial cable then transmits these measurement and control signals to a central processing unit (CPU) located in the control center, where they are displayed on a color cathode ray tube (CRT) and logged on a printer or stored on a disk.

Control of treatment processes is accomplished by coding commands into the video terminal keyboard. The microprocessor-based CPU communicates this information to all the elements of the system: the RTUs and peripherals such as bulk storage devices and printers. The CPU is capable of handling real-time commands; that is, the computer system is capable of handling all multiuser requests, system alarms, process control commands, and much more, almost on an instantaneous basis. The system operator is able to view the system data at the computer terminal.

The use of programmable controllers (PC) and programmable logic controllers (PLC), which facilitate local process control, is another aspect of the computer-based control approach. For example, the filter backwash cycle may be controlled by a PC, and the operator will be informed of the progress status of the backwash cycle. Distributed control systems (DCSs) and supervisory control and data acquisition (SCADA) systems have gained popularity in the field of treatment (plant), and water distribution systems and many water purveyors are now adopting their use.

Conventional Versus Computer-Based Systems Conventional monitoring and control instrumentation is a proven and reliable method of operating a water treatment plant. The computer-based monitoring and control system is also a proven technology; however, its use has become widespread only since the late 1970s. Both methods have advantages and disadvantages. These are listed below.

Compared to a computer-based system, a conventional monitoring and control system has the following advantages: (1) the technology is familiar; thus operation personnel do not have to be retrained, (2) troubleshooting is generally easier for operators because the system is familiar, (3) much of the equipment is standardized within the water treatment industry, and (4) the overall capital costs are generally lower than for the computer-based system. However, it also has the following disadvantages: (1) the maintenance requirements are higher, due to the use of mechanical relays and timers, (2) separate hard-wired connections are required for each measurement or control signal, (3) possible problems with electrical noise may exist if the analog signal wire comes in close contact with a source of high voltage, and (4) the incorporation of new instrumentation is more difficult and costly during plant expansion unless the initial plant design and construction are properly planned.

A computer-based system has the following advantages compared to a conventional monitoring and controls system: (1) the system is more reliable and requires less maintenance due to the use of solid-state components which are housed in sealed and noise-proof enclosures; (2) the system is more flexible due to the ability to modify the software to changing needs and the ability to incorporate new signals easily when required; (3) the system is easily expanded to accommodate future process units, in addition to accommodating water distribution data telemetered from locations remote from the water treatment plant; (4) there is a reduction in the number of hard-wired connections, thus there may be a cost savings in electrical construction due to the fact that the coaxial cables are capable of carrying much more information; (5) analog (conventional) or digital signals may be incorporated through the use of RTUs; (6) the presentation of operational data is improved through the use of a video terminal capable of color graphics; (7) the operating staff will potentially respond more effectively to impending emergency situations because they will be reacting to trends rather than critical alarms; (8) reports are generated automatically at the end of each shift, the end of each day, and the end of each month, as required by regulatory agencies; (9) the staff requirements may potentially be reduced due to the decrease in paperwork and instrumentation maintenance and the ease with which the automatic functions are incorporated; and (10) plant management and planning are improved because the presentation and integration of the plant process data are vastly improved. However, computer-based systems also have the following disadvantages: (1) should the computer system malfunction, the operator could lose both monitoring and control capabilities throughout the plant; (2) the overall capital costs may be somewhat higher than those for a conventional control system; (3) the operators must be retrained to use the computer system; (4) key plant functions must usually have conventional controls for backup operation; and (5) the computer systems are not standardized in the water treatment industry. Thus, expansion of hardware may be difficult and expensive, due to the need to match components. If the new equipment is limited to those offered by the installing manufacturer, the engineer may not be able to obtain competitive bids.

Levels of Computer-Based Monitoring and Control The size of the computer system required by the plant is not a function of the plant size. Rather, it is determined by the number of analog and digital signals that are to be fed into the computer: the number of signals increases in logarithmic fashion with an increase in the number of tasks. The computer-related tasks are data logging, report generation, alarm, plant graphic displays, analog variable displays, and manual and automatic plant control. There are six computer systems that are different combinations of these tasks:

1. Report generation
2. Data acquisition and logging/report generation/alarm indication
3. Data acquisition and logging/report generation/alarm indication/plant graphic display
4. Advanced display and data handling: data acquisition and logging/report generation/alarm indication/plant graphic displays/analog variable displays
5. Manual plant control and advanced data handling: data acquisition and logging/report generation/alarm indication/plant graphic displays/analog variable displays/manual plant control
6. Automatic plant control: data acquisition and logging/report generation/alarm indication/plant graphic displays/analog variable displays/automatic plant control.

The fifth and sixth computer systems are commonly known as DCSs or SCADA systems. The first system is a very basic computer system that only generates reports. It is not hard-wired to the plant and the data must therefore be inputted manually. Systems 2 through 6 are built around a mainframe or a powerful personal computer and may be expanded to handle many additional tasks.

The hardware costs of the last five configurations are somewhat dependent on the plant size and layout. Selection of the memory size is determined by the number of analog and digital signals that must be processed and stored by the system. The mode of control and transmission (analog or digital) also influences the size and cost of the computer system. The use of PCs, as opposed to standard relay logic control, increases the cost (PCs are normally considered as part of the system) but reduces the cost of the electrical system because there are fewer relays, circuits, and wiring. If the plant size is large and the unit processes are spread out, the use of RTUs is cost-effective since single coaxial cables may be used to transmit both the analog and digital signals. Other features that should be considered are the standard "vendor" report formats versus custom-editable report formats, vendor O&M software versus custom-editable O&M software, and the inclusion of a separate engineering or laboratory terminal.

For budgetary purposes, the estimated costs (1989 costs) for the six alternative systems for a 2-m^3/s (50-mgd) ordinary conventional process water treatment plant at an ENR index of 5000 are as follows:

- *System 1:* \$5000
- *System 2:* \$100,000
- *System 3:* \$125,000
- *System 4:* \$150,000
- *System 5:* \$600,000
- *System 6:* \$800,000–\$1,000,000

If a conventional monitoring and control system is adopted, the estimated cost for systems 1 through 4 would be approximately \$220,000. Figure 3.2.11-3 is an example of system 5.

Design of the Instrumentation and Control (I&C) System The design of the instrumentation and control system must be conducted by I&C engineers. However, the project engineer should maintain an open line of communication with the I&C engineers and perform a thorough check of the final design. Since the system must be operated and maintained by plant operators, it is beneficial to elicit input from both the operators and the owner during the design phase.

Design work usually produces the following documents in the sequence listed below. The process flow sheets (PFS) prepared by the project engineer are the basis for the P&ID diagrams.

1. Process and instrumentation diagram (P&ID)
2. Process control diagram (PCD)

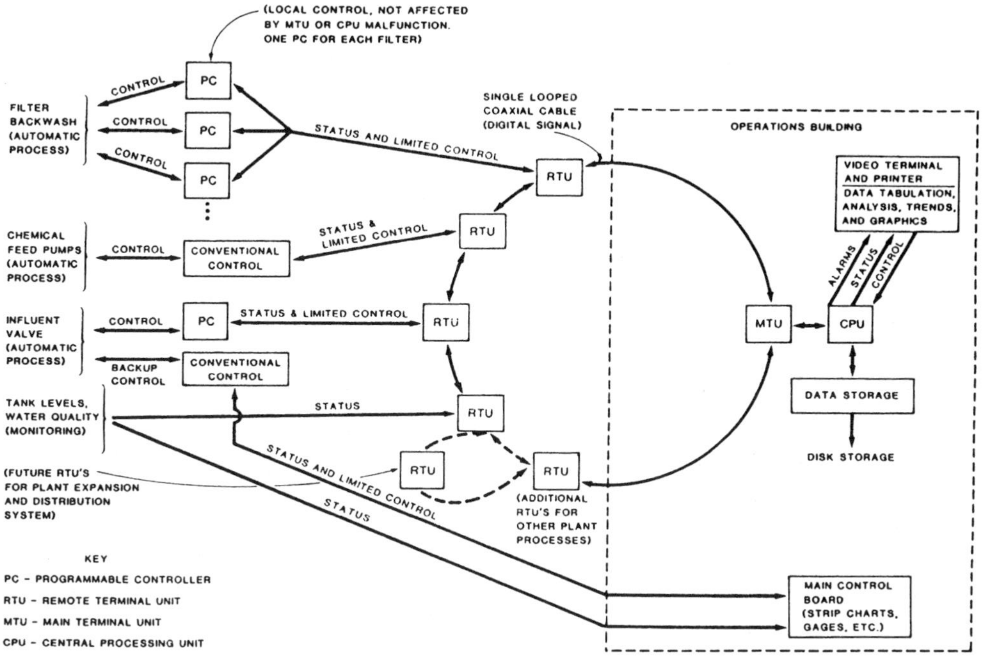

Figure 3.2.11-3 Schematic diagram of monitoring and control system (principal functions).

3. Instrumentation and input–output summaries (IIOSs)
4. Instrumentation specification sheets (ISSs)
5. Logic diagrams
6. Panel layout drawings
7. Loop interconnection drawings (LIDs)
8. Instrument installation details (IIDs)

The P&ID is a functional schematic presentation of the treatment process and the required instruments and controls. Its purpose is to illustrate the function of the treatment plant without referring to the actual hardware; the PCD specifies the monitoring and control loops.

The purpose of the IIOS is to provide continuity between the P&ID and IIOS. These documents specify both the quantity and characteristics of the analog and digital instrumentation. Similarly, the ISS provides detailed specification for each task and panel instrument. Logic diagrams for the control panels and programmable logic controls (PLCs) should be obtained by the instrumentation contractor. LIDs are used primarily in projects where the instrumentation and control system must be interfaced with existing control equipment. A typical P&ID for ordinary gravity filter is presented in Figure 3.2.11-4.

During the late twentieth century, most water treatment plant control systems use one of the three basic control systems: PLC, DCS, and a hybrid DCS system. The

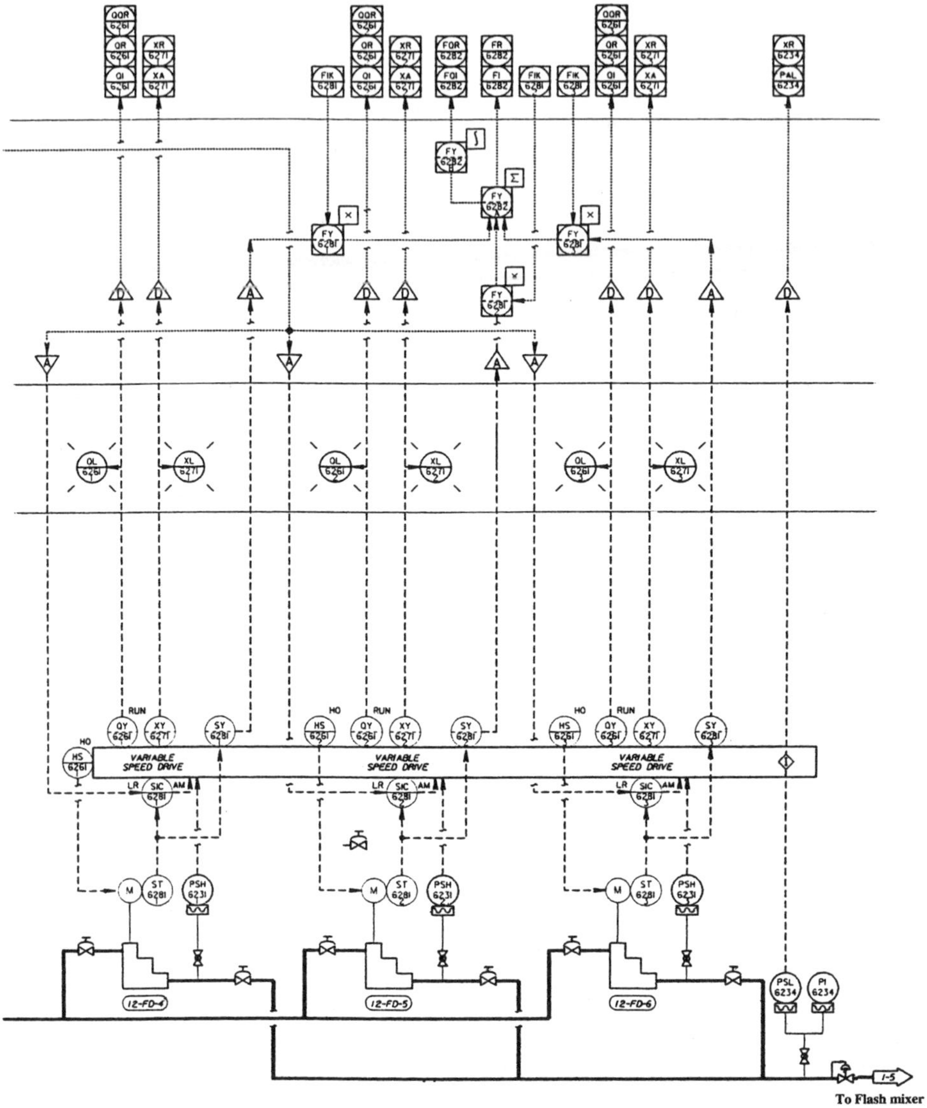

Figure 3.2.11-4 Example of P&ID drawing.

PLC-based system consists of programmable logic controllers (PLCs) networked with personal computers through a communication network (data highway). The computer workstations are the operator's window into the system. These systems usually have graphical user interface (GUI) software packages (Intellution, Factory Link) which run on the personal computer platforms. It is vital that a competent person purchase and assemble the PLCs, personal computer, communication hardware, GUI software, communication software, and system auxiliaries into an integrated system. PLCs are excellent for performing sequential logic control and are capable of providing reasonably good regulatory or closed-loop control. However, they often do not provide sophisticated control or the precision required for more demanding functions such as energy or process optimization. Ordinarily, PLC systems require a lower initial cost than that of the other two systems.

DCS (Distributed Control Systems) generally consist of process control modules (PCM), operator workstation/personal computers, a communication network, and GUI software. The DCS package is a standard product offered by one DCS manufacturer; thus there is the advantage of a single source of responsibility for both the initial purchase and ongoing long-term support services. The most reliable data acquisition and control system currently available is DCS configured with redundant PCMs and a redundant data highway. Features such as advanced control, comprehensive diagnostics, global database, fault tolerance, backup power supplies, and advanced alarming are standard features of a DCS.

DCS is a standard system and the manufacturer usually has extensive training and support services with program upgrades to keep the system relatively current. However, since the DCS system does not have a large sales volume such as the Intellution GUI for PLCs, the system may not be upgraded as frequently and therefore may lack the latest features. In general, the initial cost of DCS is higher, due to the manufacturer's higher overhead and much higher R&D costs required to develop a more sophisticated standard system.

A hybrid distributed system is a DCS that utilizes PLCs for the field controller. This hybrid system offers the advantages and flexibility of the DCS and data highway as well as single-source responsibility. This type of system is usually supplied when there is a preference for using existing PLCs or because of the owner's familiarity with programming and maintenance of PLCs. The hybrid system offers excellent sequential logic control and fair regulatory control or closed-loop control. However, troubleshooting is not as easy as with DCS because the comprehensive diagnostic function of the DCS will not extend to the PLC printed circuit card; the diagnostics will sound an alarm whenever a PLC fails. The cost of the hybrid system is usually between those of DCS and PLC systems.

The servers, workstation, and PLCs in the plant communicate with each other over a shared local area network (LAN), a real-time network and an information network. There are three basic LANs: Ethernet, token ring, and Fiber Distributed Data Interface (FDDI). However, the Ethernet is usually the network of choice, due to its throughput and reliability, ease of maintenance, and feasibility. It is important to note that the process information network (PIN) and facility information network (FIN) should be kept separate because the facility information/data traffic is much more dynamic and subject to large peaks, which can lead to a response slowdown. Also, FIN traffic may have high-bandwidth applications such as video and multimedia. Figure 3.2.11-5 is an illustration of the control and automation block diagram.

Generally, the design engineer should follow the basic design philosophy of KISS—"keep it simple, stupid." There have been several case histories where the plant operator was incapable of handling the situation when there was a sudden and complete failure of the SCADA system during the night or weekend shift. Moreover, plant op-

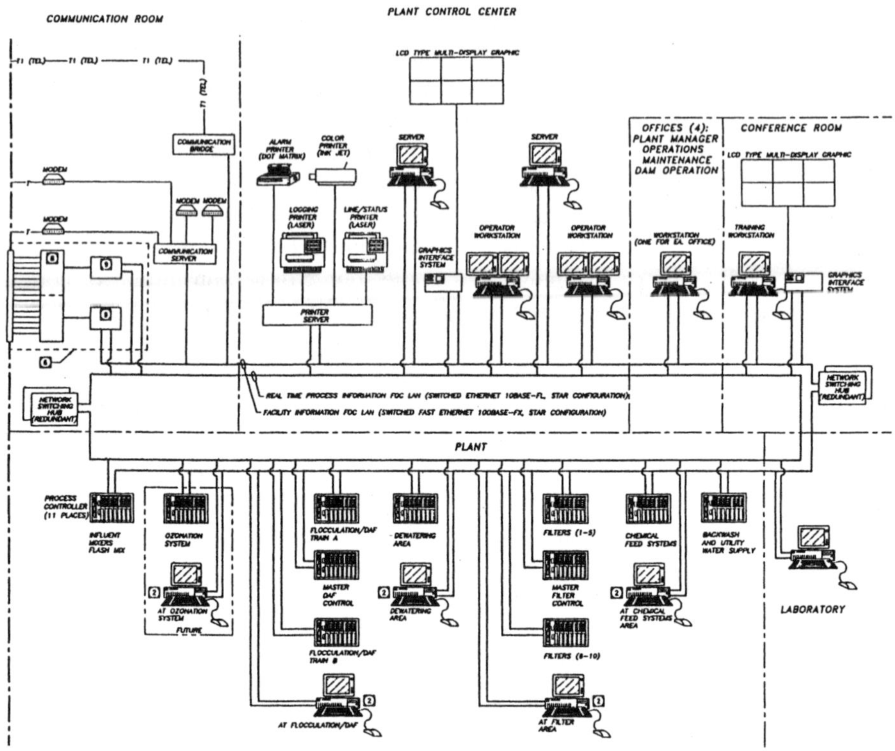

Figure 3.2.11-5 Basic design block diagram of control and automation system.

erators should not be allowed to become overly dependent on automatic control systems because this will make them ill prepared to act in an emergency situation.

Security System To assure the reliability and safety of water treatment facilities, all plants must be provided with a security system. At a minimum, the plant security system should include a perimeter fence and electric entry gates, closed-circuit television (CCTV) with a full view of the intake facility and main entry gate, and intrusion sensors on all building gates.

BIBLIOGRAPHY

Amy, J. P., and Cesarion, A. L., "Computer Applications in the Water Industry," *J. AWWA,* 85:4:129 (April 1993).

Andreas, M., Schulte, and Malm, A. P., "Integrating Hydraulic Modeling and SCADA Systems for System Planning and Control," *J. AWWA,* 85:7:62 (July 1993).

Arndt, D. J., "Making the Right Moves in SCADA Selection," *Water Eng. Manage.,* p. 25 (October 1985).

AWWA, *Automation and Instrumentation,* Manual M-2, American Water Works Association, Denver, CO, 1982.

AWWARF and Japan WWA, *Instrumentation and Computer Integration of Water Utility Operations,* AWWARF/AWWA Report, American Water Works Association, Denver, CO, 1993.

Babcock, R. H., "Instrumentation and Control," special issue, *Water and Wastewater Engineering,* 1968.

Computer Advances Research Committee, "Computers in the Water Industry, *J. AWWA,* 81:2:74 (February 1989).

Considine, D M., *Process Instruments and Controls Handbook,* 3rd ed., McGraw-Hill, New York, 1985.

Dobbs, D. M., et al., "Quick Payback Comes After PLC Installed," *Water Eng. Manage,* p. 25 (October 1988).

Eckman, D. P., *Industrial Instrumentation,* Wiley, New York, 1975.

Heywood, C. H., "SCADA Systems Today and Tomorrow," *Water Eng. Manage,* p. 18 (July 1986).

Jentgen, L. A., and Wehmeyer, M. G., "SCADA Trends and Integration Prospectives," *J. AWWA,* 86:7:44 (July 1993).

Liptak, B. G., ed., *Process Measurement Instrument Engineering Handbook,* rev. ed., Chilton Radnor, PA, 1982.

Stephen, J., Nix, and Lollins, A. G., "Expert System in Water Treatment Plant Operation," *J. AWWA,* 83:2:43 (February 1991).

WPCF, *Process Instrumentation and Control Systems,* Manual of Practice OM-6, Water Pollution Control Federation, 1978.

3.2.12 Electrical System

Purpose The modern water treatment plant cannot be operated without electrical power; almost all equipment and buildings and all control and instrumentation systems require electricity. Therefore, the plant electrical system must be reliable and safe. There are two basic standards and codes in the United States, the National Fire Protection Association (NFPA) Standard 70 and the National Electrical Code (NEC). Additionally, there are many local codes and regulations that govern the safety of the electrical system.

In general, most water treatment plants have a higher electrical load component and a smaller electrical load block. Higher load blocks consist of intake pumps, high service pumps, and ozone generators. Flocculators, sludge collectors in sedimentation tanks, various equipment and valves in the filter bank, chemical feed systems, mixers, small pumps, lighting and control systems for processes, and electricity for buildings are in the smaller load block. The electrical system must be reliable and safe, as well as efficient, clean, user friendly, and expandable.

Considerations Engineers must evaluate both the reliability of the power supply and of the power distribution at the plant when designing the electrical system of a water treatment plant.

Reliability of Power Supply Power distribution systems from power stations are generally very reliable. The utility transmission system (voltage higher than 34.5 kV) is also usually reliable. Historically, however, the power supply from utility distribution systems (15 to 25 kV) is less reliable, therefore, the engineer should first evaluate the historical data of the local power supply system to the plant. Frequency of outages, a range of outage length, and the average length of outages for last past 5 years must be investigated.

The regulatory agency's guidelines for water treatment plant systems requires either two sources of power supply from separate substations or a single supply source and an emergency power generator system. Design engineers should first contact the local power company and investigate the standard service. The engineer should then inquire about extra charges for two-line service from separate substations to the proposed plant site. If reliable service from the power company is not available or it is too expensive, installation of an on-site emergency power generator should be considered. In this situation, the power company may provide the treatment plant with rate incentives for operating this generator during the power company's peak power demand as a means of shedding power load. However, on-site emergency power generators have a high capital cost and require regular exercise and maintenance efforts. Moreover, the design of the fuel storage must comply with all local code requirements.

Reliability of Power Distribution at the Plant There are several design methods to ensure reliability of the power supply to individual equipment in a treatment plant. Higher reliability generally requires higher capital expenditure. A typical water treatment plant has four load blocks: two large loads and two small loads. If an ozonation process is included in the process train, it should be designated to the large load blocks. Large loads are typically supplied at a primary voltage of 4160 V, with 480 V of secondary supply through transformers for the smaller loads.

The power distribution within a treatment plant is always shown in a single-line (one-line) diagram using standardized symbols. Figure 3.2.12-1 depicts the standard symbols. Figure 3.2.12-2 is a single-line block diagram of a plant power system. Examples of low-voltage (secondary loads) power supply line diagrams are shown in Figure 3.2.12-3. There are three types of power distribution design schemes: radial electric, primary selective, and secondary selective systems. A radial system is the simplest and least expensive power distribution system. However, failure at any point in the system interrupts all power downstream of that point. The primary selective system is capable of isolating a fault and switching to a different path to provide uninterrupted power; this system has redundancy for the upstream part of the system, where failure causes the greatest disruption of power. The secondary selective system adds redundancy to the downstream portion of the distribution system, providing an alternative path for power to secondary loads. Another type of reliable system is the double-ended or main–tie–main arrangement. This system starts with two sources of power and two main breakers, each of which supply half the load, thus half of the system is not affected by the failure of the other system.

Design Criteria When designing a new water treatment plant, the development and evaluation of power are divided into five subissues: (1) investigation of power supply conditions at the plant site, (2) estimation of power load or power requirement, (3) evaluation of standby power requirement, (4) determination of power dis-

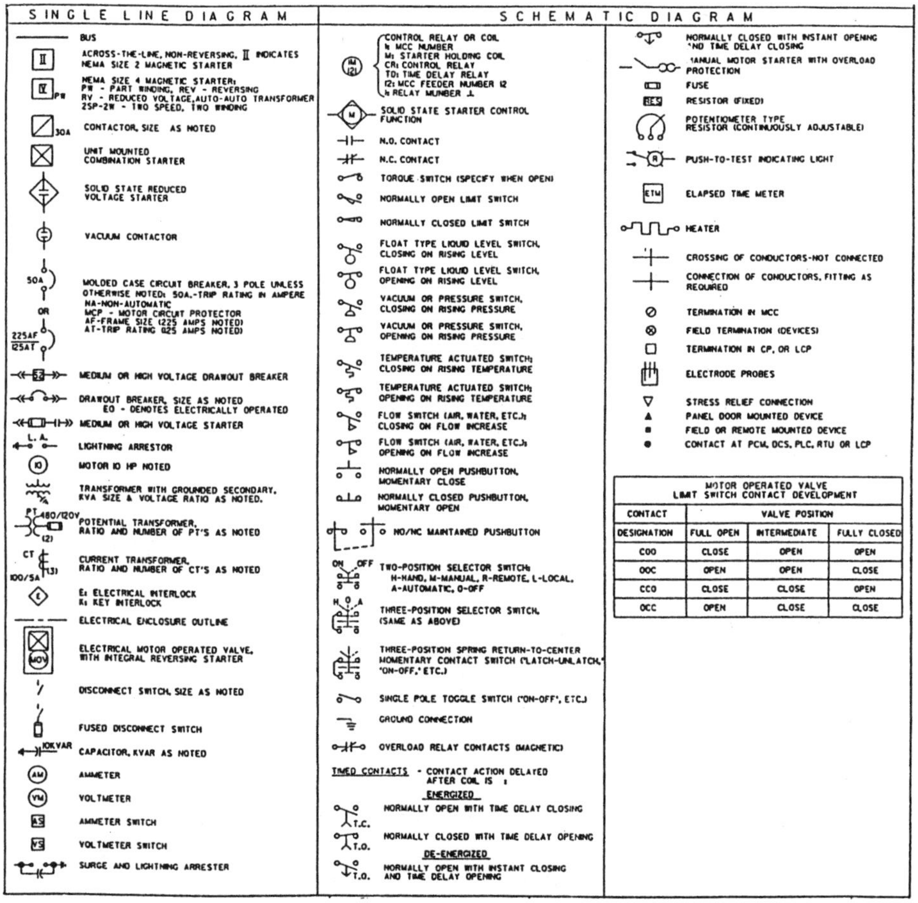

Figure 3.2.12-1 Major electrical symbols.

tribution and the required voltage, and (5) selection of the location of the main transformer within the plant site—this is a site-specific issue and is not addressed below.

Power Supply Conditions The first task is to investigate the availability of the proper type of power within a reasonable cost. The capacity and location of nearby substations, location of the main power line or nearest power pole relative to the plant service main, frequency and duration of power outage over the past several years, and power charge structure of the local power supplier (especially during peak hours) should be studied during the early stage of the project. Power supply conditions greatly affect both construction and plant operational costs, and influence the selection of equipment drive units and the size of the standby power unit (if needed).

Load The load or power requirements are dictated by plant size, type and number of unit processes, whether the treated water is distributed by gravity or by high

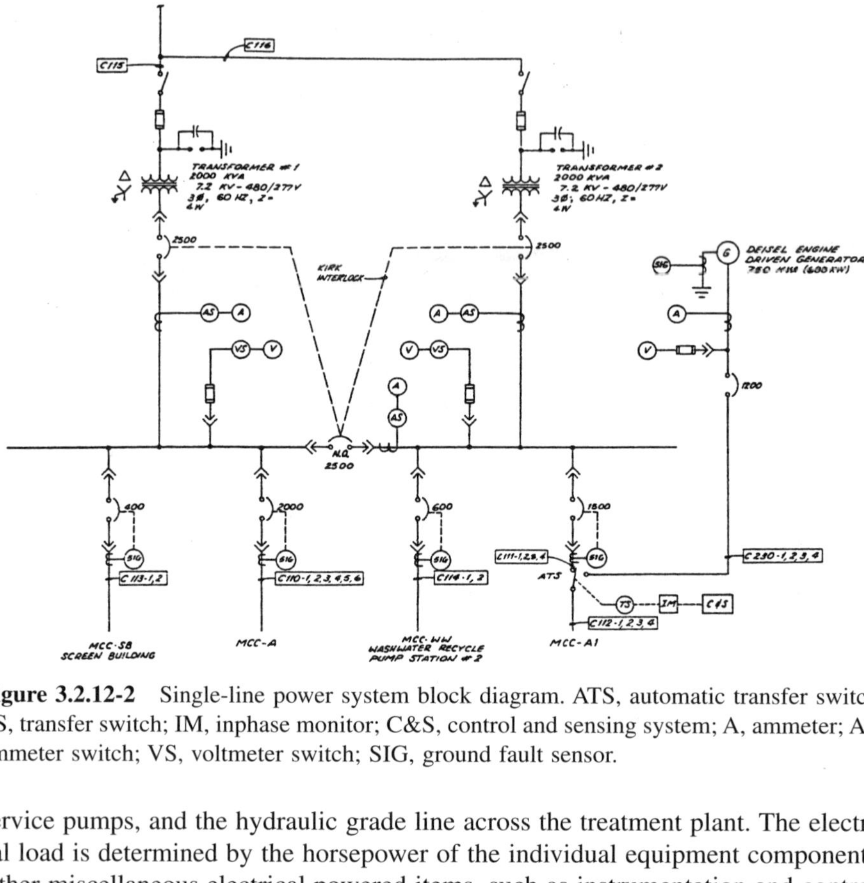

Figure 3.2.12-2 Single-line power system block diagram. ATS, automatic transfer switch; TS, transfer switch; IM, inphase monitor; C&S, control and sensing system; A, ammeter; AS, ammeter switch; VS, voltmeter switch; SIG, ground fault sensor.

service pumps, and the hydraulic grade line across the treatment plant. The electrical load is determined by the horsepower of the individual equipment components. Other miscellaneous electrical powered items, such as instrumentation and control, valve operators, chemical feeders, and lighting, are based on the actual number and scale of each unit. Table 3.2.12-1 presents an example of estimated power require-

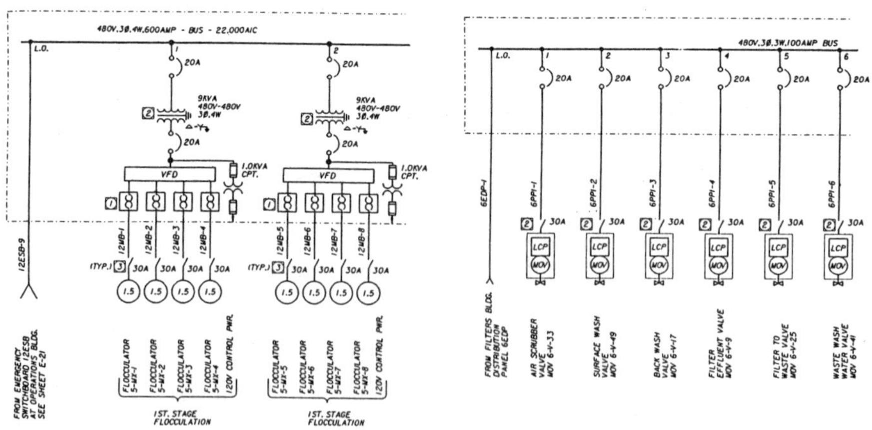

Figure 3.2.12-3 480-V single-line diagrams.

TABLE 3.2.12-1 Estimated Power Requirement of North Bay Regional Water Treatment Plant (30 to 60 mgd Ultimate)

Location/Item	Number of Units	Total Connected Load	
		hp	kW
Treated water pump station			
Lighting panel	1	—	15
Exhaust fan	1	1.5	
Backwash pumps	2	400	
Treated water pumps	9	2700 (5400 ult)	
Filters			
Lighting panel	1		10
Exhaust fan	1	1.5	
Sump pumps	4	12	
Actuated valves	16	36	—
Operations building			
Lighting panels	2	—	40
HVAC panel	1	—	30
Instrument panel	1	—	20
PAC	1	10	
Power panels	2	—	40
Chlorine monorail	1	2	
Chlorine scrubber	1	50	
Roll-up doors	3	3	—
Spare chemicals mixer	1	1	
Activated carbon mixers	2	50	—
Dust collectors	2	6	—
Metering pumps	20	16	—
Exhaust fans	4	6	—
Chemical transfer pumps	2	4	—
Actuated valves	7	10	—
Miscellaneous			
Wash-water recovery pumps	3	40	—
Sludge decant recovery pumps	2	4	—
Sludge pumps	2	1.5	
Bar screens	2	2	—
Fuel oil pump	1	1	—
Security gates	2	2	—
Sample pumps	6	2	—
Boiler pump	1	1	—
Primary A/C unit	1	1.5	—
Condenser	1	0.5	—
Floc/sed basins			
Flash mix pump	1	10	
Flocculators	36	54	—
Chain-and-flight collectors	3	6	—
Sludge cross collectors	3	3	—
Actuated valves	6	4	—

TABLE 3.2.12-1 ***(Continued)***

Location/Item	Number of Units	Total Connected Load hp	Total Connected Load kW
Generator/blower building			
Lighting panel			30
Blowers	2	150	—
Air compressors	2	20	—
Exhaust fan	1	1.5	
Ozonation building			
Lighting panel	1		20
Power panel	1		30
Ozone system (750 lb/day)	1		400
Raw water supply			
NBA pumps	3	175	
Totals		3787 (6487 ult.)	635

ments for a 30-mgd (1.32-m^3/s) plant. This plant requires 3800-kVA process motors, ozonation, and 500-kVA lighting and miscellaneous loads. Table 3.2.12-2 illustrates approximate relation between plant size and estimated power requirements.

Standby Power As mentioned previously, water treatment plants should be operated in a reliable manner; therefore, standby power is required in case of power outage unless the power to the plant is supplied by two different substations. Although this is a site-specific issue, a general list of essential equipment that should be connected to the standby power system is listed below.

- Instrumentation and control system
- Major chemical feeders (coagulants and disinfectant)
- Electrically operated major valves
- Flash mixer
- Lighting, heating, ventilating, and air-conditioning systems in the operations building
- Safety light for all areas with process units and the security system
- 50% of mechanical flocculators
- 25 to 50% of ozone generators
- 25 to 50% of intake and high-service pumps
- Minimum of one utility pump but all sump pumps
- One pump for filling the backwash tank
- All water sampling pumps and water quality analyzers
- Refrigerators, incubators, and other essential analyzers in the laboratory

TABLE 3.2.12-2 Approximate Relation Between Plant Size and Power Requirement (kVA)

Plant Size	Conventional Treatment Process	
	Without Ozonation	With Ozonation
5 to 10 mgd (0.22 to 0.44 m^3/s)	2100	2400
10 to 25 mgd (0.44 to 1.1 m^3/s)	3300	3600
25 to 50 mgd (1.1 to 2.2 m^3/s)	3400	3800
50 to 100 mgd (2.2 to 4.4 m^3/s)	5600	6500

Note: $\text{kVA} = \dfrac{\sqrt{3} \times \text{volts} \times \text{ampers}}{1000}$

Voltage: The larger the load, the highter the voltage requirements, so that large pumps and ozone generators need high-voltage service. According to Honeycutt et al., the approximate voltage requirements for certain sizes of single motors, including starter and conductors, are as follows:

For motors up to 500 hp: 480 V
For motors 500 to 5000 hp: 5 kV (2400 or 4160 V)
For motors larger than 5000 hp: 15 kV (12,470, 13,200 or 13,800 V)
For an entire plant, the approximate total horsepower values for different voltages are:
Up to 2500 hp: 480 V
From 2500 to 10,000 hp: 5 kV
Above 10,000 hp: 15 kV

If the filters are washed by direct pump wash, the wash pumps will require standby power if potential power outage is expected to last more than one day.

Economic evaluation of the various peak shaving alternatives depends on the rate structure of the utility. There are essentially four cases of peak shaving operation: (1) load shedding can be accomplished by reducing load using either a computer or microprocessor-based controller; (2) utility peak demand reduction involves the transfer of a portion of the load from the utility source to the plant's engine generator sets during times when the demand for electricity is at its highest; (3) when paralleling with the utility, the generators are directly connected to the normal distribution system—peaking power is provided directly into the system and the amount of power furnished is determined by the engine governor controls; (4) a prime power cogeneration system designates the on-site generation system as the prime source of power and only certain dedicated loads are connected to the on-site generators: The utility may be used as a standby source through the use of an automatic transfer switch; however, this alternative is more appropriate for wastewater treatment plants since methane gas produced by the sludge digestion process can drive the on-site generators.

Voltage-Drop Issue The current required to start a motor with full voltage is as high as six times the full-load running current; therefore, a protection measure should be provided. However, a 15% drop is generally considered to be tolerable. The three basic reduced-voltage starters for motor starting control measures are (1) autotransformers, (2) wye–delta starters, and (3) solid-state starters. Autotransformers can be set at 80%, 65%, and 50% of full voltage for starting (the 65% setting is commonly used). Wye–delta starters are able to control starting current to approximately two times the running current, as compared to six times. Solid-state starters provide flexible adjustment of the starting condition; therefore, there is no problem with transition current.

Miscellaneous Issues Correction of the power factor improves the power consumption of motors; therefore, electrical engineers should provide power factor capacitors. The maximum power factor is 1.0, but some existing plants have a maximum power factor of 0.9 or less. Selection of the motor is another important issue. Induction motors are generally more cost-effective than synchronous motors and are used for smaller to midsized pumps. Synchronous motors are more favorable for use in high-horsepower and slow-speed applications; however, the power factor is about 0.8. Thus correction of the power factor is essential. Installation of volts ampere reactive (VARs) is a common solution for this problem.

BIBLIOGRAPHY

Honeycutt, D. E., and Jones, J., "Electrical Systems," Chapter 21 in *Water Treatment Plant Design,* 3rd ed., AWWA and ASCE, McGraw-Hill, New York, 1997.

Square D Company, *Motor Control Fundamentals,* Square D, Oklahoma 1974.

CHAPTER 4

Subordinate Plant Facilities

4.1 CHEMICAL FEED SYSTEM

Purpose Most water treatment plants use a variety of chemical engineering processes to produce large quantities of safe drinking water within a short period of time. Therefore, the design of the chemical feed system, including handling, storing, feeding, and mixing, requires special attention. A properly designed chemical feed system can improve treatment efficiency and thereby reduce operation and maintenance costs. Three basic forms of chemical are used in water treatment: dry, liquid, and gas. It therefore follows that the basic types of chemical feeders are dry, liquid, and gas feeders. Liquid volumetric feeders are preferred because of their compact size, accuracy, and convenience in handling. Generally, chemical feeders must have a feeding capacity that is broad enough to accommodate the variations in plant flow rate and chemical dosages.

Water Treatment Chemicals Water treatment chemical should be of a high-grade food additive class or those approved by the EPA for the treatment of drinking water. The quality standards for the major water treatment chemicals are established by the American Water Works Association (AWWA). The chemicals may be classified according to their application in water treatment. The chemical names, commercially available form, and approximate unit cost (bulk purchase: 1998) are listed below. Some important characteristics of the chemicals may be found in Table 4.1-1 and Appendix 13.

COAGULANTS AND FLOCCULANTS

Aluminum sulfate (alum)	Dry or 49% liquid	\$250 per ton in dry form
Ferric chloride	35 to 45% liquid	\$250 to \$350 per ton in dry form
Ferric sulfate	Dry form	\$350 per ton
Cationic polymers	Liquid form	\$1500 per ton
Anionic polymers	Primarily dry form	\$5000 per ton
Nonionic polymers	Primarily dry form	\$5000 per ton
Lime (for softening)	Dry form	\$70 per ton
Soda ash	Dry form	\$400 per ton

DISINFECTANTS

Chlorine	Liquid form	\$250 per ton
Chloramines (Cl_2 + NH_3)		\$300 per ton for NH_3 alone
Chlorine dioxide (Cl_2 + $NaClO_2$)		\$4500 per ton for dry $NaClO_2$ alone
Ozone (on-site gas generation)		\$2000 per ton for cost of generation
Liquid oxygen (LOX)		\$60 to \$80 per ton
Sodium hypochlorite	Liquid form	\$500 to \$700 per ton Cl_2 equivalent

PH CONTROL

Lime [$Ca(OH)_2$]	Dry form	\$100 per ton
Caustic soda	50% liquid	\$600 per ton as NaOH
Soda ash	Dry form	\$400 per ton
Sulfuric acid	98% liquid	\$200 per ton

TASTE AND ODOR CONTROL

Powdered activated carbon (PAC)	Dry form	\$1300 per ton
Granular activated carbon (GAC)	Dry form	\$2100 per ton
$KMnO_4$	Dry form	\$1800 per ton or \$700 per cubic yard
Ozone (on-site gas generation)		\$2000 per ton

CORROSION CONTROL

Sodium metaphosphate	Dry form	\$1800 per ton
Zinc orthophosphate	Liquid form	\$1500 per ton
Caustic soda	50% liquid	\$600 per ton as NaOH

FLUORIDATION

Hydroflurosilicic acid	23% liquid	\$850 per ton as F
Sodium fluoride	Dry	\$400 per ton
Sodium silicofluoride	Dry	\$1300 per ton as F

SOFTENING

Lime (CaO)	Dry	\$60 per ton
Caustic soda	50% liquid	\$600 per ton
Carbon dioxide	Liquid	\$70 per ton
Sodium silicate	Viscous liquid, S.G. = 1.39	\$150 per ton

ALGAE CONTROL

Copper sulfate	Dry	\$1200 per ton

Considerations Prior to designing the chemical feed system, the engineer should be familiar with the raw water quality, the desired finished water quality, and the treatability of the water. These factors determine the types of chemical that can be employed and their dosages. The design engineer may then address the following issues: the characteristics of each chemical, local conditions, mode of delivery, availability of the chemicals, cost, local and federal safety codes, and procedures for handling the storage.

Chemical characteristics that should be analyzed are the stability (storage life), corrosivity, crystallization temperature (freezing), safety, potential hazard to health, hygroscopicity, and flammable or explosive nature. Similar to the chemical characteristics, local conditions affect the cost and the form in which the chemicals are transported and stored. The local conditions may also affect the reliability of the chemical supply. Additional design considerations include the ambient temperature, humidity, amount of snowfall, potential for flooding, distance from major freeways and rail lines, and the proximity of the plant site to residential and commercial areas.

The method of chemical transportation is an important issue because it determines the design of the receiving facility. The treatment chemicals may be delivered in the form of bags via truck or in bulk form by semitrailer, rail, or barge. If the chemicals are delivered by trucks, the access road must be wide, with an adequate turning radius. The engineer must also establish the proper grade and the type of pavement (access road), in addition to researching the weight limits of the public road and bridges.

The availability of the water treatment chemical is a very important consideration for plants located in remote areas and developing countries. For example, polymers and liquid chemicals, such as liquid alum and liquid caustic soda, are seldom available in these regions. Engineers are therefore advised to research the local availability of particular chemicals prior to designing the system.

An issue related to chemical availability is cost. Depending on the availability, the difference in cost between two chemicals may be quite significant. For example, caustic soda is generally 10 times more expensive than lime. Chemical cost is also affected by the manner in which it is purchased; bulk versus small amounts. Engineers are encouraged to study market trends in order to predict possible shortages in chemical supply, which increases their cost substantially.

With regard to the handling and storage of water treatment chemicals, the design engineer must be thoroughly familiar with the local, state, and federal safety regulations. In compliance with OSHA regulation, plants should have emergency showers, eyewashes, exhaust fans, and special clothing and gear for the safety of the workers. Additionally, chemical storage tanks must be located within a bermed area—to contain chemical spills—and acid holding tanks must be isolated from alkali chemical storage tanks by a separation berm. OSHA also demands that oxidants such as chlorine and potassium permanganate be stored in a separate room from organic chemicals such as polymers and activated carbon.

The handling and storage of treatment chemicals must be based on the nature of the chemicals. Hygroscopic chemicals such as ferric chloride, ferric sulfate, potassium permanganate, soda ash, calcium hypochlorite, and alum are generally shipped and stored in dry form and therefore require special packaging and storage to protect them from moisture. The powdered form of all substances is potentially hazardous. If the dust concentration is allowed to reach a certain level, one spark will cause an explosion. Ozone, oxygen, calcium oxide (quick lime), sodium chlorite, and activated carbon are all considered to be fire hazards; some are also explosive and toxic. They must therefore be handled and stored in strict accordance with the safety codes and rules provided by the manufacturer.

There are two other miscellaneous but important design concerns: clogging of the feed line and the mode of chemical feeding. Clogging of the chemical feed pipe is a common, yet serious problem. Alkali chemicals that include hydroxide and metal coagulants often produce heavy scale in the feed lines and diffusers. Many treatment plants dilute the chemicals downstream of the feeder to obtain better diffusion of the chemical within the process water, to minimize the feed lag time, and to avoid having the slurry settle in the feed line. However, extreme caution must be exercised when diluting metal coagulants such as alum and ferric chloride, due to their pH and solubility relationship (see Figure 3.2.3-1). In the case of alum and ferric chloride, solutions up to 1 and 5% (using ordinary filtered water), respectively, may be the maximum safe dilution. Alkaline chemicals such as lime, caustic soda, and aqua ammonia must not be diluted by carrying water unless the total hardness of the carrying water is less than 20 to 30 mg/L; otherwise, calcium carbonate scales will form rapidly, due to the softening reaction.

Regardless of the metered solution concentration of the chemicals, the diffuser pipes can still potentially become clogged as the result of instant metal hydroxide formation or softening reaction that occurs when these chemicals hit the processing water, provided that the hardness is moderate or high. For this reason, they should be designed for easy cleaning, and if the diffusers are not easily accessible, an alternative diffuser should be provided as a backup. The ideal design is to locate the feeders as close as possible to the application points and to apply these chemicals from above the water surface.

It is worth restating the importance of continuous chemical feed versus the on–off mode. The pulsation pump is not the primary choice as a feeder unless a pulsation damper can be installed to smooth out the discharge flow: noncontinuous chemical feeding will overdose one particular part of the process water while leaving the rest devoid of chemicals. Certain chemicals, such as alum, undergo instantaneous and irreversible reactions. From the view of the plant operators, chemical feed pipes should be installed at every possible location. The convenience of changing chemical feed locations will offset the initial cost of installation. If there is a possibility of future plant expansion, the design should include the space and capacity required by the additional feeders, feed lines, and appurtenances.

Overview of the Chemical Feed Systems Three types of chemical feed systems are used in water treatment: (1) liquid chemical, (2) dry chemical, and (3) gas feed. Of the three types, the liquid chemical feed system is preferred because it is clean and compact, is not labor intensive to operate, has easy automatic controls, and does not have problems with chemical dust or fumes.

Liquid Chemical Feed System The simplest, proven, and most cost-effective liquid chemical feed system has a pump capacity that is at least 10% larger than the maximum feed rate (alternative 1, Figure 4.1-1). This system contains a rotameter, flow control valves, shutoff valve(s), and a recycling pipeline; the recycling line can return the excess flow either to the tank or to the pump suction line. This type of system is not suggested for use with chemicals that coat the interior of the rotameter (glass tube). Chemicals classified under this category are caustic soda, ferric chloride, potassium permanganate, and certain types of polymer. It is recommended that rotameters not be installed in locations exposed to sunlight, since algae will grow within the glass tube.

One major drawback of this system is that there is no simple way to pace the chemical feed rate to changes in the plant flow rate. However, this problem can be avoided by using a variation of alternative 1—installing a rotodip type of feeder (alternative 2 of Figure 4.1-1). The rotodip feeder receives a 4- to 20-mA signal from the plant flow meter and is therefore capable of pacing itself to the plant flow rate. The feeder has a set of revolving cups and contains a unique feature, a wide turn-down ratio. A ratio of 1500 : 1 may be achieved by changing the revolving speed, the submergence of the cups, and the speed reduction gears. It requires minimal maintenance and can easily last more than 20 years under normal operating conditions.

The majority of modern water treatment plants use metering pumps. Most such pumps have a 150 : 1 feed range, including adjustment of the pump stroke length (15 : 1) and motor speed variation (10 : 1). The advantages of a metering pump are its simple pacing of the chemical feed rate to the plant flow rate, a very high feeding accuracy (1%), a discharge pressure up to 150 psi (1035 kPa) or more, and compact size. Refer to Figure 4.1-2 for an illustration of this second type of feed system.

A third type of chemical feed system combines a magnetic flow meter with a needle valve (for flow control) and a pressurized storage tank (Figure 4.1-3). The pressure level within the air padded tank depends on the back pressure from the processing waterline. For example, if a chemical is fed into an open channel, an air padding pressure of 10 psi (70 kPa) will be sufficient for the storage tank. This system is simple and has proved to be effective, but it has two major disadvantages: a limited turn-down ratio with an overall practical limit of 15 : 1, and a potential for clogging of the flow control valve and meters. A variation of this system uses a recycling pump rather than the pressurized storage tank, in much the same manner as the first feed system containing the rotameter. Use of a recycling pump is cost-effective since there is no need to specify a thick shell wall tank, due to the lack of internal pressure.

Since exposure to some water treatment chemicals is hazardous to human health, the plant design must conform to the local safety codes and OSHA regulations. All storage and handling facilities should therefore have eyewashes, showers, clear warning signs, berms for containments, and adequate lighting and ventilation. All storage tanks must have tags labeling their contents, an access hole, fill line, drain, overflow, discharge valve, vent, and content indicators, such as a load cell or a dif-

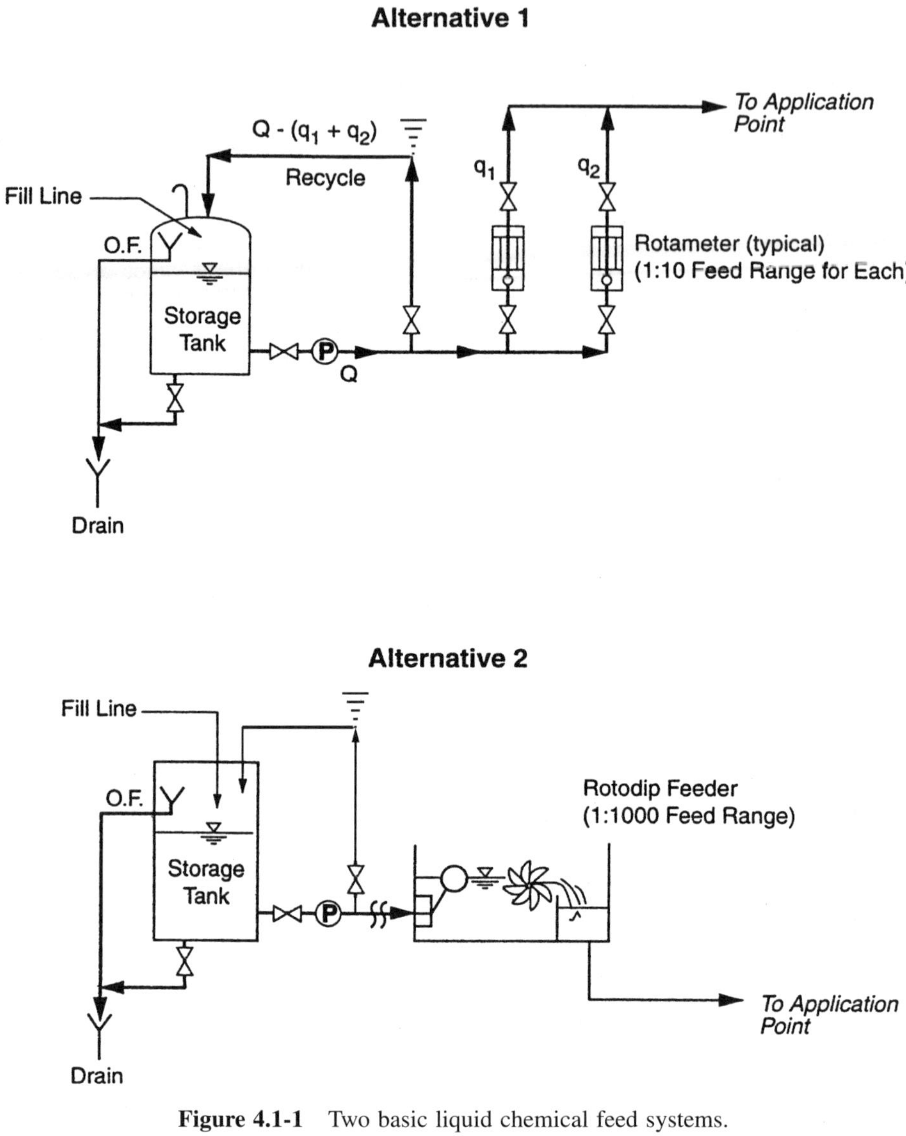

Figure 4.1-1 Two basic liquid chemical feed systems.

ferential pressure cell, or sonic-level indicators. Optionally, the tanks may contain a mixing device. The use of glass-sided gauges is not recommended, due to the potential for breakage and because they become fogged by the chemicals. Use of a bubbler tube system should be avoided, as it may introduce oxygen, carbon dioxide, or bacteria into the stored solutions and therefore degrade certain chemicals, such as caustic soda and polymers.

For any of these systems, the most important design consideration is the selection of materials that are most compatible with the chemicals to be handled—ma-

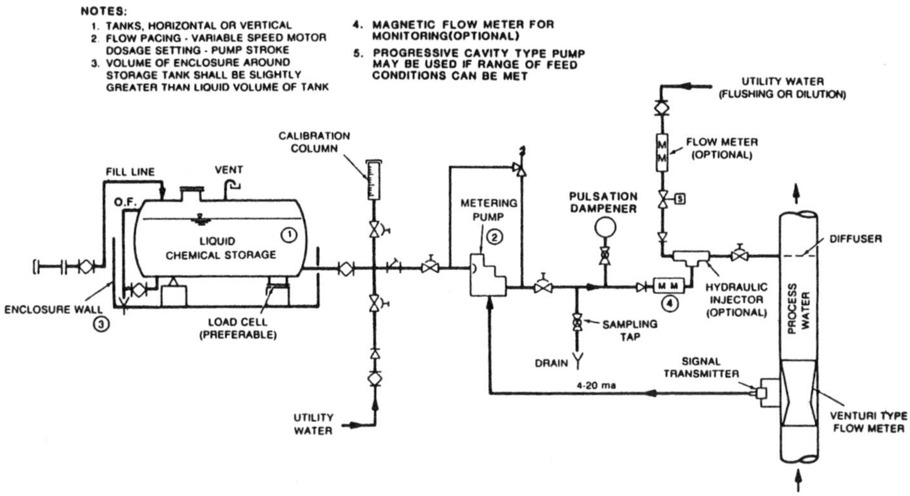

Figure 4.1-2 Liquid chemical feed system (metering pump system).

terials used in constructing the metering unit, valves, piping, and storage tank. Other design considerations include the ease of operation, reliability, safety, cost, and the method by which the chemicals are fed: It is essential to use continuous feeding and not intermittent feeding of chemicals.

Dry Chemical Feed System The dry chemical feed system is a well-established and reliable method of adding dry or solid chemicals. It has been used successfully in industrial and water treatment applications for many years. This type of system may easily be automated and problems associated with dust control can be minimized with the installation of an appropriate type of control system. Yet the dry

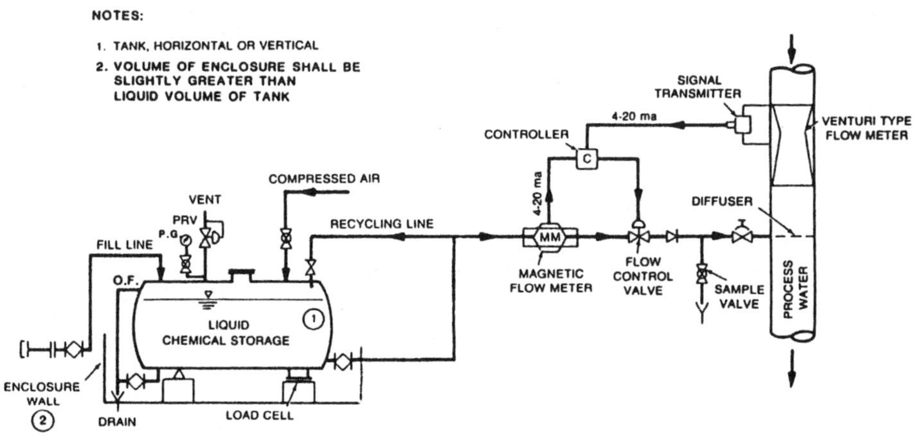

Figure 4.1-3 Flow pacing solution feed system (closed-loop control with magnetic flow meter).

chemical feed system is more labor intensive (operation and maintenance) than the liquid chemical feed system.

The dry feed system is usually not preferred for most plant operations. However, certain chemicals, such as lime, potassium permanganate, PAC, and most anionic and nonionic polymers, are available only in dry form. Moreover, a few plants can only utilize chemicals that are in dry or solid form, due to problems associated with the liquid form: freezing, high delivery cost, degradation of the liquid form of the chemicals when stored for long periods of time, or a limited selection (polymers) compared to the dry form.

The design considerations for the dry chemical feed system focus primarily on the storage silo and dissolving tank. The major concerns of the engineer are explosion-proof handling of dusty materials and the health and safety of plant operators. An appropriate type of dust collector unit and an exhaust fan must be installed in the feed rooms. Engineers should realize that all bag-type dust collectors require the dust to be completely dry. If this cannot be accomplished, a water scrubber type of collector should be specified.

For any type of solid material, especially the powdered form, the angle of the hopper bottom is an important issue. The bottom of the hopper should have an angle no greater than 30° from the vertical and should be fitted with ordinary vibrators. However, if a bin activator, manufactured by Vibra Screw, Inc., is installed in the bottom, the hopper angle may be more shallow, due to the effective control in bridging and flashing the dry material. Since the surface of the dry material stored in the silo is generally not level, it is very difficult to select a reliable inventory indicator. Currently, the best and most reliable method of keeping inventory is the installation of load cells to weigh both the silo and its contents.

In the case of hygroscopic chemicals, the storage silo should be made as hermetic as possible; this will minimize caking caused by atmospheric moisture. The silo should also have a discharge bin gate to isolate the feeder for service. This gate should be followed by a flexible connection, which isolates the vibrations of the feeder and facilitates the simple connection of the feeder to the bin outlet. If a pneumatic system is used to load the dry chemicals into a bulk storage silo, the fill connection must be of a suitable size and type so that it is compatible with the delivery trucks of the anticipated chemical suppliers. Other essential items that should be incorporated into the design are the use of elbows with a long radius in the fill line and a dust collector, which should be installed in an easily accessible location. Recently, a special type of elbow (Vortice Ell by Hammer Tek Corp.) has become available and has demonstrated very good performance.

Engineers are also advised to investigate commercially available types of packaged bulk storage which are combined with feed and mixing systems. One example is the ChemTower by suppliers such as Smith and Loveless, Advance Industries, Inc., and others. For better chemical dispersion a dissolver or slurry tank is generally required to create a chemical solution or slurry prior to application to the processing flow. Thus the dry chemical feed system (storage bins or bag storage area, dust collectors, feeder, dissolvers, duct or pipe works, and grit or waste bins) occupies a considerable amount of floor space and height.

Selection of the dissolving or slurry tank should be based on the characteristics of the chemical chosen. The necessary design considerations include the solubility of the chemical at the lowest water temperature, the time required to dissolve the chemical at the maximum feed rate, the appropriate ratio of water to chemical that creates a solution or slurry with the proper concentration, the amount of heat generated during hydration, the possibility of flow short-circuiting within the tank, and installation of overflow and drainage pipe connections that are adequate in size and in the proper location.

The feed line should provide cleanout fittings, in addition to a connection for flushing water, so that any solids or grit will be flushed out before they clog the feed pipe. The flow velocity in the feed line should be approximately 3 ft/s (0.9 m/s). If the slurry form of the chemical is to be fed to the processing water, the pipeline should avoid sharp turns and steep rises and falls because these features promote caking and increase the headloss. In other words, the design should minimize erosion and the sedimentation of slurry within the pipeline.

Figure 4.1-4 is a schematic illustration of the general dry chemical feed system. Figure 4.1-5 is a lime feed system that is installed directly above the application

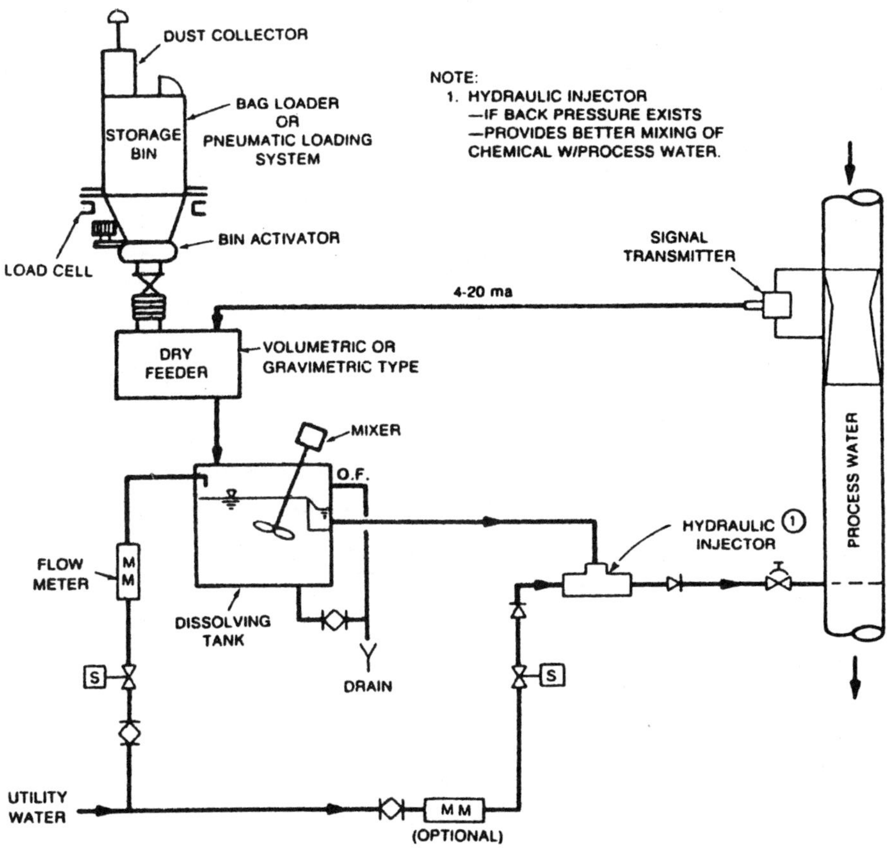

Figure 4.1-4 Dry chemical feed system with dissolving tank.

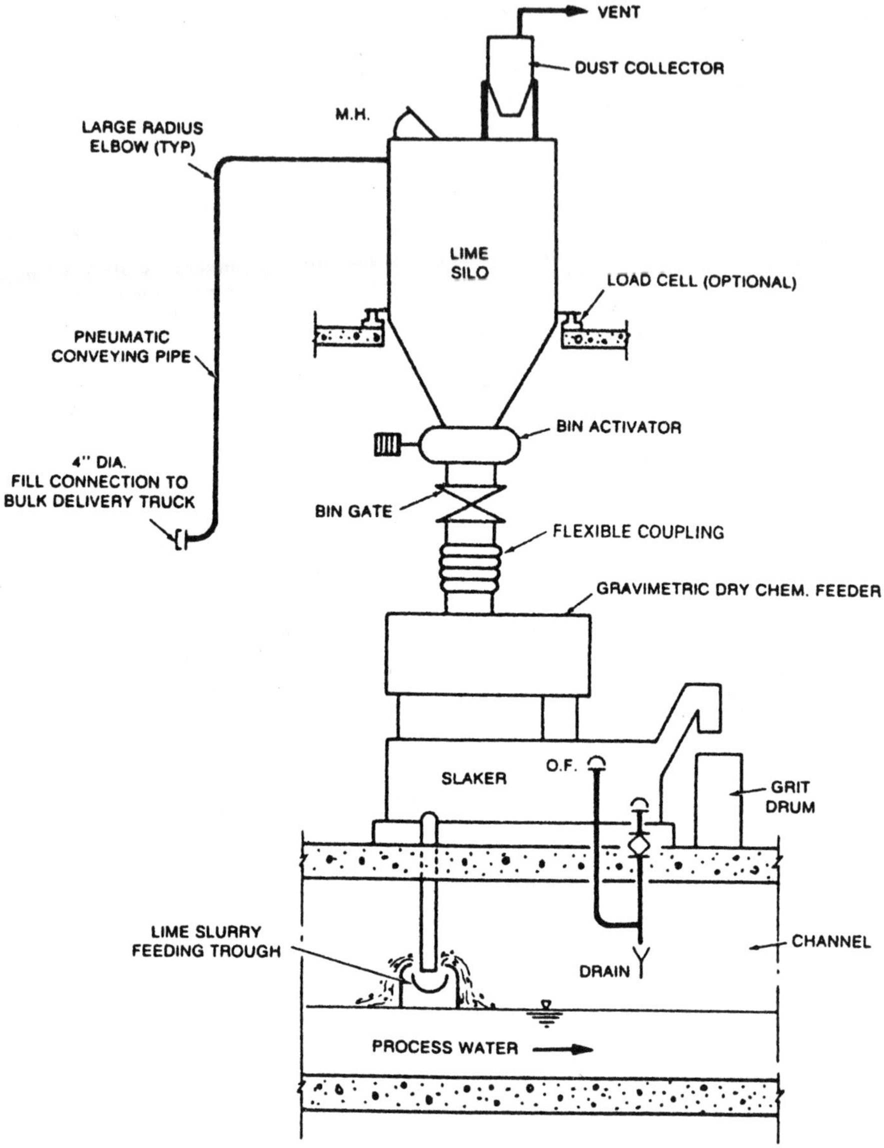

Figure 4.1-5 Lime (quick lime) feed system.

point. Figure 4.1-6 is a dry polymer feed system that may also be used as a liquid polymer feed system. This last system is proven and performs well with minimal operation and maintenance effort. However, the skid-mounted, self-contained system generally costs about $70,000. There are a few types of inexpensive dry polymer feed systems that utilize high-speed blenders without an aging tank. However, these systems are not effective, due to the high shear forces generated by the blender.

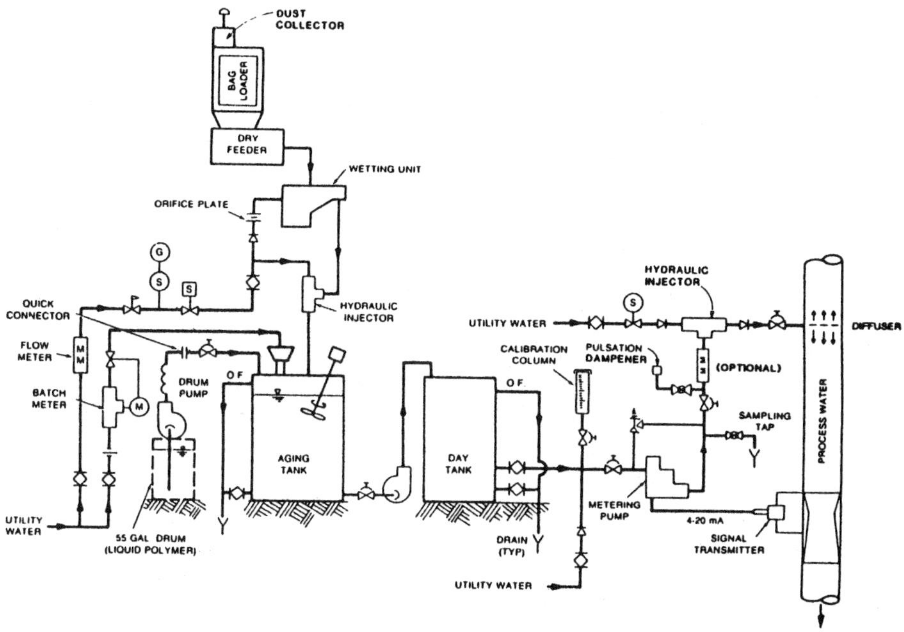

Figure 4.1-6 Nonionic/anionic polyelectrolyte feed system.

The basic types of dry chemical feeder are gravimetric and volumetric feeders. The gravimetric feeder is preferred because of its wide turn-down ratio (100 : 1), high accuracy (1%), easy automation, and ability to feed information directly to a computer. Moreover, plant operators do not have to refer to a calibration chart to set the chemical dosage. However, these units are significantly more expensive than the volumetric dry feeders. Typical gravimetric dry feeders are the belt type and loss in weight type.

Volumetric feeders include the oscillating hopper, vibrating feeders, grooved disk feeders, screw (helix) feeders, and belt feeders. The first two types should not be considered for use in the water treatment field because of their low feeding accuracy (10%). The grooved disk feeders are highly accurate but have a limited capacity. Their application is therefore limited to small plants. The screw type and belt feeders are reasonably accurate (± 1 to 3%). The feeding range of the screw feeder is 20 : 1 and 10 : 1 for the belt feeder.

The last two types are most commonly used in water treatment because of their lower cost compared to the gravimetric feeders. Yet the volumetric feeder also has a disadvantage: The feed rate setting must be calibrated frequently, due to variations in bulk density from batch to batch. This difference is the result of purchasing the same chemical from different suppliers, since the purchase of the chemicals is based on bids. The variations may also be attributed to the particle size, density, shape, and moisture content of each batch. If different types of chemicals are to be fed by means of a volumetric feeder, each chemical must have its own calibration chart, so that its dosage can be set for the feeder.

Gas Feed System With the exception of ozone and carbon dioxide, the gaseous form of a chemical is usually fed to the process water as a solution, provided that the gas is reasonably soluble in water. Although chlorine, anhydrous ammonia, and sulfur dioxide are shipped and stored in steel cylinders as liquefied gas, they are commonly made into a solution by means of a hydraulic eductor and fed to the process water as a solution. However, the liquefied gas is generally evaporated to form a gas, then injected into the process water by means of a vacuum pressure or direct gas feed system.

Gas feed systems are generally simple in mechanism and easy to operate. The solution feed vacuum type of feeder is manufactured as a proprictary type of feeder and is commonly used in chlorination, ammoniation, and dechlorination, and in feeding chlorine dioxide. This type of gas feeder shuts off the gas automatically upon a loss in vacuum, loss in operating water pressure, or when the solution stops flowing from the discharge line. Direct gas feeders are seldom used but may be utilized in situations where pressure water and/or electricity is not available. It is therefore used primarily in very small facilities such as campsites or small plants located in remote areas or in developing countries.

Both carbon dioxide and ozone, particularly ozone, have extremely low solubility in water (refer to Table 4.1-1). These gases must be fed to the process water via a rotameter, where they are released as fine gas bubbles from the bottom of the contact channels. The contact channels have a water depth ranging from 12 to 18 ft (3.5 to 5.5 m). The ozone contact channel must have a water depth of over 18 ft (5.5 m), due to the very poor solubility of ozone in water. The chlorine gas feed system is a typical gas feed system and is illustrated in Figure 3.2.9-2. The ozone feed system is discussed in detail in Section 4.8.

Required Capacity for the Chemical Feeder and Storage The feeder should be able to cover both the maximum and minimum chemical feed requirements. If the turn-down is not large enough to meet the maximum and minimum requirements, two sets of feeders may be needed to cover the entire chemical feed range adequately. The feed range may be computed by the following method (lb/hr):

$$\text{maximum feed rate} = [(\text{maximum dosage} \times 8.34) \times Q_{\text{max-day}}]$$

$$\text{average feed rate} = [(\text{average dosage} \times 8.34) \times Q_{\text{ave-day}}]$$

$$\text{minimum feed rate} = [(\text{minimum dosage} \times 8.34) \times Q_{\text{min-day}}]$$

$$\text{lb/hr} = [(\text{mg/L} \times 8.34) \times \text{mgd}] \div 24$$

NOTES:

1. Minimum plant flow occurs during winter nights or during the period of plant startup.
2. Many chemical feeders have a limited feed range of 10 : 1 to 15 : 1.
3. In the United States, the capacity of chemical feeders is commonly expressed as pounds per hour or gallons per hour.

TABLE 4.1-1 Physical Properties of Major Water Treatment Chemicals

Chemical	Common Solution Strength (%)	Specific Gravity at 60°F (Approx.)	Weight of Anhydrous Chemicals in a Gallon of Solution (lb/gal)	Freezing Temperature [°F (°C)]	Remarks	
Liquid alum	49 (35.8 Be)	1.33	5.4	4.5 (−15)	8.3% Al_2O_3 content	
Poly aluminum chloride (liquid)	10–12 Al_2O_3	1.2		10 (−12)	pH = 2.6 at 25°C	
Ferric chloride (liquid)	39 (42.6 Be)	1.42	4.6	9 (−13)	Liquid $FeCl_3$ is very corrosive	
Ferric sulfate (liquid)	50 (38 cP)	1.48	6.2	−40 (−40)	Liquid $Fe_2(SO_4)_3$ is easier to handle than dry product	
Sulfuric acid	93.2 (66 Be)	1.835	14.3	−20 (−29)	No dry product	
Fluosilicic acid	23–24	1.22	1.89 as F	−7.5 (−22)	H_2SiF_6 is very corrosive	
Zinc orthophosphate	38	1.36	0.83 as Zn	35 (2)	Based on Virchem 932	
Polyphosphate	23 as PO_4	1.22	2.35 as PO_4	50 (10)	Based on Nalco 7399	
Caustic soda	50 (50.2 Be)	1.53	6.38	50 (10)	Dry NaOH is not only hygroscopic but is also dangerous to handle	
	25 (31.5 Be)	1.28	2.65	0 (−17)		
					Vapor Pressure (psig)	
					At 120°F	At 100°F
Aqua ammonia	29.6 (26 Be)	0.897	2.2	−120 (−85)	19	7.8
	25	0.91	1.9	−70 (−57)	8.3	0.9
	20	0.927	1.5	−35 (−37)	0.1	—
Liquid catonic PE	Consider as 100% solution	1.03	8.6	27 (−3)	Based on Cat Floc T by Calgon	
Nonionic and anionic polyelectrolytes	Most nonionic and anionic PE are available as dry products. Only a few are available as liquid, and physical properties vary significantly by manufacturers.					

The quantity of water treatment chemicals that should be stored at the plant site is influenced by local weather conditions, raw water quality, and the local supply situation. Treatment plants are generally considered to be adequately stocked if they have a 15-day supply; this figure is based on the maximum dosage and the average daily flow rate. Remote areas or regions with frequent flooding or heavy snowfall should have a much larger supply.

$$\begin{aligned}\text{storage capacity (lb)} &= [(\text{maximum dosage} \times 8.34) \times Q_{\text{ave-day}}] \times 15 \text{ days} \\ &= [(\text{mg/L} \times 8.34) \times \text{mgd}] \times 15 \text{ days}\end{aligned}$$

If the maximum dosage and the maximum daily flow rate are expected to coincide for more than 1 week, the following equation may be used to calculate the storage requirement:

$$\text{storage capacity} = [(\text{maximum dosage} \times 8.34) \times Q_{\text{max-day}}] \times 15 \text{ days}$$

Specific Design Issues Due to the different nature of the water treatment chemicals, each type of feed system has a few specific design issues that must be addressed by the engineer. This section discusses special considerations associated with the (1) polymer, (2) PAC, (3) potassium permanganate, and (4) lime feed systems.

Polymer Feed System Cationic polymers are generally shipped as a solution in drums or by bulk delivery; bulk deliveries are approximately 3000 gal, but trucks with 1000-gal tanks are also available in some areas. Nonionic and anionic polymers may be shipped in either liquid or dry form. The dry form is preferred by some plants because it occupies less storage area and can be stored for a longer period of time than the maximum effective storage of 1 year for the liquid form.

The system should be designed to feed both liquid and dry polymers, as shown in Figure 4.1-6, especially in cold weather regions and remote areas. This design gives the plant operators maximum flexibility in feeding the polymer. All polymers are very viscous and slippery if spilled on wet concrete floors. Therefore, a concrete curb should be provided around the polymer feed system, in addition to hose bibs and floor drains to contain and wash away the spilled material.

Mechanical mixers with relatively low speeds should be used in preparing polymer solutions. The possibility of breaking the polymer chains is decreased if the mixers are operated at 400 rpm or less. Dilute polymer solutions (1% or less) are subject to rapid degradation. Thus, the capacity of the working solution (day tank) should be sized to require a new batch every 1 to 2 days.

Powdered Activated Carbon (PAC) Feed System There are two basic PAC feed systems.

- *Alternative 1:* dry feeder with bag-loading hopper, extension hopper, dust collector, and either a dissolving tank or a vortex mixer tank
- *Alternative 2:* PAC slurry storage tank with a metering pump

If PAC feeding is infrequent and the feed rate is less than 150 lb/h, alternative 1 should be considered because the capital investment is much less. However, if the system is used frequently and the feed rate exceeds 150 lb/h, alternative 2 is the proper choice. Two major problems have been associated with alternative 2: solidification of the PAC slurry when it is not mixed frequently and the inability of hydraulic eductors to convey the PAC slurry. The mixer for the PAC slurry tanks should be designed to prepare 1.5 lb/gal of slurry (1 lb/gal is standard).

PAC adsorbs organic compounds, including airborne gases. This characteristic may reduce its effectiveness, and engineers must incorporate special storage considerations into the plant design. A water scrubber type of dust collection system should be installed for most PACs and a bag dust collection system should be used exclusively for dry PAC. Carrying water is added to the PAC feed line so that a flow velocity of 5 ft/s (1.5 m/s) is maintained. The feed system should also have cleanouts and large radius elbows.

PAC particles are very abrasive and wet PAC is extremely corrosive. The design of the feed system should take these characteristics into account. A rotodip feeder with a hydraulic eductor system has a very successful operational history, and this system should be considered whenever applicable. A 50% moistured PAC is used exclusively in all major water treatment plants in Japan with practically no dust problems. The product has the consistency of natural table salt and is easy to handle.

Potassium Permanganate Feed System There are two basic potassium permanganate feed systems.

- *Alternative 1:* volumetric dry feed with a bag-loading hopper, extension hopper, dust collector, and dissolving tank (no metering pump)
- *Alternative 2:* a batch system consisting of two large dissolving tanks and a metering pump

Like other chemicals, the optimum dosage of potassium permanganate usually covers a wide band and a deviation of ±5% from the desired dose does not generally affect the result. For this reason, alternative 1 is preferred. Also, the floor area required by alternative 1 is usually half that required by alternative 2 because potassium permanganate has a very low solubility, 3 g per 100 mL at 10°C (50°F) and 2.8 g per 100 mL at 5°C (41°F). Alternative 2 usually requires large dissolving tanks.

Lime Feed System The most effective form of adding calcium hydroxide to process water is quick lime (CaO). Quick lime with a purity of 85% and hydrated lime [$Ca(OH)_2$] with a purity of 90% are easily obtainable. Compared to hydrated lime, both the storage and dosage requirements of quick lime are lower because 1 mol (56 g) of CaO produces 1 mol (74 g) of $Ca(OH)_2$ after hydration with water. Thus quick lime should be considered if lime is to be used frequently at a minimum dosage of 1 ton/day.

Quick lime may be shipped as pebbles or in granular form; the approximate bulk density is 60 lb/ft^3. In contrast, slaked lime has a bulk density of only 30 lb/ft^3 but is available only as a light powder. The quick lime storage silo may therefore be much smaller, and "flushing" and "bridging" problems associated with silo-stored slaked lime may be avoided. Yet quick lime must be slaked prior to use and grit must be removed from the slaker on a routine basis.

The lime feed system should consist of a storage silo with mechanical pneumatic loading, dust collector, bin activator such as those produced by Vibra Screw, bin gate, content indicator (load cells), volumetric or gravimetric dry feeder, slaker (for quick lime), or dissolving tank with mechanical mixer (for slaked lime). For optimal results, the feeder should be located directly above the application point. This arrangement will prevent "caking" within the feed pipes.

In recent years, 35 to 40% solids lime slurry has become available for water treatment for certain regions (especially in Texas). The Houston East Plant [350 mgd (15.4 m^3/s)] was the first large plant to convert successfully from the old quick lime feed system to the lime slurry system, in 1995. Since then, several large water treatment plants in both the Houston and Dallas areas have converted to the lime slurry system; these plants are satisfied with the slurry system, despite a few initial debugging problems.

The slurry system has several advantages: (1) no problems relating to lime dust; (2) much smaller storage and feeding system (overall) compared to dry lime system; (3) practically no problems with lime scale formation within the storage and feeding systems, provided that dilution of neat slurry is not made; (4) virtually no grit; and (5) much less operator attention is required by this system. The disadvantages are that there are a limited number of suppliers, the cost of the solids lime slurry is approximately 12 to 15% higher than for dry lime, and there is increased traffic by tank trailers to fill the storage tanks.

The solids lime slurry is generally stored in at least two to three tanks which have two levels of continuous slow mixing (about 35 rpm) through the use of large-diameter mixing blades (50% of the storage tank diameter). The power requirement is approximately 1 hp per 1000 gal of slurry.

Design Criteria Depending on the form of the chemical, the design parameters for the feeder, storage, feed line, and diffuser may be different. This section discusses these issues for (1) liquid chemical, (2) dry chemical, and (3) gas feed systems. The design criteria for the treatment plant access road and loading dock are also presented.

Liquid Chemical Feed System

Feeder

Type	Metering pumps, rotodip, magnetic flow meter or rotameter.
Feed range	100 : 1. This figure includes a range of 10 : 1 for the dosage setting and a range of 10 : 1 for plant flow pacing. If the feeder does not have a 100 : 1 feed range, in most cases, more than one feeder must be provided to cover the range.
Control	Ideally, the plant will have automatic flow pacing and remote dosage setting control, but this is not absolutely

	necessary. However, local manual control must be provided.
Accuracy	±1.0% is preferred, based on the maximum feed rate.
Storage	
Type	Noncorrosive material. Alternatively, a steel tank may be used if it has a protective lining. The engineer must also incorporate earthquake, freezing, safety, and aesthetic considerations into the design.
Capacity	Normally 15 days of storage, based on the maximum dosage and average daily flow rate.
Control	A content indicator should be provided for each tank and possibly another indicator for the main control board.
Feed line	Noncorrosive materials such as PVC or type 316 SS are commonly used. If pulsation pumps are used, the feed line should be sized to accommodate the peak pulsation flow rate—this rate is approximately three times higher than the rated pump capacity—regardless of whether a pulsation damper is used or not. Flushing water connections and cleanouts are also necessary items.
Diffuser	The type of diffuser depends on the type of chemical used. A perforated-pipe diffuser should be avoided if the chemical has scale-forming characteristics and types C, D, and E of Figure 3.2.3-3 should be considered. Non-scale-forming chemicals will be well dispersed by the perforated-pipe diffuser (refer to types A and B of Figure 3.2.3-3). The diffuser design is based on Figure 3.2.3-4.

Dry Chemical Feed System

Feeder	
Type	Belt feeders for large feed capacities; screw or helix feeders for small capacities; grooved disk feeders for small capacities; loss in weight type or vibration feeders in special cases.
Feed range	Each type of feeder has an inherent feed range and accuracy. The following table summarizes the data.

	Capacity	Feed Range	Accuracy
COMMON FEEDERS			
Volumetric (belt)	Minimum, 0.05 ft^3/min Maximum, 60 ft^3/min	10 : 1 normal 10 : 1 optional	±1% of set rate
Gravimetric (belt)	Minimum, 0.05 lb/min Maximum, 2,000 lb/min	100 : 1 normal	±1% of set rate

Screw (helix type) volumetric feeder	Minimum, 0.02 ft^3/h Maximum, 40 ft^3/h	20 : 1 normal	±1% by volume ±3% by weight
Grooved disk type volumetric feeder	Minimum, 0.17 ft^3/h Maximum, 17 ft^3/h	100 : 1 normal	±1% by volume ±3% by weight
	MISCELLANEOUS		
Loss in weight type (gravimetric)	Minimum, 20 lb/h Maximum, 500 lb/h	10 : 1 normal	±1% of set rate
Vibration feeder (volumetric)	Minimum, 0.7 ft^3/h Maximum 40 ft^3/h	40 : 1 normal	±10% by weight

Control — The same as liquid feeders.

Accuracy — The same as liquid feeders.

Storage — This is basically the same as the liquid feed system with respect to type, capacity, and control. Design considerations are covered in the preceding section.

Feed line — This is fundamentally the same as for the liquid feed system, except that a removable top channel with gravity flow is often used for lime slurry.

Diffuser — The design of the diffuser is the same as that for the liquid feed system. Type E of Figure 3.2.3-3 illustrates a trough type of diffuser.

Gas Feed System

Feeder

- Type — Rotameter system, venturi flow meter system.
- Feed range — 20 : 1 preferred, with a minimum of 10 : 1.
- Control — Automatic flow pacing is required; a closed-loop control with residual analyzer and plant flow meter is preferred.
- Accuracy — ±2% based on the maximum feed rate.

Storage

- Type — A high-pressure steel cylinder is standardly used in storing most liquefied gas. For large installations, a larger-capacity tank mounted on a trailer or a tank railcar (stationary storage tanks) may be used. Ozone must be generated on site.

Capacity — This is basically the same as the liquid feed system.

Control — The same as the liquid feed system.

Feed line — Fundamentally the same as the liquid feed system. However, in the case of ozone, type 316 SS is the only material that may be used.

Diffuser — Similar to the liquid feed system, but type F of Figure 3.2.3-3 must be used for carbon dioxide and ozone.

Design Criteria for the Access Road and Loading Dock The dimensions of typical delivery trucks are presented below.

	Van	Straight Body	Conventional Semitrailer
Length (ft)	15 to 20	17 to 35	55
Width (ft)	7	8	8
Height (ft/in.)	7	13/6	13/6
Floor height (ft/in)	2 to 2/8	3 to 4	4 to 4/4
Track (ft/in.)	5 to 5/4	5/10	6/6
Rear axle (ft/in.)	—	1/3 to 12	4 to 12

The allowable maximum gross weight (lb) may be calculated from $1025\,(D + 24) - 30D^2$, where D is the distance (in feet) between the first and last axle of the vehicle or combination of vehicles (AASHO). Generally, the maximum cargo weight is 18 tons. The required turning radius for a 55-ft semitrailer and truck combination is illustrated in Figure 4.1-7.

Risk Management Planning Water and wastewater treatment plants must comply with the EPA Risk Management Program (RMP) rule (40 CFR 68) by June 21, 1999 if they use, handle, or store more than 2500 lb (1135 kg) of chlorine and/or store 5000 lb (2270 kg) of sulfur dioxide at their facility. Propane has an RMP threshold of 10,000 lb (4500 kg). This rule does not apply to small facilities.

The RMP rule is catagorized under the Clean Air Act Amendments of 1990, Section 112, but is executed differently throughout the United States. The EPA has established three different program levels for facilities affected by RMP. Level 1 facilities present minimal risk to surrounding communities and therefore have reduced compliance requirements. Level 2 and 3 facilities present increasing risk, and their requirements for compliance are more complex. Facilities covered by OSHA generally must comply with the provisions of level 3.

Levels 2 and 3 have a number of similiar RMP requirements: A RMP plan must be written, available to all personnel at the facility, and updated whenever there are changes in the use of threshold chemicals. The plant must document the manner in which the chemicals are used at the facility and include the following five items:

1. A detailed hazard assessment that evaluates the possible hazards of using the chemical at the plant and how these hazards can be reduced
2. Worst-case and most-likely-case scenarios with reasonably good estimates of the population and sensitive receptors (schools, hospitals, nursing homes, and offices) within the affected area
3. A 5-year history of accidents which reports any outside releases and the extent
4. A written emergency response plan that is coordinated with the facility's local emergency planning committees if plant personnel will respond to accidental releases

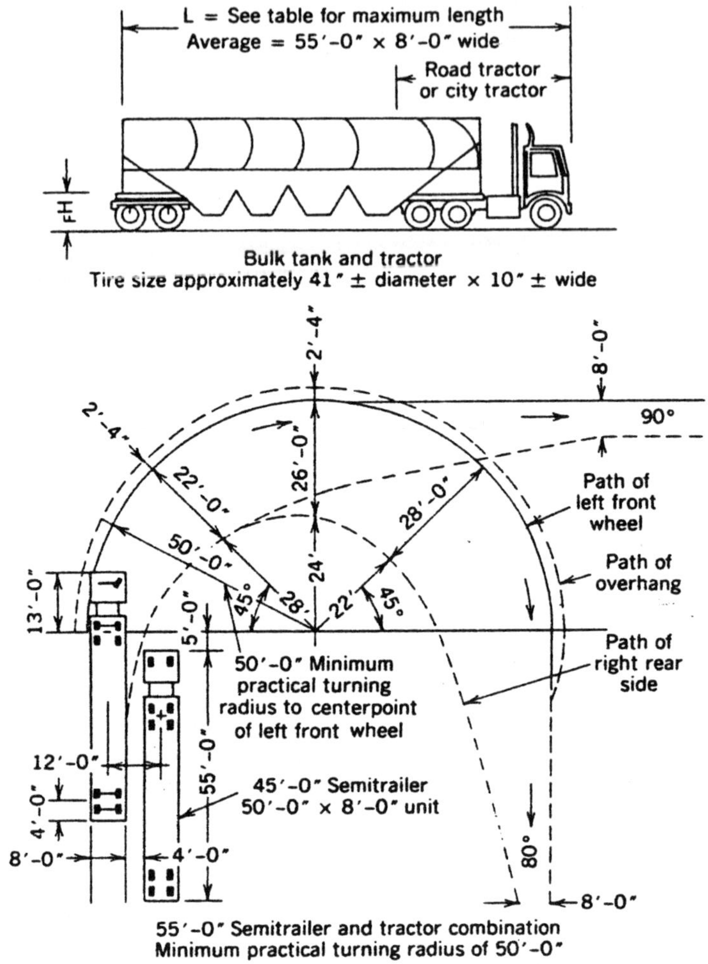

Figure 4.1-7 Turning radius required for a 55-ft semitrailer and tractor.

5. A prevention program that covers safety information, hazard review, operations and maintenance, and management practices, training, incident investigation, and compliance audits

Example Design Calculations

Example 1

Given A treatment plant with a daily average flow rate of 20 mgd is located in North Carolina and requires an alum feed and storage system. The maximum, average, and minimum alum dosages are 50, 15, and 8 mg/L, respectively.

Determine

(i) The number and capacity of each feeder
(ii) The number and capacity of the storage tank(s)
(iii) The major design considerations

Solution The maximum, average, and minimum plant flow rates should be determined first.

$$Q_{\text{max}} = 1.5 \times 20 = 30 \text{ mgd}$$
$$Q_{\text{ave}} = 20 \text{ mgd}$$
$$Q_{\text{min}} = 0.25 \times 20 = 5 \text{ mgd}$$

Commercially available liquid alum is used for operator convenience. Since liquid alum contains 5.6 lb of dry alum per gallon,

$$\text{maximum dosage rate} = 50 \times 8.34 \times 30 = 12{,}510 \text{ lb/day or } 93 \text{ gal/h}$$
$$\text{average dosage rate} = 15 \times 8.34 \times 20 = 2502 \text{ lb/day or } 18.6 \text{ gal/h}$$
$$\text{minimum dosage rate} = 8 \times 8.34 \times 5 = 334 \text{ lb/day or } 2.5 \text{ gal/h}$$
$$\text{turn-down ratio} = 93:2.5 \text{ gal/h or } 37:1$$

(i) The system should have a hydraulic diaphragm type of metering pump with a 100-gal/h capacity, a 100 : 1 turn-down unit, and a second metering pump with a capacity of 50 gal/h. A third metering pump (50 gal/h) is specified as a spare since alum is one of the most important chemicals in the treatment process. Thus, one 100-gal/h and two 50-gal/h metering pumps are specified.

(ii) The storage capacity is calculated to be $[(50 \times 8.34 \times 20) \times 15] \div 5.6 =$ 22,340 gal. Thus one 23,000-gal or two 12,000-gal fiberglass storage tanks should be provided. Freezing is not a likely problem in this location.

(iii) The major design considerations are listed below.

(a) The alum dosage should always be based on dry alum even when liquid alum is used.
(b) Alum is a very corrosive chemical. PVC, FRP, type 316 SS, or Hasteloy C SS are therefore the most appropriate types of material.
(c) Alum should be fed as a continuous flow.
(d) An effective flash mixing system is essential at the application point.
(e) To avoid clogging in both the feed lines and the diffuser, it is recommended that neat alum be fed to the process water.
(f) The alum fill line should be a type 316 SS pipe instead of PVC because liquid alum may be delivered at temperatures over 130°F (54°C). The exits of

the fill line should be above the highest liquid level in the tank because there is a tendency for the liquid alum to shoot out from the air vent at the last moment during the unloading operation; this is a function of the massive air supply from the delivery truck blower.

(g) The feed system should be designed in accordance with the safety codes and rules, including OSHA.

(h) The design should allow for ferric chloride or ferric sulfate (alternative coagulant) to be used in the alum feed system. Since ferric salts are more corrosive than alum and may contain some grit, engineers must take these issues into account when designing the feed system. Examples are the use of a hopper tank bottom to remove grit, Hasteloy C stainless steel as the metal, and Kynar for the plastic parts. Ferric chloride solution freezes at 0°C (32°F) if the solution strength is over 45%, but a 40% solution will not freeze until the temperature reaches 10°F (−12°C), approximately the same freezing point as that of commercial liquid alum (49% solution).

The concept of using an alternative coagulant is important because any treatment plant that achieves alum flocculation above a pH of 7.5 will often have a high concentration of residual aluminum, equal to or above 0.2 mg/L, in the process water. The general public and the EPA are concerned about high levels of aluminum in potable water due to its possible connection to Alzheimer's disease. It should also be noted that ferric coagulant is much more effective than alum for flocculation and clarification in cold months.

Example 2

Given A treatment plant with a maximum daily flow rate of 50 mgd (2.2 m^3/s) is located in California and requires a caustic soda feed system for pH control. The maximum, average, and minimum dosages are 10, 5, and 2 mg/L, respectively.

Determine

(i) The number and capacity of each feeder
(ii) The number and capacity of the storage tanks
(iii) The major design considerations

Solution The maximum, average, and minimum flow rates are as follows:

$$Q_{max} = 50 \text{ mgd}$$

$$Q_{ave} = 50 \div 1.5 = 33 \text{ mgd}$$

$$Q_{min} = 0.25 \times 33 = 8.3 \text{ mgd}, \qquad \text{say, 8 mgd}$$

Commercially available 50% liquid caustic soda is purchased and stored as a 25% solution to avoid crystallization during the winter. The 25% solution contains 2.7 lb of dry caustic soda per gallon.

$$\text{maximum dosage rate} = 10 \times 8.34 \times 50 = 4170 \text{ lb/day or } 64 \text{ gal/h}$$

$$\text{average dosage rate} = 5 \times 8.34 \times 33 = 1376 \text{ lb/day or } 21 \text{ gal/h}$$

$$\text{minimum dosage rate} = 2 \times 8.34 \times 8 = 133 \text{ lb/day or } 2 \text{ gal/h}$$

$$\text{turn-down ratio} = 64:2 \text{ or } 32:1$$

(i) The capacity of the pump selection is the same as for Example 1, except that a spare pump is not specified because this chemical is not absolutely essential in water treatment.

(ii) The storage capacity is calculated to be $[(10 \times 8.34 \times 34) \times 15] \div 2.7 =$ 15,753 gal of 25% solution. Provide two identical 10,000-gal tanks because a dilution step is required even during mild California winters.

(iii) The major considerations for a caustic soda feed system are as follows.

(a) The dosage is based on the dry weight and expressed in mg/L.
(b) A 50% solution of caustic soda will freeze at 55°F (13°C); however, a 25% solution will freeze at 0°F (−18°C); refer to Figure 4.1-8.
(c) Liquid caustic soda is not corrosive to steel.
(d) Caustic soda is toxic and dangerous to handle.
(e) Scale formation will occur when caustic soda is diluted with water unless the water is very soft. Also, heat is generated by dilution.
(f) Caustic soda generally costs eight times more than lime, but the cost of required dosage for pH control is approximately the same.
(g) Plant operators usually prefer to use liquid caustic soda rather than lime since it is easier to handle.
(h) The design of the feed system should be in accordance with current safety rules and regulations.

Example 3

Given A treatment plant with a daily average flow rate of 20 mgd (0.88 m^3/s) is located in Michigan. The plant requires a polymer feed system to function as a flocculation aid, as well as a filtration aid. Pilot tests confirm that an anionic or nonionic type of polymer feed system is needed. The maximum, average, and minimum dosages are 0.5, 0.2, and 0.05 mg/L, respectively.

Determine

(i) The number and capacity of each feeder
(ii) The amount of required storage
(iii) The major design considerations

Solution Commercially available anionic or nonionic polymers are primarily of the dry form; however, a limited number of these polymers are available in liquid form. The system design should allow for both dry and liquid forms to be used, even

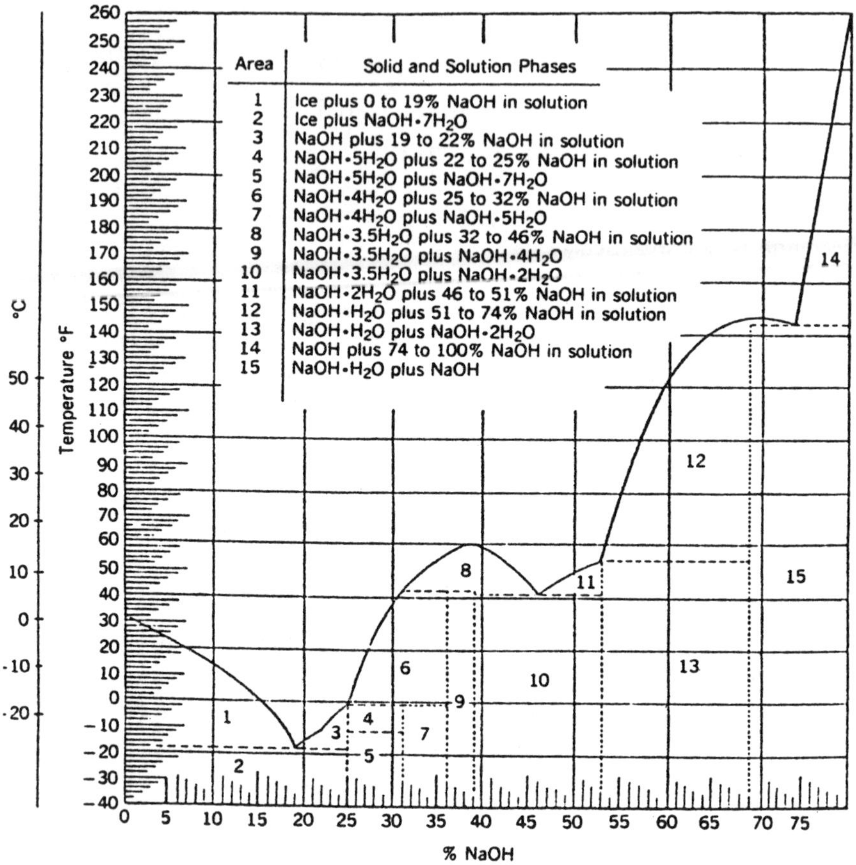

Figure 4.1-8 Graph of freezing points and solubility for a caustic soda solution. Complete solution prevails at points on or above the freezing point line. Below this line, some of the caustic soda exists as a solid, anhydrous or hydrated, depending on the temperature and composition. Numbered areas and table inset indicate the nature of the mixtures. (Courtesy of Stauffer Chemicals, New York, *Caustic Soda Bulletin,* 1965.)

though the severe winter temperatures are not favorable for liquid polymer because of the potential for freezing.

$$Q_{max} = 1.5 \times 20 = 30 \text{ mgd}$$

$$Q_{ave} = 20 \text{ mgd}$$

$$Q_{min} = 0.25 \times 20 = 5 \text{ mgd}$$

Most dry polymers are very viscous, at concentrations over 1%. Therefore, a 0.5% concentration is usually recommended; the 0.5% solution should contain 0.042 lb of dry chemical in 1 gallon of solution.

$$\text{maximum feed rate} = 0.5 \times 8.34 \times 30 = 125 \text{ lb/day}$$
$$= 125 \div 0.042 = 2976 \text{ gal/day or } 124 \text{ gal/h}$$
$$\text{average feed rate} = 0.2 \times 8.34 \times 20 = 33 \text{ lb/day or } 33 \text{ gal/h}$$
$$\text{minimum feed rate} = 0.05 \times 8.34 \times 5 = 2 \text{ lb/day or } 2 \text{ gal/h}$$
$$\text{turn-down ratio} = 124:2 \text{ or } 62:1$$

(i) A hydraulic diaphragm type of metering pump with a 150-gal/h capacity 100 : 1 turn-down, and another unit of 50-gal/h capacity should be installed.

(ii) The amount of storage is calculated to be $[(0.5 \times 8.34) \times 20] \times 15 = 1251$ lb of dry polymer. An ordinary bag of polymer weighs 30 lb. Therefore, $1251 \div 30 = 42$ bags, let us say 40 bags. Twenty bags should be stacked on a pallet for easy hauling and to protect the bags from the wet floor. Additionally, floor space for four 55-gal drums should be provided for the liquid form of the polymer.

(iii) The design considerations for the polymer are as follows:

(a) The dosage is based on the dry weight of the chemical. In the case of liquid polymer, the dosage should be determined by running jar tests, assuming that the stock solution is a 100% solution, for operator convenience.
(b) It is difficult to produce a polymer solution that forms "fish eyes" when making solution from dry polymers using ordinary dissolving tanks.
(c) Polymers that are EPA approved for the treatment of drinking water should be selected and the maximum dosages should conform with the criteria set by the EPA.
(d) Polymers act selectively; care must be taken to choose a polymer that performs best under the given conditions.
(e) The application sequence plays an extremely important role in the effectiveness of the polymer (see Table 3.2.3-1.)
(f) Never overfeed polymers; their optimum dosages are located in a very narrow range. A minute amount of overfeeding will cause the turbidity colloids to restabilize, thus resulting in a sharp increase in turbidity.
(g) Since diluted polymer solutions deteriorate quickly, (during hot weather) every batch should be used within 2 days.
(h) Some polymers are corrosive in nature.

Operation and Maintenance The major issues in operation and maintenance are (1) safety programs and training, (2) good housekeeping, and (3) good record keeping. Each of these issues is discussed in detail.

Safety Programs and Training The chemicals used in water treatment processes (Table 4.1-2) are for the most part harmful to human health. Thus the water production manager and water supply department must establish a formal safety pro-

TABLE 4.1-2 Major Water Treatment Chemicals

Name	Suitable Materials	Available Strengths	Solubility (g/100 mL)	Major Suppliers	Bulk Weight or Dry Chemical in Solution	Remarks
Aluminum sulfate (alum)	PVC, rubber, FRP, 316 SS	Dry (17% Al_2O_3) or 49% liquid alum (LA)	78 at 10°C	Allied Chemicals, Staffer Chemical	60 lb/CF (dry), 5.5 lb of dry alum/gal of 49% solution	Mostly liquid alum, S.G. = 1.33 (LA), corrosive
Ferric chloride	PVC, Hypalon Hastelloy C, FRP	35–45% solution	Infinite	Dow Chemical, Pennwalt Corporation DuPont	4.6 lb dry F.C./gal of 42% solution	S.G = 1.46 (42% solution), very corrosive
Ferric sulfate	PVC, rubber, FRP, 316 SS	Dry (68–78%) or 50% liquid	20.5 at 10°C	Stauffer Chemical, Tennessee Chemical	70 lb/CF (dry), 6.3 lb dry F.S. in 1 gal of 50% solution	Hygroscopic, S.G. = 1.5 (50% solution), corrosive
Cationic polymer	FRP, PVC, 316 SS	Mostly liquid	Infinite	Dow Chemical, Nalco Chemical, American Cyanamid	8.5–9.5 lb/gal	Eye and skin irritant, may be corrosive
Anionic polymer	FRP, PVC, 316 SS	Mostly dry	Infinite	Dow Chemical, Nalco Chemical, American Cyanamid	45–55 lb/CF	Eye and skin irritant
Nonionic polymer	FRP, PVC, 316 SS	Mostly dry	Infinite	Dow Chemical, Nalco Chemical, American Cyanamid	42–48 lb/CF	Eye and skin irritant
Chlorine gas	Steel, copper for gas; Hastelloy C, PVC, Vitor for solution	99.8% purity	0.98 at 10°C	Dow Chemical, Jones Chemical, McKeeson Chemical	d = 3.214 (air = 1)	Poisonous and very corrosive when wet
Sodium hypochlorite	Rubber, PVC, vinyl, Hastelloy C	12–15%	Infinite	Jones Chemical, Hill Brothers company Occidental	10.2 lb/gal (15%)	Highly alkaline short life, 1 gal (12.5%) = 1 lb of Cl_2

Calcium hypochlorite	Rubber, PVC, Hastelloy C	Dry form with 70% Cl_2	22% at 10°C	Delta Chemical, Jones Chemical, Pennwalt Corporation	70 lb/ft^3 (granules)	Hygroscopic
Fluosilicic acid	Rubber, PVC, Hypalon, Carp. 20	20–30% (16–24% F)	Infinite	Essex Chemical, Thompson Chemical	10.5 lb/gal (30% solution)	Corrosive, toxic, attacks glass
Sodium fluoride	Rubber, 316 SS, PVC, Hypalon	95–98% (43–44% F)	4% at 10°C, 4.1% at 25°C	Allied Chemical, Ashland Chemical, J.T. Baker Chemical	90–105 lb/CF (crystal)	Poisonous
Sodium silicofluoride	Rubber, 316 SS, PVC, Hypalon	98.5% (60% F)	0.46 at 10°C, 0.76 at 25°C	Ashland Chemical, Delta Chemical, Olin Corporation	85–105 lb/CF	Poisonous
Potassium permanganate	Steel, 316 SS FRP, PVC	97–99%	3.3 at 10°C, 5.0 at 20°C	Ashland Chemical, Carus Chemical	86–100 lb/CF	Hygroscopic, oxidant, toxic
Sodium hexameta-phosphate	316 SS, rubber, PVC, RFP	67% P_2O_5 (minmum)	Infinite	Ashland Chemical, FMC Corporation, Monsanto Company	64–100 lb/CF, glass or flake	Hygroscopic, sequestering agent
Zinc orthophosphate	RFP, PVC, 316 SS	8% zinc in sulfate type, 25% zinc in chloride type	Infinite	Technical Products Company, Monsanto Industrial Chemicals, Conray Chemical	10.5 lb/gal, S.G. = 1.27, Z_n = 0.83 lb/gal	Acidic solution, avoid skin/eye contact
Copper sulfate	Rubber, PVC, RFP, 316 SS	99%	25.2 at 15°C	Ashland Chemical, Jones Chemical, Westco Chemical	75–90 lb/CF	Poisonous, algaecide
Carbon dioxide	Steel, 316 SS PVC	99.5% liquid CO_2, 10–14% in combustion	0.35 at 10°C, 0.15 at 25°C	Union Carbide, Arco Chemical	d = 1.53 (air = 1)	Liquefied CO_2 in cylinder, recarbon-ation use

TABLE 4.1-2 *(Continued)*

Name	Suitable Materials	Available Strengths	Solubility (g/100 mL)	Major Suppliers	Bulk Weight or Dry Chemical in Solution	Remarks
Sulfur dioxide	PVC, ceramic, Hypalon, Carp. 20	100%	16.2 at 10°C, 11.3 at 20°C	Delta Chemical, Jones Chemical, Stauffer Chemical	d = 2.26 (air = 1)	Liquefied SO_2, in cylinder, dechlorination use
Air (reference)	—	—	14 mg/L at 10°C, 10 mg/L at 25°C		d = 1.29 g/L (dry)	
Ozone gas	Ceramic, Teflon 316 SS, glass	2–8% depending on generator	6.5 mg/L at 10°C, 3.5 mg/L at 25°C	On-site generation only	2.14 g/L at 0°C, d = 1.66 (air = 1)	Toxic, fire hazard, strongest oxidant
Hydrogen peroxide	Aluminum, type 5254; polyethylene; porcelain; 316 SS (fair)	35, 50, and 70% solutions	Infinite	FMC, DuPont, Inferox	S.G. = 1.29 (70% solution)	Potentially a fire/explosion hazard, eye and skin irritant
Ammonia gas	Steel, 316 SS, neoprene	99% NH_3	40 at 10°C	Delta Chemical, Monsanto Industrial Chemicals, Olin Corporation	0.77 g/L, d = 0.56 (air = 1)	Pungent, toxic, explosive
Aqua ammonia	Iron, 316 SS, PVC	30% NH_3 (26 Be)	Infinite	Allied Chemical, Hill Brothers, Jones Chemical	S.G. = 0.897	Unstable, strongly alkaline
Sodium chlorite	Glass, PVC, FRP, Hastelloy C	Dry (81%) or 25% solution	35 at 10°C	Olin Corporation, Vulcan Chemical Drew Chemical	25% solution weigh 10.25 lb/gal	Hygroscopic, combustible with organic matter

Quick lime (CaO)	Iron, concrete, PVC	82–95%	22 at 10°C	AC Industry, J.T. Baker, Kraft Chemical	60 lb/ft^3	Emits heat upon contact with water
Hydrated lime [$Ca(OH)_2$]	Iron, concrete, PVC	82–95%	22 at 10°C	AC Industry, Asher-Moore Company, Rugger Chemical	30 lb/ft^3	Dusty, irritant
Sodium hydroxide (caustic soda)	Steel, iron	Dry or 50% liquid	51 at 10°C	Stauffer Chemical, Allied Chemical, FMC Corporation	60 lb/ft^3 or 2.7 lb NaOH in 25% solution	Toxic, dangerous to handle
Sodium carbonate (soda ash)	Steel, rubber, PVC	99% pure	12.5 at 10°C	Hacros Chemical, Ashland Chemical, Stauffer Chemical	40–60 lb/CF	Hygroscopic
Sulfuric acid	Steel, PVC for 92% acid; Carp. C, Viton for <92%	60 Be; 93% H_2SO_4; 60 Be; 78% H_2SO_4; 50 Be; 62% H_2SO_4	Infinite	Allied Chemical, Ashland Chemical, Monsanto Chemical	15.2 lb/gal (66 Be), 14.2 lb/gal (60 Be)	Very corrosive, hygroscopic
Sulfur dioxide	Steel for dry gas; wet gas; glass, Carp. 20, PVC, ceramic, Viton	100% SO_2	22 at 0°C 16 at 10°C 11 at 20°C 8 at 30°C	Stauffer Chemicals, Virginia Clum	2.26 = d (air = 1.0); S.G. = 1.363 (H_2O) = 1.0	Acid gas; vapor pressure: 15 psi at 40°F, 70 psi at 100°F
Activated carbon	316 SS, FRP, rubber, Hastelloy C, epoxy paint, Saran	Power (PAC) 200 mesh size or granules (GAL); E.S.; 0.6–0.9 mm; UC; 1.6–2.4	— —	Calgon Corporation, American Novelty Company, Trans-Pacific Carbon Corporation, Calgon	13–46 lb/ft^3 (20 avg.), 25–26 lb/ft^3	Dusty, can be explosive

gram and conduct periodic hands-on training. It is essential that every treatment plant have a custom-made operation and maintenance manual for each type of chemical feed system. For example, potassium permanganate ($KMnO_4$) and ferric chloride ($FeCl_3$) must not be mixed together since toxic chlorine gas (Cl_2) is formed. Furthermore, the necessary type and number of special clothing, emergency showers, and eyewashes should be checked routinely.

Good Housekeeping Good housekeeping methods are an important issue since the chemicals are essentially hazardous in addition to being toxic. Any chemical spill should be cleaned immediately and the accumulation of gas and dust must be avoided through the frequent use of exhaust fans and a vacuum cleaner. Polymer solution spilled on the floor will result in extremely slippery conditions; all spills should therefore be washed out immediately by means of the floor drains. Operators should note that wet concrete floors are also very slippery. Appropriate hazard signs should be posted in highly visible locations at all times. Finally, regular maintenance of all mechanical equipment must be performed on time, and the chemical feed lines and diffusers should be checked periodically for clogging.

Record Keeping Record keeping is a vital task because it is required by local, state, and federal regulatory agencies. Hourly, daily, and monthly records must be kept with respect to the type and dosage of each chemical used. The chemical inventory must also be known at all times in order to maintain an adequate supply of treatment chemicals on site. Moreover, the operating performance of each piece of equipment must be logged on a daily basis and all pertinent information should be conveyed to the next shift operator through verbal communication and in writing (on the daily operational data sheet). Computer-assisted instrumentation and control can greatly assist in record keeping.

BIBLIOGRAPHY

AWWA, *Water Fluoridation: Principles and Practices,* Manual M-14, American Water Works Association, Denver, CO, 1977.

AWWA, *Water Quality and Treatment: A Handbook of Public Water Supplies,* 3rd ed., McGraw-Hill, New York, 1971.

BIF, *Chemicals Used in Treatment of Water and Wastewater,* engineering data, BIF, Providence, RI, August 1982.

Williams, R. B., and Culp, G. L., eds., *Handbook of Public Water Systems,* Van Nostrand Reinhold, New York, 1988.

4.2 HANDLING FACILITY FOR WASTE WASH WATER

Purpose Conventional water treatment processes ordinarily produce 2 to 3% of the treated water as filter waste wash water. Prior to 1965 it was common practice to discharge this waste back to the water source or other public water course.

However, due to the passage of Public Law 92-500 and the Water Pollution Control Act Amendments of 1972, this practice is now prohibited. Thus, both the waste wash water and sludge from the clarifiers are categorized as industrial wastes, and a permit must be obtained from the local regulatory agency to discharge them; the agency is governed by the National Pollutant Discharge Elimination System (NPDES).

Since the passage of Public Law 92-500, many treatment plants have been recycling the wash waste with or without clarification, and this practice has not had adverse effects on the efficiency of the process or the quality of the finished water. However, a recent EPA proposal seeks to restrict the rules governing surface water treatment by requiring a high degree of treatment—including flocculation, sedimentation, and disinfection—before recycling. Furthermore, the recycling must be directed to the head of the plant and not the filter influent.

This section describes how to design a waste-wash-water handling facility so that it meets the recent requirements set by the regulatory agencies. The design of a waste recycling system, where the waste is routed to the head of the plant without a purification process, is also discussed because this practice is still applicable to most water treatment plants outside the United States.

Considerations The quality and quantity of the wash waste, the desired quality of the recycled or discharged (treated waste), and the type and capacity of the treatment process are the major design considerations for the waste-wash-water handling facility. Alternatives such as lagooning, flocculation and sedimentation, and recycling without a purification process are also important issues and are discussed as well.

Quality and Quantity Many design engineers have chosen to mix together the waste wash water and the sludge from the basins. Yet this practice is undesirable because the solids concentrated in the sludge are allowed to disperse in the wastewater. 0.5 to 1.5% of the sludge is composed of concentrated solids: microorganisms, organic compounds, and heavy metals that have been extracted and settled from the raw water. In contrast, waste wash water has a solid concentration of 0.01 to 0.04%. If the two are mixed, the turbidity of the water increases and more chemicals are required to treat the sludge and waste mixture. This can potentially lead to the recycling of objectionable compounds and the importation of undesirable taste and odor (of the sludge) to the water. Thus the sludge from the basins should not be mixed with the filter wash waste (which is recycled) but should be discharged to the sludge handling system.

The quality of the wash waste depends on the raw water quality and pretreatment unit processes. For a conventional water purification plant using alum as the coagulant, the typical properties of the waste are

- *Turbidity:* 150 to 250 ntu
- *BOD:* 2 to 10 mg/L
- *COD:* 30 to 150 mg/L
- *pH:* 6.7 to 7.5
- *TDS:* 0.01 to 0.05%

The quantity or production rate of the wash waste is also a function of the raw water quality and the type of pretreatment processes, in addition to the efficiency of the processes. The percentage of waste to the amount of water treated is usually about 2% during the warm months and 3% during the cold. Direct filtration and in-line filtration plants generally produce higher percentages of waste, and figures greater than 10% have been reported during episodes of algal bloom and turbidity spikes in the water source. Engineers are strongly advised to recycle the waste wash water rather than discharging it after the clarification process because the cost of water is often rather significant.

Quality Goal of the Recycled or Discharged Treated Waste The quality goal of the recycled treated waste should be equal or better than the average raw water quality. However, if the waste wash water is to be discharged to a public water course, the quality of the treated waste should basically be equal to or better than the quality of the receiving body. Nonetheless, the NPDES permit may include other special requirements.

Type and Capacity of the Treatment Process The type of the waste treatment process is fundamentally the same as that of a conventional flocculation and sedimentation process with disinfection. Proprietary reactor-clarifiers, which have a large internal flow recirculation, have good process efficiency and are tolerant of shock loadings. Other advantages include the ability to thicken the sludge, easy sludge removal, and a small site requirement. Regular horizontal-flow rectangular tanks are also effective waste handling facilities, although they may require a larger site area than the proprietary types, due to their lower allowable hydraulic loading rates. The high-speed microsand settling process (ACTIFLO) should also be considered.

No matter the type of unit, the best process performance is always achieved if the units are operated on a continuous basis and not in an on-and-off fashion. This ideal condition may be attained by providing a large waste holding tank and by pumping the waste into the treatment unit at a constant flow rate. The capacity of the waste treatment facility should be based on both the maximum waste production rate and the tolerable recycling rate of the treatment plant. The practical maximum recycling rate is 20% of the plant flow rate, yet the design should be based on a recycling rate of 10% or less, so that the hydraulic loading of each process unit is not increased significantly. The sludge produced by the waste treatment facility, from the sedimentation basins, should be discharged to a sludge handling facility for further treatment.

Alternatives Under normal conditions the suspended matter in the raw water is already well coagulated and flocculated prior to filtration. Furthermore, additional effective flocculation usually occurs in the filter bed during filtration. Consequently, the backwashing wastewater contains well-flocculated particles that settle very well. Figure 4.2-1 presents the batch settling characteristics of wastewater without additional flocculation with coagulant. The figure shows that the supernatant reaches nearly 10 ntu, from an original turbidity of 210 ntu, after 30 min of settling time.

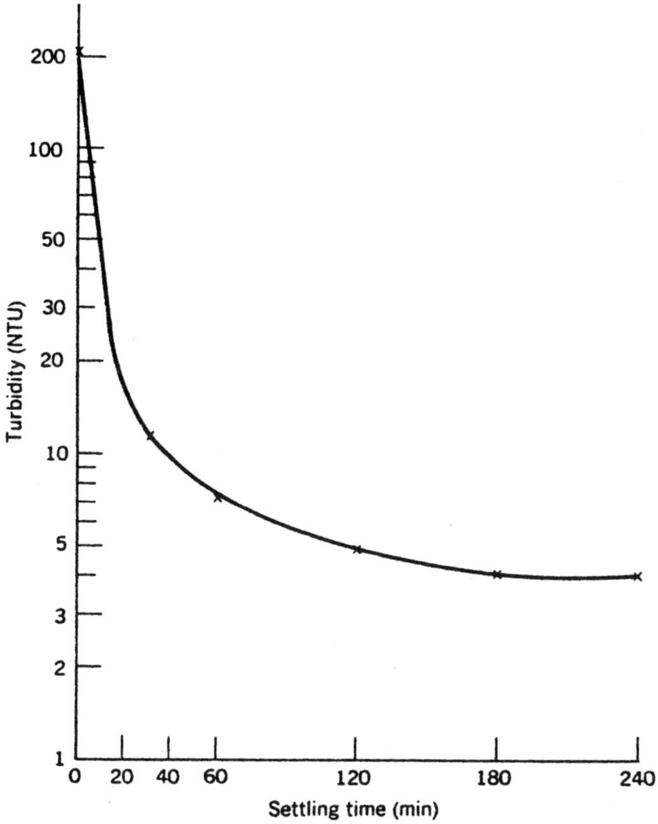

Figure 4.2-1 Settling characteristic of waste wash water.

A settling time of 1 to 2 h yields a supernatant of nearly 5 ntu. Knowing the quick-settling characteristics of the waste wash water, two basic treatment alternatives may be used: (1) lagooning with disinfection and (2) ordinary flocculation, sedimentation, and disinfection. Moreover, the wash waste may be recycled to the head of a conventional complete process without settling, as long as the waste is fresh; this practice is still a feasible alternative under certain conditions.

LAGOONING Lagooning is an effective method of clarification if the plant site has adequate space and if the filter backwash intervals are reasonably spaced. The lagoon should be large enough to hold the wash waste of approximately 10 backwashes, or a series of three or more smaller lagoons, each holding three or four filter wash wastes, may be used. All lagoons should be elongated in shape to provide the maximum distance between the inlet and outlet. This configuration minimizes the effect of the suspended sludge, which is caused by the stirring action of the incoming backwash waste. The inlet of each lagoon should have an energy dissipator. The outlet should be designed to decant as well as drain the lagoon, and should also act as an overflow facility. The sidewalls of the lagoon must be protected from erosion, rainfall, and wave action. Each lagoon should also have an adequate access road, which allows vehicles to remove accumulated sludge. Finally, a disinfection facility, together with a recycling pump, should be provided at the effluent of the lagoon. Furthermore, engineers should realize that the lagoons tend to breed mosquitos and other aquatic organisms, in addition to taste and odor problems characteristic of certain regions.

FLOCCULATION AND SEDIMENTATION PROCESSES A conventional flocculation and sedimentation pretreatment unit, with a disinfection facility, is required whenever the site does not have adequate space for lagoons or when settleability of the waste is unacceptable, unless conventional flocculation and sedimentation are implemented. As discussed previously, either the upflow type of reactor-clarifier or a conventional horizontal-flow rectangular flocculation and sedimentation basin may be used. In most cases, the preferred chemical is either 15 mg/L of alum or 2 mg/L of cationic polymer. The process will have 20 min of flocculation, followed by a surface loading of 0.8 to 2.5 gpm/ft^2 (2 to 6 m/h) and a settling time of 0.5 to 2 h, depending on the type of unit provided. A disinfection facility should be included in the design.

The conventional pretreatment process should have a backwash waste holding tank. This tank must be large enough to accommodate the water of two to three backwashes, depending on the anticipated frequency of filter washing. Floc removal should be achieved during the subsequent process (clarification), not in the holding tank. The holding tank may be square or circular rather than elongated. These shapes will prevent the floc from settling since the inflow creates a stirring effect. Experience has shown that the ideal holding tank is cylindrical in shape and has an inlet pipe that produces a tangential flow direction, thus creating a swirling or vortex flow pattern within the tank (consider the Coriolis effect for the direction of the swirling flow pattern). To be able to drain the tank completely and collect filter medium, which is very abrasive to the pump impeller, a sump must be provided at the center of the floor of the holding tank. Furthermore, an overflow facility should be included in the design. Several actual installations of this type of holding tank have performed well.

Tanks that have the transfer pump located in the center of the circular tank have been observed to be virtually free of sludge accumulation. The capacity of the transfer pump used in the flocculation and sedimentation units is dictated by the anticipated maximum filter washing frequency; the waste flowing into the holding tank must be pumped out to accommodate the next filter washing. Similarly, the rated capacity of the treatment process is determined by the maximum pumping rate of the transfer pump.

Ideally, the waste handling facility is located at a higher elevation than that of the main treatment plant and the treated water is recycled back to the head of the main plant by gravity. If this scheme cannot be adopted, recycling pumps must be installed. Yet the recycling pumps should be used only as a last resort, because the double-pumping scheme tends to have "jogging" problems from the transfer pump to the recycling pump—repeated starting and stopping of the pump at frequent intervals.

RECYCLING OF THE WASTE WITHOUT A PURIFICATION PROCESS This alternative has been applied to many existing plants and has proved to be effective provided that the recycling rate and other factors are correctly chosen. This is especially true for conventional treatment processes that must produce large and easily settleable floc. There is an abundance of clear evidence which confirms that recirculation of performed floc remarkably improves both flocculation and sedimentation; the majority of companies that design and manufacture reactor-clarifier units are exploiting floc recirculation.

The key design guides are provided by bench-scale tests. The test results obtained from a study conducted by the author are presented in Figure 4.2-2. The wastewater used in the experiment was not the supernatant but the well-stirred turbid wash water with a turbidity of 210 ntu (see Figure 4.2-1). The test results demonstrate that both flocculation and sedimentation were improved dramatically as the amount of wash waste added approached 30% of the total water volume. However, in actual treatment plants the practical blending ratio of unsettled waste wash water is up to 10%, due to the effect of hydraulic shock loading and the limited detention time of the units. Fresh settled floc (sludge) also improves the settling efficiency if the blend volume ratio is no more than 2%. However, it should be remembered that the danger associated with sludge recycling lies in the potential for recycling objectionable taste, odor, and other undesirable substances. Moreover, sludge that is over 2 to 3 days old often loses its effectiveness completely (Figure 4.2-3).

In summary, the waste wash water should be stored in a holding tank in the manner described previously and should be recycled, without clarification, at a blending rate that is 5 to 10% of the plant inflow rate. It is always a good practice to disinfect the recycling waste.

Design Criteria The design criteria for proprietary reactor-clarifiers and regular horizontal flow rectangular tanks are not discussed fully in this section because they were covered in Sections 3.2.3 through 3.2.5. However, the criteria for the alternative system of lagooning are presented, and the flocculation and sedimentation processes and disinfection facility are described briefly.

Lagooning

Number of lagoons	A minimum of two to satisfy the cleaning requirement.
Size of each lagoon	Large enough to hold the wastewater of a minimum of 10 filter backwashes.
Shape of the lagoon	Free form or rectangular. The length/width ratio should be 4 : 1 or greater.
Depth of the lagoon	A minimum of 10 ft (3 m).
Inlet	Isolation valve and an energy dissipator.
Outlet	Should function in decanting, draining, and overflow protection.
Miscellaneous	The sidewall should be protected from erosion and checked for seepage periodically. If the area is open to the public, provide a safety fencing around the lagoons.

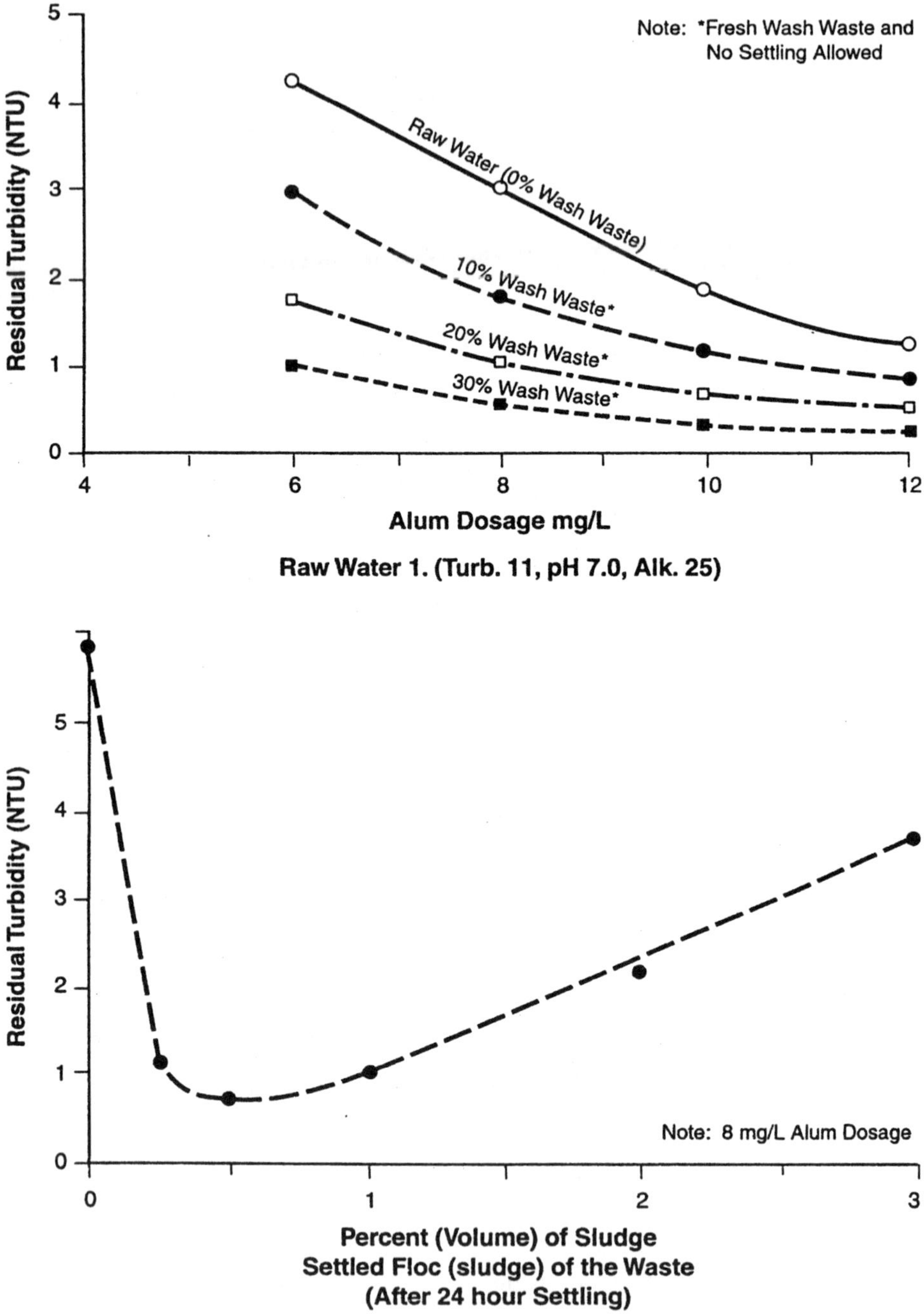

Figure 4.2-2 Improvement of flocculation by wash-waste mixing.

	Condition	Turbidity	Color	pH	Alkalinity
Raw water (reservoir)	Ⓐ 100% raw water	6	7.5	6.8	16.5
	Ⓑ Fresh wash waste	230	220	6.65	27
	Ⓒ 12 h old waste				
Filter wash waste	Ⓓ 1 day old waste				
	Ⓔ 2 day old waste				
	Ⓕ 4 day old waste				

Note: Experimental conditions Ⓑ through Ⓕ were evaluated using 10% addition of "aged" wash waste to raw water.

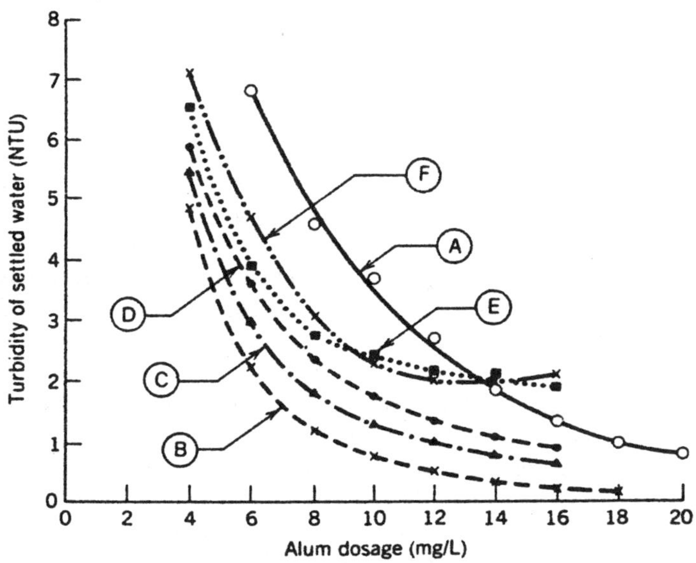

Figure 4.2-3 Settled water turbidity as a function of wash-waste admixture age.

Flocculation and Sedimentation Processes Despite the high turbidity of the waste wash water, the waste is easily flocculated and clarified. The design criteria of each unit process are fundamentally the same as presented in Sections 3.2.3 through 3.2.5. If the site area for the waste reclamation facility is very limited, a Densadag reactor-clarifier should be evaluated seriously. This unit, manufactured by Infilco Dregremont, Inc., provides a surface loading rate as high as 5 to 6 gpm/ft^2 (12.5 to 15 m/h) for the clarifier and is capable of producing an alum sludge composed of 2 to 3% solids by means of a rather high polymer dosage. The most recent superhigh rate clarification unit, ACTIFLO by Krüger, can produce settled water turbidities of 1 ntu or less with about 0.5 to 2 mg/L of cationic polymer at a surface loading rate of 20 to 25 gpm/ft^2 (50 to 62.5 m/h). Consequently, the footprint of the unit is very small.

Disinfection Facility Chlorination is the prime candidate for the disinfection of recyclable water. Since the recycling rate is generally only 2 to 3% of the plant flow rate, the THMs produced by chlorination should not be a problem unless the raw water characteristics pose a threat. Refer to Section 3.2.9 for the design criteria for the disinfection facility.

Example Design Calculations

Example 1

Given The filters in Example 2 of Section 3.2.7 are part of a 50-mgd filtration plant. The treatment plant requires a waste wash water treatment facility and a recycling facility.

Determine

(i) The type of waste reclamation facility
(ii) The treatment process
(iii) The basic design criteria for each unit process

Solution (i) A conventional pretreatment process should be selected since the treatment plant is located in a good residential area and the cost of the land does not permit the plant to purchase large extra parcels of property. Furthermore, the residents will probably not tolerate having a sludge lagoon in their neighborhood, due to aesthetic and health considerations and the potential mosquito problem.

(ii) The treatment process should have a waste holding tank located near the filter bank, waste transfer pumps on the holding tank, and conventional pretreatment units at an elevation that is several feet higher than the water level at the head of the main plant. The clarified water should be chlorinated and transported to the head of the plant via gravity flow. The head of the plant should have a flow meter in the discharge pipe line.

(iii) The six basic design criteria for the treatment process units are presented below.

DATA ON THE FILTERS

Total number of filters	Six, each with a filter bed area of 960 ft^2 (89 m^2)
Backwash condition	22 gpm/ft^2 (55 m/h) for 7 min (conservative)
Surface wash condition	3 gpm/ft^2 (7.5 m/h) for 3 min

WASTE WASH WATER HOLDING TANK

Number	One tank with a 62,800-ft^3 (1780-m^3) or 470,000-gal capacity
Size	80 ft (24.4 m) in diameter with a water depth of 12.5 ft (3.8 m) to hold three filter washings

TRANSFER PUMPS

Type	Vertical turbine pump
Number	Three identical pumps (one pump as a standby)
Capacity	1000 gpm (3.8 m^3/min) each

TREATMENT PROCESSES

Total capacity	2000 gpm (7.6 m^3/min)
Flash mixing	10 in. (25.4 cm) in-line static mixer
Flocculation tanks	Two identical baffled channel types: round-the-end type
Size of each tank	20 ft (W) × 20 ft (L) × 10 ft (SWD) or 6.1 m × 6.1 m × 3.05 m
Number of channels	Four channels with a total of 16 baffles for each tank
Volume of each	4000 ft^3 (113 m^3) or 30,000 gal
G	An average of 30 s^{-1}—tapered mixing

SEDIMENTATION TANKS

Number	Two identical tanks with tube settler modules
Size of each	20 ft (W) × 35 ft (L) × 10 ft (SWD) or 6.1 m × 10.7 m × 3 m
Tube settler	500 ft^2 (46.5 m^2) of each tank is covered by the settlers
Surface loading	2 gpm/ft^2 (5 m/h) for the area covered by the settlers
Launders	Two launders per tank, each 25 ft (7.6 m) long and a loading rate of 10 gpm/linear foot (7.5 m^3/m • h)
Sludge collector	One chain-and-flight type for each tank

TOTAL VOLUME OF THE WASTE WASH WATER

$$(22 \times 960 \times 7) + (3 \times 960 \times 3) = 156{,}480 \text{ gal/wash}$$

Since the water source is a large lake, the potential for algal bloom exists. Therefore, the worst scenario would be that each filter is washed three times a day. The average wash cycle is assumed to be 24 h for the filters. The waste holding tank should be sized to hold three backwashes because two to three filters may become clogged at the same time, requiring the filters to be washed consecutively. The total required volume is

$$156{,}480 \times 3 = 469{,}440 \text{ gal or } 62{,}800 \text{ ft}^3 \text{ } (1780 \text{ m}^3)$$

The tank should be cylindrical in shape with a maximum water depth of 12.5 ft (3.8 m). The diameter of the tank is

$$D = \left(\frac{62{,}800 \div 12.5}{0.785}\right)^{0.5} = 80 \text{ ft } (24.4 \text{ m})$$

Under the worst conditions, a total of 18 filter washes will be performed per day; thus 24 h ÷ 18 = 1.33 h or an average of 1 h 20 min between filter washings. In other words, the total volume of wash waste produced by one filter washing must be pumped out within 1.33 h. The required capacity of the transfer pump is

$$156{,}480 \div 80 = 1956 \text{ gpm } (7.4 \text{ m}^3/\text{min})$$

Make this 2000 gpm or 2.88 mgd (7.6 m^3/min).

Under normal conditions, each filter is washed once a day. Since the raw water is obtained from a large lake, it is assumed that the treatment mode of the plant (for the majority of the year) is a direct filtration simulation process. In this case, the average time between the filter washing is 24 ÷ 6 = 4 h. The required recycling pump capacity is therefore

$$156{,}480 \div 240 = 652 \text{ gpm } (2.47 \text{ m}^3/\text{min})$$

However, the pumping capacity should be roughly 50% larger (1000 gpm) since plant operators generally prefer to wash as many filters as possible during the day shift; the graveyard shift has only the absolute minimum number of operators, and in many cases it is only one or two persons.

Provide three identical transfer pumps, each with a capacity of 1000 gpm (1.44 mgd or 3.8 m^3/min), at the center of the circular tank. One of these pumps should act as a standby. The total design capacity of the treatment facility should be 2000 gpm (7.6 m^3/min). Therefore, provide two parallel units, each consisting of a 1000-gpm process train.

Since the recycling rate is rather constant, that is, only one pump is operated regardless of the water level in the holding tank, a baffled channel flocculation process is the most cost-effective method for producing good floc; it is also cost-effective to build and operate. An alternative to the baffled channels is the vertical shaft flocculator; refer to Section 3.2.4 for further information.

Each sedimentation tank should be designed with both a tube settler module and chain-and-flight collector units. The tube settler modules are used for their space-saving feature and their effectiveness. The blending ratio of the recycled wash waste should always remain below 10% of the anticipated normal operational conditions. The blending ratio is generally 5 to 8%; thus there is little concern for hydraulic shock loading to the main process units. Figure 4.2-4 illustrates the general features of the waste holding tank and the waste treatment processes.

Operational Issues An issue that is important but often neglected is the monitoring of filter media accumulation is the wastewater holding tank. Plant operators should drain the tank every 6 months and estimate the volume of accumulated sand and/or anthracite coal at the bottom of the tank. If the operator finds an excessive amount of media, it is clearly evident that the filters are losing a significant amount

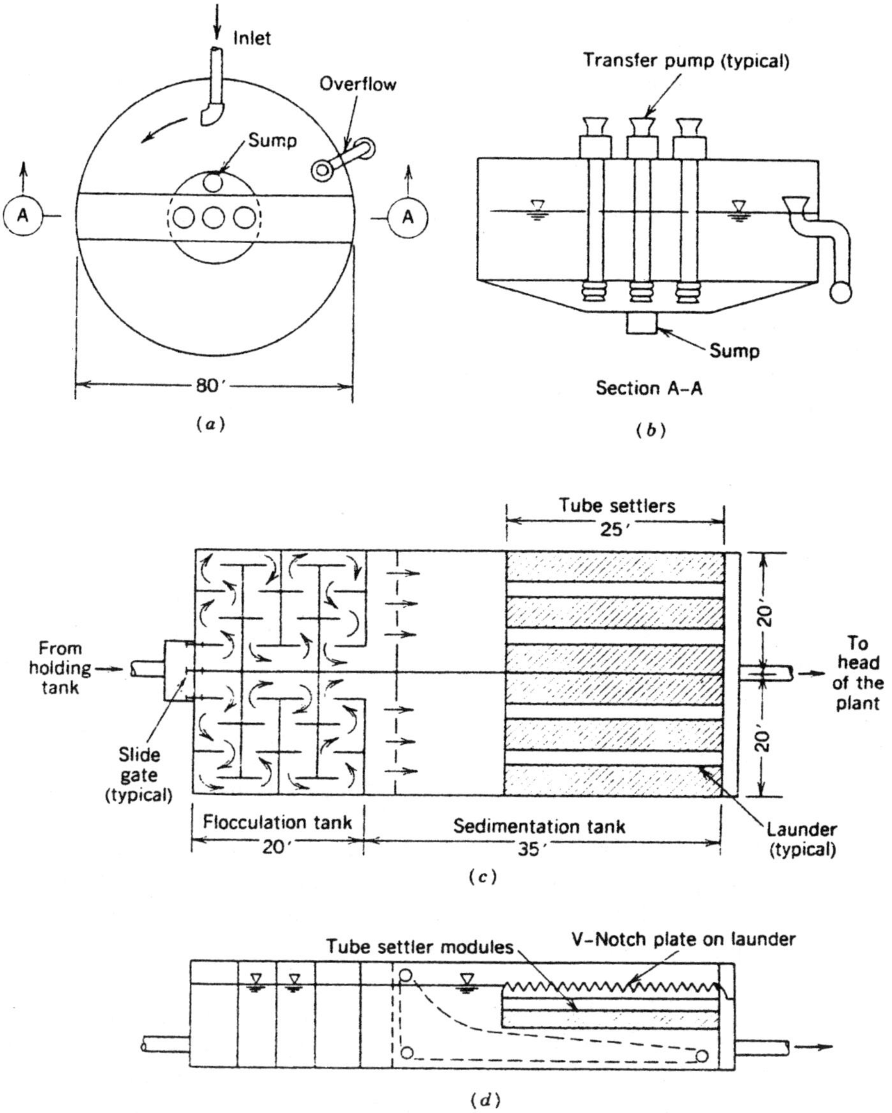

Figure 4.2-4 Waste-wash-water treatment process: (*a*) plan of holding tank; (b) section A–A; (*c*) plan; (*d*) longitudinal section.

of media during the washing process. The operator should then modify the filter washing conditions or adjust the height of the wash troughs. Accumulation of filter media should be detected at an early stage because the media are very abrasive and will erode the impeller of the transfer pumps.

With regard to the treatment of recycled water, the coagulant dosage of the main process would will not need to be adjusted if the blending ratio is less than 10%, regardless of the turbidity of the recycled water. The turbid form of wash waste contains a great deal of floc, and this generally improves the efficiency of the flocculation and sedimentation processes due to the seeding effect (Figure 4.2-2). Moreover, the dosages of other water treatment chemicals usually do not have to be readjusted as long as their blending ratios remain under 10%. Operation and maintenance of the pumps and the flocculation and sedimentation units do not require special operator considerations. Refer to Sections 3.2.4, 3.2.5, and 4.7 for further information on the operation and maintenance aspects.

BIBLIOGRAPHY

Kawamura, S., "Considerations on Improving Flocculation," *J. AWWA,* 68:6:328 (June 1976).

Kawamura, S., "Effect of Filter Wash Waste Recycling on Both Alum Flocculation and Cost of Plant Operation," *J. Jpn. WWA,* 228:28 (September 1958).

4.3 SLUDGE HANDLING AND DISPOSAL

Purpose The sludge produced by water treatment plants is categorized as chemical sludge and consists primarily of inert materials. It is a by-product of the water treatment process, and until the 1980s, treatment plants returned it to the water source downstream of the intake. Currently, federal and state legislation classify most types of sludge as pollutants and prohibit their disposal into the public water supply.

The majority of water treatment processes employ aluminum sulfate as a coagulant, and sludge is therefore commonly referred to as alum sludge. Alum sludge is difficult to dewater, and dried sludge can revert to slurry when wet and disturbed, due to its thixotropic nature. In this section we present some design guides for creating the most cost-effective and feasible sludge treatment (conditioning, thickening, and dewatering) and disposal methods.

Considerations The major considerations facing every design engineer are (1) the Water Pollution Control Act, (2) the estimation of the sludge production rate, (3) characteristics of the sludge, (4) methods of minimizing the sludge production rate, (5) dewatering methods, and (6) the ultimate disposal of the sludge. In this section we address the first five issues and present several sludge handling methods.

Water Pollution Control Act Public Law 92-500, the Water Pollution Control Act Amendment of 1972, categorizes sludge from water treatment plants as an industrial waste requiring compliance with the provisions stated in this act. Under the NPDES (National Pollutant Discharge Elimination System) provisions, a permit must be obtained to discharge materials such as sludge and waste wash water. Wide variations are found in the limitations of the permit among the various states. The parameters include the pH, BOD, turbidity, temperature, flow of discharge, and the total concentration of the following substances: suspended solids, settleable solids, iron, manganese, aluminum, chlorides, sulfate, residual chlorine, floating solids, and foam. According to this amendment, no utility is allowed to discharge waste after 1985, unless it can show good cause for not recycling the treated wastewater. Thus engineers must grapple with the problem of the ultimate disposal of sludge.

Estimation of the Sludge Production Rate It is rather difficult to estimate the sludge production rate since there are no concrete methods for predicting turbidity spikes and the chemical dosages required to treat them. However, several empirical formulas may be employed. It is recommended that engineers select parameters that are similar to the operational conditions and use the basic formulas, such as those shown below. [*Note:* In the formulas, 1.3 is the ratio between total suspended solids (mg/L) and turbidity (ntu). This ratio ranges from 1.0 to 2.0.]

Dry alum sludge production rate (lb per million gallons of water treated)
= [alum dosage (mg/L) × (0.26 × 8.34)]
+ [raw water turbidity (ntu) × 1.3 × 8.34]

Dry ferric sulfate sludge production rate (lb per million gallons of water treated)
= [ferric sulfate dosage (mg/L) × (0.54 × 8.34)]
+ [raw water turbidity (ntu) × 1.3 × 8.34]

Dry ferric chloride sludge production rate (lb per million gallons of water treated)
= [ferric chloride dosage (mg/L) × (0.66 × 8.34)]
+ [raw water turbidity (ntu) × 1.3 × 8.34]

NOTES:

1. The equations above are based on the following chemical equations:

$$Al_2(SO_4)_3 \cdot 14H_2O + 3Ca(HCO_3)_2 \rightarrow 3CaSO_4 + 2Al(OH)_3 + 6CO_2 + 14H_2O$$

$$Fe_2(SO_4)_3 + 3Ca\,(HCO_3)_2 \rightarrow 3CaSO_4 + 2Fe\,(OH)_3 + 6CO_2$$

$$2FeCl_3 + 3Ca\,(HCO_3)_2 \rightarrow 3CaCl_2 + 2Fe\,(OH)_3 + 6CO_2$$

2. According to the EPA Guidance Manual, *Enhanced Coagulation and Precipitate Softening* (PB95-100392), three waters of hydration are chemically bound to aluminum hydroxide in the residuals when the enhanced coagulation process is used. This manual shows that the dry alum sludge production rate for each million gallons of water treated by the enhanced coagulation process can be established by the equation

   8.34 × [alum dosage (mg/L) + TOC (mg/L) + (1.5 × turbidity (ntu) + A]

   where A represents additional chemical used (mg/L).
3. The sludge production rate is highly dependent on raw water quality, which is controlled by weather conditions each year. Therefore, more elaborate formulas for sludge production rate do not mean that they are better.
4. Annual sludge production rate should be computed using average figures of turbidity, alum dosage, and plant flow rate except for an expected high turbidity month.

In the case of sludge produced by lime softening, the rate of sludge production is largely a function of the hardness of the raw water and the hardness level of the softened water. The basic formula for determining the lime sludge production rate (pounds per million gallons of water treated) is

$$8.34 \times (2.0\text{Ca} + 2.6\text{Mg} + \text{SS})$$

where Ca is the calcium hardness removed as $CaCO_3$ (mg/L), Mg the magnesium hardness removed as $CaCO_3$ (mg/L), and SS the suspended solids in raw water (mg/L). In general, 2 to 3 mg/L of solids will be produced for each mg/L of hardness removed. Thus average lime softening production rate may be estimated as follows:

dry lime sludge production rate (lb per million gallon of water treated)

$$= 2.5 \times 8.34 \times \text{total hardness removed (mg/L as } CaCO_3)$$

Sludge Characteristics The characteristics of water treatment sludge differ from location to location because they are largely dependent on the raw water characteristics, the type and amount of coagulant, and other types of chemical aids used in treating the raw water. In general, dried alum sludge is composed of 30% aluminum and 50% inert solids such as sillica and calcium. A typical sample of alum sludge contains the components listed in Table 4.3-1. The typical components of lime-softened sludge are presented in Table 4.3-2.

The characteristics of lime sludge vary as a function of magnesium hydroxide content. The drying characteristics of lime sludge are similar in nature to those of alum sludge. The concentration of magnesium hydroxide may range from insignificant to 30% by weight. Lime sludge containing low levels of magnesium hydroxide may be dewatered to 50 to 60% solids through the use of drying beds. Yet these figures will be reduced to 20 and 25% solids if the sludge has higher concentrations of magnesium hydroxide. The characteristics of sludge are important to know because they greatly influence the methods of handling and disposal. For example, semidried alum and ferric sludges are thixotropic; the consistency of the sludge changes with physical force. This characteristic makes alum and iron sludge, particularly alum sludge, very difficult to handle.

The consistency of drying sludge differs primarily due to the amount of metal hydroxides and the applied polymer (coagulant) dosage. The general expression of sludge consistency is as follows:

Solids Content (% of Solids)	Sludge Consistency	
	Alum Sludge	Lime Sludge
0–10	Liquid	Liquid
10–15	Viscous liquid	Viscous liquid
15–20	Pasty	Pasty

20–25	Semisolid	Pasty to semisolid
25–30	Soft solid	Soft solid to pasty
30–50	Crumbly cake	Pasty to crumbly cake

Reduction of Sludge Production Rate and Volume To minimize capital expenditures and the operational costs of sludge handling and disposal, the first step to be considered is reduction in the sludge production rate. The sludge production rate may be cut by 30 to 80% from the amount produced by a conventional complete treatment process using alum (coagulant) if either of two methods is implemented: simultaneous use of polymer and alum (as coagulant) or adopting a direct filtration type process to replace the conventional complete treatment process. These methods should be investigated prior to design work.

The actual data from many operational water treatment plants demonstrate that if 0.1 to 1.5 mg/L of cationic polymer is applied in conjunction with alum, the alum dosage can be reduced by 30 to 50%. The combined application of 0.2 to 0.3 mg/L of anionic or nonionic polymer and alum often improves flocculation dramatically, to the extent that alum dosage may be reduced by 15 to 30%. In the case of lime softening, the use of caustic soda may decrease the production of calcium carbonate sludge by 35 to 50%. However, the cost of caustic soda often outweighs the benefits associates with a decrease in sludge production.

The generation of sludge is reduced most dramatically when the conventional complete treatment process is replaced by two-stage, direct, or in-line filtration. These processes may be used to treat raw water obtained from large lakes or reservoirs and require only 4 to 6 mg/L of alum or the combined application of 2 to 4 mg/L of alum and 0.5 to 1 mg/L of cationic polymer. In contrast, 12 to 16 mg/L of alum is required by the conventional complete process for the same raw water. Thus it is apparent that the sludge production rate may be reduced to levels that are only 10 to 20% of the conventional process. If the treatment plant is forced to adopt a conventional process due to occasional turbidity spikes or algal bloom, the plant design could possibly have a bypass channel across the basins so that plant operators have the ability to treat the water in direct filtration or in-line filtration mode for most of the year. The second step in minimizing the costs of the sludge handling and disposal processes is to decrease the volume of the sludge. The volume of the sludge may be minimized through effective dewatering techniques and by-product recovery.

It is important for engineers to select an appropriate type of clarifier and sludge handling facility. For example, the composition of alum sludge in rectangular sedimentation tanks is generally 0.3 to 1.5% solids. However, the composition is only

TABLE 4.3-1 Alum Sludge Characteristics

BOD (mg/L)	COD (mg/L)	pH	Total Solids (%)	Al_2O_3 (%)	SiO_2 and Inert (%)	Organics (%)
30–300	30–5000	6–8	0.1–4	15–40	35–70	15–25

TABLE 4.3-2 Lime Sludge Characteristics

$CaCO_3$ (%)	$Ca(OH)_2$ (%)	$Mg(OH)_2$ (%)	SiO_2 and Inert (%)
85–93	0–1	0.5–8	2–5

0.2 to 0.5% if a sludge blanket type of clarifier is used. Furthermore, basins lacking a continuous mechanical sludge removal unit usually have the sludge accumulating to heights of 10 to 12 ft (3 to 3.5 m) before the basins are cleaned. Thus the sludge is compressed to the extent that it is composed of 4 to 6% solids, whereas sludge from sedimentation tanks containing mechanical sludge removal units has a composition ranging from 0.5 to 1.5% solids, depending on the type of sludge removal unit and the scraping cycle.

Under normal conditions, approximately 0.1 to 0.5% of the water treated at a water purification plant will be wasted by the sludge withdrawal operation unless recycled. The relationship between the suspended solid content (mg/L) and the measured turbidity (ntu) is a function of the type, size, shape, and color of the suspended solids. Thus a definite relationship cannot be derived. The ratio of the total suspended solids to the turbidity usually ranges from 1 to 2, with 1.3 being an average figure. The solids content of lime softened sludge is much greater than for alum sludge; it ranges from 1 to 10%, depending on the type of clarifier unit. Sludge removed from clarifiers may be 3 to 4% solids, and 0.3 to 1.5% of the lime-softened water is used in the sludge withdrawal cycles.

Sludge Handling A certain level of handling and treatment is required prior to ultimate disposal of the sludge. The degree of handling depends on the requirements governing ultimate disposal. Mechanical dewatering systems may be used in situations where drying beds cannot be constructed, due to the lack of land or lack of capital, frequent rainfall, and/or a high required solids content. These systems include gravity thickening, vacuum filters, filter presses, centrifuges, and freeze-and-thaw techniques. Yet in regions where the weather is mild to hot and evaporation exceeds rainfall, the simplest and most cost-effective methods are sludge lagoon drying beds and the shallow-depth sludge drying beds. Properly designed and operated, these beds are capable of drying both alum and lime sludge containing 45 to 50% solids.

The drying time for sludge drying beds may be reduced drastically if the beds are aerated continuously at a rate of 1 to 2 ft^3/min per square foot (18.3 to 36.5 m/h) for the first 3 weeks of the drying cycle, until cracks appear on the surface of the sludge. One Japanese equipment manufacturer has demonstrated that the drying time for alum sludge may be shortened from 4 months (using ordinary weather drying) to 6 weeks through the implementation of sludge aeration. Rather than using compressed air, a vacuum may be applied to the sludge beds. However, this practice causes the underdrainage system to clog frequently. Similarly, sand drying beds may cut the drying time of both alum and lime sludge significantly; however, in many cases the beds must be reconstructed every 2 to 3 years, due to clogging.

In locations where severely cold temperatures are predominant for one-third of the year, freezing is a very effective method of sludge dewatering. Freezing can produce alum sludge that is 25% solids, and the sludge is no longer thixotropic in nature. The application rate of the sludge is approximately 2.5 lb/ft^2 (12 kg/m^2), and the wet sludge is usually kept below a depth of 12 in. (0.3 m). A large-scale mechanical freeze-and-thaw facility was built in Chiba [Kashiwai W.T.P., 150 mgd (6.6 m^3/s)] and Sendai, Japan, to treat alum sludge. However, widespread use of this facility has been limited by the high initial cost and high operational cost ($100 per ton).

Gravity Thickening Gravity thickening may be attained through the use of lagoons or mechanical thickening tanks, provided that a polymer is added to the liquid sludge. Alum sludge treated with 2 to 3 mg/L of cationic polymer can produce a sludge containing 3 to 5% solids through lagooning. Lime conditioning of alum sludge yields a sludge composition of 6 to 9% solids when processed by a thickener tank at a feed rate of 0.07 to 0.14 gpm/ft^2 (0.18 to 0.35 m/h) and approximately 5 to 10 lb of solids per square foot of thickener surface per day (25 to 50 kg/m^2 · d). Lime sludge may be thickened to a greater degree than alum through the use of thickeners; a loading rate of 30 to 60 lb of solids per square foot of thickener surface area per day (147 to 294 kg/m^2 · d) may yield lime sludge that is 15 to 35% solids. For alum or ferric hydroxide sludge, the standard solid loading rate for sludge thickener is 10 lb of solids per square foot per day (49 kg/m^2 · d) or 0.12 to 0.15 gpm/ft^2 (0.3 to 0.38 m/h) hydraulic loading.

Centrifugation Among the several types of commercially available units, the most applicable units are the scroll discharge, solid bowl, and basket bowl. Basket centrifuges produce alum sludge that is 10 to 11% solids without the use of polymers. However, the sludge composition can be improved to 15% solids if the sludge is preconditioned with polymer [4 to 5 lb per ton of dry solids (1.8 to 2.3 kg/ton)]. Optimum preconditioning of alum sludge calls for the application of 1 to 2 lb of polymer per ton (0.45 to 0.9 kg/ton) of solids and results in a 25 to 30% cake. Bowl centrifugation allows lime-softened sludge to be more easily dewatered than alum sludge. This technique produces a cake that is 30 to 70% solids with or without polymer preconditioning. Yet the disadvantage is that it may require a high level of maintenance and has a high operating cost.

Drum Dehydrator (Dehydrum) The drum dehydrator is a proprietary unit that has been used successfully in Japan. It transforms the sludge into pellets that are 20 to 35% solids, provided that polymer is the sludge conditioner. The shape and motion of this unit resemble those of a concrete mixer truck. Mechanically, the system is simple and therefore has the lowest operation and maintenance costs of all mechanical dewatering systems.

Filter Press (Pressure Filtration) The filter press may be used to dewater both alum and lime sludge. However, it is used to dewater alum sludge that has been preconditioned with polymer or lime or precoated with diatomaceous earth. The solid content of the cake is generally 25 to 50%, with a pressing cycle of 8 h. The life cycle of the filter cloth is normally 1.5 years. The filter press is a batch process that requires a considerable outlay of capital and has a high operational cost. Yet if the sludge is required to be over 40% dry alum solids, this process is the most reliable and quickest method available.

Belt Press The belt press is a relatively small unit requiring minimal attention and maintenance once the system is optimized. It has been used to process coagulant sludges and has yielded alum cakes that are 20 to 25% solids as long as polymer is added to the feed sludge. The Skinner Filtration Plant operated by the MWD of Southern California has been using two sets of belt presses (Winkle press: Andritz Model SMX-8-2m belt) with a processing capacity of 600 to 800 lb/h (270 to 360 kg/h) of alum sludge (dry solids base) for the past sixteen years. This plant preconditions the alum sludge, which is approximately 3% solids, with 2.5 to 3.2 lb (1.1 to 1.5 kg) of cationic polymer (Telacol) for every ton of dry solids. The product (cake) is 22 to 25% solids. The system requires only one operator for a period of about 2 to 3 h/day. The belt cloth costs $2000 per unit and lasts approximately 2000 operating hours before any major repair work is required. Other treatment plants on the East Coast of the United States report the belt press as producing cakes that are 15 to 20% solids, requiring 10 to 15 lb of polymer per ton (4.5 to 6.8 kg/ton) of solids.

Vacuum Filters Vacuum filters are used primarily in processing lime sludge. The typical yield is a cake that is 40 to 50% solids. For low-magnesium lime sludge, the loading rate is 40 to 90 lb/ft^2 per hour (195 to 440 kg/ m^2 · h), but it is only 10 to 20 lb/ft^2 per hour (49 to 98 kg/ m^2 · h) if the magnesium level is high. Vacuum filters are generally ineffective in processing alum sludge.

Table 4.3-3 summarizes the typical performance of several dewatering processes. Tables 4.3-4 and 4.3-5 show cost comparisons among various sludge dewatering systems for alum sludge and lime sludge.

Filtrate Disposal The liquid waste from all mechanical dewatering processes pose a problem because of their chemical characteristics: They contain a high concentration of polymer, alkali chemicals, aluminum, and organic compounds and a significant level of heavy metals. Recycling the liquid waste back to the head of the plant, alone or combined with the filter wash waste, is not recommended because it will adversely affect the flocculation process and will increase significantly the level of unfavorable compounds in the processing water. Moreover, Departments of

TABLE 4.3-3 Solids Content (%) of Sludge by the Dewatering Processes

	Gravity Thickening	Drying Beds	Centrifugation	Filter Press	Belt Press	Freezing
Alum sludge	2–5	30–50	10–25	25–50	20–25	25–30
Lime sludge	8–12	40–60	30–70	40–50	—	—

TABLE 4.3-4 Example of Sludge Dewatering Cost Analysis: Alum Sludge

	Belt Filter Press	Centrifuge	Filter Press	Sand Beds
Capital costs (1990–2010)				
1. Thickeners	$ 905,000	$ 905,000	$ 905,000	$ 905,000
2. Storage tank	291,000	291,000	291,000	291,000
3. Sludge pumps	133,000	133,000	133,000	133,000
4. Sludge dewatering	1,421,000	2,167,000	6,035,000	3,874,000
5. Chemical feed	64,000	64,000	64,000	64,000
6. Site work	150,000	150,000	150,000	150,000
Subtotal	$2,964,000	$3,710,000	$ 7,578,000	$ 5,417,000
10% Engineering	296,000	371,000	758,000	542,000
25% Contingency	741,000	928,000	1,894,000	1,354,000
Total capital	$4,001,000	$5,009,000	$10,230,000	$ 7,313,000
Year 2000 expansion	1,278,000	962,000	4,556,000	3,499,000
Present worth	407,000	306,000	1,452,000	1,115,000
Capital cost present worth	$4,408,000	$5,315,000	$11,682,000	$ 8,428,000
Operation and maintenance costs				
Annual cost, 1990–2000	145,900	219,000	211,500	195,000
Annual cost, 2000–2010	162,500	274,000	229,500	240,000
O&M worth	1,551,000	2,420,000	2,768,000	2,154,000
Total capital and O&M present worth	$5,959,000	$7,735,000	$14,450,000	$10,582,000

Note: Average sludge production rate, 17 tons/day; maximum sludge production rate; 32 tons/day.

Health Services in certain U.S. states discourage this practice because of uncertainty about the effect on human health of trace metal and other unfavorable compounds. Possible methods of disposal include discharge into the sanitary sewer, purification treatment, and discharge to sludge drying beds if they are on site.

Ultimate Disposal Four practical options are available for the ultimate disposal of sludge: (1) direct discharge to the sewer system, (2) hauling to a landfill, (3) on-site disposal, and (4) recovery of by-products.

Discharge to the Sewer System The easiest solution for the treatment plant is direct discharge of the sludge to a sanitary sewer. However, the waste sludge flow must be equalized prior to discharge because ordinary sewer lines are incapable of handling the flow rate (directly from the basins) and the solid content can be too thick. The equalized sludge may be pumped from the holding tank to the sewer line

TABLE 4.3-5 Example of Sludge Dewatering Cost Analysis: Lime Sludge

	Belt Filter Press	Centrifuge	Vacuum Filter	Sand Beds
Capital costs (1990–2010)				
1. Thickeners	$ 540,000	$ 540,000	$ 540,000	$ 540,000
2. Storage tank	274,000	274,000	274,000	274,000
3. Sludge pumps	143,000	143,000	143,000	143,000
4. Sludge dewatering	1,421,000	2,283,000	1,008,000	801,000
5. Chemical feed	75,000	75,000	—	75,000
6. Site work	150,000	150,000	150,000	150,000
Subtotal	$2,603,000	$3,465,000	$2,115,000	$1,983,000
10% Engineering	260,000	346,000	212,000	198,000
25% Contingency	651,000	866,000	529,000	496,000
Total capital	$3,514,000	$4,677,000	$2,856,000	$2,677,000
Year 2000 expansion	726,000	786,000	395,000	742,000
Present worth	232,000	251,000	127,000	237,000
Capital cost present worth	$3,746,000	$4,928,000	$2,983,000	$ 2,914,000
Operation and maintenance costs				
Annual cost, 1990–2000	145,900	202,600	302,000	106,000
Annual cost, 2000–2010	154,700	243,600	369,000	119,000
O&M present worth	1,529,000	2,203,000	3,315,000	1,114,000
Total capital and O&M present worth	$5,275,000	$7,131,000	$6,298,000	$4,028,000

Note: Average sludge production rate, 50 tons/day; maximum sludge production rate; 61 tons/day.

at a constant rate that does not flood the sewer system. In several cases, the discharge of waste into the sewer lines was restricted to night, due to the limited capacity of the existing sewer system. The actual effect of this practice is usually not noticeable unless sludge from the water treatment plant exceeds the total suspended solids received by the wastewater treatment plant (in the form of sewerage) by more than a few percent. On certain occasions, the alum sludge may even be beneficial to wastewater treatment processes; it may increase the removal of phosphorous or may aid in the removal of suspended solids due to the aluminum hydroxide. Water treatment plants using a high proportion of organic polymer as coagulant will yield sludge that tends to be more biodegradable. The negative effect of alum sludge on wastewater treatment plants is the reduction in methane gas (caused by excessive amounts of inert materials such as aluminum hydroxide and clay materials) being brought to the sludge digesters; increased levels of heavy metals in the sludge deters the biological treatment processes. Unless both the water and wastewater treatment plants are under the same administrative jurisdiction, a connection fee will be imposed for the direct discharge of sludge to the sewer system. Following is an example of the fee structure used in southern California in 1986:

$$\text{fee} = \$375 \times [0.0034\ (\text{gpd}) + 0.115\ (\text{ppd COD}) + 0.3595\ (\text{ppd SS})]$$

where gpd is the sludge flow (gal/day), ppd COD the pounds per day of COD, and ppd SS the pounds per day of suspended solids.

In addition to the initial connection fee, the sewer authority will assess an annual surcharge based on the established unit costs. An example is as follows:

cost per year per million gallons discharge	\$200
cost per year per 1000 lb of COD	12
cost per year per 1000 lb of SS	33

In contrast to the difficulties encountered in the United States (except in Chicago and a few other cities), the discharge of sludge to sanitary sewers is a popular method of disposal in Great Britain and the Greater Paris area in France. It has been claimed that this practice has very little effect on the wastewater treatment processes or on subsequent sludge digestion.

Hauling Sludge to Landfills Selection of a sanitary landfill that is near a city is becoming increasingly difficult. In recent years, residents have become more concerned about environmental issues and issues that may affect land value. To further complicate matters, many existing sanitary landfills are reluctant to accept liquid and semiliquid sludge in order to comply with the federal landfill guideline PL 91-512 (Solid Waste Disposal Act) and because of their operational inconvenience. The Solid Waste Disposal Act recommends that landfill sites accept water treatment sludge only if it does not contain free moisture. *Free moisture* is defined as liquid that will drain freely, by gravity, from solid material. Currently, most landfill sites that have the bottom and sidewalls lined with impervious material only accept sludge with a minimum solids content of 20 to 25%; some sites that do not have the lining demand a solids content of over 40%, due to the potential contamination of groundwaters. It is important to note that alum sludge containing less than 20% solids generally cannot easily be handled by sludge handling equipment or hauled from the plant in trucks.

Landfills generally charge a disposal fee of \$5 to \$10 for each ton of dried sludge. If the sludge is located in a remote area, the hauling charge becomes significantly higher. Design engineers should also investigate alternative landfill sites in case the nearest site becomes unavailable. If sludge hauling trucks must pass through respectable residential areas or busy commercial zones, the option of using landfills may be eliminated by public hearings. Environmental impact studies may also prevent implementation of this alternative.

On-Site Disposal On-site disposal may be considered as an alternative method of sludge disposal only if the treatment plant has an adequate site area, has an old quarry on the premises, produces very small amounts of sludge, or has a combination of these conditions. If a plant is a candidate for on-site disposal, the following issues must be evaluated carefully prior to design work: the potential for contaminating the finished water and/or groundwater, and the aesthetics and safety of the facility. The disadvantages of on-site disposal may be illustrated by the following example: Application of alum sludge to the ground surface tends to clog the soil, thus preventing the passage of air to the roots of vegetation. However, lime sludge offers some advantages: Lime sludge has been shown to improve the condition of clay soil and neutralize acidic soil.

Recovery of By-products Acidic alum recovery was first tried on a large-scale basis in Japan: 15 plants were built between 1965 and 1972. Alum recovery can generally be attained at a rate of 80% or more, through sulfuric acid extraction at a pH of 2.5. The recovered alum is nearly as effective as commercially available alum. However, the quality of the alum often does not meet water treatment chemical standards because other substances are also extracted during the acidification process: iron, manganese, heavy metals, and a wide variety of organic compounds. Consequently, the alum recovered is used in conjunction with commercially available alum. Due to the possible harmful effects of recycling and the accumulation of heavy metals, this practice is now discouraged. The recovered by-products of alum sludge have also been used in the production of low-grade bricks (in Brazil), as road stabilizer, and as aggregates for low-grade road material. In Japan, alum sludge is used as a raw material in the cement manufacturing industry, as a soil conditioner for certain types of vegetation, and in high-grade bricks.

In many locations, lime recovery is carried out by burning the lime sludge in kilns. However, very few plants now practice this alternative because of recent concerns associated with air pollution and because of economic disadvantages. Lime sludge has been used successfully in the midwestern United States to neutralize acidic soil. Lime sludge may also be used to increase the porosity of tight soils, thus allowing them to be more workable and productive for agricultural purposes.

Design Criteria The design criteria for (1) the properties of alum sludge, (2) lagoon drying beds, (3) sand drying beds, and (4) mechanical dewatering systems are now presented.

Properties of Alum Sludge

1. The solids content of wet alum sludge obtained from ordinary rectangular sedimentation tanks containing mechanical sludge collection systems is 0.5 to 1.5%.
2. Alum sludge from sludge blanket clarifiers has a solids content of 0.1 to 0.5%.
3. Sludge lagoons are used primarily in gravity thickening. The solids content of sludge processed in this manner, with 3 to 4 months of continuous sludge filling and decantation operations, may be 3 to 4%.
4. A regular sludge thickener with stirring rakes can yield alum sludge that is 4 to 6% solids with the addition of very little polymer.

5. Sludge containing 4 to 5% solids at the beginning of a drying cycle will shrink to approximately 15% of the original volume after drying; the dried sludge will thus be 45% solids.
6. The initial depth of the sludge, prior to drying, has a drastic effect on the length of the drying period. Thus the initial depth should not exceed 2 to 3 ft (0.6 to 0.9 m) if the drying cycle is established to be 3 to 4 months.
7. The supernatant must be decanted continuously during the drying cycle to facilitate the formation of cracks on the surface of the drying sludge; these cracks will speed up the drying time by exposing more surface area for evaporation.
8. Figure 4.3-1 plots the unit volume weight (bulk density) of the alum sludge at various stages of the drying cycle. This chart should be used as a reference when designing a drying bed.

Lagoon Sludge Drying Bed

Sizing guide	The required net bed area for alum sludge is estimated to be 8 lb/ft^2 (40 kg/m^2) for wet regions and 16 lb/ft^2 (80 kg/m^2) for dry regions; these figures are based on dry sludge weight.
Minimum number of lagoons	Three, but four are preferred.
Size of each lagoon	Each lagoon should be capable of holding the amount of sludge produced (average) by the treatment plant over a period of 3 to 4 months.
Depth of the liquid sludge	A maximum depth of 6 ft (1.8 m) during the filling stage; however, a depth of 4 ft (1.2 m) is preferred.
Bottom slope	A downward slope of 0.5 to 1% toward the outlet.
Shape of each lagoon	Rectangular or free form with a minimum length/width ratio of 4 : 1.
Underdrainage system	This is an optional item. The system is composed of perforated-pipe laterals with gravel packing and 1 ft (0.3 m) of sand above the gravel bed. Note that the underdrainage system tends to become clogged and is often ripped away by the sludge removal equipment.
Lining	The entire surface, including the bottom of the drying bed, should be covered by 2 to 3 in. (5 to 7.5 cm) of gunite, asphalt cement, or soil cement to facilitate easy sludge removal (by means of mechanical equipment), to prevent contamination of the groundwater, to provide protection from erosion, and for weed control.
Distance between inlet and outlet	A minimum of 100 ft (30 m) should be provided to prevent flow short-circuiting.
Inlet pipe	Isolation valve and a simple energy dissipator.

Figure 4.3-1 Unit volume weight (bulk density): (*a*) alum sludge; (*b*) lagoon sludge; (*c*) drying alum sludge.

Outlet	Multilevel [9- to 12-in. or (23- to 30-cm) interval] decantation facility, overflow protection, and complete drainage of the lagoon.
Vehicle access	A ramp should be provided for each bed to facilitate the removal of dried sludge.
Access road	A good access road, both to and around each lagoon, should be included in the design.
Flow velocity in the pipe	The flow velocity of liquid sludge should be higher than 2.5 ft/s (0.75 m/s).

Sand Drying Bed The design criteria for this type of drying bed are fundamentally the same as those listed for lagoon drying beds, with the exception of the following items:

Depth of the liquid sludge	A maximum of 18 in. (45 cm), but 12 in. (30 cm) is preferred.
Underdrain laterals	Perforated laterals with 6-in. (15-cm) diameters and 10 ft (3 m) on center.
Gravel bed	Three layers with a total depth of 14 in. (36 cm): 3-in. layer of $\frac{1}{8}$- to $\frac{3}{8}$-in. gravel; 3-in. layer of $\frac{3}{8}$- to $\frac{1}{2}$-in. gravel; and 8-in. layer of $\frac{3}{4}$- to 1.5-in. gravel.
Sand bed	Sand ranging from No. 10 to No. 30 mesh at a depth of 12 in.

NOTE:

Lime sludge is generally easier than alum sludge to dewater by the sand drying bed; this method yields a sludge that is 25% solids and is therefore spadeable [weighs 120 lb/ft^3 (1925 kg/m^3)].

Mechanical Dewatering Systems The belt press, filter press, vacuum filter, centrifugation, and freeze-and-thaw processes are the mechanical dewatering systems that are applicable to water treatment. The first two units are most frequently employed. Since these systems are proprietary items, the project engineer need only determine the functional specifications: the solids content of the dewatered sludge, floor space requirements, and limitations on power and chemical use. A pilot study must be conducted prior to writing the specifications: to confirm the effectiveness of the system and to evaluate the type and required chemical dosage for preconditioning the sludge.

Example Design Calculations

Example 1

Given A water treatment plant with a daily average flow of 20 mgd (0.88 m^3/s) is located in southern California and requires a lagoon type of sludge drying bed. The average turbidity of the raw water is 12 ntu and the average alum dosage is 10 mg/L.

Determine

(i) The required number of lagoons
(ii) The depth of each lagoon
(iii) The size of each lagoon

Solution By using the formula presented in the "Considerations" section, the sludge production rate may be calculated:

$$[(10 \times 0.26 \times 8.34) + (12 \times 1.3 \times 8.34)] \times 20$$
$$= 3035 \text{ lb (1380 kg) of dry sludge per day}$$

(i) Provide four lagoons with identical capacities. Each lagoon should be sized to hold sludge with an average of 4% solids for a duration of 3 months.

(ii) The maximum depth of the liquid sludge, during the filling stage, should be set at an average of 6 ft (1.8 m). When the drying cycle begins, continuous decantation should be initiated so that the depth of the sludge is reduced to 2 to 3 ft (0.6 to 0.9 m) prior to the actual drying process.

(iii) The required net area of one lagoon may be estimated from the sizing guide listed in the "Design Criteria" section. Since southern California is sunny for most of the year, with a warm to hot climate, use a dry solid weight of 15 lb/ft^2 (73 kg/m^2) of required area to determine the required lagoon area. Based on a 3-month supply of sludge, the average amount of dry solids is

$$(3 \times 30) \times 3035 = 273{,}150 \text{ lb (124,160 kg)}$$

The required net area for the lagoon is therefore

$$273{,}150 \text{ lb} \div 15 \text{ lb/ft}^2 = 18{,}210 \text{ ft}^2 \text{ (1700 m}^2\text{)}$$

Thus, make the lagoon rectangular in shape, 60 ft (18.3 m) wide and 300 ft (91.5 m) long, although a free shape, such as a kidney bean shape, is aesthetically better.

Check the calculations. Based on Figure 4.3-1, the volume of the 4% sludge is approximately 63.8 lb/ft^3. Therefore, the total volume of the sludge may be calculated to be

$$V = (273{,}150 \div 0.04) \div 63.8 = 107{,}034 \text{ ft}^3 \text{ (3030 m}^3\text{)}$$

Since the depth of the liquid sludge at the end of the filling cycle is 6 ft (1.8 m), the area required for 136,480 ft^3 of 4% solids sludge is

$$107{,}034 \div 6 = 17{,}840 \text{ ft}^2 \text{ (1660 m}^2\text{)}$$

which is essentially the same as 18,000 ft^2 (1675 m^2)

In summary, provide four lagoons that have a net area of approximately 72,000 ft^2 [6700 m^2 (1.7 acres)] and a depth of 6 ft (1.8 m). Other considerations, such as vehicle access, the building of dikes around the lagoons (sidewall slope of 2.5 : 1), and a vehicle access road that is 10 ft (3 m) wide at the top of the dikes, require the land to be almost twice the net area required for the lagoons [3.5 acres (1.4 ha)].

Example 2

Given A 10-mgd (0.44-m^3/s) lime softening plant is located in the midwestern United States. The plant requires that lime sludge drying beds be constructed. The average total hardness of the river water is 360 mg/L, with insignificant amounts of magnesium. The finished water hardness is, on average, 80 mg/L.

Determine

(i) The number of sludge drying beds
(ii) The filling depth of the sludge
(iii) The size of each bed

In addition, provide a sketch of the sludge bed section

Solution There are several general considerations that must be determined prior to making these four determinations. The following assumptions are made for this example.

1. For every 1 mg/L of hardness removed, 2.5 mg/L of dry solids (sludge) is produced.
2. Lime sludge containing insignificant amounts of magnesium hydroxide is easily filtered through the sand bed and dries faster than alum sludge.
3. The drying beds will yield sludge that is 50% dry solids (80 lb/ft^3 or 1280 kg/m^3). Sludge that is 25% solids contains 72 lb of dry solid per cubic foot of sludge (1156 kg/m^3); this sludge is spadeable and is also easily handled by mechanical sludge removal equipment.
4. Assume that the average daily flow is 67% of the plant capacity of 10 mgd.
5. The hardness removed is averaged to be 360 − 80 = 280 mg/L.

The average sludge production rate (dry sludge) is 39,115 lb/day (17.780 kg/d):

$$[(2.5 \times 280) \times 8.34] \times 6.7 = 39{,}115 \text{ lb/day}$$

Now assume that the process unit yields a liquid lime sludge that is 5% solids with a unit weight of 65 lb/ft^3 (1044 kg/m^3). The volume of the 5% solids sludge produced each day may be calculated as

$$V = (39{,}115 \div 0.05) \div 65 = 12{,}035 \text{ ft}^3\text{/day } (341 \text{ m}^3\text{/d})$$

The volume of the 25% solids sludge is 39,115 ÷ 72 = 543 ft^3/day (15.4 m^3/d)

(i) A total of eight sludge beds is needed. Each bed should be able to hold the 25% solids sludge produced over an average of 15 days, so that a total of 4 months storage is provided.

(ii) The maximum depth of accumulated sludge during the filling cycle is 12 in. (0.3 m).

(iii) The size of each bed must be calculated as follows. The estimated volume of the semidried sludge (25% solids) for 15 days of storage is

$$V = 543 \times 15 = 8145 \text{ ft}^3 \ (230 \text{ m}^3)$$

The depth of the bed is limited to 1 ft (0.3 m). If the width of each bed is designated to be 40 ft (12.2 m), the required length is

$$(8145 \div 1) \div 40 = 203 \text{ ft}, \qquad \text{say, } 200 \text{ ft } (61 \text{ m})$$

Therefore, the drying bed is 40 ft (W) × 200 ft (L) or 12.2 m (W) × 61 m (L) per sand bed. Figure 4.3-2 is a sketch of the designed sludge bed.

NOTES:

1. The sludge beds tend to become clogged and the beds are easily disturbed by the heavy equipment used during sludge removal.
2. The sludge lagoon drying bed is the most popular and practical method of sludge handling.
3. The design criteria for the sludge lagoons are basically the same for lime sludge and alum sludge.
4. A conventional lime sludge lagoon system has a minimum of three lagoons, each having a depth of 10 ft (3 m) and the capacity to hold nearly a 1-year production of wet sludge. The land requirement is approximately 1 acre per 100 mg/L hardness removed (4050 m^2 per 100 mg/L) for each million gallons treated (3785 m^3). A decantation facility is essential for proper sludge drying.

Operation and Maintenance Plant operators must pursue three major issues: (1) reduction of the sludge production rate, (2) withdrawal of highly concentrated sludge from the process, and (3) optimization of the sludge dewatering operation.

Reduction of the Sludge Production Rate To decrease the sludge production rate, operators must select the appropriate coagulant, optimize the chemical dosage, and possibly optimize the flocculation process. Since 26% of the added alum produces aluminum hydroxide (i.e., becomes part of the sludge), the primary goal of the operator should be to reduce the alum dosage without degrading the finished water quality. The operator should first check to see if the polymers are being em-

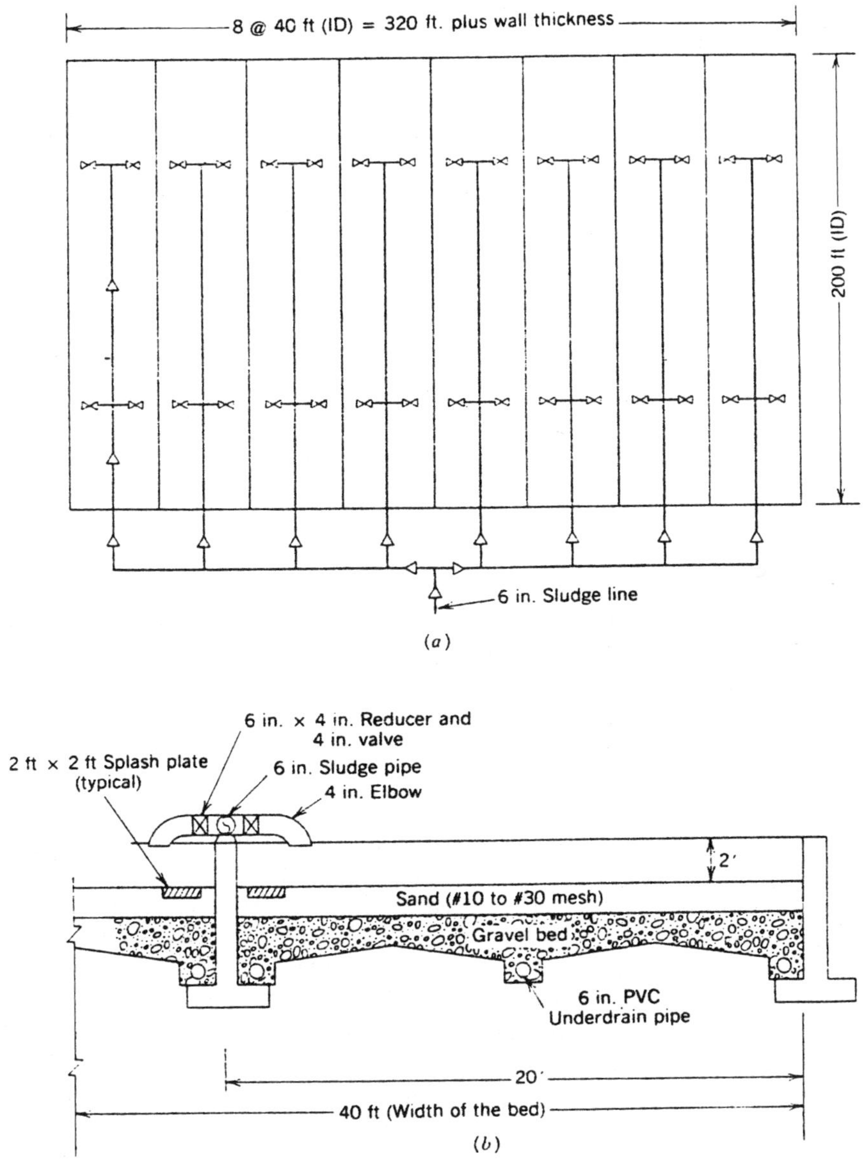

Figure 4.3-2 Lime sludge drying bed: (*a*) plan; (*b*) partial cross section.

ployed effectively. Since the cost of most cationic polymers is approximately six times higher than alum, there will be no difference in the total chemical cost if 1 part of polymer is used to replace 6 parts of alum. There are many case histories which prove that a cationic polymer dosage of 0.5 mg/L, together with alum, reduces the alum dosage by 3 to 5 mg/L when compared to flocculation by alum alone. The use of anionic and nonionic polymers as flocculation aids can also reduce the required alum dosage, as illustrated in Table A10-1. In the case of direct or two-stage filtration, the use of cationic polymers virtually replaces alum as the coagulant. Since there is a minimal amount of aluminum hydroxide in the sludge, there is a reduction in sludge production and the sludge is more easily handled.

The adoption of a direct filtration process during periods of good raw water quality is an alternative method of drastically reducing sludge production. Many water treatment plants that draw their water supply from large reservoirs or lakes should be able to implement this mode of treatment, but only at certain times of the year. Rather than feeding 15 mg/L of alum to raw water containing 3 ntu of turbidity to form settleable floc, the direct filtration process requires an alum dosage of only 6 mg/L, or a combination of alum (3 mg/L) and cationic polymer (0.5 mg/L) may be used to coagulate the turbidity. The flocculation and filtration processes occur subsequently in the filter bed. To reiterate, there are many actual case histories of successful treatment of water by this mode; technical articles on this subject may be found elsewhere.

The third item that the operator must be aware of is the effect of feeding lime with alum at the head of the plant. Many old textbooks often illustrate this practice without regard to the alkalinity and pH of the raw water, and a number of treatment plants consider this practice as a given. Unless the alkalinity of the raw water is below 15 mg/L and the required alum dosage exceeds 15 mg/L, there is no need to add extra alkalinity. Flocculation in a pH range of 5 to 7 is actually ideal, except for the potential for corrosion; however, the pH may be adjusted in the later stages. If lime is not added with alum to the head of the plant, both residual alum in the treated water and sludge production will be reduced. All cases described in this section require bench-scale tests to be performed to verify and obtain the optimum chemical dosage.

Withdrawal of Thick Sludge If the sludge delivered to the dewatering system is thick, the process is more cost-effective and more effective in concentrating the solids, thus allowing the ultimate disposal operation to be more efficient. For example, the operator may adjust the sludge collection cycle of the clarifier by observing the consistency of the liquid sludge, provided that the facility allows the operator to do so. Excessive sludge withdrawal will result in an extremely low solids content. Thus operators should adjust plant operations so that the sludge will have a solids content of at least 0.5%; alum sludge obtained from the clarifier preferably has a solids content of 1%. A word of caution: Do not allow the sludge to remain in the clarifier for a long period of time in an attempt to increase sludge concentration, because the sludge may become septic and may impart an objectionable taste and odor to the treated water.

Although it is common practice to discharge the sludge from the clarifiers to the filter waste-wash holding tank, it is highly discouraged for three reasons: (1) the sludge becomes diluted, (2) unfavorable substances such as organic compounds, microorganisms, heavy metals, and taste- and odor-causing compounds may potentially be recycled back to the treated water, and (3) the EPA's upcoming rule will prohibit this practice. Sludge from the clarifiers should be directed to either the sludge beds or the sewer system.

Optimization of Sludge Thickening and Disposal Operations There are many areas in which the plant operator can implement change; however, these depend on the type of thickening system. Plant operators have the ability to optimize the sludge loading rate to the thickening process, the thickness of the feed sludge, the belt speed, and the pressure level to the filter press. However, the efficiency of the sludge thickening process may be greatly improved by adding an appropriate type and proper dosage of polymer to the liquid sludge, regardless of the type of sludge being processed. The EPA's Filter Wash Waste Recycling Rule (effective in the year 2003) prohibits discharge of sludge from the clarifiers to the filter wash waste if the water is recycled.

In regions where lime is readily available (at a reasonable cost) and where the cost of personnel is also reasonable, the application of an additional 15% (or more) of lime to alum sludge raises the pH of the sludge to 12. Experience has shown that this practice dramatically improves the dewatering characteristics of the alum sludge.

BIBLIOGRAPHY

AWWA, *Water Quality and Treatment,* 3rd ed., McGraw-Hill, New York, 1971.

AWWA Committee Report, "Research Needs for Alum Sludge Discharge," *J. AWWA,* 79:6:99 (June 1987).

Bishop, M. M. et al., "Mechanical Dewatering of Alum Solids and Acidified Solids: An Evaluation," *J. AWWA,* 83:9:50 (September 1991).

Bishop, M. M., et al., "Testing of Alum Recovery for Solids Reduction and Reuse," *J. AWWA,* 79:6:76 (June 1987).

Clark, D. K., and Anderson, C. A., "Shared Residual Processing Will Benefit Adjacent Agencies," *J. AWWA,* 89:6:48 (June 1997).

Cornwell, D. A., and Lee, R. G., "Water Stream Recycling: Effect on Water Quality," *J. AWWA,* 86:11:50 (November 1994).

Cornwell, D. A., et al., "An Overview of Liquid Ion Exchange with Emphasis on Alum Recovery," *J. AWWA,* 71:12:74 (December 1979).

Culp, R. L., et al., "Is Alum Sludge Advantageous in Wastewater Treatment?" *Water Waste Eng.,* p. 16 (July 1979).

Dulin, B. E., and Knocke, W. R., "The Impact of Incorporated Organic Matter on Dewatering Characteristics of Aluminum Hydroxide Sludges," *J. AWWA,* 81:5:74 (May 1989).

Garabeck, R. J., et al., "Silvicultural Application of Alum Sludge," *J. AWWA,* 79:6:84 (June 1987).

Geertsema, W. S., et al., "Long-Term Effects of Sludge Application to Land," *J. AWWA,* 86:11:64 (November 1994).

Japan WWA, "Design Criteria on Handling Facilities for Water Treatment Wastes," *J. Jpn. WWA,* 489:103 (June 1975).

Johnson, G., Buchanan, G. G., and Newkirk, D. D., "Optimizing Belt Press Dewatering at the Skinner Filtration Plant," *J. AWWA,* 84:11:47 (November 1992).

Knocke, W. R., "Effects of Coagulation on Sludge Thickening and Dewatering," *J. AWWA,* 79:6:89 (June 1987).

Koorse, S., "The Role of Residuals Disposal Laws in Treatment Plant Design," *J. AWWA,* 85:10:57 (October 1993).

Lucas, J. B., et al., "Effects of Alum Sludge on Plant Growth," *J. AWWA,* 86:11:75 (November 1994).

Martel, J. C., "A Pilot Scale Study of Alum Sludge Dewatering in Freezing Bed," *J. AWWA,* 83:12:51 (December 1991).

Novak, J. R., and Bandak, N., "The Effect of Shear on Dewatering of Water Treatment Residuals," *J. AWWA,* 86:11:84 (November 1994).

Novak, J. T., et al., "Use of Polymers for Chemical Sludge Dewatering on Sand Beds," *J. AWWA,* 69:2:106 (February 1977).

Peters, G. H., Baumann, E. R., and Larson, M. A., "Effect of Various Parameters on the Thickening or Softening of Plant Sludges," *J. AWWA,* 81:3:74 (March 1989).

Thompson, P. L., and Paulson, W. L., "Dewaterability of Alum and Ferric Coagulation Sludges," *J. AWWA,* 90:4:164 (April 1998).

"Water Treatment Plant Sludges: An Update on the State of the Art," Parts 1 and 2, *J. AWWA,* 70:9:498 and 70:10:548 (September and October 1978).

Wilheim, J. T., et al., "Freeze Treatment of Alum Sludge," *J. AWWA,* 68:6:312 (June 1976).

4.4 INTAKE SYSTEM

Purpose A water treatment plant ceases to function when the intake system fails to supply water. If the plant is relying on a single intake system, malfunction of this unit results in interruption of the water supply to the community. Therefore, the intake must be located in an easily accessible location and designed and built to supply a specified quantity of the best available quality of water; the source should be reliable, that is, no interruption under any conditions. The intake system is often combined with the grit chamber and raw water pumping station. However, in this section we discuss only the intake system.

The intake structure is constructed at the water source for the sole purpose of extracting water for water treatment and water supply. The capacity of the intake should be the maximum daily demand anticipated for the next 50 years; the permits required to build the intake and the cost of construction are both difficult and costly tasks. If an intake is twice the size required to satisfy present needs, it would cost an additional 20 to 30%. However, a well-designed intake system, for middle-sized to large plants, may cost 15 to 20% of the total cost of constructing the entire water treatment plant.

Considerations The essential factors for the intake system are reliability, safety, and minimal operation and maintenance costs. Thus the design engineer must conduct extensive studies and use ingenuity in selecting the site and designing the intake works. Numerous studies must be performed prior to selection of the intake site: water rights, the quality of the water source, climatic conditions, fluctuations in the flow rate and water level, regulations set by departments of fish and game, water navigation, geographical and geological information, and economic issues.

The intake may be located on a river, lake, or reservoir or may be designed to extract groundwater. The basis for site selection varies according to the water source.

River Intakes Site selection for the intake system should be based on (1) obtaining the best quality water through the implementation of procedures that avoid polluting the water source; (2) forecasting possible changes in the flow and course of the river; (3) minimizing the effects of icing, flooding, floating debris, river navigation, and surges in flow; (4) providing easy access to the intake for maintenance and repair; (5) providing adequate maneuvering space for vehicles; (6) allowing for future additions to the facility; (7) maintaining a safe quantity of water flow during periods of drought; (8) minimizing the effects of the facility on aquatic life; and (9) obtaining the best or at least good, geological conditions.

If the intake site is located at a bend in the river, the outside bank of the curve should be selected since it offers better conditions. The inside bank will probably be troublesome because of the shallow water depth and the presence of sand bars and accumulated debris. Most river intake structures are constructed as shore intakes: an intake system and raw water pumping station combined. Yet several alternatives are available; submerged intake, intake tower, suspended intake, siphon intake, floating intake, and movable intake. The last two types are worth evaluating if the water level of the river bed tends to fluctuate widely, if there is a frequent flooding problem, and if the intake capacity is rather small—less than 5 mgd (0.2 m^3/s).

The design of the intake system should include multilevel intake ports and bar screens or racks. Additionally, engineers must provide protection against ice, floating debris, boats, and barges, and the design must meet the requirements established by the department of fish and game. The infiltration galleries or pipes are a unique type of intake system. This system is a gravel-packed gallery containing many openings (perforated pipes). It may be constructed under the river bed parallel to the shore or in a gravel-packed well located on the river bed. The advantages are that it avoids problems associated with ice formation, and water quality is substantially better than that of the river water itself. The disadvantages are the potential reduction in yield due to clogging by the sand and gravel layer, and its limited capacity.

Lake and Reservoir Intakes Tower intakes and submerged intakes are two typical types of lake and reservoir intakes. Submerged intakes are generally less expensive than intake towers to build. These units also have no aesthetic impact and do not interfere with river navigation. Submerged inlets that are properly designed and located will have minimal problems with ice as long as the water depth exceeds 20 to 30 ft (6 to 9 m). The characteristics of tower intakes are described in the "Design Criteria" section below.

An alternative to a tower and submerged intakes is a shore shaft. The shore shaft is the part of the intake system that connects to the submerged inlet structure by means of a pipe or conduit. It functions as a screen chamber as well as a pump suction well. Thus the shore shaft should be located at a depth that allows for the safe withdrawal of the required maximum flow rate when the lake or reservoir is at its lowest level.

Site selection for the lake and reservoir intake system should be based on the considerations listed for river intakes and the following items: (1) the stratification characteristics of the water due to seasonal changes in water temperature; (2) the water quality of minor streams, algal growth characteristics, and the growth cycle of other microorganisms; (3) shore conditions, wind direction, and velocity; (4) the conditions of the catchment area, including potential sources of pollution; (5) conditions conducive to siltation in the reservoir; (6) the ultimate purpose of the reservoir—multipurpose, such as hydroelectric power, irrigation, water supply, recreation—and the impact of joint users on the inlet location; (7) the level of sanctioned recreational activities and the degree to which body contact sports will be allowed; and (8) the flood level. In addition to these considerations, the design engineer must also contend with issues such as a cooperative design arrangement with the joint users (if the intake is for multipurpose use), the possible incorporation of an artificial water mixing device to destratify the water in the intake design, and special ice-breaking measures (air agitation) at the intake location for cold weather regions. Refer to Figures 4.4-1 and 4.4-2 for various types of intake structures.

Groundwater Intakes The term *groundwater* includes water in artesian wells, deep infiltration galleries or pipes, shallow wells, and deep wells. In all cases, the quality and quantity of the water should be investigated by a test well or other methods. The first step of the design project is to find a reliable and suitable aquifier to function as the source of groundwater. The initial study must also investigate (1) the relationship between the level of the nearby river and the groundwater level of the site during the dry season; (2) the quality of the groundwater during the dry season if the wells are close to the seashore; (3) the quality, yield, and geological condition of wells and infiltration galleries in the immediate vicinity; (4) the temperature and quality of the water and if artesian wells and springs are targeted as the potential sources of water, the yield; and (5) the effect of industrial wastes and the possible seepage of wastewater (nitrates) into the aquifer. Site selection for groundwater intakes must also consider the potential intrusion of seawater, the effects of having the groundwater level being drawn down by a cluster of wells, and the potential migration of pollutants from remote dump sites harboring industrial toxic wastes.

After a well site has been selected, a test well should be drilled and pumping tests must be performed to obtain a safe yield from the test well and to check the quality of the well water. A safe yield is assumed to be approximately 60 to 70% of the maximum yield obtained during the pumping test. Since the quality of the groundwater is generally very good, all conventional treatment processes except chlorination may be eliminated. Thus the design of the water supply well must be in strict accordance with all requirements dictated by the regulatory agencies.

Design Criteria In this section we present the design criteria for the intake tower, shore intake, intake crib, intake pipe, and infiltration gallery.

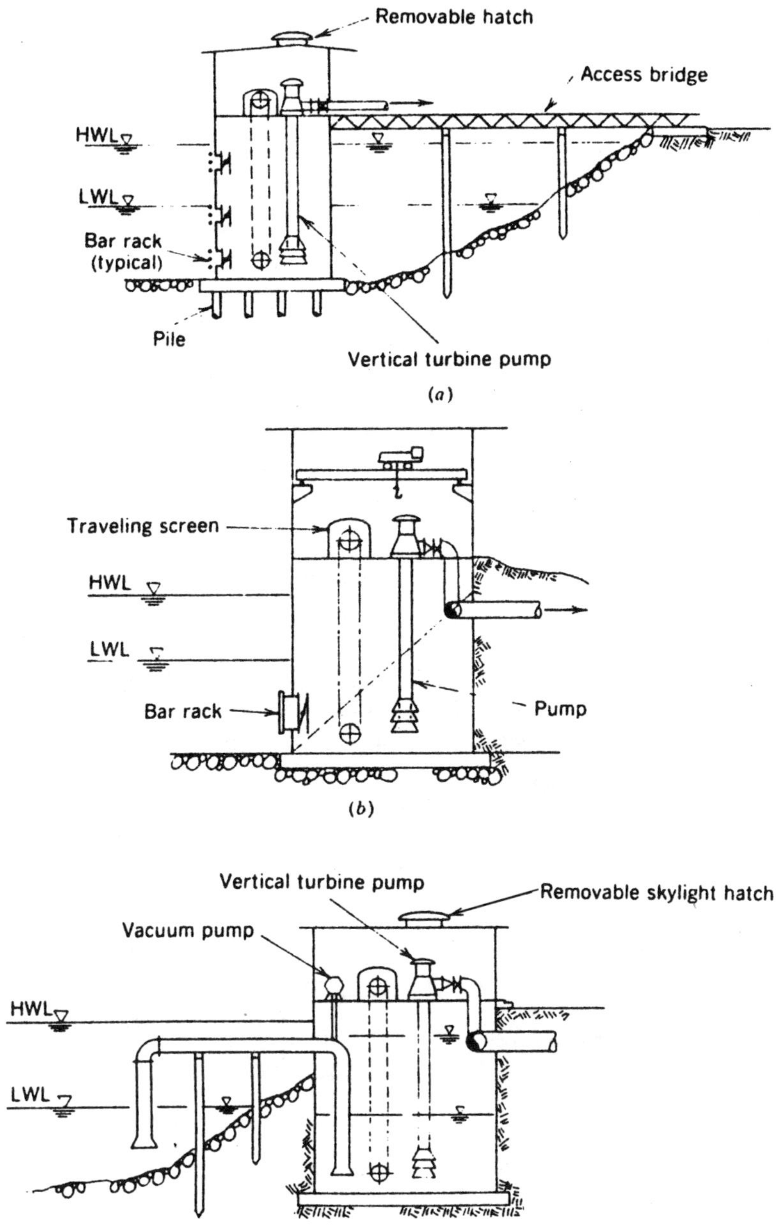

Figure 4.4-1 Various types of intake structures: (*a*) intake tower; (*b*) shore intake; (*c*) siphon well intake.

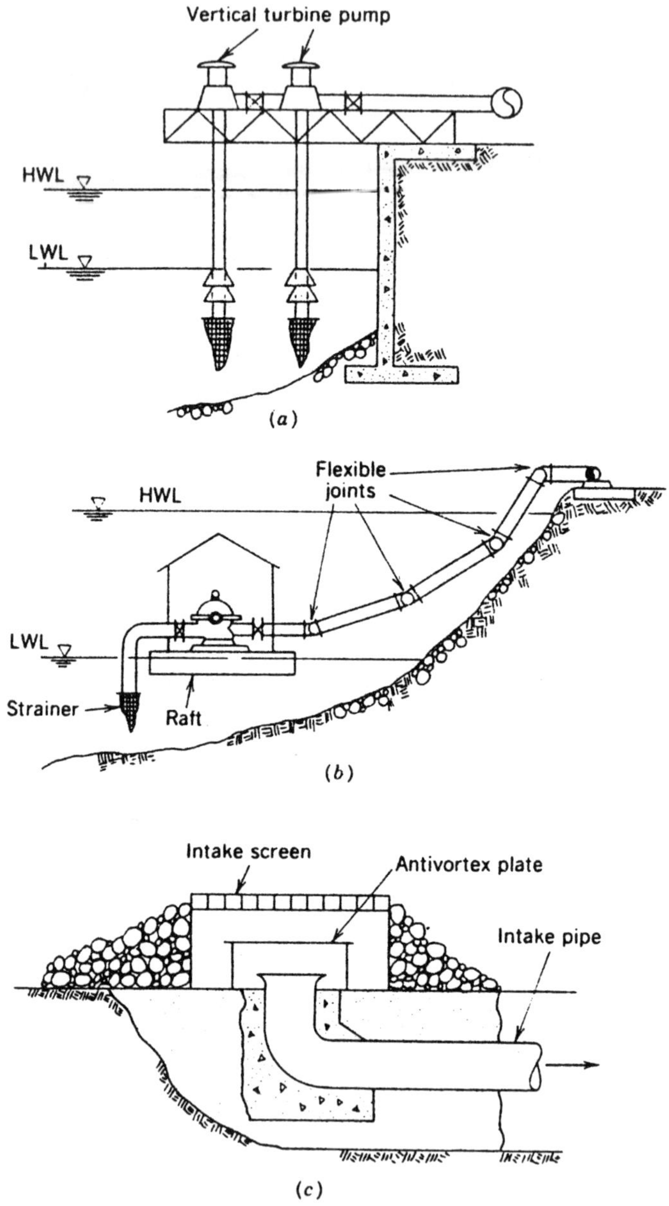

Figure 4.4-2 Various types of intake: (*a*) suspended intake; (*b*) floating intake; (*c*) intake crib.

Intake Tower

Location	As close to the shore as possible but at a minimum water depth of 10 ft (3 m), with the exception of small intakes.
Shape and size	The standard cross section of the tower should be circular or oval. In the case of a river intake, the major axis of the oval must be parallel to the flow direction. The top of the tower should be at least 5 ft (1.5 m) above the highest water level and the access bridge to the tower must also have the same clearance. The interior diameter of the tower must be large enough to be able to install and service the intake gates and pumps (if pumps are installed).
Structure	The materials used to construct the tower should be strong and durable, such as reinforced concrete, and should be built on a solid foundation so that it may resist the force of flooding.
Intake ports	Gated intake ports must be provided at various depths; the lowest of the ports should be approximately 2 ft (0.6 m) above the bottom. The vertical interval of the ports should be 10- to 15-ft (3 to 4.5 m). The flow velocity through the Gross area of the ports, at the same elevation, should generally not exceed 1 ft/s (0.3 m/s). In regions where the hazards of icing are to be expected, a velocity of 0.5 ft/s (0.15 m/s) or less is desirable. Sluice gates or butterfly valves must be specified for the intake ports on the interior or exterior of the tower. The gates are used for isolation, but butterfly valves are preferred since the sluice gates have a tendency to freeze up if they are not used on a regular basis.
Bar screen	Bar screens (racks) should be provided at each port exterior to the valve. The bars should be composed of $\frac{1}{2}$- to $\frac{3}{4}$-in. (13- to 19-mm)-diameter steel spaced 2 to 3 in. (5 to 7.5 cm) on center. Under normal conditions, the flow velocity through the net opening area of the bar screens must not exceed 2 ft/s (0.6 m/s). In certain cases the velocity is limited to under 0.35 ft/s (0.1 m/s) by fish and game authorities to protect small fish from being sucked in.
Fine screen	If screens are not installed downstream of the tower, it will be necessary to install fine screens to remove small floating objects and to protect fish. In most cases, the clear openings of these screens should be approximately $\frac{3}{16}$ to $\frac{3}{8}$ in. (5 to 9.5 mm), and the maximum flow velocity through the net openings in the screens should be

no more than 2 ft/s (0.6 m/s) at the maximum designed flow rate. The use of an automatic hydraulic cleaning device is highly recommended. Moreover, in regions with a very cold climate, the intake tower and screens should be protected from frazil and anchor ice formation. The two methods that are often used are compressed air and steam injection.

Shore Intake

Location	Minimum water depth of 6 ft (1.8 m).
Type	Typical shore intake—siphon well, suspended, or floating type—depending on the situation.
Structure	This is determined by the type of intake but is basically the same as for the intake tower.
Intake bay	The cross section of the intake opening should be capable of accommodating a maximum flow velocity of 1.5 ft/s (0.45 m/s). The flow velocity should be reduced to below 1 ft/s (0.3 m/s) if an unusual amount of debris or icing is to be expected.
Bar screen	These should be provided at the intake bay and set approximately 60° from the horizontal. The bar should be composed of $\frac{1}{2}$- to $\frac{3}{4}$-in. steel spaced 2 to 3 in. (5 to 7.5 cm) on center. Under normal conditions, the flow velocity through the screens should not exceed 2 ft/s (0.6 m/s). The bar screen has a trapezoidal cross section, which prevents the accumulation of solids, with reciprocating action, that is, a climbing screen inclined 80° to the horizontal.
Fine screen	Fine screens are often installed just downstream of the bar screen or may be installed in the grit chamber. The clear openings in the screen must be $\frac{3}{16}$ to $\frac{3}{8}$ in. (5 to 9.5 mm), and under normal conditions, the maximum flow velocity through the net opening should occur at minimum submergence. As with the tower intake, in regions with cold weather, the fine screens must be protected from icing hazards through the use of compressed air or steam injection.

Intake Crib

Location	Deeper than 10 ft (3 m) from the surface and in a location where it will not be buried by sediment, washed away, or hampered by problems associated with ice formation.
Structure	In areas where the water depth exceeds 10 ft (3 m), the top

of the intake should be 3 ft (1 m) above the river bottom. However, if the water depth is less than 10 ft, the crib must be buried anywhere from 1 to 3 ft (0.3 to 1 m) below the river bed. Cribs are usually polygonal and built with a firm frame of lumber or reinforced concrete.

All sides of the crib should be protected by riprap or a concrete slab. Crib ports must be sized to provide a maximum velocity of 0.25 to 0.5 ft/s (0.08 to 0.15 m/s). The crib should surround a bellmouthed pipe that is connected to the intake conduit.

Intake Pipe or Conduit

Size	To minimize the accumulation of sediments, the pipe or conduit must be capable of providing a minimum flow velocity of 3 to 4 ft/s (0.9 to 1.2 m/s).
Protection	If the pipe or conduit crosses a river or lake to connect with the shore shaft, it should have 3 to 4 ft (0.9 to 1.2 m) of cover over the top of the crown. Additionally, crushed rocks should be placed above the trench; a rule of thumb is to use 1 yd^3 of crushed rock per linear foot of pipe or conduit (2.5 m^3/m).
Slope	To avoid creating an air blockage in the line, the pipe or conduit must be laid on a continuously rising or falling grade.
Permit	A permit must be obtained from the appropriate authorities prior to the design phase if the intale pipe or conduit is designed to cross the bottom and bank of a river, lake, or reservoir.

Infiltration Gallery

Direction	At right angle to the river or parallel with the river flow, depending on the underflow pattern, the anticipated difficulties in construction, and the cost of constructing the gallery.
Depth	The common depth is generally 15 ft (4.5 m) below the bottom of the river or lake. Yet the actual depth should be determined through hydrogeological studies.
Orifice size	The diameter of each perforation is generally $\frac{3}{8}$ to $\frac{3}{4}$ in. (10 to 20 mm). The number of orifices is 2 or 3 per square foot (20 to 30 per square meter) of the collector surface.
Length	The length of the gallery is based on the assumption that the flow velocity through each orifice, at maximum intake flow rate, must be 0.1 ft/s (0.03 m/s).

Structure	The gallery must be constructed with perforated reinforced-concrete pipe or conduit. A stainless steel well strainer may be used for small-scale installation.
Slope	The gallery can be laid on level ground, but a slight slope of 500 : 1 will minimize the problem of air blockage.
Velocity	The flow velocity must not exceed 3 ft/s (0.9 m/s) at the outlet of the gallery.
Backfilling	The collector is packed with cobblestone, gravel, and sand similar to that used in filter bed gradation. The thickness of each layer must be a minimum of 20 in. (0.5 m). The trench of the gallery collector is backfilled with the specified sand, or native sand may be laid on top of the gravel and sand packing.
Junction well	This feature must be provided at the junction points of the gallery and at the end of the gallery line. The diameter of the junction well must be a minimum of 3 ft (1 m) to facilitate the inspection of the gallery.

Shallow Wells and Deep Wells The design and construction of the wells are addressed in many books. Therefore, no attempt is made to discuss this subject (see the Bibliography). The regulatory agencies and the *Ten States Standards* provide detailed instructions on the design of the wells.

Operation and Maintenance Under normal conditions, a properly designed and constructed intake system requires very little in the way of operation and maintenance efforts. The usual operational tasks include regular maintenance of mechanical equipment—traveling screen, intake pumps, and appurtenances—and the removal of large floatables collected by the bar racks. If the intake is provided with multilevel ports, the valves of each port must be operated on a regular basis. There have been too many cases where the sluice gates attached to various levels of intake ports became inoperable because they were not operated periodically for maintenance and severely corroded valve operating stems and the gates.

Floods are a potential problem for the intake system because logs and other large objects may jam the intake bar racks. In regions of cold weather, ice floes driven by the wind may potentially block the ports of the shore intake. Operators must keep careful watch for these problems and appropriate actions must be taken before any damage occurs to the intake or before there is a drastic reduction in the capacity. After each flood or earthquake it is vital to send a diver down to examine the condition of the underwater intake structures to determine if the ports have become blocked by sand migration or if the foundation of the intake has been eroded or damaged.

When the intake is located in a large reservoir the plant operator and laboratory personnel should periodically check the quality of water obtained from various depths of the lake. If the quality is better at a particular level, the intake level should be adjusted accordingly. This practice reduces the chemical cost of water treatment and saves a considerable amount of extra labor by avoiding conditions associated with algal bloom and taste and odor problems.

In some cases, the raw water must be conveyed to the treatment plant over a long distance by means of a pipeline or conduit. One of the problems associated with this practice is the growth of slimy organisms or freshwater muscles in the interior of the pipeline. The growth of these organisms may dramatically reduce the carrying capacity of the raw water aqueduct. A common solution is to chlorinate the water at the intake site, leaving approximately 0.5 mg/L of free residual for 15 to 30 min. Chlorination may be continuous or a shock treatment, depending on the situation; operators must give special consideration to the potential for THM formation. Regardless of the potential for THM formation, it is a good practice for engineers to incorporate a chlorine feed system, including a chlorine diffuser, during the design phase. If no such facility is provided, plant operators should consider adding a chlorination system before the situation becomes uncontrollable. In regions such as the Great Lakes area, zebra mussel infestation has caused great problems. Potassium permanganate is a good control chemical (dosage rate as low as 0.3 to 0.5 mg/L) since there is not THM formation by this chemical.

BIBLIOGRAPHY

Anderson, K. E., *Water Well Handbook,* 4th ed., Missouri Water Well and Pump Contractors Association, Belle, MO, 1971.

AWWA, *Groundwater,* AWWA Manual M21, American Water Works Association, Denver, CO, 1973.

Davis, C. V., ed., *Handbook of Applied Hydraulics,* McGraw-Hill, New York, 1969.

DHV Consulting Engineers, *Shallow Wells,* DHV, Anersfoot, The Netherlands, 1979.

Fair, G. M., et al., *Water and Wastewater Engineering,* Wiley, New York, 1966.

Gibson, U. P., *Water Well Manual,* Premier Press, Berkeley, CA, 1971.

Japan WWA, *Design Criteria for Waterwork Facilities,* Japan Water Works Association, Tokyo, 1969.

National Water Well Association, *Design and Construction of Water Wells,* Van Nostrand Reinhold, New York, 1988.

National Water Well Association, Committee on Water Well Standards, *Water Well Specifications,* Premier Press, Berkeley, CA, 1981.

Richardson, W. H., "Intake Construction for Large Lakes and Rivers," *J. AWWA,* 61:8:365 (August 1969).

Urquhart, L. C., ed., *Civil Engineering Handbook,* McGraw-Hill, New York, 1959.

4.5 GRIT CHAMBER

Purpose *Grit* is defined as a combination of silt, sand, gravel, shells, and other abrasive materials. The grit chamber is a plain sedimentation tank that removes grit by simple gravity sedimentation. The purpose of this chamber is to protect moving mechanical equipment (such as pumps and mixers) from the abrasive substances and to prevent the accumulation of grit in the raw water line and the pretreatment processes, including the ozone contact tanks and basins. The grit chamber should be situated upstream of the raw water pumps and as close to the intake structure as possible. This arrangement prevents problems associated with siltation and simplifies grit handling. As described in Section 4.4, the grit chamber and inlet structure are often constructed at the intake site, and screens are often installed at the inlet of the grit chamber. A hydraulic scale model conducted by the author clearly demonstrated that when the screen is installed properly, it functions as an inlet diffuser wall.

Engineers should note that the grit chambers for water treatment processes are distinctly different from those used in sewage treatment in regard to one function, the settling of undesirable compounds. Sewage plants are designed primarily to remove the large inert compounds of suspended matter. Thus, most of the organic decomposable solids are permitted to flow through the tank. The tank design therefore does not need to provide good flow characteristics for settling and the tank is often aerated to produce a spiral flow pattern. For this reason it is obvious that engineers should not use the blueprints of a sewage treatment grit chamber when designing a water grit chamber.

Considerations The four basic considerations in designing a grit chamber are (1) the location of the tank, (2) the number of required tanks, (3) the shape of each tank, and (4) the size of the grit to be removed. The first two issues are discussed only briefly. The most desirable location of the grit chamber is at the intake site since this will maximize the purpose of the tank. A single divided tank or two separate tanks are generally used to meet the minimum operational requirements. This arrangement allows one tank or side to be drained for cleaning or repair while allowing the plant to remain operational. In situations where sand carryover is not a major problem, a single tank with a bypass channel or pipe is considered to be satisfactory.

Grit chambers are available in two basic shapes: rectangular and square. The rectangular tank is identical to the rectangular horizontal flow sedimentation tank. The square tank, also known as a *detritus tank,* is commonly used in the treatment of sewage (see Figure 3.2.5-9). The square tank is satisfactory for the purpose of wastewater treatment but does not perform well as a water treatment grit chamber. The main problem with the detritus tank is a function of its physical characteristics. The short tank length and the uneven inflow to the tank cause severe flow short-circuiting. Figure 4.5-1 illustrates the result of a tracer test using a square tank. The graph clearly demonstrates the poor flow characteristics in the tank. To provide good flow characteristics for grit settling, the chamber should be rectangular with a contracted inlet (Figure 4.5-2).

If there is a significant amount of grit carryover, the grit may be removed by chain-and-flight or traveling bridge scrapers since these will remove the grit continuously. If these conditions do not exist or are not expected to occur, a gutter should be provided at the center of the tank bottom (longitudinally) to facilitate man-

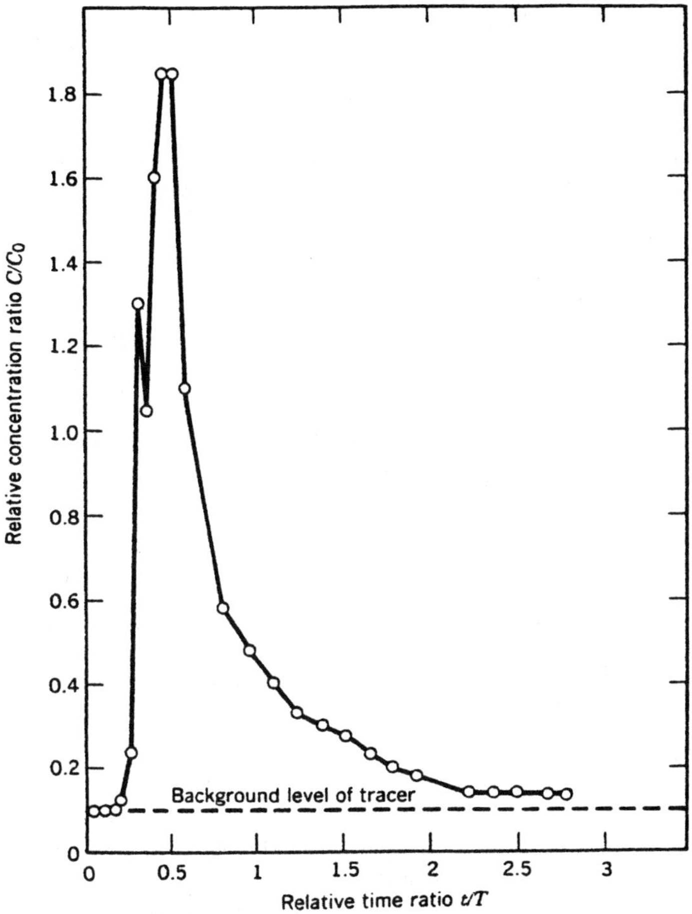

Figure 4.5-1 Grit basin tracer curves for a squire tank.

ual cleaning of grit (by hosing) and removal by a grit pump or a hydraulic eductor system. The lower portion of the tank may be longitudinally sloped 100 : 1 toward the sump and transversely sloped 1 : 50, or steeper, toward the center gutter.

A screen may be constructed at the downstream portion of the inlet diffuser wall at an angle of 70° from the horizontal. Since the screen acts as a diffuser wall and the direction of flow (just past the screen) is perpendicular to the screen angle, an angle of less than 70° would direct the flow to the tank bottom and scour the sediment (Figure 4.5-2). It is also important to provide isolation valves at both ends of the tank and to install a weir type of effluent so that a minimum water depth can be maintained in the tank. The diffuser wall should have a total orifice area of approximately 15% of the cross-sectional area of the tank. These criteria are based on the hydraulic scale model test of a rectangular grit chamber conducted by the author to optimize the diffuser wall design and screen angles (see the Bibliography).

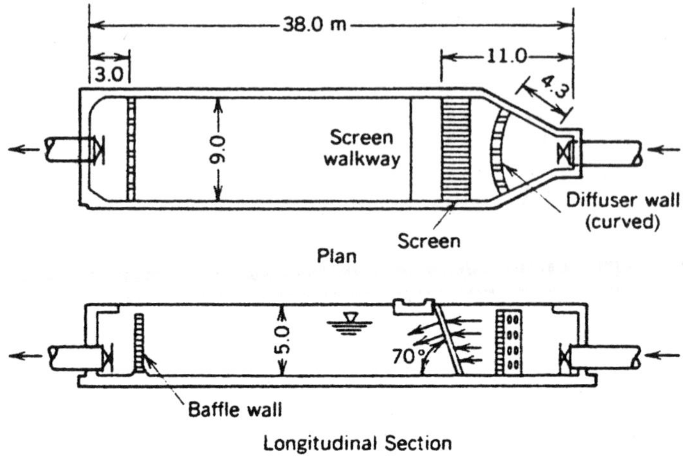

Figure 4.5-2 Grit chamber.

Generally, the grit to be removed is sand or silt with diameters exceeding 0.1 mm or larger than No. 100 mesh (0.15 mm). Grit any smaller than this does not pose a major threat to the pumps and pipelines. Figure 4.5-2 is a grit chamber designed to treat 35 mgd (1.5 m^3/s) of flow and has been performing well since 1955; the design is based on the hydraulic scale model study.

Design Criteria

Number of tanks	Two tanks (one with a bypass line may be used as an alternative)
Water depth	10 to 13 ft (3 to 4 m) with a grit remover; 11.5 to 16 ft (3.5 to 5 m) without a grit remover
Length/width ratio	Minimum of 4 : 1
Length/water depth ratio	Minimum of 6 : 1
Mean flow velocity	10 to 15 ft/min (3 to 4.5 m/min)
Detention time	6 to 15 min
Surface loading	4 to 10 gpm/ft^2 (10 to 25 m/h)

NOTE:

If a gravity flow system is planned, the highest water level in the tank must be lower than the lowest water level at the intake.

Operation and Maintenance A properly designed grit chamber does not require much operator attention. The inlet diffuser vanes of square tanks, such as the detritus tank, require periodic adjustment to maintain uniform inlet flow conditions. Yet it is virtually impossible to create a good inflow condition based solely on the adjustment of the diffuser vane. Plant operators must observe the conditions of the grit chamber on a daily basis. However, more attention should be given to the intake and the grit chamber after heavy rainfall, and tasks such as trash removal, adjustments in the grit removing cycle, and the quality of the incoming raw water must be monitored before the water is permitted to flow to the treatment processes. If the plant design does not include an automatic screen cleaning system, the plant operator may be quite busy removing debris from the screen. Plant operators should also consider installing a turbidimeter in the grit tank for early detection of turbidity spikes if the design has failed to provide one.

Regardless of the grit chamber conditions, the interior of the tanks should be inspected on an annual basis during periods of low water demand. All required maintenance should also be performed at this time. It is always good practice to collect grit samples from the tank because sieve analysis of the samples provides operators with valuable information on tank performance. An example design calculation of a grit chamber design is shown in Example 2 of Section 3.2.5.

BIBLIOGRAPHY

Degremont, *Water Treatment Handbook,* 5th ed., Halstead Press, New York, 1979.

Kawamura, S., "Hydraulic Scale Model Simulation of the Sedimentation Process," *J. AWWA,* 73:7:372 (July 1981).

Metcalf & Eddy, *Wastewater Engineering: Treatment, Disposal, Reuse,* 2nd ed., McGraw-Hill, New York, 1979.

Montgomery, J. M., Consulting Engineers, *Water Treatment: Principles and Design,* Wiley, New York, 1985.

4.6 OPERATIONS BUILDING

Purpose The operations building of many of the older, small to midsized plants were designed with little attention to architectural and human considerations. More emphasis should be placed on designing a functional and pleasing environment since this building is considered to be the human-oriented structure of the entire water treatment plant and often dictates the theme of all the surrounding structures.

Considerations The three basic issues that must be addressed are the (1) functional, (2) architectural, and (3) structural considerations. The issues range from building configuration and exterior finish to the design loads.

Functional Issues The operations building should have an administrative zone, operational management zone, product quality control zone (laboratory), and a mechanical/workshop zone. Additionally, the chemical feed system is often included in the operations building. The administrative zone is composed of a lobby with an area for the receptionist or secretary, a storage room for records and office supplies, toilet facilities for visitors, a conference room, and offices for the water production manager, chief operator, and supervisor.

The zone of operational management minimally contains the control room with its monitors and data loggers, a process control laboratory, and a lunchroom. The laboratory should at least have a jar tester, turbidimeter, pH meter, and a water sampling sink. A large meeting room suitable for training operators and accommodating a large group (30 people) of visitors is preferably included in the design. By providing an accordion type door, part of this meeting room may be sectioned off for use by shift operators and maintenance personnel who are stationed at the facility.

A lunchroom with vending area is an essential part of modern treatment plants, for the morale and well-being of the plant personnel. Locker room facilities, with showers, must also be provided for both men and women operators. Modern supervisory and control systems produce a large quantity of records, thus an adequate room must be provided for the storage of operational records and computer supplies. The Safe Drinking Water Act was recently amended to increase the number of water sampling and analysis items from the present 49 to 183 items by the year 2000. Based on this knowledge, all major treatment plants should have a rather large laboratory zone unless the water samples are sent to an outside laboratory.

The laboratory should have four distinct areas based on their specialty: general chemistry lab, instrumentation lab, bacteriological lab, and an office for the water quality control superintendent. One of the laboratories must have a sampling sink with sampling taps from the head of the plant, settled water, filtered water, finished water, and other necessary phases of the processing water. A minimum total lab space (floor area) of 1500 to 2000 ft^2 (140 to 190 m^2) is required to process the drastic increase in future water analysis.

The fourth zone in the operations building is the mechanical equipment room and the machine (maintenance) shop. The function of this zone is to house the mechanical equipment—the heating, air conditioning, and ventilation—necessary for the entire operations building and the workshop for the process equipment. Engineers must design one room in this zone with a separate air-conditioning system. This room does not need to be large, generally 200 ft^2 (19 m^2), but it must have a dust-free environment so that it may be used as an electronics and computer workshop.

The chemical storage and feeding system may be designed as part of the operations building. It is usually located at the rear of the administrative building, behind the administrative and operational management zones for reasons of safety and traffic control. The advantages of having a separate building for the chemical feed systems are improved and convenient access for delivery trucks, shorter chemical feedlines and suction pipelines for the liquid chemical metering pumps, better housekeeping, better climate control, and safety—in case of a severe chemical leak.

Architectural Issues There are eight major architectural issues in the design of the operations facility. Some of these are common sense. However, many are a function of owner preference and state and federal laws.

Environmental Issues This category includes the topography of the site, weather conditions, the type and the quality of the surrounding neighborhood, and the view of the process units from the control room. Most of these data will have been collected during the plant site evaluation phase.

SPACE REQUIREMENTS The size, type, and number of rooms are strongly influenced by owner preference and real needs. The limits on the occupancy rate affects the layout of the office, process, and maintenance zones. Two areas within the administrative zone—the toilet/locker room and the laboratory—require larger than average space. Handicap toilet facilities and separate locker rooms for both men and women are mandated by law, and the need for a larger laboratory space is a result of the Safe Drinking Water Amendment.

Based on past experience, mid-sized plants [30 to 60 mgd (1.3 to 2.6 m^3/s) capacity] should have the following major rooms and functional sizes:

Lobby/reception area	18 ft × 20 ft	360 ft^2
Control room	15 ft × 20 ft	300 ft^2
Conference room	15 ft × 20 ft	300 ft^2
Offices		
Superintendent	15 ft × 15 ft	225 ft^2
Plant supervisor	15 ft × 15 ft	225 ft^2
Supervisor	12 ft × 15 ft	180 ft^2
Chemist	12 ft × 15 ft	180 ft^2
Maintenance	12 ft × 15 ft	180 ft^2
Lunchroom	12 ft × 12 ft	144 ft^2
Laboratory	25 ft × 50 ft (minimum)	1250 ft^2
Toilet/locker room	16 ft × 32 ft	512 ft^2
HVAC equipment room	15 ft × 16 ft	240 ft^2
Electrical equipment room	12 ft × 16 ft	192 ft^2
Machine shop	25 ft × 30 ft	750 ft^2
Warehouse	24 ft × 30 ft	720 ft^2
Chemical feed room	40 ft × 62 ft	2480 ft^2
Chlorine feed room	15 ft × 20 ft	300 ft^2
Chlorine storage room	15 ft × 44 ft	660 ft^2
		9018 ft^2 (840 m^2)

NOTE: 1 m = 3.28 ft, 1 m^2 = 10.76 ft^2.

The other rooms should be sized according to the type of equipment, usage, and the requirements set by the manufacturer.

CONFIGURATION OF THE ROOMS Factors such as the location of the visitor parking lot, location of plant facilities, and the view of the treatment plant often influence the configurations of the rooms in the operations building. In many cases, the plant superintendent and control room operators demand that their rooms have a view of the main treatment process units or the plant entrance. Thus the following is a list of several design criteria that may be used to arrange the space within the operations building.

1. The control room should be adjacent to the lobby. This will allow operators to monitor and control visitor entry.
2. Ideally, the control room should be located in the interior of the building to avoid exposing computer equipment to the effects of direct sunlight and fluctuations in temperature. Moreover, the computer terminals are easier to view if they are shielded from window glare.
3. Access to the conference room, offices, and so on, should be convenient for employees.
4. Visitor access to the toilet facilities must be simple. A separate toilet facility should be provided for all visitors, and this room should be kept clean at all times.
5. Separate rooms should be provided for meetings and lunch breaks.
6. The lunchroom should have the same appliances and storage space as those found in a small residential kitchen.
7. If the treatment plant is located in a residential area, the operations building should be a one-story structure, to minimize the visual impact to the neighborhood. However, the height of the building is a function of the situation and owner preference.

EXTERIOR FINISH The exterior of the building should be attractive, durable, and require minimal or no maintenance. The walls should be constructed of a material that may be used for both the interior and exterior of the building; concrete or masonry meets these requirements. Masonry is relatively inexpensive and is more attractive than concrete. The roof should be pitched to minimize the potential for leaking and also to create an architectural feature. Many options exist for roofing material: clay tile, shingle, or metal. Finally, the windows and window wall units should be specified to be insulated glass. Engineers may even wish to stipulate that all rooms requiring climate control have tinted windows.

INTERIOR FINISH The materials selected for the interior of the building should be both durable and attractive. However, owner preference often strongly influences the selection. The following is a typical room finish schedule.

	Floor	Wall	Ceiling
Lobby and corridors	Quarry tile	Wallpaper	Acoustic tile
Control room	Sheet vinyl or tile	Wallpaper	Acoustic tile
Offices	Carpeting or sheet vinyl	Wallpaper	Acoustic tile

Toilet/locker rooms	Mosaic tile	Ceramic tile	Painted gypsum board
Laboratory	Sheet vinyl	Painted gypsum board	Acoustic tile
Electric room	Concrete	Painted gypsum board	Painted gypsum board
Process room	Concrete	Sealed masonry	Acoustic tile or painted gypsum board

BUILDING CODE The building code divides the operations building into an occupancy zone and a factory zone. The codes also dictate certain design features based on the occupancy rate and safety considerations; these considerations include:

- Number of exits
- Fire-rated corridor walls
- Fire-rated doors and windows
- Area separations
- Fire-rated ceilings
- Flame spread classifications
- Smoke detectors

The factory zone must be separated from the occupancy zone by a fire-rated wall as required by the Uniform Building Code (UBC). The heating, ventilation, and air conditioning room (HVAC) requires a fire-rated wall containing sound-insulating material. Additional issues are the amount of insulation in the walls and ceiling. Refer to the UBC for the specific requirements related to each of these considerations.

BUILDING ARRANGEMENT It is in the best interest of all parties if the building arrangement issue is developed by a qualified architect who is familiar with the functions of a water treatment plant or an architect who is guided by a project engineer. Figure 4.6-1 is a sample building arrangement.

Structural Issues The structural system of the operations building is affected directly by the occupancy groups in the building. Each occupancy group has a maximum allowable floor area; refer to the UBC for the maximum allowable floor area for the five different types of construction. Note that the washdown and durable materials used in the construction of water treatment plants often give the operations facility type II or III building characteristics.

The goal of the structural design for hydraulic structures is to produce strong units that will remain serviceable and leak-free for the life of the plant. This design should follow the recommendations of the UBC and ACI-350, *Concrete Sanitary Engineering Structures*. Regardless of the specific application, the design and de-

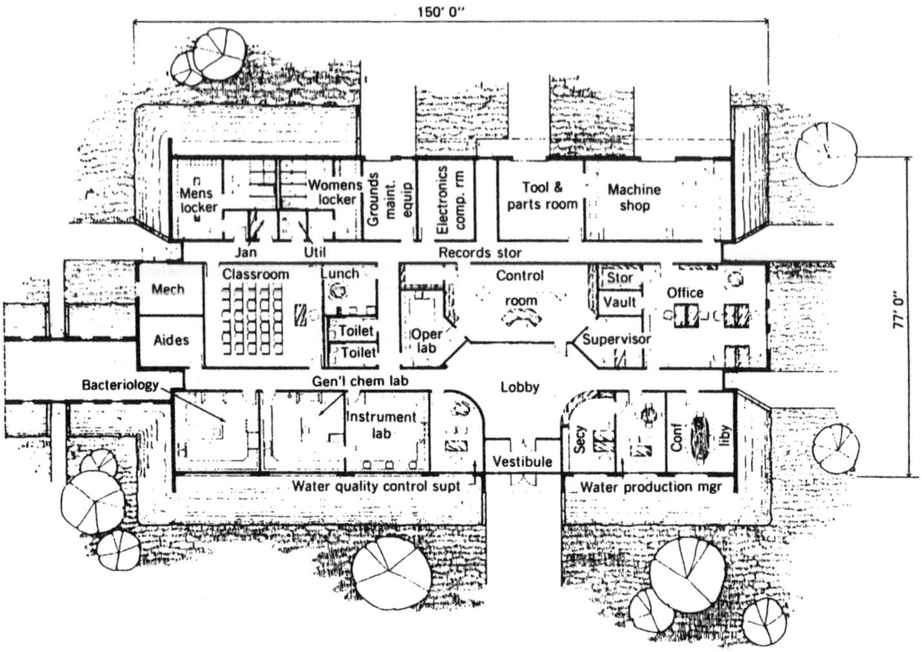

Figure 4.6-1 Building floor plan.

tails of the structure must minimize cracking due to shrinkage and limit direct cracking at construction joints and expansion–contraction joints. Since geotechnical conditions of the plant site also affect the structural design of the plant facilities, recommendations should be obtained from a geotechnical consultant in regard to the subgrade and groundwater level of the property. Advice should also be obtained on shoring, earth pressures, sheet piles, and so on.

Design Criteria In this section we present very general design criteria for the materials and the design loads used in constructing an operations building. Specific recommendations cannot be presented because adjustments must be made according to the local and specific conditions of each plant site.

Construction Materials

Concrete	4000 psi (280 kgf/cm^2) compressive strength
Reinforcing steel	ASTM A615, grade 60
Structural steel	ASTM A36
Masonry	ASTM C90, grade N-1, lightweight block, 1500 psi (105 kgf/cm^2) combined compressive strength
Aluminum	Alloy 6061-T6
Stainless steel	Type 304

Design Loads

Office, laboratory, and control room	100 lb/ft^2 (490 kgf/m^2)
Areas with process units and mechanical equipment	150 lb/ft^2 (730 kgf/m^2)
Areas with heavy equipment	Equipment weight plus 200 lb/ft^2 (980 kgf/m^2)
Electrical control room	250 lb/ft^2 (1220 kgf/m^2)
Traffic loading	HS-20 (AASHTO)
Roof loads (UBC requirement)	20 lb/ft^2 or 97 kgf/m^2 (basic)
Wind loads (UBC requirement)	100 mph (161 km/h)
Seismic loads	Recommendation of a geotechnical expert

4.7 PUMPING SYSTEM

Purpose Although pumping is not a water treatment process per se, it is a major ancillary item. The water treatment design engineer is responsible for selecting and specifying the appropriate type of pump, for setting the system requirements, writing the pump specifications, and developing all other necessary information. Since the pumping system is usually not the design engineer's area of expertise, he or she is strongly advised to refer to the literature published by pump manufacturers and to consult with mechanical engineers to obtain more detailed information and data. This section presents only basic guidelines for pump selection, designing a pumping facility, preparing pump specifications, and other special considerations.

Considerations Several items must be addressed when designing a pump system. These items, in addition to those used in selecting pump drive units and determining the number of pumps and standby generators, are covered briefly in this section.

Type of Fluid The primary concerns that must be addressed during pump selection are both the chemical and physical characteristics of the fluid to be pumped. The main characteristics that strongly influence pump selection and system design are viscosity, temperature, density, corrosivity, chemical stability, lubricating property, and the type and amount of suspended material (in water).

System–Head Curve The design engineer should establish the system–head curve by analyzing the pumping system; the preliminary design of the system should include a pump layout, piping scheme, and instrumentation diagram. A clear and accurate curve of the pumping equipment characteristics, in the expected range of operation, may be obtained from the manufacturer. Engineers should remember that hydraulic headloss is a function of the flow rate and the physical characteristics of the piping system. Therefore, a clear piping scheme will reduce the chance of additional headloss and ultimately minimize the operational cost of pumping. The details for establishing the system–head curve may be found in most engineering and hydraulic textbooks.

Potential System Modifications In the majority of cases, the pumping system designed during the initial design phase is later expanded or modified. If future modification of the pumping station is required by the master plan, the pumping system should be designed to accommodate them. This will allow the original design to still be effective and will not hinder existing plant operations during construction.

Operational Modes Important considerations for designing a pumping system are the degree of flow or head fluctuation and the mode of operation, such as continuous or intermittent. The operational modes dictate the number of pumps and their capacities and determine if booster pumps are required in a flow path.

Required Margins Engineers must specify a pumping unit that has adequate margins so that the equipment will be able to withstand infrequent, short-lived, abnormal conditions: fluctuation in pressure, dips in electric voltage and frequency, hydraulic surges, or the loss of cooling water. With respect to small electrical disturbances, the margins may be added to the pump head and capacity. Yet the pumps should only be specified to be 15 to 20% over the design points because oversized pumps may develop mechanical and hydraulic problems.

Pump Selection Different types of applications require specific types of pump. Pump selection is based on factors such as the fluid characteristics, turn-down ratio, discharge pressure and system requirements, availability of space, layout, life, energy and pump costs, code requirements, and the materials used in construction.

Reciprocating pumps, such as the plunger or diaphragm type, should be selected for liquid chemical metering and injection applications. These types of pumps deliver a constant capacity of liquid over a wide range of system–head variations. However, the capacity of these pumps is rather small and the pumps are characterized by a pulsating flow.

Positive-displacement pumps, also known as rotary pumps, may be used for both liquid and gas pumping. Although this type of pump does not exhibit a pulsating type of flow and may also be used as a slurry or metering pump, its application is limited to a low to medium pressure range.

Centrifugal pumps are used for a wide variety of hydraulic head and over a wide range of capacity requirements. However, in water treatment applications they are generally used in situations requiring low to medium capacity with medium to high pressure. These pumps require priming unless installed below the water surface in a dry well.

Axial flow pumps are best suited for applications requiring low hydraulic head and high-flow conditions. The mixed flow type is used in situations with intermediate conditions, between the axial flow and centrifugal pump. Refer to Figure 4.7-1 for the pump characteristics.

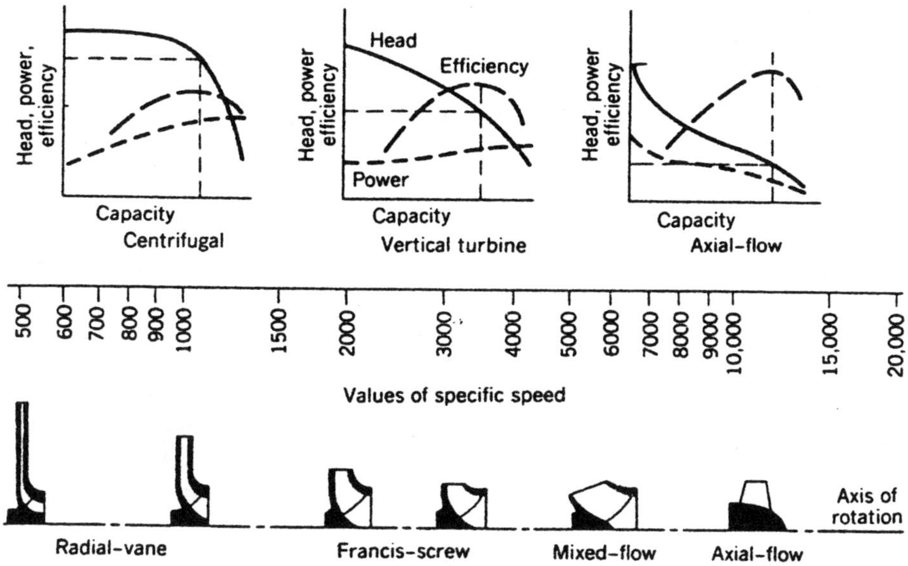

Figure 4.7-1 Type characteristics and specific speed scale for various impeller designs. (Adapted, with permission, from R. P. Walker, *Pump Selection,* Butterworth-Heinemann, Stoneham, MA, 1972.)

Vertical turbine pumps require much less space than the centrifugal type of pump and are also self-priming. Yet this type of pump generally requires more head room. Thus, if the pumps are installed in a housing, a removable skylight or hatch must be provided directly above each pump to facilitate repairs. Nonetheless, vertical pumps such as turbine pumps, axial flow pumps, and mixed flow pumps are often installed without a housing. Rather than using a pump well construction, vertical turbine pumps may use a "canned" suction type of design in the suction pipeline, to help minimize the construction costs of a pumping station.

For further information of the various aspects of pumps, refer to the *Hydraulic Institute Standards,* 14th edition and the more updated versions published in 1997 and 1998 (see the Bibliography).

Drive Selection The available energy source, the cost of operation, and the various local conditions all affect selection of the pump drive unit. Pumps may be driven by a variety of means—electric motor, internal combustion engine, or steam—and may have either constant- or variable-speed drives. The internal combustion engine can be the drive unit due to local conditions and to avoid increasing the energy demand during peak hours of electrical consumption. An electric motor may be selected as part of the dual drive, to act as a redundancy drive. Yet the majority of pumps are driven by squirrel-cage induction motors, due to their versatility and availability.

Other miscellaneous types include synchronous motors, used in applications requiring a high horsepower rating and where a power factor correction is required;

wound-rotor motors, which are used occasionally in applications requiring variable-speed drives; and reduced-voltage starting and low-in-rush current motors. The last two types may be used as high-power motors in situations where an initial transient reduction in-line voltage would adversely affect other equipment.

Number of Pumps and Standby Generators Reliability considerations often tend to favor the use of multiple-pump schemes. A three-pump system employing identical electric motor–driven pumps, each having a capacity that is 50% of the maximum demand, is commonly selected. This type of system requires only that the power supply run two of the pumps at any one time and maintain the third pump as a standby. If frequent power outages are anticipated, an engine-driven pump or standby diesel generator should be incorporated into the pumping system design. If a standby generator is selected, the system will require voltage starting and the engine generator should be capable of supplying a starting current that is at least three times the normal full-load current. However, a smaller emergency generator will be sufficient if sequence starting is used, since each pump may be started in a stepwise fashion.

The size of the pumping system is also an important factor in determining the number of pumps. Smaller systems may opt to utilize two to three pumps of the same capacity in order to make the pumping system more flexible. This scheme will enable the system to meet the water demands, will facilitate easier maintenance, and will allow interchangeability of parts. However, larger installations experiencing high peak demands would benefit from a two-bank system. In this scheme, one bank of pumps (constant-speed drives) supplies the base flow to satisfy the normal demand. The second bank supplies the extra flow necessary to satisfy the peak demands. The pumps of the second bank may be driven by internal combustion engines, such as diesel engines, so that the expensive peak power costs (including the potential penalty charge) may be avoided.

Pump Specifications The two basic types of pump specifications are those concerned with pump construction and those related to performance. The first specifies details such as the type of design, the design method, and pump construction. The performance specifications dictate the performance standards for the pump and contain a few details related to pump construction, such as the pump construction material.

Since the owner is primarily responsible for performance of the pump, it is more than likely that the plant owner, not the manufacturer, will be held legally responsible if pump malfunctions develop, provided that the specifications describe the pump design and construction. Thus, most water treatment plant projects use the pump performance specifications when constructing the pump system. The project engineer can always seek help from pump manufacturers when developing the clarifying specification. However, the engineer should remember that each supplier is partial to its own product. It is always a good practice to include a series of alternatives, in addition to the basic specifications, because it allows pump suppliers to present their best offers and provides the owner with the advantage of obtaining bids from several manufacturers. The pump specifications should refer to the quality codes and standards of the materials to be used. Industrial standards such as ANSI and ASTM are good sources for basic requirements relating to metallurgy, dimensions, tolerance, and flange details.

It is important to note that a number of design engineers have problems with selecting the appropriate type of metal for pump construction. In the majority of cases, stainless steel is used, with excellent results. However, if the plant is pumping anaerobic sludge or deep well water or pumping out the bottom portion of deep lakes, type 316 stainless steel will become severely pitted under these anaerobic conditions. Therefore, other types of stainless steel, such as type 301 or 304, should be specified with manufacturer assistance. Another area of concern is the problem of abrasion when stainless steels of the same type rub together. As a rule of thumb, the hardness of the two adjacent metals should have a difference of 50 Bhn on the Brinell scale of hardness.

The bidding documents are usually divided into two major parts, the technical specifications and general information (commercial term). The technical specifications should include the following items:

1. Scope of work
2. Work not included in the document: for example, foundations, anchor bolts, labor
3. Ratings and service conditions such as the characteristics of the liquid, temperature of the liquid, flow rate, total dynamic head, preferred speed range, efficiency at the design point, load conditions, NPSH, overpressure, runout, and operating conditions
4. Design and construction brief detailing the codes, standards, material used, type of casing, stage arrangement, balancing, nozzle orientation, type of support, vents and drains, type of bearings, shaft seals, baseplates, connecting pipes, and instruments
5. The lubricating oil system
6. Details of the drive unit, such as motor voltage, power supply and regulation, local panel requirements, wiring standards, terminal boxes, and electrical devices; and the type of fuel, number of cylinders, cooling system, speed governing, starting requirements, couplings, and exhaust muffler for the internal combustion drives
7. Cleaning and painting
8. Performance testing
9. Drawings and supporting data
10. Special tools
11. Criteria for evaluation

The general information should include the name and address of the owner, the applicable codes and standards, information on the location and condition of the plant site, and a schedule that explains the drawing requirements, bid dates, manufacturing time, and equipment delivery. The clause concerning the plant site should specify, among other things, the elevation of the plant site, the range of ambient temperatures, and humidity. It is recommended that the general information also clarify the key people associated with the project: for example, owner, engineer, and seller.

Performance testing is generally not an important consideration for small pumps: those with a discharge size that is less than 6 in. (15 cm). Yet it is vital for larger pumps because a loss of efficiency of 1 to 2% from the specified value results in a significant increase in the annual power cost. It is very important that the owner and supplier agree on the method of testing and whether testing of the pump should be witnessed.

Special Considerations In addition to the considerations mentioned previously, the design engineer should address five other issues: (1) the pump starting conditions, (2) the use of high-speed versus low-speed pumps, (3) the method used in controlling hydraulic surcharge, (4) the design of the pump suction well, and (5) selection of the sludge pump.

Pump Starting Conditions The discharge valve of centrifugal pumps must be closed at a particular time during startup to prevent hydraulic surges. Under these circumstances, the pumping unit will not require additional horsepower. As long as an adequate net positive suction head is provided, this practice also removes the concern with regard to cavitation. This is also true for mixed flow pumps, up to an approximate specific speed of 5000.

Propeller pumps, including most types of mixed flow pump, require very high horsepower at startup. Thus the pump discharge valve should be sufficiently opened after the drive unit has operated at full speed for a few seconds. If the discharge valve is allowed to remain in a closed position, the water in the bowls will become heated, thereby allowing steam to displace the water in the column. For this reason, valve control during pump startup must be evaluated carefully.

High-Speed Versus Low-Speed Pumps The pumping requirement for a particular flow rate at a specific total dynamic head may be met by two different methods. The first is to use the combination of a small impeller, capable of high speeds such as 3600 rpm, with a high-speed motor. The second alternative is to use a slower-speed drive (800 to 1200 rpm) with a larger impeller. Although pump selection is influenced by owner preference and plant conditions, the high-speed pumps are cheaper to manufacture and the owner therefore has the option to select the lowest bidder. Despite their lower cost, high-speed pumps are generally maintenance intensive, due to problems associated with wear, erosion, misalignment, and vibration. High-speed units also tend to produce a high-pitched noise. It is the responsibility of the engineer to evaluate the various offers carefully, including the economic life span of the pump and the associated maintenance costs.

Hydraulic Surge Control Two types of hydraulic surge have been associated with the pumping system. The first occurs during pump startup and in small plants; this type of surge may be eliminated by installing a surge control valve between the pump and the check valve (Figure 4.7-2). The second type, the water hammer, is produced if the pump is suddenly stopped by the operator or if all the pumps suddenly fail due to a power outage. If the facility is expected to have a significant problem with water hammer due to power outages, a surge tank should be installed to alleviate this problem. Figure 4.7-2*b* illustrates surge tank installation for a hydraulic power generation system. However, the basic idea of the surge tank is the same for the pumping system.

A special technique used in controlling hydraulic surges is surge analysis. Although the technique requires both knowledge and experience to be used effectively, a rule of thumb is to close the surge control valve (during pump startup) and to open the valve (during pump topping) after the shock wave has reflected. The required time may be calculated from the formula $2L/a$, where a is the velocity of the shock wave (4000 to 5000 ft/s) and L is the length of the pipeline in feet. Computer analysis of the surge is often implemented using software developed by the University of Kentucky.

Design of the Pump Suction Well A critical part of the pumping system design, for the engineer, is the pump well, because the pump impeller requires a particular type of flow condition to exist. The engineer must verify that flow conditions are uniform and ascertain that the design of the pump suction well will provide the required type of flow. The pump well may be an open pit or large pipe. Detailed design criteria for optimum pump suction configurations may be found in the *Hydraulic Institute Standards* (15th edition, 1997). Figures 4.7-3 and 4.7-4 illustrate some of these design guides.

Sludge Pumps Three types of pump are best suited for sludge pumping: the positive-displacement type, such as the plunger and progressive cavity pumps, the centrifugal type, and the torque flow type. Regardless of the type of pump chosen, the sludge piping should be no smaller than 6 in. (15 cm) in diameter and the velocity should be 5 to 6 ft/s (1.5 to 1.8 m/s), to prevent clogging problems.

The plunger type of pump has a satisfactory history of operation. These pumps are self-priming and the pulsating action tends to concentrate the sludge in the hoppers (located prior to the pumps) and resuspends solids in the pipelines during low flow velocity. The main disadvantages of the plunger pumps are their bulky size and cost. For these reasons, this type of pump is rarely used today.

The use of progressive cavity pumps, such as the Moyno pump, has been particularly successful in pumping concentrated sludge. The Moyno pumps are progressive cavity pumps and are self-priming at suction lifts up to approximately 20 ft (6 m). However, the pump is usually installed under flooded suction conditions to prevent the rubber stator from burning out.

Specially designed centrifugal pumps, such as screw feed, bladeless, and torque flow, have also been used successfully. The torque flow type of pump, such as the Wemco pump, has fully recessed impellers and is very effective in pumping sludge but has very low pump efficiency. Engineers should refer to *Wastewater Engineering* (2nd edition) by Metcalf & Eddy, Inc., for more detailed information on sludge pumps.

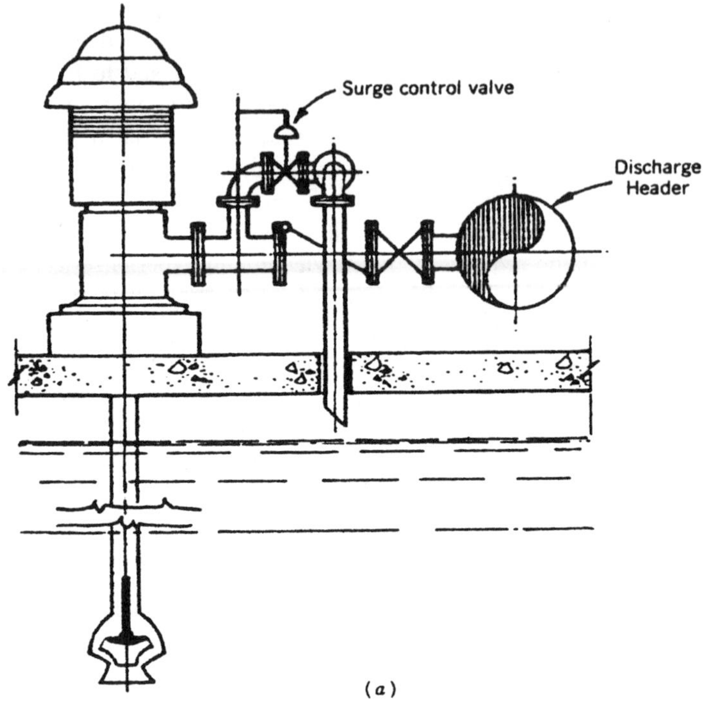

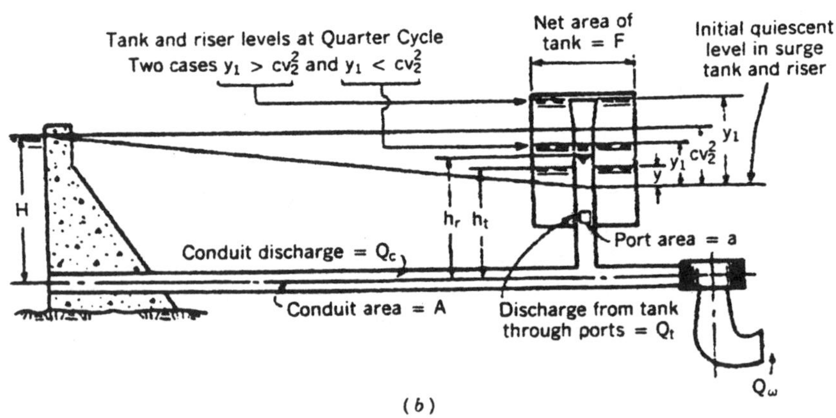

Figure 4.7-2 Surge control valve and surge tank. (*a*) Surge control valve. (Adapted from R. P. Walker, *Pump Selection,* Ann Arbor Science Publishers, Ann Arbor, MI, 1973.) (*b*) Surge tank installation. (Adapted from *Handbook of Applied Hydraulics,* C. V. Davis, ed., McGraw-Hill, New York, 1952.)

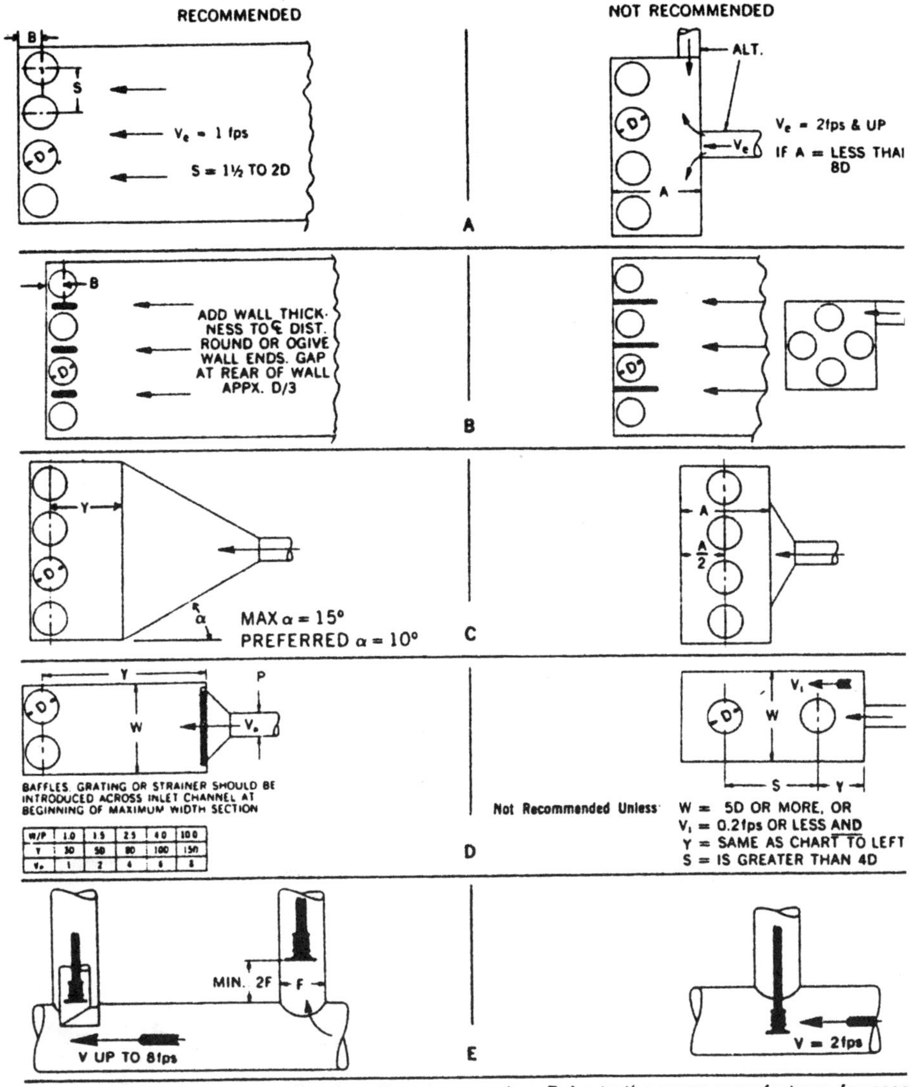

The Dimension D is generally the diameter of the suction bell measured at the inlet. This dimension may vary depending upon pump design. Refer to the pump manufacturer for specific dimensions.

Figure 4.7-3 Multiple pump pits. (Adapted from *Hydraulic Institute Standards,* 14th ed., 1983.)

Operation and Maintenance As described earlier, the starting and stopping of the pump requires special consideration. These procedures are dependent on the pump type and the associated accessories. Pumps with low hydraulic head, mixed flow, and of the propeller type should be started with the discharge valve wide open against a check valve to prevent backflow. However, pumps with medium and high

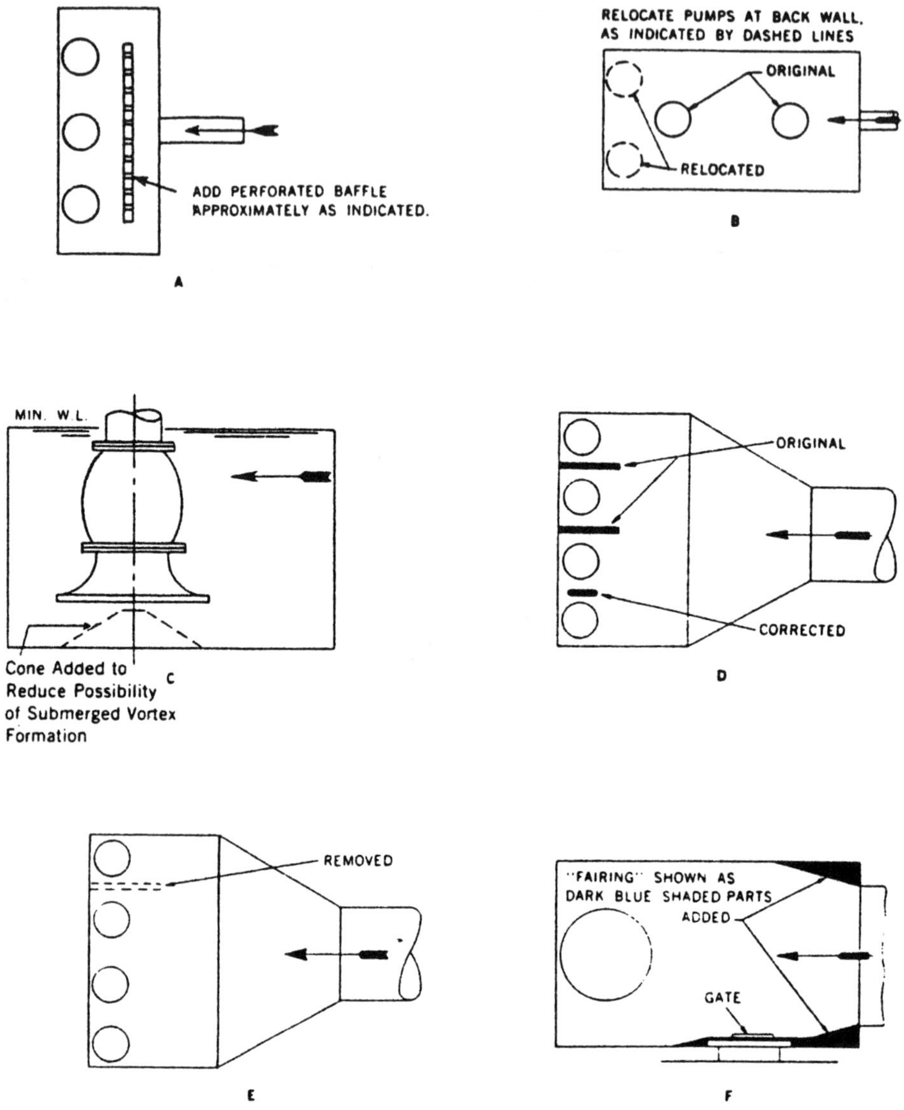

Figure 4.7-4. Correction of existing sumps (part one). (Adapted from *Hydraulic Institute Standards,* 14th ed., 1983.)

hydraulic head should be started against a closed discharge valve to reduce the starting load on the motor. If the discharge line is already pressurized by another pump, a check valve will be equivalent to a closed valve. It should be remembered that whenever a pump is started against a closed discharge valve the recirculation bypass line must be opened to prevent overheating.

The pump stopping procedure is also dependent on pump type and accessories. The plant operator is therefore advised to follow strictly both the operational manual provided by the manufacturer and the plant operation and maintenance manual. The major areas of maintenance concern inspection and replacement. Three types of inspection must be practiced: hourly and daily, semiannual, and annual. Hourly and daily inspections require the operator to note the normal sounds of the pump and the bearing temperature, and check for leaks in the stuffing box. Semiannual procedures involve inspection of the pumps, stuffing box, packing, and alignment and the refilling of the lubrication oil and grease. Annual inspection requires that all parts of the pump assembly be checked and cleaned thoroughly, and the connecting pipes be realigned with the pump. Finally, the pumping system should be overhauled completely whenever necessary, usually every 2 to 4 years.

Record keeping is also a very important issue in the operation and maintenance of a pumping system. Each pump should have its own complete record on file. This record should note the maintenance costs, the number and type of repairs, and the total operating hours. Experience has shown that photographs of the internal parts of the pump, taken during semiannual and annual inspection, prove to be very valuable.

BIBLIOGRAPHY

Allis-Chalmers, *Reference Data,* Allis-Chalmers Pump, Inc., Cincinnati, OH, 1970.

Colt Industries, *Hydraulic Handbook,* Fairbanks Morse Pump, KS, 1975.

Davis, C. V. , Handbook of Applied Hydraulics, 3rd ed., McGraw-Hill, New York, 1969.

Hydraulic Institute, *American National Standard for Centrifugal/Vertical Pumps,* Hydraulic Institute, Parsippany, NJ, 1997.

Hydraulic Institute, *American Standard for Pump Intake Design,* Hydraulic Institute, Parsippany, NJ, 1998.

Hydraulic Institute, *Hydraulic Institute Standards,* 14th ed., Hydraulic Institute, Cleveland, OH, 1983.

Karassik, I. J., et al., *Pump Handbook,* McGraw-Hill, New York, 1970.

Langteau, R. R., "Clarification Control in Pumps," *J. WPCF,* 38:4:585 (April 1966).

Metcalf & Eddy, *Wastewater Engineering: Treatment, Disposal, Reuse,* 2nd ed., McGraw-Hill, New York, 1979.

Shaw, G. V., and Loomis, A. W., *Cameron Hydraulic Data,* Compressed Air Magazine Co., Ingersoll-Rand, Phillipsburg, NJ, 1970.

4.8 OZONATION SYSTEM

Purpose The ozonation process recently gained popularity as a treatment alternative in the United States after the regulatory agencies placed restrictions on disinfection by-products, especially the acceptable levels of trihalomethane. Ozonation is an established and proven disinfection alternative as well as a preoxidant for the control of THM and HAA_5 precursors. Ozonation also has the added benefits of (1) oxidation and volatilization of organics; (2) control of algae and associated taste- and odor-producing compounds; (3) destabilization (microflocculation) of certain types of turbidity; (4) removal of color-causing compounds; (5) oxidation of iron and manganese; (6) very short disinfection times for all pathogens, including *Cryptosporidium;* (7) partial oxidation of organics for subsequent removal by microorganisms; and (8) methyl tertiary butyl ether (MTBE) removal. In this section we discuss the design considerations for the ozonation system and present some design issues with respect to ozone generation and the ozone contact tanks.

Considerations Ozonation is considered to be a relatively new treatment process in the United States. However, the process has a history of almost 100 years of practical experience. Thus basic knowledge should first be obtained from the extensive source of published articles and operational experiences prior to conducting design works. The second step, provided that there is both the time and budget, is to perform a pilot study; this will furnish the engineer with valuable design guides.

The design issues are (1) selection of a feed gas system; (2) preparation of the feed gas system; (3) selection of the ozone generator; (4) design of the ozone contact basin; (5) destruction of off-gas ozone; (6) control of DBPs by ozonation, particularly bromate; (7) control of assimilable organic carbon (AOC) levels due to ozonation in order to minimize after growth of microbes in the water distribution system; and (8) safety. The issue of safety will also be discussed briefly since ozone is both a toxic and a flammable gas. The two most practical articles (published) on this subject so far are the *Design Manual on Municipal Wastewater Disinfection* (EPA/625/1-82/021) written by the EPA and "Ozonation of Drinking Water" by Domenic Grasso [*Ozone Sci. Eng.,* 9:109 (1987)]. There are also many valuable technical papers on the ozonation system as listed in the Bibliography.

Feed Gas Selection Ozone many be generated from air, oxygen-enriched air, or oxygen. The concentration of ozone produced by air is 1.5 to 2.5% by weight. The ozone concentration is increased to 3 to 5% if high-purity oxygen is processed by the same low-frequency ozone generators. However, 8 to 12% ozone can be generated with modern medium-frequency generators containing ceramic dielectrodes.

Feed Gas Treatment Both air and pure oxygen must be treated prior to being fed to the ozone generator; this maximizes the ozone production rate and minimizes maintenance work on the ozonator. The pretreatment process includes the removal of dust, moisture, oil, and in some instances, nitrogen gas: Dust reduces the efficiency of ozone production; oil is capable of fouling the dielectric; and nitrogen has the ability to damage the ozone generator because of the production of nitric acid within the unit.

If the gas contains excessive moisture, the life span of the dielectric of the ozone generator becomes reduced and the power required to produce a specified level of ozone is increased. Thus the removal of moisture is the most important of the gas pretreatments. This task is achieved by means of a dryer, which uses a combination of pressure, temperature, and desiccant.

The traditional ozonation system for application to water treatment is generally composed of a precompressor with a 5-μm paper filter, main compressor, aftercooler [less than 95°F or (35°C)], oil coalescer (if required), refrigerant dryer (optional), heat-reactivated desiccant dryer with activated alumina and molecular sieves or silica gel, 1-μm filter, hygrometer, gas flow meter, and pressure-regulating valve.

The efficiency of ozone generation is a function of the feed gas treatment. The capability of the feed gas treatment may be maximized by reducing the moisture content of the feed gas to below 0.00596 lb of water per 1000 ft^3 of air (0.14 g/m^3), with a dewpoint of −40°F (−40°C). This dewpoint can only be attained if 99.71% of the moisture is removed. Yet gas containing this moisture content is still considered to be in poor condition for the ozone generator. A dewpoint of −58°F (−50°C) is judged to be marginal for the generator. The desired feed gas humidity is achieved at a dewpoint of −76°F (−60°C); at this temperature the feed gas contains only 0.00051 lb of water per 1000 ft^3 of air (0.01 g/m^3) with 99.98% of the moisture removed.

The feed gas compressing process generally removes approximately 60% of the moisture content of the air; an additional 20% is removed by the refrigerant dryer and the desiccant dryer removes the remaining moisture so that a total of 99.98% is eliminated. Excessive moisture content in the feed gas (air) not only reduces ozone production and increases generator maintenance, but also damages the internal components of the generator. Since the pretreatment of the gas is vital to the efficient production of ozone, sufficient monitoring procedures and control devices should be implemented in the process train. Serious consideration should be given to duplicating critical elements such as the desiccant dryer. Table 4.8-1 lists the moisture content of air at various ambient temperatures.

The most popular air feed gas systems used in ozone generation systems are classified by their operating pressure. The system generally used is the low-pressure system. It operates at a pressure of 10 to 40 psig (70 to 275 kPa). High-pressure systems operate at pressures of 70 to 100 psig (480 to 690 kPa) and decrease the gas pressure prior to the ozone generator. These high-pressure systems are typically employed in small to medium-sized applications.

As discussed, air feeding to the ozone generator is a complex, costly, and precisely controlled pretreatment process that is maintenance intensive. However, a great majority of the feed gas treatment process may be eliminated if high-purity oxygen such as liquefied oxygen (LOX) is used to generate ozone. LOX is presently available at a very reasonable price ($3.00 per cubic foot or $0.04 per pound). On-site generation of oxygen may also be achieved by means of cryogenic, pressure-swing adsorption (PSA), or vacuum swing adsorption (VSA) processes. However, these on-site oxygen generation processes require complex mechanical and control units. These units also create a somewhat negative visual impact and emit a certain degree of noise. Thus design engineers should seriously evaluate the use of LOX and the oxygen feed ozone generation system for a plant of any size, especially for small to midsized treatment plants. This particular type of process offers several advantages for both the design engineer and the plant owner: The design engineer benefits from the simple overall design scheme, especially instrumentation and control, and a required floor space that is approximately half that of the air feed system. The owner benefits from the potentially significant savings in both capital and operation and maintenance costs. Consequently, most new ozonation facilities designed after

TABLE 4.8-1 Moisture Content of Air

Air Temperature °C	Air Temperature °F	Moisture Weight (lb H_2O/lb air)	Weight Volume[a] (lb/ft^3)	Moisture Content[b] (lb H_2O Per 1000 ft^3)
−80	−112	0.0000003168	0.07526	0.00002384
−75	−103	0.0000007713	0.07526	0.00005805
−70	−94	0.000001640	0.07526	0.0001234
−65	−85	0.000003342	0.07526	0.0002515
−60	−76	0.000006743	0.07526	0.0005074
−55	−67	0.00001311	0.07526	0.0009866
−50	−58	0.00002464	0.07526	0.001854
−45	−49	0.00004455	0.07526	0.003352
−40	−40	0.00007925	0.07526	0.005964
−35	−31	0.0001381	0.07526	0.010393
−30	−22	0.0002344	0.07526	0.017640
−25	−13	0.0003903	0.07526	0.029373
−20	−4	0.0006731	0.07526	0.050665
−15	5	0.001020	0.07526	0.07676
−10	14	0.001606	0.07526	0.1209
−5	23	0.002485	0.07526	0.1870
0	32	0.003788	0.07526	0.2851
5	41	0.005421	0.07526	0.4080
10	50	0.007658	0.07526	0.5763
15	59	0.01069	0.07526	0.8045
20	68	0.01475	0.07526	1.1100
25	77	0.02016	0.07526	1.5171
30	86	0.02731	0.07526	2.0552
35	95	0.03673	0.07526	2.7642
40	104	0.04911	0.07526	3.6959

Source: Adapted from EPA's *Design Manual,* EPA/625/1-86/02, October 1986.

[a]Air weight/volume is corrected to 68°F and 1 atm pressure. Air weight/volume = 1205 g/m^3.

[b]g/m^3 = lb/1000 ft^3 × 16.012.

1990 employ LOX as the feed gas. Figure 4.8-1 is a schematic illustration of the Lox feed system, VSA feed system, and classic air feed system for ozone generation.

Ozone Generator Ozone is generated by passing a high-voltage alternating current (6 to 20 kV) across a dielectric discharge gap through which oxygen or oxygen-bearing gas is injected. There are currently three basic types of ozone generator for use in water treatment applications: (1) low-frequency units with variable voltage; (2) medium-frequency units with constant voltage, variable voltage, or frequency control; and (3) high-frequency units. These units are tube generators, in contrast to the plate generators used exclusively in small ozone generators.

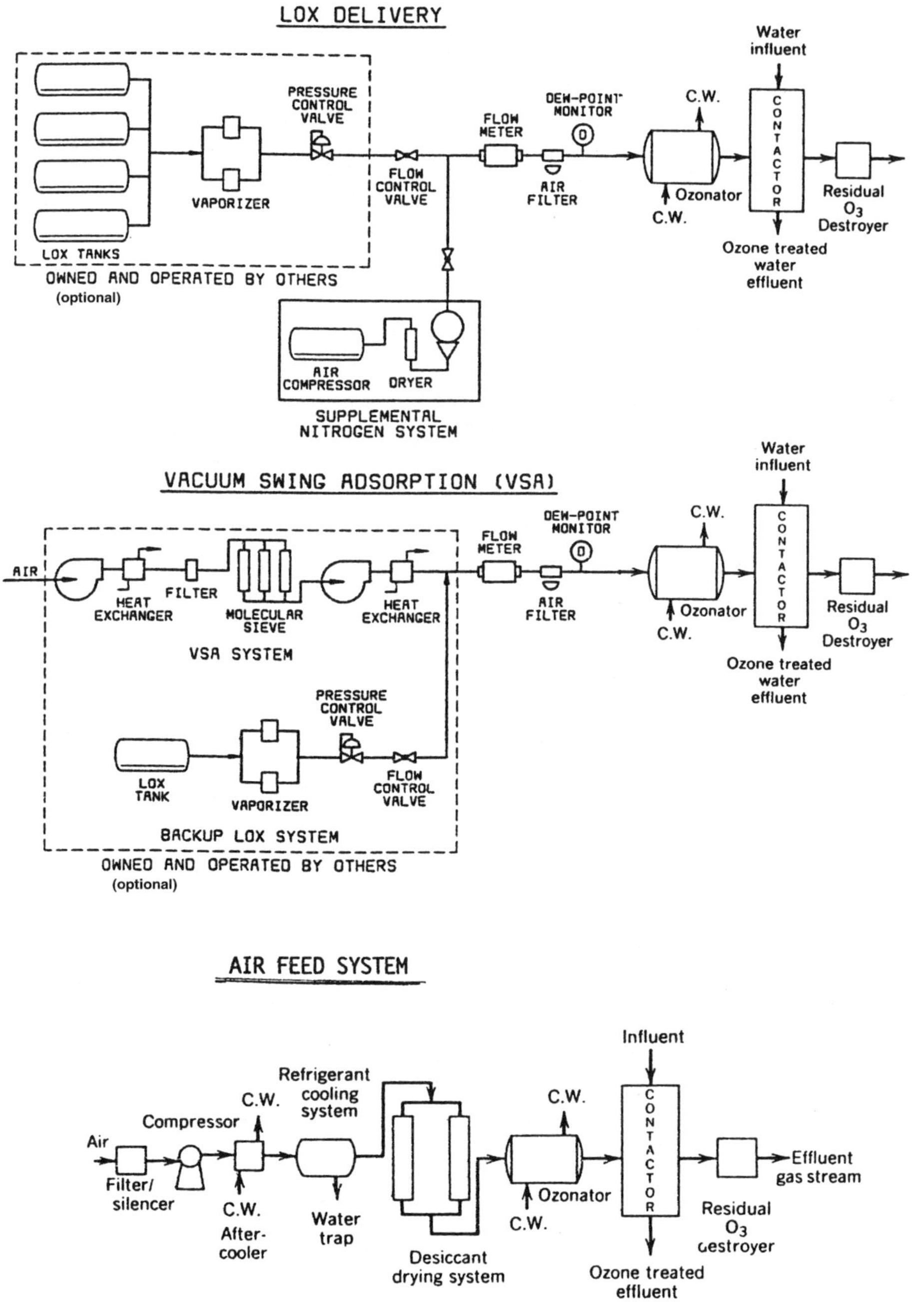

Figure 4.8-1 Examples of three different feed gas systems.

Low-frequency units have a frequency of 50 to 60 Hz, single-phase power, and a peak voltage of 10 to 20 kV. The performance of these units is proven. Medium-frequency units operate at a frequency of 200 to 1000 Hz, are three phase, and have a peak voltage of 8 to 14 kV; their performance is proven effective, thus they are widely accepted in the industry today. Another ozone generator is a high-frequency unit operating at 600 to 2000 Hz, with three phases, and a peak voltage of 8 to 10 kV. However, they are manufactured by a very limited number of companies. Figure 4.8-2 provides a general concept of ozone-generating systems.

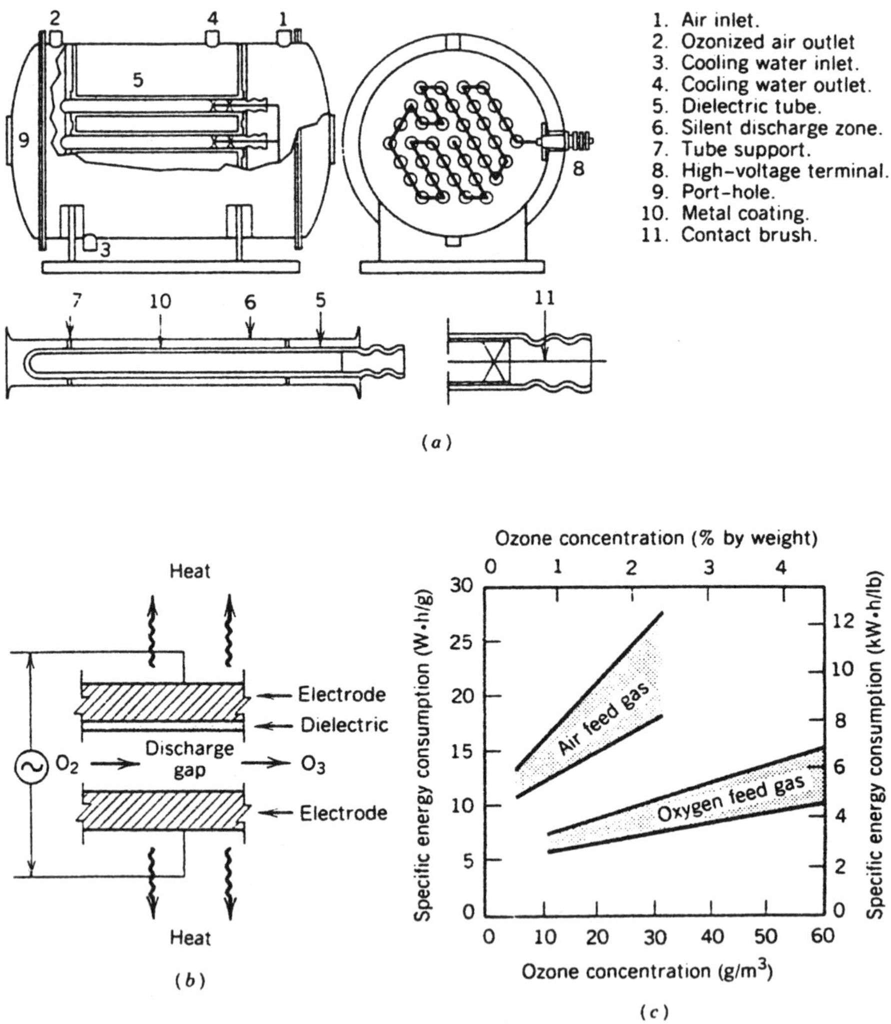

Figure 4.8-2 General concepts on ozone generation: (*a*) horizontal tube, voltage-controlled, water-cooled ozonator; (*b*) corona cell arrangement for ozone production; (*c*) energy consumption versus ozone production for an ozone generator.

Much credence should not be given to the sales pitch of the manufacturers. The most practical ozone generators are either low- or medium-frequency units, primarily because of their years of operational experience, reliability, and the large number of suppliers from which to choose. Units requiring lower voltages, including many medium-frequency units, generate less heat, thus minimizing the cooling water requirements and possibly increasing the life span of the dielectric. Medium-frequency units have a better power to production turn-down ratio than the low-frequency generators: Low-frequency units may achieve a production of 20% at 20% power if the airflow rate is reduced, whereas medium-frequency units may achieve a 10% production rate at 10% power. Yet medium-frequency units contain more electronic components than do low-frequency units; failure of any one of these components will cause the power supply to malfunction and terminate ozone production.

Selection between low- and medium-frequency ozone generators should therefore be based on the following considerations: (1) reliability and maintenance, (2) energy cost differential, (3) turn-down ratio, (4) cooling water temperature, (5) benefits attained by using oxygen-enriched feed gas, and (6) owner preference. Table 4.8-2 briefly summarizes the characteristics of the two types of ozone generator.

In the overall production of ozone, electrical power is a critical element because the greater the power consumption, the greater the heat generation. Gas flow also affects the generation of heat; a gas flow that is too low will cause the generator to heat up, whereas an excessively high gas flow rate will decrease ozone production. Since 90 to 95% of the supplied power is converted to heat, the cooling system is a very important issue with respect to the generator. Fortunately, manufacturers are highly competitive in

TABLE 4.8-2 Brief Comparison of Two Types of Ozone Generator

	Low-Frequency Unit	Medium-Frequency Unit
Electronics complexity	Lower	Higher
Power requirement to produce 1 lb of ozone	Specific energy of 4.5–5.5 kWh; a total of 12 kWh with air feed but 8–10 kWh with O_2	Specific energy of 4–5 kWh; a total of 5–6 kWh with O_2
Ozone concentration (weight %)	1–1.5 with air, 2–3 with oxygen (can be 5–6 with O_2)	2–3 with air, 5–6 with glass dielectrodes and O_2, 8–12 with ceramic dielectrodes and O_2
Capital cost	Lower for smaller-capacity units, higher for larger-capacity units ($370/lb of ozone for 2000-lb/day unit)	Higher for smaller-capacity units, lower for large-capacity units ($280/lb of ozone for 2000-lb/day unit)
Operation and maintenance cost	Higher, $0.8–$1.2/lb of ozone	Lower, $0.8–$1.0/lb of ozone
Reliability	Excellent	Good to very good
Effect of generator cooling temperature	Moderate	Sensitive

their cooling designs. The cooling water temperature should generally be 68°F (20°C) or less and approximately 0.75 to 1 gal (3 to 4 L) of cooling water is required for each gram of ozone generated (0.23 to 0.3 gpm of water for each pound/day of ozone). Refer to Figure 4.8-3 for the various conditions that affect the rate of ozone production. Figure 4.8-4 is an example of the bid prices of both low- and medium-frequency ozone generators in the United States in recent years. Figure 4.8-5 is a breakdown of the construction costs. However, if all ancillary items such as a new power substation are included, the total cost can be more than two times higher than that indicated in the figure.

After the type of ozone generator unit is selected, the design engineer must establish the following key design issues:

- Size of the ozone generator
- Number of generators
- Cost of energy
- Availability and cost of oxygen

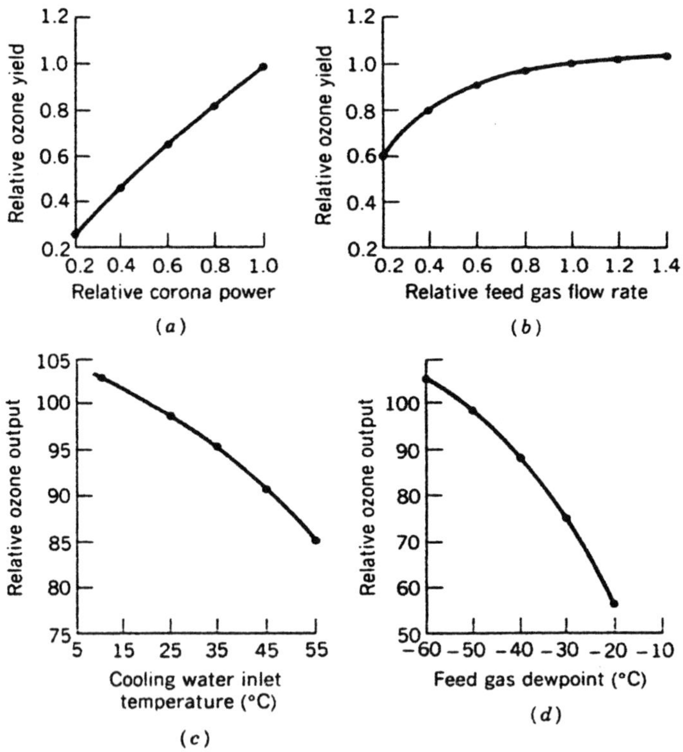

Figure 4.8-3 Various conditions that affect ozone production rate: (*a*) ozone yield versus corona power for a constant feed gas flow rate; (*b*) ozone yield versus gas flow rate for constant corona power; (*c*) effect of cooling water on ozone yield; (*d*) effect of moisture on ozone yield. (Adapted, with permission, from R. P. Rice and A. Netzer, *Handbook of Ozone Technology and Applications,* Vol. 1, Butterworth-Heinemann, Stoneham, MA, 1982.)

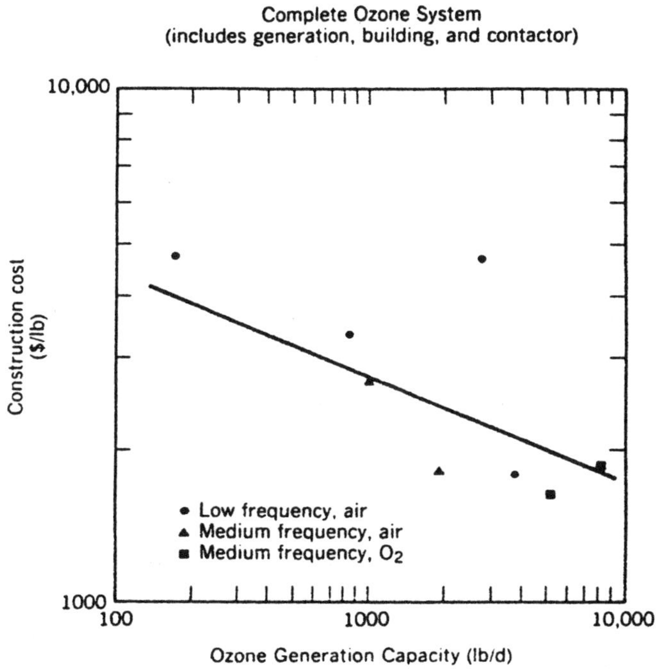

Figure 4.8-4 Construction cost of ozonation system.

- Type of feed gas treatment system
- Reliability of each component
- Operation and maintenance costs
- Ozone contactor design
- Destruction of the off-gas
- Use of catalysts such as UV and hydrogen peroxide

The major ozone generators currently are Emery-Trailigas (ET), Ozonia (formerly IDI and ABB) Schmidding-Werke, PCI, Mitsubishi, Sumitomo, and Fuji.

Ozone Contact Tank As in any other water treatment process unit, the effective mixing of ozone and process water is critical in maximizing the performance of the ozonation system. Due to the extremely low solubility of ozone in water, a special ozone contact tank must be designed. Table 4.8-3 lists the solubility of ozone in water. The contact tank is an effective means of transferring the ozone from the gas bubble to the bulk of the process water: The net ozone content in the gas bubbles is usually only 5 to 10%. This effective transfer of ozone to water is a critical step in the ozonation process.

Several types of contactor have been developed: (1) diffused bubbles (concurrent and countercurrent), (2) positive pressure injection (U-tube), (3) negative pres-

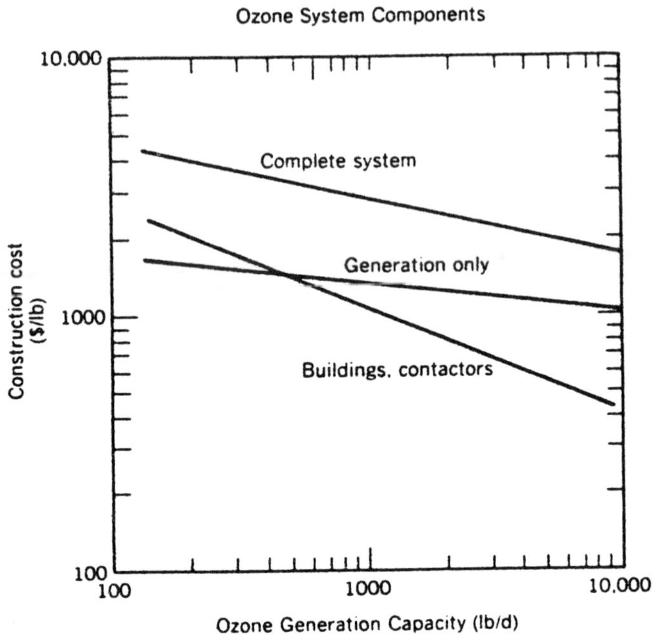

Figure 4.8-5 Breakdown of ozonation system construction cost.

sure (Venturi tube), (4) turbine mixer tank, (5) static mixer, and (6) packed tower. The countercurrent bubble contactor is considered to be the most efficient and cost-effective of the alternatives for use in metropolitan water treatment plants and were designed in the manner illustrated in Figure 4.8-6 until the early 1990s. However, an extensive ($13 million) study on the advanced ozone oxidation process conducted by the Metropolitan Water District of Southern California (MWD) using a 5.5-mgd (0.24-m^3/s) pilot facility, as well as special studies using three computer models, including the Finite Element Analysis (FEA) hydrodynamic model, revealed that a severe flow short-circuiting problem existed for the traditional ozone contact tank design. The effective contact tank design created in the 1995 study by the MWD, illustrated in Figure 4.8-7, was called the baseline contact geometry by the MWD; a T_{10}/T of 0.75 can be achieved with this design. Due to the rapid reaction kinetics of ozone, a hydraulic detention time (HDT) of 5 to 10 min is considered to be practical for regular oxidation and disinfection purposes. However, inactivation of *Cryptosporidium* would require 15 to 20 min of HDT even though the EPA will not release the required $C \times t$ value or the MCL until mid-2002.

The ozone contact tanks have several common design features: The tanks must be enclosed completely; they are composed of concrete and are located primarily outside; and each tank has several cells and is capable of handling 50% of the maximum daily flow of the plant. Additionally, each cell should have a drain to allow periodic dewatering and cleaning; stainless steel hatches should be provided to facilitate inspection and maintenance; a slight negative pressure [2 in. (5 cm) of water] should be maintained in the tank to ensure that there is no leakage of ozone

TABLE 4.8-3 Solubility of Ozone in Water

Water Temperature (°C)	Henry's Constant (atm/mol)	Ozone Concentration[a]		Ozone Solubility (mg/L)
		mg/L	ppm vol.	
0	1,940	12.07	6,044	8.31
5	2,180	12.07	6,044	7.39
10	2,480	12.07	6,044	6.50
15	2,880	12.07	6,044	5.60
20	3,760	12.07	6,044	4.29
25	4,570	12.07	6,044	3.53
30	5,980	12.07	6,044	2.70
0	1,940	18.11	9,069	12.47
5	2,180	18.11	9,069	11.09
10	2,480	18.11	9,069	9.75
15	2,880	18.11	9,069	8.40
20	3,760	18.11	9,069	6.43
25	4,570	18.11	9,069	5.29
30	5,980	18.11	9,069	4.04
0	1,940	24.14	12,088	16.62
5	2,180	24.14	12,088	14.79
10	2,480	24.14	12,088	13.00
15	2,880	24.14	12,088	11.19
20	3,760	24.14	12,088	8.57
25	4,570	24.14	12,088	7.05
30	5,980	24.14	12,088	5.39
0	1,940	36.21	18,132	24.92
5	2,180	36.21	18,132	22.18
10	2,480	36.21	18,132	19.50
15	2,880	36.21	18,132	16.79
20	3,760	36.21	18,132	12.86
25	4,570	36.21	18,132	10.58
30	5,980	36.21	18,132	8.09

Source: Adapted from EPA's *Design Manual,* EPA/625/1-86/02, October 1986.

[a]The concentration of the ozone gas is determined at a standard temperature of 68°F (20°C) and a standard pressure of 1 atm (101 kPa).

above the water surface; all pipe (type 316L SS) connections should be welded or flanged inside the tank and a check valve should be installed in the gas feed line prior to the contactor to prevent the backflow of water into the gas line; the water depth in the tank should be 18 to 20 ft (5.5 to 6 m) to provide a minimum ozone transfer efficiency of 95%. Finally, if the ozone concentration in the feed gas is 10% or above, a different system, such as the sidestream pump injection method or a hydraulic eductor and in-line static mixer, may be considered.

Off-Gas Destruction The ozone contact tank typically has a 90 to 95% transfer of ozone to the water; the remaining 5 to 10% results as off-gas. The ozone content of the off-gas must be reduced to levels below the OSHA and local Air Quality

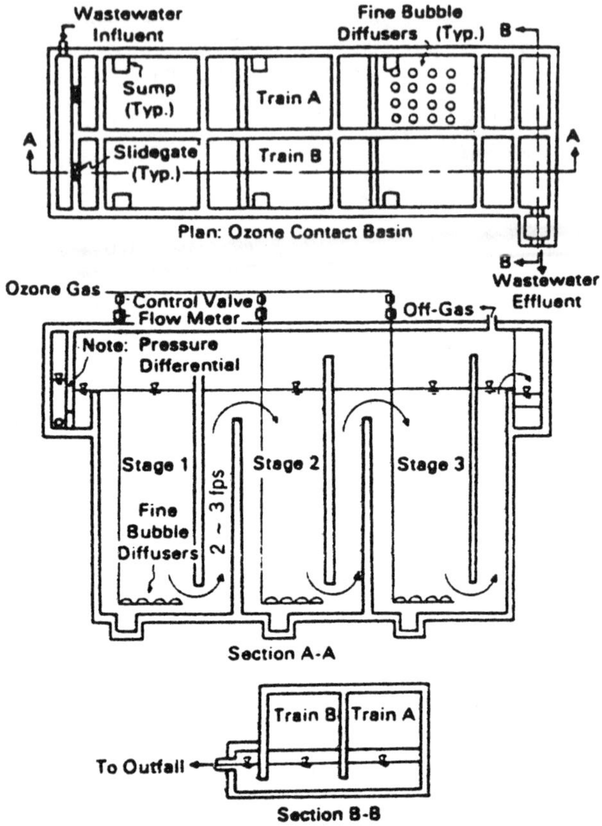

Figure 4.8-6 Schematic of a three-stage bubble diffuser ozone contact basin. (Adapted from EPA Design Manual, EPA/625/1-86/02, October 1986.)

Management District (AQMD) standards prior to venting into the atmosphere. According to OSHA, the maximum allowable ambient concentration of ozone is 0.0002 g/m^3 (0.1 ppm by volume) for an 8-h working day. The off-gas normally contains levels greater than 1 g/m^3 (500 ppm by volume). Three basic methods of destruction have been used: thermal destruction, thermal destruction with a catalyst, and catalytic destruction.

Thermal destruction is used almost exclusively for the air feed gas system; the oxygen feed gas system supplies such a high content of oxygen in the off-gas that use of thermal destruction is considered to be a fire hazard. Temperatures ranging from 570 to 660°F (300 to 350°C) for a period of 3 s are required to obtain 99% ozone destruction by thermal destruction. Due to the high temperature and the large power consumption of the unit, a heat recovery unit is usually provided. A demister, a stainless steel wire mesh, should also be provided upstream of the ozone destructor to minimize the accumulation of foam on the heating elements. The heat loss of the thermal destructor is approximately 30%.

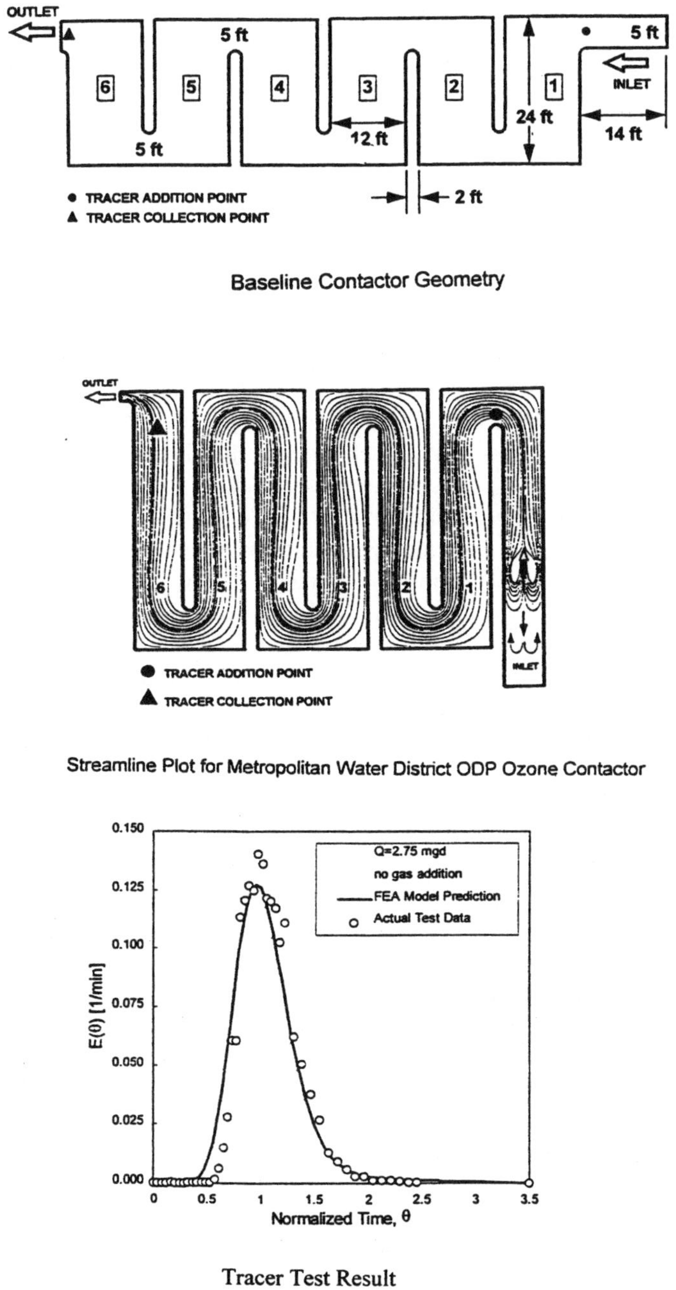

Figure 4.8-7 Concept of a recommended six-stage bubble diffuser ozone contact tank. (Adapted from Metropolitan Water District of Southern California, Report 1105.)

The system of thermal destruction with catalyst consumes less energy than thermal destruction alone. Thus this system has become more popular in recent years. The catalyst may be metal (platinum or palladium), metal oxides (aluminum oxide or manganese oxide), or hydroxides and peroxides. However, the hydroxides and peroxides are used only for pilot-scale studies. The advantage of the metal catalyst system is its lower operating cost. The metal oxide catalyst system can operate at 85 to 120°F (30 to 50°C); this is much lower than the thermal unit. Yet the disadvantage is its sensitivity to chlorine, chlorine derivatives, sulfides, and nitrogen; these chemicals may react with and destroy the catalyst. Activated carbon adsorption has been employed, but the powdery carbon produced by the reactivation process is potentially explosive. Thus this method has limited application.

In certain cases the off-gas is recycled back to the feed gas after conditioning or is injected into the filter waste backwash water, which is discharged to a holding lagoon. It should be noted that these alternatives are used very rarely because of the need to maintain an additional process and the lack of high reliability for the off-gas treatment.

By-products Control The major products of ozonation are brominated compounds, including bromate, assimilable organic carbon (AOC), and aldehydes. The levels of these compounds depend on the quality of raw water and level of ozone dosage. A large increase in the level of AOC after ozonation is a concern because this condition promotes the regrowth of bacteria in the water distribution system when there is a regular level of chlorine residual in the finished water; AOC level increases two to five times that prior to ozonation have been reported.

To control the major ozonation by-products (except brominated compounds), granular activated carbon (GAC) is placed in filter cells to make them biological filters or bacterial active filters (BAFs). Microorganisms that flourish in the upper portion of the GAC bed remove AOC, aldehydes, as well as TOC and DBP precursors, thereby reducing these compounds to the levels regulated. Regular dual-media filters and sand filters also act as biological filters after prefilter ozonation. However, the regular filter beds lose the capability of biofiltration for several hours when backwashed with chlorinated water, whereas the GAC bed is virtually unaffected. Thus it is recommended that engineers consider the use of GAC beds after ozonation. The GAC filter bed is discussed further in Section 7.7.

Safety The issue of safety must be addressed because ozone is both a toxic gas and a fire hazard. Moreover, if the system uses oxygen as the feed gas, the situation becomes more dangerous. Although the ozonation system may be less hazardous than the chlorination system because it can be shut down if an ozone leak develops, it may also be more dangerous because the system must use high-voltage electrical power to generate the ozone.

The American Industrial Hygiene Association (AIHA), OSHA, and other associations recommend the following permissible levels of ozone in the air: "Workers will not be exposed to ozone concentrations in excess of a time weighted average of 0.2 mg/m^3 (0.1 ppm by volume) for eight hours or more per workday, and that no worker be exposed to a ceiling concentration of ozone in excess of 0.6 mg/m^3

(0.3 ppm by volume) for more than 10 minutes." Local Air Quality Management Districts may have more stringent criteria that are part of the restrictions on smog.

A few other safety considerations include the installation of self-contained breathing apparatuses in the event of a severe ozone leak, the installation of eyewashes, and the availability of comprehensive safety and operation and maintenance manuals, which may be consulted during an emergency situation. Ideally, the ozone generation building and ozone contact tanks are isolated from the operations building for obvious safety reasons.

Design Criteria The design criteria for the ozonation system are numerous. They range from the ozone generator, to the contact tank, to the destruction unit. The requirements for these and many more issues are presented below. Additionally, an example of a detailed design is presented. (*Note:* 1 psi = 0.0703 kg/cm^2 = 6.895 kPa.)

Ozone dosage	1.5 to 3 mg/L (normal) depending on the purpose or 0.3 to 1.0 mg/L of ozone per mg/L of TOC
Number of ozone generators	Minimum of two, preferably three; one always acts as a standby
Ozone generator	
Minimum production	10 to 20% of rated capacity
Maximum production	75% of rated capacity if a continuous run
Cooling water temperature	Less than 75°F (24°C) at the inlet
Vessel construction	Pressure vessel [15 psig (104 kPa)] constructed with type 304L or 316L stainless steel with Hypalon or Teflon gaskets
Type of generator	Low frequency with variable voltage, medium frequency with frequency control, or another type, depending on the selection
Compressor pressure	10 to 40 psig if heat-reactive desiccant; 80 to 100 psig if pressure swing desiccant
Number	Minimum of two; one acts as standby
Type	Liquid ring, <1000 lb/day (455 kg/d) ozone production; centrifugal, >1000 lb/day (455 kg/d) ozone production; piston type, oil-free, or oil lubricated with oil removal filters, <100 lb/day—small plant
Filters	
Before compressor	Regular filter with silencer
Before dryer	3 to 5 μm
Before generator	0.3 μm
Pressure drop	0.5 psi when clean, 2 psi for coalescing filter
Moisture removal	
Refrigerant dryer	41°F (5°C) dewpoint (80% removal from air)
Desiccant dryer	−76°F (−60°C) dewpoint (99.98% removal from air)

NOTES: Ambient air contains approximately 2.7 g of water per kilogram of air. A dewpoint of −58°F (−50°C) is marginal for the desiccant dryer. Air with a dewpoint of −76°F (−60°C) contains 6.7 mg of water per kilogram of air.

Desiccant regeneration cycle (heat reactivation type)	
Minimum cycle time	8 h
Design cycle time	16 h
Oxygen supply alternative: Liquid oxygen (LOX)	+98% purity, 4400-gal (16.5-m^3) tank truck, 13,000-gal (50-m^3) standard-tank-size storage tank
Pressure swing adsorption (PSA)	15 to 30 tons/d gaseous oxygen (GOX) production, 90 to 92% purity, 30% turndown
Vacuum swing adsorption (VSA)	30 to 90 tons/day GOX production capacity, 90 to 94% purity, 30% turndown
Cryogenic distillation	100 tons/day GOX as well as LOX production, +98% purity, 50% turndown, high capital cost
Ozone contact tanks	
Number of tanks	Minimum of two
Transfer efficiency	Minimum of 95% if possible; may range from 90 to 95%
Detention time (HDT)	5 to 20 min (depending on application)
Stage of contact	2 to 10 (depending on application)
Water depth	18 to 20 ft (5.5 to 6.1 m)
Submergence of diffuser	16 to 18 ft (4.9 to 5.5 m)
Freeboard	4 to 6 ft (1.2 to 1.8 m) to allow deposition of foam
Ozone diffuser	
Material	Type 304L stainless steel, glass, ceramic, or Teflon
Bubble size	2 to 5 mm
Gas flow	0.5 to 4 ft^3/min (14 to 113 L/min), depending on the diffuser: rod diffuser (2.5 in. × 24 in. L), 4 ft^3/min maximum (113 L/min); disk diffuser (7 in. diameter), 1.25 ft^3/min maximum (35 L/min); disk diffuser (9 in. diameter), 1.8 ft^3/min maximum (51 L/min)
Headloss	Maximum of 0.5 psi
Permeability	8 to 20 ft^3/min per square foot (225 to 565 L/min^2)
Porosity	35 to 45%
Residual ozone (ozonated water) monitor	
Type	UV ozone monitor (continuous monitoring) such as PCI Model HC or potassium iodide monitoring

Ozone destruction unit	
Type	Heat catalyst unit (most popular)
Temperature	80 to 100°F (27 to 38°C)
Catalyst	Metal (platinum) or metal oxides: these catalysts are proprietary items
Size of ozonation system building (estimated)	2 to 4 ft^2/lb of O_3 production (0.4 to 0.8 m^2/kg)
Overall power requirement	10 to 12 kWh/lb of O_3 (air feed) (22 to 26 kWh/kg); 8 to 10 kWh/lb of O_3 (O_2 feed) (17.5 to 22 kWh/kg)

Figure 4.8-1 is a schematic presentation of a typical ozonation system.

Example Design Calculations

Example 1

Given The ozone concentration of the gas flowing from the ozone generator to the tank is customarily expressed as "percentage of ozone by weight" rather than the weight of ozone in grams per cubic meter of gas. Assume that the ozone concentration is 20 g/m^3.

Determine The percentage weight of ozone for the air feed ozone generation system and the oxygen feed ozone generation system. Assume that the gases are all at 20°C (68°F) and under 1 atm of pressure (STP conditions).

Solution

$$\text{Gram molecular weight of air} = 29 \text{ g/mol}$$

$$\text{Gram molecular weight of oxygen} = 32 \text{ g/mol}$$

$$\text{Gram molecular weight of ozone} = 48 \text{ g/mol}$$

(i) For the air feed ozone generation system the density of air is

$$\frac{\text{MP}}{\text{RT}} = \frac{29}{24.053} = 1.2056 \text{ g/L}$$

$$= 1205.6 \text{ g/m}^3 \quad \text{(see Appendix 7)}$$

and the density of ozone is

$$\frac{\text{MP}}{\text{RT}} = \frac{48}{24.053} = 1.9956 \text{ g/L}$$

$$= 1995.6 \text{ g/m}^3$$

The volume of 20 g of ozone is

$$\frac{gRT}{MP} = \frac{20 \times 24.053}{48} = 10.02 \text{ L}$$

$$= 1\% \text{ of } 1 \text{ m}^3$$

Since the volume of 20 g of ozone is 1% of m^3 of the gas, the weight of the gas is

$$1205.6 + [(1995.6 - 1205.6) \times 0.01] = 1213.5 \text{ g/m}^3$$

Thus the percent weight is

$$(20 \div 1213.5) \times 100 = 1.648\%, \qquad \text{say, } 1.65\%$$

(ii) For the oxygen feed ozone generation system the density of oxygen is

$$\frac{MP}{RT} = \frac{32}{24.053} = 1.3304 \text{ g/L}$$

$$= 1330.4 \text{ g/m}^3$$

Knowing that the volume of 20 g of ozone is 1% of the gas, the weight of the gas is

$$1330.4 + [(1995.6 - 1330.4) \times 0.01] = 1337.05 \text{ g/m}^3$$

The percent weight is

$$(20 \div 1337.05) \times 100 = 1.496\%, \qquad \text{say, } 1.5\%$$

Example 2

Given The ozone concentration of the gas from an air feed ozone generator is measured by the potassium iodide wet-chemistry method as 25.6 mg/L.

Determine The percent weight of ozone at STP conditions.

Solution

$$25.6 \text{ mg/L} = 25.6 \text{ g/m}^3$$

$$\text{Weight of air} = 1205.6 \text{ g/m}^3 \quad \text{from Example 1}$$

$$\text{Weight of ozone} = 1995.6 \text{ g/m}^3 \quad \text{from Example 1}$$

$$\text{Volume of 25.6 mg/L of ozone} = \frac{\text{gRT}}{\text{MP}} = \frac{25.6 \times 24.053}{48} = 12.83 \text{ L}$$

$$= 1.28\% \text{ of the gas}$$

The weight of the gas is calculated to be

$$1205.6 + [(1995.6 - 1205.6) \times 0.0128] = 1215.7 \text{ g/m}^3$$

Thus the percent weight of the ozone is

$$(25.6 \div 1215.7) \times 100 = 2.1\%$$

Example 3

Given An air feed ozone generator is operating at an airflow rate of 300 standard cubic feet per minute (scfm; 8.5 m^3/min); the standard conditions are 20°C (68°F) and 1 atm (14.7 psi) of pressure. The ozone concentration is measured to be 1.65% weight by the ozone monitor. The plant water flow rate is 20 mgd (0.88 m^3/s).

Determine

(i) The ozone production rate lb/day
(ii) The ozone dosage in mg/L

Solution (i) The conversion factor for g/L to lb/ft^3 is 0.0753. The ozone production rate is calculated to be

$$\frac{1.65 \text{ lb O}_3}{100 \text{ lb air}} \times 0.0753 \times (300 \times 1440) = 537 \text{ lb/day (244 kg/d)}$$

(ii) The ozone dosage is

$$537 \div (8.34 \times 20) = 3.22 \text{ mg/L}$$

Example 4 Design the ozonation system of a 20-mgd (average) water treatment plant using the design criteria presented in the "Design Criteria" section.

Solution *Step 1: Ozonation Production Rates*

$$\text{Maximum ozone production rate} = 3 \times 8.34 \times 30 = 751 \text{ lb/day (340 kg/d)}$$

$$\text{Average ozone production rate} = 3 \times 8.34 \times 20 = 500 \text{ lb/day (227 kg/d)}$$

$$\text{Minimum ozone production rate} = 1.5 \times 8.34 \times 5 = 63 \text{ lb/day (29 kg/d)}$$

Step 2: Type, Size, and Number of Ozone Generators A medium-frequency ozone generator is selected primarily because of its 10 : 1 turn-down ratio, lower power consumption, and its ability to enrich the feed air should the plant need to double the ozone output at a future date. Generally, any ozone generator should be operated for extended periods of time at approximately 75% of the rated capacity. The rated maximum daily production of ozone should be 751 ÷ 0.75 or 1000 lb/day. Three generators, each capable of 335 lb/day (150 kg/d), should therefore be provided. This type of generator can accommodate a turn-down ratio of 10 : 1 so that the minimum production rate is approximately 35 lb/day (16 kg/d). Thus a rate of 63 lb/day (28 kg/d) is easily achieved. Since two of the generators can satisfy the required 500 lb/day (227 kg/d), under normal conditions, one unit should be left on standby.

Step 3: Size and Number of Air Compressors Since the maximum ozone concentration is 2.5% by weight for most air feed medium-frequency generators, the air flow at maximum condition is

$$1000 \text{ lb/day} \div \left(0.0753 \times 1440 \times \frac{2.5}{100}\right) = 369 \text{ scfm or } 370 \text{ scfm } (10.5 \text{ m}^3\text{/min})$$

Add 60 scfm to this number for purge airflow to the dryers. Thus the maximum airflow may be calculated as

$$370 + 60 = 430 \text{ scfm } (12 \text{ m}^3\text{/min})$$

The minimum ozone concentration is approximately 1% by weight when this type of generator is employed. Thus the minimum airflow rate is

$$75 \text{ lb/day} \div \left(0.0753 \times 1440 \times \frac{1}{100}\right) = 69 \text{ scfm or } 70 \text{ scfm } (2 \text{ m}^3\text{/min})$$

If 2.5% by weight ozone is produced on a continuous basis, the minimum airflow rate would be

$$75 \text{ lb/day} \div \left(0.0753 \times 1440 \times \frac{2.5}{100}\right) = 28 \text{ scfm } (0.8 \text{ m}^3\text{/min})$$

This value is the smaller of the two. Add 22 scfm (0.6 m^3/min.), for purge airflow to the dryer. Therefore, the minimum airflow is

$$28 + 22 = 50 \text{ scfm } (1.4 \text{ m}^3\text{/min})$$

Based on these calculations, the compressors may be either medium [35 to 50 psig (243 to 345 kPa)] or low [10 to 30 psig (69 to 207 kPa)], depending on the type of drying unit selected. Provide:

- One 50-scfm compressor (1.4 m^3/min)
- Two 100-scfm compressors (2.8 m^3/min)
- Two 150-scfm compressors (4.2 m^3/min)

Step 4: Size and Number of Desiccant Dryers The design basis is as follows:

1. Operating pressure of 20 to 25 psig for low-pressure systems and 50 psig for medium-pressure systems
2. Adjustable site elevation for the pressure
3. Inlet air temperatures: 77°F (25°C) with after-cooler and 40°F (4.4°C) with refrigerated dryer
4. A required maximum dewpoint temperature for the air flowing to the ozone generator of −76°F (−60°C)
5. Designed desiccant dryer cycle time of 16 h; minimum drying time of 8 h

Operation and Maintenance The ozonation system is both very sensitive and expensive compared to other types of water treatment equipment. Plant operators must be very conscientious in maintaining optimum operational conditions, for reasons of efficiency and safety. The operational conditions of the ozonation system must be adjusted by the plant operator whenever the plant flow rate changes. This will provide the system with optimal operational conditions, thus facilitating maximum process efficiency and minimal operation and maintenance costs. Included in the optimization process is the determination of the optimum ozone dosage; a 0.5-mg/L reduction in dosage translates into a significant savings in power consumption. The regular performance of several other tasks also directly affects the efficiency of the ozonation system: checking the temperature of the cooling water, checking the dewpoint temperature at the desiccant dryer unit, checking and calibrating the ozone concentration monitors, checking the off-gas concentration, and adjusting the ozone dosage. The plant operation and maintenance manuals, as well as manuals furnished by the equipment supplier, should describe these very important items in detail.

A common problem for many large metropolitan areas is air pollution due to smog. In these situations, the local AQMD sets the maximum allowable ozone concentration in the air. Thus the efficiency of the off-gas destruction unit should be maintained at a level high enough to meet the stringent limits. Moreover, the exhausting air from the ozone contact tanks and the building housing the ozone generator must be monitored closely for excessive emission of ozone. With regard to the safety aspects of the ozone system, strict adherence to the safety codes and regulations released by OSHA, the American National Standards Institute/American Society for Testing and Materials (ANSI/ASTM), the American Conference of Governmental Industries (ACGI) and the American Industrial Hygiene Association (AIHA) is absolutely necessary.

BIBLIOGRAPHY

Chang, S. D., and Singer, P. C., "The Impact of Ozonation on Particle Stability and Removal of TOC and THM Precursors," *J. AWWA,* 83:3:71 (March 1991).

Cipparone, L. A., et al., "Ozonation and BDOC Removal: Effect on Water Quality," *J. AWWA,* 89:2:84 (February 1997).

Dowbiggin, W. B., and Singer, P. C., "Effect of Natural Organic Matter and Calcium on Ozone-Induced Particle Destabilization," *J. AWWA,* 81:6:77 (June 1989).

Edwards, M., et al., "Effects of Ozonation on Coagulation of NOM Using Polymer Alone and Polymer/Metal Salt Mixtures," *J. AWWA,* 86:1:105 (January 1994).

Ferguson, D. W., et al., "Applying Ozone for Organic Control and Disinfection: Utility Perspective," *J. AWWA,* 83:5:32 (May 1991).

Ferguson, D. W., et al., "Comparing Peroxone and Ozone for the Control of Taste and Odor Compounds, Disinfection By-products and Microorganisms," *J. AWWA,* 82:4:181 (April 1990).

Jacangelo, J. G., et al., "Ozonation: Assessing Its Role in the Formation and Control of D.B.Ps," *J. AWWA,* 81:8:94 (August 1989).

Katzenelson, E., et al., "Inactivation Kinetics of Viruses and Bacteria in Water by Use of Ozone," *J. AWWA,* 66:12:725 (December 1974).

Krasner, S. W., et al., "Formation and Control of Bromate During Ozonation of Water Containing Bromide," *J. AWWA,* 85:1:73 (January 1993).

LePage, W. L., "A Treatment Plant Operator Assesses Ozonation," *J. AWWA,* 77:8:41 (August 1985).

Liang, S., et al., "Oxidation of MTBE by Ozone and Peroxone Processes," *J. AWWA,* 91:6:104 (June 1999).

Malorey, S. W., et al., "Ozone-GAC Following Conventional U.S. Drinking Water Treatment," *J. AWWA,* 77:8:66 (August 1985).

Metropolitan Water District of Southern California, *Oxidation Retrofit Program: Design Report,* Report 1105, November 1995.

Monk, R. D. F., et al., "Prepurchasing Ozone Equipment," *J. AWWA,* 77:8:49 (August 1985).

Rakness, K. L., "Ozone Disinfection," Chapter 6 in *EPA Manual: Municipal Wastewater Disinfection,* EPA/625/1-86-021, U.S. Environmental Protection Agency, Washington, DC, October 1986.

Rice, R. G., "Uses of Ozone in Drinking Water Treatment," *J. AWWA,* 73:1:44 (January 1981).

Rice, R. G., and Netzer, A., *Handbook of Ozone Technology and Applications,* Vol. 1, Ann Arbor Science, Ann Arbor, MI, 1982.

Rice, R. G., and Netzer, A., *Handbook of Ozone Technology and Applications,* Vol. 2, Butterworth, Boston, 1984.

Schylz, C. R., and Hawkins, R. A., "Evaluating a High Efficiency Ozone Injection Contactor," *J. AWWA,* 87:8:85 (August 1995).

Weinberg, H. S., et al., "Formation and Removal of Aldehydes in Plants That Use Ozone," *J. AWWA,* 85:5:72 (May 1993).

CHAPTER 5

Design of Plant Components

5.1 FLOW MEASUREMENT

Purpose Flow measurement is the most important variable measured in a treatment plant. The three principal applications for flow measurement are establishing the flow rate of the process and determining the liquid chemical and gas flow rates of various subordinate processes. Differential pressure producers, direct discharge measurement, positive volume displacement measurement, and flow velocity–area measurement are the four basic types of measurement commonly used in water treatment plants.

Considerations Several questions must be considered to select the most appropriate type of flow measurement equipment.

1. Is liquid or gas flow being measured?
2. Is the flow occurring in a pipe or in an open channel?
3. What is the magnitude of the flow rate?
4. What is the range of flow variation?
5. Is the liquid being measured clean, or does it contain suspended solids or air bubbles?
6. What is the accuracy requirement?
7. What is the allowable headloss by the flow meter?
8. Is the flow corrosive?
9. What types of flow meters are available to the region?
10. What types of postinstallation service is available to the area?

Types of Flow Meters Flow meters commonly used in the field of water treatment are the venturi meter, orifice meter, propeller meter, magnetic flow meter, ultrasonic flow meter, vortex meter, rotameter (variable-area meter), flumes, and weirs. Liquid chemical flow is almost exclusively measured by positive displacement pumps rather than flow meters, with the exception of the rotameter which may be used in certain situations.

Differential Pressure-Producing Meters Differential pressure–producing meters currently on the market are the venturi, Dall tube, Hershel venturi, universal venturi, and venturi inserts (Figure 5.1-1). The accuracy of these flow tubes is ±1% for a flow range of 10 : 1. However, since the square-root extractor is used before the 4- to 20-mA signal is transmitted, the range of accurate flow measurement is restricted to no more than 5 : 1. The headloss across a venturi flow meter is relatively small, ranging from 3 to 10% of the differential, depending on the ratio of the throat diameter to the inlet diameter (beta ratio). These flow tubes should be installed at an elevation that will sustain the meter continuously at full pipe flow and ideally will not allow negative pressure to form at the throat during the maximum flow rate.

Orifice plates are another type of differential pressure producer. Both concentric and eccentric orifice plates are available. Their accuracy is ±2% for a flow range of 4 : 1. The orifice plate yields a high headloss ranging from 40 to 90% of the dif-

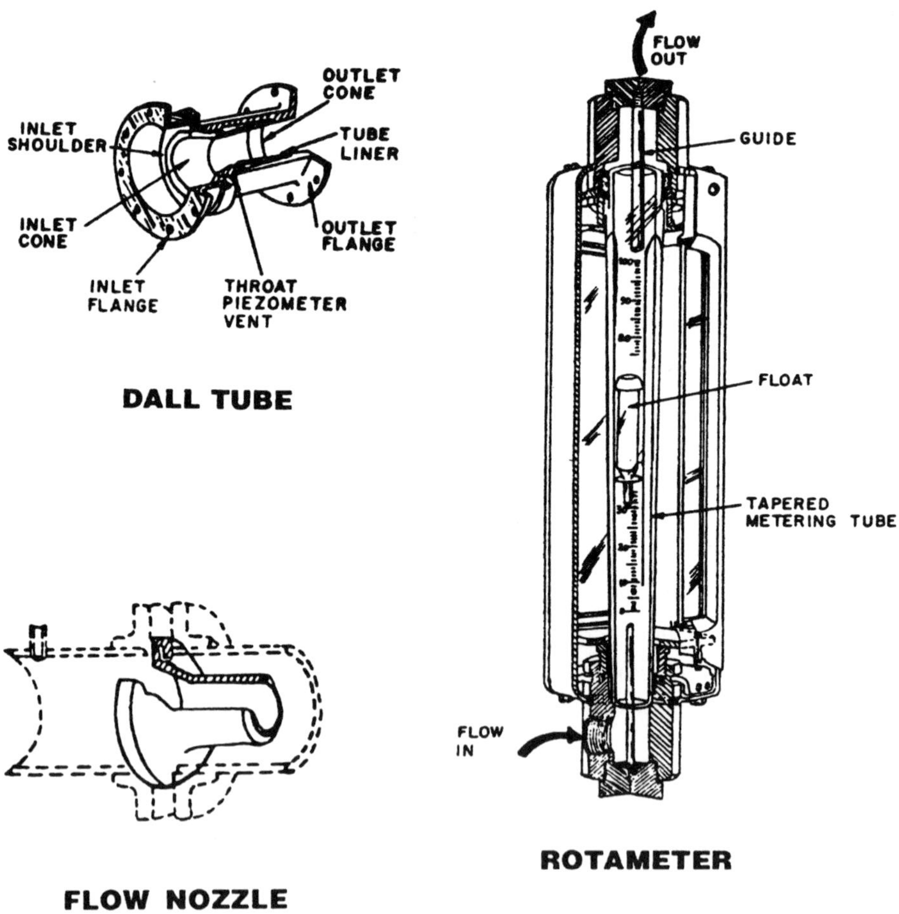

Figure 5.1-1 Common flow meters.

ferential; these values depend on the ratio of the orifice diameter to the pipe diameter. This type of flow measurement equipment is suitable for measuring both gas and liquid flow.

A third type of differential producer is the flow tube (flow nozzle). The throat ratio determines the headloss of the flow tube; headloss ranges from 10 to 20% of the differential. However, its application is primarily in sewage works (Figure 5.1-1). To assure uniform flow through the meter, a straight pipe with a length equivalent to at least 10 pipe diameters should ideally be provided upstream of all differential flow meters. Yet the meter manufacturer may only allow for a straight pipe that is five pipe diameters in length in conjunction with a flow straightener. Figure 5.1-2 presents the headloss characteristics of several venturi flow meters and their piping requirements.

Propeller and Turbine Meters The Sparling, Rockwell, Badger, and Neptune meters have been on the market for decades. These types of meter have a magnetic coupling that senses the revolutions of the spinning rotor and sends out electronic pulses. The accuracy is 1 to 2% for a flow range of 10 : 1, provided that there is a uniform and straight pipe of at least five pipe diameters and straightening vanes ahead of the meter. The minimum flow velocity should generally be 0.7 ft/s (0.2 m/s) and the maximum flow velocity must not exceed 6 ft/s (1.8 m/s). The propeller and turbine meters should be employed in situations where the water is relatively clean and noncorrosive and for liquids that are not too viscous since they are subject to wear, particularly in cases where the water contains suspended solids such as silt. Thus the meter must be installed in a readily accessible location to facilitate easy maintenance. Main-line meters are available in sizes up to 24 in. (0.6 m) in diameter; the saddle type, up to 72 in. (1.8 m); and the intake meters, up to 78 in. (2 m) in size.

Magnetic Flow Meters The magnetic flow meters operate by having a conductor, such as water flowing in a magnetic field, create a voltage that is proportional to the velocity of relative motion. Fixed magnetic coils wrapped around the pipe produce a magnetic field that is perpendicular to the flow of liquid. The induced voltage is detected by two electrodes embedded in the pipe. The accuracy of this type of meter is ±1% for a flow velocity range of 3 to 30 ft/s (0.9 to 9.1 m/s) and within ±2% if the range is 1 to 3 ft/s (0.3 to 0.9 m/s). Yet this type of direct discharge meter is relatively expensive and may require periodic calibration to correct drifting of the signal.

The magnetic flow meter may be used for measuring most any type of liquid, including the measurement of sludge flow. Sizes up to 48 in. (1.05 m) are readily available, and larger sizes may be custom manufactured. Since there is no flow restriction inside the meter, there is essentially no headloss within the unit. However, the size of the meter should be one size smaller than the size of the line, so that the flow velocity requirements can be met. The two fundamental requirements are that a uniform and straight pipe of at least five pipe diameters be placed ahead of the meter and that the unit should receive a full pipe flow.

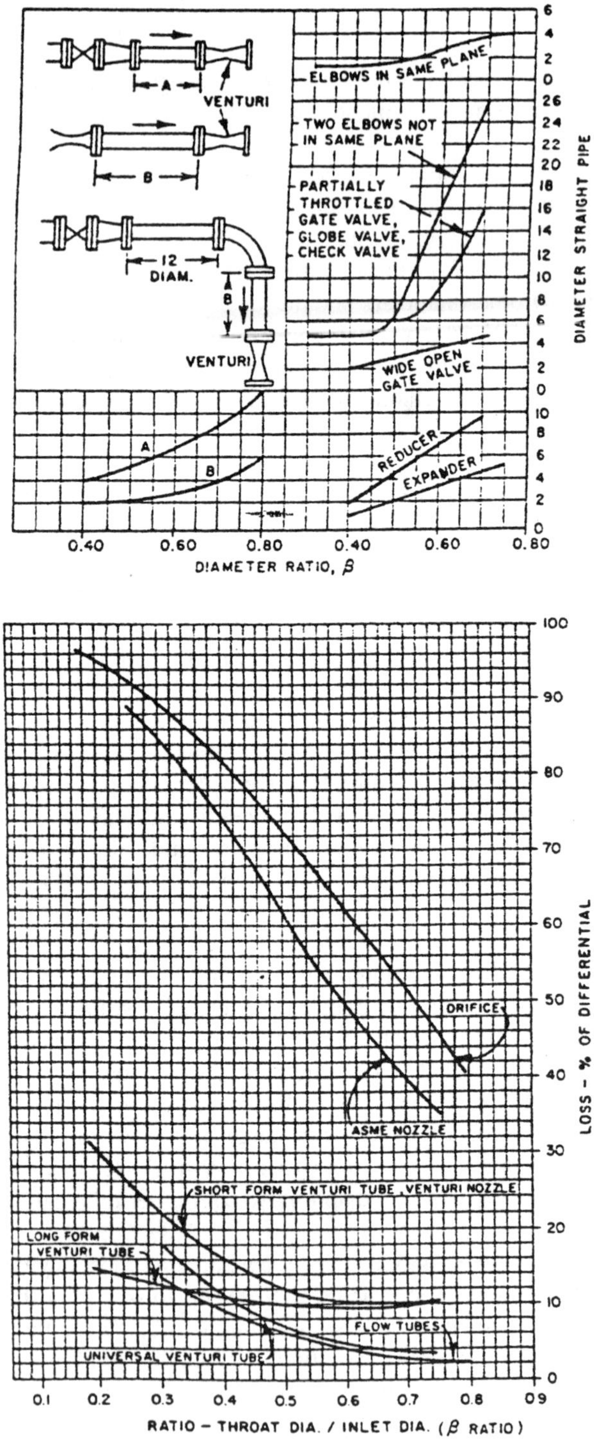

Figure 5.1-2 Piping requirement and headloss for venturi flow meters. (Adapted from BIF Catalog, 1968.)

Ultrasonic Flow Meters The flow measurement of the ultrasonic flow meter is based on the time required for an ultrasonic pulse to diagonally traverse a pipe or a channel against the liquid flow. Application of this type of direct discharge meter should be to clean water so that it measures the transit time or time of flight and not the reflection of the pulse off particles or air bubbles (Doppler effect). The flow velocity is proportional to the difference in the time of the two pulses. The accuracy is ±1% for a flow velocity ranging from 1 to 25 ft/s (0.3 to 7.6 m/s), but the meter reading is greatly affected by a change in the fluid composition. Thus this type of meter may only be employed for water that is free of particles and air bubbles as much as possible. In the case of large pipes or a channel, more than one pair of sensors is required.

Vortex Meters When a liquid flow hits a bluff object, eddies and vortices are formed; this is known as *vortex shedding*. The frequency at which the vortices are generated is proportional to the velocity of the liquid flow. The resulting pressure changes may be detected by probes mounted on the body of the bluff. This information is then relayed to an indicating recorder by a 4- to 20-mA signal. The manufacturers claim that this type of meter has an accuracy of ±1% for a flow range of 12 : 1. The headloss is approximately two times the velocity head through the meter. The cost of the vortex meter is moderate. However, this type of direct discharge meter does not function well with viscous, dirty, abrasive, or corrosive fluids.

Rotameters (Variable-Area Flow Meters) Rotameters consist of a tempered glass tube containing a freely moving float. This glass tube is vertically mounted with the wide side up (Figure 5.1-1). Rotameters may be used in measuring both gas and liquid flow rates. The flow rate of the particular gas or liquid is indicated when the float reaches a position of equilibrium between its weight and the upward force of the medium. The flow range for this type of meter is approximately 10 : 1, with an accuracy of ±2% of full scale. Yet the accuracy is highly dependent on the skill of the person reading the meter.

A rotameter may be applied in situations with very low flow rates, 0.1 to 140 gal/h (0.4 to 530 L/h) for water and 1 to 520 standard cubic feet per hour (0.028 to 14.7 m^3/h) for air, and the accuracy of the meter is not affected by the piping configuration immediately upstream of the rotameter. Basically, these meters are controlled manually and read or measured visually.

Flumes (Positive Volume Displacement Type) Flumes are used for open-channel flow. The Parshall flume and Palmer–Bowles are the types used predominantly in wastewater treatment. These meters are used to measure small to very large flow rates [0.02 to 3000 mgd (0.9 L/s to 130 m^3/s)] and are applicable to both clean and dirty water. The accuracy of the flumes is ±5% over a range of 20 : 1, but the range may be expanded to 35 : 1 with some sacrifice in accuracy. It is important that engineers provide a sufficient fall downstream of the flume to avoid the effects of the backwater, because the level measurement is conducted at a specific point in the flume by either an ultrasonic level detector or a float located directly above the point. The flow rate may also be determined in a stilling well by a bubbler or float.

Weirs Several types of weirs—for example, the V-notch, rectangular, or trapezoidal—are currently available for use in the field of water treatment. The head above the crest of the weir is usually measured by a float, bubbler system, acoustic meter, hook gauge mounted above the weir, or staff board mounted on the sidewall of the channel. The practical flow range of the weirs is approximately 30 : 1 and the accuracy is ±3%. Rectangular and trapezoidal weirs have a head–flow power relationship of approximately 1.5, while the V-notch weirs have a power relation of 2.5 (see Appendix 14 for the formulas). Therefore, any inaccuracy in the measurement of the hydraulic head will greatly affect the accuracy of the flow measurement, especially in the case of the V-notch weirs. Yet the V-notch weir plates may be used for a much wider flow range; an extreme case would be 3000 : 1.

Although the weirs are easy to manufacture and are cost-effective, the main disadvantages are their high headloss and the tendency for settleable solids and large debris to collect on the upstream side of the weir. Figure 5.1-3 illustrates flow curves for Parshall flumes and a 90° V-notch weir. The equipment manufacturers can provide the engineer with detailed information on the various types of flow meter currently available, and all engineers are advised to review the brochures prior to selection. Virtually all types of meters, with the exception of the rotameter, require an adequate length of uniform and straight conditions upstream of their location. Most of the flow measurement meters are capable of sending a standard 4- to 20-mA signal.

Example Design Calculations

Example 1

Given A water treatment plant has a daily maximum, average, and minimum design flow rate of 22, 15, and 4 mgd, respectively. However, this plant is designed to double its capacity in approximately 15 years.

Determine The appropriate size and elevation of the venturi tube, which is connected to a plant inlet tank containing a water depth of 10 ft.

NOTES:

Conversion factors from English units to the metric system are as follows: 1 mgd = 0.044 m^3/s; 1 ft = 0.305 m; 1 in. = 2.54 cm.

2. The selection guides for the venturi meter are as follows:

(a) Select a meter size that will allow for an inlet flow velocity of 6 to 10 ft/sec at the average daily flow rate.

(b) Select a proper beta ratio for the Venturi tube so that a differential pressure of approximately 80 in. will be produced at the average daily flow.

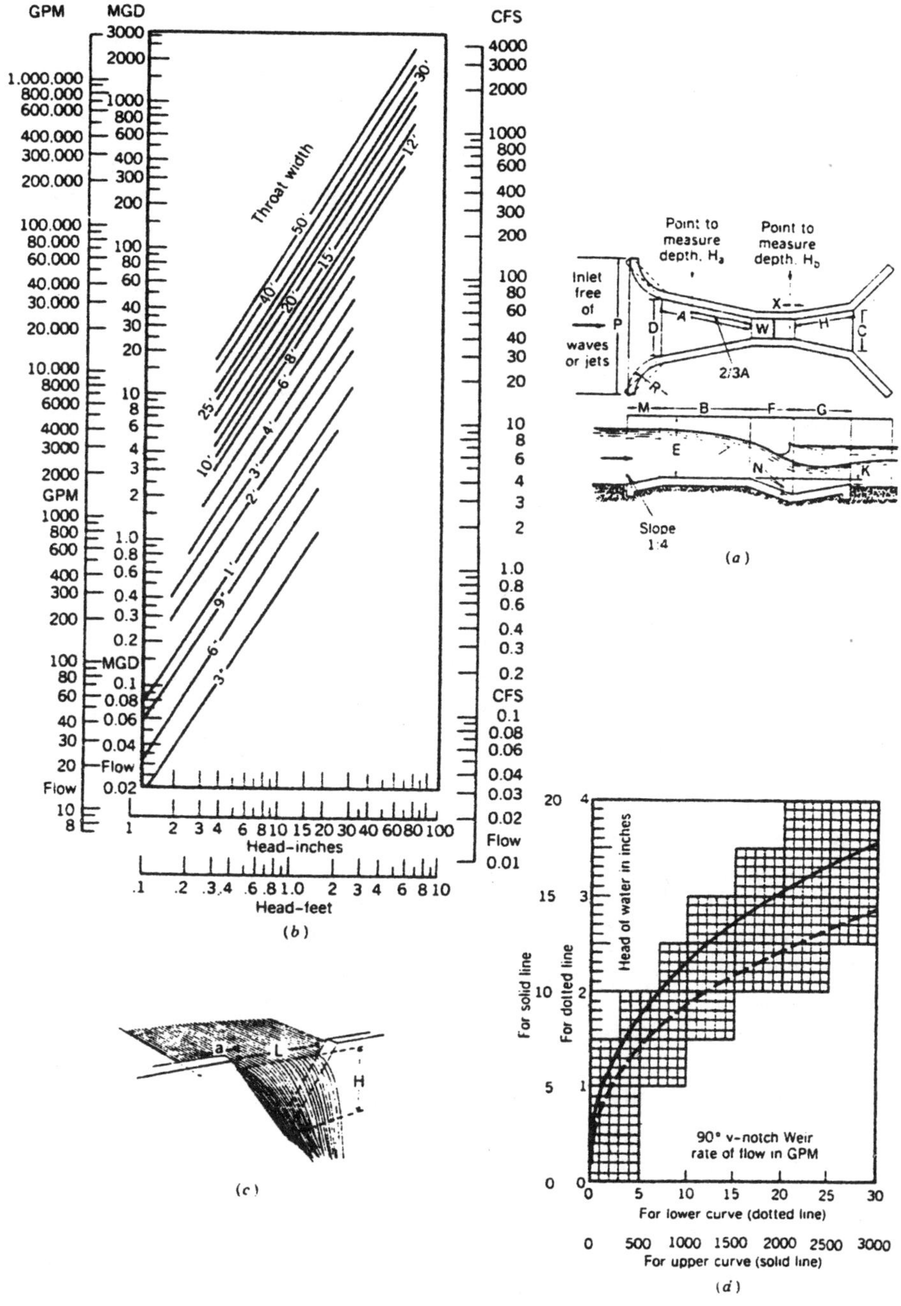

Figure 5.1-3 (*a*) Parshall flume; (*b*) flow curves for Parshall flume; (*c*) V-notched weir; (*d*) flow curves for 90° V-notched weir.

(c) Check the magnitude of the differential at the daily maximum and daily minimum design flows. The minimum differential should be at least 3 in. and the negative pressure should be no more than 2 ft at the throat of the venturi (at the differential value) during the maximum flow rate.

Solution The plant influent pipe should be sized for a future flow rate of 30 mgd (daily average flow rate). Thus the line size should be 36 in. in diameter ($v =$ 6.6 ft/sec). Yet the flow velocity is only 3.3 ft/sec (1 m/s) at 15 mgd; this is the average daily flow rate of the initial stage. The Venturi tube selection table (Appendix 12) indicates that none of the 36-in. tubes will satisfy the criteria; a type A throat (beta ratio A) of 36 in. is characterized by a 120-in. differential at 16 mgd, but the differential is 240 in. (20 ft) at 22 mgd; that is, the centerline of the Venturi tube must be installed approximately 20 ft lower than the water level in the inlet tank to avoid producing a significant amount of negative pressure at the throat of the Venturi tube. The differential at the minimum design flow rate of 4 mgd is

$$120 \text{ in.} \div (16/4)^2 = 7.5 \text{ in.} > 3 \text{ in.} \qquad \text{OK}$$

If the site condition allows this type of configuration for the plant inlet pipe, without raising the cost significantly, a Venturi tube with a Type A throat of 36 in. should be selected for the initial plant. When the plant capacity is doubled (at a future date), the Venturi may be replaced with a Venturi tube with a Type C throat, which would produce a differential of 80 in. at a daily flow rate of 32 mgd, 160 in. (13.3 ft) at 45 mgd (maximum daily flow), and 5 in. at 8 mgd (minimum daily flow rate).

Alternatively, the pipe size may be reduced from 36 in. to 20 in. with a minimum pipe length of 30 ft, if the influent line cannot be installed greater than 20 ft below the water level of the tank due to the site condition or cost. Under these circumstances, a 20-in. venturi tube ($v =$ 10.5 ft/sec at 15 mgd) should be selected: The Venturi tube selection table indicates that the type D (20 in.) Venturi tube will give a differential of 80 in. at a flow rate of 16 mgd. The flow rate at 15 mgd is

$$80 \text{ in.} \div (16/15)^2 = 70.3 \text{ in. (6 ft)}$$

The differential at the daily maximum flow of 22 mgd is

$$80 \text{ in.} \times (22/16)^2 = 151 \text{ in. (12.6 ft)}$$

and the differential at 4 mgd is 5 in. A minimum of 17 ft of uniform straight pipe and a straightener should be provided upstream of the Venturi tube (Figure 5.1-4).

When the plant capacity is doubled, the 20-in. line (containing the Venturi tube) is replaced by a 36-in. line and a 36-in. Type C Venturi tube. This alternative requires that the centerline elevation be approximately 14 ft below the water level in the inlet tank; this is easily achieved in most cases.

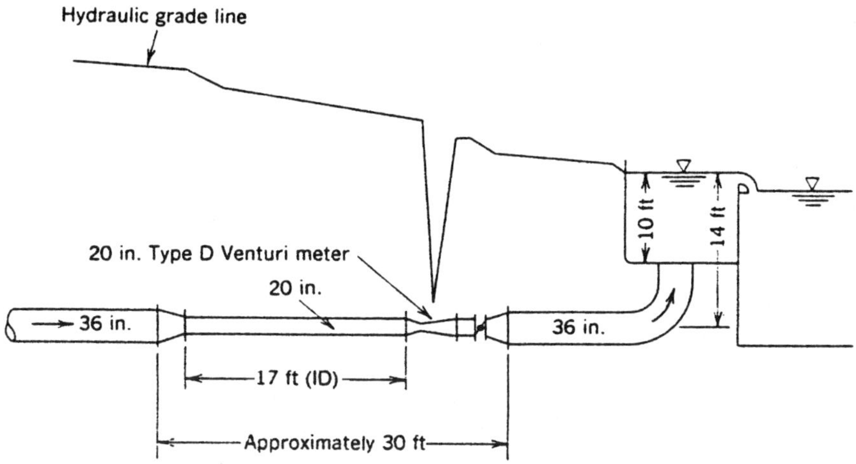

Figure 5.1-4 An example of Venturi flow tube Installation.

BIBLIOGRAPHY

Andrew, W. G., and Williams, H. B., *Applied Instrumentation in the Process Industries,* 2nd ed., Vols. 1 and 2, Gulf Publishing, Houston, TX, 1980.

Babcock, R. H., "Instrumentation and Control in Water Supply and Wastewater Disposal," *Water Wastewater Eng.,* (1968).

Department of Civil Engineering, "Instrumentation, Control, and Automation for Water Supply and Wastewater Treatment Systems," *Proceedings of the 9th Sanitary Engineering Conference,* University of Illinois, 1967.

Knudsen, D. I., ed., *Process Instrumentation and Control System,* Manual of Practice OM-6, Water Pollution Control Federation, 1982.

5.2 LEVEL MEASUREMENT

Purpose Level measurement is an essential item in plant operations because the plant operator must be aware of the water level of all the process units, the levels of all chemical storage tanks and silos, and the pressure of water or compressed air lines—that is, the water level in the distribution mains and the utility lines. There are several means of measuring liquid levels; via a float, pressure elements, bubbler system, or an ultrasonic system. Methods used in measuring dry, powdery materials include ultrasonic systems, photocell systems, rotary paddle switches, diaphragm units, and wire strain gauge systems. However, it is often very difficult to measure the level of powdery materials held in storage bins accurately, due to the concavity of the surface or irregular surface pattern. The most accurate and reliable method of monitoring any dry or liquid material in storage bins or tanks is to use load cells to measure the total weight (tank plus content).

In the remainder of this section we describe some features of the (1) float system, (2) pressure elements, (3) bubbler tube system, and (4) ultrasonic level detector. Other methods of level measurement are not commonly employed in water treatment and are therefore not discussed.

Float System In the past, the most common method of level measurement of open tanks or channels was the float-operated transmitter. This is a simple and reasonably accurate system. Nevertheless, the system is very unsightly, and the installation process is very time consuming and expensive, due to the need for a stilling well and a collection of wires, wheels, and tackles. The float system also requires periodic maintenance to assure friction-free motion of the float and cable assembly.

Pressure Elements Level measurement via pressure elements is very common in the field of water treatment. This method determines liquid levels by measuring the pressure produced by the head or by measuring the level above the pressure measuring tap. A pressure transducer connected to the pressure elements measures the water pressure at the base of the tank and reads the liquid level directly if properly calibrated for the liquid.

NOTE: Conversion factors from the English unit system to metric and SI systems are as follows: 1 psi − 0.0703 kg/cm^2 = 6.895 kPa.

The bourdon tube, bellow element, diaphragm element, and manometer are examples of pressure element measurers. The bourdon tube has both helical and spiral units and is suited for high-pressure (water depth included) measurements. The pressure range of the spiral units is commonly 0 to 15 and 0 to 200 psi (0 to 1 and 0 to 14 kg/cm^2). The helical unit is typically used for measuring much higher pressures: 0 to 200, 0 to 400, and 0 to 6000 psi (0 to 14, 0 to 28, and 0 to 420 kg/cm^2) or higher.

Bellow units measure pressures in the intermediate range. The units are commonly manufactured to measure water levels (depth) of 0 to 100 in. of water (0 to 2.5 m) and a pressure range of 0 to 5 and 0 to 30 psi (0 to 0.35 and 0 to 2 kg/cm^2). The diaphragm type of pressure element is particularly suited for obtaining measurements within a small range in the low-pressure zone. The available units can cover a range of 0 to 10 and 0 to 20 inches (0 to 0.25 and 0 to 0.5 m), and 0 to 40 and 0 to 140 inches of water (0 to 1 and 0 to 3.6 m).

The use of manometers is limited to pilot studies or temporary use; when the prime meter is out of service. Pressure elements often require the following conditions to yield accurate readings: zero suppression, zero elevation, or differential pressure for pressurized tanks (D/P cell).

Bubbler Tube System The bubbler tube system is used to measure the back pressure of the hydrostatic head. This is equivalent to the hydrostatic head of the liquid at the bottom of the bubbler tube. This type of level measurement is currently the most widely used method applied to tanks. The bubbler tube system has a tube placed inside a tank which runs from the top of the tank and opens approximately 3 in. (7.5 cm) from the bottom. During operation, compressed air is supplied to the submerged bubbler tube via a regulator (pressure regulator, differential pressure regulator) or a purge rotameter. The bubble purge rates are very low and as long as the bubbles escape periodically, the system is functioning properly. The purge rate is generally 0.5 ft^3 of free air per hour (14 L/h).

The source of the compressed air may be a small air compressor, a compressed air tank, or a nitrogen gas cylinder. If electrical power cannot be provided, for whatever reason, the use of a compressed air cylinder is the most economical method; a standard scuba diving compressed air tank can supply 1 month of air and is easily refilled. The end of the bubbler tube must be set at a point of zero elevation or zero level measurement. For example, if the end of the bubbler tube is located 12 ft (3.6 m) below the surface of the water, the bubbler tube would indicate a water level of 12 ft (3.6 m). If the water level drops by 5 ft (1.5 m), the measured depth would be 7 ft (2.1 m). This system detects the water depth (pressure) by means of an appropriately calibrated gauge or manometer.

The bubbler system has the advantages of (1) simple design, (2) easy accessibility and little concern for the corrosion of the pressure sensing device, and (3) the ability to be installed at the bottom of the tank where accessibility is blocked or impossible. Figure 5.2-1 shows a bubbler system used in open-tank level measurement.

Ultrasonic Level Detector The ultrasonic system is used to monitor either the water level in a tank or dry material stored in a storage bin open to the atmosphere. The height of the water/material is measured by means of an acoustic pulse; the ultrasonic transmitter and receiver units are located above the maximum level of the

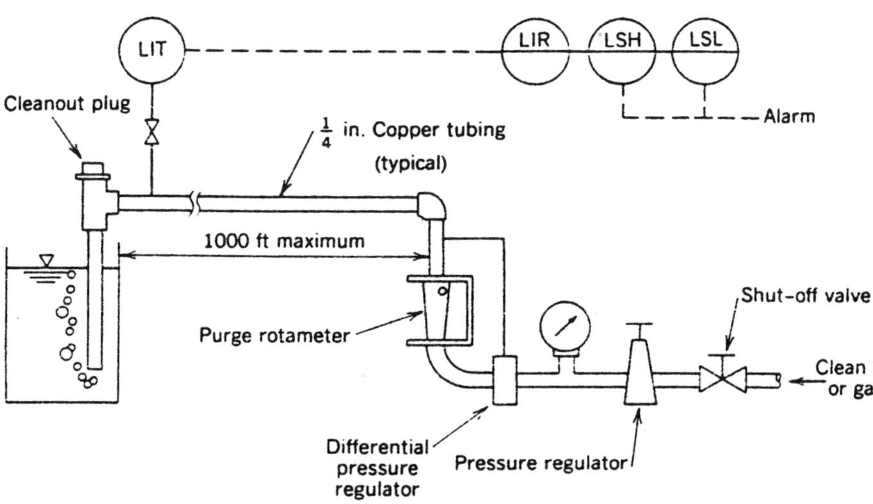

Figure 5.2-1 Bubbler tube system of tank level measurement.

object. The transmitter generates a pulse down through and perpendicular to the liquid level or surface of the dry material. The time elapsed between pulse generation and the detection of the reflected pulse energy is a function of the speed of sound through air. Since the speed of sound in air is dependent on temperature (0.18% per degree Celsius), use of a temperature correction factor is necessary for accurate measurement. Prior to selecting the level measuring device, the design engineer should be familiar with the following considerations: the required pressure range, required accuracy, equipment reliability, environmental conditions of the plant site, and operations and maintenance concerns.

BIBLIOGRAPHY

Andrew, W. G., and Williams, H. G., *Applied Instrumentation in the Process Industries,* 2nd ed., Vols. 1 and 2, Gulf Publishing, Houston, TX, 1980.

Babcock, R. H., "Instrumentation and Control in Water Supply and Wastewater Disposal," *Water Wastewater Eng.,* (1968).

Department of Civil Engineering, "Instrumentation, Control, and Automation for Water Supply and Wastewater Treatment Systems," *Proceedings of the 9th Sanitary Engineering Conference,* University of Illinois, 1967.

Knudsen, D. I., ed., *Process Instrumentation and Control System,* Manual of Practice OM-6, Water Pollution Control Federation, 1982.

5.3 VALVE SELECTION

Purpose The flow of liquid or gas in pipes is regulated by valves, whereas gates are used to regulate the flow of water from reservoirs, tanks, or channels. Valves are used much more frequently because they are capable of performing a multitude of important functions. Their primary functions include shutoff, throttle, prevention of backflow, or a combination of these functions.

To be able to select an appropriate type of valve, all design engineers must have a basic knowledge of the various types of commonly used valves, the characteristics of the fluid that is to be regulated, and the hydraulic characteristics of the piping system in which the valve is to be installed. Valves are manufactured out of a wide variety of materials to meet a wide range of requirements, and the engineer is responsible for specifying the appropriate valve type as well as valve materials. Engineers should also be aware that all large [over 12 in. (0.3 m)] valves, including gates, require adjustment after installation; without the necessary adjustments, most valves will leak to some degree. Design engineers should therefore include the final valve specifications.

Considerations Items that must be considered during valve selection are (1) the type of fluid or gas that is to be regulated, (2) the temperature of the system, (3) the flow range, (4) the pressure of the system, (5) valve function, (6) valve location,

(7) the type of valve operator, and (8) the reliability and cost of the valve. Each of these items is discussed.

Type of Fluid or Gas The characteristics of the liquid or gas are important when selecting the appropriate type of valve. If the liquid or gas is corrosive in nature, noncorrosive materials must be specified for all internal parts of the valve; type 18-8 stainless steel is typically selected. Valves used in ozone gas lines must be composed of type 316 stainless steel, and Teflon seats, not rubber, must be used. If a chemical slurry is used in the treatment process, the valve must not have internal recesses which may become filled with slurry solids. Finally, if abrasive matter is present in the liquid, the fluid passages must be composed of materials that are resistant to this type of erosion.

Temperature Although temperature is not a major issue when selecting the major valves of a water treatment process, it becomes a very important consideration when valves are used in conjunction with auxiliary equipment such as heating boilers and certain types of chemical feed systems—those that handle exothermic chemicals such as caustic soda and sulfuric acid. In these types of situation, valves ordinarily used in the water treatment process should not be employed because, due to thermal distortion, they do not perform well at operating temperatures above 150°F (65°C) unless special metal parts are specified.

Flow Range Flow range is an important issue when selecting throttling valves. Control of the flow range is extremely important because most throttling valves have a limited range in which they yield satisfactory performance. Unlike the throttling valve, flow range is not an important consideration for multijet valves. These valves are designed and manufactured to service a wide range of pressures and flow control without cavitation.

For the purpose of simple shutoff, the flow range is usually not a design consideration. However, most valves are unsuitable for service when water velocity exceeds 35 ft/s (10.5 m/s), based on the valve port area, and the engineer must therefore specify special instructions for valve construction.

Pressure The maximum differential pressure across the valve and the normal and the extreme line pressures should be known prior to valve selection. Extremely high pressures often result from hydraulic surges in the piping system, and selection of the valve operator should therefore be based on the maximum line pressure.

Valve Function Each type of valve has an optimal function. The valve function may be isolation of a line, drainage of a tank, prevention of backflow, reduction in pressure, or flow modulation. Thus the design engineer should specify the most appropriate type of valve for each application.

Valve Location Valves may be located in a variety of locations: in a valve vault, a pipe gallery, in the wall at the entrance of a tank, at the exit of a pipeline, buried in the ground, or submerged in the water at all times. There are also differences in valve construction for buried and regular types of valve. Furthermore, valves installed in the tank wall require only one flange rather than a double flange. Engineers must select the proper type of valve and specify the conditions for valve installation.

Valve Operator The design engineer must specify whether a valve is to be operated manually or by power. If a manually operated valve is selected, the type of operator (i.e., a wheel or a square nut with key) and the orientation of both the operator and stem supports must be specified. Certain types of valves, such as flow modulating, automatic sequencing, and remotely operated, necessitate the use of powered operators because manual power is insufficient.

Powered operators may be energized by means of electricity, compressed air, water, or oil. However, the first two options are most commonly used in waterworks. Strictly from the standpoint of reliability and maintenance, the use of air actuators is preferred during a power failure or during repair work because of its simplicity. However, an electrical motor offers better control over the modulating speed and is easier to incorporate into the instrumentation and control system.

Reliability and Cost The degree of valve reliability may play an important role in the control system, yet may be independent of the application. If a high degree of reliability is essential to the treatment process, the cost of the valve is unimportant. However, the relative costs of the various sizes and types of valve should generally be compared for each application. A good cost comparison study will include the cost of the valve itself as well as the projected maintenance costs and the cost of replacing equipment when necessary.

Types of Valves There are five basic types of valves: (1) slide, (2) rotary, (3) globe, (4) swing, and (5) multijet. The slide valve has a sliding disk that travels perpendicular to the flow direction. An example is a gate valve. Rotary valves operate by having a plug or disk move in a rotary fashion. The butterfly, ball, plug, and cone valves are classified as rotary type valves. Globe valves are used extensively in ordinary home plumbing fixtures. They have disks or plugs that move parallel to the flow direction. Globe valves do not cavitate under most household usage conditions.

The swing valve is comprised of a swing check valve, which prevents reverse flow. The swing check valve is fundamentally a combination of a rotary and a globe valve. The multijet valve, also known as a sleeve valve, is composed of two concentric pipes. Both the inner and outer pipes are covered with a multitude of small orifices. The inner pipe slides back and forth like a telescoping pipe, thus aligning or staggering the holes. An alternative type has perforation only in the outer sleeve. A third type of multijet valve has multiple orifices on the slide gate disk. The multijet valves became available in the early 1970s. They are used exclusively to reduce high pressure and to control flow rate without causing cavitation (under most conditions). Table 5.3-1 summarizes valve characteristics and the main valve applications commonly used in water treatment.

Flow Control Valve Considerations The design engineer must carefully select the most appropriate type of valve for each application; the use of an improper type of valve may result in cavitation, corrosion, or erosion and thereby degrade the structural integrity of the valve itself. Valve selection may be conducted in two steps: selection of the proper type of valve, followed by sizing.

Valve Selection To select the most appropriate type of valve for a given application, the design engineer must evaluate the pressure drop characteristics and the working range of the valves. Figure 5.3-1 presents examples of valve rangeability. Figure 5.3-2 shows the flow characteristics of certain types of valves when throttled. Engineers should not confuse rangeability with turn-down. *Rangeability* is the ratio between the maximum controllable flow and the minimum controllable flow rate. *Turn-down* is a ratio of the normal maximum flow rate versus the minimum controllable flow rate.

If a valve is to be used in a situation requiring water pressure control, valves characterized by equal percentage flow, such as the ball and butterfly valves, should be selected for ordinary cases—where there is a normal pressure drop of at least 15% but less than 30%. If a higher pressure drop, such as 50%, is expected, a valve with linear characteristics (plug or multijet/sleeve valve) should be specified. In situations requiring a single valve to handle a pressure drop of over 50%, the sleeve valves perform effectively in most applications without cavitation. For applications requiring the control of liquid levels, a valve with linear characteristics, such as the plug valve, is the most appropriate type in most cases.

Equal-percentage valves are most appropriate under the following conditions: (1) the process is fast acting; (2) the situation requires high rangeability; (3) the dynamics of the system are not well known; and (4) in the case of heat exchangers, an increase in the production rate requires a correlative and greater increase in the volume of the heating or cooling medium.

Linear valves should be used (1) for slow-acting processes and (2) when there is a system pressure drop of over 40% across the valve. Quick-opening valves should be used (1) for on–off control and (2) when maximum valve capacity must be obtained immediately. The quick-opening valve exhibits linear characteristics for approximately one-fourth of its travel from shutoff. Beyond this point, the quick-opening valve is useful only in an on–off function.

Valve Sizing The sizing of control valves involves the application of basic sizing formulas, evaluation of the process fluid characteristics, and use of basic data for the valves. Some design engineers fail to take the size of the control valves into consideration and end up with the wrong-size valve. Nevertheless, system control by these valves is achieved to an acceptable degree most of the time; this is possible because of the high rangeability of the control valves and the flexibility of the controller mode and because the available pressure drops are often greater than the design points. However, optimum system control may be achieved only if the control valve is capable of handling both the minimum flow and a flow that is over 100% (120%) of the maximum flow rate.

Another factor that should be considered when sizing the control valves is the capacity factor (Cv). Cv is defined as the number of gallons of water [at room tem-

TABLE 5.3-1 Characteristics and Applications of Common Valves

Type	Standard Size	Application	Rated Pressure (psi)	Remarks
Gate valves ANSI/AWWA C 509-87 for gate valves AWWA C 501-87 for sluice gate	2–48 in. (5–120 cm)	• On/off control but does not have a driptight shut-off • Not suitable for flow rate control by throttling	150 (10.5 kg/cm^2)	• This class includes wedge, double disk gate valves, and sluice gates • Debris deposits in the seats will cause valve closure problems • Full size openings allow pipe cleaning pigs to pass through • The valves often become inoperable when left in an open or closed position for extended periods of time • Double disk type should only be installed vertically • Slow opening and closing speeds minimize water hammer • Requires a large space for installation
Butterfly valves ANSI/AWWA C 504–87 "A" valves have a rated maximum line flow velocity of 8 fps; "B" valves, up to 16 fps	4–90 in. (10–230 cm)	• On/off control with drip-tight shut-off • Limited degree of flow control by throttling	75 and 150 (5.3–10.5 kg/cm^2)	• Easy to operate and at a low cost • Minimal maintenance requirements • Requires a small space for installation • Quick closing causes water hammer problem • Applicable to both air and liquid • Water should not contain any large debris • Does not allow the pipe cleaning pig to pass through

Ball valves ANSI/AWWA C 507–85	$\frac{1}{2}$–48 in. (1.3–120 cm)	• Driptight shut-off • Limited degree of flow control by throttling	Up to 300 available (21 kg/cm^2)	• Full size openings allow pipe cleaning pigs to pass through • Very little head loss when fully opened. • The size of the valves is often necked down to $\frac{7}{10}$ of the size of the connecting line • Services clean water as well as slurry • Large size valves are heavy
Cone valves	6–24 in. (15–60 cm)	• On/off control with non-driptight closure • Limited degree of flow control by throttling	Up to 300 available (21 kg/cm^2)	• Full size openings allow the pipe cleaning pigs to pass through • Operation of the valve begins to lift the plug off the seat before rotation of the cone • The valve operator is rather complicated and expensive • Should not be installed in an upside-down position • The port consists of a tapered plug or a full-area circular port
Plug valves	$\frac{1}{2}$–36 in. (1.3–90 cm)	• On/off control with driptight closure • Often used as a flow control valve • Mild throttling is acceptable	120–170, depending on the size (8.4–12 kg/cm^2)	• Plug valves are suitable for service of dirty water and slurry as well as clean water • Lubricated and eccentric plug valves are available • Lubricated type comes with cylindrical or tapered plugs with a diamond shaped port for throttling application • Lubricated types are easily operated, provided that they are frequently used; however, the plug type tends to "freeze" if is not operated for an extended period of time

TABLE 5.3-1 *(Continued)*

Type	Standard Size	Application	Rated Pressure (psi)	Remarks
				• Lubricated types require periodic lubrication with grease • Nonlubricated types are easier to operate and require less maintenance • Eccentric plug valves have rectangular ports whose area equals 80% of the pipe area • Eccentric valves may be installed in any position; however, the shaft must be kept horizontal and the plug must remain in the upper part of the valve body when the valve is opened
Globe valves	1–16 in. (2.5–40 cm)	• On/off with a high pressure drop and driptight closure • Accurate throttling and flow control	150 but up to 300 available (10.5–21 kg/cm^2)	• Disk globe, angle valve, Y-valve, needle valve, and altitude valves are in this class • Application to clean water only • The disk design is used for manual control and the needle valves are used for accurate and automatic control • Most all household valves are disk globe valves
Check valves ANSI/AWWA 506–78 and ANSI/AWWA 508–82	2–60 in. (5–150 cm)	• Prevents reversal of flow	150 (10.5 kg/cm^2)	• Lift, swing, and ball checks are the basic types of check valve • Internal spring-loaded types are called "silent" checks • Lift checks are often used with globe valves for both liquid and gas. Angle lift check

				valves are used for vertical line; horizontal lift types are used for horizontal lines • Swing checks are often used with gate valves and may be installed in either a horizontal or vertical position • Ball checks are used for air vent, vacuum valves, and chemical metering pumps—on both the suction and discharge sides • Headloss across the check valves is significant
Diaphragm valves	$\frac{1}{2}$–24 in. (1.3–60 cm)	• Driptight shut-off for corrosive chemicals, slurry, and sewage	150 but up to 300 available (10.5–21 kg/cm^2)	• The valves are used in lime slurry and PAC slurry lines • Intermittent operation may cause leakage • Poor flow control characteristics • Has a low cost but a high capacity
Multijet valves	8–72 in. (20–180 cm)	• Flow control with a high pressure drop across the valve without cavitation	150 (10.5 kg/cm^2)	• Both in-line and submerged discharge types are available • Has a history of good performance; emits low noise and has minimal maintenance requirements • Limited number of valve manufacturers • These valves are expensive

Notes: (1) ANSI/AWWA C 540–87 standard for power-activated devices (operators) for valves and sluice gates.
(2) ANSI/AWWA C 550–81 standard for protective interior coating for valves and hydrants.

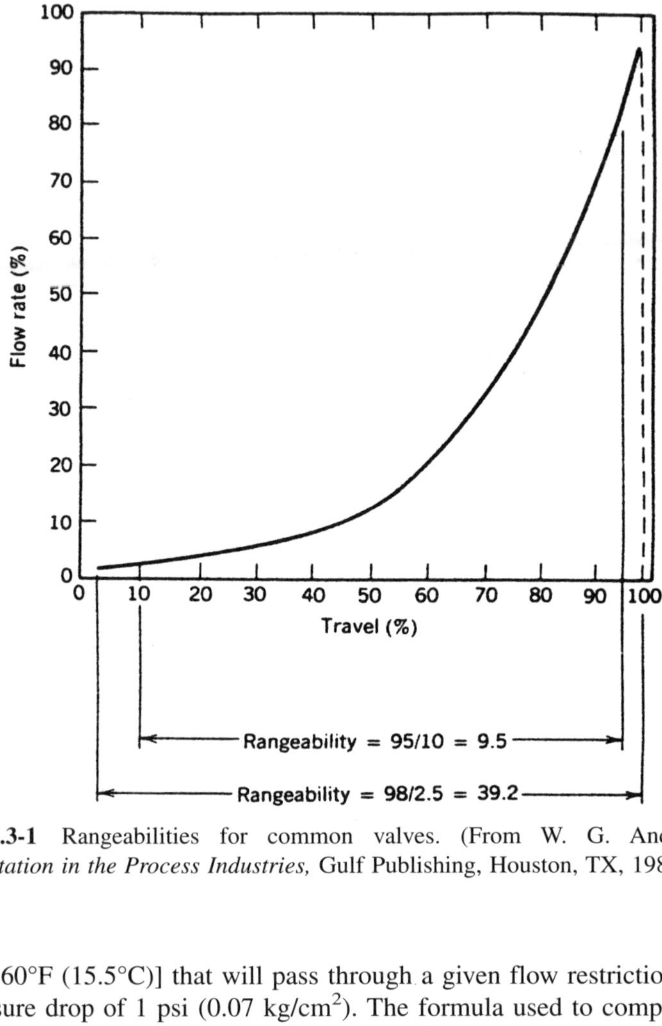

Figure 5.3-1 Rangeabilities for common valves. (From W. G. Andrew, *Applied Instrumentation in the Process Industries,* Gulf Publishing, Houston, TX, 1980.)

perature, 60°F (15.5°C)] that will pass through a given flow restriction per minute at a pressure drop of 1 psi (0.07 kg/cm^2). The formula used to compute Cv is established by the Fluid Controls Institute, Inc. (FCI), and the formula for noncompressive fluids such as water is

$$\text{Cv} = Q\left(\frac{G}{\Delta P}\right)^{0.5} \quad \text{(volume based)}$$

where Cv is the capacity factor (gpm), Q the flow rate of the liquid (gpm), G the specific gravity of the liquid (1.0 for water), and ΔP the differential pressure drop across the valve (psig).

$$\text{Cv}\phi = \frac{\text{Cv}}{D^2}$$

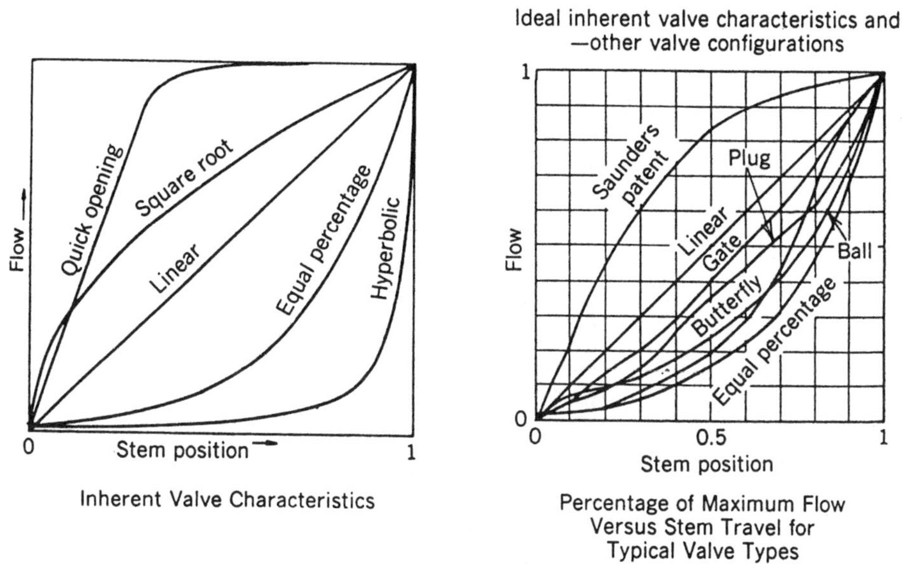

Figure 5.3-2 Flow characteristics for common valve. (From W. G. Andrew, *Applied Intrumentation in the Process Industries,* Gulf Publishing, Houston, TX, 1980.)

where Cvϕ is the capacity factor through a 1-in. (2.5-cm) valve for a given valve angle with a pressure drop of 1 psig (0.07 kg/cm^2) and D is the pipe diameter in inches. Table 5.3-2 compares the capacity factor (gpm) for several types of valves. Most major valve manufacturers will provide the Cv values of their products.

The control valve is generally selected to open only 25 to 60% at the designed flow rate. The following general guides should be used in valve selection:

1. To select the appropriate size valve, begin by comparing the value of Cv at both maximum and minimum flow conditions.
2. Next, select a valve size that is capable of handling the maximum Cv when the valve is 85 to 90% open, and the minimum Cv when the valve is opened only 10 to 15%.

Cavitation Considerations *Cavitation* is defined as the formation and subsequent collapse of vapor-filled cavities resulting from the dynamic action in a liquid. The cavities may be bubbles, vapor-filled pockets, or a combination of the two. Dissolved gases are often liberated shortly before vaporization begins.

Cavitation causes noise and vibration, and if it is severe, both the disk and body of the valve will become pitted. Under these unfavorable conditions, cavitation may become so severe that the butterfly valve (flow modulator) must be replaced every few months. The design engineer must make sure that the valve selected for flow control does not cavitate under critical conditions, that is, when the valve is at a

TABLE 5.3-2 Capacity Comparison of Valve Body Types

	Cv Factor Comparison for Various Valve Types by Size (in.)										
Valve Type	$\frac{1}{2}$	$\frac{3}{4}$	1	1.5	2	3	4	6	8	10	12
Single-port globe	5.0	6.4	11.0	25.0	48.0	95	147	324	436	987	1180
Double-port globe			13.0	32.0	53.0	121	190	438	765	1320	1760
Split body		6.4	11.0	25.0	48.0	93	140	307			
Angle			16.5	38.2	68.7	152	293	492	766	1030	
Sauders	4.0	8.5	8.5	31.0	55.0	180	305	515	1130	1450	2350
Cage	6.5	14.0	21.4	38.0	67.2	150	235	460			
Butterfly (60°)			15.9	38.8	68.0	152	280	638	1163	1843	2679
Vee-ball					99.0	254	536	1040	1700	2690	3930

high throttling position. The cavitation index (σ) is calculated from the following equation:

$$\sigma = \frac{H_d - H_v}{\Delta H + v^2/_{2g}}$$

where σ is the cavitation index, H_d the static head at the valve exit (absolute feet), H_v the vapor pressure of the liquid, ΔH the pressure drop across the valve (ft), and v the velocity through the pipe, which is directly connected to the valve (ft/s).

NOTES:

1. The barometric pressures, expressed as feet of water column, at given ground elevations are as follows:

33.9 ft at sea level	10.332 m at sea level
32.8 ft at 1000 ft elevation	9.997 m at 305 m
31.6 ft at 2000 ft elevation	9.631 m at 610 m
30.5 ft at 3000 ft elevation	9.296 m at 914 m
29.4 ft at 4000 ft elevation	8.961 m at 1220 m
28.3 ft at 5000 ft elevation	8.625 m at 1524 m

2. The absolute water vapor pressures, expressed as feet of water column, are as follows:

40°F (4.4°C)	0.28 ft (0.085 m)
50°F (10°C)	0.41 ft (0.125 m)
60°F (15.6°C)	0.59 ft (0.180 m)
70°F (21.1°C)	0.83 ft (0.253 m)
80°F (26.7°C)	1.18 ft (0.360 m)
90°F (32.2°C)	1.61 ft (0.491 m)
100°F (32.8°C)	2.21 ft (0.674 m)

The criteria for valve cavitation are as follows:

$\sigma > 2.5$–3	In general, cavitation will not occur and no particular noise will develop.
$2.5 < \sigma > 1.5$	There is a development of noise; "hammer blows" and a steady roar persist.
$\sigma < 1.5$	Under these circumstances, the pipe and valve are apt to vibrate and cavitation develops. When σ reaches 0.5, cavitation will be fully developed.

Figure 5.3-3 illustrates a multijet (sleeve) valve. This type of valve is designed and used as a throttling valve in situations with a high pressure drop in the system. The multijet valve does not develop cavitation until the cavitation index approaches 0.2. This figure also compares the cavitation index of various types of multijet valves.

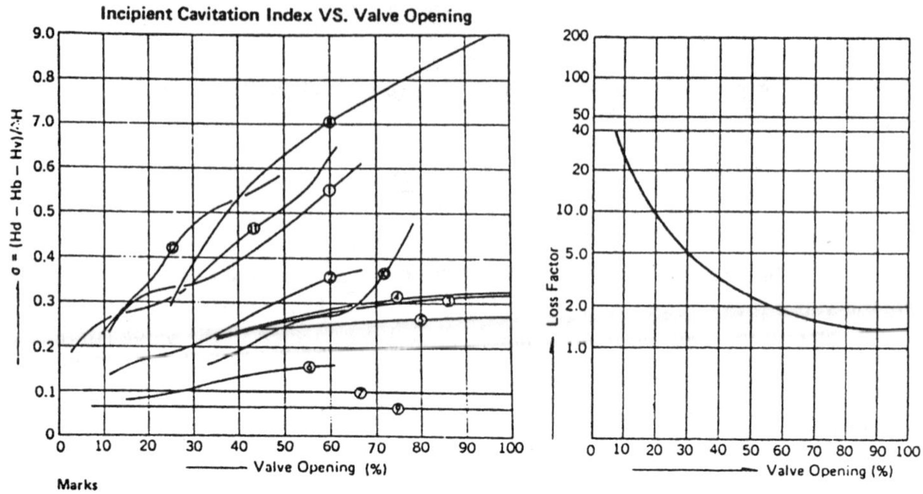

Marks

(1) Conical Plug Valve With Skeleton Plug.
(2) Conical Plug Valve With Skirted Plug.
(3) Globe Valve With Serrated Disc Ring.
(4) Globe Valve with V-Ported Sleeve.
(5) Globe Valve With Perforated Sleeve.
(6) Needle Valve Diacharging Into Expansion Tank.
(7) In-Line Multiple Jet Sleeve.
(8) Globe Linear Valve.
(9) Vertical Multi-jet Sleeve Valve.
(10) Lubricated Plug Valve
(11) Butterfly Valve
(12) Gate Valve

$$\text{Cavitation index } (\sigma) = \frac{Hd + Hb - Hv}{\Delta H}$$

where Hd = Downstream pressure (feet)
Hb = Barometric pressure (feet)
Hv = Vapor pressure (feet)
ΔH = Head drop across valve (feet)

Loss-Factor Table

Note: The loss factor varies with the number of ports, so please consult us before designing piping.

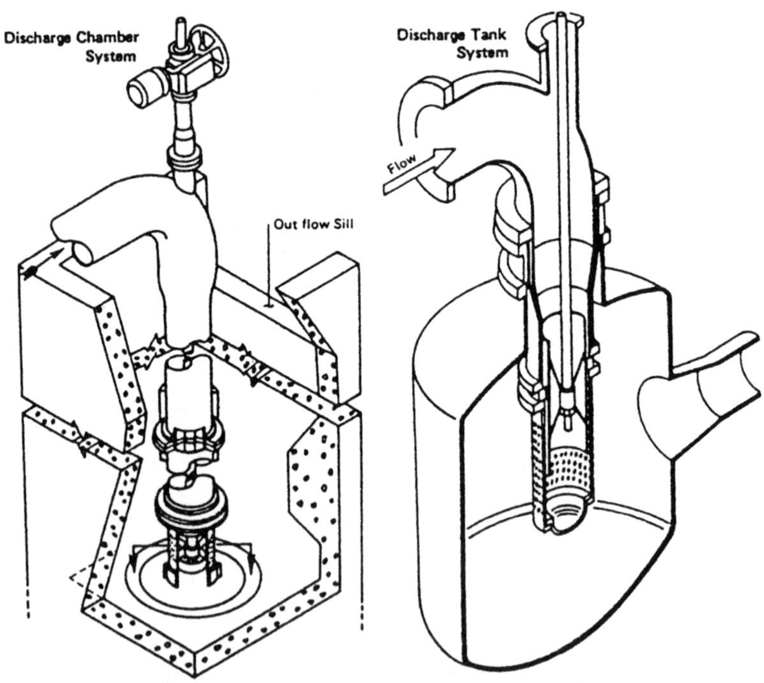

Figure 5.3-3 Multijet (sleeve) valve and cavitation index of values. (From Bailey Valve brochure.)

Example Design Calculations

Example 1

Given A 36-in. (0.9-m) pipeline connects a water reservoir to a water treatment plant. The difference in static water level is 30 ft (9.1 m) A butterfly valve is selected to control the planned flow rate from 31.5 to 5 mgd (1.38 to 0.22 m^3/s. The pipeline elevation, with the venturi flow tube and the flow control valve, is 15 ft (4.6 m) below the water surface at the head of the plant. Assume that the headloss in the line is 14 ft (4.3 m) at the maximum flow rate and 0.5 ft (0.15 m) at the minimum flow rate. The plant elevation is 1000 ft (305 m) above sea level and the ambient temperature is 80°F (27°C).

Determine The proper-size flow control valve, and evaluate the cavitation situation

Solution *Step 1.* Select a 36-in. (0.9-m) butterfly valve, which is the same size as the pipeline. The capacity factor Cv at the maximum flow rate of 31.5 mgd (1.4 m^3/s) is

$$\mathrm{Cv} = Q\left(\frac{G}{\Delta P}\right)^{0\ 5}$$

$$= (31.5 \times 695)\left[\frac{1}{(45 - 14 - 15) \div 2.31}\right]^{0.5} = 8318$$

$$\mathrm{Cv}\sigma = \frac{\mathrm{Cv}}{D^2} = 8318 \div 36^2 = 6.4$$

Based on the graph in Figure 5.3-4, the disk angle of the valve is approximately 35°. For the purpose of flow control, the recommended disk angle of the butterfly valve is generally 15 to 60°, and 35° is therefore an appropriate valve.

The Cv at the minimum flow rate of 5 mgd (0.22 m^3/s) is

$$\mathrm{Cv} = (5 \times 695)\left[\frac{1}{(30 - 0.5) \div 2.31}\right]^{0.5} = 972$$

$$\mathrm{Cv}\sigma = \frac{\mathrm{Cv}}{D^2} = 972 \div 36^2 = 0.75$$

In this case the disk angle is approximately 9°, which is a very throttled position. This condition is unacceptable since it is less than the recommended angle of 15°.

Step 2. Since the rule of thumb is to size the flow control valve one size smaller than the size of the line, begin by selecting a 24-in. (0.6-m) valve. The placement of a pipe size reducer and an increaser on the ends of the 24-in. valve will produce

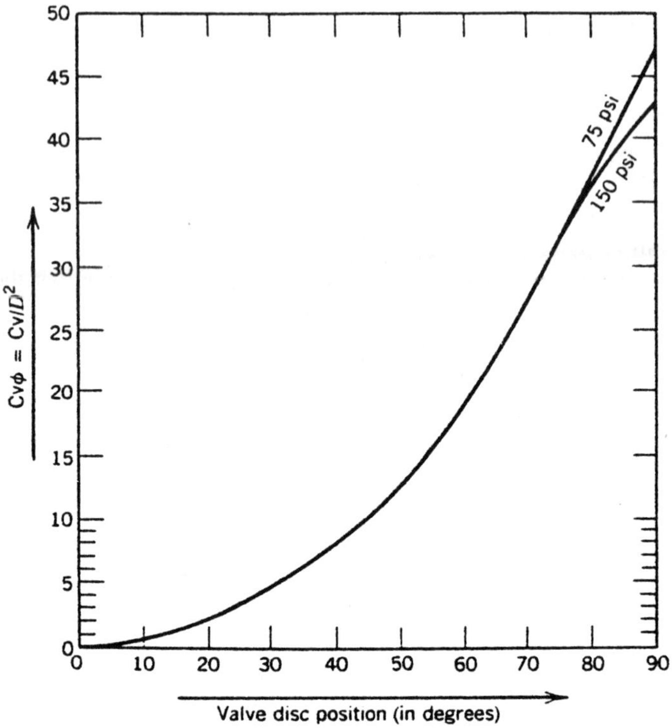

Figure 5.3-4 Unit capacity factor (Cv/D^2) versus approximate disk angle of butterfly valve.

extra headloss. Although the headloss from the reducer is negligible, the increaser will yield a headloss of approximately 1 ft (0.3 m). The pressure drop across the valve is therefore 45 − 14 − 16 = 15 ft or 6.5 psi (0.45 kg/cm^2):

$$\text{Cv} = (31.5 \times 695)\left(\frac{1}{6.5}\right)^{0.5} = 8587$$

$$\text{Cv}\sigma = 14.9$$

The disk angle of the valve is 53°, which is within the accepted range.

The pressure drop across the valve for the minimum flow rate is practically the same as in the preceding trial.

$$\text{Cv}\phi = 972 \div 24^2 = 1.7$$

The valve is open 16°. This valve is on the borderline, but still within the proper range. Now check the cavitation index (σ) when the 24-in. (0.6-m) valve is in the full throttle position. The hydraulic head at the valve exit is 15 − 1 = 14 ft (4.3 m) and the barometric pressure at an elevation of 1000 ft is 32.8 ft; thus

$$H_d = 14 \text{ ft} + 32.8 \text{ ft} = 46.8 \text{ ft } (14.3 \text{ m})$$

The water vapor pressure at 80°F is 1.18 ft (0.36 m); the flow velocity through the valve inlet is 2.46 ft/s at 5 mgd; and the pressure drop across the valve is calculated to be 45 − 14 = 31 ft (9.45 m):

$$\sigma = \frac{H_d - H_v}{\Delta H + v^2/2g} = \frac{46.8 - 1.18}{31 + [(2.46)^2 \div 64.4]} = 1.47$$

This value is definitely in the cavitation range. The important decision-making factors are how often and how long this 5 mgd (0.22 m^3/s) flow requirement occurs. If this low-flow condition is infrequent and short in duration, the 24-in. (0.6-m) butterfly valve may be used with the knowledge that the cavitation condition exists and that the valve disk will cavitate every time. However, a plug valve or even a multijet valve should be considered if this low-flow condition occurs frequently and if the duration of the 5-mgd (0.22-m^3/s) flow is expected to last throughout most of the winter season.

BIBLIOGRAPHY

Andrew, W. G., and Williams, H. B., *Applied Instrumentation in the Process Industries,* 2nd ed., Vols. 1 and 2, Gulf Publishing, Houston, TX, 1980.

Benjes, H. H., et al., "How to Select Valves for Water and Wastewater," *Water Wastewater Eng.,* 5:62 (May 1968).

Daneker, J. R., "Sizing Butterfly Valves," Parts 1 and 2, *Water Wastes Eng. Ind.,* O-10 (July 1970) and E-12 (September 1970).

Edwards, J. A., "Valves, Pipes, and Fittings," *Pollut. Eng.,* p. 22 (December 1974).

Hutchinson, J. W., *ISA Handbook of Control Valves,* 2nd ed., Instrument Society of America, Research Triangle Park, NC, 1976.

5.4 PIPING SYSTEM

Purpose There are six basic piping systems in a water treatment plant: (1) raw water and finished waste distribution mains; (2) plant yard piping that connects the unit processes; (3) plant utility, including the fire hydrant lines; (4) chemical lines; (5) sewer lines; and (6) miscellaneous pipings, such as drainage and irrigation lines. The cost of the total plant piping system generally runs 5 to 7% of the total plant construction cost depending on the layout, site conditions, and winter temperatures. Careful attention must be paid to the design of the piping system because if the pipes in the main processes fail, the entire plant will be shut down for an indeterminate length of time.

Considerations The design considerations for the piping system are the function of the specifics of the system. However, all piping systems share a few common issues: The pipe strength must be able to resist internal pressure, handling, and earth and traffic loads; the pipe characteristics must enable the pipe to withstand corrosion and abrasion and expansion and contraction of the pipeline (if the line is exposed to atmospheric conditions); engineers must select the appropriate pipe support, bedding, and backfill conditions; the design must account for the potential for pipe failure at the connection point to the basins due to subsidence of a massive structure; and the composition of the pipe must not give rise to any adverse effects on the health of consumers.

Raw Water Line Depending on the site specifics, the raw water line may be either a gravity flow or a pressure line. Factors that can adversely affect the functioning of the water line are the growth of clams and slime in the interior of pipes; accumulation of sediment within the pipeline; air blockage at high points of the pipeline; presence of a vacuum condition within the pipeline since this can lead to collapse of the pipeline; creation of a thrust force by the water flow, especially during flow surges; and the expansion and contraction of exposed pipeline resulting from changes in ambient temperatures.

Many case histories can be found to illustrate the problem concerning the growth of aquatic organisms in the raw water lines and how the carrying capacity of such pipelines can be drastically reduced. If there is a possibility that these conditions could arise, a temporary or permanent potassium permanganate or chlorine feed system may be incorporated into the piping system design, depending on the anticipated severity of the problem and DBP formation.

Since the majority of pipelines seldom have a constant pipeline grade from the intake to the plant, the air release and air vacuum valves should be installed at the high points of the pipeline and a blow-off assembly should be provided at the low points. If they must pass through areas with poor ground conditions, the pipelines must be extensively supported, for example, by the use of pile supports. Alternatively, the design may allow an adequate amount of flexibility to occur at the pipe joints.

Whenever a pipeline must cross a geological fault, the design engineer is faced with additional problems. The recommendations of geotechnical experts must be included in the design of the pipeline. Yet there is one basic design rule: The design should allow the pipeline to undergo a sufficient degree of movement; therefore, flexible joints should be installed on either side of the fault and an articulated configuration (Z-shaped pattern) should be used. Figure 5.4-1 presents two types of flexible joints that can endure a certain degree of pipe displacement. The Dresser coupling provides for a deflection of approximately 6°; Flex-Tend by EBBA Iron and TAI-Flex by Taisei-kiko can tolerate a deflection of 15° and an expansion of 4 in. (10 cm).

Plant Yard Piping The plant yard piping is a low-pressure system. The main design concerns are the flow velocity in the pipeline, allowable headloss, connections to the structures, physical strength of the pipes (in the case of deep installations), corrosion of the pipes, and cost of the system. Flow velocity is an important consideration because of its effect on the fragile alum floc contained in the process water. If the flow is too fast, the alum floc will be destroyed; if the flow velocity is

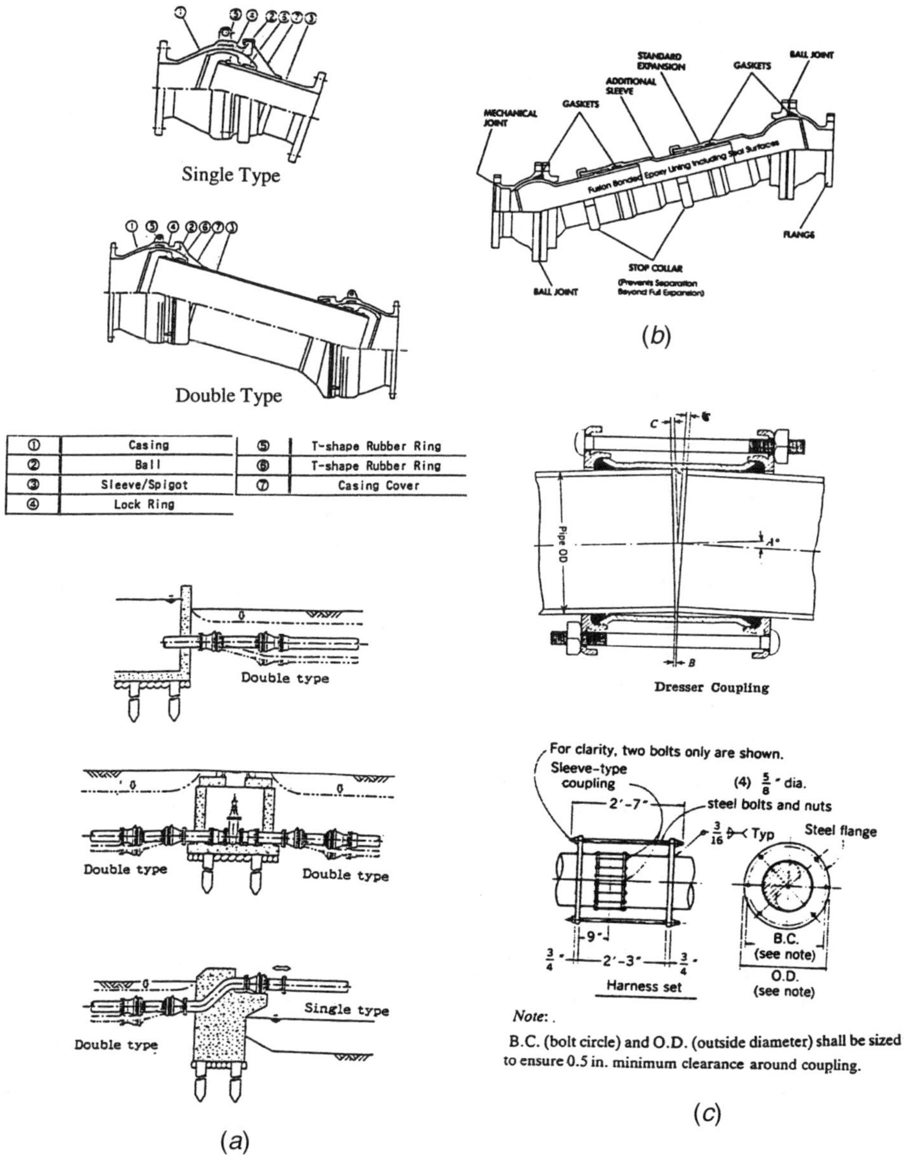

Figure 5.4-1 Flexible pipe joints and applications: (*a*) Tai-Flex joints (courtesy of Taisi Kiko); (*b*) Flex-Tend joint (courtesy of EBBA Iron); (*c*) Dresser coupling.

too slow, sludge will settle in the pipelines. The suggested flow velocity for each connecting pipeline (among treatment unit processes) is listed in Table 3.2.2-1.

Another important design issue is the pipe connection to the structures and tanks. Many pipe breakdowns occur within 6 ft (1.8 m) of the structure wall, due to differential settlement between the buried pipeline and the structure. The solution is either to allow a rotating action to occur at the point where the pipe meets the wall through the use of a bell and spigot type of joint or to allow the pipeline to flex a couple of degrees by installing one or two pipe couplings in the pipeline adjacent to the wall. Nonetheless, sleeve couplings have a limited degree of allowable deflection; if the angle of deflection is greater than 3 to 4°, water may begin to leak from the pipe. Figure 5.4-2 presents several examples of these designs. The Flex-Tend joint (Figure 5.4-1) is capable of tolerating three types of movement—rotation, deflection, and stretching or contracting—in one assembly, thus eliminating the need for a joint harness assembly.

The yard piping may have to be laid at a depth that is much greater than the standard; the standard depth is dictated by vehicle wheel loading. Additionally, the yard pipings are buried in twos and threes in a wide-open trench. In both these cases, the thickness of the pipe is determined by the earth load and not by the internal water pressure; thus the engineer must be careful when specifying the pipe thickness. Furthermore, the design engineer should specify the proper bedding conditions and details regarding backfill to avoid pipeline failure.

There are cases when large trench excavations are required, due to a maze of piping or because the pipes must be laid alongside large buried structures—to which they must connect. These types of yard piping conditions should be considered as wide trench or embankment conditions. Under these circumstances, the pipes may be protected from high external loading through the use of three basic methods: (1) reinforced-concrete encasement, (2) reinforced-concrete cradles, and (3) heavy-walled pipe. Due to the generally congested nature of the yard pipings, the most common choice is the installation of heavy-walled pipes that are specifically designed for embankment conditions.

Based on Marston's formula, the external loading resulting from backfill on the pipe under trench and embankment conditions is as follows. For the trench condition:

$$W_d = C_d W B_d^2$$

where W_d is the fill load (pounds per linear foot of pipe), C_d the load coefficient (a function of H/B_d and K_μ; refer to Appendix 19), W the unit weight of fill material (lb/ft^3), B_d the width of the trench at the level of the top of the pipe (ft). For the embankment conditions:

$$W_c = C_c W B_c{}^2$$

where W_c is the load per unit length of pipe, C_c the load coefficient (a function of H/B_c and $T_{sd}P$; refer to Appendix 20), W the unit weight of fill material, B_c the external diameter of the pipe (ft).

Pipe corrosion may occur internally and/or externally. The use of cement-lined cast iron and steel pipes provides adequate protection against internal corrosion since the process water of most water treatment plants is generally not highly corrosive. However, certain soil characteristics are very corrosive to the pipes. For this reason, the geotechnical study of the site must include soil resistivity measurements. If the soil is "hot," the engineer should obtain expert recommendations. In situations where

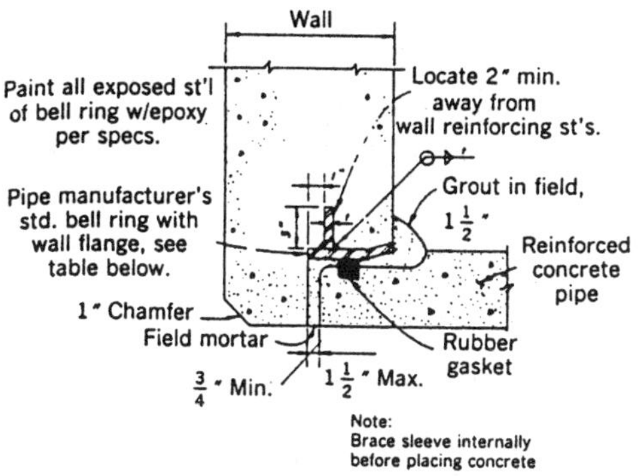

Bell ring and wall flange plate thickness

Pipe size	Steel plate
30" & under	$\frac{1}{4}$"
31" to 60"	$\frac{5}{16}$"
61" & over	$\frac{3}{8}$"

Reinforced concrete pressure pipe terminating in wall

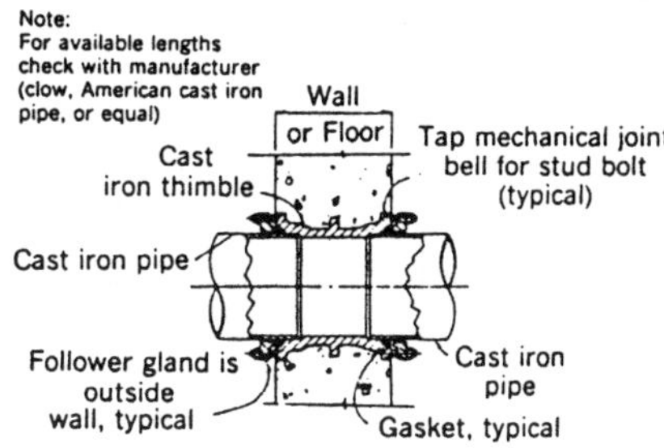

Mechanical joint cast iron thimble

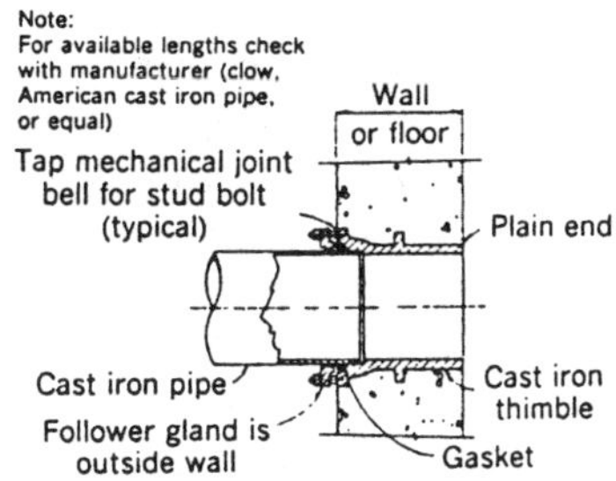

Mechanical joint and plain end Cast iron thimble

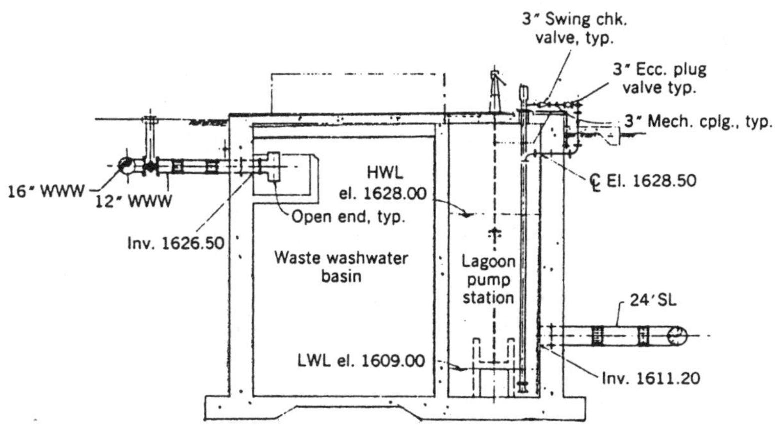

Figure 5.4-2 Typical methods of pipe connections to a structure.

corrosion poses a serious threat, the use of cathodic protection (as a backup to the protective coatings) should be considered seriously for steel and cast iron pipes.

Since the internal pipe pressure is relatively insignificant for yard pipings, reinforced concrete pipes and steel cylinder concrete pipes should also be evaluated along with the cast iron or cement-lined and cement-coated steel pipes. This recommendation considers the problem of corrosion, the need for flexibility at the pipe joints, the pressure imparted by the heavy earth, and the cost of the system.

Plant Utility and Water Distribution Lines These pipelines are high-pressure lines that must normally withstand pressures of 75 psi (5.3 kg/cm^2) or more. The economic velocity of the distribution main is approximately 5 ft/s (1.5 m/s). However, a flow velocity of about 7 ft/s (2.1 m/s) is quite common for utility lines carrying potable water, as well as utility water (nonpotable) and lines servicing fire hydrants.

There are four specific issues that greatly affect the design of the piping system: (1) extreme caution must be exercised when installing pipelines near existing sewer lines, (2) cross connections with the chlorine injector actuation line and other utility lines (not always potable water) must be avoided, (3) a joint harness must be specified when using a Dresser type pipe coupling so that the pipe will not slip out from the coupling, and (4) pipe material that will not degrade the quality of the potable water must be selected.

In most states, the Department of Health Services requires the water main and the sewer line to be physically separated. The basic rules set by regulatory agencies are as follows. The horizontal separation between the water line and the drain or sewer line should be 10 ft (3 m) if the water main must cross a sewer line or storm drain. The water line should be laid at an elevation that allows the bottom of the water main to be 18 in. (0.45 m) above the top of the drain or sewer.

Regulatory agencies also dictate the type of cross connections that may be used between potable and nonpotable water lines; this includes steam condensate, the cooling water from the engine jacket or other heat exchanger, and pipe connections to chemical tanks or injectors. If a direct connection is absolutely necessary, an approved type of backflow prevention valve must be placed in the pipeline or an air gap of at least 6 in. (0.15 m) must be provided in all connections to chemical storage tanks. The installation of surge tanks is often required at the beginning of the distribution main whenever high-service pumps are used to pump potable water into a lengthy distribution main. The surge tanks function to protect the pipeline and pumps.

Dresser couplings are commonly used in the high-service pump discharge lines. Whenever sleeve couplings such as the Dresser coupling are used in high-pressure lines, the joint harness should be capable of resisting the unbalanced thrusts at the elbows or to tolerate the line pull caused by, among others, hydraulic thrusts. A mechanical type of coupling such as the Victaulic coupling does not require that a harness be installed; however, it does not allow the pipes to undergo much movement.

The recent EPA Drinking Water Quality Standards set stringent rules with respect to the allowable maximum concentration level of many items. Certain types of pipe can release undesirable components, such as pipe material or coating material, into the potable water. Of major concern are asbestos cement pipes and pipes coated with bituminous material. Although lead pipes were used primarily in servicing residential homes in the past, engineers should be aware that small lead pipes pose a threat to the health and safety of the public. All new pipe designs should avoid specifying the aforementioned types of pipe when establishing the potable water supply line.

Chemical Feed Lines The main concerns surrounding chemical feed lines are corrosion, abrasion, formation of scale, and heat resistance. When plastic pipes are exposed to the sun, they can potentially deteriorate as a result of ultraviolet rays and weathering. Moreover, these pipes should have a proper type of pipe support to minimize pipe sagging between the supports, especially when plastic pipes are used.

Selection of an appropriate type of pipe material depends on the type and solution strength of each chemical that will flow through the pipe; refer to Table 4.1-1 and Appendix 13. Any chemical that is fed in slurry form requires a minimum flow velocity of 5 ft/s (1.5 m/s), and if the chemical feed line is exposed to the atmosphere, the pipes must be insulated from freezing and direct sunlight; certain chemical solutions tend to crystallize when the ambient temperature is well above 32°F (0°C). When employing scale-forming chemical solutions, the pipe design must have a minimum number of sharp turns and should provide cleanouts and a water flushing line.

If acid, alkali, oxidizing, and organic compound chemical lines are installed in a pipe chase, double-walled chemical feed pipes must be used, as dictated by local OSHA regulations or other codes. This practice will minimize the potential for contamination of soil and fire or explosion in the event that a pipe ruptures. Similarly, from a safety standpoint, chlorine should be fed to the application point as chlorine gas (under negative pressure) rather than as a strong solution under pressure, unless the feed line is short and double-walled pipe is used.

Miscellaneous Pipings Included in the category of miscellaneous pipings are sewer, drain, storm drain, and irrigation pipelines. If these pipelines are designed to enter any building, the uniform plumbing codes must be adhered to strictly. Should these pipelines be laid outside, the following materials may be used, depending on the conditions of the site: cast iron, salt-glazed vitrified clay, concrete, or PVC. With respect to sewer lines, the standard sewer line design should be followed when establishing details such as the size, flow velocity, minimum slopes, and access manhole requirement (refer to the *Ten States Standards for Sewage Works*).

Piping Materials The most commonly used materials for water supply are cast iron, steel, concrete, and PVC. In the past, asbestos cement pipes were employed extensively; however, in June 1989 health concerns over the detrimental effects of the asbestos fibers caused the EPA to ban their use.

Cast iron pipes have a certain resistance to corrosion. The standard sizes range from 2 to 24 in. (5 to 60 cm) in diameter. Most larger cast iron pipes are cement lined. The Hazen–William C value may be as high as 145 for new pipes larger than 8 in. (0.2 m) in diameter. The disadvantages of this type of pipe are its slightly brittle nature, its heaviness, and its rather high cost, particularly for large pipes. Yet ductile cast iron pipes may be manufactured as large as 78 in. (2 m) in diameter due to their ductility. However, the length of large pipes must be limited to reduce the weight of the pipes, and this is achieved by using a large number of pipe joints. Mechanical joints are generally used in these types of circumstances rather than the bell and spigot and the flange joints normally associated with cast iron pipes (ANSI/AWWA C110 to C151-86 for cast iron and ductile iron pipes and fitting).

Steel pipes are commonly used when large and relatively small pipelines as well as nonstandard large fittings are required. In most cases, both the interior and exterior of these pipes are lined with cement for protection against corrosion. Both longitudinal and spiral welded steel pipes are available for use. The joints between the pipes are usually constructed by welding or through the use of a bell and spigot with rubber gasket, sealed flanges, sleeve couplings, or mechanical couplings. The advantages of steel pipe include thinner walls, lighter weight, and in general, lower cost when large-diameter pipes are specified. Yet this type of pipe also has several disadvantages: its inability to carry high external loads, its tendency to collapse when subjected to negative pressures, and the additional expense of providing corrosion protection measures such as cathodic protection (ANSI/AWWA C200 to C213-85, steel water pipe standards).

Concrete pipes are available in various forms: precast, reinforced concrete, prestressed with steel bars, and steel cylinder pipes. These pipes have low maintenance costs, are resistant to corrosion under normal conditions, and are capable of withstanding external loads. However, the concrete pipes also have some undesirable features, such as a short pipe length due to the weight, free lime being leached from the concrete, tendency to crack under pressure and cause leaking, permeability of the concrete, and corrosion in the presence of strong acids (ASTM C14, C76, C361 and AWWA C300 to C303, specifications for concrete to modified concrete pipes).

Plastic pipes are used primarily as chemical solution, drain, sewer, and irrigation lines. The standard specifications for plastic and glass fiber–reinforced pipes may be found in ANSI/AWWA C900 to C950-81, and the thickness design may be found in the ANSI/AWWA standards piping handbook or the reference data and/or catalogs provided by the pipe manufacturer. The recommended pipe trench dimensions and details concerning pipe support, bedding, compact fill around the pipes, and backfill may also be found in the aforementioned publications and various other publications.

BIBLIOGRAPHY

ACPA, *Concrete Pipe Design Manual,* American Concrete Pipe Association, Vienna, VA, 1987.

AWWA, *PVC Pipe: Design and Installation,* Manual M-23, American Water Works Association, Denver, CO, 1980.

AWWA, *Steel Pipe: A Guide for Design and Installation,* Manual M-11, American Water Works Association, Denver, CO, 1985.

Bealey, M., and Lemons, J. D., eds., *Concrete Pipe and the Soil–Structure System,* STP 630, American Society for Testing and Materials, Philadelphia, 1977.

IAPMO, *Uniform Plumbing Code,* 18th ed., International Association of Plumbing and Mechanical Officials, Los Angeles, 1988.

King, R. C., and Crocker, S., *Piping Handbook,* 5th ed., McGraw-Hill, New York, 1973.

5.5 STANDBY POWER SUPPLY

Purpose All water treatment plants should be operated in a reliable and cost-effective manner. Power outages will undoubtedly occur for a number of reasons, including damage to the power generation system or power supply system as the result of accidents such as earthquakes, fires, floods, tornadoes, hurricanes, lightning, or even sabotage. Certain equipment, those of the monitoring and control systems, require a continuous power supply. This may be generated by a standby emergency power supply unit.

The standby unit also offers a monetary benefit for water treatment plants using high-service pumping stations with large electric motors. Variations in water demand affect pumping requirements, which in turn affect the power demand. Since the rate structure or amount the utility charges for power is based in part on anticipated peak demand, a standby power unit may be used to reduce the base rate by peak shaving.

Considerations

Power Outage It is important to investigate the reliability of the local power supply system over a period of at least 5 to10 years. Issues that should be studied prior to the design work are the frequency of the power outages and the duration of each episode. The type and scope of the standby power supply system are dependent on several factors: (1) the frequency of power outages, (2) the duration of outages, (3) the amount of essential equipment, (4) the total power load required by the essential equipment, (5) the types of fuel available for the power generator, (6) fuel cost, and (7) the availability of alternative power supplies from other substations in the vicinity of the plant.

Water treatment plants using raw water pumps will have a different-sized standby power supply system than that of plants receiving raw water from the source via gravity. The standby power system is installed in substantially greater numbers for the latter type of treatment plant. This is because the raw water keeps flowing into the plant regardless of whether the power is on or off. Thus the treatment process units must be capable of processing the water based on standby power. Although the plant flow rate and power consumption can be reduced, the plant flow may not be allowed to be reduced if the water source is shared by more than two or three users, even in the event of a power failure. Such a reduction would result in a significant increase in pressure or a massive overflow at the transmission line. In most cases, the daily flow cannot be changed without giving advanced 24-hr notice to the agency controlling the source. For any given situation, the minimum capacity of the standby unit should be capable of providing a power supply that will operate 25% of the designed total capacity of the plant. Yet specific local conditions as well as the characteristics of the water supply system can dictate the minimum supply quantity.

The following is a list of equipment whose functions must essentially be maintained during a power outage:

1. All equipment associated with the instrumentation system and the major units of the control system.
2. The computer system; all data stored in the RAM of the system will be lost during the time between the loss of power and the startup of the standby generator; thus it is crucial to have an uninterrupted power supply.
3. The emergency lighting of all site structures.
4. Electrically operated valves.
5. Major chemical feeders such as coagulant and disinfectant feeders.

6. Chlorinator booster pumps, leak detectors, and emergency fans.
7. Flash mixer.
8. Twenty-five to 50% of the mechanical flocculators.
9. Sampling pumps.
10. Sump pumps.
11. A designated number of raw water and treated water pumps; less than 50% of the units (for each type) must generally be functioning.
12. Certain laboratory equipment: refrigerators, incubators, compressors, vacuum pumps, room lights, and other essential analyzers.
13. The hot-water circulating pumps of the heating system.

Depending on the duration of the power outage, other major process units, such as the filter backwash equipment, may or may not require the services of a standby power unit. If a reliable power supply system can be obtained from other electrical substations in the vicinity, the treatment plant may opt to use a dual power supply system rather than a standby generator.

Peak Power Shaving The economic evaluation of the various peak shaving alternatives is influenced by the rate structures of the utility companies. The four basic treatment plant operational modes used to achieve peak shaving are:

1. *Load shedding.* Significant load shedding may be accomplished by reducing the load through the implementation of a computer or microprocessor-based controller.
2. *Reduction of the peak utility demand.* This method involves the transfer of part of the power load from the utility to the engine generator supply during periods of high electricity demand (peak).
3. *Paralleling with utility.* In this scenario, the power generator is connected directly to the normal distribution system. Energy required during peak power consumption is supplied directly to the system and the amount of power furnished is determined by the engine governor control.
4. *Prime power cogeneration.* A prime power system is a system in which the on-site generator is the prime source of power. Only certain dedicated loads may be connected to the on-site generator, and the utility may be used as a standby source through the use of an automatic transfer switch.

Natural gas or diesel engine generators are generally used to supply the standby power necessary to maintain the essential equipment during power outages; under these circumstances, the second peak shaving method is normally considered. The startup and shutdown functions of the engine generator should be fully automatic. However, a time delay of approximately 60 s should be provided before the generator reaches the proper speed, to differentiate extremely short-term power outages resulting from lightning and other causes from long-term power failures.

CHAPTER 6

Elements of Detailed Designs

6.1 PLANT HYDRAULICS

Purpose One of the major design efforts of a water treatment plant is the in-plant hydraulics. Since each water treatment unit process is functionally dependent on hydraulic factors, a hydraulic analysis must be conducted during the design phase because it plays a crucial role in assuring overall plant function and the efficiency of each treatment process.

Considerations Obtaining an even flow distribution to parallel process units is an extremely important hydraulic consideration. It is rather amazing to find that a number of existing treatment plants have problems with uneven flow distribution: among sedimentation basins, between the treatment process modules, and even among the filters. This has resulted in the presence of significant variations in flow distribution. Differences as high as 47% have been observed among process units. A sample computation for sizing the inlet channel may be found in Example 1 of Section 3.2.3 and "Choice of Flocculation Tank Inlets" in Section 3.2.4.

Another common hydraulic problem is the occurrence of severe flow short-circuiting in the tanks of process units such as flocculation and sedimentation as well as the clearwell and disinfection contact tank. Flow short-circuiting in sedimentation tanks has been studied extensively. These studies show that only 20 to 30% of the mean detention time (tank volume divided by the flow rate) may be considered to be the actual flow-through time. Figure 3.2.5-3 and Appendix 9 present some examples of flow short-circuiting resulting from density flow and improper design of the inlet diffuser wall.

Many college engineering courses teach that selection of the pipings connecting two separate tanks should be based on the most economical size that is capable of carrying a particular flow rate from point A to point B using computer analysis. However, practical and operational water treatment experience indicates that the proper size is determined by two factors: the required flow velocity, as dictated by the process, and the extreme operational conditions of the plant. For instance, all pipes or channels designed to transport floc to the sedimentation tank must be sized neither to break the fragile floc nor allow floc settling. If the pipe is used to carry sludge, a minimum flow velocity of 2 ft/s (0.6 m/s) must be maintained to prevent the sludge from settling.

Another hydraulic consideration is the maximum and minimum plant flow rate. Should one side of a module be taken off line, the remaining side is required to carry nearly twice the designed flow rate and an extremely high flow will occur. Thus all pipes and channels should generally be sized at 150% of the designed flow rate (maximum daily flow rate). If this scheme is followed, the output of the treatment plant may easily be increased by 50% (i.e., filtration rates of 4 to 6 gpm/ft^2) at some future date. The raw water main and the main distribution lines are generally sized to handle future plant flow rate, in anticipation of plant capacity expansion after construction of the original plant.

The required available headloss across an ordinary conventional water treatment plant averages approximately 16.5 ft (5 m). However, an advanced treatment plant with preozonation, postozonation, and GAC adsorption processes requires an available head of approximately 25 ft (7.5 m): available head is defined as the difference between the water surface elevation in the tank at the head of the plant where the plant influent flows in, and the high water level in the clearwell at the designed plant flow rate.

The flow hydraulics through granular media beds, including filter beds, have been studied extensively and may be found in various textbooks and technical journals. However, it is important to note that most inexperienced design engineers tend to spend many hours calculating the headloss through a filter bed at various filtration rates and water temperatures, determining the results up to two decimal points. Yet this entire process is quite unnecessary; the engineer actually only needs to calculate a conservative approximate figure, which may be assumed from similar existing filter beds or pilot study data. Furthermore, design engineers should realize that the headloss across the filter is a function of many factors that come into play after the media is in place: changes in media gradation after numerous backwashings, changes in media size due to attrition loss and the loss of the fine grains, formation of mud balls within the bed, and changes in the depth of the filter bed resulting from media loss.

Generally, the filtration cycle requires a filter design with over 8 ft (2.4 m) of available headloss across the bed. Consequently, a difference of ± 3 in. (75 mm) in the initial headloss does not have a significant impact on the filtration process from an operational standpoint. The same analogy may be applied to computations on backwash hydraulics and filter bed expansion. The calculation of headloss through the filter underdrain orifices is another example where practical experience saves the expenditure of a lot of unnecessary time and effort. The engineer should not spend time evaluating the orifice coefficient when performing hydraulic calculations because most filter underdrains are covered with a gravel bed. Consequently, over 30% of the orifices do not have a clean orifice area. Thus the most conservative coefficient (0.65) should be used for the calculations.

The hydraulics of water troughs, gullets, channels, conduits, orifices, weirs, pipes, and fitting are not discussed in this book because they may be found in a number of other textbooks.

Example Design Calculations

Example 1

Given A 10-mgd (0.44-m^3/s) water treatment plant has two flocculation/sedimentation tanks in parallel and four filters. A 20-in. (0.5-m) venturi flow meter and a

20-in. butterfly valve are located just upstream of the head of the plant. Each flocculation tank has two 14-in. (0.36-m) butterfly valves (as the inlet), and each of the two baffle walls (one per tank) produces a 0.1-ft (0.3-cm) headloss across the wall. Each sedimentation tank is equipped with two launders that are 35 ft (10.7 m) long with a grand total of 280 V-notches (90°) to decant the settled water. Each filter has an inlet weir that is 4 ft (1.2 m) in width and a 14-in. (0.36-m) inlet butterfly valve as shown in Figure 6.1-1. The required water elevation in the filters is elevation 100.

Determine The hydraulic grade line between the filter and the inlet to the 20-in. (0.5-m) Venturi flow meter at a flow rate of 10 mgd (0.44 m^3/s)

Solution *Step 1.* Determine the water level control points in the process train. In this case, the filter inlet weir and the V-notched weirs in the sedimentation tanks are the level control points. Both weirs should be set at an elevation that will allow a free-fall condition (not submerged) to be maintained, resulting in a uniform flow rate at each weir.

Step 2. Compute the headloss or required head over the weirs between the level control points. Use a computation sheet (tabulation form as shown here) so that the values may easily be checked. Refer to Appendix 14 for the proper K factors and the basic hydraulic formulas that are applicable to this problem.

(a) *Between the Filters and Settled Water Channel* Refer to Table 6.1-1a for the calculations.

(b) *Between the Settled Water Conduit and the Head of the Plant* The proper size of each launder in the sedimentation tank can be determined quickly by using Figure 3.2.7-16. Since each launder carries 2.5 mgd (1740 gpm), the width should be 1.5 ft (0.46 m) and the height should be specified as 1.67 ft (0.5 m). Set the bottom elevation of all launders at an elevation of 104.75. Each V-notch on the launders should have a free discharge flow. Thus the elevation of the V-notch should be set to give 106.50 water elevation in the tank. Since each tank has a total of 280 of a 3-inch height V-notched weirs, the flow rate of each 90° V-notched weir is

$$(10 \times 695) \div 560 = 12.4 \text{ gpm or } 0.0277 \text{ ft}^3/\text{s}$$

Since 0.0277 $ft^3/s = 2.54H^{2.5}$, $H = 0.164$ ft or 1.97 in.

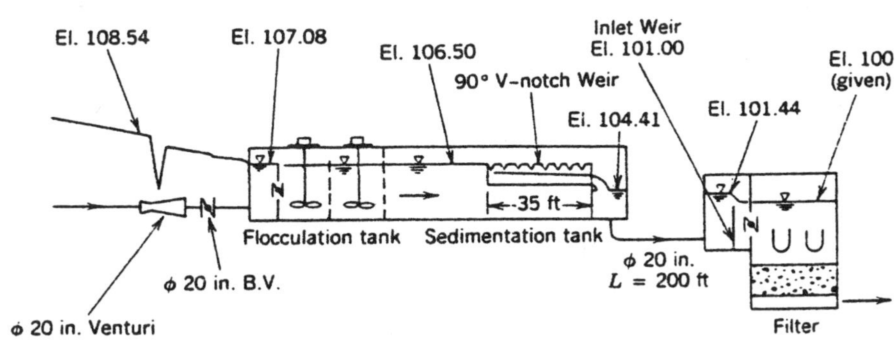

Figure 6.1-1 Process train and hydraulic grade line.

TABLE 6.1-1 Calculations for Example 1

(a) Between the Filters and the Settled Water Conduit

Item	Q	Diameter	v	$v^2/2g$	K	Friction Loss L	Friction Loss Δh	L.O.H.	Water Elevation
Water level in the filter cell									100
14 in. BV	2.5 mgd	14 in.	3.62	0.2	0.35	—	—	0.07 ft	100.07
Inlet hole	2.5 mgd	14 in.	3.62	0.2	1.6 (C = 0.8)	—	—	0.32 ft	100.39
Inlet weir*	2.5 mgd = 3.97 cfs = 3.33 × 4 $H^{1.5}$, where H = 0.45 ft							0.45 ft	101.44
Filter inlet channel								Negligible	101.44
20 in. pipe exit	10 mgd	20 in.	7.09	0.78	1.0	—	—	0.78 ft	102.22
Pipeline	10 mgd	20 in.	—	—	—	200 ft	0.9 ft/100 ft (C = 120)	1.8 ft	104.02
Pipe inlet	10 mgd	20 in.	7.09	0.78	0.5	—	—	0.39 ft	104.41
Water level in the settled water conduit								Negligible	104.41

*Set the crest of the weir at 101.00 to ensure a free discharge condition.

(b) Between the Settled Water Conduit and the Head of the Plant

Item	Q	Diameter	v	$v^2/2g$	K	L.O.H.	Water Elevation
Water level in the settled water conduit							104.41
Water level over the V-notched weirs* on the launders						0.164 ft	106.50
Sedimentation tank						Negligible	106.50
Diffuser wall (sedimentation tank inlet)						0.1 ft (given)	106.60
Baffle wall (flocculation tank)						0.1 ft (given)	106.70
Inlet BV	2.5 mgd	14 in.	3.62	0.2	0.3	0.06 ft	106.76
Inlet hole	2.5 mgd	14 in.	3.62	0.2	1.6	0.32 ft	107.08
Inlet channel						Negligible	107.08

*The crest of the V-notch is set at an elevation of 106.34.

(c) Inlet of the Venturi Meter to the Inlet Channel to the Flocculation Tanks

Item	Q	Diameter	v	$v^2/2g$	K	Friction Loss L	Friction Loss Δh	L.O.H.	Water Elevation
Water level in the inlet cell									107.08
20 in. Pipe exit	10 mgd	20 in.	7.1 fps	0.78 ft	1.0	—	—	0.78 ft	107.86
20 in. BV (full open)	10 mgd	20 in.	7.1 fps	0.78 ft	0.25	—	—	0.20 ft	108.06
20 in. Venturi†	10 mgd	20 in.	—	—	—	—	—	0.34 ft	108.40
Pipeline	10 mgd	20 in.	—	—	—	15 ft	0.9 ft/100 ft (C = 120)	0.14	108.54

†The Venturi differential is 80 in. (6.67 ft) at 10 mgd and the headloss across the Venturi is approximately 5% of the differential.

(c) *Inlet of the Venturi Meter to the Inlet Channel to Flocculation Tanks.* See Table 6.1-1c.

Refer to earlier chapters for examples of hydraulic calculations on baffled walls in basins, density flow and flow stability (Froude number) of sedimentation tanks, and the basic hydraulics of filter structures. An example of the hydraulic grade line across a plant is illustrated in Figure 6.1-2.

BIBLIOGRAPHY

Benefield, L. D., et al., *Treatment Plant Hydraulics for Environmental Engineers,* Prentice Hall, Englewood Cliffs, NJ, 1984.

Chao, J. L., and Trussell, R. R., "Hydraulic Design of Flow Distribution Channels," *J. Environ. Eng. Div. ASCE,* 106:2:321 (February 1980).

Davis, C. V., *Handbook of Applied Hydraulics,* 3rd ed., McGraw-Hill, New York, 1970.

Hudson, J. E., *Water Clarification Processes: Practical Design and Evaluation,* Van Nostrand Reinhold, New York, 1981.

Huisman, L., *Sedimentation and Floatation,* Delft University of Technology, Delft, The Netherlands, 1973.

Hydraulic Institute, *Engineering Data Book,* Hydraulic Institute, Cleveland, OH, 1979.

Karassick, I. J., et al., *Pump Handbook,* McGraw-Hill, New York, 1976.

King, H. W., Krutzsch, W. C., Fraser, W. H., and Messina, J. P., *Handbook of Hydraulics,* McGraw-Hill, New York, 1976.

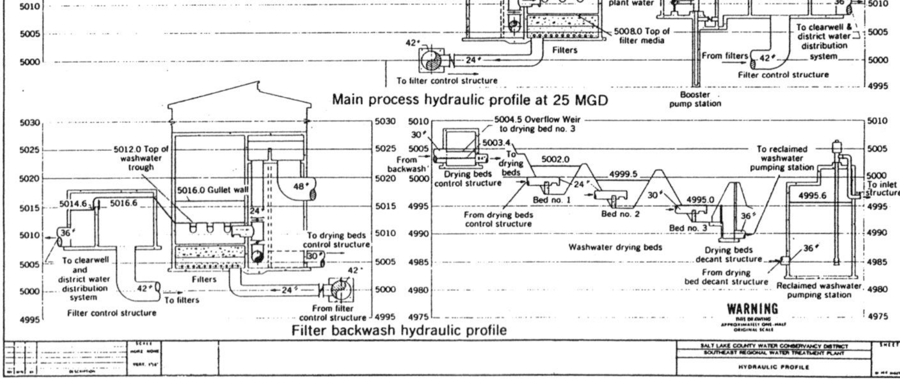

Figure 6.1-2 Example of hydraulic profile.

Miller, D. S., *Internal Flow Systems,* Vol. 5 in the BHRA Fluid Engineering Series, BHRA Fluid Engineering, London, 1978.

Montgomery, J. M., Consulting Engineers, *Water Treatment: Principles and Design,* Wiley, New York, 1985.

Shaw, G. V., and Loomis, A. W., *Cameron Hydraulic Data,* Compressed Air Magazine Co., Ingersoll-Rand, Phillipsburg, NJ, 1965.

Yao, K. M., "Hydraulic Control for Flow Distribution," *Proc. Sanitary Eng. Div. ASCE,* 2:275(April 1972).

6.2 SPECIFICATIONS

Clear and precise specifications containing drawings are a mandatory part of any modern project. Vague or erroneous specifications result in inferior-quality products, in addition to a large number of change orders and delays in the project. Moreover, they are a potential source of lawsuits. General guidance on the various types of engineering specifications are available from several agencies, including the Construction Specification Institute, American Consulting Engineers Council, American Society of Civil Engineers, National Society of Professional Engineers, American Institute of Architects, and the General Services Administration.

Project engineers should write the specifications in a manner that is both clear and convenient for the contractors since the specifications are written for the contractors who must furnish all the materials and equipment, build the structures, perform the civil works, provide and assemble the components, and finish the electrical work. For instance, a system such as ozonation or chlorination contains many subsystems; thus the system specifications should include all the components and subsystems under one heading. This frees the contractor from having to search through several other divisions and sections to find the specifications for each subsystem. If the specifications are not organized in this manner, the contractor will waste precious time when formulating a bidding price for the project and may even overlook some components and subsystems.

Finally, project engineers must be aware that the specifications of certain projects must be tailored to meet the particular needs of the system. Although the master and standard specifications certainly save time, minor adjustments are always required in most treatment plant constructions.

Contents of Specifications The specifications are generally comprised of two major parts: the administrative requirements and the detailed requirements of the work to be performed. Some major items that must be considered by the engineer are the size of the project, the duration of construction, and the nature of the project—whether it is a privately or publicly financed project. The outline of the specifications should contain the following items:

- Bid advertisement (invitation to bid)
- Information to bidders

- Form of the proposal
- Form of the contract agreement
- Bond forms
- Contract revisions
- Standard specifications
- Master specifications
- Special provisions/conditions
- General provisions
- Technical specifications

Bid Advertisement It is standard practice for public agencies to publicly advertise all public works projects. Sealed bids are received through advertising in various news media for a required period of time. After the bids are opened, publicly read aloud, tabulated, and evaluated, the lowest bidder is selected.

Information to Bidders It is customary to issue a set of drawings and specifications to all prospective bidders for a reasonable charge.

Form of the Proposal In most cases, the proposals are accompanied by a guarantee in the form of a certified check or surety bond. This practice ensures that the successful bidder will enter into a contract; the guarantee is returned when the bid is awarded. However, if the lowest bidder fails to execute the contract, the certified check is forfeited as liquidated damages and the obligations of the surety (under the bond) are enforced as compensation to the owner for the cost of awarding the contract to the next bidder or for the additional cost of re-advertising. If the project is financed by private owners, it is not necessary to obtain competitive bids, and negotiations are often conducted with selected contractors.

Types of Contracts For both public and privately financed works, the two basic types of contract are the unit process and lump sum. Negotiated and special contracts are also possible.

Unit Price Contract The unit price contract should be used when it is impossible to define the exact limits of the various work items encompassed in the contract and/or bid documents. For payment purposes, the work is broken down into major segments with respect to the type of work and trade involved. Each segment is designated as a payment item (estimated quantities) and is listed in the proposal. Bidders are requested to submit a written bid for each unit. For example, a specific number of cubic yards of concrete will be bid at a certain unit price per cubic yard. The total bid is the sum (in dollars) of all items listed in the proposal and is the basis by which all received bids are compared to determine the lowest bidder. Payment to the contractor is determined by the actual quantity measured for each item incorporated into the work at the contract unit price.

Lump-Sum Contract The lump-sum contract is selected when it is possible to accurately delineate, through drawings, the limits of the work encompassed in the contract. With this type of contract, the bidder makes quantity takeoffs and presents them as the basis for the bid. However, it is essential that the bidder be provided with drawings and specifications that are extremely detailed with respect to all features and requirements of the work. Payment is made in accordance with the lump-sum bid to cover all work and services required by the drawings and specifications.

Negotiated Contract Private work contracts often take the form of negotiated contracts. Several different payment methods are available: lump sum or unit price, fixed fee plus reimbursable costs (with a ceiling price), fixed fee plus reimbursable costs (no ceiling), reimbursable costs plus a percentage of the cost, and a construction-management contract. Furthermore, incentives such as a bonus for economic construction and/or completion ahead of schedule may be added. Conversely, the contractor is penalized for inefficiency and/or failure to meet the completion date.

Special Contract The use of a special contract may be applicable in specific circumstances such as the procurement and installation of highly specialized equipment. In this case, the owner simply invites proposals from a select group of prequalified contractors without advertising the project. However, contracting procedures are generally followed for contracts procured through public bidding.

Contract Revisions There are four basic types of contract revision: (1) addenda, (2) stipulations, (3) change orders, and (4) supplementary agreements.

Addenda Revisions to contract documents, made before bids are received from contractors, are classified as addenda. The revisions are usually changes to the contract drawings and specifications resulting from errors, omissions, or the need for clarification. Bidders should be notified of any changes in the bid opening dates by means of an addendum. All addenda should be delivered sufficiently in advance of the bid opening date to all persons who have picked up the bid documents so that they may adjust their proposals. In turn, the bidders must acknowledge receipt of all addenda; otherwise, their bids will not be accepted.

Stipulations Stipulations are issued to the bidder when the contract is executed. They contain modifications of the contract terms as proposed by the owner.

Change Orders This is a written order to the contractor authorizing a change in work from the original drawings and specifications. It usually increases or decreases the quantity of work to be performed. Thus a payment adjustment should be made accordingly. The change order must be approved by the owner and signed by the engineer. Good design work will have a total adjusted payment of only 2 to 3% of the total contract price resulting from change orders. If the total cost of the changes exceeds a specified percentage (such as 20%) of the total contract price, a supplementary agreement, acceptable to both the owner and the contractor, should be executed before proceeding with the affected work.

Supplementary Agreements This written agreement encompasses all modifying work that is beyond the scope and terms of the contract or work changes that are within the scope of the contract yet exceed a set percentage of the original price. The supplementary agreement must be signed by both the owner and the contractor.

Standard Specifications The standard specifications are often preestablished by public agencies to ensure uniformity in administrative procedures and in the quality of the facilities. These criteria list the specific requirements for all materials and work quality. The specifications are updated or amended periodically and published. Although in some instances the standard specifications may be used without any changes, they generally require minor modifications and additions. The assembled modifications and additions are referred to as supplementary specifications, special previsions, or special conditions.

Master Specifications Many consulting engineering firms provide master specifications that are used for projects owned by either private or local agencies. These specifications are written in-house and pertain to general items. When the master specifications are applied to a project, the engineer should delete requirements that are irrelevant to the particular project. This practice helps to reduce the time required to write the contract specifications, as well as serving as a checklist, thereby minimizing errors and omissions.

General Provisions The general provisions define the rights and responsibilities of both the owner and the contractor to the construction contract, surety, and requirements, as well as delineating the legal relationships and the authority of the engineer. The general provisions are usually comprised of the following sections:

- Definitions and abbreviations
- Bidding requirements
- Contract and subcontract procedures
- Scope of the work
- Control of the work
- Legal and public relations
- Damage claims
- Workers' compensation insurance
- Comprehensive general liability of the contractor
- Laws, ordinances, and regulations
- Responsibility for work
- Explosives
- Sanitary provisions
- Public safety and convenience
- Property damage

- Public utilities
- Prosecution and progress
- Time of completion
- Suspension of work
- Unavoidable delays
- Annulment and default of contract
- Liquidated damages
- Time extensions
- Measurement and payment
- Payment
- Termination of contractor responsibility
- Guaranty against defective work

These sections are often referred to as *boilerplate.*

Technical Specifications The common types of technical specifications pertain to (1) materials and work quality, (2) material procurement, and (3) performance. Each of these is discussed.

Materials and Work Quality Specifications The material and work quality specifications are included in most construction projects. They cover general and specific conditions related to the performance of the work, material requirements, construction details, quantity measurements, and the criteria used for determining payment.

Materials Procurement Specifications These are usually written for special situations such as accelerated construction schedules, for projects requiring the use of materials that have a long lead time, and for projects involving several general construction contracts. This type of technical specification ensures that the materials are available to the job site on time. Furthermore, it ensures uniformity in the materials. The technical specifications should clearly delineate all elements concerning materials, work quality, and the mode and time of delivery. Examples are filter media (i.e., anthracite and filter sand), filter underdrains, special valves, and certain precast concrete items.

Performance Specifications The performance specifications are divided into two categories. One type specifies the water treatment unit process or an entire water treatment plant. These generally include a basic site plan, the maximum required hydraulic loading rate for each unit process for the product water, a guaranteed water quality limit (required) from each unit process, the time of completion, and other miscellaneous basic requirements. This type of performance specification is commonly employed in most European and Asian countries.

The second type of performance specification concerns the prepurchase of equipment by the owner prior to signing a construction contract; this is often practiced in the United States. Certain types of equipment must be prepurchased due to their long lead time or because the owner is determined to obtain a particular type of equipment that may not be purchased through competitive bidding procedures. The performance specifications should therefore define the quality, function, and all requirements pertaining to the water treatment equipment, in addition to the required tests, affidavits, and other supporting evidence of compliance. It is vital that the test conditions and procedures be clearly defined to prevent conflict and/or future lawsuits.

Construction Specification Institute (CSI) The purpose of the CSI is to organize and improve construction specifications. It was established by architects, engineers, and certain manufacturers. The CSI recommends that companies use the CSI format for all water treatment projects. Their format consists of 16 divisions, with several sections per division pertaining to specific construction activities. The CSI format is widely issued in the construction of buildings and public works and for product specifications released by manufacturers. The CSI also publishes a *Manual of Practice,* which contains a great deal of information on writing specifications. Instruction documents and project manuals such as *Specification Language, Specification Writing and Production, Performance Specifications, Methods of Specifying,* and *Bidding Requirements* are valuable reference items for all project engineers.

BIBLIOGRAPHY

Abdallah, E. T., Preparing Specifications for Design-Bid-Build Project, ASCE Manual, Jan 2000.

Goldbloom, J., and White, J. J., "Specifications," in Section 3 of *Standard Handbook for Civil Engineers,* 3rd ed., F. S. Merritt, ed., McGraw-Hill, New York, 1983.

White, J. J., "Specifications," in Section 3 of *Standard Handbook for Civil Engineers,* Hicks, T. G. editor. McGraw-Hill, New York, 1972.

6.3 FINAL COST ESTIMATES

Cost estimations for a project is generally required in three stages: feasibility, preliminary design, and the design completion phase. Each of these phases would be expected to have a respective level of accuracy of ± 45, ± 30, and $\pm 7\%$. Cost estimates conducted during the early stages of a project are often used to prepare funding for the project budget through the use of cost estimation curves and by adjusting the cost to a geographical area as described in Section 2.4. In this section we discuss the final cost estimate for bid evaluation.

The final cost estimate is not merely a simple summary of the anticipated costs of various items based on quantity takeoffs, extensive cost data, and statistics. Many other factors are an integral part of the final cost estimate: the complexity of the project, site and climatic conditions, job site locations, judgment of the cost estimator with respect to economic trends, dynamism of the construction industry, and the extent to which the project attracts contractors. Professional cost estimators generally require 2 to 5 work-hours per sheet of construction drawings to complete an estimate.

If the project engineer is required to make the final construction estimate, he or she should be aware of three characteristic weaknesses of most engineers: (1) the tendency to devote too much time to the design aspects and not enough to cost estimation, (2) the tendency to focus on direct costs and to neglect the necessary indirect costs incurred by the contractor, and (3) the omission or insufficient allotment of funds to cover specific business costs—risks, escalation, and legal requirements—incurred by the contractor.

Development of Strategy and Data The first step in preparing a construction cost estimate is to develop the strategy and data used in the estimating process. One method involves dividing this task into four basic components: (1) site conditions, (2) classification of the work and quantity takeoffs, (3) construction program, and (4) unit cost development.

Site Conditions The topographical, geological, cultural, climatic, and socioeconomic conditions of the site all affect project cost directly. Site conditions such as the topography, nature of the soil, physical characteristics of the soil, underground water table, construction costs, the source of the imported materials, and the available disposal sites (for the excess soil) relate directly to the total design and construction costs. Furthermore, items such as site access, working conditions, productivity of the local labor, and cost of materials and labor (in the vicinity of the site area) should not be overlooked.

If the project is in a foreign country, the cost estimator should review all applicable regulations with respect to taxes, import duties, port and freight charges, currency conversion rates, local inflation rate, and limitations on money transferred between countries. Other factors, such as specific laws or regulations governing construction by foreign contractors and the employment of local labor, must also be evaluated.

Classification of Work and Quantity Takeoffs Once the design plans and specifications are completed, the engineer should divide the project into a series of specific bid items. Each of these items should then be broken down into the most logical and practical individual categories of work so that payments may be proportioned as the work progresses. Since each bid item generally includes a number of separate and distinct construction operations or classes of work, a separate cost estimate must be prepared for each item. The quantity takeoffs are divided further to define the scope or size of each construction operation based on the site condition and the engineer's judgment. The quantities (numerical figures) are then tabulated using the appropriate units: feet, yards, cubic yards, or days. Be aware, however, that certain projects benefit from having the items expressed as a lump sum.

Construction Program Having executed the classification of work and quantity takeoffs, the cost estimator must now develop a realistic construction program. Items that must be tabulated are the required type and level of laborers and professionals, required type and number of construction equipment, and the frequency of use and duration of need (anticipated) for such equipment and the crews to operate them. The cost estimator should establish a preliminary but fairly accurate overall construction progress schedule, including sequential and concurrent construction operations. The schedule should provide a reasonable time for mobilization and preparation at the beginning of the project, in addition to demobilization and completion of the work.

Unit Cost Development In anticipating the direct costs of a project, the cost estimator must develop the unit cost for (1) direct labor, (2) equipment operations, (3) consumable materials, and (4) nonconsumable materials. The unit cost for direct labor is determined by establishing a schedule and dividing the project into various classifications, such as crafts, skills, and labor. The availability, quality, and rates for the various trade classifications, in addition to the hourly rates of payment for regular work hours, must be designated for each classification. Various taxes and fringe benefits should be taken into account when determining the hourly rates. Direct equipment operating costs include the cost of fuel, oil, tires, all equipment parts, and labor—operators, maintenance, and repair personnel.

Consumable materials are defined as items that are used during the construction operation; they are not part of the permanent installation. This category includes items such as small tools and accessories. Depending on the type of construction, the unit cost for consumable materials may be expressed as cost per hour of use or as a percentage of the direct labor cost. Nonconsumable goods are defined as materials and equipment that are installed on a permanent basis. A schedule of these items should be prepared early in the estimating work because a considerable amount of time will be spent on obtaining quotes from all suppliers and manufacturers, including transportation or freight charges.

Direct Costs Direct costs encompass direct labor, the operation of construction equipment, consumable and nonconsumable materials, and subcontracted services. If the project contains several construction operations, these costs must be estimated separately for each operation and the costs then totaled to yield the cost for each bid item. Later, the direct costs of each bid item are summarized.

Indirect Costs The principal indirect cost items are administrative and overhead labor; transportation and equipment overhead; general office expenses; staff expenses; expenses incurred in establishing an on-site project office, in drawing up the construction plan and shop drawings; expenses associated with the operation and maintenance of the facilities; payroll taxes and burden; equipment mobilization and demobilization; equipment ownership; insurance and bond; taxes; and licenses and fees. Items such as depreciation, interest on investment, major repair and storage costs, overhead, insurance, and taxes are categorized as equipment ownership costs. If equipment is rented, rental charges are substituted for ownership costs.

Estimate Summary The estimate summary should be comprised of direct and indirect costs, makeup, and cost escalation adjustment. *Makeup* is defined as the allowance granted to the contractor for profit and contingencies. The time schedule, working conditions, and uncertain ground and climatic conditions, in addition to the economic situation, all affect the amount of the makeup; the amount is generally expressed as a percentage of the total cost and is approximately 20% for many water treatment construction projects. The percentage will be higher for small projects, and vice versa.

The final cost estimate described above is a general approach that is applicable to almost any type of construction project, including the construction of water treatment plants. Table 6.3-1 is an example of a cost estimate summary for a 50-mgd (2.2-m^3/s) simple direct filtration plant. Table 6.3-2 is another example of a cost estimate summary for a 40-mgd (1.75-m^3/s) conventional process plant with ozonation and filters with GAC beds. It should be noted, however, that both cost estimates do not include the contingency costs (about 20% of total cost), bids evaluation, or construction management services (CMS) by engineers (3 to 6% of total cost, depending on the contract). Actual cost estimates for the construction of water treatment plants should include these two items. Table 6.3-3 lists the ENR Index of major U.S. cities (1999).

Whether it is regarded as an art, science, or equal parts of good data, good judgment, and good fortune, cost estimators rely increasingly on computers to help prepare the final cost estimate; since computers are more accurate and efficient, they reduce project costs. Approximately 170 different types of computerized cost estimating software are currently available on the market. The popular packages are distributed by Contractors Management (CMS) of San Diego, California; Management Computer Controls (MCC) of Memphis, Tennessee; Timberline Software Corporation and Bidtek, Inc., of Wilsonville, Oregon; and G-2, Inc., of Boise, Idaho.

BIBLIOGRAPHY

Dickson, R. D., in *Water Treatment Plant Design: Estimating Water Supply System Costs,* R. L. Sank, ed., Ann Arbor Science, Ann Arbor, MI, 1978.

Guttman, D. L., et al., *Computer Cost Models for Potable Water Treatment Plants,* EPA/600/2-78-181, U.S. Environmental Protection Agency, Washington, DC, September 1978.

Hopper, T. W., "Estimating Construction Costs," *J. Consult. Eng.,* p. 94 (August 1968).

Montgomery, J. M., Consulting Engineers, *Water Treatment-Principles and Design,* Wiley, New York, 1985.

6.4 SUPPLEMENTAL STUDIES

Designing a water treatment plant requires the combined talents of experts from various disciplines. In many cases, the entire plant design cannot be handled by one design firm since some areas are simply beyond the capability of the firm or because it is more cost-effective to subcontract them. These areas include (1) the geotechnical/geological study, (2) cold weather design, (3) corrosion study, (4) acoustic study, (5) surveying, and (6) landscaping.

TABLE 6.3-1 Detailed Cost Estimate of a 50-mgd Direct Filtration Plant

Utah Valley Water Filtrations Plant
Detailed Construction Cost Estimate Summary, 1977

Description	Estimate	Adjusted Estimate[a]	($)%
Earthwork, grading, paving	$ 381,699	$ 343,529	4.8
Yard piping	1,225,249	1,102,724	15.3
Landscaping and irrigation	68,779	68,779	1.0
Operations building	844,730	824,938	11.5
Pumping station	722,961	702,206	9.7
Meter vault 1	297,612	286,484	4.0
Meter vault 2	83,497	77,865	1.0
Valve vault 1	61,927	57,665	0.8
Valve vault 2	71,639	66,597	1.0
Valve vault 3	219,241	203,525	2.8
Valve vault 4	47,871	45,312	0.6
Filters	1,735,277	1,628,577	22.5
Filter waste washwater reclamation basin	328,034	321,342	4.5
Filter waste washwater basin	169,746	157,571	2.2
Flocculation basins	374,815	349,080	4.8
Chemical storage and activated carbon tanks	226,435	218,832	3.0
Treated water reservoir	465,491	426,906	6.0
Retaining wall	76,328	68,695	1.0
L.P. gas tanks	14,330	12,897	0.2
Testing and disinfecting	20,000	20,000	0.3
Fencing and gates	25,675	23,107	0.3
Overflow structure to Salt Lake aqueduct	43,848	39,463	0.5
Sludge supernatant pumping station	16,620	15,412	0.2
Move on and off	150,000	150,000	2.0
Subtotal (1)	7,671,824	7,211,506	100
Contractor's overhead and profit (20% of subtotal)	1,534,703	1,442,301	
Subtotal	9,206,527	8,653,807	
Electrical, instrumentation, telemetry (approx. 17.5% of subtotal)	1,341,189	1,341,189	
Total	$10,547,716	$9,994,996	

[a]Adjusted downward by 10% to account for lower Salt Lake City area costs for earthwork, paving, fencing, yard piping, concrete, reinforcement steel, miscellaneous metalwork, structural steel, waterstops and certain other construction materials, and associated labor.

Notes: (1) No cost for the construction management services (CMS) by the engineer is included.
(2) No contingency cost is included.

TABLE 6.3-2 Detailed Cost Estimate of a 40-mgd Conventional Treatment Process Including Ozonations and GAC Filter Beds

North Bay Regional Water Treatment Plant, California
Detailed Construction Cost Estimate Summary, 1988

Description	Estimate	%
Site works	$ 3,270,000	11.3
Yard piping	4,500,000	15.6
Flocculation/sedimentation basins	2,990,000	10.4
Filters	3,180,000	11.0
Blowers and building	510,000	1.8
Ozone generators with building	2,060,000	7.1
Preozonation tanks	910,000	3.2
Postozonation tanks	840,000	2.9
High-service pump station	1,580,000	5.5
Clearwell	1,000,000	3.5
Operations building	3,350,000	11.6
Wash-waste reclamation building	410,000	1.4
Wash-waste recycling facility	470,000	1.6
Chemical feed system	1,040,000	3.6
Chemical feed pipe trenches	380,000	1.3
Intake structures	460,000	1.6
Miscellaneous structures	1,140,000	4.0
Miscellaneous	780,000	2.7
Total	$28,870,000	100

Notes: (1) The cost is rounded up below $1000.
(2) No contractor's overhead and profit are included.
(3) No contingency and CMS costs by the engineers are included.

Geotechnical/Geological Study The geotechnical/geological study and recommendations may be obtained from reputable local geotechnical consultants for a reasonable fee. A local firm or a large firm with a local office is generally preferred simply because it is assumed to be most familiar with local conditions and the history of events. These firms usually provide such services as test borings of several or more strategic plant site locations and laboratory analysis of test samples, and furnish a report on their findings. Additionally, the firm should provide recommendations on the plant design, especially with respect to the structural design, the design of the foundation, and the construction aspects of the civil engineering work. Subcontracting firms are expected to supply the following information: in situ moisture, density, a laboratory maximum density and optimum moisture curve, direct shear tests, moisture sensitivity test, consolidation tests, resistivity, pH, and an analysis of the soluble sulfate and chloride content of the soil.

The design recommendation should also address (1) seismic design parameters; (2) structural design parameters such as the type and size of foundations, the various anticipated earth pressures for subterranean walls, and the design of slabs on various grades; (3) design data on earthwork, such as cut, fill, and compaction re-

TABLE 6.3-3 Example of ENR Construction Cost Index

COST INDEXES BY CITY

1913=100 CITY	CONSTRUCTION COST JULY '99 INDEX	CONSTRUCTION COST % CHG YEAR	BUILDING COST JULY '99 INDEX	BUILDING COST % CHG. YEAR	COMMON LABOR JULY '99 INDEX	COMMON LABOR % CHG. YEAR	SKILLED LABOR JULY '99 INDEX	SKILLED LABOR % CHG. YEAR	MATERIALS JULY '99 INDEX	MATERIALS % CHG. YEAR
ATLANTA	3822.76	+4.5	2789.81	+5.7	6563.16	+4.1	3849.85	+6.3	2143.16	+ 5.2
BALTIMORE	4496.88	+0.2	3061.75	+0.3	8442.11	0.0	4671.47	0.0	2078.85	+0.8
BIRMINGHAM	4245.00	+1.3	2649.57	+3.1	7852.63	0.0	3658.86	+2.0	2033.87	+4.4
BOSTON	6994.40	+4.1	3832.58	+4.8	15131.58	+3.6	6819.82	+4.0	2007.10	+6.6
CHICAGO	7192.96	+5.0	3915.94	+5.0	15168.42	+4.6	6553.15	+4.1	2304.77	+ 6.6
CINCINNATI	5872.81	+4.7	3220.27	ı3.3	11789.47	+5.4	4839.04	+3.9	2246.46	+2.5
CLEVELAND	6461.51	+2.1	3583.05	+1.2	13350.00	+2.9	5782.58	+2.6	2239.53	–0.9
DALLAS	3945.14	–0.1	2668.00	–0.5	6742.11	+1.6	3385.59	+2.5	2230.87	–3.1
DENVER	4429.43	–1.0	2823.00	–1.3	8255.26	0.0	4033.03	+0.5	2084.57	– 3.4
DETROIT	7042.05	+3.5	3956.43	+4.3	14742.11	+3.7	6630.63	+5.1	2322.65	+2.8
KANSAS CITY	6016.54	+1.8	3385.37	+3.6	12052.63	+2.0	5135.14	+5.3	2317.01	+1.2
LOS ANGELES	6823.97	+2.0	3589.02	–0.1	14457.89	+3.2	5952.55	+1.6	2145.11	– 2.9
MINNEAPOLIS	6846.46	+2.9	3577.58	+2.3	14531.58	+4.0	5936.94	+4.7	2136.22	– 1.5
NEW ORLEANS	3926.00	–2.6	2600.60	–3.9	6842.11	0.0	3358.56	0.0	2138.71	– 7.4
NEW YORK	9301.94	+7.0	5093.38	+5.5	20597.37	+5.9	9534.53	+2.7	2378.93	+12.8
PHILADELPHIA	7467.87	+2.1	4083.58	+0.9	15736.84	+3.5	6839.64	+3.2	2399.79	– 2.8
PITTSBURGII	6087.39	+2.2	3514.25	+1.5	12497.37	+3.4	5733.33	+3.8	2158.70	– 2.0
ST. LOUIS	6801.36	+3.3	3488.24	+5.5	14552.63	3.1	5841.44	+6.1	2050.58	+4.4
SAN FRANCISCO	6818.39	+0.8	3900.26	+1.5	14410.53	+1.8	6739.94	+4.0	2165.14	– 3.1
SEATTLE	6932.16	+2.8	3456.31	+0.4	14810.53	+4.0	5670.87	+2.3	2103.48	– 2.4
MONTREAL	6359.59	+4.8	3700.16	+8.6	12378.95	0.0	5387.39	0.0	2670.31	+21.4
TORONTO	8051.54	+1.3	4099.67	+1.4	16913.16	+2.0	6521.92	+3.2	2620.23	– 1.3

Source: Engineering News-Record, July 12, 1999.

quirements, and the type of equipment required for excavation and compaction; (4) corrosivity of the soil, complemented by recommendations on the major underground pipeline design and the required type of cement for concrete structures set in the ground; (5) a prediction as to the groundwater level during plant construction and after construction is completed; and (6) magnitude of settling (for structures) after completion of construction.

Cold Weather Design If the water treatment plant is located in a region with extremely cold winter weather conditions, the plant design requires many special considerations. For instance, all major unit processes should either be housed or buried in the ground (sedimentation tanks and the clearwell) to prevent freezing of the process water. Certain items in the building and construction design should be specifically outlined, especially the roofing system and the vapor barriers (these should be designed to reduce condensation) otherwise, the doors will freeze in place. Heat traces should be conducted for all outdoor equipment and chemical feed pipes; these items should also be insulated. Drain pipes, especially filter waste wash and sludge pipes, should be protected from freezing and be capable of draining tanks or lagoons that may be frozen solid. The exit of these pipes must be located at the bottom of the tank, beneath the anticipated thickness of surface ice. Furthermore, the capacity of the tank should be large enough to accommodate a required volume of water or sludge below the ice caps. Another important design consideration concerns pipelines. All pipes transporting liquid must be buried below the local frost depth so that the piping system functions on a continuous basis and to avoid pipe ruptures which result from freezing of the water.

The design and construction of buildings and treatment plants in cold weather regions are unique tasks requiring very special knowledge. Thus projects located in arctic regions should employ project engineers who are licensed as cold weather design specialists (in the engineering field). The engineer, as well as the architect, mechanical engineer, and other project staff members, should be either registered as cold weather design specialists or hired/selected only if qualified to design and construct structures for cold weather regions.

Corrosion Studies Corrosion is a common concern in the field of water treatment. Corrosion of pipes due to the carrying water is rarely a serious problem because all steel and cast iron pipes, with the exception of pipes less than 4 in. (10 cm) in diameter, are now cement lined and coated. Moreover, in the United States, the pH of the filtered water is now commonly adjusted prior to distribution to the consumer, thereby significantly reducing internal corrosion in the main distribution lines. Nevertheless, certain sites, primarily marshy areas, can have very corrosive soils. Under these circumstances, the issue of corrosion must be given special consideration during design of the yard pipings and concrete structures. There are also a few cases where iron and steel pipes (buried) corroded severely due to stray currents emanating from an electric railroad track or as the result of galvanic action along the pipelines. A corrosion expert should be involved in the project if pipe corrosion is anticipated to be a major problem.

Acoustic Studies If a water treatment plant is located in a quiet residential area, the noise generated by certain water treatment equipment may be a significant issue. Noise emanating from air compressors, blowers, high-service pumps, and engine-driven power generators is loud and disturbing to both plant operators and neighboring residents. OSHA recommends that the maximum noise level be maintained below 80 decibels (dB) for the health and safety of operators, and this criterion is generally much more stringent with respect to neighboring communities. The design engineer should consult an acoustic control specialist for recommendations on how to solve the noise problem properly in a cost-effective manner.

Surveying Although most engineering firms are capable of surveying, an outside firm of surveying specialists should be used if it is the more cost-effective alternative. Whenever the job site is remote or does not have easy access, or if it is difficult to conduct surveying using regular surveying techniques, an aerial surveying specialist should be chosen.

Landscaping Proper landscaping not only makes a plant site attractive but reduces maintenance and irrigation costs. Experience has shown that reputable local landscaping professionals can do a good job at a reasonable cost.

CHAPTER 7

Specific Water Treatment Processes

7.1 LIME–SODA ASH SOFTENING

7.1.1 General Discussion

The main purpose of water softening by means of lime–soda ash and lime softening is to reduce the levels of calcium and magnesium in the process water to a total hardness of approximately 80 mg/L (as $CaCO_3$) as well as reducing the magnesium hardness to approximately 40 mg/L (as $CaCO_3$). Over 1000 municipal water softening plants currently operate in the United States, the majority located in the Midwest and Florida.

Prior to the mid-twentieth century, the primary problem of the domestic consumer was the high consumption of soap by hard water. However, due to the widespread use of synthetic detergents, this is no longer a major issue today. Aside from the more immediate consumer concerns, the additional benefits of lime softening are quite substantial, including the removal of heavy metals, metallic elements, and organic compounds, as well as effectively killing bacteria, viruses, and algae. Lime softening also improves water quality with respect to pipe corrosion, boiler feed, and cooling waters. The spontaneous removal of elements such as iron, manganese, lead, mercury, and chromium is particularly attractive to purveyors of potable water, while the reduction of silica and total dissolved solids is beneficial to industries.

Since 1990, special attention has been given to the excess lime softening process because of its effectiveness in controlling THMs and HAA_5 by removing natural organic matter (NOM) prior to chlorination. The softening process is termed *enhanced softening* by the EPA, and TOC removal can be as high as 80% at pH 10.5 or higher with 10 to 15 mg/L of ferric sulfate (Figure 7.1.1-1).

Water hardness is defined as the amount of divalent metallic cations in the water and is expressed in mg/L as $CaCO_3$. The major divalent metallic cations that contribute to water hardness are calcium and magnesium. A few major European countries are adopting their own scales of water hardness, and this confusing situation is compounded by the fact that many industries in the United States still use the old expression of hardness: grains per gallon. Table 7.1.1-1 presents the relationship between the expressions of hardness.

Engineers should note that hard water is not currently known to adversely affecting human health in any significant manner. Quite a few studies performed during the 1970s in the United States, Canada, United Kingdom, and Japan have demon-

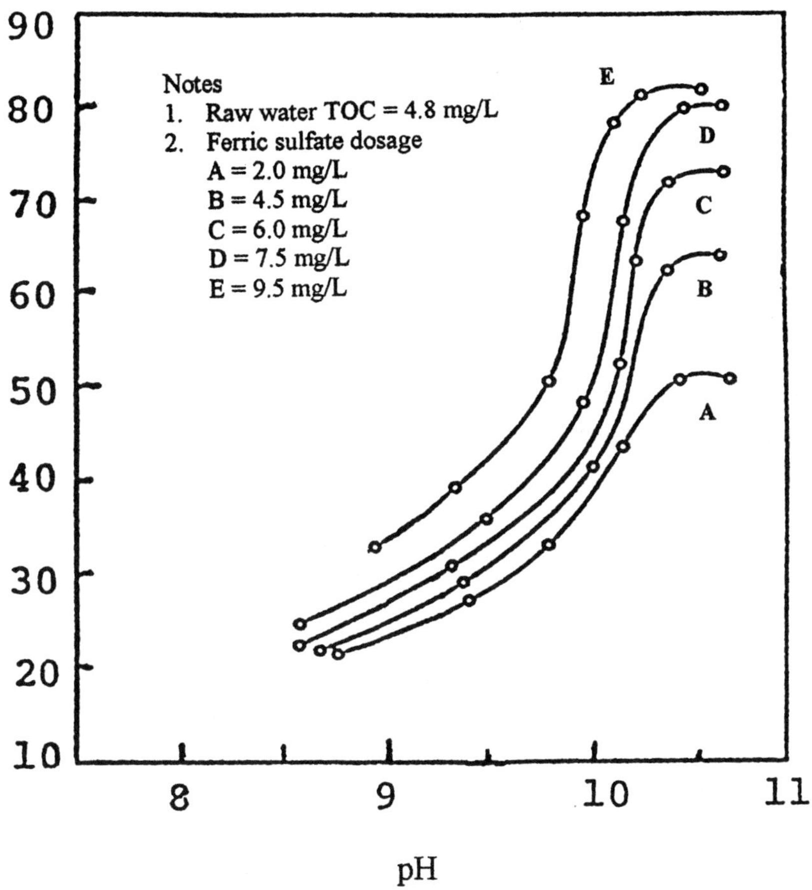

Figure 7.1.1-1 Enhanced softening process for TOC removal. [Data from S. R. Qasim, et al., *J. Environ. Eng. Div. ASCE,* 118(3) (May–June 1992).]

strated an inverse correlation between the incidence of cardiovascular disease and the hardness level of drinking water. Yet conflicting studies have also been reported.

Lime softening, as well as other processes such as ion exchange, electrodialysis, nanofiltration, reverse osmosis, distillation, and freezing may be employed to soften hard water. However, only the process of lime softening is discussed in this chapter. Yet engineers should be aware that there are a few technical articles that demonstrate that the use of magnesium carbonate, in conjunction with lime softening, improves the calcium carbonate sludge characteristics.

7.1.2 Basic Chemical Reactions

In this section we present the basic chemical reactions of (1) lime and lime–soda ash softening, (2) caustic soda softening, and (3) water stabilization of softened water.

TABLE 7.1.1-1 Relationships Among Hardness Units

United States	mg/L as $CaCO_3$, which is 0.02 meq/L
	1 grain of $CaCO_3$/gal = 17.1 mg/L as $CaCO_3$
Germany	Equivalent to 10 mg/L of lime expressed as CaO
	1 German degree = 17.86 mg/L as $CaCO_3$
France	1 French degree = 10 mg/L as $CaCO_3$

The level of water hardness is generally classified as follows:

0–75 mg/L	Soft water
75–150 mg/L	Moderately hard water
150–300 mg/L	Hard water
Over 300 mg/L	Very hard water

Lime and Lime–Soda Ash Softening

$$CaO + H_2O \rightarrow Ca(OH)_2 \quad (1)$$

$$CO_2 + Ca(OH)_2 \rightarrow CaCO_3 + H_2O \quad (2)$$

$$Ca(HCO_3)_2 + Ca(OH)_2 \rightarrow 2CaCO_3 + 2H_2O \qquad pH \geq 9.5 \quad (3)$$

$$Mg(HCO_3)_2 + Ca(OH)_2 \rightarrow CaCO_3 + MgCO_3 + 2H_2O \qquad pH \geq 9.5 \quad (4)$$

$$MgCO_3 + Ca(OH)_2 \rightarrow Mg(OH)_2 + CaCO_3{}^* \qquad pH \geq 11 \quad (5)$$

*plus $Ca(OH)_2$ due to an excess lime dosage

$$MgSO_4 + Ca(OH)_2 \rightarrow Mg(OH)_2 + CaSO_4 \quad (6)$$

$$CaSO_4 + Na_2CO_3 \rightarrow CaCO_3 + Na_2SO_4 \quad (7)$$

$$CaCl_2 + Na_2CO_3 \rightarrow CaCO_3 + 2NaCl \quad (8)$$

Caustic Soda Softening

$$CO_2 + 2NaOH \rightarrow Na_2CO_3 + H_2O \quad (9)$$

$$Ca(HCO_3)_2 + 2NaOH \rightarrow CaCO_3 + 2Na_2CO_3 + 2H_2O \quad (10)$$

$$Mg(HCO_3)_2 + 4NaOH \rightarrow Mg(OH)_2 + 2Na_2CO_3 + 2H_2O \quad (11)$$

$$MgSO_4 + 2NaOH \rightarrow Mg(OH)_2 + Na_2SO_4 \quad (12)$$

$$CaSO_4 + 2NaOH \rightarrow Ca(OH)_2 + Na_2SO_4 \quad (13)$$

$$CaCl_2 + 2NaOH \rightarrow Ca(OH)_2 + 2NaCl \quad (14)$$

NOTE: Reactions (13) and (14) need to add CO_2 to form $CaCO_3$ precipitate.

Water Stabilization of Softened Water

$$2Ca(OH)_2 + 3CO_2 \rightarrow CaCO_3 + Ca(HCO_3)_2 + H_2O \qquad pH = 8.8 \qquad (15)$$

$$CaCO_3 + CO_2 + H_2O \rightarrow Ca(HCO_3)_2 \qquad pH \leq 8.3 \qquad (16)$$

$$2CaCO_3 + H_2SO_4 \rightarrow Ca(HCO_3)_2 + CaSO_4 \qquad pH \leq 8.3 \qquad (17)$$

$$2CaCO_3 + 2HCl \rightarrow Ca(HCO_3)_2 + CaCl_2 \qquad pH \leq 8.3 \qquad (18)$$

7.1.3 Softening Process Alternatives

There are five alternatives to the softening process. These depend on the type of hardness, degree of softening, operational convenience, degree to which the production of lime sludge must be reduced, and desired savings in chemical cost. The alternatives include (1) partial lime softening, (2) excess lime softening, (3) lime–soda ash softening, (4) caustic soda softening, and (5) softening with both lime and caustic soda.

Alternative 1: Partial Lime Softening This process removes only calcium carbonate. Lime (alone) is fed to the process water to raise the pH to approximately 9.5 [reaction (3)]. The softened water is then stabilized by recarbonation [reactions (15) and (16)] or through the addition of acid [reactions (17) and (18)] to reduce the pH to approximately 8.8 or below, thus minimizing the occurrence of heavy scaling within the pipes and valves.

Alternative 2: Excess Lime Softening This alternative process is used in removing the carbonate form of calcium and magnesium. An excess amount of lime is applied to the process water at a minimum pH of 10.6, preferably 11 to 11.3 [reactions (3) through (5)] since magnesium hydroxide will not adequately form unless the pH of the water is raised above 11. Following the formation of magnesium hydroxide, the process water should be recarbonated to reduce the pH to 8.7 to 8.8. This step will convert the hydroxide form to carbonate and bicarbonate because the water has a calcium hydroxide concentration of 40 to 50 mg/L at pH 11 and the water is unstable.

At pH 8.8, approximately 10% of the alkalinity is carbonate and 90% is in the bicarbonate form. Calcium bicarbonate is dissolved in the water, yet calcium carbonate precipitates out and must be removed by means of sedimentation or filtration to minimize its effect of hardening the softened water. The sedimentation process generally follows the recarbonation process because the filtration process receives a rather excessive amount of calcium carbonate precipitate. When clear water such as underground water is softened by means of this process, a portion of the raw water is blended to the softened water (called *split treatment*). The carbon dioxide and bicarbonates (in the raw water) react with the calcium hydroxide in the softened water to yield the calcium carbonate precipitate.

Alternative 3: Lime–Soda Ash Softening This alternative should be used to remove both carbonate and noncarbonate hardness. Noncarbonate hardness is typically produced by calcium sulfate and calcium chloride; soda ash must be used to remove these compounds [reactions (6) through (8)]. The pH of the softening water should be a minimum of 10.6; therefore, recarbonation or acidification is required to stabilize the water prior to filtration.

Alternative 4: Caustic Soda Softening Caustic soda may be used as a substitute for lime in alternatives 1 through 3. The advantages of caustic soda are its decreased sludge production, the elimination of problems associated with chemical dust, and the option of utilizing simpler storage and feed systems. Another advantage is the production of Na_2CO_3 as shown in reactions (9) through (11). Therefore, this process removes noncarbonate hardness without feeding soda ash. However, caustic soda also has several major disadvantages; the cost of caustic soda is four to six times higher than that of lime; it is a potential safety hazard to operators should there be a massive leak; and freezing problems occur for 50% solutions at 55°F (13°C).

Alternative 5: Softening with Both Lime and Caustic Soda This last alternative is a variation of alternative 4; it is used to reduce the overall chemical cost and to reduce the capital cost of the lime feed system. The chemical reaction involved in this alternative is as follows:

$$2Ca(HCO_3)_2 + Ca(OH)_2 + 2NaOH \rightarrow 3CaCO_3 + Na_2CO_3 + 4H_2O$$

7.1.4 Overall Softening Treatment Process

All five water softening alternatives basically have the same overall water treatment process as the conventional complete water process: flash mixing, flocculation, sedimentation, recarbonation, second sedimentation, filtration, and chlorination. A mechanical mixer should not be used as the flash mixer in the softening process because the mixing blades will be coated by calcium carbonate, resulting in a large ball of $CaCO_3$. The appropriate type of flash mixing is achieved by hydraulic mixing through the use of baffled channel or water fall from weirs. If the softened water is stabilized through acid rather than carbon dioxide, the recarbonation and second sedimentation processes are deleted. However, the addition of acid increases either the sulfate or chloride concentration and slightly increases water hardness due to the dissolving of $CaCO_3$ by acids, whereas the use of carbon dioxide does not.

The normal lime softening process can manage occasional turbidity spikes of up to 500 to 1000 ntu because the large dosage of lime (for softening) also acts as a coagulant. Nevertheless, turbidity removal and filter performance will both improve if metal coagulants such as sodium aluminate or ferric sulfate are fed in conjunction with lime. If the turbidity frequently exceeds 1000 ntu, engineers should consider installing an additional clarification process (alum flocculation and sedimentation) upstream of the regular lime softening process train.

Design engineers should note that recycling lime sludge to the head of the softening process train always improves the efficiency of the softening process; recycling gives a seeding effect in the formation of calcium carbonate floc. The solids contact and sludge recirculation types of reactor clarifier are frequently used in lime softening plants precisely because of this feature and are proven to be quite effective in this capacity (Figure 7.1.4-1). If the carbon dioxide content of the raw water exceeds 10 mg/L, an aeration process such as a coke tray aerator should be con-

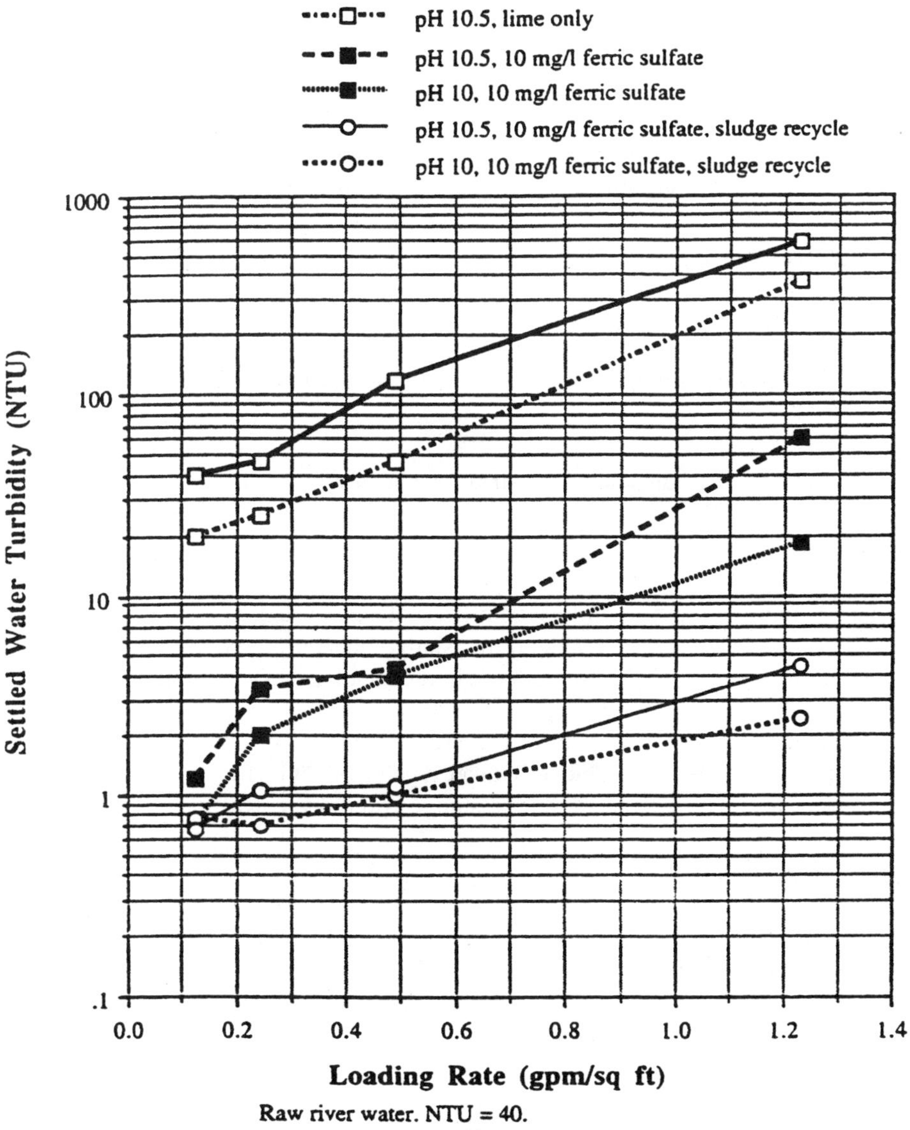

Figure 7.1.4-1 Effect of pH, ferric and sludge recycle on settling velocities.

sidered for use in gas stripping. This scheme reduces the dosage of lime or caustic soda required for the removal of carbon dioxide [reactions (2) and (9)], as well as reducing the volume of sludge production.

7.1.5 Design Criteria

The design criteria for (1) flash mixing, (2) flocculation, (3) sedimentation, (4) recarbonation, (5) filtration, (6) chlorination, and (7) sludge handling and disposal (for lime–soda ash softening) are presented below.

Flash Mixing

Mixing energy (G)	300 to 500 s^{-1}
Mixing time	10 to 30 s

Flocculation

Mixing energy (G)	5 to 50 s^{-1}
Mixing time	30 to 40 min
Flow velocity through ports	0.5 to 1.2 ft/s (0.15 to 0.36 m/s)

NOTE: Adequate mixing time is very important because of the slow reaction rate.

Sedimentation

Surface loading	0.75 to 1 gpm/ft^2, rectangular tank (1.9 to 2.5 m/h); 1 to 1.75 gpm/ft^2, sludge blanket type (2.5 to 4.4 m/h)
Detention time	Minimum of 2 h, rectangular tank; minimum of 1 h, sludge blanket type
Weir loading	15 gpm/ft, rectangular tank (11 m^3/m · h); 20 gpm/ft, sludge blanket type (15 m^3/m · h)
Total water loss	$<$ 3% due to sludge withdrawal

Recarbonation

Carbon dioxide diffusion tank	3 min minimum contact time
Recarbonation tank	20 min minimum detention time
Water depth	Minimum of 12 ft (3.7 m)
pH of water after recarbonation	8.7 to 8.8

NOTE: The use of liquid carbon dioxide has become more commonly practiced for recarbonation.

Filtration (Dual Media Bed)

Filtration rate	4 to 6 gpm/ft^2 (10 to 15 m/h)
Anthracite coal	
Depth	20 in. (0.5 m)
E.S.	1.0 to 1.1 mm
U.C.	$<$1.4
S.G.	1.67 to 1.7
Sand	
Depth	10 in. (0.25 m)
E.S.	0.55 to 0.6 mm
U.C.	$<$1.5
S.G.	$>$2.63
Backwash rate	20 gpm/ft^2 average (50 m/h)

NOTES:

1. A surface washing system must be installed because it plays an essential role in maintaining the top portion of the filter bed free of mud balls and mud pans; these form as the result of the cementing action of the calcium carbonate precipitates.
2. The design criteria for the surface wash system are identical to those used in designing regular filters (refer to Section 3.2.7).
3. Based on the bed expansion, use of filter backwash (alone) is not an effective method.

Chlorination Although the $C \times t$ rule issued by the EPA does not acknowledge its role, the softening process is always characterized by a good disinfecting capability; the level of THM and HAA formation due to subsequent chlorination is usually also quite low because of effective removal of TOC by the softening process. The required chlorine dosage is defined as the minimum amount required to maintain a chlorine level of 0.5 mg/L throughout the distribution system to fight post-contamination. This restriction is imposed because under high-pH conditions, residual chlorine is in the hypochlorite form. The disinfecting power of this ion is much less effective than hypochlorous acid. Under a pH range of 8.6 to 8.7, approximately 90% of the chlorine residual is in hypochlorite form. Thus the chlorine disinfecting power increases with a decrease in pH after recarbonation or acid neutralization.

Sludge Handling and Disposal

Sludge production rate	2 to 3.5 lb of solids for each pound of hardness removed (refer to Section 4.3)
Sludge dewatering	Sludge lagoon type of drying bed or a mechanical dewatering system (refer to Section 4.3)

TABLE 7.1.5-1 pH versus Hydroxide Alkalinity

pH	9.7	10	10.2	10.4	10.5	10.6	10.8	11.0
OH (mg/L as $CaCO_3$	2–3	4–5	8–9	12–13	14–16	17–20	26–30	41–50

NOTE: Lime recovery by lime sludge recalcification has been practiced to some degree. However, due to recent laws against air pollution, most urban areas severely limit the use of this process.

Table 7.1.5-1 lists the approximate levels of hydroxide alkalinity at various pH levels (of the water). Figure 7.1.5-1 illustrates the relationship between carbonate and bicarbonate ions at various pH levels that are common to the recarbonation process.

7.1.6 Operational Considerations

The lime–soda ash softening process is based on a stoichiometric chemical reaction between the elements of hardness and the softening chemicals. A properly trained operational staff and a basic chemical laboratory are therefore essential in ensuring

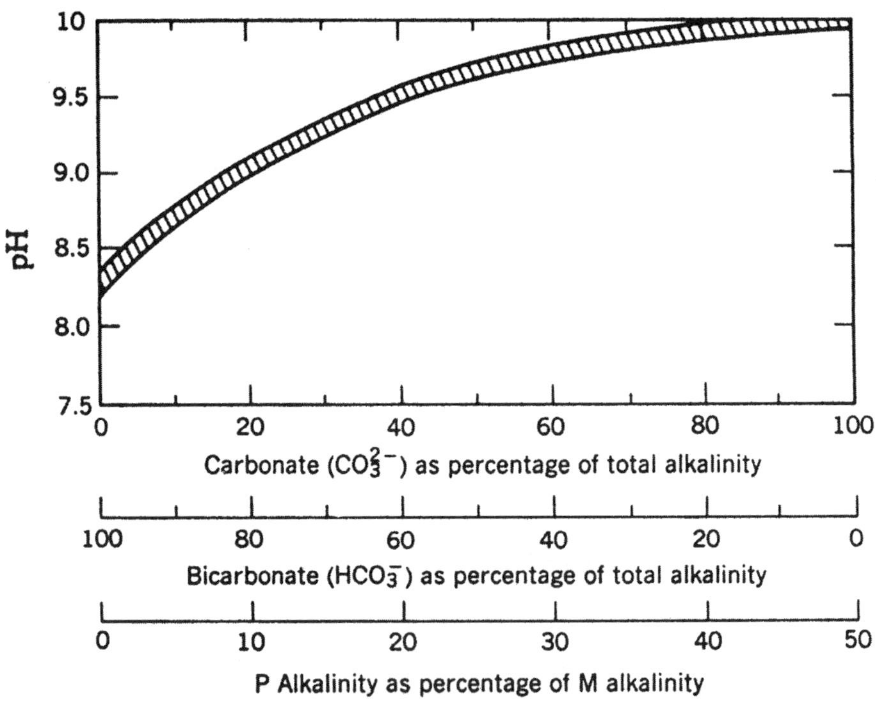

Figure 7.1.5-1 Relationships between pH and carbonate and bicarbonate ions. (Adapted from *The Nalco Water Handbook*, F. N. Kemmer, ed., McGraw-Hill, New York, 1979.)

TABLE 7.1.6-1 Molecular Weights of Major Items

CaO	$Ca(OH)_2$	Na_2CO_3	$CaCO_3$	$NaOH$	CO_2	Ca	Mg
56	74	106	100	40	44	40	24.3

successful water softening. One of the most important operational parameters is proper chemical dosage. The molecular weights of the major items involved in the softening process are listed in Table 7.1.6-1.

Chemical Dosages When softening water containing carbonate hardness, the following general formula will determine the lime dosage (the chemical feed details are discussed in Section 4.1):

$$\text{lime (CaO in mg/L)} = (A + B + C) \times D/E$$

where

$A = CO_2$ as CO_2 (mg/L) $\times$ 56/44

$B = HCO_3$ alkalinity as $CaCO_3$ (mg/L) $\times$ 56/100

$C =$ Mg as Mg (mg/L) $\times$ 56/24.3

$D =$ excess lime required to raise pH to 11 to remove magnesium; it is generally 10 to 20% of $A + B + C$: thus $D = 1.1$ to 1.2

$E =$ purity of the quick lime (0.88 to 0.95); therefore, $E = 0.88$ to 0.95

For slaked lime [$Ca(OH)_2$]:

$$\text{slaked lime } [Ca(OH)_2] = (A' + B' + C') \times D/E$$

where

$A' = CO_2$ as CO_2 (mg/L) $\times$ 74/44

$B' = HCO_3$ alkalinity as $CaCO_3$ $\times$ 74/100

$C' =$ Mg as Mg (mg/L) $\times$ 74/24.3

$D =$ excess lime: $D = 1.1 - 1.2$

$E =$ purity of the lime: generally, $E = 0.93 - 0.95$

The soda ash dosage for noncarbonate hardness is determined as follows:

$$\text{soda ash } (Na_2CO_3 \text{ in mg/L}) = (NH - L) \times 106/100$$

where NH is the noncarbonate hardness as $CaCO_3$ (mg/L) and L is the noncarbonate hardness remaining in the softened water (mg/L).

Since commercially available soda ash is nearly 100% pure, a purity correction factor is usually not necessary.

If caustic soda is employed instead of lime, in the same manner as described above, refer to reactions (9) through (14) and use a molecular weight of 40 (for NaOH) rather than 56 (for CaO).

The carbon dioxide dosage for recarbonation is

$$CO_2(\text{mg/L}) = \{[\text{HA} + (\text{CA} \times \text{R})] \times 44/100\} \times \text{F}$$

where HA is the hydroxide alkalinity as $CaCO_3$ (mg/L), CA the carbonate alkalinity as $CaCO_3$ (mg/L), R the ratio used to convert CO_3 to HCO_3 (generally ranges from 0 to 0.2), and F the excess CO_2 dosage factor: usually F = 1.2. The pH of the recarbonated water must be approximately 8.8 in order to facilitate the formation of HCO_3 and $CaCO_3$: these compounds are vital in removing excess calcium hardness by means of settling and filtration.

The treatment plant operator should remember the basic relationships of hydroxide, carbonate, and bicarbonate alkalinity:

1. When the phenolphthalein alkalinity (PA) is zero, the water does not contain hydroxide or carbonate alkalinity. Methyl orange alkalinity (MA) is the bicarbonate form.
2. If PA $<$ 0.5MA, there is no hydroxide alkalinity. The carbonate alkalinity is 2PA and the bicarbonate alkalinity is MA $-$ 2PA.
3. When PA $=$ 0.5MA, both the hydroxide and bicarbonate alkalinities are zero, and MA is all in carbonate form.
4. When PA $>$ 0.5MA, the bicarbonate alkalinity is zero. The hydroxide alkalinity is 2PA $-$ MA and the carbonate alkalinity is 2(MA $-$ PA).
5. When PA $=$ MA, both the carbonate and bicarbonate alkalinities are zero and the hydroxide alkalinity is equal to either PA or MA.

It is also worth remembering that the pH for phenolphthalein endpoint titration is 8.3.

Sludge Handling and Disposal This issue also has great bearing on the operations of the softening plant. In Section 4.3 we present a detailed discussion of this topic.

7.1.7 Example Calculations

Example 1

Given The raw water contains the following: 250 mg/L of bicarbonate hardness, 30 mg/L of magnesium, 75 mg/L of noncarbonate hardness, and 15 mg/L of carbon dioxide. Assume that the softened water delivered to the consumer has a pH of 8.7 and a total hardness of 85 mg/L, of which 35 mg/L is carbonate hardness.

Determine The required chemical dosages, using commercially available chemicals, based on a plant flow rate of 15 mgd (0.7 m^3/s)

Solution

$$\text{Dosage of quick lime (mg/L)} = (A + B + C) \times D/E$$

where

$$A = 15 \times 56/44 = 19 \text{ mg/L}$$
$$B = (250 - 35) \times 56/100 = 120 \text{ mg/L}$$
$$C = 30 \times 56/24.3 = 69 \text{ mg/L}$$
$$D = 1.15 \text{ (15\% excess lime)}$$
$$E = 0.88 \text{ (88\% purity is assumed)}$$

Thus

$$\text{CaO (mg/L)} = (19 + 120 + 69) \times (1.15/0.88) = 272 \text{ mg/L}$$
$$\text{dosage of soda ash (mg/L)} = (\text{NH} - \text{L}) \times 106/100$$

where NH = 75 and L = 35 mg/L. Thus

$$Na_2CO_3 = 40 \times (106/100) = 42.2 \text{ mg/L}, \qquad \text{say, 43 mg/L}$$

It is assumed that the water qualities of the softened water and recarbonated water are the same as those listed in Table 7.1.7-1.

Recarbonation is achieved by means of liquid carbon dioxide, since it is nearly 100% pure, is easily handled, is easily fed, and has low maintenance and capital costs when compared to flue gas; flue gas is attained by burning oil or natural gas and only yields approximately 12% CO_2.

The carbon dioxide dosage is determined from Table 7.1.7-1. Since hydroxide alkalinity of 43 mg/L and a carbonate alkalinity of 2 mg/L must be converted to bicarbonate alkalinity,

$$CO_2 \text{ (mg/L)} = \{[\text{HA} + (\text{CA} \times \text{R})] \times (44/100)\} \times \text{F}$$

where HA = 43 mg/L and CA × R = 2 mg/L or 32 × 0.063 = 2. Thus

$$CO_2 = [(43 + 2) \times 0.44] \times 1.2 = 24 \text{ mg/L}$$

TABLE 7.1.7-1 Changes in Water Characteristics During the Softening Process (Assumed)

Characteristic	Raw Water	Softened Water	Recarbonated Water
CO_2	15	0	0
pH	7.5	11	8.7
Mg	30	3	3
Total hardness	325	110	85
Noncarbonate hardness	75	35	35
Total alkalinity (MA)	250	75	50
P alkalinity (PA)	0	59	15
HCO_3 alkalinity	250	0	20
CO_3 alkalinity	0	32	30
OH alkalinity	0	43	0

Notes: (1) MA represents methyl orange alkalinity.
(2) PA is phenolphthalein alkalinity.
(3) Magnesium cannot be completely removed by this process. Therefore, a residual of 3 mg/L is assumed to be present in the softened water.

SUMMARY OF CHEMICAL DOSAGES

$$\text{Lime (CaO) dosage} = 272 \times 8.34 \times 15 = 34{,}027 \text{ lb/day}$$

$$\text{Soda ash dosage} = 43 \times 8.34 \times 15 = 5380 \text{ lb/day}$$

$$\text{Carbon dioxide dosage} = 24 \times 8.34 \times 15 = 3002 \text{ lb/day}$$

BIBLIOGRAPHY

AWWA/ASCE, *Water Treatment Plant Design,* American Water Works Association, New York, 1969.

Cox, C. R., *Operation and Control of Water Treatment Processes,* World Health Organization, Geneva, 1964.

Craun, G. F., and McCabe, L. J., "Problems Associated with Metals in Drinking Water," *J. AWWA,* 67:11:593 (November 1975).

Graveland, A., et al., "Developments in Water Softening by Means of Pellet Reactors," *J. AWWA,* 75:12:619 (December 1983).

Hess, J. S., "Lime and Caustic Soda Softening at Fremont, Ohio," *J. AWWA,* 60:8:980 (August 1968).

Judkins, J. F., and Wynn, R. H., "Crystal-Speed Conditioning of Lime Softening Sludge," *J. AWWA,* 64:5:306 (May 1972).

Kemmer, F. N., ed., Nalco Chemical Co., *The Nalco Water Handbook,* McGraw-Hill, New York, 1979.

National Research Council, *Drinking Water and Health: Water Hardness and Health,* National Academy of Sciences, Washington, DC, 1977, p. 349.

National Research Council, *Drinking Water and Health: Water Hardness and Health,* Vol. 3, National Academy of Sciences, Washington, DC, 1980, p. 20.

Reh, C. W., in *Water Treatment Plant Design: Lime-Soda Softening Processes,* R. L. Sanks, ed., Ann Arbor Science, Ann Arbor, MI, 1982.

Riehl, M. L., *Hoover's Water Supply and Treatment,* 10th ed., National Lime Association, Washington, DC, 1970.

Schroeder, J. A., "Municipal Drinking Water and Cardiovascular Death Rates," *J. Am. Med. Res.,* 195:2:125 (February 1966).

Singhal, A. K., "Conventional Lime–Soda Ash Softening vs. Split Treatment," *J. AWWA,* 69:3:158 (March 1977).

Snoeyink, V. L., and Jenkins, D., *Water Chemistry,* Wiley, New York, 1980.

Thompson, C. G., et al, "Magnesium Carbonate: A Recycled Coagulant," Parts 1 and 2, *J. AWWA,* 64:1:11 (January 1972) and 64:2:93 (February 1972).

Van der Veen, C., and Graveland, A., "Central Softening by Crystallization in a Fluidized-Bed Process," *J. AWWA,* 80:6:51 (June 1988).

Winton, E. F., and McCabe, L. J., "Studies Relating to Water Mineralization and Health," *J. AWWA,* 62:1:26 (January 1970).

Wood, F. O., "Selecting a Softening Process," *J. AWWA,* 64:12:820 (December 1972).

Wyness, D. K., "The Helical Flow Reactor Clarifier," *J. AWWA,* 71:10:580 (October 1979).

7.2 IRON AND MANGANESE REMOVAL

Iron and manganese are minerals commonly found in soil, where they exist in the form of oxides: namely, insoluble ferric oxide and manganese oxide (very insoluble). However, when water contains carbon dioxide, or in acidic water, ferric iron is reduced to the ferrous form under anaerobic conditions; this ion *is* soluble in water. Under the same conditions, the dioxide form of manganese is reduced from a valence of 4 to a valence of 2 and is also soluble in water. Consequently, a significant amount of iron and manganese is often found in deep well water and the hypolimnion water of stratified lakes and reservoirs. Iron and manganese may occasionally be found in groundwater containing hydrogen sulfide. Organically bound iron and manganese may also be found in groundwater containing humic acid.

The iron and manganese concentrations of surface water seldom exceed 1 mg/L. The iron content of groundwater may reach levels up to 10 mg/L under conditions of low alkalinity (less than 50 mg/L) and up to 2 mg/L of manganese. The presence of iron and manganese in potable water is not known to cause health problems. These minerals are associated primarily with aesthetic factors, such as unpleasant taste and odor, and domestic problems, such as staining of laundry and fixtures, and can potentially be a serious problem for many industries—for example, staining of the product and/or formation of mineral deposits. Another potential problem is the growth of iron bacteria within the distribution main. The Secondary Drinking Water Standards (EPA) recommends MCL levels of 0.3 mg/L iron and 0.05 mg/L manganese.

7.2.1 Alternative Methods

Several alternatives exist for the removal of iron and manganese in the field of water treatment. The basic methods are (1) oxidation followed by clarification and filtration, (2) ion exchange, (3) stabilization by means of a sequestering agent, and (4) lime softening. Each of these schemes is discussed. A case history is also presented at the end of the section.

Oxidation Oxidation may be achieved through (1) aeration, (2) chlorination, (3) chlorine dioxide, (4) potassium permanganate, or (5) ozonation.

Aeration Aeration can be used to fulfill two purposes: transfer of oxygen to water and the removal of volatile gases.

$$4Fe\,(HCO_3)_2 + O_2 + 2H_2O \rightarrow 4Fe\,(OH)_3 + 8CO_2$$

$$2MnSO_4 + 2Ca(OH)_2 + O_2 \rightarrow 2MnO_2 + 2CaSO_4 + 2H_2O$$

The air diffusion type of aeration uses air diffusers to supply air to the process water at a depth of 12 to 15 ft (3 to 4.5 m). The air/water volume ratio is 0.75 : 1.0. This type of unit has an average oxygen transfer efficiency of only 5 to 10% and is therefore not considered to be effective.

The coke tray type of contact aerator is a more effective method of oxidation. This aerator consists of a series of trays that are each 12 to 18 in. deep (0.3 to 0.45 m). The bottoms of the trays are perforated and coke, crushed stone, limestone, or plastic medium is placed in each tray as the contact material; the loading rate is 15 to 20 gpm/ft^2 (37 to 50 m/h). Theoretically, 1 mg of oxygen oxidizes 7 mg of divalent iron and 3.4 mg of divalent manganese. Organically bound iron cannot be oxidized through aeration.

The rate of oxidation is a function of the pH of the process water: the higher the pH, the better the result. To complete the process of oxidation within 15 min, the pH of the water should be higher than 7.5, preferably 8. Manganese oxidizes very slowly and is not effectively oxidized at pH values below 9.5. In fact, it would take over 1 h to oxidize manganese at pH 9.5.

Alum flocculation and sedimentation are generally required after aeration. These processes are then followed by filtration, especially in cases where the iron concentration of the raw water exceeds 5 mg/L.

Chlorination Chlorine is often used in the oxidation of divalent iron and manganese because, unlike aeration, it has a faster oxidation rate and is capable of oxidizing organically bound iron.

$$2Fe(HCO_3)_2 + Ca(HCO_3)_2 + Cl_2 \rightarrow 2Fe(OH)_3 + CaCl_2 + 6CO_2$$

$$Mn(HCO_3)_2 + Ca(HCO_3)_2 + Cl_2 \rightarrow MnO_2 + CaCl_2 + 4CO_2 + 2H_2O$$

The equations above show that 1 mg/L of chlorine oxidizes 1.58 mg/L of iron and that 0.78 mg/L of manganese is oxidized by 1 mg/L of chlorine.

The rates of the oxidation reactions are pH dependent. For iron to be oxidized within 15 to 30 min, the pH of the process water should be approximately 8.0 to 8.3, preferably 8.5. Under these same conditions, manganese will be oxidized within 2 to 3 h. When the raw water contains ammonia, chloramines are formed upon addition of chlorine. Consequently, the oxidation rate is reduced. If THMs and HAA_5 formation is discouraged, chlorination cannot be employed as an oxidant.

Generally, the standard process train for iron and manganese removal is alum flocculation, followed by clarification and filtration; the practice of prechlorination leaves a minimum level of 0.5 mg/L of available free chlorine. This process train quickly coats the filter medium grains with a layer of manganese dioxide and ferric hydroxides or ferric oxides. Once the filter grains have been coated, the filter bed is capable of effectively removing manganese and iron by means of continuous prechlorination.

Chlorine Dioxide Chlorine dioxide is a strong oxidant that effectively oxidizes organically complexed iron and manganese. The pH level necessary for oxidation is a very important factor in the reaction rate and should be a minimum of 7.0.

$$Fe(HCO_3)_2 + NaHCO_3 + ClO_2 \rightarrow Fe(OH)_3 + NaClO_2 + 3CO_2$$

$$Mn(HCO_3)_2 + 2NaHCO_3 + 2ClO_2 \rightarrow MnO_2 + 2NaClO_2 + 4CO_2 + 2H_2O$$

Theoretically, 1 mg/L of chlorine dioxide will oxidize 0.83 mg/L of iron, and the same amount of chlorine dioxide will oxidize 0.41 mg/L of manganese.

Engineers must remember that many regulatory agencies are limiting the use of chlorine dioxide due to the potential toxicity of the chlorite and chlorate ions formed during the production of chlorine dioxide. Another important issue is the operational cost. The chlorine dioxide process is approximately three times as expensive as chlorination.

Potassium Permanganate Potassium permanganate is a strong oxidant. The reaction time is therefore fast in a wide range of pH. The oxidation time varies from as short as 5 min for both iron and manganese, provided that the pH is over 7.0.

$$3Fe(HCO_3)_2 + KMnO_4 + 7H_2O \rightarrow 3Fe(OH)_3 + MnO_2 + KHCO_3 + 5H_2CO_3$$

$$3Mn(HCO_3)_2 + 2KMnO_4 + 2H_2O \rightarrow 5MnO_2 + 2KHCO_3 + 4H_2CO_3$$

The reactions above indicate that 1 mg/L of potassium permanganate oxidizes 1.06 mg/L of iron and 0.52 mg/L of manganese. If both iron and manganese are present in the raw water, the oxidation process may be more cost-effective if chlorine is added prior to feeding potassium permanganate; the chlorine oxidizes iron rather easily, leaving the more expensive potassium permanganate to oxidize the manganese.

In the field of water treatment, it is well recognized that the addition of potassium permanganate to the head of a conventional complete treatment process helps to improve flocculation and sedimentation. This phenomenon is due to the seeding effect of the oxidized iron and manganese particles. Under these conditions, the media grains in the upper portion of the filter bed soon become coated with manganese oxide and ferric hydroxide in the same manner as the prechlorination process.

Ozonation The use of ozone for the sole purpose of oxidizing ferrous iron and manganese is very unlikely due to the availability of other cost-effective methods. Yet ozonation has been practiced successfully in Europe for the removal of both iron and manganese by means of a conventional complete treatment process coupled with preozonation. Experience has shown that excessive ozone oxidation of manganese forms permanganate and the water in the clearwell becomes pink in color.

$$2Fe(HCO_3)_2 + O_3 + 2H_2O \rightarrow 2Fe(OH)_3 + O_2 + 4CO_2 + H_2O$$

$$Mn(HCO_3)_2 + O_3 + 2H_2O \rightarrow MnO_2 + O_2 + 2CO_2 + 3H_2O$$

Theoretically, 1 mg/L of ozone will oxidize 2.3 mg/L of iron and 1.15 mg/L of manganese.

Ion Exchange When the raw water contains less than 0.5 mg/L of iron and manganese, either a cation or hydrogen cation type of ion exchange unit may be employed to remove these elements; these units are also capable of removing hardness. Yet if the processing water contains any amount of dissolved oxygen, there is always the danger of fouling the ion exchange material and clogging the exchanger bed. Thus it is vital that the process be run in the absence of oxygen to prevent the elements from oxidizing. The ion exchange process is generally limited to the processing of industrial water.

Manganese Zeolite Filtration Green sand impregnated with manganese (glauconite) is commonly known as zeolite. Each grain of the sand is coated with iron and manganese oxides. This type of filter medium is a form of ion exchange commonly used in industries. In this process, potassium permanganate is fed to the filter influent on a continuous basis, oxidizing soluble iron and manganese, and continuously regenerating the filter medium. It is very important that a proper dosage of potassium permanganate be applied to the process water, since an excessive dosage yields an effluent that is pink in color, while an inadequate dosage results in the leakage of manganese into the filter effluent.

In the case of groundwater treatment, zeolite is placed in pressure filter cells rather than gravity filters, to preserve the discharge pressure of the well pump. This scheme eliminates the problem of repumping and allows one filter to be backwashed by using the pressured filtered water from the remaining filters. The filtration rate of this process usually ranges from 3 to 4 gpm/ft^2 (7.5 to 10 m/h). Manganese zeolite filtration is generally limited to water containing a maximum level of 1 mg/L of iron or manganese. However, it has been reported that zeolite has the capacity to remove 0.09 lb/ft^3 (1.5 kg/m^3) of either iron or manganese and to regenerate 0.18 lb/ft^3 (2.9 kg/m^3) of filter medium by potassium permanganate.

Sequestering Process The purpose of the sequestering process is to hold both iron and manganese in solution by means of sequestering chemicals. This process is generally applicable to water containing less than 2 mg/L of iron and manganese; both elements must be present in their bicarbonate form. The sequestering agents include compounds such as sodium silicate, trisodium phosphate, hexametaphosphate, and zinc orthophosphate.

Proper dosage of the sequestering chemicals should be calculated on the basis of the manufacturer's suggested dosage. Approximately 2 mg/L of hexametaphosphate may be used for 1 mg/L of iron. However, polyphosphate dosages are limited to less than 5 mg/L due to its tendency to stimulate biological growth in the distribution system. In this case, an adequate amount of residual chlorine must be present in the process water so that the growth of the microorganisms can be controlled.

Engineers should note that the sequestering process is effective only on cold water: the sequestering agents lose their dispersing properties if the water is either highly heated or boiled. For this reason, this process is seldom employed in treating medium-sized to large domestic water supply systems. Yet small water systems are often financially incapable of implementing oxidation and filtration or lime softening processes, and the much simpler and less costly (capital) sequestering process may therefore be the appropriate choice for treating their well water.

Lime Softening As described in Section 7.1, iron and manganese are effectively removed by the softening process provided that the pH of the process water is above 9.5. Based on the pH and solubility relationships, 83% of the iron precipitates at a pH of 8.4, but 92 and 100% of the iron precipitates at the respective pH values of 8.8 and 9.6. Manganese is more difficult to remove: No manganese will precipitate out at a pH of 8.8; however, a pH of 9.4 and 9.8 will remove 98 and 100% of the manganese, respectively.

7.2.2 Case History

A water treatment plant located in the northern part of the United States treats shallow groundwater containing approximately 5 to 5.5 mg/L of ferrous ion and 0.6 mg/L of manganese. The water has a total hardness of 200 mg/L and a pH of approximately 7.1. A sludge blanket type of reactor clarifier and standard dual-media filters are used to treat this water. The basic treatment process is partial softening with lime. This scheme automatically reduces the iron level of the filtered water to nearly 0.1 mg/L and also reduces the manganese level to below 0.05 mg/L.

The plant conducted an experimental study to evaluate the feasibility of reducing the required lime dosage as well as the feasibility of reducing the sludge production rate based on the oxidation of ferrous iron through application of an appropriate amount of chlorine. The results were very interesting and proved to be educational for those people designing iron removal systems under similar conditions. Table 7.2.2-1 presents some of the results. These results indicate that the chlorinated water must have a pH of at least 8.2 if the treatment process is expected to produce filtered water containing less than 0.2 mg/L iron within a reasonable

TABLE 7.2.2-1 Main Results of the Iron Removal Test

Lime dosage—range (mg/L)	45–55	55–65	65–75	75–85	85–95
pH of the settled water	7.8–7.9	7.9–8.2	8.3–8.6	8.7–8.9	8.9–9.1
Turbidity of the filtered water	0.7–2.2	0.1–1.2	0.05–0.2	0.05–0.12	0.05–0.15
Iron (mg/L)					
Settled water	3.9–4.1	3.1–3.9	2.6–3.1	2.1–2.5	2.0–2.5
Filtered water	0.5–0.8	0.1–0.5	0.01–0.2	0.01–0.05	0.01–0.05
Filter run length (h)	10–16	12–22	26–30	24–34	24–30

Notes: (1) Approximately 5 mg/L of chlorine were fed with lime for all cases.
(2) The hardness of the filtered water increased in proportion to the lime dosages for all cases where the pH was less than 8.5. However, a 5–10 mg/L reduction in hardness only occurred in cases where the pH exceeded 8.5.
(3) The water temperature ranged from 6 to 7°C.
(4) The filtration rate was approximately 2 gpm/ft^2.

period of time: a filter run length of 24 h. The increase in filtered water hardness at pH values below 8.5, after the addition of lime, is the obvious result of the conversion of most of the lime (calcium hydroxide) to calcium bicarbonate; the calcium bicarbonate is completely soluble in water at lower pH conditions.

Bench-scale studies were also conducted at the plant to evaluate the effectiveness of anionic polymer for improving the settleability of iron hydroxide prior to filtration. Some of the test results are listed in Table 7.2.2-2. The results listed in

TABLE 7.2.2-2 Jar Test Results for the Polymer Evaluation Tests

Beaker Number	Cl_2 Dosage	$KMnO_4$ Dosage	Polymer Dosage	Floc Size	Settled Water	
					Turbidity	Iron
1	0	5	0	Small	5.8	4.5
2	0	5	0.2	Medium to large	3.5	2.3
3	2.5	2.5	0.2	Large	2.7	2.1
4	5	0	0	Small	7.0	4.7
5	5	0	0.2	Medium to large	3	1.6
6	5	0	0.2	Large	2	1.4

Notes: (1) Iron content of the raw water was 5 mg/L.
(2) The jar test was conducted at $G \times t = 3 \times 10^4$, with a total mixing time of 10 min and a settling time of 20 min.
(3) Polymer (Nalco 7763) was added approximately 1 min after the oxidant was applied to the beakers, with the exception of beaker 5 and 6.
(4) The polymer was added 2.5 min after chlorine application for beaker 5 and 5 min after the addition of chlorine for beaker 6.
(5) The pH was adjusted to 8.2 by NaOH prior to the addition of oxidant.
(6) The temperature of the raw water was 5.6°C.
(7) The chemical dosages are all in mg/L. The unit of iron in water is also mg/L.

the table demonstrate two major issues. The first is an obvious improvement in both the iron hydroxide characteristics and iron removal (via sedimentation) through the addition of anionic polymer in concentrations as low as 0.2 mg/L. The second issue is the time lag between the addition of oxidant and polymer. With respect to the polymer, a time delay of 5 min significantly improves the removal of both turbidity and iron by means of sedimentation.

BIBLIOGRAPHY

Adams, R. B., "Manganese Removal by Oxidation with Potassium Permanganate," *J. AWWA,* 52:2:219 (February 1960).

AWWA, *Water Treatment Plant Design,* American Water Works Association, New York, 1969.

AWWA Committee Report, "Research Needs for the Treatment of Iron and Mn," *J. AWWA,* 79:9:119 (September 1987).

Boatby, J. R., "Optimum Mn Removal and Washwater Recovery at Filtration Plant in Brazil," *J. AWWA,* 80:12:71 (December 1984).

Conneley, E. J., "Removal of Iron and Manganese," *J. AWWA,* 50:5:697 (May 1958).

Culp, R. L., Wesner, M. R., Culp, G. L., *Handbook of Public Water Systems,* Van Nostrand Reinhold, New York, 1986.

Jenkins, S. R., et al., "Effective Manganese Removal Using Lime as an Additive," *J. AWWA,* 76:9:82 (September 1984).

Knocke, W. R., et al., "Examining the Reactions Between Soluble Iron, DOC, and Alternative Oxidants During Conventional Treatment," *J. AWWA,* 86:1:117 (January 1994).

Knocke, W. R., et al., "Soluble Mn Removal on Oxide-Coated Filter Media," *J. AWWA,* 80:12:65 (December 1988).

Knocke, W. R., et al., "Using Alternative Oxidants to Remove Dissolved Manganese from Water Laden with Organics," *J. AWWA,* 79:3:75 (March 1987).

Montgomery, J. M., Consulting Engineers, *Water Treatment: Principles and Design,* Wiley, New York, 1985.

Robinson, R. B. and Ronk, S. K., "The Treatability of Mn by Sodium Silicate and Chlorine," *J. AWWA,* 79:11:64 (November 1987).

Robinson, R. B., et al., "Iron and Manganese Sequestration Facilities Using Sodium Silicate," *J. AWWA,* 84:2:77 (February 1992).

Sanks, R. L., *Water Treatment Plant Design,* Ann Arbor Science, Ann Arbor, MI, 1978.

Victoreen, H. T., "Controlling Corrosion by Controlling Bacteria Growth," *J. AWWA,* 76:3:87 (March 1984).

Vitaraghavan, T., et al., "Removing Mn from Water at Fredericton, N.B., Canada," *J. AWWA,* 79:8:43 (August 1987).

Welch, W. A., "Potassium Permanganate in Water Treatment," *J. AWWA,* 55:6:735 (June 1963).

Willey, B. F., et al., "Iron Removal with Potassium Permanganate," *J. AWWA,* 55:6:729 (June 1963).

Wong, J. M., "Chlorination Filtration for Fe and Mn Removal," *J. AWWA,* 76:1:76 (January 1984).

7.3 TASTE AND ODOR CONTROL

In the water supply industry, it is not unusual for water purveyors to receive customer complaints regarding the objectionable taste and odor of the potable water. These complaints are more frequent when the source is surface water rather than groundwater, because surface water is more likely to be affected by microorganisms and organic substances as well as manufactured wastes. Some consumers also consider the odor of the residual chlorine in tap water to be offensive.

The most frequent causes of taste and odor in the water supply are algae and *Actinomycetes* (during certain seasons). The second is decaying vegetation. Taste and odor problems are occasionally the result of hydrogen sulfide, agricultural runoff, industrial chemical spills, illegally discharged industrial chemicals, and pollution by sewage. The current (1999) drinking water quality standards set the threshold odor number (TON) to less than 3. It is generally difficult to control taste and odor problems once they have developed. The plant design should therefore provide the operator with a certain amount of flexibility, that is, several available options. It is very important to allow the use of alternative chemicals, as well as supplying alternative feed points, so that the location and type of chemical can be changed in accordance with changes in raw water quality. Moreover, the intake design should have provisions that allow the intake to withdraw water from selected depths in the lake or reservoir; this is an essential design factor.

7.3.1 Major Taste- and Odor-Producing Substances

The detection and intensity of taste and odor are very subjective issues: Some people are very sensitive, whereas others have a very high level of tolerance. One method that has been developed in recent years is flavor profile analysis (FPA). FPA is designed to reduce or eliminate the subjective nature of this problem by using a panel of people who are trained to analyze odors associated with processed water.

Most odor problems are the result of algae or *Actinomycetes*. Table 7.3.1-1 lists the major odor-producing compounds and the responsible species of algae. Among the typical odor-producing compounds, geosmin and 2-methylisoborneol (MIB) often impart objectionable odor at very low concentrations (10 ng/L or 0.01 μg/L). Therefore, the acceptable level of these compounds is considered to be approximately 8 ng/L.

The taste of the potable water can generally be attributed to the presence of small amounts of metals, such as iron and copper, and phenolic compounds. For example, 0.05 to 0.1 mg/L of dissolved iron, 2 to 5 mg/L of divalent copper, and 4 to 9 mg/L of dissolved zinc are considered to be taste thresholds. In the case of phenol, the threshold level is 1 mg/L; it is 0.002 g/L for chlorophenol (chlorophenol is formed by the chlorination of phenol-bearing water). Generally, tastes and odors are objectionable to the consumer. However, under normal conditions, they are not a threat to the health of the consumer, provided that water treatment plants are operated properly.

Refer to Table 7.3.1-2 for a list of taste- and odor-producing algae and their characteristic features. Table 7.3.1-3 presents the troublesome algae and the recommended copper sulfate and chlorine dosages that should be used in their control.

TABLE 7.3.1-1 Taste- and Odor-Producing Algae and Characteristics of the Taste and Odor

Algal Genus	Algal Group	Odor When Algae Are: Moderate	Odor When Algae Are: Abundant	Taste	Tongue Sensation
Anabaena	Blue-green	Grassy, nasturtium, musty	Septic	—	—
Anacystis	Blue-green	Grassy	Septic	Sweet	—
Aphanizomenon	Blue-green	Grassy, nasturtium, musty	Septic	Sweet	Dry
Asterionella	Diatom	Geranium, spicy	Fishy	—	—
Ceratium	Flagellate	Fishy	Septic	Bitter	—
Dinobryon	Flagellate	Violet	Fishy	—	Slick
Oscillatoria	Blue-green	Grassy	Musty, spicy	—	—
Scenedesmus	Green	—	Grassy	—	—
Spirogyra	Green	—	Grassy	—	—
Synura	Flagellate	Cucumber, muskmelon, spicy	Fishy	Bitter	Dry, metallic slick
Tabellaria	Diatom	Geranium	Fishy	—	—
Ulothrix	Green	—	Grassy	—	—
Volvox	Flagellate	Fishy	Fishy	—	—

Source: Adapted from Palmer (1962).

7.3.2 Control Measures

Three basic taste and odor control measures are used in the treatment of potable water: (1) prevention at the source, (2) removal at the treatment plant, and (3) control within the distribution system (refer to Section 7.3.4). The fundamentals of each of these measures are discussed.

Preventive Measures Methods used in prevention at the source include (1) reservoir mixing, (2) aquatic plant control, and (3) water pollution control.

Reservoir Mixing Most lakes and reservoirs stratify during warm seasons, frequently causing the bottom layer (hypolimnion) to be oxygen deficient. Under such anaerobic conditions, both hydrogen sulfide and the degradation of organic substances result in the emission of objectionable odors. Experience gained during the last half of the twentieth century has shown that artificial mixing of lakes by means of mechanical mixers, submersible pumps, or air-lift pumps significantly improves water quality, thereby reducing the burden on water treatment plants in controlling taste and odor problems, as well as iron and manganese. The design engineer should therefore evaluate the feasibility of installing a reservoir mixing device as a preventive measure for taste and odor.

TABLE 7.3.1-2 Odor-Producing Algae and Chemical Structures of Odors

Compound	Structure	Associated Organisms
Methylisoborneol (MIB)	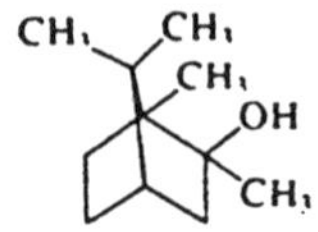	*Actinomycetes* *Oscillatoria curviceps* *Oscillatoria tenuis*
Geosmin	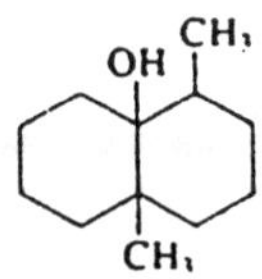	*Actinomycetes* *Sympioca muscoum* *Oscillatoria tenuis* *Oscillatoria simplicissima* *Anabaena scheremetievi*
Mucidone	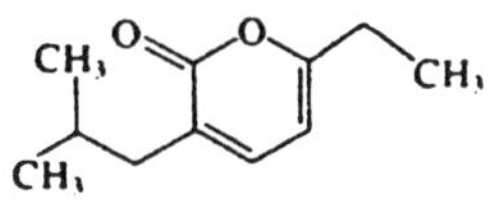	*Actinomycetes*
Isobutyl mercaptan	$(CH_3)_2{>}CHCH_2{-}SH$	*Microcystis flos-aquae*
N-Butyl mercaptan	$CH_3(CH_2)_3{-}SH$	*Microcystis flos-aquae* *Oscillatoria chalybea*
Isopropyl mercaptan	$(CH_3)_2{>}CHSH$	*Microcystis flos-aquae*
Dimethyl disulfide	$CH_3{-}S{-}S{-}CH_3$	*Microcystis flos-aquae* *Oscillatoria chalybea*
Dimethyl sulfide	$CH_3{-}S{-}CH_3$	*Oscillatoria chalybea* *Anabaena*
Methyl mercaptan	CH_3SH	*Microcystis flos-aquae* *Oscillatoria chalybea*

Source: Adapted from J.M. Montgomery, Consulting Engineers, *Water Treatment: Principles and Design,* Wiley, New York, 1985.

Aquatic Plant Control The control of marginal aquatic plants is a very effective method in preserving the (good) water quality of a reservoir. However, this task is not easily achieved when the lake or reservoir is large. One method of controlling marginal plants, including algae, is to drastically change the water level several times a year. When the water level is lowered by 4 to 5 feet (1.2 to 1.5 m), the roots of the marginal plants become exposed to sun and dry up.

The most common method of controlling algae growth is the use of copper sulfate. The required dosage varies with the type of algae and ranges from 0.05 to 0.8 mg/L. Sensitive species such as blue-green algae, which frequently cause intense taste and odor, may be treated with lower dosages. However, green algae, which occasionally causes taste and odor problems, require much higher dosages. Yet dosages higher than 0.8 mg/L are not recommended since they are lethal to most types of fish. Potassium permanganate is an alternative to copper sulfate. Its effective dosage ranges from 0.4 to 4 m/L. Nonetheless, factors such as low solubility (5 g/100 mL at 20°C), higher cost ($2000 per ton), and the development of a purple color generally discourage engineers and plant operators from using potassium permanganate.

TABLE 7.3.1-3 Troublesome Algae and Required Dosages of Control Chemicals

	Organism	Trouble	Copper Sulfate (mg/L)	Chlorine (mg/L)
Algae				
Diatoms	*Asterionella, Synedra, Tabellaria*	Odor: aromatic to fishy	0.1–0.5	0.5–1.0
	Fragillaria, Navicula	Turbidity	0.1–0.3	—
	Melosira	Turbidity	0.2	2.0
Grass-green	*Eudorina,*[a] *Pandorina*[a]	Odor: fishy	2–10	—
	Volvox[a]	Odor: fishy	0.25	0.3–1.0
	Chara, Cladophora	Turbidity, scum	0.1–0.5	—
	Coelastrum, Spirogyra	Turbidity, scum	0.1–0.3	1.0–1.5
Blue-green	*Anabaena, Aphanizomenon*	Odor: moldy, grassy, vile	0.1–0.5	0.5–1.0
	Clathrocystis, Coelosphaerium	Odor: grassy, vile	0.1–0.3	0.5–1.0
	Oscillatoria	Turbidity	0.2–0.5	1.1
Golden or yellow-brown	*Cryptomonas*[b]	Odor: aromatic	0.2–0.5	—
	Dinobryon	Odor: aromatic to fishy	0.2	0.3–1.0
	Mallomonas	Odor: aromatic	0.2–0.5	—
	Synura	Taste: cucumber	0.1–0.3	0.3–1.0
	Uroglenopsis	Odor: fishy. Taste: oily	0.1–0.2	0.3–1.0
Dinoflagellates	*Ceratium*	Odor: fishy, vile	0.2–0.3	0.3–1.0
	Glenodinium	Odor: fishy	0.2–0.5	—
	Peridinium	Odor: fishy	0.5–2.0	—
Filamentous bacteria	*Beggiatoa* (sulfur)	Odor: decayed. Pipe growths	5.0	—
	Crenothrix (iron)	Odor: decayed. Pipe growths	0.3–0.5	0.5
Crustacea	*Cyclops*	*c*	—	1.0–3.0
	Daphnia	*c*	2.0	1.0–3.0
Miscellaneous	*Chironomus* (bloodworm)	*c*	—	15–50
	Craspedacusta (jellyfish)	*c*	0.3	—

Source: Adapted from J.M. Montogomery, Consulting Engineers.

[a]These organisms are classified also as flagellate protozoa.

[b]Classification uncertain.

[c]These organisms are individually visible and cause consumer complaints.

Other methods that have been tried in the past include PAC—to create a "black-out" condition—and the application of chlorine compounds. Carbon treatment is usually limited to small reservoirs and is a temporary measure because the carbon settles out rather quickly unless a certain degree of agitation is provided in the reservoir. The effective dosage of PAC ranges from 2 to 10 mg/L. Chlorine compounds such as calcium hypochlorite and sodium hypochlorite may be used to control algae with 0.2 to 1 mg/L of residual free chlorine. Yet their effectiveness is quickly lost with exposure to sunlight and waterborne organic materials.

Groundwater Source Management Groundwater is frequently contaminated by pollutants such as gasoline, industrial solvents, and a wide range of volatile organics. The prevention of groundwater contamination has just begun in recent years, but this issue will most likely continue being a source of engineering concern for several more years, since today's problems are generally the result of practices and politics that have been ongoing for the past 10 to 50 years.

Reservoir Management In regions that experience the four seasons, most lakes and reservoirs undergo thermal stratification. Under these conditions, a stagnation zone (hypolimnion) is formed and the water below 25 ft (7.5 m) is nearly stagnant during both the warm and cold months. Since the stagnation zone has little to no dissolved oxygen, the sulfates and nitrates serve as a source of oxygen for biochemical oxidation by anaerobic bacteria and the sulfate ion is reduced to sulfide ion to form hydrogen sulfide. The level of hydrogen sulfide formed in this manner may be as high as 5 to 6 mg/L. However, the well water in some areas may have hydrogen sulfide levels as high as 20 to 30 mg/L. Conversely, the water of impoundments overturns twice a year—spring overturning and autumn overturning—due to the change in water temperature. This phenomenon brings the hydrogen sulfide, as well as other odor-producing compounds and unfavorable elements, to the surface.

Positive and effective control of hydrogen sulfide formation may be attained by mixing the entire body of lake water, thus preventing the formation of a stagnation zone within the reservoir. This artificial destratification control method has proved to be effective for most reservoirs. Mixing may be achieved through the use of mechanical mixers, pumps, or diffused air pipes. However, the mixing is not always effective since the shape, length, and depth of the lake, as well as the design and location of the mixer, strongly affect effective mixing.

Removal at the Treatment Plant If the preventive measures are ineffective or impractical, the problem of taste and odor must be controlled at the treatment plant. The three basic measures are (1) aeration, (2) oxidation, and (3) adsorption.

Aeration Aeration is a practical solution in controlling taste and odor problems caused by volatile compounds such as hydrogen sulfide. It is generally not the best method in controlling geosmin and MIB, the most frequent taste- and odor-producing compounds. Examples of the aeration process include diffused mechanical nozzle spraying, multiple tray cascading, and packed tower type.

Oxidation In most cases, oxidation is the most practical and effective method of controlling taste and odor problems. Oxidation can be potassium permanganate oxidation, ozonation, ozone and hydrogen peroxide oxidation, chlorine dioxide, or chlorination. The most frequently used oxidizing chemical is potassium permanganate. The dosage ranges from 0.5 to 5 mg/L: however, taste and odor are commonly reduced to an acceptable level of 3 TON through the use of 0.3 to 0.6 mg/L.

Ozonation is also an effective method of oxidation. This process actually changes the characteristics of the odor and flavor, in addition to reducing the level of the odor-producing compounds. The effective dosage of ozone ranges from 1 to 5 mg/L, with 2 mg/L being the average effective dosage. With this method, the key to successful taste and odor control is to provide alternative chemical feed points and an alternative chemical feed system in case it becomes necessary to use an alternative oxidant. Several pilot studies conducted in recent years indicate that the concurrent use of ozone and hydrogen peroxide may or may not be superior to the use of ozone alone. Thus bench test or pilot studies should be conducted to evaluate the effectiveness of this method.

Chlorine dioxide has been used effectively in the past. However, its use is now discouraged by both the EPA and the departments of health services in certain states due to the potential harmful effects of chlorite on human health. Similarly, chlorination is also an effective method of taste and odor control, but its use must be carefully evaluated due to the possible formation of DBPs and chlorophenol (chlorophenol is produced when phenol is oxidized in water). Generally, if any taste and odor problems are anticipated, the water treatment plant must have a potassium permanganate and a chlorine feed system. The design of these chemical feed systems is covered in Section 4.1. It is also essential to conduct bench-scale studies to help select the optimum type and dosage of oxidant prior to plant design.

Figure 7.3.2-1 illustrates the relation between MIB concentration and ozone dosage with a contact time of 2.5, 5, and 10 min. Figure 7.3.2-2 presents the changes in flavor at various ozone dosages and contact times. The required ozone dosage to remove MIB is specific to individual water. However, a general idea of the dosage and removal rate can be obtained from Figure 7.3.2-3.

Adsorption The two basic and effective adsorption processes in removing taste and odor are the addition of PAC to the process water and the use of GAC adsorption beds (to filter the water). The use of PAC adsorption is most appropriate in situations where moderate taste and odor problems are infrequent: midsized to small plants. The two basic types of PAC storage and feed systems are dry carbon storage and dry feeding. The rule of thumb dictates that if the hourly feed rate of PAC is less than 150 lb, the PAC dry feeder should be used in combination with a solution mixing tank or a vortex mixer. If the hourly feed rate exceeds 150 lb and there is a need for frequent PAC addition, the use of a slurry feed system should be considered. The PAC dosage varies from 1 to 50 mg/L. However, based on past experience, 25 mg/L should be considered to be the maximum allowable dosage, regardless of the type of feeding (dry or slurry). Refer to Section 4.1 for the design consideration of the PAC feed system.

Whenever a moderate to severe chronic taste and odor problem exists, the use of GAC adsorption beds should be considered. GAC adsorption is similar to ordinary filters; however, the empty bed contact time (EBCT) is a very important design issue. The EBCT values typically range from 3 to 10 minutes (mostly 3 min) for the purpose of taste and odor removal; the filtration rate of the GAC adsorption

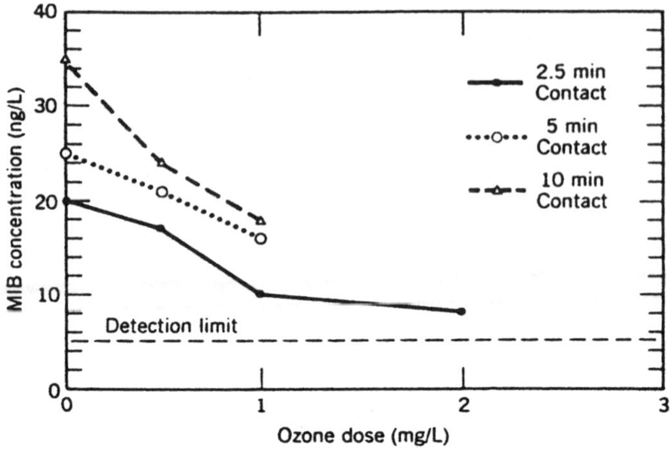

Figure 7.3.2-1 Effect of ozone contact time on MIB removal.

beds ranges from 3 to 6 gpm/ft^2 (7.5 to 15 m/h). The standard size of the medium is approximately the same as that of rapid sand filter beds, and the depth of the layer is approximately twice the size. Operational experience gained from European facilities indicates that the GAC beds need to be regenerated every 4 to 5 years. Further details on the GAC adsorption process are given in Section 7.7.

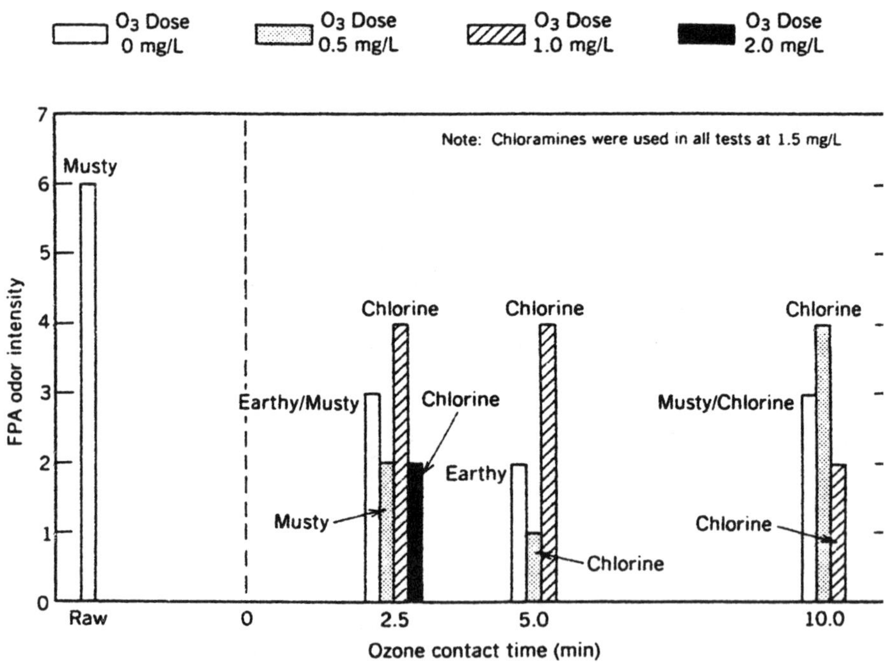

Figure 7.3.2-2 Changes of flavor and odor intensity by ozonation.

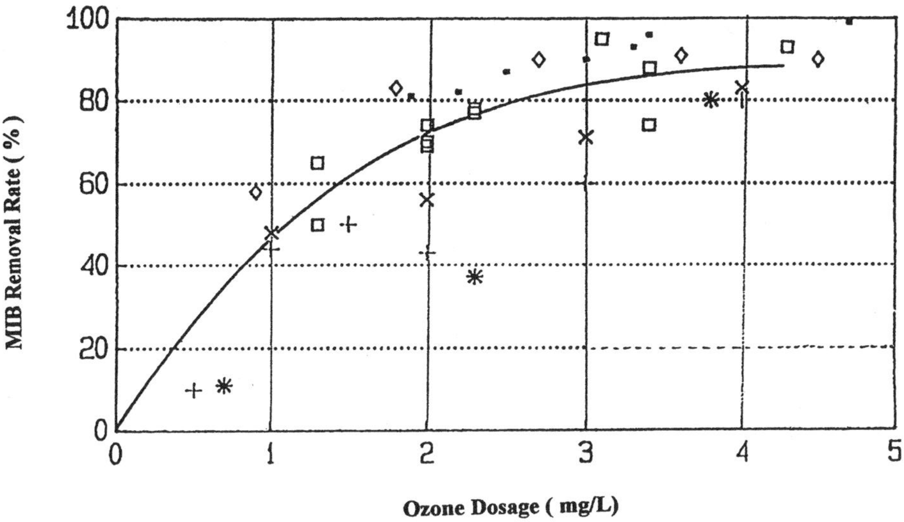

Figure 7.3.2-3 MIB removal by ozone for several surface waters.

Figure 7.3.2-4 illustrates the effect of the EBCT on the removal of several different types of odor based on a pilot study. It is very important to run a bench-scale or pilot test so that the appropriate type and size of GAC may be selected and to find the optimum EBCT for dominant taste and odor compounds. Water treatment plants that require frequent taste and odor removal by activated carbon should compare the cost and ease of operation and maintenance, as well as the level of maintenance work, of PAC versus GAC. Figure 7.3.2-5 presents an example cost comparison between PAC and GAC for taste and odor control. The figure is created based on the following assumptions: an average plant flow rate of 114 mgd (5 m^3/s), a filtration (adsorption) rate of 5 gpm/ft^2 (12.5 m/h), cost of PAC at \$1000 per ton and \$1700 per ton for GAC. The capital costs for a PAC storage and feeding system and GAC facilities with a regeneration cost of GAC are not included; thus the comparison is based only on the cost of material. This figure shows the initial placement cost of GAC for 3, 4, and 5 minutes of EBCT and an averaged annual cost of replacement (every 3 years). The cost of PAC is based on the duration of PAC feeding from 0 to 365 days (12 months) at a feed rate of 5, 10, 15, and 20 mg/L. A general idea for the cost of PAC and GAC under selected conditions can be obtained; for instance, the graph shows the annual cost of GAC when replaced every 3 years (this is considered normal) for 3 min of EBCT is approximately \$270,000. The same cost for PAC can be found where the four diagonal lines of PAC dosages, indicated as 5, 10, 15, and 20 mg/L, intersect the vertical line of 3 min of EBCT. Thus if the PAC dosage is 5 mg/L, a total of 3 months per year costs the same as GAC with 3 min of EBCT. Also, PAC dosages of 10 mg/L for 45 days, 15 mg/L for 1 month, and about 20 days at 20 mg/L yield the same cost as GAC with 3 min. If the days of PAC feeding exceed these numbers of days, the cost of PAC is higher. Since PAC ends up in the sludge and is disposed, there is no recovery cost. However, GAC can

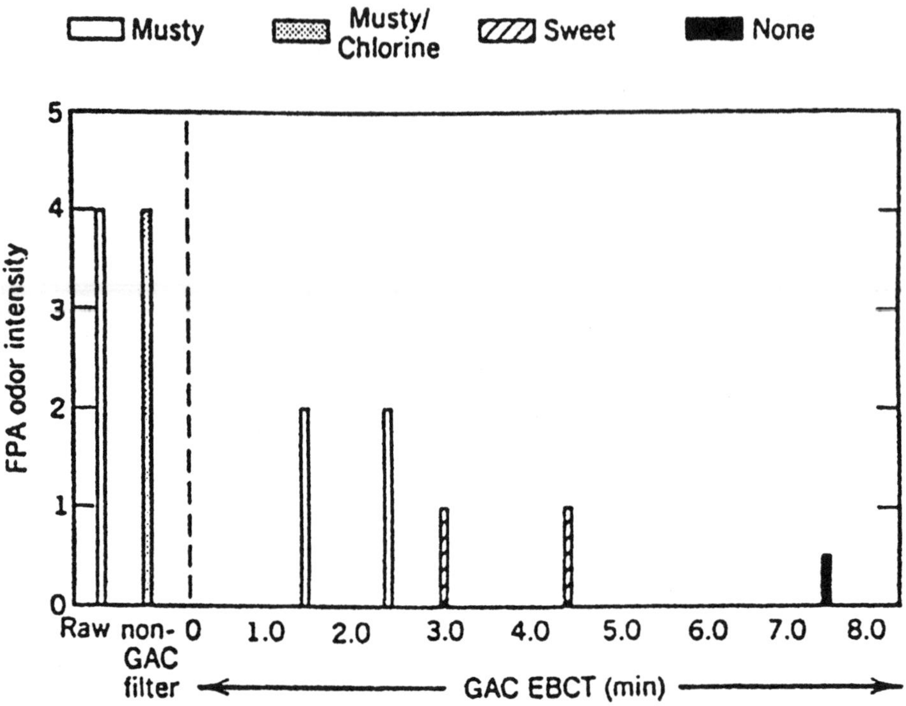

Figure 7.3.2-4 Odor removal by GAC adsorption process.

be reactivated, and approximately 85% of the original GAC is reused. It should be noted that the PAC feeding and storage facility is bulky and dirty, and PAC slurry tends to solidify in the holding tank, feed line, and even the sludge in the clarifier will solidify if the dosage is high. The effective PAC feeding rate for most taste and odor control situations is 10 to 20 mg/L when MIB and geosmin are the cause of the problem, since the removal rate of these compounds by PAC is about 50% (Table 7.3.2-1). Table 7.3.2-1 presents the results of bench-scale tests conducted to evaluate the effectiveness of taste and odor removal by PAC and four other typical oxidants, spiking 100 ng/L each of MIB and geosmin to filtered water.

7.3.3 Hydrogen Sulfide Removal

The removal of hydrogen sulfide is discussed separately from the removal of other odor producing compounds because of its unique characteristics. In natural water, under normal conditions, the sulfur species exists in five stable forms: HSO_4^-, SO_4^{2-}, S^0, H_2S, and HS^-. Other species such as thiosulfate, polysulfide, and polythionate also exist in the natural water. However, they are not thermodynamically stable. The following relationship exists among the sulfur species. In water with a pH of 8 or below, a normal pH range for natural water, H_2S and HS^- are the dominant forms of sulfur. At pH 7, approximately 80% of the sulfur is in the form of hydrogen sul-

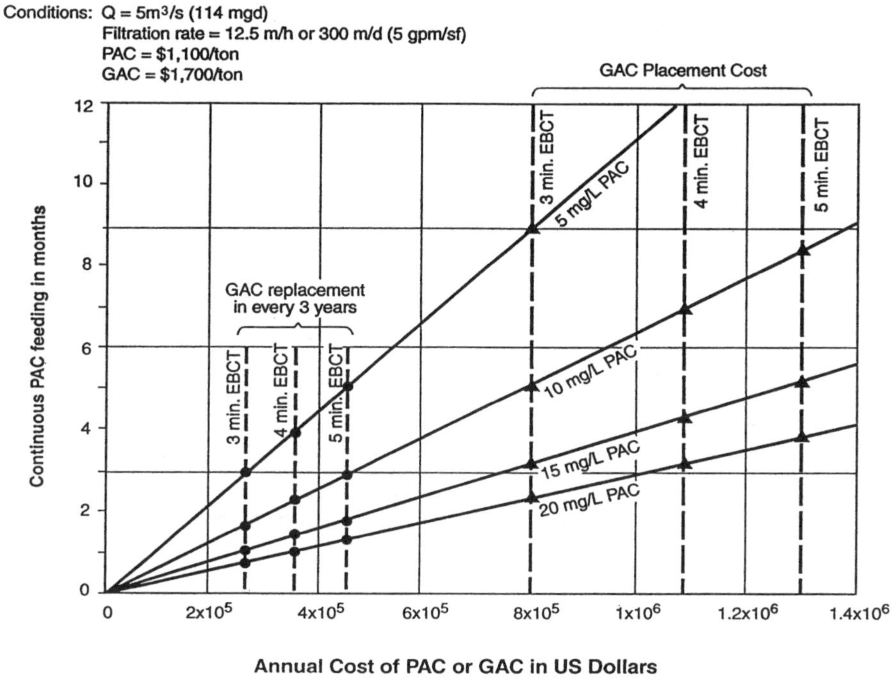

Figure 7.3.2-5 Choice of PAC or GAC for taste and odor control.

fide and almost 100% at pH 6. However, at pH levels of 8 and above, the reduced sulfur exists in the water as HS^- and SO_4^{2-} ions, and the amount of free H_2S is very small and insignificant.

The most common method of removing hydrogen sulfide from water is oxidation. Adsorption by activated carbon is also an effective method but rather expensive. The oxidation process used in the field of water treatment includes (1) aeration, (2) chlorination, (3) ozonation, and (4) potassium permanganate oxidation.

Aeration There are three basic types of aeration: spraying, diffused air, and multiple tray. The aeration process is both practical and satisfactory if the level of total sulfides (expressed as H_2S) is less than 3 to 4 mg/L. It is essential to adjust the pH of the water to below 7, preferably 6, prior to aeration. Under the proper pH conditions, aeration may liberate CO_2 and other volatile organic substances. The effective air/water flow rate ratio should be in the range 80 to 120 : 1, and the hydraulic loading for a tray aerator—the type most commonly used in well water treatment—is 7 to 15 gpm/ft^2 (17.5 to 37.5 m/h). For diffused air aeration, the power requirement for the blower is approximately 1 kW per each million gallons of water treated (1 kW per 3,785 m^3/d).

$$H_2S + \tfrac{1}{2}O_2 \rightarrow H_2O + S^0$$

TABLE 7.3.2-1 An Example of Geosmin and MIB Removal Rate by PAC, $KMnO_4$, Cl_2, H_2O_2, and O_3

Chemical	Feed Rate (mg/L)	Removal Rate (%)	
		Geosmin	MIB
Powdered Activated Carbon (PAC)	10	40	62
Powdered Activated Carbon	25	52	65
Potassium Permanganate ($KMnO_4$)	0.8	42	28
Chlorine (Cl_2)	2	45	33
Hydrogen Peroxide (H_2O_2)	1	50	72
Ozone (O_3)	2.5	94	77
Ozone and Hydrogen Peroxide	2.5 and 0.5	97	95

Note: 100 ng/L each of Geosmin and MIB were spiked to filtered water for these tests.

Chlorination Chlorination is often used to oxidize hydrogen sulfide in well water. The reaction is

$$4Cl_2 + H_2S + 4H_2O \rightarrow H_2SO_4 + 8HCl$$

Therefore, 8.3 parts of chlorine are required to oxidize 1 part of hydrogen sulfide. Factors that affect this reaction include pH, temperature, and reaction time. The oxidation rate causes the pH to increase sharply to 6.5 to 7.0.

Ozonation The ozonation process is seldom used to oxidize hydrogen sulfide. The efficiency of this process is affected strongly by the level of pH; the pH of the water should be maintained at 7 or less. The reaction is

$$H_2S + 4O_3 \rightarrow H_2SO_4 + 4O_2$$

The ratio of ozone to hydrogen sulfide is 5.7 : 1.

Potassium Permanganate Oxidation Potassium permanganate is a common oxidant that is often used in the water treatment process.

$$4KMnO_4 + 3H_2S \rightarrow S^0 + 2K_2SO_4 + 3MnO + MnO_2 + 3H_2O$$

This reaction illustrates that 6.2 parts of potassium permanganate are required to oxidize 1 part of hydrogen sulfide. Once again, pH control plays an important role in maintaining the pH in the range 6.5 to 7.

Elemental sulfur exists in water in a virtually insoluble form and thus may be removed by filtration. However, sulfur may also exist in water in colloidal form, as a milky white emulsion. In this form, the sulfur may not be removed satisfactorily by ordinary filtration processes and the odor problem will remain. The use of sodium pyrosulfite or sulfur oxide has been recommended by Monscvitz in treating this colloidal form of sulfur.

7.3.4 Control in the Distribution System

Taste and odor problems may also occur in the distribution system, primarily as the result of corrosion of pipe material and/or the growth of iron bacteria such as *Crenothrix* and *Leptothrix* within the water main. Additionally, in cases where the sulfate content of the water is high and the water is allowed to stand in dead-end mains, taste and odor problems may be compounded by sulfate-reducing bacteria. These problems can be rectified only if design engineers minimize the use of dead-end mains and provide blow-off and cleanout assemblies at strategic points in the distribution system. Furthermore, maintenance crews must keep the water lines clean by periodical flushing of deposits and microorganisms that accumulate within them. In cases where the interior of the pipe is badly corroded, the main must be cleaned by the "pig" and relined with new cement.

Water treatment plants can also minimize taste and odor problems by maintaining an adequate level of residual chlorine to combat bacteria in the distribution system and taking anticorrosion measures—either adjusting the pH of the water based on the saturation index or through the addition of a corrosion inhibitor to the finished water.

BIBLIOGRAPHY

Ansleme, C. I. H., et al., "Effects of Ozonation on Taste and Odor," *J. AWWA,* 80:10:45 (October 1988).

AWWA, *Water Treatment Plant Design,* American Water Works Association, New York, 1969.

AWWA Committee Report, "Research on Taste and Odors," *J. AWWA,* 62:1:59 (January 1970).

AWWA Committee Report, "Artificial Destratification in Reservoirs," *J. AWWA,* 63:9:597 (September 1971).

Bernhardt, H., and Clasen, J., "Flocculation of Micro-organisms," *J. Water SRT-Aqua,* 40:2:76–87 (1991).

Biederman, W. J., et al., "Destratification Using Air," *J. AWWA,* 63:7:462 (July 1971).

Craig, S. L., et al., "Optimizing the Removal of Geosmin and MIB by Powdered Activated Carbon," *J. AWWA,* 80:3:73 (March 1988).

Doughtery, J. D., and Morris, R. L., "Studies on the Removal of Actinomycetes Musty Tastes and Odors in Water Supply," *J. AWWA,* 59:10:1320 (October 1967).

Glase, W. H., et al., "Evaluating Oxidants for Removal of Model Taste and Odor Compounds," *J. AWWA,* 82:5:79 (May 1990).

Khiari, D., et al., "Sensory GC Analysis of Decaying Vegetation and Septic Odors," *J. AWWA,* 89:4:150 (April 1997).

Knoppert, P. L., et al., "Destratification Experiment at Rotterdam," *J. AWWA,* 62:7:448 (July 1970).

Krasner, S. W., et al., "Taste and Odors: The Flavor Profile Method," *J. AWWA,* 77:3:34 (March 1985).

Lalezary, S., et al., "Air Stripping of Taste and Odor Compounds from Water," *J. AWWA,* 76:3:83 (March 1984).

Lalezary, S., et al., "Evaluating Activated Carbons for Removing Low Concentrations of Taste and Odor Producing Organics," *J. AWWA,* 78:11:76 (November 1986).

Lalezary, S., et al., "Oxidation of Five Earthy-Musty Taste and Odor Compounds," *J. AWWA,* 78:3:2 (March 1986).

Laverty, G. L., et al., "Quality Improvements by Reservoir Aerations," *J. AWWA,* 62:11:711 (November 1970).

Meadows, M. D., "Reservoir Management," *J. AWWA,* 79:2:26 (February 1987).

Means, E. G., "An Early Warning System for Taste and Odor Control," *J. AWWA,* 78:3:77 (March 1986).

Medsker, L. L., et al., "Odorous Compounds in Natural Water," Environ. Sci. Technol., 2:6:461 (June 1968).

Monscvitz, J. T., and Ainsworth, L. D., "Treatment for Hydrogen Polysulfide," *J. AWWA,* 66:9:537 (September 1974).

Montgomery, J. M., Consulting Engineers, *Water Treatment: Principles and Design,* Wiley, New York, 1985.

Mouchet, P., and Bonnelye, V., "Solving Algae Problems: French Expertise and World-wide," *J. Water SRT-Aqua,* 3:125–141 (1998).

Nelson, M. K., "Sulfide Odor Control," *J. WPCF,* p. 1285 (October 1963).

Nielson, B. J., "Reaeration Dynamics of Reservoir Destratification," *J. AWWA,* 66:10:617 (October 1974).

Palmer, C. M,. "Algae in Water Supplies," in *Water Supply,* U.S. Department of Health, Education, and Welfare, Washington, DC, 1962.

Powell, S. T., "Removal of Hydrogen Sulfide from Well Water," *J. AWWA,* 40:12:1277 (December 1948).

Raman, R. K., "Controlling Algae in Water Supply Impoundments," *J. AWWA,* 77:8:41 (August 1985).

Rashash, D. M. C., et al., "FPA of Selected Odorous Compounds," *J. AWWA,* 89:4:131 (April 1997).

Schiller, B., "Vacuum Degasification of Water for Taste and Odor Control," *J. AWWA,* 47:2:124 (February 1955).

Silvey, J. K., "Growth and Odor: Production Studies," *J. AWWA,* 64:1:35 (January 1972).

Steichen, J. M., et al., "The Effect of Lake Destratification on Water Quality," *J. AWWA,* 71:4:219 (April 1979).

Suttet, I. H., "An Evaluation of Activated Carbon for Drinking Water Treatment: A National Academy of Sciences Report," *J. AWWA,* 72:1:41 (January 1980).

Symons, J. M., et al., "Impoundment Destratification of Raw Water Quality Control Using Either Mechanical or Diffused Air Pumping," *J. AWWA,* 59:10:1268 (October 1967).

Symons, J. M., et al., "Mixing of Water Supply Reservoirs for Quality Control," *J. AWWA,* 62:5:322 (May 1970).

Teerick, J. R., et al., "Artificial Destratification in Reservoirs of the California State Water Project," *J. AWWA,* 61:9:436 (September 1969).

Vajdic, A. H., "Gamma-Ray Treatment of Taste and Odors," *J. AWWA,* 63:7:459 (July 1971).
Walker, G. S., et al., "Chlorine Dioxide for Taste and Odor Control," *J. AWWA,* 78:3:84 (March 1986).

7.4 DISINFECTION BY-PRODUCT CONTROL: THMS, HAA_5, AND VOCS

One of the dilemmas facing the water industry today is how to minimize the potential DBP risks while maintaining protection from microbial contamination using adequate disinfectants. If the level of applied disinfectant increases, the microbial risk decreases but the DBP risk increases. Trihalomethanes (THMs) and haloacetic acids (HAA_5) are major by-products of chlorine disinfection. Synthetic volatile organic compounds (VOCs) are industrial wastes frequently found in many groundwaters. These substances are categorized by the EPA as a threat to public health since most are suspected of being carcinogenic compounds. Since THMs, HAA_5, and VOCs are organic contaminants that predominantly cause problems in the field of water supply for both treatment practices and cost, among others, the control methods are discussed briefly.

7.4.1 THMs

THMs are formed by chlorination of natural organic matter (NOM), often described as an organic precursor, by free residual chlorine. Organic precursors are generally humic substances and fluvic acids. Chloramines are also capable of producing THMs but at such low levels that this process is usually considered to be a non-THM-forming process.

In 1979, the EPA established a maximum contaminant level (MCL) of 0.1 mg/L for the total trihalomethane (TTHM) concentration, but Stage 1 of the D/DBP Rule has lowered it to 0.08 mg/L. TTHM is defined as the sum of trihalomethane ($CHCl_3$), tribromomethane ($CHBr_3$), bromodichloromethane ($CHBrCl_2$), and dibromochloromethane ($CHBr_2Cl$).

To establish THM control methods and to comply with THM regulation, the characteristics of THM formation under different conditions must be understood. There are four basic definitions for THM:

1. *Instantaneous THM.* Instantaneous THM is the THM concentration at the moment of sampling. Compliance with the MCL of THM is determined by this value.
2. *Terminal THM.* THM concentration at the farthest end of the distribution main is know as the terminal THM. If a bench-scale test is used to evaluate terminal THM, the test should simulate the actual chlorine contact time in the treatment plant, plus the anticipated detention time in the distribution system. Furthermore, the temperature and pH of the sample water should be identical to the actual water treatment and supply conditions.

3. *THM formation potential.* The THM formation potential is determined by subtracting the instantaneous THM concentration from the terminal THM.
4. *Maximum total THM potential.* This value represents the maximum THM concentration level under the most favorable conditions for THM formation. The test condition as defined by the EPA requires either a period of 7 days at a water temperature of 77°F (25°C) with a minimum of 0.2 mg/L of free chlorine residual, or an initial chlorine concentration of 5 mg/L with a pH of 9 to 9.5 for a period of 7 days.

The rate of THM formation and the terminal THM concentration depend on six major factors: (1) the amount of organic precursors, (2) level of free residual chlorine, (3) water temperature, (4) water pH, (5) bromide concentration in the water, and (6) chlorine contact time.

Precursors If organic precursors are not present in the water, there will be no THM formation. Thus the concentration and type of precursors have a direct impact on THM formation. The purpose of the maximum total THM potential test is to measure the quantity of organic precursors available to react with free residual chlorine.

Level of Free Chlorine Residual Since THM is a by-product of chlorination, the amount of chlorine residual directly influences the level of THM formation. Therefore, the initial mixing conditions and the design of the contact tank also affect the rate of THM formation.

Water Temperature Water temperature directly influences the rate of THM formation: a higher temperature induces a faster rate and therefore results in a higher rate of formation. In many cases, THM levels during the summer months, when the water temperature is 77°F (25°C), is nearly twice as great as THM levels during the winter season [water temperature is generally 39°F (4°C)]. Thus no special THM control measures are necessary during the winter months. Although water temperature is a major factor in the rate of THM formation, the type and concentration of organic precursors also greatly affect this phenomenon.

pH of the Water The rate of THM formation and terminal THM are greatly affected by the pH of the water. The higher the pH, the faster the reaction rate and the higher the THM levels. This phenomenon is believed to be due to pH-induced changes in the functional groups of the precursor molecules.

Bromide Concentration The rate of reaction between bromide and the precursor molecules is faster than that of chlorine and the precursors: the level of THM increases within a shorter period of time. If bromide is present in the process water, the concentration of terminal THM is higher than the chlorine compounds. The mechanism involved in this reaction is that the bromide ion is oxidized by free residual chlorine to form bromine, which is capable of quickly reacting with precursors to form THMs. In coastal regions the bromide ions are often supplied to surface water, as well as groundwater, from seawater. Yet bromide ions may also be present in water obtained from many other sources, such as in regions that were below sea level during prehistoric times.

Chlorine Contact Time The THM level is a function of chlorine contact time: The longer the contact time, the higher the level of THM concentration. This phenomenon is particularly distinct during the first several hours but generally tapers off after 2 to 3 days.

7.4.2 THM Control Measures

THM control measures should be implemented in two basic types of situations: the design of a new water treatment plant and during the modification of an existing plant (due to high THM levels). The THM control strategies are the same for both cases and the basic steps are as follows:

Step 1. Look for alternative local water supplies such as groundwater. If alternative supplies do not exist or are not economically feasible, the second step should be evaluated.

Step 2. This step contains several options, which are modifications of the conventional treatment technique.

- (a) Improve flocculation and sedimentation to increase the removal of organic precursors (enhanced coagulation).
- (b) Change the points of chlorination to shorten the contact time.
- (c) Substitute potassium permanganate for prechlorination and use small amounts of chlorine dioxide as preoxidant.
- (d) Use chloramines as an alternative disinfectant. This option should include the addition of ammonia 10 to 20 min after chlorine application and observation of the THM formation level.
- (e) Add PAC to remove precursors and THM (seasonal use).

If any of these options appears feasible after bench-scale test evaluation, a pilot-scale test should be conducted (or even testing in an actual plant) to confirm the effectiveness of the option and to establish the anticipated costs. Should any of the second-step options fail, a more drastic and expensive treatment technique should then be evaluated (step 3).

Step 3. The alternatives in this step of the THM control strategies are as follows:

- (a) Use preozonation to oxidize and remove organic precursors.
- (b) Strip THMs through the use of an aeration tower.
- (c) Provide a GAC adsorption bed to remove THMs.
- (d) Utilize a membrane process, either nanofiltration (NF) or reverse osmosis (RO), to remove THMs.

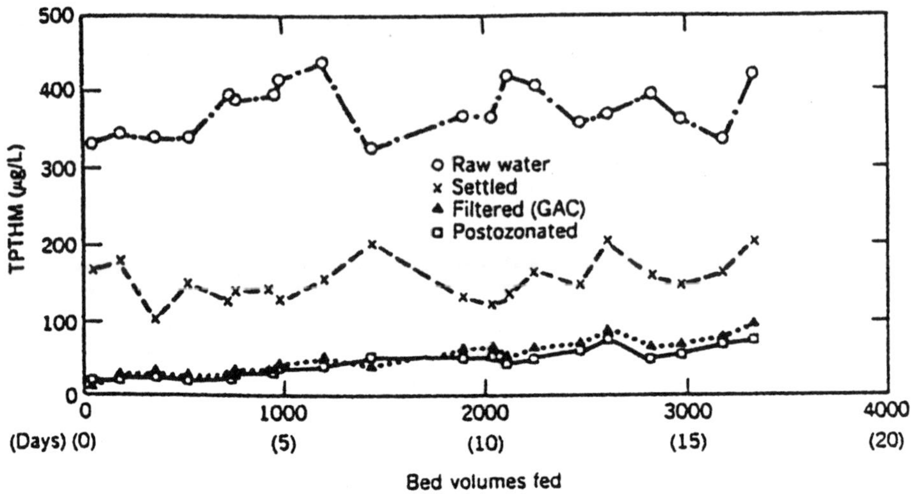

Figure 7.4.2-1 Total potential trihalomethane (TPTHM) reduction through pilot plant processes.

(e) Change to a water supply source containing a minimal level of precursors.

Use of ozone and the GAC adsorption bed have been reported to be an effective means of controlling THM levels since the mid-1970s; technical articles concerning this subject may be found elsewhere. Figure 7.4.2-1 is an example of the effective control of THMs using a combination of GAC adsorption bed and postozonation. In this case, the THMs were removed by GAC, and postozonation was necessary to combat significant leakage of microorganisms from the GAC bed; the GAC bed is literally a nest of microorganisms, due to the abundant supply of nutrients.

7.4.3 HAA_5

Haloacetic acids (HAAs) consist of nine compounds. However, the MCL for HAA is applicable only to the first five compounds, so it is expressed as HAA_5: monochloroacetic acid, dichloroacetic acid, trichloroacetic acid, monobromoacetic acid, dibromoacetic acid, and chloral hydrate.

Under Stage 1 of the D/DBP Rule for HAA_5, the MCL is 0.06 mg/L. The MCLs for both HAA_5 and THMs are regulated on a running annual average based on quarterly samples. The best available technology to meet Stage 1 MCL includes enhanced coagulation (enhanced softening) or granular activated carbon adsorption with an empty bed contact time of 10 min. Like THMs, HAA_5 formation is higher with elevated temperature and TOC concentration in water. The GAC process removes HAA_5 rather effectively in contrast to THMs, which have a low affinity for GAC. However, the control strategy for HAA_5 is basically the same as for THM control.

7.4.4 Operational Issues

The plant operator must run the treatment plant in a manner that produces a processed water that is in compliance with the MCL of the THM standard. Refer to Figure 7.4.4-1 for the appropriate steps. The first of these steps is the quarterly sampling requirement. The quarters end on March 31, June 30, September 30, and December 31. The THM levels of these samples must be submitted to the regulatory agency 30 days after receipt of the analysis. The sampling requirements and the conditions for reduced sampling frequency are different for surface-water and groundwater supplies.

Quarterly Sampling for Surface Water Supplies Each treatment plant is required to obtain and analyze a minimum of four samples per quarter. The samples must be collected from the distribution system within 24 h of each other, and at least 25% must be collected from a location that reflects the maximum resident time within the system. The remaining samples are collected from representative locations in the distribution system based on the person served, the water source, and the treatment methods used.

Compliance with the MCL of THM and HAA standards is based on a running average of the four quarterly averages. If the instantaneous TTHM concentration never exceeds 0.08 and 0.06 mg/L for HAA_5 after one year of compliance monitoring, a request to decrease the monitoring frequency can be made to the primary agency. If the agency agrees to a reduced sampling frequency of one sample per quarter, this sample should be collected at a point (in the distribution system) reflecting the maximum resident time.

Quarterly Sampling for Groundwater Supplies The quality of groundwater is generally more consistent and has a lower concentration of precursors than surface water. For these reasons, the monitoring requirement for THMs and HAA are less stringent than for surface water supplies. Distribution systems that use groundwater as their sole source of water may have a monitoring frequency as low as one maximum total THM and HAA potential test per year per aquifer. This sample must be collected from a location in the distribution system with the longest residence time. However, the primary agency will agree to this reduced frequency only if the utility can substantiate that the maximum total THM and HAA_5 potentials are

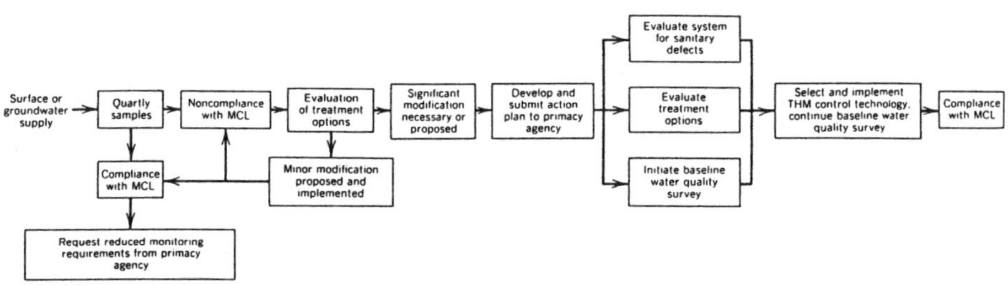

Figure 7.4.4-1 Steps required to comply with MCL of the THM regulation.

less than 0.08 and 0.06 mg/L during periods of maximum total THM and HAA formation (high temperatures and pH) and/or during periods of highest total organic compound (TOC) concentrations.

7.4.5 VOCs

It has been estimated that approximately 20% of all groundwater supplies in the United States are contaminated by synthetic volatile organic compounds (VOCs). This constitutes a real threat to public health since approximately 80% of public water supplies depend on groundwater and 95% of rural areas use groundwater to satisfy domestic water needs. Volatile organic chemicals are widely employed in industrial, agricultural, commercial, and domestic activities. Yet it is believed that the majority of VOC contamination (of groundwater) results from improper disposal, leaks, or spills of industrial wastes. In some instances, VOCs may enter the potable water as a by-product of chlorination: Certain joint compounds used in reservoir membrane liners and certain types of plastic pipe or plastic pipe liner react with chlorine to yield carbon tetrachloride.

Once the VOCs enter an aquifier, they can be transported over great distances due to their low affinity for soil, the lack of opportunity to evaporate, and the minimal chance for biodegradation. Recent federal and state studies reveal that the VOC most frequently found in the highest concentrations is trichloroethylene (TCE); tetrachloroethylene (PCE) ranks second. The most common VOCs and their sources are discussed briefly.

Trichloroethylene (TCE) TCE is a nonflammable liquid with an odor resembling that of chloroform. Its primary application is as a degreasing solvent in the metal industry. Some household products, such as spot removers, rug cleaners, dry cleaning liquid, and air fresheners, also contain TCE.

Tetrachloroethylene (PCE) Tetrachloroethylene is also known as perchlorethylene and is therefore commonly abbreviated as PCE to distinguish it from trichloroethylene (TCE). PCE is a colorless, nonflammable liquid with an etherlike odor. It is used primarily as a commercial dry cleaning solvent. Other usages vary from rug and upholstery cleaner; stain, lipstick, and rust remover; degreaser (metal industry); textile scouring; to vegetable fumigant.

Carbon Tetrachloride (Tetrachloromethane) This VOC is a colorless, nonflammable, heavy liquid used as a fire extinguisher; in the cleaning of cloth; as a solvent for oil or fat; and as a fumigant. Its application in consumer goods was banned in 1970 and ceased being used as an aerosol propellant in 1978.

1,1,1-Trichloroethane (Methychloroform) This compound is a nonflammable liquid that is virtually insoluble in water. It is used primarily for cold type metal cleaning. Other applications include the cleaning of plastic molds and drains, leather tanning, degreasing of septic tanks, and as a shoe polish.

1,2-Dichloroethane This chemical is an intermediary in the manufacture of vinyl chloride monomers and chlorinated solvents such as TCE and PCE. It is used as a solvent in cleaning textiles, PVC processing equipment, and so on. It is also used in the manufacturing of paint, adhesives, and other similar substances.

Vinyl Chloride (Chlorethylene) This compound is a colorless, flammable gas that is soluble in alcohol, ether, carbon tetrachloride, and benzene; it is also slightly soluble in water. Vinyl chloride is used primarily in the manufacturing of plastic and PVC resins. Since plastic and PVC products are widely consumed in the modern world, vinyl chloride is used extensively by many industries.

Benzene (Benzol) Benzene is a colorless and highly flammable liquid, possessing a characteristic odor, which is fairly soluble in water. Derived originally from coal, benzene has been used in the manufacture of a wide variety of goods: medicinal chemicals, dyes, artificial leather, linoleum, oil cloth, varnishes, lacquers, and numerous other organic compounds and goods.

7.4.6 VOC Control Measures

The VOC control measures can be classified into two basic schemes: (1) management and (2) treatment. Each of these is discussed.

Management Control Measures The management alternatives are (1) finding a new source, (2) blending contaminated water with water of a good quality, (3) containment of pollutants, and (4) elimination of the pollution source. The first alternative is generally not feasible since it requires a significant expenditure of capital or is simply not practical. The blending alternative may be attractive to some water purveyors: those who own multiple wells or own both groundwater and surface water supplies. Yet this second alternative is limited by the lack of system flexibility, insufficient dilution capability, and lack of quality water sources. The third and fourth alternatives are theoretically good but are very difficult to implement because there is no easy way of locating the source of contamination, predicting the migration of pollutants, or stopping the migration.

Treatment Control Measures The treatment control strategies are (1) stripping by aeration, (2) removal by adsorption, and (3) a combination of aeration and adsorption.

Air-Stripping Method The process of aeration transfers volatile substances from water to air or from air to water. The former process is defined as air stripping. Air stripping has been used to strip carbon dioxide, hydrogen sulfide, and certain taste- and odor-producing compounds from water. It has also been used successfully in the removal of VOCs. However, air stripping has certain limitations: The process is temperature dependent and is therefore not applicable to treatment plants located in cold climates; its exhaust pollutes the air; and a high air/water ratio is required to remove VOCs effectively. Other minor drawbacks include the potential degradation of water quality by airborne particulates: the oxidation of the reduced form of metallic inorganics, thereby promoting floc formation; the increased potential for corrosion-related problems due to the increase in dissolved oxygen; and increased biological growth within the unit.

Three basic types of air-stripping methods have been tested and applied to actual treatment plants: (1) diffused air, (2) spray, and (3) packed tower. According to Hess et al. (1983) (Figure 7.4.6-1), if over 90% of the VOCs must be removed, the packed tower type of air stripping is the only method capable of meeting this requirement.

When designing air-stripping units, five basic design considerations (minimum) must be addressed:

1. Characteristics of the VOCs to be removed
2. Temperature of the water and air
3. Air/water ratio
4. Contact time
5. Required surface area for mass transfer

Generally, an air/water ratio [(ft^3/min)/(ft^3/min)] ranging from 20 : 1 to 25 : 1 is required to obtain over 90% removal of both TCE and PCE. Based on the results of several pilot-scale studies, this air/water ratio must be at least 20 : 1 for THM re-

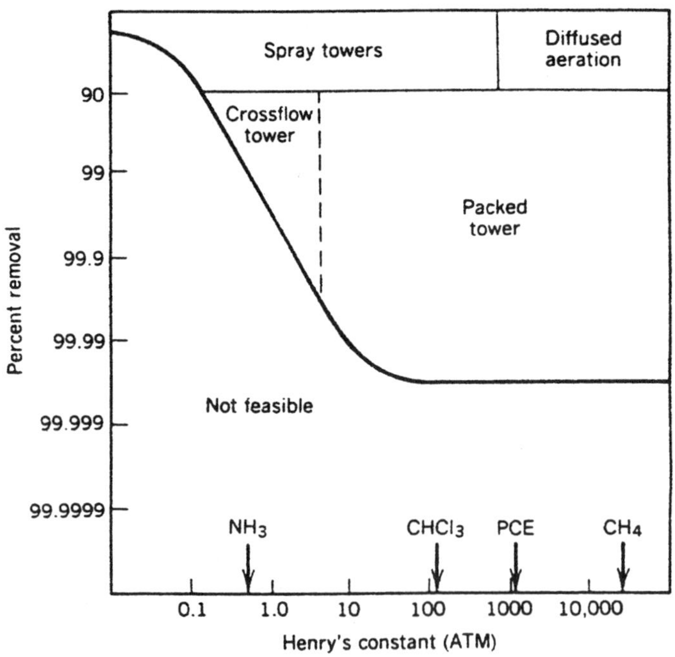

Figure 7.4.6-1 Selection of a feasible aeration process for VOCs. (Adapted from A. F. Hess et al., AWWA Research Foundation Report, 1983.)

moval. The basic formula for designing a packed tower was developed by chemical engineers prior to the mid-twentieth century. However, modifications of this basic design (for special water treatment applications) have been published by several people in the water treatment industry after 1980; these articles are listed in the Bibliography at the end of this section.

Adsorption Methods The three basic methods of adsorption used in the removal of VOCs are PAC, synthetic resin adsorption, and GAC. PAC adsorption is not very promising in the removal of VOCs, due to the nature of the medium: Compounds with high molecular weights tend to adsorb more readily to PAC, but VOCs are low-molecular-weight compounds that do not readily adsorb to PAC unless very high dosages are used. Synthetic resins such as Amersort XE-340, manufactured by Rohm & Hass, were designed to remove nonpolar organic compounds with low molecular weights. This alternative has had some promising results.

Currently, the most effective adsorption process for VOCs is the GAC bed. Results obtained from pilot studies indicate that this type of bed effectively removes PCE and other VOCs. The prime factors that affect the efficiency of adsorption are solubility and affinity. Thus a crucial consideration for design engineers is the potential desorption of VOCs, THM, and HAA from the GAC bed. The desorption process can result from competitive adsorption and from reversal of the adsorption process due to a decreased influent concentration. Since the adsorption rate is a function of water temperature, higher temperatures improve the adsorption rate; pH does not affect the adsorption rate of VOCs.

Combination of Aeration and Adsorption This alternative is an effective method of VOC removal. Aeration reduces the loading of contaminants to the GAC bed, thereby decreasing the frequency of regeneration for the bed.

BIBLIOGRAPHY

Aieta, E. M., et al., "Advanced Oxidation Processes for Treating Groundwater Contaminated with TCE and PCE: Pilot Scale Evaluation," *J. AWWA,* 80:5:64 (May 1988).

Arora, H., et al., "DBP Occurrence Survey," *J. AWWA,* 89:6:60 (June 1997).

AWWA, "Organic Chemical Contaminants in Ground Water: Transport and Removal," *AWWA Seminar Proceedings,* No. 20156, American Water Works Association, Denver, CO, June 1981.

AWWA Research Foundation, *Advanced Oxidation Process for Control of Off-Gas Emissions from VOC Stripping,* AWWA Research Fund/AWWA, *J. AWWA,* 78:2:62 (February 1986).

AWWA Research Foundation, *Occurrence and Removal of VOC from Drinking Water,* Cooperative Research Report, American Water Works Association, Denver, CO, 1983.

Bilello, L. J., and Singley, J. E., "Removal of Trihalomethanes by Packed-Column and Diffused Aeration," *J. AWWA,* 78:2:62 (February 1986).

Black, B. D., et al., "Reducing Cancer Risks by Improving Organic Carbon Removal," *J. AWWA,* 88:6:40 (June 1996).

Carlson, M., et al., "Controlling DBPs with Monochloramine," *J. AWWA,* 90:2:95 (February 1998).

Crittenden, J. C., et al., "Design Considerations for GAC Treatment of Organic Chemicals," *J. AWWA,* 79:1:74 (January 1987).

Crittenden, J. C., et al., "Using GAC to Remove VOCs from Air Stripper Off-Gas," *J. AWWA,* 80:5:73 (May 1988).

Crozes, G., et al., "Enhanced Coagulation: Its Effect on NOM Removal and Chemical Costs," *J. AWWA,* 87:1:78 (January 1995).

Ferguson, D. W., et al., "Comparing Peroxone and Ozone for Control of Taste and Odor Compounds, Disinfection By-products and Microorganisms," *J. AWWA,* 82:4:181 (April 1990).

Glaze, W. H., and Kang, J., "Advanced Oxidation Process for Treating Groundwater Contaminated with TCE and PCE: Laboratory Studies," *J. AWWA,* 80:5:57 (May 1988).

Glaze, W. H., et al., "Determining Health Risks Associated with Disinfectants and Disinfection By-products: Research Needs," *J. AWWA,* 85:3:53 (March 1993).

Glaze, W. H., et al., "Evaluation of Ozonation By-products from Two California Surface Waters," *J. AWWA,* 81:8:66 (August 1989).

Gordon, G., et al., "Minimizing Chlorite Ion and Chlorate Ion in Drinking Water Treatment with Chlorine Dioxide," *J. AWWA,* 82:4:160 (April 1990).

Hand, D. W., "Design and Economic Evaluation of a Treatment of VOCs from a Contaminated Groundwater," *J. AWWA,* 78:9:87 (September 1986).

Hand, D. W., et al., "Design and Evaluation of an Air-Stripping Tower for Removing VOCs from Groundwater," *J. AWWA,* 78:9:87 (September 1986).

Hand, D. W., et al., "Destruction of DBP Precursors with Catalytic Oxidation," *J. AWWA,* 87:6:84 (June 1995).

Hand, D. W., et al., "Simplified Model for Design of Fixed-Bed Adsorption Systems," *J. Environ. Eng.,* 110:4:440 (April 1984).

Hess, A. F., et al., *Control Strategy—Aeration Treatment Technique: Occurrence and Removal of VOCs from Drinking Water,* AWWA Research Foundation Report, American Water Works Association, Denver, CO, 1983.

Jacangelo, J. G., et al., "Ozonation: Assessing Its Role in the Formation and Control of Disinfection By-products," *J. AWWA,* 81:8:94 (August 1989).

Jacangelo, J. G., et al., "Selected Processes for Removing NOM: An Overview," *J. AWWA,* 87:1:64 (January 1995).

Jacangelo, J. G., et al., "UF With Pretreatment for Removing DBP Precursors," *J. AWWA,* 87:3:100 (March 1995).

Kavanaugh, M. C., and Trussell, R. R., "Design of Aeration Towers to Strip Volatile Contaminants from Drinking Water," *J. AWWA,* 72:12:684 (December 1980).

Krasner, S. W., et al., "The Occurrence of Disinfection By-products in U.S. Drinking Water," *J. AWWA,* 81:8:41 (August 1989).

Lykins, B. W., Jr., et al., "GAC for Controlling THM's," *J. AWWA,* 80:5:85 (May 1988).

Means, E. G., et al., "D/DBP Regulation: Issue and Ramifications," *J. AWWA,* 85:2:68 (February 1993).

Milter, R. J., et al., Disinfection By-product Formation and Control by Ozonation and Biotreatment," *J. AWWA,* 84:11:53 (November 1992).

Najm, I. N., et al., "Effect of Bromide and NOM on By-product Formation," *J. AWWA,* 87:1:106 (January 1995).

Najm, I. N., et al., "Evaluating Surrogates for Disinfection By-products," *J. AWWA,* 86:6:98 (June 1994).

Perry, R. H., and Green, D. W., *Chemical Engineering Handbook,* 6th ed., McGraw-Hill, New York, 1986.

Pontius, F. W., "D/DBP Rule to Set Tight Standards," *J. AWWA,* 85:11:22 (November 1993).

Pourmoghaddas, H., et al., "Effect of Bromide Ion on Formation of HAA_5 During Chlorination," *J. AWWA,* 85:1:82 (January 1993).

Ram, N. M., et al., *Significance and Treatment of VOC in Water Supplies,* Lewis Publishers, Chelsea, MI, 1990.

Siddiquis, M. S., et al., "Factors Affecting DBP Formation During Ozone–Bromide Reaction," *J. AWWA,* 85:1:63 (January 1993).

Singer, P. C., et al., "Correlations Between Trihalomethanes and Total Organic Halides Formed During Water Treatment," *J. AWWA,* 81:8:61 (August 1989).

Singley, J. E., *Trace Organics Removal by Air Stripping,* AWWA Research Foundation Report, American Water Works Association, Denver, CO, 1980.

Snoeyink, V., "Adsorption as a Treatment Process for Organic Contaminant Removal from Groundwater," *Proceedings of the AWWA Seminar,* American Water Works Association, Denver, CO, 1981.

Symons, J. M., *Treatment Techniques for Controlling THM's in Drinking Water,* EPA/600/2-81-156, U.S. Environmental Protection Agency, Washington, DC, September 1981.

Symons, J. M., et al., "An Advanced Oxidation Process for DBP Control," *J. AWWA,* 87:11:66 (November 1995).

Symons, J. M., et al., "Precursor Control in Water Containing Bromide," *J. AWWA,* 86:6:48 (June 1994).

Tan, L., et al., "Comparing Ozonation and Membrane Separation for Color Removal and Disinfection By-Product," *J. AWWA,* 83:5:74 (May 1991).

Thompson, J. D., et al., "Enhanced Softening: Factors Influencing BBP Precursor Removal," *J. AWWA,* 89:6:94 (June 1997).

Umphries, M. D., et al., "A Study of Trihalomethane Removal by Packed-Tower Aeration," *J. AWWA,* 75:8:414 (August 1983).

Vrijenhoek, E. M., et al., "Removing Particles and THM Precursors by Enhanced Coagulation," *J. AWWA,* 90:4:139 (April 1998).

Weinberg, H. S., at al., "Formation and Removal of Aldehydes in Plants That Use Ozone," *J. AWWA,* 85:5:72 (May 1993).

Westerhoff, P., et al., "NOM's Role in Bromide and Bromine Formation During Ozonation," *J. AWWA,* 90:2:82 (February 1998).

7.5 FLUORIDATION AND FLUORIDE REMOVAL

During the early twentieth century, studies confirmed that drinking water containing fluoride ion concentrations in excess of 1.5 mg/L caused dental fluorosis (mottled enamel of human teeth) and that the severity of mottling increased with an increase in the fluoride level. Other studies found that dental caries were less prevalent in people with mottled enamel and concluded that it was desirable for the public water supply to have a fluoride ion concentration of approximately 1 mg/L to promote optimal dental health.

The fluoride concentration of potable water should be maintained at an optimal level based on the figure set by the local health department; fluoridation of the public water supply is determined by public vote in many states. Consequently, design engineers must design either a fluoride feed system to maintain an optimal level of fluoride or a process that removes fluoride when the water supply contains excessive levels of fluoride.

7.5.1 Fluoridation

General Considerations As mentioned previously, the recommended level of fluoride in potable water is 1 mg/L. The California Department of Health Services suggests that this figure be an annual average at water temperatures near 15°C; the recommended fluoride concentration falls to 0.8 mg/L during hot months (summer) and rises to 1.2 mg/L during cold seasons. The reason for adjusting the levels is that the average water intake by consumers varies with the seasons—more water is consumed during warm seasons than the winter months.

Fluoride is not found in the free state; it always exits in combination with other elements. All fluoride compounds dissociate to yield fluoride ion when added to the process water. The most frequently used fluoride compounds are sodium fluoride (NaF), sodium silicofluoride (Na_2SiF_6), and hydrofluosilicic acid (H_2SiF_6). These three compounds are chosen because of their availability, reasonable cost, and safety. The solubility of the first two compounds is very low and the maximum solution concentration is 4% for sodium fluoride and only 0.76% for sodium silicofluoride at 25°C. Hydrofluosilicic acid is available in liquid form and its solubility is therefore infinite. However, the concentration of commercially available solutions are generally only 23%. Refer to Table 7.5.1-1 for a summary of the chemical characteristics and the major components of the feed system.

Both sodium fluoride and sodium silicofluoride are available in powered form or fine crystal. Thus problems associated with dust must be addressed during preparation of the solution. Plant operators must wear the proper type of protective gear and use an exhaust fan when handling storage bags because the dust is harmful to human health. Furthermore, if the hardness of the dissolving water exceeds 75 mg/L (as $CaCO_3$), it must be softened prior to preparing the fluoride solution or the fluoride ion will react with calcium and magnesium; the resulting precipitate is capable of clogging the feeders and the piping system.

Hydrofluosilicic acid is much easier to handle than the other two compounds but is a very corrosive chemical (strong acid). Moreover, if the process water has low alkalinity, the application of a very small amount of this chemical, to produce a dosage of 1 mg/L, can significantly depress the pH of the water. The design of these chemical feed systems should follow the guidelines presented in Section 4.1.

TABLE 7.5.1-1 Fluoride Compounds and Feed System

Item	Na_2SiF_6	NaF	H_2SiF_6
Commercially available form	Dry	Dry	Liquid
Commerical purity (%)	98–99	90–98	23–24
Molecular weight (MW)	188.05	42.0	144.08
Specific gravity	—	—	1.220 at 17.5°C (10.17 lb/gal)
Percentage of fluoride in MW	60.7	45.25	79.2
Solubility (g/100 mL at 25°C)	0.762	4.05	Infinite
Freezing point of the solution	0°C	0°C	−22°C
Shipping containers	100 lb bags or 125–400 lb drums or in bulk form		55 gal drums or in bulk form
Approximate cost ($/lb)	0.25	0.45	0.3
Feed system components	• Dry chemical storage • Water softener if total hardness of tap water exceeds 75 mg/L as $CaCO_3$ • Two solution tanks for a batch system • A scale to monitor daily dosage • A reliable metering pump • A dust collector for dry feeder or solution preparation area		• Liquid storage tanks such as steel tank with Neoprene lining • A day tank with a scale to monitor daily usage • A reliable metering pump

Example Design Calculations

Example 1

Given A 20-mgd (0.88-m^3/s) water treatment plant in Oklahoma requires a new fluoridation system. The water supply consistently has a natural fluoride concentration of 0.3 to 0.4 mg/L. Hydrofluosilicic acid will be used for fluoridation because of its easy handling and safety (to the operator).

Determine

(i) The required type and capacity of the feeder
(ii) The capacity of the day tank
(iii) The capacity of the storage tank
(iv) The application point

Solution (i) Since the commercially available products is a 23% solution and the percentage of fluorine in the hydrofluosilicic acid is 79.2%, based on the molecular weight derived from the chemical formula, the fluorine content (%) of the product is

$$79.2 \times 0.23 = 18.2\%$$

The dosages are

maximum dosage as fluoride ion

$$= (1 - 0.3) \times 8.34 \times 20 = 116.8 \text{ lb/day } (53 \text{ kg/d})$$

minimum dosage as fluoride ion

$$= (1 - 0.4) \times 8.34 \times (20 \times 0.25) = 25 \text{ lb/day } (0.97 \text{ kg/d})$$

Knowing that the specific gravity of the solutions is 1.220 at 17.5°C, the weight of the solution is

$$8.34 \times 1.22 = 10.17 \text{ lb/gal}$$

Thus the actual chemical dosages in gallons per hour are

$$\text{maximum dosage of 23\% solution} = 116.8 \text{ lb/day} \div (10.17 \times 0.182)$$
$$= 63 \text{ gal/day or } 2.6 \text{ gal/day } (9.84 \text{ L/h})$$
$$\text{minimum dosage of 23\% solution} = 25 \text{ lb/day} \div (10.17 \times 0.182)$$
$$= 13.5 \text{ gal/day or } 0.56 \text{ gal/day } (2.1 \text{ L/h})$$

The required capacity for the metering pump is therefore 3 gal/h (11.4 L/h). Since the turn-down ratio of any metering pump is 10:1, the minimum dosage is 0.3 gal/h (1.1 L/h).

The type of pump that should be specified is a diaphragm–plunger combination pump, because it is best for handling very corrosive solutions. The pump should have Hastelloy C stainless steel parts, a polyethylene or neoprene diaphragm, and PVC pipings.

Two identical pumps are provided, with one acting as a standby. The metering pumps are preferably paced to the plant flow rate.

(ii) The maximum amount of chemical used per day is 63 gal (240 L). Therefore, provide a day tank with a minimum capacity of 80 gal (300 L). This tank should be set on a weighing scale to facilitate the monitoring of daily use.

(iii) Bulk shipments of hydrofluosilicic acid are most cost-effective and are safer: Operators do not have to handle the 55-gal (210-L) drums frequently. Since the capacity of a tank (trailer) is 4000 gal (15 m^3), a storage tank with a 5000-gal (19-m^3) capacity should be installed. This storage tank should be fabricated from steel, for protection from tornadoes and potential vandalism, and lined with neoprene rubber to protect the steel from corrosion. Based on the average dosage, the 4000-gal supply provides approximately 100 days' worth of storage. The tank is installed outside for reasons of safety. A heating system is not provided since the ambient temperatures in this area does not fall below $-8°F$ ($-22°C$), which is the freezing point of this chemical.

(iv) The application point for hydrofluosilicic acid is at the inlet to the clearwell. A diffuser with a static mixer is provided for good mixing. This location is chosen because application of fluoride, as well as lime (for pH control), at the head of the plant results in the loss of some of the added fluoride due to alum flocculation. Refer to Section 7.5.2.

7.5.2 Fluoride Removal

The two alternative methods for removing excess fluoride from the water supply are (1) chemical precipitation and (2) the process of ion exchange.

Chemical Precipitation The fluoride content of the process water can be reduced to an acceptable level by means of either conventional alum flocculation and sedimentation, provided that a large alum dosage (200 to 300 mg/L) is used, or a lime softening process. Results obtained from bench-scale tests indicate that fluoride concentrations up to 3.5 mg/L (raw water) may be reduced to 1 mg/L by employing alum dosages up to 350 mg/L. Lime softening also effectively reduces fluoride to a safe level by producing calcium fluoride precipitates.

Ion Exchange Process There are presently only three types of ion exchange material that may have practical applications in defluoridation: (1) tricalcium phosphate, including bone char or bone meal; (2) activated alumina; and (3) ion exchange resins. The effectiveness of these materials has been proven by actual plant-scale tests and/or pilot-scale tests.

Bone Char and Bone Meal These two materials are very similar in that they are both comprised of processed bone material. Their only difference is the slight variation in the manufacturing process and products. An example of defluoridation by ion exchange using bone char is as follows:

$$Ca(PO_4)_6 \cdot CaCO_3 + 2F^- \rightarrow Ca(PO_4)_6 \cdot CaF_2 + CO_3^{2-}$$

Bone char, expressed as $Ca_3(PO_4)_2$, can be regenerated through the application of 1% caustic soda.

Activated Alumina (Al_2O_3) Fluoride removal by means of activated alumina has been demonstrated by an actual plant (Gila Bend, Arizona) and several pilot-scale studies. In each of these cases, granular activated alumina (filter sand size)—approximately 5 ft (1.5 m) deep—was used to reduce fluoride from an initial level of 3 mg/L to less than 0.5 mg/L using a filtration rate of 5 to 7 gpm/ft^2 (12.5 to 17.5 m/h). These studies also demonstrated that it was necessary to adjust the pH of the raw water to 5.5 prior to filtration to maximize the efficiency of fluoride removal.

Regeneration of the bed was achieved through the addition of 1 % caustic soda solution.

Ion Exchange Resin Process Fluoride can be removed from process water through the use of ordinary cation exchange and anion exchange resin columns arranged in series. The first column exchanges its polarized hydrogen with sodium to form the equivalent acid:

$$2NaF + H_2R^+ \rightarrow H_2F_2 + Na_2R^+$$

The hydrogen fluoride is subsequently removed by the second column by the following reaction:

$$2R^- + H_2F_2 \rightarrow 2RHF$$

Design engineers should note that one problem does exist for all three types of ion exchange: the disposal of the concentrated fluoride. These processes cannot be implemented unless proper disposal methods are used or an appropriate site is found.

As a final note, it must be stressed that both bench- and pilot-scale tests must be conducted prior to the design of fluoride removal process. These procedures are of extreme importance because defluoridation is affected significantly by the presence of other ions in the raw water for these ions compete with the fluoride removal reaction.

BIBLIOGRAPHY

AWWA, *Water Fluoridation: Principles and Practice,* Manual M4, American Water Works Association, Denver, CO, 1977.

Harmon, J. A., and Kalichman, S. G., "Defluoridation of Drinking Water in Southern California," *J. AWWA,* 57:2:245 (February 1965).

Kreft, P. H., "Removal of Fluoride by Actived Alumina," M.S. thesis, Loyola University, Los Angeles, 1986.

Maier, F. J., "Defluoridation of Municipal Water Supplies," *J. AWWA,* 45:8:879 (August 1953).

Raboski, J. B., and Miller, J. P., "Fluoride Removal by Lime Precipitation and Alum and Polyelectrolyte Coagulation," *Proceedings of the 29th Purdue Industrial Waste Conference,* Purdue University, Lafayette, IN, 1974, p. 669.

Savinelli, E. A., "Defluoridation of Water with Activated Alumina," *J. AWWA,* 50:1:33 (January 1958).

Sawyer, C. N., *Chemistry for Sanitary Engineers,* McGraw-Hill, New York, 1960.

Smith, H. R., and Smith, L. C., "Bone Contact Removes Fluoride," *Water Works Eng.,* 90:5:600 (May 1937).

Williams, R. B., and Culp, G. L., *Handbook of Public Water Systems,* Van Nostrand Reinhold, New York, 1986.

Wu, Y. C., and Nitya, A., "Water Defluoridation and Activated Alumina," *J. Environ. Eng. Div. ASCE,* 105:2:359 (1979).

7.6 CORROSION CONTROL

Businesses in the industry of water treatment and supply are hampered by the serious problem of corrosion (iron pipes and metals) and degradation of concrete. Corrosion protection is therefore a major design consideration for design engineers. Painting, coating, proper maintenance of the cathodic protection system (for metal structures), and the continuous feeding of corrosion inhibitors to the distribution system are the major tasks of the plant operation personnel. Corrosion is far from being a simple phenomenon. It is a combination of many factors: electrochemical, physical, chemical, biological, and metallurgical. Since corrosion and corrosion control are subjects which on their own could fill a book, only a brief discussion is presented here.

7.6.1 Corrosion of Metals

The electrode potential is the potential difference between a metal and a solution of its ions. It is expressed by the *Nernst equation:*

$$E = \frac{RT}{nF} \ln \frac{P}{P'} = \frac{RT}{nF} \ln \frac{C}{C'}$$

where

n = valency of the metal ions

T = absolute temperature

R = gas constant

P = pressure at which the metal dissolves

P' = osmotic pressure of the solution

C = a constant

C' = activity of the metal ions in the solution

F = Faraday number

The electrode potential is standard for a standard solution of its ions and is expressed as E_0, thus simplifying the Nernst equation to give the general equation defining the electrode potential of a metal at 25°C:

$$E = E_0 + \frac{0.058}{n} \log C'$$

In the case of iron, this equation becomes

$$E = -0.44 \text{ patient } 0.029 \log \text{Fe}^{2+}$$

Table 7.6.1-1 presents the oxidation reactions and potential for the major types of common metal at 25°C. Metals with positive electrode potentials are classified as noble metals (cathodic); those with negative electrode potential are base metals (anodic). The corrosion of iron is caused primarily by electrochemical cell action. The presence or absence of oxygen alters the rate of corrosion. Other factors that influence the corrosion of iron are the total mineral concentration, the presence of dissolved gases, the temperature of the water, the surface condition of the iron, the flow velocity across its surface (pipe flow), and biological action.

General Corrosion of Metals General corrosion of metals is induced by galvanic action. A good example is the case where two dissimilar metals are connected and exposed to water: One metal becomes cathodic and the other anodic, thus setting up a galvanic cell. For instance, when a steel and a copper pipe are connected (which is often the case in most household plumbing), the steel pipe becomes the anode. Since metal loss occurs at the anode, the steel pipe corrodes. As shown in Table 7.6.1-1, there is a 0.96-V difference in electrode potential between steel and copper, and the corrosion is therefore severe. The corrosion rate of metals that are close to one another in the table is less than that of metals that are widely separated.

Just as in the case where dissimilar metals are exposed to water of a uniform concentration, a galvanic current is also produced when a single metal is exposed to water of varying concentrations (ionic strengths). The portion of the metal located in the more concentrated solution becomes the anode and therefore corrodes. This process is known as concentration cell corrosion.

As long as a treatment plant is characterized by poor design and/or has deposits of debris on the metal surface, concentration cell corrosion may occur at any plant

TABLE 7.6.1-1 Oxidation Potentials Against Normal Hydrogen Electrode at 25°C

Metal	Electrode Reactions	Equilibrium Potential (V)
Magnesium	$Mg = Mg^{2+} + 2e^-$	−2.34
Beryllium	$Be = Be^{2+} + 2e^-$	−1.70
Aluminum	$Al = Al^{3+} + 3e^-$	−1.67
Manganese	$Mn = Mn^{2+} + 2e^-$	−1.05
Zinc	$Zn = Zn^{2+} + 2e^-$	−0.762
Chrome	$Cr = Cr^{3+} + 3e^-$	−0.71
Iron	$Fe = Fe^{2+} + 2e^-$	−0.440
Nickel	$Ni = Ni^{2+} + 2e^-$	−0.250
Lead	$Pb = Pb^{2+} + 2e^-$	−0.126
Hydrogen	$H_2 = 2H^+ + 2e^-$	−0.000 by convention
Copper	$Cu = Cu^{2+} + 2e^-$	+0.345
Copper	$Cu = Cu^+ + e^-$	+0.522
Silver	$Ag = Ag^- + e^-$	+0.800
Platinum	$Pt = Pt^{2+} + 2e^-$	+1.2 approx.
Gold	$Au = Au^{3+} + 3e^-$	+1.42
Gold	$Au = Au^+ + e^-$	+1.68

Source: Adapted from Degremont, *Water Treatment Handbook,* 5th ed., Halsted Press, New York, 1979.

site. These factors create a localized concentration of a specific element such as oxygen, sulfate, or chloride, which is significantly different from other portions of the water. For instance, the oxygen concentration beneath the deposits of debris is less than the surrounding water; thus this portion of the metal becomes anodic and acts as the starting point for corrosion. This type of corrosion mechanism is called an *oxygen differential cell*. Coexisting with most oxygen cells are the corresponding sulfate and chloride concentration cells. These coexist because chloride and sulfate ions penetrate the deposit and become concentrated. The corrosion rate is influenced by the surface area ratio of cathodic and anodic metals: The larger the cathode area, the higher the corrosion rate, and vice versa. For example, a large anode will produce a general corrosion condition, whereas a small anode yields a severe pitting type of corrosion.

As mentioned previously, the surface condition of the metal greatly influences the conditions for corrosion. When a metal surface has deposits of debris or when alum floc or lime particles create nonaerated zones (after precipitation) beneath the deposits, corrosion begins in these anodic regions. Furthermore, if bacteria are growing under the deposits, they create an oxygen-deficient (reducing) condition, which intensifies the corrosion condition. Corrosion of stainless steel often occurs under this type of condition. Anaerobic microbes also play a role in corrosion. The presence of sulfate-reducing bacteria within the deposit creates a very aggressive condition, which further accelerates metal loss. Some anaerobic bacteria, such as iron bacteria (i.e., *Lepthothrix*), actually consume ferrous iron and thereby increase the rate of corrosion.

Secondary Corrosion Factors Secondary factors that influence corrosion are (1) the concentration of dissolved solids in the water, (2) the level of dissolved gases (in water), (3) water temperature, (4) stress corrosion, (5) fatigue, and (6) impingement attacks. Each of these is described briefly.

Dissolved Solids The effect of the dissolved solids content on water corrosivity is a complex issue. Both the species and the concentration of ions are important factors. Some species, such as carbonate and bicarbonate, reduce corrosion, whereas chloride, sulfate, bromide, and nitrate ions markedly accelerate corrosion. In regard to concentration, less than 1 mg/L of copper ion can increase the rate of corrosion significantly: Copper ions are deposited on the metal and act as an anode, causing a severe pitting type of corrosion to occur. Household aluminum utensils have been known to corrode as the result of copper ions.

Dissolved Gases Dissolved gases such as free carbon dioxide, oxygen, hydrogen sulfide, and ammonia strongly influence the corrosion of metals, especially iron and copper.

Water Temperature Higher water temperatures accelerate the rate of corrosion by increasing the rate of cathodic reaction. The chemical reaction rate generally doubles for every 15°F (8°C) increase in temperature.

Stress Corrosion Metals that are under tensile or bending stress, due to external loads or internal stress locked into the metal during fabrication, will exhibit stress corrosion when subjected to a corrosive environment.

Fatigue Corrosion Fatigue corrosion occurs when a metal undergoes repeated stress in a corrosive environment.

Impingement Attacks Cavitation, erosion, abrasion by waterborne grit and other solids, and erosion of the protective coating on metal surfaces are examples of impingement attacks.

7.6.2 Corrosion Control for Metals

Five basic measures exist for corrosion control: (1) the selection of construction materials, coatings/linings, and insulation; (2) application of corrosion inhibitors; (3) installation of cathodic protection systems; (4) installation of sacrificial anodes; and (5) pH control by alkali chemicals.

Design Efforts When corrosion problems are expected for a treatment plant, the design engineer may consider using corrosion-resistant materials such as stainless steel, concrete, plastic, or copper as a substitute for carbon steel. If these materials cannot be implemented, special coatings and/or linings should be applied to the metal to separate it from the corrosive water or environment. Paints, plastic sheeting, rubber, and cement are often used in this capacity. Galvanic corrosion resulting from the connection of dissimilar metals must be avoided by insulation. Nonconductive materials such as plastic or synthetic rubber (Figure 7.6.2-1) should be inserted between the metals.

Application of Corrosion Inhibitors Despite the best design efforts, most water treatment systems include a system that applies chemical corrosion inhibitors to the process water. The inhibitors control corrosion by stopping the anodic and/or cathodic reaction. Chromate, orthophosphate, nitrate, and silicate are anodic inhibitors; calcium carbonate, polyphosphate, and zinc are typical cathodic inhibitors. There are some chemicals that act as both anodic and cathodic inhibitors: mercaptobenzothiazole, benzotriazole, and tolytriazole. These chemicals form an organic film of amines of the surface of the metal, thereby inhibiting corrosion (similar to painting). The corrosion inhibitors most commonly used in the field of water treatment are polyphosphate (cathodic inhibitor requiring a dosage of approximately 2 mg/L), zinc orthophosphate (anodic inhibitor requiring a dosage of 1 mg/L or less), and sodium silicate (anodic inhibitor requiring a dosage of 8 to 10 mg/L as silica).

Cathodic Protection Cathodic protection is used to lower the electrode potential of metals and is used to protect buried pipes or equipment exposed to corrosive water, including seawater. Cathodic protection is generally effective in applying an electrical potential of 0.85 to 1 V.

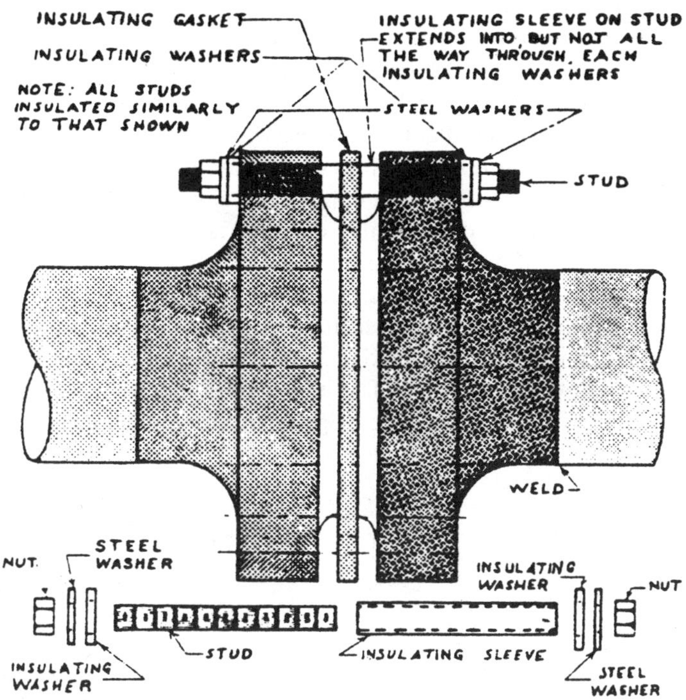

Figure 7.6.2-1 Suggested insulation flange assembly. (Adapted from A. W. Peabody, *Control of Pipeline Corrosion,* NACE, Houston, TX, 1967.)

Sacrificial Anodes Sacrificial anodes reduce galvanic cell corrosion for steel or iron structures. They are composed of zinc or magnesium and become anodic under corrosive conditions, thereby supplying electrons to cathodic surfaces.

pH Control by Alkali Chemicals The most common method of corrosion control for municipal water distribution systems is pH control of water. This method provides a slightly positive (+0.2) Langelier Index or a Stability Index (Ryznar Index) of 6 or below. The chemical generally fed to the process water is lime since it provides calcium and alkalinity, in addition to raising the pH. However, lime particles tend to settle out in the distribution main, creating a condition conducive to concentration cell corrosion. Thus caustic soda is often employed as a substitute for lime, due to its ease of handling and because it eliminates the precipitation of lime particles.

$$\text{Langelier Index (LI)} = \text{pH} - \text{pH}_s$$

where pH is the pH of the sample water and pH_s is the saturation pH.

$$\text{Stability Index (SI)} = 2\text{pH}_s - \text{pH}$$

Figure 7.6.2-2 and Appendix 17 are graphical determinations of pH and LI. SI can be determined by simple calculation after LI is obtained. Table 7.6.2-1 illustrates the relation between corrosion characteristics and LI and SI values.

As a means of corrosion control, some treatment plants feed lime ahead of filtration (pH ≥ 8) in the belief that this scheme will eliminate the after-precipitation of undissolved lime in the distribution system. In actuality, this practice increases the finished water concentration of aluminum; the alum floc in the filter influent is dissolved by the lime due to the solubility characteristics of the floc (refer to Figure 3.2.3-1), and the level of residual aluminum is often 0.3 mg/L or more.

Plant operators should realize that an adjustment in pH based on LI and SI does not mean that corrosion does not occur. Indeed, in many cases, iron structures became corroded, although the water had a slightly positive Langelier Index. This situation often occurs with soft water. The best way to monitor metal corrosion and to optimize corrosion inhibitor dosage or pH control by alkali chemicals is to conduct a continuous corrosion coupon test.

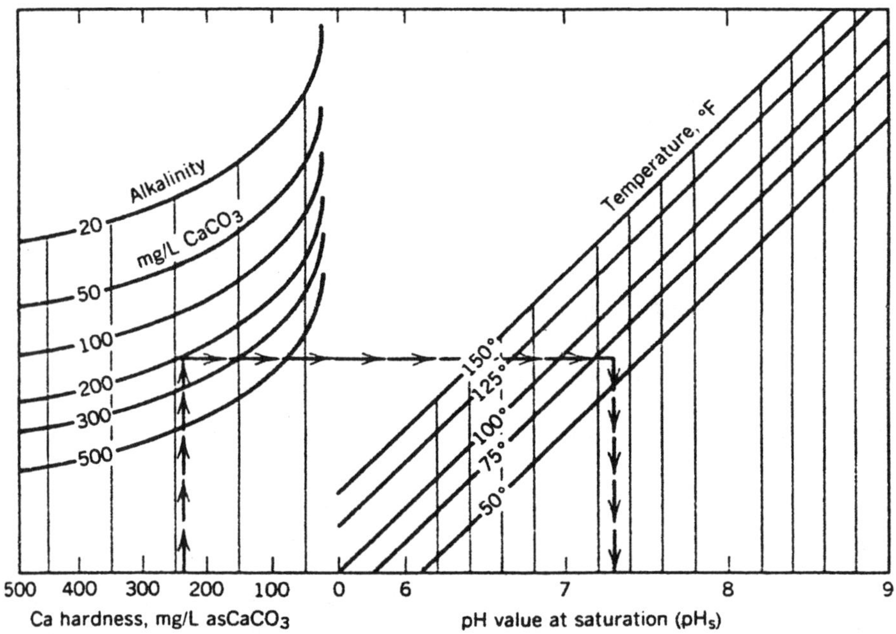

Example:

Raw water calcium = 240 mg/L
Alkalinity = 190 mg/L
pH = 6.8
Temperature = 70°F
From chart: pH_s = 7.3
Langelier index = 6.8 − 7.3 = −0.5
Stability index = (2 × 7.3) − 6.8 = 7.8

Figure 7.6.2-2 Graphical determination of pH_s and Langelier Index. (Adapted from *The Nalco Water Handbook,* F. N. Kemmer, ed., McGraw-Hill, New York, 1979.)

TABLE 7.6.2-1 Corrosion Characteristics Versus LI and SI

	LI	SI
Medium to heavy scaling	+0.5 to +1.0	4–5
Slight scale formation	+0.2 to +0.3	5–6
Equilibrium	0	6–7
Slightly corrosive	−0.2 to −0.3	7–7.5
Medium to heavy corrosion	−0.5 to −1.0	7.5–8.5

7.6.3 Corrosion Coupon Test

A number of tools have been developed to measure corrosion by indirect means. One method is the corrosion coupon test: Figure 7.6.3-1 presents a typical installation of the corrosion coupon test assembly. The major steps of this test are as follows:

1. The pretest weights of the metal specimens are recorded prior to placing the metals into the system for a period of 30 to 100 days.
2. Following removal, the specimens are cleaned, reweighed, and evaluated. The loss of weight and type of corrosion (general, pitting, etc.) are observed and recorded.

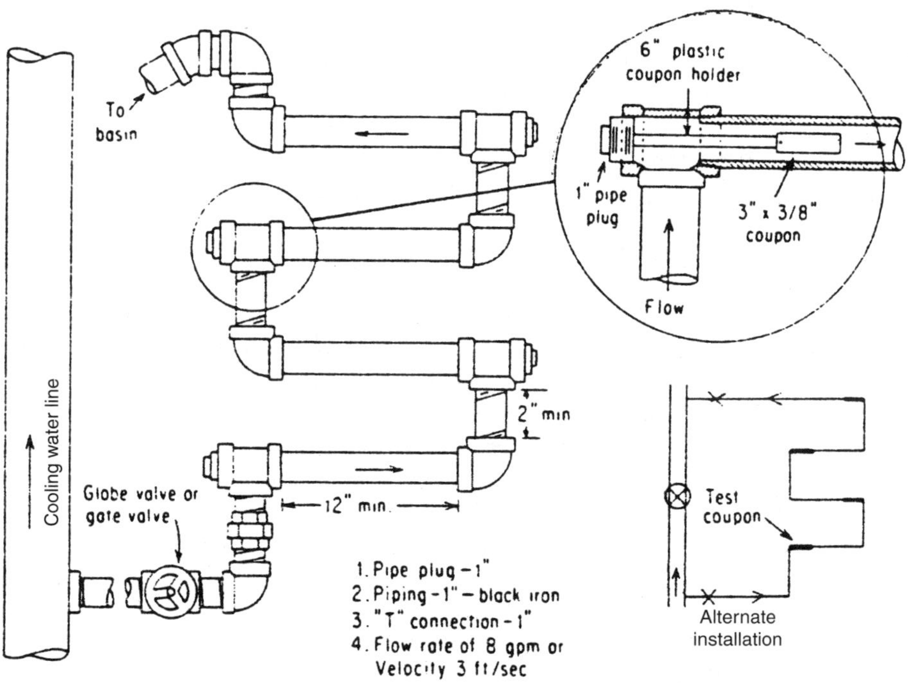

Figure 7.6.3-1 Coupon test assembly. (Adapted from *The Nalco Water Handbook,* F. N. Kemmer, ed., McGraw-Hill, New York, 1979.)

3. The corrosion rate is expressed as metal loss in mils per year, a mil being 0.0001 in. One mil/yr is equivalent to 0.025 mm/yr.

The corrosion coupon test is occasionally run for visual observation only. The results of the coupon test can vary widely depending on the condition of the coupon: whether the metal was sandblasted versus polished, the location of the coupon within the test system, the length of exposure (duration of the test), the type of metallurgy, and the quality of the water.

7.6.4 Corrosion (Degradation) of Concrete

Today, almost all concrete structures are composed of steel bar–reinforced concrete. Theoretically, the steel bars do not corrode, due to the positive potential in the concrete, that is, until the covering concrete [approximately 2 in. (5 cm) thick] degrades. The electrode potential of the steel bars (in concrete) is approximately +0.1 V and is referred to as the normal hydrogen electrode for concrete at a pH of approximately 11.6. Yet, in reality, corrosion of the reinforcing steel bars is an important concern for engineers since degradation of concrete or the formation of cracks in the concrete can be attributed to three causes: (1) physical causes, (2) chemical attack, and (3) bacterial action.

Physical Causes The degradation of concrete will be accelerated if the concrete is excessively permeable. This characteristic results from using the improper-size gravel; the wrong mixing proportion of cement, sand, and gravel; and poor concrete casting techniques. If the concrete is poorly constructed—that is, has cracks and cavities, and insufficient thickness of concrete over the steel bars, or badly constructed expansion joints that leak water—these defects will contribute to its degradation. The last of the physical causes is erosion resulting from heavy vehicle traffic, scouring by sand or grit, or a high velocity of water flow [over 15 ft/sec or (4.6 m/s)].

Chemical Attacks Carbon dioxide, strong acids, ammonia, sulfate, and strong alkali chemicals all attack concrete. Of the five, carbon dioxide has the least effect because the high lime residue of the cement simply converts the carbon dioxide into calcium carbonate. However, if a concrete surface is exposed continuously to water containing over 15 mg/L of CO_2, its surface will begin to degrade. In or near large cities, concrete corrosion due to exhaust gas from industries and cars has become a serious problem.

Strong acids such as sulfuric, hydrochloric, nitric, and phosphoric, as well as organic acids, are highly destructive to concrete. The only effective method of protecting concrete from the action of these acids is to line or coat it with a suitable type of material, such as plastic membrane, rubber, or paint.

Ammonia degrades concrete by replacing calcium (in concrete) with ammonia. This loss of calcium from cement causes the concrete to corrode rapidly. In recent years, many water treatment plants began feeding ammonia (in addition to chlorine) to the process water. The hydrochloric acid produced by chlorine and ammonia is capable of degrading concrete, provided that the water is soft and has very little alkalinity, although its dosage is so small that it will not quickly attack concrete except at the ammonia feed points.

Certain regions have water supplies containing a high level of sulfates. Since sulfates are capable of reacting with aluminates—an ingredient of cement—and cause the original volume to swell by a factor of 2 to 2.5, they may be considered to be the primary cause of concrete deterioration in water treatment structures. A number of existing plants, especially those located in regions with a high sulfate content in the water, exhibit severe degradation of their concrete surfaces, especially in areas continuously exposed to cascading water and flowing water: Concrete troughs in sedimentation basins and filters, and the sidewalls of channels and basins. Engineers are therefore strongly advised to consider a dense and high-strength concrete with type 5 cement when designing plants for these regions (sulfate content over 100 mg/L). Table 7.6.4-1 lists the available types of cement and their major ingredients.

Concrete can also be deteriorated by strong alkali chemicals such as caustic soda, potassium hydroxide, and soda ash because the aluminum oxides in the cement (see Table 7.6.4-1) tend to become soluble when in contact with water with a high pH: The strong alkali chemicals react with the sodium molecule in the molecular structure. Thus engineers are advised either to select cement with a low aluminum content (type 2 or type 4) or to coat the concrete surface with an appropriate type of paint. If general-purpose cement is specified, a PVC membrane should be used, provided that the liquid pH exceeds 12.

Bacterial Action The third cause of concrete degradation is bacterial action. A typical example is crown corrosion of concrete sewer pipes. In this situation, the concrete is deteriorated by sulfuric acid formed by anaerobic bacteria, which consume waterborne sulfate. This phenomenon rarely occurs in water treatment plants.

7.6.5 Special Construction and Design Considerations

The construction supervising engineer must give special consideration to the contact between steel reinforcing bars and iron pipes passing through concrete walls. Absolutely no direct contact should exist between these two elements, since con-

TABLE 7.6.4-1 Major Ingredients of Portland Cement

Composition (%)	Type 1 General Purpose	Type 2 Modified	Type 3 High Early Strength	Type 4 Low Heat	Type 5 Sulfate Resisting
$3CaOSiO_2$	53	47	58	26	38
$2CaOSiO_2$	24	32	16	54	43
$3CaAl_2O_2$	8	3	8	2	4
$4CaOAl_2O_3Fe_2O_3$	8	12	8	12	8
Total	93	94	90	94	93

tact will produce galvanic cell corrosion in the iron pipe. The same type of problem will occur whenever steel members are embedded in the reinforced concrete structure. If the end of a steel member, which is either exposed to water or buried in the ground, must be welded to reinforcing steel bars (in concrete), it must be well coated or cathodically protected from galvanic cell corrosion. Design engineers should be very careful when designing these types of contacts and are advised to make notations on the drawings.

Additional design considerations include the use of insulation flanges in cases where dissimilar metal pipes must be connected and in situations where the pipeline exits a concrete wall to enter the ground or a water tank (see Figure 7.6.2-1). Furthermore, stray current electrolysis type of corrosion may occur in areas near direct current (dc) transit systems. It is also very important to specify a dense and high-strength concrete [compressive strength of over 5000 psi (350 kg/cm^2)] reducing the water/cement ratio to less than 0.45. Recent case histories have shown a clear superiority of the high-strength reinforced concrete structures against both corrosion and erosion. It is also known that overuse of form release agents have an obvious effect on the rate of concrete corrosion. For further information, refer to *Control of Pipeline Corrosion* (Peabody, 1967), released by the National Association of Pipe Corrosion (AAPC) Engineers.

BIBLIOGRAPHY

Degremont, *Water Treatment Handbook,* 5th ed., Halsted Press, New York, 1979.

Edwards, M., et al., "Effect of Selected Anions on Copper Corrosion," *J. AWWA,* 86:12:73 (December 1994).

Edwards, M., et al., "The Pitting Corrosion of Copper," *J. AWWA,* 86:7:74 (July 1994).

Fair, G. M., et al., *Water Supply and Wastewater Disposal,* Wiley, New York, 1954.

Good, R. J., et al., *Coping with Deterioration of Concrete Surfaces in Adelaide's Early Water Filtration Plants,* Report by the Engineering and Water Supply Department, Adelaide, South Australia.

Hatch, G. B., and Rice, O., "Influence of Water Composition on the Corrosion of Steel," *J. AWWA,* 51:6:719 (June 1959).

Holm, T. R., and Schock, M. R., "Potential Effects of Polyphosphate Water Treatment Products on Lead Solubility in Plumbing Systems," *J. AWWA,* 83:7:76 (July 1991).

Joint Task Group on Calcium Carbonate Saturation, "Suggested Methods for Calculating and Interpreting Calcium Carbonate Saturation Index," *J. AWWA,* 82:7:71 (July 1990).

Kemmer, F. N., ed., Nalco Chemical Co., *The Nalco Water Handbook,* McGraw-Hill, New York, 1979.

Korshin, G. V., et al., "Influence of NOM on Copper Corrosion," *J. AWWA,* 88:7:36 (July 1996).

Larson, T. E., "Chemical Control of Corrosion," *J. AWWA,* 58:3:354 (March 1966).

Montgomery, J. M., Consulting Engineers, *Water Treatment: Principle and Design,* Wiley, New York, 1985.

Peabody, A. W., *Control of Pipeline Corrosion,* National Association of Corrosive Engineers, Houston, TX, 1967.

Pisigan, R. A., Jr., and Singley, J. E., "Effects of Water Quality Parameters on Corrosion of Galvanized Steel," *J. AWWA,* 77:11:76 (November 1985).

Pisigan, R. A., Jr., and Singley, J. B., "Influence of Buffer Capacity, Chlorine Residual, and Flow Rate on Corrosion of Mild Steel and Copper," *J. AWWA,* 79:2:62 (February 1987).

Reiber, S. H., et al., "Corrosion Monitoring and Control in the Pacific Northwest," *J. AWWA,* 79:2:71 (February 1987).

Rothberg, M. R., et al., "Computerized Corrosion Control," *J. AWWA,* 85:7:26 (July 1993).

Singley, J. E., "Electrochemical Nature of Lead Contamination," *J. AWWA,* 86:7:91 (July 1994).

Stone, A., et al., "The Effects of Short Term Changes in Water Quality on Copper and Zinc Corrosion Rates," *J. AWWA,* 79:2:75 (February 1987).

Strum, W., "Corrosion Studies," *Public Works,* p.78 (1957).

Strum, W., "Investigation on the Corrosive Behavior of Water," *J. Sanit. Eng. Div. ASCE,* 86:27 (1960).

7.7 GRANULAR ACTIVATED CARBON ADSORPTION

7.7.1 Purpose

Granular activated carbon (GAC) has recently been used as a substitute for granular filter medium or as an additional process in the conventional treatment process, for the removal of organic compounds, including disinfection by-products, those producing taste and odor, pesticides, and other synthetic organic compounds. GAC has been proven effective in removing these various organic compounds and even organic mercury. The GAC filter bed is capable of functioning in the same manner as ordinary filters (removal of suspended matter) but also removes organic compounds. However, the Europeans use GAC adsorption beds for the removal of taste and odor, other trace organic compounds, and total organic carbon, by installing GAC contactors downstream of ordinary filters (in most cases) preceded by a preozonation process.

In recent years, the GAC adsorption process has suddenly gained a great deal of attention in the United States as the result of the 1986 Amendments to the Safe Drinking Water Act. These amendments set the maximum contaminant levels (MCLs) for over 50 organic compounds, of which 26 are synthetic organic compounds. The water treatment industry in the United States has recently begun to take large steps in improving the aesthetic quality of drinking water, by reducing color, objectionable tastes and odors, in addition to making it safer. The GAC adsorption process is certainly capable of meeting these objectives. Yet the operation and maintenance costs are significantly higher, due primarily to the need to regenerate the GAC.

7.7.2 Characteristics of GAC

GAC may be manufactured from a variety of material: wood, nut shells, coal, peat, or petroleum residues. However, GAC used in water treatment is usually produced from bituminous or lignite coal through slow furnace heating under anaerobic conditions. The absence of oxygen ensures that the coal does not burn but is transformed into a porous carbon material. This product is then activated by exposure to a mixture of steam and air at a temperature of 1500°F (815°C). The activation process oxidizes the surface of the carbon pores, thereby allowing the surface to attract and hold organic compounds.

According to the *Calgon Bulletin* (Calgon, 1976), the physical properties of GAC are as follows:

Total surface area of GAC	890 to 900 m^2/g
Bulk density of wetted GAC	30 lb/ft^3 (480 kg/m^3)
Bulk density of dry GAC	28 lb/ft^3 (450 kg/m^3)
Particle density	1.4 to 1.5 g/cm^3
Effective size	0.8 to 0.9 mm for Filtrasorb 100, 0.55 to 0.65 mm for Filtrasorb 200
Uniformity coefficient	1.9 for Filtrasorb 100, 1.7 for Filtrasorb 200
Iodine number	850 to 875
Abrasion number	70 to 86
Moisture as packed	0.5 to 2%
Cost of drying GAC	\$1.0 to \$1.2/lb (\$2.2 to \$2.6/kg) or \$760 to \$905/yd^3 (\$1000 to \$1185/m^3)

Figure 7.7.2-1*a* illustrates the hydraulic characteristics of a GAC bed. The optimum backwash rate is approximately 30 to 40% of the bed expansion rate, due to

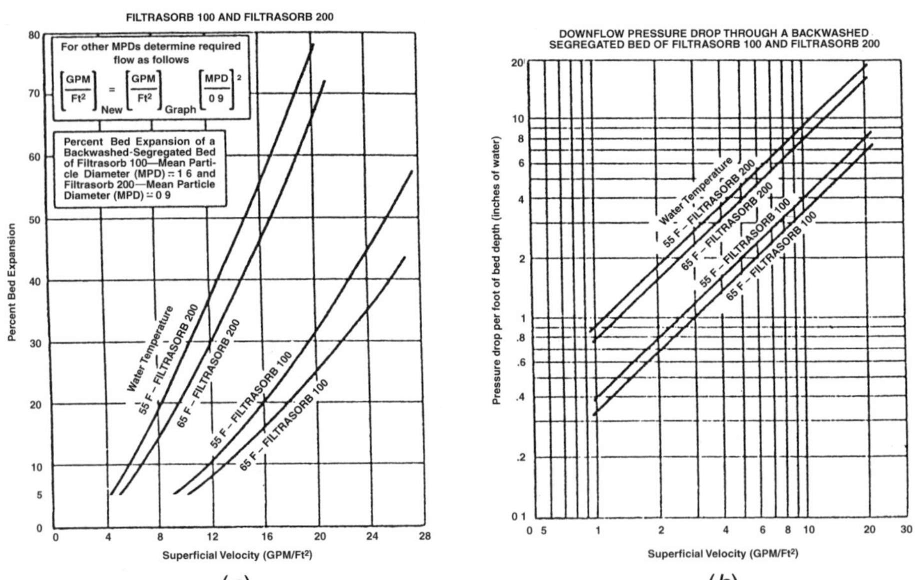

Figure 7.7.2-1 Hydraulic characteristics of typical GAC medium beds: (*a*) backwash rate versus GAC bed expansion rate; (*b*) headloss versus backwash rates. (Adapted from *Calgon Bulletin 20-60,* 1976.)

the light specific gravity of the material. The GAC beds must be backwashed periodically to remove suspended matter and microorganisms, even though the mass transfer zone is destroyed completely by backwashing. Figure 7.7.2-1*b* depicts the headloss across a clean GAC bed at various filtration rates.

7.7.3 Design Considerations

Design engineers are advised to conduct the following sequence of activities whenever GAC contactors are being considered as part of a treatment process train:

1. Survey intensively all available literature on the subject.
2. Collect all available water analysis data on the subject water. Identify the possible options for removing the organic compounds.
3. Procure a fresh sample of water and conduct a complete water analysis centered on the organic compounds.
4. Reevaluate the preliminary options and establish both the carbon usage rate and the pilot testing conditions.
5. Conduct a GAC pilot study.
6. Establish the design parameters of the process after analyzing the pilot test data.

The major design considerations for GAC contactors are:

1. The designed flow rate and available headloss for this process
2. The nature and amount of all organic compounds in the water supply
3. The MCLs of the regulated organic compounds
4. The overall treatment process train and each unit process
5. The type of GAC
6. The size of GAC
7. The filtration rate (superficial velocity)
8. The empty bed contact time (EBCT)
9. The type of contact units
10. The location of the GAC contactor in the process train
11. The regeneration frequency and the method of regeneration

Designed Flow Rate The designed flow rate should allow for the present required flow rate as well as future expansion of the facility (if needed).

Nature and Amount of All Organic Compounds Both the physical and chemical properties of the organic compounds strongly affect the type and level of pretreatment prior to the GAC adsorption bed. Alum coagulation and flocculation with clarifier are effective schemes for high-molecular-weight organic compounds; the majority of highly volatile, low-molecular-weight compounds may possibly be removed by air stripping upstream of the GAC contact beds.

MCL of Regulated Organic Compounds Refer to Section 2.4 for a detailed discussion.

Overall Treatment Process Train and Each Unit Process Refer to Section 2.4, especially Figure 2.4.5-1*b*.

Type of GAC The pore size of the GAC is an important factor in the removal of organic substances. For example, GAC with large pore volumes may be effective in removing waterborne compounds of high molecular weight. Moreover, activated carbon adsorbs inorganic ions in a preferential order:

$$H^+ > Al^{3+} > Ca^{2+} > Li^+ > Na^+ \qquad \text{for cations}$$

$$NO_3^- > Cl^- \qquad \text{for anions}$$

Engineers are therefore advised to consult the literature provided by manufacturers prior to selecting the GAC medium.

GAC BED AS FILTRATION/ADSORPTION PROCESS A recent design of filters used after preozonation utlizes GAC media instead of traditional sand and anthracite coal to remove carcinogenic ozonation by-products and reduce the high level of AOC in the ozonated water to the allowable level. Furthermore, the replacement of existing regular filter media with GAC and GAC capping over existing filter beds has been evaluated and implemented in quite a few existing plants, not because of preozonation, but to control taste and odor or remove SOCs found in agricultural, urban, and industrial wastes/pollution.

The removal of ozonation by-products by a GAC bed largely depends on the microorganism growth within the bed. A filter bed composed of biological activated carbon filters is called a *bacterial active filter* (BAF) or *biofilter*. Regular sand and anthracite filter beds can also become BAF with preozonation to some degree, but the concentration of biomass is much less than for a GAC bed. Moreover, the biomass in regular filter beds is easily washed out during backwash with residual chlorine, in contrast to a GAC bed, which is capable of holding most of the biomass. It is important to realize that most TOC removal is achieved in the top 6 to 10 in. (15 to 20 cm) of the bed (about 1.5 min of EBCT). Therefore, it is worthwhile to consider capping regular filter beds with GAC. If the filter is equipped with a surface wash system, surface washing should be applied for less than 2 to 3 min to retain the biomass.

There are three important design issues that must be addressed when using GAC capping and GAC with a layer (barrier) of sand or garnet at the bottom of the filter bed. The first is to select the proper size and specific gravity of GAC so that both the GAC bed and sand or anthracite layers are fluidized to the same degree during backwash; that is, both GAC and other types of media in a filter cell must be hydraulically matched during the filter washing period. Otherwise, GAC is either washed out or becomes dirty and clogged with suspended matter and biomass.

The second design issue is to provide a proper *L/d* ratio across the filter bed. Since the filter bed does not act as an adsorption bed but as a filtration and adsorption bed, the turbidity of the filter effluent should be less than 0.1 ntu as a goal. The third issue is the size and gradation of the GAC grains. For the adsorption process, GAC of small size and with a high uniformity coefficient is preferable. The filtration process, however, requires a larger size with a small uniformity coefficient; this is essential for the longer filter run and to achieve good effluent quality. Modern filter beds, including GAC, are composed of media with an effective size of 1.2 to 1.5 mm, with a uniformity coefficient of less than 1.4, and have a bed depth of 6 to 7 ft (1.8 to 2.1 m) with or without a sand barrier at the bottom.

Since a GAC bed is loaded with biomass such as bacteria, protozoans, rotifers and even nematodes—which may potentially break through the GAC bed—it is strongly recommended that a thin [10 in. (25 cm)] sand barrier (0.55 to 0.60 mm effective size) be placed at the bottom of the bed. It is important to recognize that coal-based GAC is preferred to wood-based GAC for filter media and TOC removal. The three filter bed design issues are discussed thoroughly in Section 3.2.7.

There are three major operational issues for the GAC filter: (1) to control the biomass, (2) to avoid feeding certain chemicals prior to the filter, and (3) to prevent anaerobic conditions within the GAC bed. Ozone treatment of a high-NOM (TOC) water produces a high concentration of biomass during a warm-water period. Despite regular air-scouring wash, initial loss of head increases with time, thereby unreasonably shortening the filter run length. A practical solution to this problem is to add chlorine to the backwash water (0.3 to 0.5 mg/L residual) until the initial headloss drops to a normal level. A few plants always backwash GAC filters with chloraminated water to control biomass without sacrificing the adsorption capacity of GAC to any significant degree.

Caution must be exercised when selecting the type of chemical that is to be fed upstream of the GAC filter bed. Polymer feeding as a filter aid is common practice for regular high-rate filters, but this practice potentially decreases the adsorption capacity of GAC and also creates a possible release of monomerous acrylamide. All anionic and nonionic types of polymers are made from polymerized acrylamide. Microorganisms in the biomass can potentially break the polymer chains and produce monomerous products; monomerous acrylamide is a known carcinogen. Potassium permanganate is often fed to control iron and manganese as well as taste and odor. If this is practiced for a long period of time, the GAC grains become coated with manganese oxide, thus the adsorption capacity is reduced or even lost.

The third operational issue is unique to the GAC bed. A GAC bed that is left as a standby (without draining the water) for a long period will have an anaerobic condition within the bed. When this GAC filter is put back on line after a long standby period, there will be a bad taste and odor problem, and soluble iron and manganese will be released from the bed. The taste and odor problem is the result of decaying organic matter and hydrogen sulfide. The reduced form of iron and manganese released from a GAC filter will produce a high color when chlorine is fed to the water during the final disinfection stage. Therefore, if a GAC bed is not used for more than 2 to 3 weeks (especially during summer months), it is recommened that the water be drained and the bed be kept in an aerobic condition by applying air scouring at certain intervals.

Size of GAC The size of the GAC strongly influences both the depth of the bed and the headloss across the bed given a specific flow rate. Coarse GAC has definite advantages for application to deep-bed gravity contactors under high filtration rates. However, smaller GAC is characterized by a higher rate of adsorption, given an equal level of activity with a shallower bed, since it has a greater active surface area. Calgon Carbon Corp., one of the major GAC producers, provides various size ranges of GAC, as shown below.

	F-100	F-200	F-300	F-400	F-820	F-816	F-614
Effective size (mm)	0.85	0.63	0.95	0.64	1.0	1.4	1.8
UC	1.9	1.7	2.3	1.6	1.5	1.4	1.4
Iodine number	850	850	900	1000	850	850	850

NOTE: Both effective size and uniformity coefficient (UC) are approximate numbers. F-100 through F-400 are products used primarily for adsorption purposes. The filtration/adsorption use GAC are F-820, F-816, and F-614.

Filtration Rate Generally, the most common filtration rate is 2 to 5 gpm/ft^2 (5 to 12.5 m/h), although higher rates are also used.

EBCT A practical EBCT is one that ranges from 7.5 to 10 min. The shorter EBCT may be used when removing synthetic organic compounds (SOCs) since these are easily adsorbed, but the longer contact time must be used in removing the more soluble and less adsorbable organic compounds.

Figure 7.7.3-1 illustrates the EBCT versus the THM breakthrough at EBCT values of 5, 7.5, and 15 min. The graph shows that the EBCT should be a minimum of 15 min, given a reasonable regeneration cycle (a 6-month to 2-year cycle). It is very interesting to see that increasing the EBCT from 7.5 to 15 min, increases the carbon life significantly. Similar results have been obtained from a number of pilot tests, including the test presented in an EPA publication entitled *Treatment Techniques for Controlling THM's in Drinking Water* (1981).

Most of the design parameters (1 through 8) may be evaluated by a GAC pilot study, provided that the project has both the time and the budget to do so. The pilot study should test a variety of pretreatment options, such as ozonation, aeration, alum coagulation, sedimentation, and filtration. Since the actual raw water usually contains a unique combination of suspended solids, organic compounds, and inorganic compounds indigenous to the region, the only way to properly establish most of the design criteria is to conduct a pilot-scale study on the plant site.

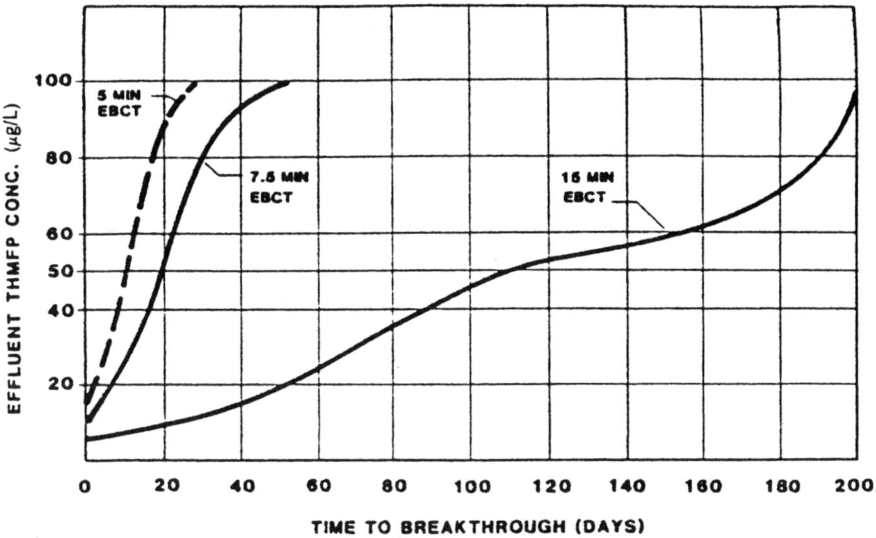

Figure 7.7.3-1 EBCT versus THM breakthrough against time. (Reprinted from *AWWA Seminar Proceedings, Granular Activated Carbon Installations: Conception to Operation,* by permission. Copyright © 1987, American Water Works Association.)

Type of Contact Unit The configuration of the GAC bed may be upflow, downflow in parallel trains, or downflow in series trains. The upflow configuration is not suitable for application in a potable water treatment process unless the regular filter is located downstream of the upflow GAC bed. Otherwise, the GAC grains will be present in the finished water due to the escape of fine GAC particles during the latter stages of the filtration cycle: the result of bed expansion due to the development of headloss across the bed.

Location of the GAC Contactor in the Process Train The GAC contact bed can be placed upstream of the flocculation process, before the ordinary filters, or downstream of the regular filters, depending on the nature of the substance to be removed and the result of the pilot study. The GAC contact bed is often used as both a filtration and adsorption bed. Figure 7.7.3-2 is a cross section of a GAC adsorption bed designed for the city of Cincinnati.

7.7.4 GAC Regeneration

GAC is a very expensive product ($2000 per ton); thus the exhausted medium cannot be discarded like spent PAC. Fortunately, GAC is fairly easy to regenerate. The need to regenerate the GAC ranges from 6 months to 5 years, depending on the amount, type of organic contaminants and level of biological activities in the GAC bed. The two basic alternatives for GAC regeneration are off-site regeneration and on-site regeneration. The on-site alternative is generally not cost-effective unless the carbon ex-

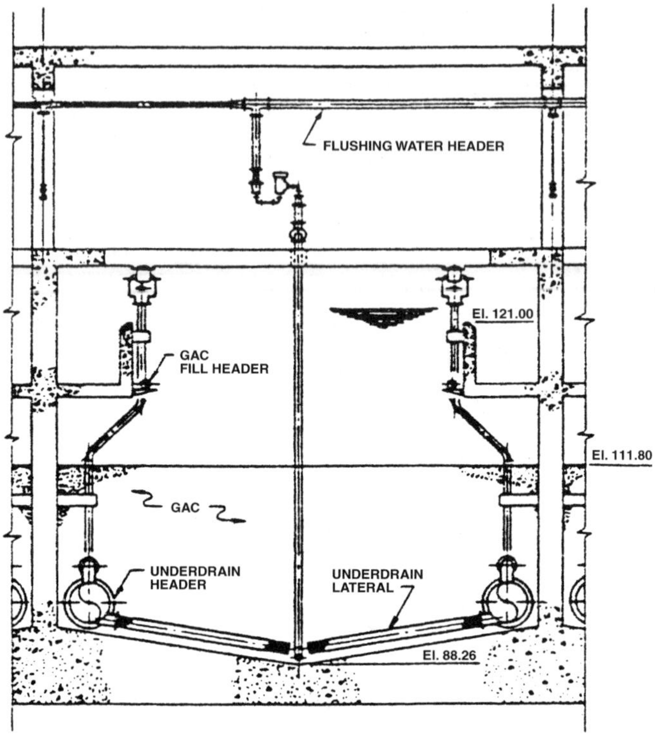

Figure 7.7.3-2 Cross section of a GAC adsorption bed. (Reprinted from *AWWA Seminar Proceedings, Granular Activated Carbon Installations: Conception to Operation,* by permission. Copyright © 1987, American Water Works Association.)

haustion rate is very large (over 5 tons/day). Moreover, the additional air pollution from the furnace off-gas can pose a great problem in large metropolitan areas, which already have poor-quality air. The off-site regeneration area can be any site outside urban areas where air pollution is not a major issue. In most cases, the GAC manufacturer is capable of regenerating the spent GAC at its facility on a contractual basis.

The three methods used in GAC regeneration are (1) regeneration by steam, (2) thermal regeneration, and (3) chemical regeneration. Since only an outline of these methods is presented, refer to the *Carbon Adsorption Handbook* edited by Cheremisinoff and Ellerbusch (1978) and/or the manufacturer for further details.

Steam Regeneration This method is limited to carbon that has been saturated with a limited type and level of volatile compounds. For further details, refer to the *Carbon Adsorption Handbook.*

Thermal Regeneration This is a pyrolytic process that burns off adsorbed compounds. The dewatered GAC is heated to approximately 1500°F (815°C) under controlled conditions: steam is used to prevent the carbon from igniting. The GAC is then quenched with water for reuse. Approximately 10 to 15% of the original GAC is generally lost during this entire process. The types of furnace commonly used in thermal regeneration of GAC are multiple hearth, fluidized bed, and rotary kiln. All of these are very expensive.

Both spent and regenerated GAC is transferred in slurry form by means of hydraulic eductors. The slurry is transported through stainless steel piping for most of the distance; stainless steel screw conveyers are used to feed the spent GAC to the furnace. Figure 7.7.4-1 is an example of the off-gas handling system.

Chemical Regeneration This method uses a solvent to extract adsorbed compounds at a temperature of approximately 210°F (100°C) by means of pH. The process was developed by the French firm Degremont and its application is very limited at present.

7.7.5 Capital and Operational Costs

The capital cost of a GAC adsorption system can be estimated based on the total volume of GAC medium required for the system. The capital cost includes the GAC medium, reactor cells, pipings, pumping, flow control, electrical, and concrete works. The estimated capital cost of the system is shown below.

Volume of GAC (ft^3)	2,500	5,000	10,000	25,000	50,000	100,000
Cost ($ million)	0.48	0.84	1.40	3.90	7.80	13.0

The costs above do not include storage of the GAC medium or the spent GAC reactivation system because they are site specific. The total cost is generally 2.5 to

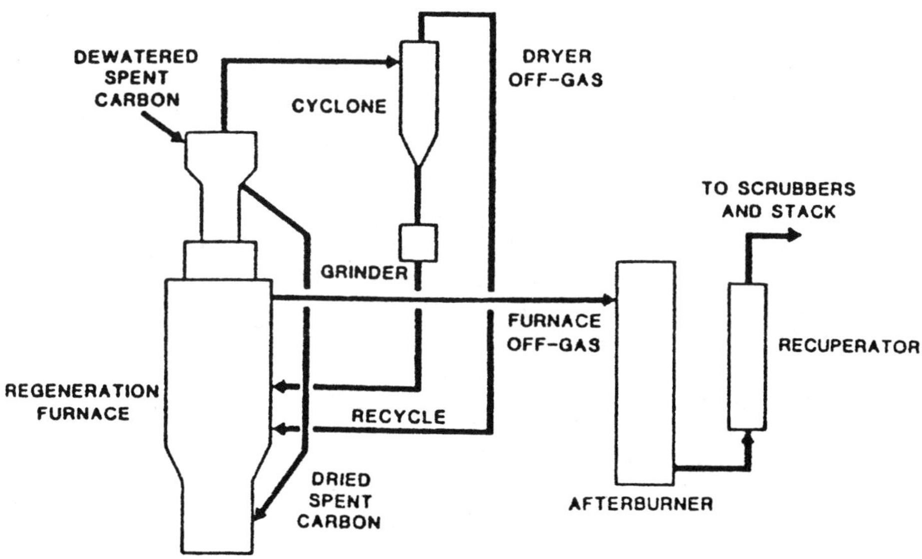

Figure 7.7.4-1 Off-gas control system. (Adapted from *AWWA Seminar Proceedings, Granular Activated Carbon Installations: Conception to Operation,* by permission. Copyright © 1987, American Water Works Association.)

3.5 times higher than the cost shown above, depending on the size of the facility and other factors, such as if the storage and the reactivation facilities are included. Operational costs include labor, fuel, power, electrical and control, virgin GAC makeup, miscellaneous indirects, and maintenance costs. The operational cost is usually estimated based on the GAC exhaustion rate.

GAC exhaustion rate (lb/day)	1000	2500	5000	10,000	25,000	50,000	100,000
Cost ($/lb)							
On-site regeneration	2.1	1.3	0.9	0.72	0.65	0.61	0.58
Off-site regeneration	2.7	2.65	2.6	2.6	2.6	2.6	2.6
Throwaway	4.5	4.45	4.43	4.4	4.4	4.38	4.35

BIBLIOGRAPHY

Adams, J. Q., and Clark, R. M., "Cost Estimates for GAC Treatment Systems," *J. AWWA,* 81:1:34 (January 1989).

AWWA, *AWWA Seminar Proceedings—GAC Installations: Conception to Operation,* annual conference in Kansas City, American Water Works Association, Denver, CO, 1987.

Bablon, G. P., "Developing a Sand–GAC Filter to Achieve High Rate Biological Filtration in France," *J. AWWA,* 80:12:47 (December 1988).

Calgon Corp., *Calgon Bulletin, 20-60,* Calgon, Pittsburgh, PA, 1976.

Carlson, M. A., et al., "Comparing Two GAC for Adsorption and Biostabilization," *J. AWWA,* 86:3:91 (March 1994).

Cheremisinoff, P. N., and Ellerbusch, F., ed., *Carbon Adsorption Handbook,* Ann Arbor Science, Ann Arbor, MI, 1978.

Crittenden, J. C., et al., "Design Considerations for GAC Treatment of Organic Chemicals," *J. AWWA,* 77:1:132 (January 1987).

Culp, R. L., and Clark, R. H., "Granular Activated Carbon Installations," *J. AWWA,* 75:8:398 (August 1983).

EPA, *Treatment Techniques for Controlling THM's in Drinking Water,* EPA/600/2-81-156 U.S. Environmental Protection Agency, Washington, DC, September 1981.

Grease, S. L., et al., "GAC Filter Adsorbers," *J. AWWA,* 79:12:64 (December 1987).

Hand, D. W., et al., "Designing Fixed-Bed Adsorbers to Remove Mixtures of Organics," *J. AWWA,* 81:1:67 (January 1989).

Hassler, J. W., *Purification with Activated Carbon,* Chemical Publishing, New York, 1974.

Hooper, S. M., et al., "Improving GAC Performance by Optimized Coagulation," *J. AWWA,* 88:8:107 (August 1996).

Hozalski, R. M., et al., "TOC Removal in Biological Filters," *J. AWWA,* 87:12:40 (December 1995).

Hyde, R. A., et al., "GAC as Sand Replacement in Rapid Gravity Filters," *J. AWWA,* 79:12:33 (December 1987).

Lykins, B. W., et al., "Granular Activated Carbon for Controlling THMs," *J. AWWA,* 80:5:85 (May 1988).

McCarty, P. L., et al., "Operational Experiences with GAC Adsorbers at Water Factory 21," *J. AWWA,* 71:11:741 (November 1979).

Oxenford, J. L., and Jykins B.W., Jr., "Conference Summary: Practical Use of GAC," *J. AWWA,* 83:1:58 (January 1991).

Quinn, J. E., and Snoeyink, V. L., "Removal of Total Organic Halogen by GAC Adsorbers," *J. AWWA,* 72:8:483 (August 1980).

Roberts, P. V., and Summers, R. S., "Performance of GAC for Total Organic Carbon Removal," *J. AWWA,* 74:2:113 (February 1982).

Suffett, I. H., *Activated Carbon Adsorption of Organics from the Aqueous Phase,* Vols. 1 and 2, Ann Arbor Science, Ann Arbor, MI, 1980.

Suffett, J. J., "An Evaluation of Activated Carbon for Drinking Water Treatment: A National Academy of Sciences Report," *J. AWWA,* 72:1:41 (January 1980).

Summers, R. S., et al., "Bench Scale Evaluation of GAC for NOM Control," *J. AWWA,* 87:8:69 (1995).

Symons, J. J., "A History of the Attempted Federal Regulation Requiring GAC Adsorption of Water Treatment," *J. AWWA,* 76:8:34 (August 1984).

Waer, M. A., et al., "Carbon Regeneration: Dependence on Time and Temperature," *J. AWWA,* 84:3:82 (March 1992).

Wang, J. Z., et al., "Biofiltration Performance: Relationship to Biomass," *J. AWWA,* 87:12:55 (December 1995).

Weber W. J., Jr., and Jocellah, A. M., "Removing Humic Substances by Chemical Treatment and Adsorption," *J. AWWA,* 77:1:132 (January 1985).

Wiesner, M. R., et al., "Optimizing the Placement of GAC Filtration Units," *J. AWWA,* 79:12:39 (December 1987).

7.8 ION EXCHANGE AND MEMBRANE SEPARATION PROCESSES

The ion exchange and membrane separation processes are not ordinarily used in municipal water treatment practices. However, membrane filtration process has gained great attention because it has the potential to achieve the highest removal rate of NOM and DBP precursors and complete removal of *Giardia* cysts and *Cryptosporidium* oocysts. These processes are the products of highly specialized chemical engineering. Consequently, water treatment engineers not involved in the detailed design of these processes—the design, fabrication, and operational aspects—are advised to seek the assistance of specialized equipment manufacturers and resin suppliers.

Provided that the project has both the budget and the time, water treatment engineers should run a pilot test on one of the processes to establish the design and operational parameters (for a specific water supply). Although engineers should understand the basic concept and important characteristics of these processes in order to write performance specifications and possibly P&ID diagrams, it is very important that engineers obtain assistance and advice from equipment manufacturers since they possess a great deal more knowledge and experience.

In this section we discuss some fundamental features of ion exchange processes and five membrane processes: (1) microfiltration, (2) ultrafiltration, (3) nanofiltration, (4) reverse osmosis, and (5) electrodialysis.

7.8.1 Ion Exchange

The process of ion exchange removes undesirable ions from raw water by exchanging them with a number of desirable ions; the desirable ions are stored in a solid material (ion exchanger). Since the ion exchanger has a limited capacity for exchange, it must be washed with a regenerating solution to regain its function. This cycle generally entails backwashing, regeneration, rinsing, and service. The most primitive ion exchangers are natural soil and sodium aluminosilicates. The first synthetic ion exchanger was composed of processed natural greensand (zeolite). However, insoluble synthetic plastic (resin) is used almost exclusively today. Most commercial resins are composed of synthetic plastic materials such as copolymers of styrene and divinyl benzene and are expected to survive 5 to 10 years of continuous use.

The resins are classified as either cation exchangers—that is, resins with negatively charged sites that take up positively charge ions—or anion exchangers. The resin has a porous and permeable structure; thus the whole resin particle participates in the exchange process. Resins are typically spherical in shape and range from 20 to 40 mesh (0.8 to 0.4 mm). Cation resins generally have an exchange capacity of approximately 2 milliequivalents per dry gram (mEq/g). The typical strong base anion resins have a 1.3 mEq/g exchange capacity.

In the field of water treatment, the ion exchange process is used for water softening, the removal of fluoride or nitrate, and the production of demineralized water for laboratory use. The resinous ion exchange process is the only process, with the exception of evaporation, that has the potential to remove minerals from water completely, thus yielding ultrahigh-purity water.

Characteristics Whether ion exchange is performed in fixed beds, continuous loops, pulsed beds, or by countercurrent extraction, it is still a batch process with respect to exhaustion and regeneration of the medium. Resins have a discrete exchange capacity and must be regenerated when exhausted. The cost of the chemicals per unit volume of water treated is almost directly proportional to the amount of minerals removed.

The reaction for the removal of water hardness is

$$\left.\begin{matrix} Ca^{2+} \\ Mg^{2+} \end{matrix}\right\} + Na_2R \rightarrow \left.\begin{matrix} Ca \\ Mg \end{matrix}\right\} R + 2Na^+$$

where R is the exchange resin. A more comprehensive expression of ion exchange, with a hydrogen cycle (H-form exchange), used in treating a typical water containing a variety of ions is as follows:

$$\left.\begin{matrix} Ca^{2+} \\ Mg^{2+} \\ Fe^{2+} \\ 2Na^+ \\ 2NH_4^+ \end{matrix}\right\} + H_2R \rightarrow \left.\begin{matrix} Ca \\ Mg \\ Fe \\ Na_2 \\ (NH_4)_2 \end{matrix}\right\} R + 2H^+$$

A typical anion exchange resin, with a chloride cycle, used to remove nitrate ions is

$$NO_3^- + ClR \rightarrow NO_3R + Cl^-$$

Ion exchange resins are characterized by selectivity: the preferential removal of certain ions. Selectivity primarily varies with the type of resin, ionic strength of the solution, relative amounts of different ions, and water temperature. This selectivity is a very important characteristic in the regeneration process. For example, in the sodium cycle the exchanger has a preference for calcium ion over sodium at concentrations of 1000 mg/L or below. However, at 100,000 mg/L the exchanger prefers sodium ions over calcium; thus it can be regenerated with brine.

The order of ion selectivity (in ordinary water) is listed below. For cations:

$$Fe^{3+} > Al^{3+} > Pb^{2+} > Ba^{2+} > Sr^{2+} > Cd^{2+} > Zn^{2+} > Cu^{2+}$$
$$> Fe^{2+} > Mn^{2+} > Ca^{2+} > Mg^{2+} > K^+ > NH_4^+ > H^+ > Li^+$$

For anions:

$$CrO_4^{2-} > SO_4^{2-} > SO_3^{2-} > HPO_4^{2-} > CNS^- > CNO^- > NO_3^-$$
$$> NO_2^- > BR^- > Cl^- > CN^- > HCO_3^- > HSiO_3^- > OH^- > F^-$$

The ion exchange units can be used to deionize water simply by allowing the water to flow continuously through a cationic resin column (hydrogen cycle) and a strong base anionic resin column (hydroxide cycle).

$$\begin{array}{ccccc} \left.\begin{array}{l} Ca\,(HCO_3)_2 \\ Mg(HCO_3)_2 \\ CaSO_4 \\ MgSO_4 \\ CaCl_2 \\ MgCl_2 \\ NaCl \\ SiO_2 \cdot H_2O \end{array}\right\} & + HR & \rightarrow & \left.\begin{array}{l} CO_2 \\ H_2O \\ H_2SO_4 \\ HCl \\ SiO_2 \cdot H_2O \end{array}\right\} & \rightarrow ROH \rightarrow H_2O \\ & \downarrow & & & \downarrow \\ & \left.\begin{array}{l} Ca \\ Mg \\ Na \end{array}\right\} R & & & R\left\{\begin{array}{l} SO_4 \\ Cl \\ HSiO_2 \end{array}\right. \end{array}$$

The important issues that must be addressed when designing an ion exchange process are reduction of chemical costs, reduction of waste products, and improvement of the processed water. However, it is vital that the water be pretreated by means of a proper process such as alum flocculation and sedimentation (possibly with filtration) prior to ion exchange (reverse osmosis or lime softening) to reduce the ionic loading to the exchanger system and to remove particulates such as iron, manganese, and organic substances which may foul the resins.

Tables 7.8.1-1 and 7.8.1-2 can help in selecting the most appropriate type of ion exchange process and alternative. The major U.S. suppliers of resin are Dow

TABLE 7.8.1-1 Basic Types of Demineralizer and Resins Used[a]

Scheme	Removal: Cations	Removal: Strong anions	Removal: Weak anions
1 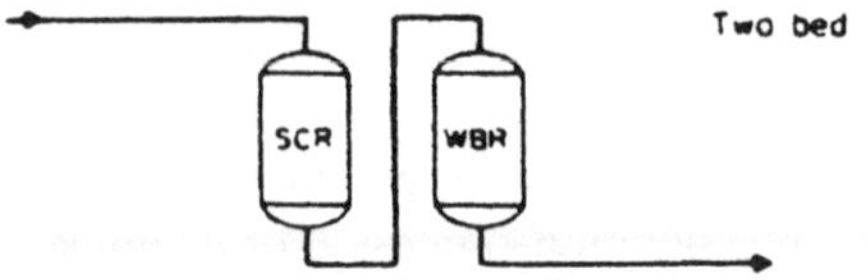	By SCR	By WBR	None
2 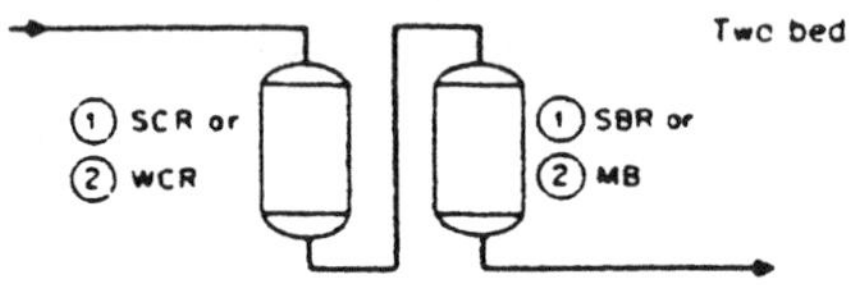	1. By SCR 2. By WCR + SCR in MB	← By SBR → ← By SBR in MB →	
3 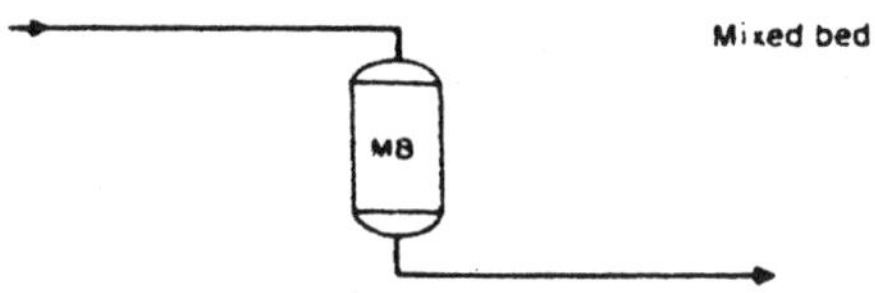	By SCR	← By SBR →	
4 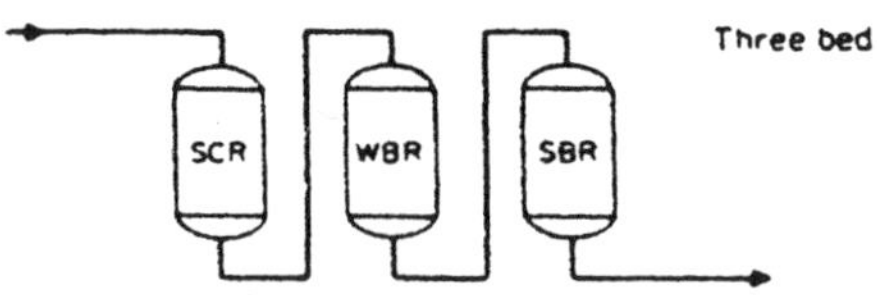	By SCR	By WBR	By SBR
5 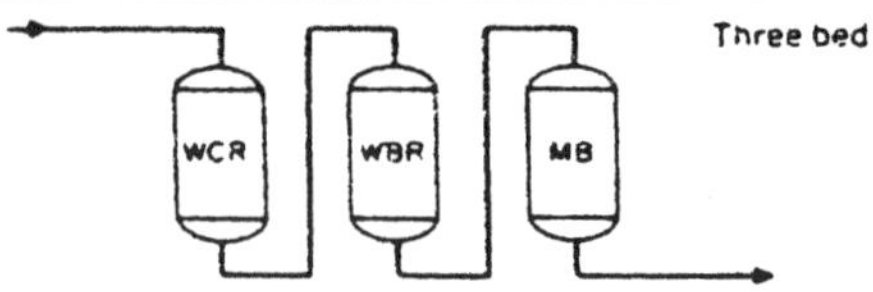	By WCR + SCR in MB	By WBR	By SBR in MB
6 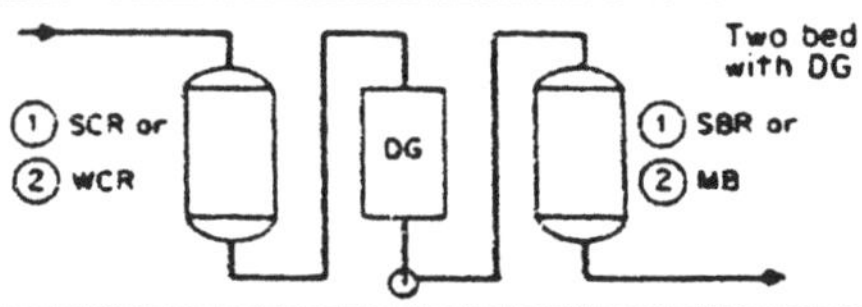	1. By SCR 2. By WCR + SCR in MB	By SBR By SBR in MB	By DG + SBR By DG + SBR in MB
7 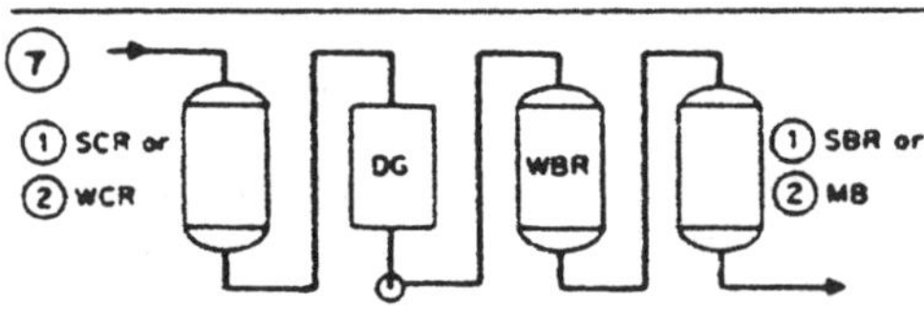	1. By SCR 2. By WCR + SCR in MB	By WBR By WBR	By DG + SBR By DG + SBR in MB

Source: Adapted from *The Nalco Water Handbook,* F.N. Kemmer, ed., McGraw-Hill, New York, 1979.

[a]SCR, strong cation resin; WCR, weak cation resin; SBR, strong base resin; WBR, weak base resin; DG, degasifier; MB, mixed bed.

TABLE 7.8.1-2 Ion Exchange Dimineralization Unit Selection Chart

Flow rate gpm	Amount of Impurity to be Removed, mg/L			Resins Required			Units to be Used				
	FMA	CO_2	SiO_2	C	WB	SB		C	DG	A	MB
1. Any	Any	None	None	×	×	—	(1)	×	—	×	—
							(2)	—	—	—	×
2. Any	Any	Any	None	×	×	—	(1)	×	×	×	—
							(2)	—	×	—	×
3. 0–20	Any	Any	Any	×	—	×	(1)	×	—	×	—
							(2)	—	—	—	×
4. 20–50	Any	0–50	Any	×	—	×	(1)	×	—	×	—
							(2)	—	—	—	×
5. Over 100	0–100	Over 100	Any	×	—	×	(1)	×	×	×	—
							(2)	×	×	—	×
6. Over 100	Over 200	Over 100	Any	×	×	×	(1)	×	×	×	—
							(2)	×	×	×	×

Source: Adapted from *The Nalco Water Handbood,* F.N. Kemmer, ed., McGraw-Hill, New York, 1979.

Notes: (1) C, cation; A, anion; DG, degasifier; MB, mixed bed; WB, weak base; SB, strong base; FMA, free mineral acidity (SO_4 + Cl + NO_3).

(2) Numbers in parentheses: (1) multibed plant, (2) mixed bed for 1 μS effluent.

(3) Intermediate ranges (50–100 gpm flow, 50–100 mg/L CO_2, 100–200 mg/L FMA) require careful evaluation for best balance in capital and operating costs.

Chemical, Rohm & Haas, and Zeolite. Many water treatment equipment manufacturers are also capable of producing ion exchange process units.

7.8.2 Membrane Separation Process

In the field of water treatment, in-depth filtration with granular medium is a regular filtration system capable of removing particles that are approximately 0.1 μm or above in size, provided that chemical coagulation pretreatment is performed. In response to the EPA's recent emphasis on safe drinking water, engineers have begun to reevaluate all the best available technology (BAT), including the membrane separation process. Membrane filtration, sometimes called a pressure-driven membrane process, has distinct advantages over the conventional treatment process: (1) removal of suspended solids, with no coagulant, up to about 200 ntu turbidity; (2) reliable production of good filtered water; (3) very high "log removal" of *Giardia*- and *Cryptosporidium*-sized particles; (4) much less space (footprint) required than for the conventional treatment process; (5) easy integration into the automatic control system; (6) minimum labor requirement so that it can maintain unattended operation most of the time; (7) chemical-free backwash water that can often be discharged to local water bodies; and (8) long-term compliance with drinking water regulations.

The membrane separation process uses semipermeable membranes to separate impurities from water. Semipermeable membranes are selectively permeable to water and certain solutes and play an extremely important role in nature: Nutrients and waste products are transferred across the semipermeable membrane of cells. Similarly, the membrane separation process used in the field of public water supply uses a driving force to pass the water through the membrane, leaving the impurities behind as a concentrate. The degree of purification is dictated by the type of membrane, the type and level of driving force, and the characteristics of the water (Table 7.8.2-1). However, the process does have several associated problems: (1) membrane fouling (fouling by bacteria, chlorine residual, and cationic polymer for certain types of membranes), (2) requires treatment of chemically washed waste before disposal, and (3) need for pretreatment of poor-quality raw water, which is almost identical to that in the conventional treatment process, including ozonation and activated carbon adsorption for taste and odor control.

Despite the problems, the number of membrane filtration plants has grown dramatically each year and membrane systems are replacing package filtration plants in some small water systems. Low-pressure membrane filtration provides a physical barrier to turbidity and pathogens; thus regulatory agencies in all U.S. states are now allowing the installation of this system provided that a pilot study is conducted. The economically applicable size of membrane filtration plants is no longer small and a few 23 mgd (1 m^3/s) plants will be on line at the end of the twentieth century in the United States.

The development of the submerged microfiltration process (Zenon's ZeeWeed and CMF-S by US Filter) now makes it possible to design plant capacities of 50 to 100 mgd (2.2 to 4.4 m^3/s). The cost of the membrane filtration system has decreased steadily since 1990, and the capital cost of the microfiltration system without appurtenances is approximately \$0.25 per gallon per day (\$65/$m^3 \cdot d$) at the end of the twentieth century; in comparison, the cost of regular high-rate filters are approximately 50% of this figure.

There are several major suppliers of membrane filtration systems: US Filter (Memcor), Infilco Degremont (Aquasource), Asahi Chemical Ind., Co., Ltd. (Pall Corp), Mitsubishi Rayon Co., Ltd., Leopold/PWT (FYNE System), Hydranautics/Nitto Denko Corp., Osmonics Corp., and Zenon Environmental Inc.. Dupont's Permasep is a well-known RO system. With regard to desalination sys-

TABLE 7.8.2-1 RO Membrane Characteristics (Based on 90% Rejection)

	Type Membrane		
	Triacetate Hollow Fibers	Polyamide Hollow Fibers	Cellulose Acetate Spiral Wound
Flux at 400 psi, gpd/sq ft	1.5	1.0	15–18
Back pressure, psi	75	50	0
pH range	4–7.5	4–11	4–6.5
Maximum temperature, °F	86	85	85
Cl_2, maximum mg/L	1.0	0.1	1.0
Bio-resistance	Good	Excellent	Fair
Backflushing	Ineffective	Ineffective	Effective

Source: Adapted from *The Nalco Water Handbook,* F.N. Kemmer, ed., McGraw-Hill, New York, 1979.

tems, Ionics Inc., Dupont De Nemours Inc., Culligan International Co., Osmonics Inc., and US Filter are major suppliers.

One special characteristic of membrane filtration is the great influence of water temperature on the production capacity (flux). Wintertime [39°F (4°C)] flux can be about half of summer [72°F (22°C)] condition under the same applied pressure. The department of health services in each state sets the allowable flux rate for membrane filters; for instance, the state of California limits the flux to 0.7 gpm/m^2 (159 L/h · m^2), whereas the state of Wisconsin has a limit of 0.5 gpm/m^2 (115 L/h · m^2). The most important aspect of the membrane filters is the integrity of the system; thus the system should have a reliable method of checking integrity. Table 7.8.2-2 briefly compares the characteristics and applications of these processes, and Figure 7.8.2-1 indicates the size ranges of suspended matter in raw water and selected processes.

Microfiltration (MF) Microfilter is the most popular type of membrane filter and the US Filter (Memtec) system has dominated the water treatment field in the United States during recent years. The purpose of microfiltration is to remove rather large particles, those over approximately 0.5 μm in diameter. One application of MF is in the field of sanitary engineering: Millipore filters are used to isolate coliform bacteria from water to collect suspended solids from water samples. Another application is in the removal of airborne particulates for air-feed ozone generators.

The MF membranes may be manufactured from a variety of material, including cellulose acetate. The pore size of these membranes is generally larger than 0.1 μm and a pore size of 0.2 μm is frequently employed for water treatment. Thus removal of all colloidal matter (generally, 0.01 to 1 μm in size) and dissolved materials cannot be expected. Yet the membrane is capable of swelling and changing its characteristics, thereby allowing it to retain certain types of colloidal and soluble substances.

Whenever the MF membrane becomes clogged, the filtration rate drops below a practical level, thereby necessitating backwashing and chemical cleaning. A self-cleaning cartridge filter with 500-μm openings is generally installed upstream of membrane filters. Consequently, the water should be pretreated and have most of its suspended solids removed prior to the membrane filtration process. Backwashing of the membranes is normally done by air-scouring wash every 20 min for a minute or so. Moreover, chemical cleaning of the membrane should be performed every 2 to 3 months, depending on the raw water quality, by caustic soda or acid with detergent when the pressure exceeds 15 psi (1 kg/cm^2).

As discussed previously, the new larger capacity of MF modules and the more reasonable cost of the system enables engineers to design 100-mgd (4.4-m^3/s) MF systems beyond the year 2000. Figure 7.8.2-2 is an artist's concept of a 30-mgd (1.3-m^3/s) MF system with six cells of a submerged type MF filter system.

Ultrafiltration (UF) Ultrafiltration uses membranes with pore sizes that are significantly smaller than 0.1 μm and generally require a driving pressure ranging from 10 to 90 psi (69 to 620 kPa or 0.7 to 6 kg/cm^2). For water treatment application, a "loose" membrane is used so that 10 to 40 psi (0.7 to 2.8 kg/cm^2) pressure is required even though their pore size is 0.01 to 0.02 μm. UF is capable of removing

TABLE 7.8.2-2 Comparison Between Membrance Separation and Ion Exchange Processes

Process	Suitable Water	Pore Size or Resin Size	Molecular Weight Cutoff (daltons)	Driving Force	Removal Objects	Notes
Microfiltration (MF)	500-μm self-cleaning cartridge filter	0.1–0.2 μm, 0.2 μm is common	300,000	10–20 psig (0.7–1.4 kg/cm^2)	Particulates and microbial	Batch process, removal of particles over 0.5 μm
Ultrafiltration (UF)	200 to 500-μm self-cleaning cartridge filter	0.003–0.01 μm, 0.01 μm is common	50,000	10–40 psig (0.7–7.8 kg/cm^2)	Molecular size compounds, particulates and microbial	Batch process, liquid-solid separation
Nanofiltration (NF)	Regular filter effluent or MF filtrate	0.001–0.005 μm	200–400	75–150 psig (5.3–10.6 kg/cm^2)	NOM, including color, virus, Ca, Mg	Batch process, DBPs control and softening
Reverse osmosis (RO)	Filtered water, 100–36,000 mg/L salts	<1 nm	—	>200 psig (>14 kg/cm^2)	Ionized salt ions and colloidal matter	Continuous process, 90–95% inorganic salts and 95–99% organic matter
Electrodialysis (ED)	Filtered water, 500–8000 mg/L salts	<1 nm	—	Dc electric current, 0.27–0.36 kW/lb, salts (0.6–0.8 kW/kg)	Ionized salt ions	Continuous process, complete removal of salts is not possible
Ion exchange (IX)	Settled or filtered water, 50–1,000 mg/L salts	<1 nm	—	<7 psig (<0.5 kg/cm^2)	Ionized ions, cationic and/or anionic	Batch process, complete removal of salts is possible

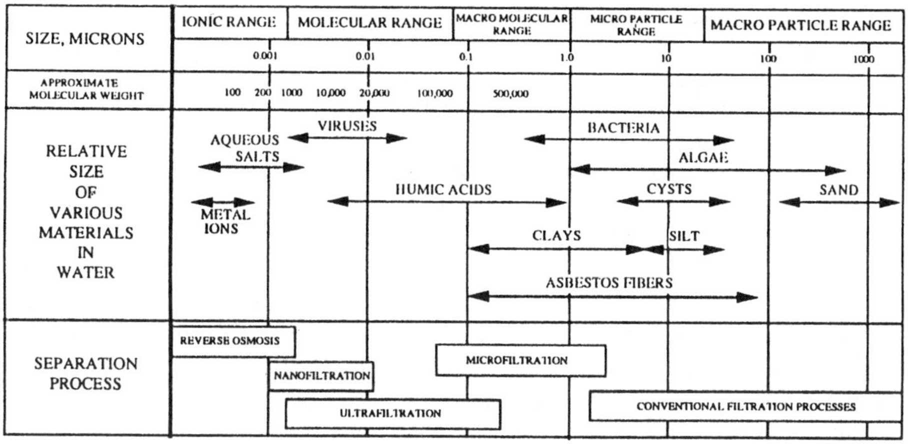

Figure 7.8.2-1 Size ranges of suspended matter in water and corresponding membrane processes. [Reprinted from *Journal American Water Works Association,* Vol. 81, No. 11 (November 1989), by permission. Copyright © 1989, American Water Work Association.]

colloids, bacteria, viruses, and high-molecular-weight organic compounds. The membranes are consequently susceptible to clogging. However, most types of UF membranes can be backwashed.

UF membranes are usually composed of two layers: an extremely thin (1 to 5 μm) and fine porous skin over a thicker (25 to 50 μm) and more porous substrate. The flow rate of the filtration process is reduced by the suspended matter and by the concentration of polarization, resulting in a localized increase in rejected impu-

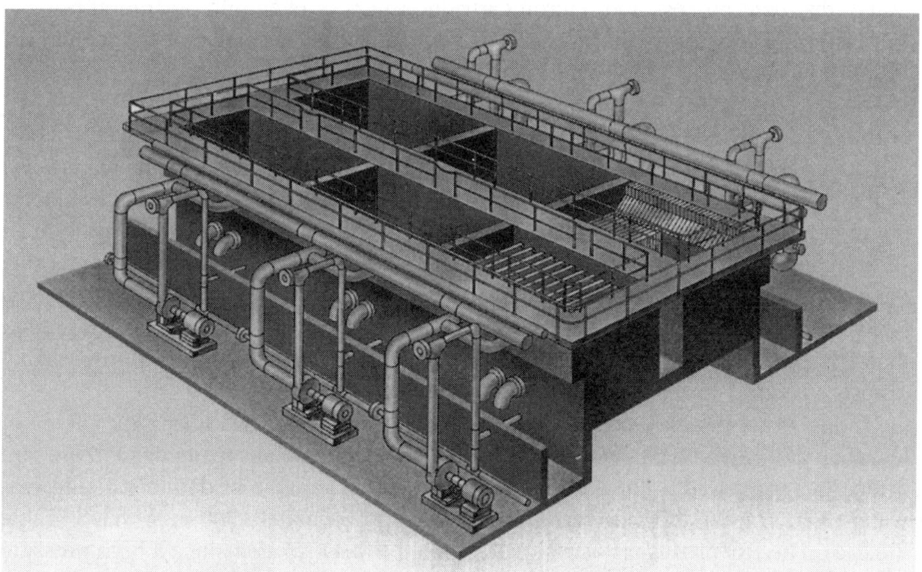

Figure 7.8.2-2 Concept of 30-mgd MF system. (Courtesy of U.S. Filter.)

rities. This increase in the density and viscosity of the solution at the surface of the membrane causes a reduction in the flux (flow) rate: The water production rate is approximately 50 gal/day per square foot (2 m^3/day · m^2) of membrane surface area and the energy requirement ranges from 0.4 to 2 Wh/gal (0.1 to 0.5 kWh/m^3) of the product water, depending on the dead-end and cross-flow filtration.

UF membranes may be flat with a filter press type of support, flat and spirally wound, tubular with a supporting cylinder (inside), or a tubular (hollow) fiber with an external diameter of 0.5 to 1.5 mm that does not have a supporting member (the raw water flows inside the tubes). This last configuration requires prefiltering through membrane filters with a pore size of 10 to 20 μm to prevent the membranes from becoming quickly clogged. The UF process has been used in the medical field for the separation of blood plasma and also has a number of industrial applications: concentration of macromolecular suspensions (i.e., polyvinyl alcohol), recovery of certain substances from wastes, and concentration of emulsions.

Nanofiltration (NF) The pore size of this membrane is between those of UF and RO. Nanofiltration membranes for potable water applications have a pore size of 0.001 to 0.002 μm and a molecular weight cutoff (MWCO) of approximately 200 daltons. Nanofiltration is capable of removing DBP precursors as well as calcium and magnesium ions, and water softening can be achieved more cost-effectively than with the lime–soda ash softening process. Permeate recovery is 80 to 90%, and 50 to 70% removal of total dissolved solids can be achieved. Since TOC removal is effectively achieved by the NF process, enhanced coagulation is not necessary to control DBPs. Also, microbial removal is 100%. Therefore, NF is capable of replacing ozonation, GAC adsorption, and enhanced coagulation. However, the NF membrane has a biofouling problem; thus effective pretreatment such as a conventional treatment process and micron cartridge filters or even microfiltration prior to NF, can be very helpful. Table 7.8.2-3 shows typical performance of the membrane filtration process.

Reverse Osmosis (RO) When pure water flows through a semipermeable membrane (separating the two waters) into water with a high salt content, the water level in the salted water compartment rises until the osmotic pressures on either side of the membrane are equalized: This is osmosis. In reverse osmosis, an osmotic pressure greater than that of the salted water is applied to the compartment containing the salted water. This then forces pure water to flow back across the membrane. The pressure required depends on the difference in salt concentration but is often greater than 300 psig (2070 kPa or 21 kg/cm^2). The rule of thumb is to apply approximately 10 psig (70 kPa) for each 1000 mg/L difference in TDS.

The RO process may be used in cases where the water has a salinity (TDS) of less than 500 mg/L to produce a 25-mg/L TDS product water using a small amount of energy (compared to the EO process). RO is also capable of demineralizing seawater (36,000 mg/L TDS) to water containing approximately 300 mg/L TDS (Table 7.8.2-4). The RO process is occasionally called *hyperfiltration* since a high pressure is required to drive the system. The major differences between RO and nanofiltration, which also applies a fairly high pressure, are the magnitude of the applied pres-

TABLE 7.8.2-3 Typical Performance of Membrane Filtration Processes (Quality of Permiates)

Parameter	Raw Water	Microfilter (Memcor)	Ultrafilter (Aquasource)	Nanofilter (Osmonics)
Turbidity (ntu)	1.3	0.05	0.06	0.01
Temperature (°C)	6	6	6	6
Particles (>3 μm/100 mL)	4500	40	57	9
TOC (mg/L)	2.6	2.2	2.2	1.2
UV_{254}	0.087	0.071	0.073	0.017
Iron (mg/L)	0.093	0.011	0.012	0.011
Manganese (mg/L)	0.064	0.032	0.033	0.036
HPC (cfu/mL)	54	—	570	64
THMs				
Chlorine residual (mg/L)	0.3	0.2	0.2	0.2
Chloroform (μg/L)	66	61	60	14
Dibromochloromethane (μg/L)	1.3	1.3	1.3	2.5
Bromodichloromethane (μg/L)	11	11	11	5.5
Total trihalomethanes (μg/L)	79	74	72	22
HAA_5				
Chlorine residual (mg/L)	0.3	0.2	0.2	0.2
Bromochloroacetic acid (μg/L)	3	3	3	1
Bromodichloroacetic acid (μg/L)	2	2	2	0
Chlorodibromoacetic acid (μg/L)	0	0	0	0
Dichloroacetic acid (μg/L)	25	25	25	4
Trichloroacetic acid (μg/L)	21	18	18	2
Total HAA_5	46	43	43	7

sure and the size of the substances that can be removed. Nanofiltration and ultrafiltration are essentially incapable of removing solutes of low to intermediate molecular weight. Reverse osmosis can remove ionic species with low molecular weight, as well as all types of colloids, bacteria, and viruses. The ratio between the amount of demineralized water and the amount of raw water fed to the system is defined as the conversion factor and is expressed as a percentage.

Two types of RO membranes are currently being marketed: cellulose acetate and aromatic polyamide. Membranes composed of cellulose acetate yield a high flow rate per unit of surface area. They are available in tube shape, as spirally wound flat sheets, and as hollow fibers. Conversely, polyamide membranes have a low rate of flow and are

TABLE 7.8.2-4 Typical Performance of RO Process

(*a*) Demineralization (Softening) by RO Process

	Constituent (mg/L)	
	Raw	Finished
Hardness, as $CaCO_3$	380	20
Alkalinity, as $CaCO_3$	215	16
Total electrolyte, as $CaCO_3$	445	29
Silica, as SiO_2	25	3
pH	7.2	6.0
CO_2	25	25

(*b*) Performance of a B-10 Module in Treating Seawater

Operating conditions: P = 56 bar, conversion factor Y = 25%, acidification H_2SO_4 = 810 psig

	Feedwater (mg/L)	Purified Water (mg/L)	Salt Passage (%)
Ca^{2+}	329	1.5	0.5
Mg^{2+}	1,031	3	0.3
Na^+	9,419	77	0.8
K^+	355	3	0.8
Sr^{2+}	13	0.2	—
So_4^{2-}	2,200	1	0.05
PO_4^{3-}	9	0	—
HCO_3^-	68	26	38
Cl^-	15,825	120	38
F^-	0	0	—
NO_3^-	3.4	0	—
SiO_2	0	0	—
Total salinity	32,680	185	0.6
Conductivity (μS/cm)	37,250	390	1
pH	6.3	6.3	—

Source: (*a*) Adapted from *The Nalco Water Handbook,* F.N. Kemmer, ed., McGraw-Hill, New York, 1979. (*b*), Adapted from Degremont, *Water Treatment Handbook,* 5th ed., Halsted Press, New York, 1979.

marketed as hollow fiber to provide the maximum surface area per unit volume. The advantage of the polyamide membrane is its excellent resistance to chemical degradation and biological attacks, thereby providing it with a longer service life (3 to 5 years) than that of cellulose acetate membranes (2 to 3 years); refer to Table 7.8.2-1. However, the polyamide material in RO membranes becomes irreversibly damaged when exposed to oxidizing biocides and most cationic biocides used in water treatment.

Since RO is a continuous process, the membranes must be replaced every 3 to 5 years. However, unlike ED, there are a number of membrane and RO process unit manufacturers to choose from. Moreover, the operational costs associated with demineralization are lower for RO than for ED and ion exchange; it is also uniquely characterized by a constant cost production factor.

Electrodialysis (ED) The semipermeable dialysis membrane is impermeable to water but permeable to either cations or anions under different driving forces. The driving force may be a concentration gradient (simple dialysis), pressure (piezodialysis), or an electrical potential (electrodialysis). The most commonly used membranes are composed of hydrated cellophane, and selectivity of a given membrane is largely a function of pore size. Although dialysis is rarely used in the field of water treatment and water reclamation, a few industries, such as the textile industry, use this process to reclaim caustic soda from mercerizing baths.

The electrical potential is the driving force in the ED process. Membranes having anion and cation exchange properties are stacked alternately in a press with a narrow water passage between them. The channels are made as narrow as possible to minimize electrical consumption. When direct current is applied to the electrodes (located on either side of the stacked membrane), the anions migrate to the anode, and vice versa. Since the cation exchange membrane only permits cations to pass through, and only anions for the anion exchange membrane, alternating concentration and dilutions are created in the various (alternating) compartments of the membrane stack (Figure 7.8.2-3).

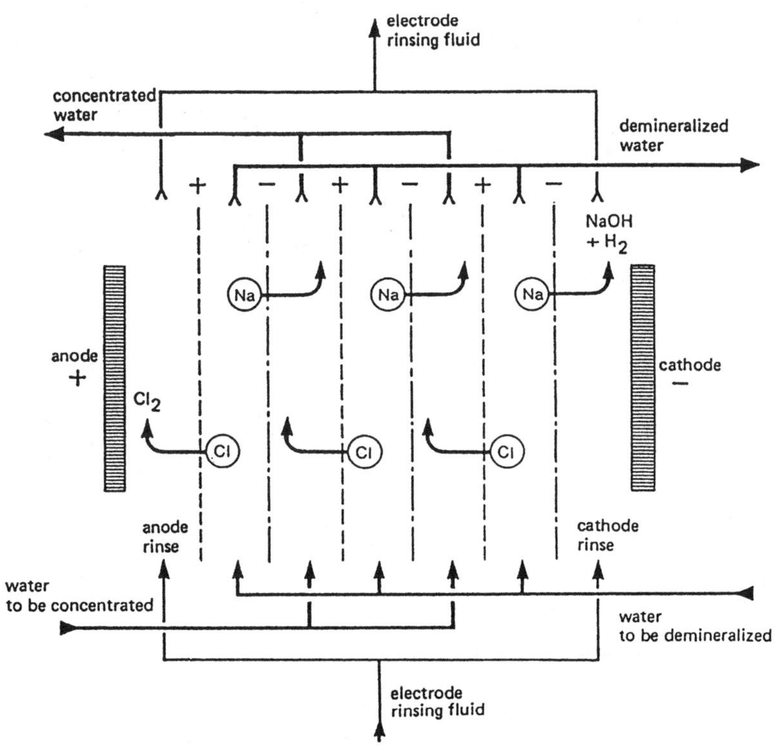

Figure 7.8.2-3 Concept of electrodialysis.

Unlike ion exchange, ED is a continuous process where the amount of salt removed from the water is proportional to the applied electrical current. However, application of an excessive amount of current results in electrolysis, forming hydrogen and oxygen gas, and thereby reduces overall efficiency. The efficiency of ED may also be reduced by concentration polarization and by the formation of scale and deposits on the membrane surface; these phenomena increase the electrical current requirement. For example, calcium carbonate and magnesium hydroxide scales will develop at the anion exchange membrane, due to the high pH resulting from the hydroxide ions. Although improvements in the stack design, as well as in the membranes, have helped to reduce both polarization and scale formation, they have not eliminated these problems.

If the raw water contains nonionized molecules such as organic compounds and colloids, they remain partially demineralized in the water. It is therefore necessary to pretreat the raw water to remove turbidity, organic compounds, and iron, prior to being applied to the ED process; otherwise, they will foul the membranes. The power consumption of the ED process is approximately 0.27 to 0.35 kWh per pound of salt removed (0.6 to 0.8 kWh/kg).

The ED process has three major limitations: (1) it is incapable of producing highly demineralized water; (2) the cost of the process increases rapidly with the salinity of the feed water, with the maximum salinity limit for economical treatment being 8000 mg/L; and (3) the number of equipment manufacturers is very limited: Ionix, Inc., Aqua-Chem Inc., and Dow Chemical USA are the only major suppliers in the United States. Figure 7.8.2-3 presents the basic principle of the electrodialysis system.

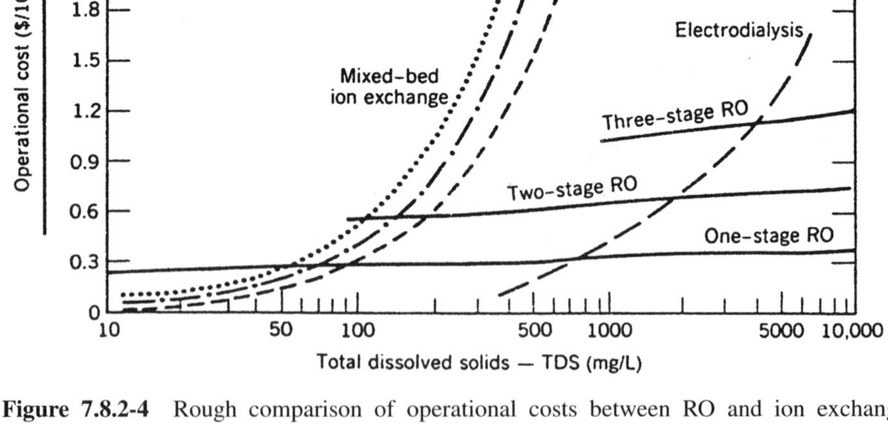

Figure 7.8.2-4 Rough comparison of operational costs between RO and ion exchange processes.

Figure 7.8.2-4 is an example of the demineralization costs associated with ion exchange, electrodialysis, and the process of reverse osmosis, excluding the capital amortization cost.

BIBLIOGRAPHY

Adham, S. S., et al., "Characteristics and Costs of MF and UF Plants," *J. AWWA,* 88:5:22 (May 1996).

Adham, S. S., Jacangelo, J. G., and Laine, J. M., "Low Pressure Membranes: Assessing Integrity," *J. AWWA,* 87:3:62 (March 1995).

Applegate, L. E., "Post Treatment of Reverse Osmosis Product Waters," *J. AWWA,* 78:5:59 (May 1986).

Bergman, R. A., "Cost of Membrane Softening in Florida," *J. AWWA,* 88:5:32 (May 1996).

Bersillon, J. L., et al., "Ultrafiltration Applied to Drinking Water Treatment: Case of a Small System," *Proceedings of the AWWA Annual Convention,* Los Angeles, 1989.

Chellam, S., et al., "Effect of Pretreatment on Surface Water Nanofiltration," *J. AWWA,* 89:10:77 (October 1997).

Chellam, S., et al., "Estimating Cost for Integrated Membrane Systems," *J. AWWA,* 90:11:83 (November 1998).

Cicek, N., et al., "Recycling Municipal Wastewater Using a Membrane Bioreactor," *J. AWWA,* 90:11:105 (November 1998).

Clifford, D., and Liu, X., "Ion Exchange for Nitrate Removal," *J. AWWA,* 85:4:135 (April 1993).

Degremont, *Water Treatment Handbook,* 5th ed., Halsted Press, New York, 1979.

Escobar, I. C., and Randall, A. A., "Influence of NF on Distribution System Biostability," *J. AWWA,* 91:6:76 (June 1991).

Fair, G. M., et al., *Water Supply and Wastewater Disposal,* Wiley, New York, 1954.

Fu, P., et al., "Selecting Membranes for Removing NOM and DBP Precursors," *J. AWWA,* 86:12:55 (December 1994).

Gere, A. R., "Microfiltration Operating Costs," *J. AWWA,* 89:10:40 (October 1997).

Haas, R. D., "Microfiltration Beats Conventional Treatment," *J. AWWA,* 89:10:10 (October 1997).

Jacangelo, J. G., et al., "Assessing Hollow-Fiber Ultrafiltration for Particulate Removal," *J. AWWA,* 81:11:68 (November 1989).

Jacangelo, J. G., et al., "Low Pressure Membrane Filtration for Removing Giardia and Microbial Indicators," *J. AWWA,* 83:9:97 (September 1991).

Jacangelo, J. G., et al., "Mechanism of *Cryptosporidium, Giardia,* and MS2 Virus Removal by MF and UF," *J. AWWA,* 87:9:107 (September 1995).

Jacangelo, J. G.,et al., "The Membrane Treatment," *Civil Eng.,* pp. 40–45 (September 1998).

Jurcand, V. L., et al., "Coagulation Pretreatment for Ultrafiltration of a Surface Water," *J. AWWA,* 82:12:76 (December 1990).

Karimi, A. A., et al., "Microfiltration Goes Hollywood," *J. AWWA,* 91:6:90 (June 1999).

Kemmer, F. N., ed., Nalco Chemical Co., *The Nalco Water Handbook,* McGraw-Hill, New York, 1979.

TABLE 7.9-1 Effective Processes for the Removal of Inorganic Contaminants

Contaminant	Applicable Effective Treatment Processes: Chemical Coagulation and Sedimentation		Ion Exchange	Reverse Osmosis	Remarks
Arsenic Ar^{3+} and Ar^{5+}	$Fe_2(SO_4)_3$ Alum Lime	(pH 6–8) (pH 6–7) (pH 10.5–11.3)	—	—	IX process (Na and H cycles) are not effective
Barium Ba^{2+}	Lime	(pH 10–11)	IX (Na cycle) is effective	—	Alum and ferric coagulant are not effective
Cadmium Cd^{2+}	Lime $Fe_2(SO_4)_3$	(pH 8.5–11.3) (pH 8–9)	—	—	Alum coagulation is not effective
Chromium Cr^{3+} and Cr^{6+}	Lime $Fe_2(SO_4)_3$ Alum	(pH 10–11.3) (pH 7–9) (pH 7–8.5)	—	—	Cr^{6+} can be removed by $FeSO_4$ (pH 7–7.9) but not by $Fe_2(SO_4)_3$
Fluoride F^-	Lime	(pH 9.5–11.3)	IX by activated alumina is good	—	Bone char is also good
Lead Pb	$Fe_2(SO_4)_3$ Alum Lime	(pH 7–9) (pH 6–9) (pH 9.5–11.3)	—	—	
Inorganic Hg	$Fe_2(SO_4)_3$ Lime	(pH 7–8) (pH >11)	—	—	Both GAC and PAC (>50 mg/L) are effective for mercury
Organic Hg	—	—	—	—	
Nitrate N^-	—	—	IX resin process is effective	—	Both strong base IX and weak base IX are good
Selenium Se^{4+} and Se^{6+}	—	—	Cation IX is good	RO is effective	Coagulation by metal coagulants is not very good
Silver Ag	$Fe_2(SO_4)_3$ Alum Lime	(pH 6–9) (pH 6–8) (pH >10)	—	—	
Asbestos fibers	$Fe_2(SO_4)_3$ Alum	(pH 7–9) (pH 6–7.5)	—	—	Filtration should be followed by coagulation and sedimentation

Lauch, R. P., and Guter, G. A., "Ion Exchange for Removal of Nitrate from Well Water," *J. AWWA,* 78:5:52 (May 1986).

Martin, H. J., and Miller, G. R., "A Zero Discharge Steam Electric Power Generating Station," *J. AWWA,* 75:5:52 (May 1986).

Metcalf, P. R., and Schroeder, E. D., "Dentrification Using a Membrane-Immobilized Biofilm," *J. AWWA,* 87:3:77 (March 1995).

Montgomery, J. M., Consulting Engineers, *Water Treatment: Principles and Design,* Wiley, New York, 1985.

Mulford, L. A., et al., "NF Performance at Full and Pilot Scale," *J. AWWA,* 91:6:64 (June 1991).

Pointius, F., "Regulatory Compliance Using Membrane Processes," *J. AWWA,* 88:5:12 (May 1996).

Pointius, F., et al., "Regulation Governing Membrane Concentrate Disposal," *J. AWWA,* 88:5:44 (May 1996).

Reinhard, M. R., et al., "Removing Trace Organics by RO Using Cellulose Acetate and Polyamide Membranes," *J. AWWA,* 78:4:163 (April 1986).

Sinsigalli, P. D., and McNutt, J. L., "Industrial Use of Reverse Osmosis," *J. AWWA,* 78:5:47 (May 1986).

Snoeyink, V. L., et al., "Strong-Acid Ion Exchange for Removing Barium, Radium and Hardness," *J. AWWA,* 79:8:66 (August 1987).

Snyder, D. W., et al., "Weak-Acid Ion Exchange for Removing Barium, Radium, and Hardness," *J. AWWA,* 78:9:98 (September 1996).

Wiesner, M. R., et al., "Cost Estimates for Membrane Filtration and Conventional Treatment," *J. AWWA,* 86:12:33 (December 1994).

7.9 INORGANICS REMOVAL PROCESSES

In 1985 the EPA released its recommendations on the maximum containment level (MCL) for 10 inorganic contaminants: arsenic, barium, cadmium, chromium, fluoride, lead, mercury, nitrate, selenium, and silver. This section briefly covers the applicable and tested processes used to remove these contaminants to levels below the MCL.

In 1978, the EPA published a manual that presented the recommended treatment techniques for meeting the interim primary drinking water standards; these are listed in Table 7.9-1. Since the EPA did not present effective methods for removing asbestos fibers, it has been added to the table.

BIBLIOGRAPHY

EPA, *Manual of Drinking Water Treatment Techniques for Meeting the Interim Primary Drinking Water Regulation,* EPA-600/8-77-005, U.S. EPA, Cincinnati, OH, 1978.

CHAPTER 8

Management of Procurement and Construction Phases

The design phase of a project is completed once the final plans and specifications of the project are presented to the owner and the required regulatory agencies. As part of the design phase, the design engineer should compile a complete set of project files, including all design calculations and correspondence, since these documents may prove to be essential for the project in the latter stages and particularly in the case of future lawsuits.

Upon completing the design phase, the engineer must begin managing the procurement and construction phases. The sequence of events is (1) bid advertising, (2) bid opening, (3) bid evaluation and award of contract, and (4) the construction phase. There are generally three principal parties involved in the construction phase: the owner, the designer, and the bidder.

8.1 PROCUREMENT PHASE (BID PHASE)

As stated previously, the procurement phase is composed of bid advertising, bid opening, and bid evaluation and award of contract.

8.1.1 Bid Advertising

During bid advertising the project engineer and the resident engineer (inspector) should review the contract documents together to ensure that the constructability of the structures and important field consideration have been incorporated into the plant design and specifications. Since the period between bid advertising and bid opening is generally a minimum of 6 to 8 weeks (unless the project is very small)—to allow the prospective contractors to perform the necessary activities—the engineers have time to review the documents carefully. If any omissions or errors are found during this period, the project engineer should issue an addendum to the plans or specifications to all bidders a minimum of 10 days before bid opening. The project manager should also look ahead and meet with the resident engineer to develop an agenda and a list of key subjects that must be discussed with the contractor at the preconstruction meeting, which is held with the successful bidder.

During the bidding phase, the project and resident engineers should perform the items listed below.

1. Arrange for bid advertising and select a bid opening date. Log the issuance of bid documents and release the project data to construction trade papers.
2. Keep an accurate log of all sets of contract documents issued: for example, name of the firm, address of each bidder, name of the person acquiring the documents for the firm, date of pickup, and deposit fee received.
3. If addenda are issued to the bidding documents, a copy of the addenda should be mailed to each holder of a set of documents by certified mail with a return receipt requested.
4. If the demand for the documents exceeds the supply prepared, exact reproductions of the original documents must be supplied to the requesting firms.

For obvious reasons, it is vital that all bidders receive identical sets of bidding documents and that the same information be provided to all bidders whenever they request clarification by phone, e-mail, or mail. To protect the owner against financial loss, which may occur because the bidder later refuses or is unable to sign the contract after submitting a bid, the requirement for a bid bond (security) should be included in the bidding documents. The bid bond should be 5 to 10% of the contractor's bid price and may be in the form of a certified or cashier's check and, occasionally, a negotiable security. The bid securities are returned to the bidders after all bids are opened and the contract is awarded.

All public works contracts require the contractor selected to submit performance and payment (labor and materials) bonds once the contract is awarded. The performance bond acts as a surety to cover costs associated with completion of the contract should the contractor fail to execute the contractual requirements. Under the payment bond, if the contractor becomes insolvent, the surety guarantees payment of all legitimate bills (labor or the acquisition of material) resulting from performance of the contract.

8.1.2 Bid Opening

The bid-opening procedure generally proceeds as follows:

1. Receipt of sealed bids at the designated time and place
2. Confirmation that all bids are responsive
3. Acceptance and logging of each bidder's name and the bid amount of all responsive bids
4. Summary of all line-item prices for all unit-price bids

After all the bids have been opened, the nonresponsive or informal bids are rejected. The project manager then identifies the apparent low bidder, which after careful bid review and evaluation is subject to confirmation.

8.1.3 Bid Evaluation and Award of Contract

Following the bid opening session, all bids are tabulated and each bidder is checked for financial responsibility, licensing, integrity, and reputation. The remaining tasks for the project and resident engineers are:

1. Analyze the bids and compare them with the engineers' estimate.
2. Make recommendations with regard to awarding the contract.
3. Prepare notification of award of contract and procure bid bonds from the three lowest bidders.
4. Remove copies of payment and performance bonds and certificate of insurance from the file prepared on the selected contractor and have the contractor submit the payment and performance bond with an actual insurance policy.
5. Have the owner sign the contract and return all other bid bonds.

8.2 CONSTRUCTION PHASE

There are several types of contract for a construction project. Fisk (1982) describes the following types as the most commonly used: (1) traditional architect/engineer (A/E) contract, (2) design/construction manager (D/CM) contract, (3) professional construction manager (PCM) contract, and (4) design–build (turnkey) contract.

8.2.1 Traditional Architect/Engineer Contract

The traditional A/E contract limits the engineering services provided during construction to occasional visits to the construction site and a few other functions, such as review and approval of shop drawings, interpretation of plans and specifications during construction, evaluation of claims by contractors, and review of payment to the contractor.

8.2.2 Design/Construction Manager Contract

The D/CM contract is similar to the A/E contract except that the project manager of the A/E firm is responsible to the owner during both design and construction. The project manager must satisfy all project needs, including scheduling, cost control, quality control, prepurchasing, and coordination of all work. The responsibility of the firm does not end until the owner's final acceptance of the project.

8.2.3 Professional Construction Manager Contract

Under the PCM contract, the professional construction management firm only acts as a representative of the owner during the entire duration of the project but is not involved in the design or construction of the treatment plant. Under this type of contract, the construction manager operates as an organization, not as an individual, and is responsible for the total project time, cost and quality control, and supervision; the construction manager also has control over the functions of the A/E firm and the contractor.

8.2.4 Design–Build Contract

The design–build contract is commonly called *turnkey construction*. The owner contracts with a single firm that plans, designs, and constructs the entire project with its own in-house capabilities. In addition, the turnkey contractor often offers the owner financial incentives. The major advantages of turnkey construction are (1) the elimination of claims by the contractor against the owner for errors or omissions in the plans and specifications, and (2) the ability to design construction since the design of each independent phase of the project is completed without having to wait for completion of the overall project design, which is often referred to as *fast track*. The disadvantage of turnkey contracts is that there is no competitive bidding process for plant construction. Thus turnkey construction contracts are illegal in the United States for all publicly funded construction projects, although it is commonly used to build water treatment plants in the rest of the world and is applicable to any project in the United States provided that the project uses only private funds. In recent years, the design–build–operate (DBO) contract has become popular worldwide, including the United States, due to the increasing number of privatized projects.

If a design–build contract is chosen, the firm should be selected on the basis of individual expertise, previous designing experience in the particular type of project, financial status, and the reputation of the firm. Although it appears that this concept is beneficial and saves a great deal of cost to the owner, case histories have proved that this is not necessarily true. Without competitive bidding procedures, the construction costs tend to be higher, and the quality of the design and specifications is not necessarily superior.

8.3 CONSTRUCTION MANAGEMENT

Construction involves the combined skills of engineering, science, and organization, in addition to the use of educated guesses and calculated risks. Numerous complex details must be carefully coordinated among various parties: the owner, engineers, architects, general and specialty contractors, manufacturers, material suppliers, equipment distributors, governmental agencies, utilities, labor, and so on. Construction contractors are private enterprises that engage in various types of construction projects, including water supply. The prime objectives of the contractor are (1) to complete the construction project to the satisfaction of the owner, within the set budget and on schedule, and (2) to make a profit. Since a contractor's goal is to complete the project as quickly and as economically as possible, there is always the possibility of poor-quality work on the project.

Engineers, on the other hand, tend to believe that if their design is correct and the plans and specifications are prepared carefully, field construction will proceed without problems. However, many case histories have proved this idea to be far from true and engineers therefore need to ensure that all work and construction practices are equal to, or better than, the standards dictated in the construction contract documents. For these reasons it is not surprising that owners are reluctant to place the responsibility for overseeing the quality of work on the project solely in the hands of the contractor.

The construction management program is a vital tool in completing the construction phase of a project in a satisfactory manner, for it meets the requirements of both the owner and the engineers. As a matter of fact, many federal construction contracts require the resident inspector, who is employed by the contractor, to meet the contractor quality control requirements. Ideally, construction management is handled by the firm that designed and prepared the plans and specifications, since it is intimately familiar with the plant site and design details and therefore capable of effectively administering the construction phase, providing quality control for the owner, securing the public safety, and maintaining the professional reputation of both the design firm and the contractor.

Construction administration and construction quality control should be characterized by continuous on-site inspection by one or more experienced and competent engineers who are technically qualified. These inspectors are often know as resident engineers. The resident engineer is responsible for ensuring that all the design details shown on the plans and specifications, and on the shop drawings approved by the design engineer, are constructed in strict accordance with the requirements. The resident engineer must also ensure that the work quality and construction practices are equal to or better than the standards dictated in the construction documents.

One important restriction is that both the resident project representative and the inspector are not authorized to change the plans and specifications, nor are they allowed to implement their own interpretations of the documents. If errors or problems associated with construction practice arise during construction, they should immediately be brought to the attention of the project manager for further clarification. Whenever there are obvious errors or if the changes vastly improve the situation or save time and material without sacrificing quality, these changes should be implemented by having the owner sign a formal change order (consent). The change order generally involves an adjustment in the construction fee.

The construction operation is a dynamic process requiring maximum efficiency, speed, and economy for the contractor. Therefore, the resident engineer(s) should make quick, yet appropriate decisions on all minor technical problems so as not to delay the operation of the contractor; resident engineers have the authority to approve materials and work. As in other fields, computerized project management programs are available for tasks such as scheduling, budgeting, and the plotting of job progress. Use of this type of software is recommended because the time required to perform these tasks is accelerated to a short period of time.

8.4 RESPONSIBILITY OF RESIDENT ENGINEERS

The function of the resident engineer is discussed in Section 8.3. However, the major responsibilities are as follows:

1. Be thoroughly familiar with the plans and specifications.
2. When the work does not meet the requirements, notify the contractor of the construction site immediately and record the infraction and date of notification in the daily log. If the contractor chooses to ignore the notice, advise the project manager of the inaction immediately.
3. Write a daily report that includes the daily activities, the items inspected, any instructions given, any agreements made with the construction team, the weather conditions, and other pertinent items. The daily report and diary/log books have legal importance in the event of contract disputes.
4. Perform all on-site testing specified in the contract documents. The tests should be performed in an expeditious and careful manner. All test failures must be reported to the contractor without delay.
5. Perform all inspections in a timely manner.
6. Discover unacceptable work in its early stages and report it to the contractor as soon as possible.
7. All problems beyond the capability of the resident engineer should be reported to the project manager of the engineering firm.
8. Do not be forced into making hasty decisions; carefully evaluate the situation and the possible consequence of the decision.
9. Be prepared to stand behind any decision made in regard to the contractor's work.
10. Be concerned with the safety of the public and all persons associated with the construction project and attempt to be objective in all situations (control emotions).

8.5 PROTESTS AND CLAIMS

Protests are defined as disputes arising from the issuance of a contract change order by the engineer against the objection of the contractor. When the engineering firm receives a letter of protest from the contractor, it should be examined carefully before acknowledging its receipt to assure that the basic requirements of the specification are included in the letter. The project manager and the owner should then review the merits of the protest and issue a letter advising the contractor of his rights under the contract to file a legal claim. One of the most important items that is considered before presenting any legal claim is documentation. When negotiations between the owner and the contractor fail, the data documented by the resident engineer(s) and the contractor become valuable in the litigation or arbitration procedures.

8.6 PROJECT CLOSEOUT

The main activities during project closeout are final inspection, acceptance of the plant for the owner, and final payment to the contractor. However, many important issues remain: review of the record (*as built*) drawings, transfer of the guarantee, obtaining the operation and maintenance instructions for the equipment, return of bonds, composing the plant operation and maintenance manual, and compiling the punch list items. The punch list is a detailed checkoff list that enumerates all items still requiring completion or correction before the work will be accepted and a certificate of completion is issued.

In some cases, the project is accepted as being "substantially complete" where only minor items remain to be completed or the "beneficial use" of the facility is accepted by the owner. Beneficial use (occupancy) is defined as the owner's use of the facility prior to its full completion. The date of beneficial use is the beginning of the 1-year guarantee period and total responsibility for proper maintenance and operation of the facility is transferred to the owner. Even if the certificate of substantial completion is filed with the punch list, the owner will not release the retained funds until all remaining deficiencies are corrected and a waiver of liens is completed. Final payment signifies the release of all retainers held during the project; this generally amounts to 5 to 10% of the total project cost.

Liquidated Damages If the contractor fails to complete the project by the date specified, he is financially liable to the owner for a preagreed sum for each day beyond the original finish date. This amount of money represents the opportunity cost of the owner resulting from such delays, and in the case of water treatment plants, this is usually the loss of income from the sale of potable water to customers.

Cleanup The contractor is responsible for cleaning up the construction site at the end of the job. The work will not be accepted until all temporary utilities, construction access roads, security fences (for material storage), field offices, stockpiles, surplus materials, and scraps are removed from the plant site. Site cleanup is the obligation of the contractor and must be conducted at the contractor's expense. The contractor should not be permitted to bury any waste or rubbish on the plant site unless the owner issues written approval.

8.7 MAJOR CLOSEOUT ACTIVITIES

The major closeout activities are as follows:

1. Perform closeout inspections as presented in the punch list.
2. Prepare clear and concise documentation of the closeout inspection of each item on the punch list.
3. Implement partial reduction of the field office inspection staff.
4. Conduct a final inspection with qualified engineers: a final punch list must be developed for any outstanding deficiencies requiring correction.
5. Prepare the record drawing with the contractor.
6. Obtain the following items from the contractor before accepting the project:
   (a) Guarantees
   (b) Certification of inspection

(c) Operating manuals and instructions for all equipment
(d) Keying schedules
(e) Spare parts that are specified to be received
(f) Record drawings
(g) Bonds (guarantee bonds, maintenance bonds, etc.)
(h) Certificate of inspection and compliance issued by local agencies
(i) Waivers of liens
(j) Consent of surety for final payment

7. Receive the request for final progress payment from the contractor.
8. Check that all remaining work requiring completion is fixed properly.
9. Obtain signatures from the A/E project manager, the contractor, and the owner on the certificate of completion. File this certificate with the office of the County Recorder.
10. Notify the owner that the project is complete and ready for operation.
11. Arrange for final payment and release of the retainer by the owner.
12. Clean up and close the field office.

8.8 CONSTRUCTION PHASE CHECKLIST

The project manager and resident engineer are responsible for the following major tasks during the construction phase:

1. Approve the cost breakdown documents submitted by the low bidder.
2. Identify and list the contractor's key personnel and the phone numbers at which they may be contacted.
3. Finalize the quality control and inspection plans.
4. Hold preconstruction meetings with the owner and with the contractor and owner.
5. Arrange for surveys, test labs (i.e., determine concrete strength), and mill and factory inspections.
6. Arrange for the construction documentation tools (photos, diary, field log).
7. Issue a notice to proceed.
8. Review the mobilization and construction schedule as submitted by the contractor.
9. Provide the necessary construction management service for on-site quality assurance of materials and work; the scope of this service depends on the type of construction contract.
10. Prepare daily construction reports and a diary/field log. Transmit these to the project manager.
11. Prepare and distribute monthly reports.
12. Review and process shop drawings with the transmittals.

13. Receive, log, and transmit all submittals from the contractor.
14. Review the record drawings and update them on a monthly basis.
15. Confirm the amount of each payment.
16. Process all progress payments.
17. Process the relocation of retainer fees.
18. Estimate the proposed changes and the extra work required by these changes.
19. Prepare, log, and process all change orders.
20. Document all extra work and work order protests.
21. Process and document time delays and requests for time extensions.
22. Assist in negotiations related to disputed work.
23. Administer the owner's acceptance of "beneficial use."
24. Compile a punch list and arrange for final inspection and job acceptance.
25. Prepare a notice for project completion, final payment, and final billing. Advise the owner that a 1-year guarantee bond is required from the contractor prior to occupation.
26. Finalize the record (as built) drawings.
27. Prepare a construction contract summary report and file this in the construction management project document file.
28. Inspect the project prior to the end of the warranty; prepare a report on the findings and make it part of the project management file.
29. Evaluate the performance of the construction management team.
30. File all pertinent documents for permanent record.

BIBLIOGRAPHY

Fisk, E. R., *Construction Project Administration,* 2nd ed., Wiley, New York, 1982.

Montgomery, J. M., Consulting Engineers, *Project Control Manual,* J.M. Montgomery Consulting Engineers, Pasadena, CA, 1976.

O'Brien, J. J., *CPM in Construction Management,* 3rd ed., McGraw-Hill, New York, 1983.

PMI, *Project Management Seminar/Symposium: Innovation,* Project Management Institute, Drexel Hill, PA, 1984.

PMI, *Research Goals with Project Management,* Project Management Institute, Drexel Hill, PA, 1987.

CHAPTER 9

Operations and Maintenance Manual and Operator Training

Since water treatment plants generally require a period of 2 years to be constructed, adequate time exists for the preparation of the operation and maintenance manual. This manual must be provided to the plant personnel at least 1 to 2 months before plant startup. Training sessions for the plant operators must begin during the final month of plant construction, when a substantial portion of the plant is completed. Thus the owner should either hire or assign the key positions of the operational staff by this date. All new members of the plant operating staff—especially the superintendent, chief operator, and maintenance supervisor—are advised to visit the construction site frequently, beginning in the early stages of construction, so that they are familiar with the location, elevation, and method of installation of all pipes, weirs, valves, and other items that are not visible after plant construction is completed. Most of these structures are either buried in the ground or embedded in the concrete walls and slab once the plant is finished.

Today, all municipal and public works officials, as well as directors of private water purveyors, are aware of the new and stringent drinking water quality standards set by the EPA and the states, and the requirements imposed on the water treatment process. In the midst of all these complex interactions of needs and concerns are the public officials and the directors of the private sector, who have identified the needs of their communities for improved and more sophisticated water treatment and control. Furthermore, the public is demanding safe drinking water. Thus millions of dollars have been funded for the construction of technologically sophisticated, complex, and expensive water treatment plants and the modification of existing plants to meet the new requirements. Yet use of high-quality equipment and computer-based instrumentation and control does not guarantee the continuous production of safe water; the plant must be operated by qualified and well-trained personnel. Thus the need for a well-organized and practical plant operations and maintenance manual, as well as systematic and highly skilled on-site training, is obvious. Unfortunate disasters such as Three Mile Island clearly demonstrate this need.

9.1 OPERATIONS AND MAINTENANCE MANUAL

Too many operations and maintenance (O&M) manuals are relegated to occupying shelf space on the bookshelves of water treatment plants. They are seldom used as a reference because they are either abstracts of textbooks or "boilerplate" that has little practical value. Although the EPA issued guidelines for composing an O&M manual in 1974, the key to creating a useful manual is to be practical and specific. An O&M manual should have practical features such as discussions of the basic principles used in the plant; talk-through presentations of the process layouts; detailed descriptions of each process component; step-by-step operating instructions for each process and process component; discussions of the variables that arise during process control and process status evaluation; descriptions and catalog cuts of all major equipment; tabulated troubleshooting guides listing the probable causes and solutions for problems commonly encountered in the field; and emergency response procedures.

The format of the O&M manual can be any style; however, the manual should at least include the following subjects

1. Technical literature (equipment)
2. Process operating manual
3. Preventive operation procedures
4. Standard operating procedures
5. Maintenance servicing schedules
6. Emergency response procedures
7. Policy formulation procedures

The process operating manual listed above (point 2) is the heart of an O&M manual. This section should include the following items:

1. Process description
2. Design criteria
3. Operating procedures
4. Troubleshooting guides
5. Emergency response program
6. Servicing and maintenance requirements

Photographs of the control systems and the major equipment of each process train and a narrative instruction—which provides information on how to start up, shut down, and troubleshoot respective pieces of equipment—have been found to be very effective in educating plant personnel. Use of such documents (videocassettes) provides even newcomers to the plant with a visual direction on how to operate and maintain any system. The use of schematic illustrations is always helpful in creating a better understanding of the electrical, electronic control, and extensive valving schemes.

One of the most popular and effective O&M manuals is that prepared by television and automobile manufacturers. Their concise manuals are almost always only a few pages long: a summary of the key items supported by a detailed description of each function. The plant O&M manual can also be arranged in a similar manner. This type of format allows the plant operator to obtain key information from a small pocket guide that summarizes the key subjects. If this concept is adopted, the O&M manual may be organized as shown below.

Main O&M Manual

1. Purpose of the treatment process
2. Process descriptions

TABLE 9.1-1 Operations Checklist (Example)

OPERATIONS CHECKLIST

Date:________________

Operator:______________

Item Description	Equipment No.	Status

3. Component descriptions
4. Detailed operating procedures
5. Performance evaluation and process control
6. Troubleshooting guide
7. Maintenance and servicing schedule
8. Emergency response program

TABLE 9.1-2 Equipment Data Form

EQUIPMENT DATA FORM

Plant No._____ Equipment Description_____

Equipment No._____ Component of_____

Supplier Address

Manufacturer_____
Model No._____
Serial No._____
Installation Date_____
Component Cost_____

Recommended Spare Parts

Capacity_____
Speed_____
TDH_____
Frame_____
Type_____

Comments_____

DRIVE MOTOR

Supplier Address

Manufacturer_____
Model No._____
Serial No._____
Installation Date_____
Component Cost_____

Horsepower_____	DC/AC_____	Type_____
Frame_____	Ins. Class_____	°C Rise_____
Speed_____	Phase_____	Enclosure_____
Amperes_____	Cycles_____	Condition_____
Voltage_____	S.F._____	

Comments_____

Operator Pocket Guide

1. Flow schematics
2. Concise operating procedures
3. Troubleshooting guides
4. Sample problems, including emergency response

TABLE 9.1-3 Problem Identification Form

PROBLEM IDENTIFICATION FORM

Observer

PROBLEM STATEMENT

Observer: ______________________ Date: __________

Subject: ______________________

Description of Problem: (Attach sketches or additional sheets if needed.) ______________________

____

Suggested Solution

____

Submit one copy each to:

______________ Plant Superintendent

______________ Project Coordinator

______________ Sup. San. Engineer

Project Manager

RESOLUTION OF PROBLEM (Decision /Corrective Actions)

____

DISTRIBUTION
Sup. San. Engineer
Project Coordinator
Plant Superintendent

______________ Plant Superintendent (JMM/OCI)

____

Maintenance Pocket Guide

1. Maintenance procedures
2. Servicing schedules
3. Schematics of major processes and control loops

The O&M manual should additionally contain the recommended types of forms used for records, logs, and reports. Today, data logging by means of computers is almost the standard method of producing the plant operation log and report. Therefore, an additional operation log sheet is not necessary. However, it is recommended that a few additional forms be distributed to the operators so that they can familiarize themselves with each piece of equipment, its location, and its physical condition (e.g., the need for painting). Overreliance on a computerized system will make plant operators ill prepared for any emergencies resulting from failure of the computer. Tables 9.1-1 through 9.1-3 are examples of this type of form.

9.2 OPERATOR TRAINING BEFORE PLANT STARTUP

There are two key requirements for successful plant startup: (1) good operation and proper maintenance and (2) familiarity with all plant facilities and a working knowledge of all water treatment processes. The goal of the operator training sessions (prior to startup) is to educate the plant operators in these two essential items. The training sessions should last a minimum of three working days with the O&M manual as the textbook. The project engineer and the heads of the mechanical, electrical, instrumentation, and control disciplines are assigned as instructors for each area of expertise. Representatives of major equipment manufacturers should also be invited to participate in the training sessions. Full use of audiovisual aids such as videotapes, color slides, and a chalkboard is highly recommended.

BIBLIOGRAPHY

EPA, *Considerations for Preparation of Operation and Maintenance Manuals,* EPA/430/9-74-001, U.S. Environmental Protection Agency, Washington, DC, 1974.

CHAPTER 10

Plant Startup and Follow-Up Services

10.1 STARTUP

Even if the plant is constructed exactly according to the plans and specifications, plant operators must always debug a few items during plant startup. Moreover, certain processes, such as ordinary granular medium filters and sludge blanket reactor-clarifiers, require a period of maturation in order to function properly. Plant startup is a hectic time for plant operators. Many equipment and process control units require adjustment. Therefore, serious problems and complications are most likely to arise during this period. The plant owner is strongly advised to contract the services of a qualified professional consulting firm so that problems often associated with plant and process startup can be avoided, thereby quickly placing the plant into service with fewer complications; the firm may also troubleshoot and optimize process efficiency. The most appropriate choice would be to hire the engineering firm that conducted the design, as well as construction management, simply because it is the most familiar with the plant, provided that it is qualified and experienced in doing so.

Optimization of coagulant dosage, adjustment of mixing conditions for flocculation as well as ozonation and disinfection, optimization of filter wash conditions, ensuring the proper pump operation sequence, and careful analysis and control of the finished water quality prior to delivery to the customer are items that are effectively executed by the consulting firm. Plant startup should be initiated at 20 to 25% of plant capacity, continuing until the majority of debugging is complete and the plant is steadily producing good-quality water, and before delivering the water to the public. It is important that the plant manager create a solid plan for startup and has all available plant operational staff participate. It is very beneficial if well-experienced plant operators from other plants and consulting engineers are invited to participate during this phase.

10.2 FEEDBACK

Feedback from the owner (client) with respect to the quality of the design, the performance of the engineers, and the performance and efficiency of the facility (water treatment plant) is absolutely necessary for maintaining quality engineering services. Most owners are reluctant to voice their dissatisfaction or their constructive opinions unless outrageous mistakes have been made. Instead, the unhappy client never awards another job to the engineering firm and the engineers have no idea what caused the dissatisfaction. Thus, from the standpoint of both business and engineering, it is imperative that engineers ask the client "How have we been doing?"

TABLE 10.2-1 Project Feedback Report Form (Example)

PROJECT FEEDBACK REPORT

FROM: ______________________

TO: ______________________ Division of Human Resources and Technology

		CHECK APPROPRIATE SPACE
PROJECT	______________________	
JOB NO.	______________________	____ General Design Problem
CLIENT	______________________	____ Specification Problem
OBSERVER	______________________	____ Standard Detail
DATE INITIATED	______________________	____ Special Detail
SPEC. SECT. & PAGE	______________________	____ Special Legal Problem
DRAWING SHEET NO.	______________________	____ Other

CASE HISTORY OR PROBLEM: (Attach sketches or additional information)

____

SUGGESTED IMPROVEMENT:

____

ACTION TAKEN IN THIS CASE:

____

and "How can we improve our service?" (feedback). It is a cold fact that the performance of the engineers and the firm will neither improve nor progress unless proper feedback is obtained from the client.

Generally, once all the debugging has been performed and the water treatment plant has been operating in a satisfactory manner, the engineers who were assigned to the project begin work on the next project. Seldom do engineers have a chance to view the operational plant unless they are asked by the client to remedy significant problems. However, these problems are often minor in nature: problems with mechanical equipment, problems associated with the electrical or control systems, or problems with the civil works or architectural items. Engineers from these various disciplines are capable of handling these types of problem; thus the project engineer and project manager are generally not involved or even supplied with the details of the problem. This is especially true for busy consulting engineers. Consequently, some engineers continue making the same mistakes over and over. For this reason, a project engineer should visit each plant that he or she has designed at least once a year for the first 5 years. This allows the engineer (potentially) to obtain feedback from the plant operators. Furthermore, the engineer should compose a memorandum describing the findings during the plant visitation, including the owner's criticisms and comments (Table 10.2-1). The memo should be filed in the project file and should also be disseminated among the key engineers of the project team; this allows every key engineer to benefit from the findings.

When an error of a serious nature is discovered, some project engineers find it very difficult to disclose this information to their boss and other engineers and the error tends to be covered up. This is not in the best interest of the engineer or the firm. All grave errors should be reported immediately to both the chief engineer and the legal department of the firm. Numerous cases have needlessly ended up in court, costing the engineering firm 10 or even 100 times more than if the problem had quickly been remedied; the latter case usually costs less than $50,000. Any delay in corrective action not only aggravates and angers the client but can potentially lead to a lawsuit.

10.3 OPERATOR TRAINING SERVICES

Both the EPA and public officials recognize the importance of operator training for wastewater treatment plants in improving plant effectiveness and, more important, in cutting operation and maintenance costs. The EPA requires that all publicly funded plants provide operator training services. Due to the requirements imposed by both the states and the EPA on water treatment plants, many administrators have begun to feel the same need for operators of public facilities, including water treatment plants. This concept is not a new idea: Operator training has been used for a number of years to help plant operators obtain their operator certification (required by all states).

The operator training program has been offered as a state-sponsored short school program to help operators pass the certification examination, as a correspondence course, as in-plant practical technical assistance, and as in-plant training. Of these programs, the in-plant training is undoubtedly the most comprehensive and effective, since it is a combination of hands-on and classroom instruction conducted at the treatment plant. This type of program provides operators with a sound understanding of the water treatment process and plant management, as well as useful information with lasting value. Several well-qualified consulting firms and professional organizations offer this type of training service.

The operator training program generally teaches plant operators how to analyze plant processes and provides instruction in mathematics, chemistry, and biology. Additional topics include housekeeping, safety equipment inspection techniques, and emergency procedures. In addition to the operator program, several other types of specific instruction are available in categories such as management, maintenance, and laboratory procedures. Management training is intended for the plant management staff and focuses on human resources development, quality control, quality assurance, process analysis, and writing skills. Maintenance training provides instruction in equipment performance standards, electrical and control system wiring diagrams, the scheduling of preventive and corrective maintenance, maintenance management programming, mechanical and electrical troubleshooting, and emergency services.

The laboratory training program is designed to provide detailed and functional training for plant operators in standard testing for routine operations and on-site specific process control, including tasks such as jar testing, core sampling of filter bed, and filter medium analysis. Laboratory training should not be limited to plant operators; the laboratory staff should be provided with on-site instruction as well as supplemental training at a state-certified laboratory in *Giardia* and virus monitoring; the use of gas chromatography/mass spectrometer equipment for the analysis of organic pollutants; adsorption and emission spectroscopy, inductively coupled plasma, and other techniques used in heavy metal analysis; and techniques used in conducting asbestos fiber counts with the transmission electron microscope. Since the tracer test for evaluating the tank detention time has become an important issue with respect to disinfection criteria, it is essential that plant personnel be trained to perform the test and be allowed to practice the technique.

10.4 PLANT OPERATION AND MANAGEMENT SERVICES

Before any type of training program can be implemented, the owner must hire the water treatment personnel. The entire plant operation and management is often conducted by a private professional organization. The primary reasons for this arrangement are cost-effective plant operation, better process control, and the steady production of safe drinking water by a well-educated and well-trained (licensed) professional staff. The cost-effectiveness and effective performance provided by qualified private professional firms are both well known and proven in many sectors of society. All major water treatment plants in Paris are operated and managed by a private professional firm (those in London have also implemented this option), and this concept has also gained popularity in the United States. Several consulting firms and private professional organizations offer this type of service.

One of the newest approaches to effective plant management is the comprehensive maintenance program using a computerized maintenance management system. One type, the Computer Assisted Management System for Environmental Operations (CAMEO), developed by Envirotech Operating Services, generates data on issues such as equipment age and number of breakdowns. The master report also provides information on equipment cost and repair. By using this type of system, an organized maintenance program, including preventive maintenance work and prioritized work orders, can easily be established. The effective use of thermographic analysis, vibration detection, and analysis systems such as the Snapshot system can identify problems in major equipment before the problems become serious, thus avoiding costly and major repairs. These types of assistance make overall management of a water treatment plant more efficient and economical and are provided by a few professional service firms.

Another special type of service that may become essential for water purveyors is that of a complete laboratory which performs tasks such as sample collection, analysis, data evaluation, and report preparation. As a result of the new EPA regulation, based on the Safe Drinking Water Act, very specific requirements are imposed on all public water systems in regard to sampling: The analysis of over 100 and over 180 regulated contaminants was required by the early 1990s and the year 2000, respectively, and the report must follow a specific format. It is obvious that most small to midsized public water supply agencies are presently incapable of handling the analysis of all contaminants, due to the need for special sophisticated analyzers and well-qualified scientists/chemists. In the author's opinion, one of the best solutions is either to have the state use the services of certified laboratories or to have each water purveyor enter a service contract with a certified private laboratory that will handle these requirements. Several certified laboratories presently operate in major metropolitan regions and are capable of meeting these new needs for a reasonable fee.

APPENDIX 1

Abbreviations

AAPC	American Association of Pipe Corrosion
AASHTO	American Association of the State Highway Transportation Officials
ACGI	American Conference of Governmental Industries
A/E	architect/engineer
AIHA	American Industrial Hygiene Association
ANSI	American National Standards Institute
AOC	assimilable organic carbon
AQMD	Air Quality Management District
ASCE	American Society of Civil Engineers
ASTM	American Society for Testing and Materials
ASU	areal standard units
AWWA	American Water Works Association
BAF	bacterial or biologically active filter
BAT	best available technology
BDOC	biodegradable dissolved organic carbon
BOM	biodegradable organic matter
°C	Celsius or Centigrade, degree
CADD	computer-aided design and drafting
CCTV	closed-circuit television
CMS	construction management service
COD	chemical oxygen demand
CPM	critical path method
CPU	central processing unit
CRT	cathode ray tube
CSI	Construction Specification Institute
CSO	combined sewer overflow
CSSE	Conference of State Sanitary Engineers
$C \times t$	product of concentration (mg/L) and time (min)
Cv	capacity factor for valves
DAF	dissolved air flotation
DBO	design, build, and operate
DBP	disinfectant by-product
dc	direct current

D/CM	design/construction management
DCS	distributed control system
DHS	Department of Health Services
DNA	deoxyribonucleic acid
DO	dissolved oxygen
EBCT	empty bed contact time
ED	electrodialysis
EIS	environmental impact study or statement
ENR	*Engineering News-Record*
ENR Index	*Engineering News-Record* Construction Index
EPA	U.S. Environmental Protection Agency
ES	Effective size (Hazen)
ESWTR	Enhanced Surface Water Treatment Rule
°F	Fahrenheit degree
FCI	Fluid Controls Institute
FDDI	fiber-distributed data interface
FEA	finite element analysis
FIN	facility information network
FPA	flavor profile analysis
Fr	Froude number
ft	foot or feet
ft^3/h	cubic foot per hour
ft^3/min	cubic foot per minute
ft^3/s	cubic foot per second
g	gram
g	acceleration of gravity
G	velocity gradient or mixing intensity (s^{-1})
$G \times t$	product of G (s^{-1}) and time t (s)
GAC	granular activated carbon
GOX	gaseous oxygen
gph	gallons per hour
gpm	gallons per minute
gpm/ft	gallons per minute per linear foot
gpm/ft^2 or gpm/sf	gallons per minute per square foot
GUI	graphical user interface
G value	velocity gradient
HAA	haloacetic acids
HDT	hydraulic detention time
HPC	heterotrophic plate count
HVAC	heating, ventilation, and air-conditioning
ICR	Information Collection Rule
ID	inside diameter
IEEE	Institute of Electrical and Electronic Engineers
in.	inch
IOC	inorganic chemical

ISA	Instrument Society of America
ISPM	integrated system of project management
Jtu	Jackson turbidity unit
K	kelvin (absolute temperature)
kg/cm^2	kilogram per square centimeter
kPa	kilopascal
kW	kilowatt
kWh	kilowatthour
LAN	local area network
lb	pound
LCR	Lead and Copper Rule
LI	Langelier Index
LID	loop interconnection diagram
LOH	loss of head
LOX	liquid oxygen
μg/L	micrograms per liter
μ Ω/cm	microhms per centimeter
μm	micrometer (10^{-6})
μg/L	micrograms per liter
m	meter
m^3/s	cubic meters per second
MA	methyl orange alkalinity
max	maximum
MCL	maximum contaminant level
MCLG	maximum contaminant level goal
mEq/g	milliequivalent per gram
MF	microfiltration
mg	milligram
mg/L	milligrams per liter
mgd	million gallons per day
m/h	meter per hour
MIB	2-methylisoborneol
min	minutes
Min	minimum
mL	milliliter
ML/d	megaliters per day
mm	millimeter
MPa	megapascal
MPN	most probable number
MTBE	methyl tertiary butyl ether
MWCO	molecular weight cutoff
$mW \cdot s/cm^2$	milliwatts per second per square centimeter (ultraviolet light)
MX	3-chloro-4(dichloromethyl)-5-hydroxy-2[5*H*]-furanone
N	newton ($kg \cdot m/s^2$)
NAS	National Academy of Sciences

NEC	National Electrical Code
NEMA	National Electrical Manufacturers Association
NF	nanofiltration
NFPA	National Fire Protection Association
ng	nanogram
ng/L	nanograms per liter
NIPDWR	National Interim Primary Drinking Water Regulations
nm	nanometer (10^{-9})
NOM	natural organic matter
NPDES	National Pollutant Discharge Elimination System
NPDWR	National Primary Drinking Water Regulations
ntu	nepherometric turbidity unit
Ω	ohm
O&M	operation and maintenance
OSHA	Occupational Safety and Health Administration
PA	phenolphthalein alkalinity
PAC	powdered activated carbon
P&ID	process and instrumentation diagram
PC	programmable controller
PCD	process control diagram
PCE	tetrachloroethylene (perchloroethylene)
pC/L	picocuries per liter
PCM	process control module
PCM	professional construction management
PD	process diagram
PERT	program evaluation and review technique
PIN	process information network
PLC	programmable logic controller
POADMA	polydiallyl dimethyl ammonium
ppd	pounds per day
ppm	parts per million
PSA	pressure-swing adsorption
psi	pound per square inch
PTB	pitched turbine blade
PVC	polyvinyl chloride
RCML	recommended maximum contaminant level
Re	Reynolds number
RFP	request for proposal
RFQ	request for qualification
RMP	risk management planning
RO	reverse osmosis
RTU	remote terminal unit
s	seconds
SCADA	supervisory control and data acquisition
SDWA	Safe Drinking Water Act

SI	Stability Index
SI unit	international system of units
SMCLs	Secondary Maximum Contaminant Levels
SOCs	synthetic organic chemicals
SUVA	specific UV absorbance
SWTR	Surface Water Treatment Rule
TCE	trichloroethylene
TCR	total coliform rule
TDH	total dynamic head
TDS	total dissolved solids
THMs	trihalomethanes
TOC	total organic carbon
TON	threshold odor number
TTHMs	total trihalomethanes
UBC	Uniform Building Code
UF	ultrafiltration
UFRV	unit filter run volume
UV_{254}	ultraviolet radiation of 254 nm
VE	value engineering
VFD	variable-frequency drive
VOCs	volatile organic compounds
VSA	vacuum swing adsorption
W	watt
WHO	World Health Organization

APPENDIX 2

Basic Physical Tables and Greek Alphabet

Systems of Basic Units

Designation	Dimensions	Systems: English (FPS)	Systems: Metric (MKS)	Systems: International (SI)
Length	(*L*)	foot (ft)	meter (m)	meter (m)
Mass	(*M*)	pound (lb)	kilogram (kg)	kilogram (kg)
Time	(*t*)	second (sec)	second (s)	second (s)
Electric current	(*A*)	ampere (A)	ampere (A)	ampere (A)
Temperature	(*T*)	degree Fahrenheit (°F)	degree Celsius (°C)	degree Kelvin (°K)
Luminous intensity	(*I*)	candela (cd)	candela (cd)	candela (cd)

Decimal Multiples and Fractions of SI Units

Factor	Prefix	Symbol	Factor	Prefix	Symbol
10^{1}	deka	D (da)	10^{-1}	deci	d
10^{2}	hecto	h	10^{-2}	centi	c
10^{3}	kilo	k	10^{-3}	milli	m
10^{6}	mega	M	10^{-6}	micro	μ
10^{9}	giga	G	10^{-9}	nano	n
10^{12}	tera	T	10^{-12}	pico	p
10^{15}	femta	F	10^{-15}	femto	f
10^{18}	atta	A	10^{-18}	atto	a

Systems of Derived Units, Geometry, Mass

Designation	Dimensions	FPS	MKS	SI
Area	$(L)^2$	ft^2	m^2	m^2
Static moment of area	$(L)^3$	ft^3	m^3	m^3
Moment of inertia of area	$(L)^4$	ft^4	m^4	m^4
Product of inertia of area	$(L)^4$	ft^4	m^4	m^4
Polar moment of inertia of area	$(L)^4$	ft^4	m^4	m^4
Volume	$(L)^3$	ft^3	m^3	m^3
Static moment of volume	$(L)^4$	ft^4	m^4	m^4
Moment of inertia of volume	$(L)^5$	ft^5	m^5	m^5
Product of inertia of volume	$(L)^5$	ft^5	m^5	m^5
Polar moment of inertia of volume	$(L^5$	ft^5	m^5	m^5
Mass[a] $= M = W/g$	(M)	lb	kg	kg
Static moment of mass	$(M)(L)$	lb-ft	kg-m	kg-m
Moment of inertia of mass	$(M)(L)^2$	lb-ft^2	kg-m^2	kg-m^2
Product of inertia of mass	$(M)(L)^2$	lb-ft^2	kg-m^2	kg-m^2
Polar moment of inertia of mass	$(M)(L)^2$	lb-ft^2	kg-m^2	kg-m^2
Curvature of a curve	$(L)^{-1}$	1/ft	1/m	1/m
Torsion of a curve	$(L)^{-1}$	1/ft	1/m	1/m
Plane angle[b]	(R)	rad	rad	rad
Solid angle[b]	(S)	sr	sr	sr

[a]In the English system (FPS) and in the metric system (MKS), the mass M is a derived unit, defined as the weight W divided by the acceleration due to gravity g. lb = pound mass, kg = kilogram mass.

[b]The unit of plane angle called *radian* (rad) and the unit of solid angle called *steradian* (sr) are supplemental units.

Alphabet Table

Greek Letter	Greek Name	English Equivalent
Α α	Alpha	(a)
Β β	Beta	(b)
Γ γ	Gamma	(g)
Δ δ	Delta	(d)
Ε ε	Epsilon	(e)
Ζ ζ	Zeta	(z)
Η η	Eta	(ā)
Θ θ	Theta	(th)
Ι ι	Iota	(ē)
Κ κ	Kappa	(k)
Λ λ	Lambda	(l)
Μ μ	Mu	(m)
Ν ν	Nu	(n)
Ξ ξ	Xi	(ks)
Ο ο	Omicron	(o)
Π π	Pi	(p)
Ρ ρ	Rho	(r)
Σ σ ς	Sigma	(s) 6
Τ τ	Tau	(t)
Υ υ	Upsilon	(ü, o͞o)
Φ φ	Phi	(f)
Χ χ	Chi	(H)
Ψ ψ	Psi	(ps)
Ω ω	Omega	(ō)

Russian Letter	English Equivalent
А а	(ä)
Б б	(b)
В в	(v)
Г г	(g)
Д д	(d)
Е е	(ye)
Ж ж	(zh)
З з	(z)
И и	(i, ē)
Й й	(ē) 7
К к	(k)
Л л	(l)
М м	(m)
Н н	(n)
О о	(ô, o)
П п	(p)
Р р	(r)
С с	(s)
Т т	(t)
У у	(o͞o)
Ф ф	(f)
Х х	(kh)
Ц ц	(ts)
Ч ч	(ch)
Ш ш	(sh)
Щ щ	(shch)
Ъ ъ	8
Ы ы	(ë)
Ь ь	9
Э э	(e)
Ю ю	(ū)
Я я	(ya)

APPENDIX 3

Metric (SI) System and Selected Useful Conversion Factors

PHYSICAL QUANTITY	CUSTOMARY ("British" or "Imperial")	METRIC†† spelled out	METRIC†† symbolic	RECIPROCAL ***
DISTANCE	1 foot	0.3048 meter	0.3048 m	3.281
	1 yard	0.9144 meter	0.9144 m	1.094
	1 mile	1.609 kilometer	1.609 km	0.6215
AREA	1 square inch	6.452 square centimeter	6.452 cm^2	0.155
	1 square foot	0.0929 square meter	0.0929 m^2	10.76
		or 929 square centimeters	929 cm^2	0.001076
	1 square yard	0.836 square meter	0.836 m^2	1.196
	1 acre	4,047 square meters	4,047 m^2	0.000247
		or 0.4047 hectare	0.4047 h	2.47
	1 square mile	2.590 square kilometers	2.590 km^2	0.386
		or 259.0 hectares	259.0 h	0.00386
VOLUME	1 cubic inch	16.39 cubic centimeters	16.39 cm^3	0.0610
	1 pint (liquid)	473.2 cubic centimeters	473.2 cm^3	0.002113
	1 quart	946.4 cubic centimeters	946.4 cm^3	0.001057
		or 0.9464 liter	0.946 l	1.057
	1 gallon	3.785 liters	3.785 l	0.2642
	1 cubic foot	0.0283 cubic meter	0.0283 m^3	35.3
	1 cubic yard	0.765 cubic meter	0.765 m^3	1.308
	1 acre-foot	0.1233 hectare-meter	0.1233 h m	8.11
		or 1.233 megaliter	1.233 Ml	0.811
		or 1,233 cubic meters	1,233 m^3	0.000811
TIME	1 minute	60 seconds	60 s	0.01667
	1 hour	3.6 kiloseconds	3.6 ks	0.2778
	1 day	86.4 kiloseconds	86.4 ks	0.01157
		or 0.0864 megasecond	0.0864 Ms	11.57
	1 week	0.6048 megasecond	0.6048 Ms	1.6534
	1 dozen days	1.0368 megaseconds	1.0368 Ms	0.96451
	1 month (30 days)	2.592 megaseconds	2.592 Ms	0.3858
	1 year (365¼ days)	31.56 megaseconds	31.56 Ms	0.03169
VELOCITY	1 foot per hour	84.6 micrometer per second	84.6 μm/s	0.01182
	1 foot per minute	5.080 millimeters per second	5.080 mm/s	0.1969
	1 foot per second	0.3048 meter per second	0.3048 m/s	3.281
	1 mile per hour	0.4470 meter per second	0.4470 m/s	2.237
FLOW	1 gallon per minute	63.08 milliliter per second	63.08 ml/s	0.01585
	1 cubic foot per minute	0.4719 liter per second	0.4719 l/s	2.119
	1 cubic foot per second	28.32 liters per second	28.32 l/s	0.0353
	1 million gallons per day	43.812 liters per second	43.812 l/s	0.022824
		or 0.043812 kiloliter per second	0.043812 kl/s	22.824
MASS	1 ounce mass	28.35 grams	28.35 g	0.03527
	1 pound mass	453.6 grams	453.6 g	0.002205
		or 0.4536 kilogram	0.4536 kg	2.205
	1 ton mass	0.907 megagram	0.907 Mg	1.102
	(short, 2000 pounds mass)	or 0.907 metric ton	0.907 t	1.102
		or 0.907 tonne	0.907 t	1.102
FORCE	1 ounce force	0.2780 newton	0.2780 N	3.597
	1 pound force	4.448 newtons	4.448 N	0.2248
	1 ton force (2,000 pounds force)	8.897 kilonewtons	8.897 kN	0.11240
PRESSURE	1 pound per square foot	47.88 newtons per square meter	47.88 N/m^2	0.02089
	1 pound per square inch	6.895 kilonewtons per square meter	6.895 kN/m^2	0.11240
	1 millimeter of mercury	133.3 newtons per square meter	133.3 N/m^2	0.0075
	1 foot of water	2.989 kilonewtons per square meter	2.989 kN/m^2	0.3346
	1 atmosphere	0.1013 meganewton per square meter	0.1013 MN/m^2	9.87
ENERGY	1 watt-second	1.000 joule	1.000 J	1.000
	1 foot-poundforce	1.356 joule	1.356 J	0.7375
	1 BTU	1.055 kilojoule	1.055 kJ	0.948
	1 watt-hour	3.60 kilojoules	3.60 kJ	0.2778
	1 horsepower-hour	2.684 megajoules	2.684 MJ	0.3726
	1 kilowatt-hour	3.60 megajoules	3.60 MJ	0.2778
POWER	1 horsepower	745.7 watts	745.7 W	0.001341
		or 0.7457 kilowatt	0.7457 kW	1.341
	1 joule per second	1.000 watt	1.000 W	1.000
	1 BTU per hour	0.293 joule per second	0.293 J/s	3.41
TEMPERATURE	1 degree Fahrenheit	5/9 degree Celsius for each Fahrenheit degree above or below 32 °F	$(5/9 \times (T_F - 32))$ °C	1.8 degree Fahrenheit for each Celsius degree, plus 32 °F

Source: J. J. Tuma, *Handbook of Physical Calculations,* McGraw-Hill, New York, 1976.

APPENDIX 3 *(Continued)*

SPECIAL COMPOUND UNITS	BTU per cubic foot	37.30 joules per liter	37.30 J/l	0.0268
	BTU per pound of mass	2.328 joules per gram	2.328 J/g	0.4296
	BTU per square foot per hour	3.158 joules per square meter	3.158 J/m^2	0.3167
	1 pound per thousand cubic feet per day	185.4 grams per cubic centimeter per second	185.4 g/cm^{3}s	0.00539
		or 0.1854 kilogram per cubic meter per second	0.1854 kg/m^{3}s	5.39
	1 cubic foot per minute per thousand cubic feet	16.67 microliters per liter per second	16.67 μl/l s	0.0600
		or 16.67 x 10^{-6} per second*	16.67 x 10^{-6} s^{-1}	60,000
	1 cubic foot per second per acre	69.98 liter per second per hectare	69.98 l/s h	0.01429
		or 6.998 micrometers per second†	6.998 μm/s	0.1429
	1 gallon per acre	9.35 milliliters per hectare	9.35 ml/h	0.1070
		or 9.35 centimeters**	9.35 cm**	0.1070
	1 gallon per day per linear foot	143.7 microliters per second per meter	143.7 μl/s m	0.00696
		or 0.1437 square millimeters per second	0.1437 mm^2/s	6.96
	1 gallon per day per square foot	0.4715 milliliter per square meter per second	0.4715 ml/m^{2}s	2.12
		or 0.4715 micrometers per second	0.4715 μm/s	2.12
	1 pound mass per acre	1.122 kilogram per hectare	1.122 kg/h	0.892
		or 0.1122 gram per square meter	0.1122 g/m^2	8.92
	1 pound mass per cubic foot	16.02 grams per liter	16.02 g/l	0.0624
	1 pound mass per linear foot	1.488 kilogram per meter	1.488 kg/m	0.672
	1 pound per horsepower-hour	0.1690 milligram per joule	0.1690 mg/J	5.917

*i.e., 16.67 parts in 1,000,000, per second
†i.e., 6.998 micrometers depth change per second
**i.e., 9.35 centimeters depth

***Multiply quantity known in metric units by this number to get British equivalent.
††Multiply quantity known in British units by this number to get metric equivalent.

Quantities and SI Units

Unit Type	Quantity	Unit	Symbol	Expressions in Terms of Other Units
Base units	Length	meter	m	
	Mass	kilogram	kg	
	Time	second	s	
	Electric current	ampere	A	
	Thermodynamic temperature	kelvin	K	
	Amount of substance	mole	mol	
	Luminous intensity	candela	cd	
Supplementary units	Plane angle	radian	rad	
	Solid angle	steradian	sr	
Derived units having special names	Frequency	hertz	Hz	s^{-1}
	Force	newton	N	$kg \cdot m/s^2$
	Pressure, Stress	pascal	Pa	N/m^2
	Energy, Work, Quantity of heat	joule	J	$N \cdot m$
	Power	watt	W	J/s
	Electric charge	coulomb	C	$A \cdot s$
	Electric potential Potential difference Electromotive force	volt	V	W/A
	Electric resistance	ohm	Ω	V/A
	Electric conductance	siemens	S	A/V
	Electric capacitance	farad	F	C/V
	Magnetic flux	weber	Wb	$V \cdot s$
	Inductance	henry	H	Wb/A
	Magnetic flux density	tesla	T	Wb/m^2
	Luminous flux	lumen	lm	$cd \cdot sr$
	Illuminance	lux	lx	lm/m^2
Some other derived units	Area	square meter		m^2
	Volume	cubic meter		m^3
	Velocity—angular	radian per second		rad/s
	Velocity—linear	meter per second		m/s
	Acceleration—angular	radian per second squared		rad/s^2
	Acceleration—linear	meter per second squared		m/s^2
	Density (mass per unit volume)	kilogram per cubic meter		kg/m^3
	Moment of force and torque	newton meter		$N \cdot m$

QUANTITIES AND SI UNITS *(Continued)*

Unit Type	Quantity	Unit	Symbol	Expressions in Terms of Other Units
Some other derived units (cont'd)	Viscosity—dynamic	pascal second		Pa·s
	Thermal conductivity	watt per meter kelvin		W/m·K
	Thermal flux density or irradiance	watt per square meter		W/m^2
	Thermal capacity or entropy	joule per kelvin		J/K
	Permeability	henry per meter		H/m
	Permittivity	farad per meter		F/m
	Magnetic field strength	ampere per meter		A/m
	Luminance	candela per square meter		cd/m^2
	Molar entropy	joule per mole kelvin		J/mol·K

METRIC (SI) CONVERSION AND USEFUL CONVERSION FACTORS

Length:	1 mil	$= 2.54 \times 10^{-3}$ cm
	1 in.	= 2.54 cm
	1 ft	= 30.48 cm
	1 yd	= 0.914 m
	1 mile	= 1.61 km
Area:	1 ft^2	= 0.093 m^2
	1 yd^2	= 0.836 m^2
	1 $mile^2$	= 2.59 km^2
	1 ha	= 10,000 m^2 = 2.47 acres
Volume:	1 ft^3	= 28.32 L = 7.48 gal
	1 gal	= 3.785 L
	1 yd^3	= 0.7646 m^3
	1 acre-ft	= 0.326 million gallons
		= 1235 m^3
	1 m^3	= 35.31 ft^3 = 264 gal
Weight:	1 dalton	$= 1.65 \times 10^{-24}$ grain
	1 lb	= 453.6 g
	1 ton (short)	= 2000 lb = 907.18 kg
Temperature:	(32°F) × 5/9 = °C	
Feeding rate:	1 mg/L	= 8.34 lb/million gallons
Flow rate:	1 ft^3/s	= 448.8 gpm
		= 0.6463 mgd
	1 ft^3/min	= 28.32 L/min
	1 mgd	= 1.547 ft^3/s
		= 3785 m^3/d
		= 3.785 ML/d
	1 m^3/s	= 22.826 mgd
	1 gpm/ft^2	= 2.5 m/h = 60 m/d
Pressure:	1 psi (lbf/in^2)	= 0.0703 kgf/cm^2 (kg/cm^2)
	1 psig	= 14.7 lb/in^2 = 2.31 ft H_2O
	1 kPa (N/m^2)	= 0.145 psi
	1 MPa	= 145 psi
	1 bar	= 1.0197 kgf/cm^2
		= 14.5 psi
Power:	1 hp	= 550 ft-lb/s
		= 42.44 Btu/min
	1 hp	= 0.7457 kW
Gravitational acceleration		= 32.2 ft/s^2 = 9.807 m/s^2

The following table shows the number of pounds of chemical to add to 1 gallon of water to make % solutions

% solution	lb/gal	% solution	lb/gal	% solution	lb/gal
0.1	0.008	3.0	0.258	20.0	2.20
0.2	0.017	5.0	0.440	25.0	2.76
0.5	0.042	6.0	0.553	30.0	3.56
1.0	0.084	10.0	0.927	40.0	5.52
2.0	0.170	15.0	1.473	50.0	8.33

APPENDIX 4

Temperature Conversions

Equivalent Temperature Readings for Fahrenheit and Celsius Scales

$$°F = \tfrac{9}{5}\,°C + 32°, \quad °C = \tfrac{5}{9}\,(°F - 32°)$$

Fahrenheit deg	Centigrade deg	Fahrenheit deg	Centigrade deg	Fahrenheit deg	Centigrade deg	Fahrenheit deg	Centigrade deg	Fahrenheit deg	Centigrade deg	Fahrenheit deg	Centigrade deg	Fahrenheit deg	Centigrade deg	Fahrenheit deg	Centigrade deg
-459.4	-273	-21.	-29.4	17.6	-8.	56.	13.3	95.	35.	134.	56.7	172.4	78.	211.	99.4
-436.	-260.	-20.2	-29.	18.	-7.8	57.	13.9	96.	35.6	134.6	57.	173.	78.3	212.	100.
-418.	-250.	-20.	-28.9	19.	-7.2	57.2	14.	96.8	36.	135.	57.2	174.	78.9	213.	100.6
-400.	-240.	-19.	-28.3	19.4	-7.	58.	14.4	97.	36.1	136.	57.8	174.2	79.	213.8	101.
-382.	-230.	-18.4	-28.	20.	-6.7	59.	15.	98.	36.7	136.4	58.	175.	79.4	214.	101.1
-364.	-220.	-18.	-27.8	21.	-6.1	60.	15.6	98.6	37.	137.	58.3	176.	80.	215.	101.7
-346.	-210.	-17.	-27.2	21.2	-6.	60.8	16.	99.	37.2	138.	58.9	177.	80.6	215.6	102.
-328.	-200.	-16.6	-27.	22.	-5.6	61.	16.1	100.	37.8	138.2	59.	177.8	81.	216.	102.2
-310.	-190.	-16.	-26.7	23.	-5.	62.	16.7	100.4	38.	139.	59.4	178.	81.1	217.	102.8
-292.	-180.	-15.	-26.1	24.	-4.4	62.6	17.	101.	38.3	140.	60.	179.	81.7	217.4	103.
-274.	-170.	-14.8	-26.	24.8	-4.	63.	17.2	102.	38.9	141.	60.6	179.6	82.	218.	103.3
-256.	-160.	-14.	-25.6	25.	-3.9	64.	17.8	102.2	39.	141.8	61.	180.	82.2	219.	103.9
-238.	-150.	-13.	-25.	26.	-3.3	64.4	18.	103.	39.4	142.	61.1	181.	82.8	219.2	104.
-220.	-140	-12.	-24.4	26.6	-3.	65.	18.3	104.	40.	143.	61.7	181.4	83.	220.	104.4
-202.	-130.	-11.2	-24.	27.	-2.8	66.	18.9	105.	40.6	143.6	62.	182.	83.3	221.	105.
-184.	-120.	-11.	-23.9	28.	-2.2	66.2	19.	105.8	41.	144.	62.2	183.	83.9	222.	105.6
-166.	-110.	-10.	-23.3	28.4	-2.	67.	19.4	106.	41.1	145.	62.8	183.2	84.	222.8	106.
-148.	-100.	-9.4	-23.	29.	-1.7	68.	20.	107.	41.7	145.4	63.	184.	84.4	223.	106.1
-139.	-95.	-9.	-22.8	30.	-1.1	69.	20.6	107.6	42.	146.	63.3	185.	85.	224.	106.7
-130.	-90.	-8.	-22.2	30.2	-1.	69.8	21.	108.	42.2	147.	63.9	186.	85.6	224.6	107.
-121.	-85.	-7.6	-22.	31.	-0.6	70.	21.1	109.	42.8	147.2	64.	186.8	86.	225.	107.2
-112.	-80.	-7.	-21.7	32.	0.	71.	21.7	109.4	43.	148.	64.4	187.	86.1	226.	107.8
-103.	-75	-6.	-21.1	33.	+0.6	71.6	22.	110.	43.3	149.	65.	188.	86.7	226.4	108.
-94.	-70.	-5.8	-21.	33.8	1.	72.	22.2	111.	43.9	150.	65.6	188.6	87.	227.	108.3
-85.	-65.	-5.	-20.6	34.	1.1	73.	22.8	111.2	44.	150.8	66.	189.	87.2	228.	108.9
-76.	-60.	-4.	-20.	35.	1.7	73.4	23.	112.	44.4	151.	66.1	190.	87.8	228.2	109.
-67.	-55.	-3.	-19.4	35.6	2.	74.	23.3	113.	45.	152.	66.7	190.4	88.	229.	109.4
-58.	-50.	-2.2	-19.	36.	2.2	75.	23.9	114.	45.6	152.6	67.	191.	88.3	230.	110.
-49.	-45.	-2.	-18.9	37.	2.8	75.2	24.	114.8	46.	153.	67.2	192.	88.9	231.	110.6
-40.	-40.	-1.	-18.3	37.4	3.	76.	24.4	115.	46.1	154.	67.8	192.2	89.	231.8	111.
-39.	-39.4	-0.4	-18.	38.	3.3	77.	25.	116.	46.7	154.4	68.	193.	89.4	232.	111.1
-38.2	-39.	0.	-17.8	39.	3.7	78.	25.6	116.6	47.	155.	68.3	194.	90.	233.	111.7
-38.	-38.9	+1.	-17.2	39.2	4.	78.8	26.	117.	47.2	156.	68.9	195.	90.6	233.6	112.
-37.	-38.3	1.4	-17.	40.	4.4	79.	26.1	118.	47.8	156.2	69.	195.8	91.	234.	112.3
-36.4	-38.	2.	-16.7	41.	5.	80.	26.7	118.4	48.	157.	69.4	196.	91.1	235.	112.8
-36.	-37.8	3.	-16.1	42.	5.6	80.6	27.	119.	48.3	158.	70.	197.	91.7	235.4	113.
-35.	-37.2	3.2	-16.	42.8	6.	81.	27.2	120.	48.9	159.	70.6	197.6	92.	236.	113.3
-34.6	-37.	4.	-15.6	43.	6.1	82.	27.8	120.2	49.	159.8	71.	198.	92.2	237.	113.9
-34.	-36.7	5.	-15.	44.	6.7	82.4	28.	121.	49.4	160.	71.1	199.	92.8	237.2	114.
-33.	-36.1	6.	-14.4	44.6	7.	83.	28.3	122.	50.	161.	71.7	199.4	93.	238.	114.4
-32.8	-36.	6.8	-14.	45.	7.2	84.	28.9	123.	50.6	161.6	72.	200.	93.3	239.	115.
-32.	-35.6	7.	-13.9	46.	7.8	84.2	29.	123.8	51.	162.	72.2	201.	93.9	240.	115.6
-31.	-35.	8.	-13.3	46.4	8.	85.	29.4	124.	51.1	163.	72.8	201.2	94.	240.8	116.
-30.	-34.4	8.6	-13.	47.	8.3	86.	30.	125.	51.7	163.4	73.	202.	94.4	241.	116.1
-29.2	-34.	9.	-12.8	48.	8.9	87.	30.6	125.6	52.	164.	73.3	203.	95.	242.	116.7
-29.	-33.9	10.	-12.2	48.2	9.	87.8	31.	126.	52.2	165.	73.9	204.	95.6	242.6	117.
-28.	-33.3	10.4	-12.	49.	9.4	88.	31.1	127.	52.8	165.2	74.	204.8	96.	243.	117.2
-27.4	-33.	11.	-11.7	50.	10.	89.	31.7	127.4	53.	166.	74.4	205.	96.1	244.	117.8
-27.	-32.8	12.	-11.1	51.	10.6	89.6	32.	128.	53.3	167.	75.	206.	96.7	244.4	118.
-26.	-32.2	12.2	-11.	51.8	11.	90.	32.2	129.	53.9	168.	75.6	206.6	97.	245.	118.3
-25.6	-32.	13.	-10.6	52.	11.1	91.	32.8	129.2	54.	168.8	76.	207.	97.2	246.	118.9
-25.	-31.9	14.	-10.	53.	11.7	91.4	33.	130.	54.4	169.	76.1	208.	97.8	246.2	119.
-24.	-31.7	15.	-9.4	53.6	12.	92.	33.3	131.	55.	170.	76.7	208.4	98.	247.	119.4
-23.8	-31.1	15.8	-9.	54.	12.2	93.	33.9	132.	55.6	170.6	77.	209.	98.3	248.	120.
-23.	-30.	16.	-8.9	55.	12.8	93.2	34.	132.8	56.	171.	77.2	210.	98.9	249.	120.6
-22.	-30.	17.	-8.3	55.4	13.	94.	34.4	133.	56.1	172.	77.8	210.2	99.	249.8	121.

APPENDIX 5

Periodic Table of the Elements

KEY TO CHART: Atomic Number → **50** +2 +4 ← Oxidation States; Symbol → Sn; Atomic Weight → 118.69; -18-18-4 ← Electron Configuration

Transition Elements — Group 8 — Transition Elements

1a	2a	3b	4b	5b	6b	7b	8	8	8	1b	2b	3a	4a	5a	6a	7a	0	Orbit
1 +1 −1 H 1.00797 1																	**2** 0 He 4.0026 2	K
3 +1 Li 6.939 2-1	**4** +2 Be 9.0122 2-2											**5** +3 B 10.811 2-3	**6** +2 +4 −4 C 12.01115 2-4	**7** +1 +2 +3 +4 +5 −1 −2 −3 N 14.0067 2-5	**8** −2 O 15.9994 2-6	**9** −1 F 18.9984 2-7	**10** 0 Ne 20.183 2-8	K-L
11 +1 Na 22.9898 2-8-1	**12** +2 Mg 24.312 2-8-2											**13** +3 Al 26.9815 2-8-3	**14** +2 +4 −4 Si 28.086 2-8-4	**15** +3 +5 −3 P 30.9738 2-8-5	**16** +4 +6 −2 S 32.064 2-8-6	**17** +1 +5 +7 −1 Cl 35.453 2-8-7	**18** 0 Ar 39.948 2-8-8	K-L-M
19 +1 K 39.102 -8-8-1	**20** +2 Ca 40.08 -8-8-2	**21** +3 Sc 44.956 -8-9-2	**22** +2 +3 +4 Ti 47.90 -8-10-2	**23** +2 +3 +4 +5 V 50.942 -8-11-2	**24** +2 +3 +6 Cr 51.996 -8-13-1	**25** +2 +3 +4 +7 Mn 54.9380 -8-13-2	**26** +2 +3 Fe 55.847 -8-14-2	**27** +2 +3 Co 58.9332 -8-15-2	**28** +2 +3 Ni 58.71 -8-16-2	**29** +1 +2 Cu 63.546 -8-18-1	**30** +2 Zn 65.37 -8-18-2	**31** +3 Ga 69.72 -8-18-3	**32** +2 +4 Ge 72.59 -8-18-4	**33** +3 +5 −3 As 74.9216 -8-18-5	**34** +4 +6 −2 Se 78.96 -8-18-6	**35** +1 +5 −1 Br 79.904 -8-18-7	**36** 0 Kr 83.80 -8-18-8	-L-M-N
37 +1 Rb 85.47 -18-8-1	**38** +2 Sr 87.62 -18-8-2	**39** +3 Y 88.905 -18-9-2	**40** +4 Zr 91.22 -18-10-2	**41** +3 +5 Nb 92.906 -18-12-1	**42** +6 Mo 95.94 -18-13-1	**43** +4 +6 +7 Tc (97) -18-13-2	**44** +3 Ru 101.07 -18-15-1	**45** +3 Rh 102.905 -18-16-1	**46** +2 +4 Pd 106.4 -18-18-0	**47** +1 Ag 107.868 -18-18-1	**48** +2 Cd 112.40 -18-18-2	**49** +3 In 114.82 -18-18-3	**50** +2 +4 Sn 118.69 -18-18-4	**51** +3 +5 −3 Sb 121.75 -18-18-5	**52** +4 +6 −2 Te 127.60 -18-18-6	**53** +1 +5 +7 −1 I 126.9044 -18-18-7	**54** 0 Xe 131.30 -18-18-8	-M-N-O
55 +1 Cs 132.905 -18-8-1	**56** +2 Ba 137.34 -18-8-2	**57*** +3 La 138.91 -18-9-2	**72** +4 Hf 178.49 -32-10-2	**73** +5 Ta 180.948 -32-11-2	**74** +6 W 183.85 -32-12-2	**75** +4 +6 +7 Re 186.2 -32-13-2	**76** +3 +4 Os 190.2 -32-14-2	**77** +3 +4 Ir 192.2 -32-15-2	**78** +2 +4 Pt 195.09 -32-16-2	**79** +1 +3 Au 196.967 -32-18-1	**80** +1 +2 Hg 200.59 -32-18-2	**81** +1 +3 Tl 204.37 -32-18-3	**82** +2 +4 Pb 207.19 -32-18-4	**83** +3 +5 Bi 208.980 -32-18-5	**84** +2 +4 Po (210) -32-18-6	**85** At (209) -32-18-7	**86** 0 Rn (222) -32-18-8	-N-O-P
87 +1 Fr (223) -18-8-1	**88** +2 Ra (226) -18-8-2	**89**** +3 Ac (227) -18-9-2																-O-P-Q

															Orbit
*Lanthanides	**58** +3 +4 Ce 140.12 -19-9-2	**59** +3 Pr 140.907 -20-9-2	**60** +3 Nd 144.24 -22-8-2	**61** +3 Pm (145) -23-8-2	**62** +2 +3 Sm 150.35 -24-8-2	**63** +2 +3 Eu 151.96 -25-8-2	**64** +3 Gd 157.25 -25-9-2	**65** +3 Tb 158.924 -26-9-2	**66** +3 Dy 162.50 -28-8-2	**67** +3 Ho 164.930 -29-8-2	**68** +3 Er 167.26 -30-8-2	**69** +3 Tm 168.934 -31-8-2	**70** +2 +3 Yb 173.04 -32-8-2	**71** +3 Lu 174.97 -32-9-2	-N-O-P
**Actinides	**90** +4 Th 232.038 -19-9-2	**91** +5 +4 Pa (231) -20-9-2	**92** +3 +4 +5 +6 U 238.03 -21-9-2	**93** +3 +4 +5 +6 Np (237) -22-9-2	**94** +3 +4 +5 +6 Pu (244) -23-9-2	**95** +3 +4 +5 +6 Am (243) -24-9-2	**96** +3 Cm (247) -25-9-2	**97** +3 +4 Bk (247) -26-9-2	**98** +3 Cf (251) -28-8-2	**99** Es (254) -29-8-2	**100** Fm (257) -30-8-2	**101** Md (256) -31-8-2	**102** (254) -32-8-2	**103** Lw	-O-P-Q

Numbers in parentheses are mass numbers of most stable isotope of that element.

APPENDIX 6

Basic Properties of Air and Water

COMPONENTS OF ATMOSPHERIC AIR (APPROXIMATE)

Constituent	Content (%) by Volume	Content (ppm) by Volume
N_2	78.08	—
O_2	20.95	—
CO_2	0.33	—
Ar	—	18.18
He	—	5.24
CH_4	—	2.0
H_2	—	0.5
NO_2	—	0.5

MASS DENSITY AND VISCOCITY OF AIR

Temperature (°C)	Density (kg/m^3)	Kinematic Viscosity (m^2/s)
−20	1.392	1.152×10^{-5}
−10	1.340	1.245×10^{-5}
0	1.293	1.319×10^{-5}
10	1.247	1.412×10^{-5}
20	1.201	1.486×10^{-5}
30	1.160	1.607×10^{-5}
40	1.129	1.691×10^{-5}

NOTE: For conversion to ft^2/s, use 1 m^2/s = 10.764 ft^2/s; for conversion to lb/ft^3, use 1 kg/m^3 = 0.06243 lb/ft^3

Atmospheric Pressure

1 atm = 10.1325 N/cm^2 = 1.0332 kgf/cm^2 = 14.69 lbf/in. (psi)

1 atm = 760 mm mercury/cm^2 @ 0° C = 406.8 in. water/$in.^2$

Notes:

Bar (b) = 10 N/cm^2 = 1.0197 kgf/cm^2

Torr is the pressure of 1 mm of mercury per square centimeter at latitude 45, sea level. 1 torr = 1.333×10^{-2} N/cm^2 = 1.3596×10^{-3} kgf/cm^2

MASS DENSITY OF WATER

	0°C	5°C	10°C	15°C	20°C	25°C	30°C
ρ (kg/m^3)	999.9	1000.0	999.7	999.1	998.2	997.1	995.7

KINEMATIC VISCOSITY OF WATER (ν): $\nu = \mu/\rho$

	0°C	5°C	10°C	15°C	20°C	25°C	30°C
(centistokes)							
cm^2/s ($\times 10^{-2}$)	1.792	1.519	1.31	1.146	1.011	0.898	0.803
m^2/s ($\times 10^{-6}$)	1.792	1.519	1.31	1.146	1.011	0.898	0.803
ft^2/s ($\times 10^{-5}$)	1.926	1.633	1.408	1.232	1.087	0.965	0.864

NOTE: To convert centistokes to ft^2/s, multiply by 1.075×10^{-3}.

ABSOLUTE VISCOSITY OF WATER (μ)

	0°C	5°C	10°C	15°C	20°C	25°C	30°C
(centipoises)							
$g/cm \cdot s$ ($\times 10^{-2}$)	1.792	1.519	1.310	1.145	1.009	0.895	0.800
$kg/m \cdot s$ ($\times 10^{-3}$)	1.792	1.519	1.310	1.145	1.009	0.895	0.800
$lb \cdot s/ft^2$ ($\times 10^{-5}$)	3.74	3.17	2.73	2.39	2.11	1.87	1.67
$newton \cdot s/m^2$ ($\times 10^{-3}$)	1.83	1.55	1.336	1.17	1.009	0.913	0.816

NOTES: To convert centipoises to $lb \cdot s/ft^2$, multiply by 2.088×10^{-5}. Newton = $kg \cdot m/s^2 = 10^{-5}$ dyne.

TABLE A6-1 Density and Viscosity of Water

Temperature (°C)	Density ρ, γ (g/cm^3), also δ	Absolute viscosity, μ (cP)[a]	Kinematic viscosity, ν (cSt)[b]
0	0.99987	1.7921	1.7923
1	0.99993	1.7320	1.7321
2	0.99997	1.6740	1.6741
3	0.99999	1.6193	1.6193
4	1.00000	1.5676	1.5676
5	0.99999	1.5188	1.5188
6	0.99997	1.4726	1.4726
7	0.99993	1.4288	1.4288
8	0.99988	1.3872	1.3874
9	0.99981	1.3476	1.3479
10	0.99973	1.3097	1.3101
11	0.99963	1.2735	1.2740
12	0.99952	1.2390	1.2396
13	0.99940	1.2061	1.2068
14	0.99927	1.1748	1.1756
15	0.99913	1.1447	1.1457
16	0.99897	1.1156	1.1168
17	0.99880	1.0876	1.0888
18	0.99862	1.0603	1.0618
19	0.99843	1.0340	1.0356
20	0.99823	1.0087	1.0105
21	0.99802	0.9843	0.9863
22	0.99780	0.9608	0.9629
23	0.99757	0.9380	0.9403
24	0.99733	0.9161	0.9186
25	0.99707	0.8949	0.8975
26	0.99681	0.8746	0.8774
27	0.99654	0.8551	0.8581
28	0.99626	0.8363	0.8394
29	0.99597	0.8181	0.8214
30	0.99568	0.8004	0.8039
31	0.99537	0.7834	0.7870
32	0.99505	0.7670	0.7708
33	0.99473	0.7511	0.7551
34	0.99440	0.7357	0.7398
35	0.99406	0.7208	0.7251
36	0.99371	0.7064	0.7109
37	0.99336	0.6925	0.6971
38	0.99299	0.6791	0.6839
39	0.99262	0.6661	0.6711

[a] 1 centipoise = 10^{-2} (gram mass)/(cm)(s).
[b] 1 centistoke = 10^{-2} cm^2/s.

APPENDIX 7

Gas Law

Generalized gas law:

$$PV = nRT$$

where

P = atmospheric pressure (P = 1 as standard pressure)
V = volume of a gas in liter
n = number of moles of gas
R = molar gas constant
T = absolute temperature (K) (0°C = 273.15K)

Under standard temperature (20°C) and pressure (1 atm), R and n are given as follows:

$$R = \frac{PV}{nT} = \frac{1 \text{ atm} \times 22.4 \text{ L}}{1 \text{ mol} \times 273.15 \text{ K}} = 0.0821 \frac{\text{L}}{\text{K} \cdot \text{mol}} = 0.0821 \text{ L/K} \cdot \text{mol}$$

$$n = \frac{g}{M}$$

where g is the mass of the gas (g) and M is the molecular weight of the gas. Under standard temperature and pressure (STP), the density, volume, and molecular weight of a gas are given as follows:

Density (g/L) of gas:

$$d = \frac{g}{V} = \frac{MP}{RT} \text{ or } d = \frac{M}{24.053}$$

Volume (L) of gas:

$$V = \frac{gRT}{MP} \text{ or } V = \frac{g \times 24.053}{M}$$

Molecular weight (g/mol) of gas:

$$M = \frac{gRT}{PV} \text{ or } M = \frac{g \times 24.053}{V}$$

APPENDIX 8

Ozone Concentration Conversion

English Unit Equivalents for Ozone Concentration

Standard Pressure .... 14.696 psi (101 kPa)
Standard Temperature .... 68°F (20°C)
Gram Molecular wt. of air .... 29 g/mole
Gram Molecular wt. of oxygen .... 32 g/mole
Gram Molecular wt. of ozone .... 48 g/mole
Molar Volume .... 0.08205 L/mole/°K

Metric	English					
Wt Ozone to Vol Gas (g/m³)	Percent Weight		Percent Volume (%)	PPM Weight		PPM
	Air (%)	Oxygen (%)		Air (ppm)	Oxygen (ppm)	Volume (ppm)
2	0.17	0.15	0.10	1,658	1,503	1,002
4	0.33	0.30	0.20	3,313	3,004	2,004
6	0.50	0.45	0.30	4,967	4,503	3,007
8	0.66	0.60	0.40	6,618	6,001	4,009
10	0.83	0.75	0.50	8,267	7,498	5,011
12	0.99	0.90	0.60	9,914	8,993	6,013
14	1.16	1.05	0.70	11,559	10,486	7,015
16	1.32	1.20	0.80	13,201	11,978	8,018
18	1.48	1.35	0.90	14,842	13,469	9,020
20	1.65	1.50	1.00	16,480	14,958	10,022
22	1.81	1.64	1.10	18,116	16,446	11,024
24	1.98	1.79	1.20	19,750	17,932	12,026
26	2.14	1.94	1.30	21,382	19,417	13,029
28	2.30	2.09	1.40	23,012	20,900	14,031
30	2.46	2.24	1.50	24,640	22,381	15,033
32	2.63	2.39	1.60	26,265	23,862	16,035
34	2.79	2.53	1.70	27,889	25,340	17,038
36	2.95	2.68	1.80	29,510	26,818	18,040
38	3.11	2.83	1.90	31,129	28,294	19,042
40	3.27	2.98	2.00	32,746	29,768	20,044
42	3.44	3.12	2.10	34,362	31,241	21,046
44	3.60	3.27	2.20	35,974	32,712	22,049
46	3.76	3.42	2.31	37,585	34,182	23,051
48	3.92	3.57	2.41	39,194	35,651	24,053
50	4.08	3.71	2.51	40,801	37,118	25,055

Source: Adapted from U.S. EPA *Disinfection Manual.*

APPENDIX 9

Tracer Test

TRACER TEST STUDY

Tracer tests are commonly performed by injecting a slug dose of dye, radioactive substance, or a salt solution into the inlet of a basin or tank, then measuring the concentration of the tracer in the effluent at various time intervals. The effluent monitoring should be carried on until substantially all the tracer has passed through the basin. The type of tracer that can be used for water treatment and water supply systems is limited to lithium chloride and fluoride because of public health safety concerns.

Chlorides such as sodium chloride, which is readily available as well as cost-effective, is not an effective tracer because of its high specific gravity as a solution, and it produces a density flow in the tank or basin. Often, lithium chloride is used as a tracer since it is safe to human health at tracer dosage, safe to handle, and easy to put into solution due to its high solubility (63 g per100 m/L in near freezing water, exothermic reaction). However, lithium chloride is rather expensive ($3 per pound). Lithium chloride is available from Lithco, a subsidiary of FMC Corp., Gastonia, North Carolina (phone: 704-868-5350). Commonly available fluoride compounds are sodium fluoride, sodium silicofluoride, and hydrofluosilicic acid. The first two are available in dry form, but the latter is available only as a liquid. The dry compounds have very low solubility in water (4 and 0.76%, respectively, at 25°C) and dust from these compounds can be hazardous to health. Hydrofluosilicic acid is available as a liquid (23% solution) and has infinite solubility, but is extremely acidic and therefore creates a safe handling problem. Although the cost of these fluoride chemicals is about 10 to 30% of that for lithium chloride, the cost benefit is often offset by the safety concerns and troublesome preparation of the tracer solution. One important characteristic of an effective tracer is that it cannot be removed or consumed by the treatment process. Fluoride can be removed by a high dosage of alum flocculation and to a significant degree by the lime softening process. Lithium, however, is much more stable than fluoride.

TRACER TEST METHODS

The two basic methods of tracer addition are the step-dose method and the slug-dose method. In the step-dose method the tracer is dosed at a constant rate until the concentration at the desired endpoint reaches a steady-state level. In the slug-dose method a large and instantaneous dose of tracer is added at an entrance to a tank where adequate mixing is provided, and the tracer is recorded as it passes through the tank by taking samples at the effluent of the tank at certain intervals.

AMOUNT OF TRACER TO BE USED

Regardless of the type of tracer, the amount of the tracer chemical to be fed should be adequate so that the residual of the tracer in the water samples at the outlet of the tank is sufficiently high to monitor the flow through time very clearly. As a rule of thumb, the dose for lithium chloride is about 2 to 3 mg/L of lithium at peak concentration with the slug-dose method. When fluoride compound is used, the maximum concentration of fluoride ion in the sample water should be 1 to 1.5 mg/L in order not to exceed the MCL of fluoride of the drinking water standards. Often, the background level of fluoride ion in most waters is on the order of 0.5 mg/L, and therefore not much additional fluoride can be applied. The background level of lithium in most waters for water supply is negligibly small and generally does not fluctuate.

It should be realized that the flow pattern in a tank or basin is never an ideal plug flow and that substantial degrees of flow short-circuiting, dead zone, and mixing may exist. Therefore, the amount of tracer chemical required is often significantly greater than the amount computed based on the assumption that the flow in a tank is plug flow. When a tracer test (using the slug-dose method) is conducted for a flocculation tank with flocculators, the applied tracer chemical is dispersed in the entire flocculation tank. Therefore, the amount of tracer to be added is based on the entire flocculation tank volume and desired lithium ion concentration in the sampled water.

SAMPLING

Sampling at the outlet of the tank should include the water before tracer addition in order to establish the background. After adding the tracer, sampling should begin at 5% of the computed mean detention time and sampling should be continued until the tracer level drops back to the background level. For the slug-dose method, the testing period is usually 2 to 2.5 times longer than the computed detention time. The frequency of sampling depends on the test method, the anticipated detention time, the cost of tracer analysis, and the total budget of the tracer study. For the slug-dose method, where the calculated detention time is less than 30 min, 2.5 to 3-min sampling intervals are preferable and should be carried out until the tracer concentration decreases to a level that is about 25% higher than the background level. After this point, the sampling frequency can be every 5 to 10 min until the tracer level is no longer detected. The initial sampling intervals can be 5 to 10 min when sedimentation tanks or clearwells are tested since their mean detention time is normally 2 to 5 hours.

In the case of the step-dose method, the sampling should also be started at 5% of the computed mean-time detention time of the tank, and the same sampling frequency as the slug-dose method should be applied. However, the sampling should be terminated after the measured residual concentration reaches a steady-state value. Notes of the field test should include the test method, time of tracer addition, time of each sampling, plant flow rate during the test, water level (if it is fluctuating), water temperature, and ambient temperature. In addition, the time at which the tracer feed is stopped should also be recorded if the step-dose method is used.

DATA EVALUATION

Slug-Dose Method The test results should be tabulated and the results organized by the values of C/C_0 versus t/T, where C_0 the background concentration, C the tracer concentration in sample water (concentration in sample minus background concentration), t the time of sample collection, and T the calculated detention time (the volume of water in the tank divided by the flow rate). The results are then plotted on a graph as shown in Figure A9-1, with the values of C/C_0 on the vertical axis and t/T on the horizontal axis. If desired, the actual sampling time after the tracer injection can be used instead of t/T on the horizontal axis. Case 1 in the figure is a sedimentation tank with an intermediate diffuser wall at the middle of the tank length and an inlet diffuser wall. The samples were collected at three points along the tank effluent. The graph shows that three different levels of peaks, different shapes (tracer peak at location A is a flat shape), and the peak of all three lines are located at different t/T scales. The results show that the flow in the tank is not uniform: The flow is skewed to the west side of the tank and a flow stagnation occurs on the eastern side of the wall, with an average flow through time of about 50% of the calculated detention time. The skewed flow is due to the flow direction in the tank. The diffuser walls have too much opening (15% of cross section), so that they do not provide a controlled headloss to redistribute the flow into the tank. Case 2 shows the tracer curves for two common rectangular sedimentation tanks constructed without effective influent diffuser walls. Samples were taken only at the middle point of the tank effluent and therefore evaluation of the flow characteristics in the tank is limited. However, it is obvious that the actual flow through time is only about 30% of the computed detention time.

Step-Dose Method Table A9-1 presents the results of the step-dose tracer test for a poorly compartmentalized flocculation tank that has $T = 30$ minutes using lithium chloride as a tracer. Here the background lithium concentration was 0.01 mg/L, and a constant dosage of lithium chlorine (3 mg/L as lithium) was added to the inlet of the tank. Figure A9-2 is the plot from Table A9-1 with the C/C_0 figures plotted against the sampling time on the horizontal axis. The figure shows a severe flow short-circuiting; it took only 6 min for the first tracer to travel to the outlet of the tank. A C/C_0 value of 0.5 corresponds to $t = 11$ min, so that one-half of the total flow is flowing out after 11 min of detention time. After 30 min, almost 90% of flow has passed through the tank. If the disinfectant contact time (t for $C \times t$) is determined through this tracer test, t_{10} should be used as the value for t. The t_{10}

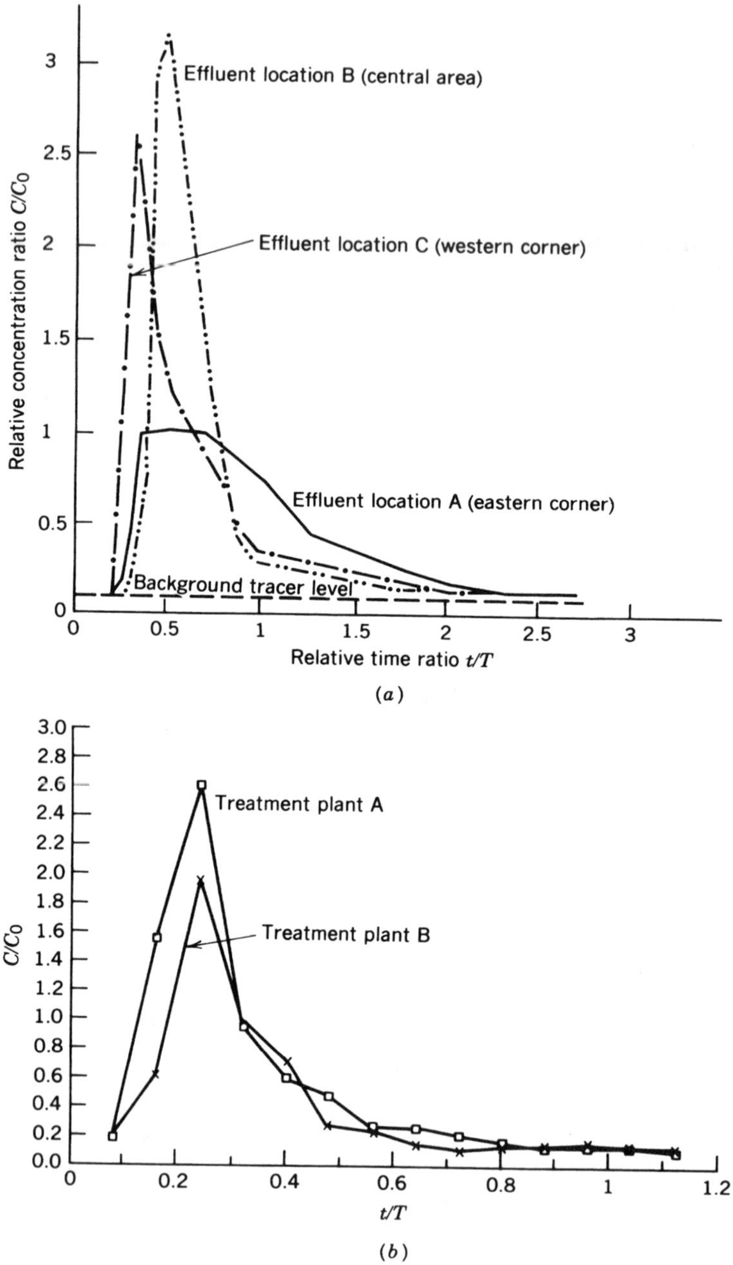

Figure A9-1 Tracer curves of conventional sedimentation basins (slug-dose method).

TABLE A9-1 Flocculation Tank Tracer Test Data

Time (A) (min)	Total Lithium Level (mg/L)	C^a (mg/L)	$C/C_0{}^b$
0	0.01	0	0
3	0.01	0	0
6	0.31	0	0.1
9	1.03	1.02	0.34
12	1.76	1.75	0.58
15	1.91	1.90	0.63
18	2.24	2.23	0.73
21	2.15	2.14	0.71
24	2.68	2.67	0.89
27	2.72	2.71	0.90
30	2.66	2.65	0.88
33	2.69	2.68	0.90
36	2.73	2.72	0.91
39	2.84	2.83	0.94
42	2.86	2.85	0.95
45	2.85	2.84	0.95

[a]Net increase of lithium concentration (0.02 mg/L is background level).
[b]C_0 = applied tracer dosage = 3 mg/L.

value is obtained from Figure A9-2 at $C/C_0 = 0.1$ by first drawing a line to the horizontal axis to give the time t. In this particular case, t_{10} is 5 min.

EXAMPLE CALCULATION

A tracer test is planned for one of three flocculation tanks (flocculator system) which has a 20-min mean detention time (calculated time) and also for a chlorine contact tank (effluent weir control), which has a 15-min contact time at a normal plant flow rate of 30 mgd (1.32 m^3/s). Select the test method and determine the amount of chemical required as well as the application method.

Tracer Test for the Flocculation Tank

Set the test conditions as follows:

Method	Slug-dose method.
Tracer	Lithium chloride.
Level of tracer in sampled water	Approximately 3 mg/L as lithium ion.
Flocculation tank	Tracer test for only one of the three tanks.
Tracer feeding	Prepare a 30% LiCl solution on site using several plastic buckets, and pour this into the flocculation tank inlet where turbulence is greatest.

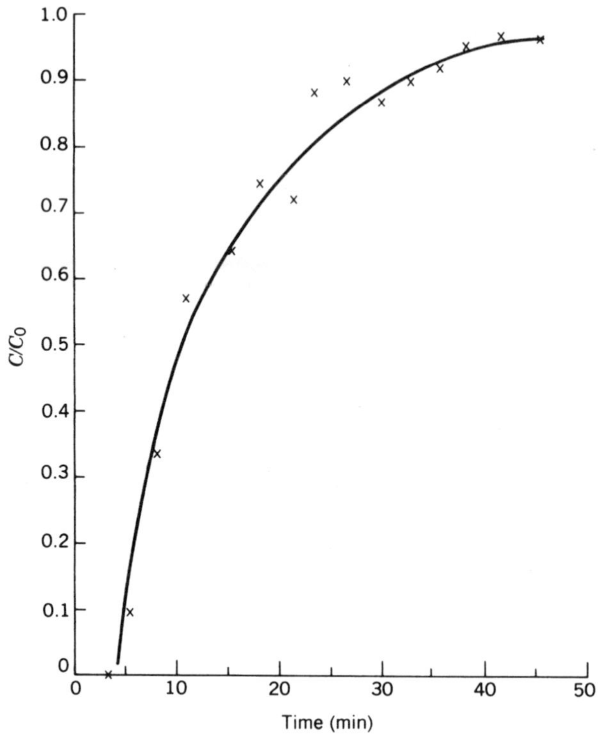

Figure A9-2 Tracer curve of a poorly compartmentalized flocculation tank (step-dose method).

We now calculate the amount of lithium chloride required for the flocculation tank. When we treat three tanks with a detention time of 20 min each at 30 mgd, the volume of one tank is

$$[(30 \div 3) \times 1.547] \times (60 \times 20) = 18{,}564 \text{ ft}^3 \quad (526 \text{ m}^3)$$

The amount of chemical required to obtain 3 mg/L residual should be computed on the entire volume of the tank, since the chemical will be completely dispersed into the entire tank by the mechanical flocculators.

$$\frac{3}{1{,}000{,}000} = \frac{x}{18{,}564 \times 62.4}$$

$$x = 3.48 \text{ lb } (1.6 \text{ kg})$$

The molecular weight of lithium chloride is 42.5 and the molecular weight of lithium is only 7 or 16.5% of the chemical, assuming almost 100% purity. The 3 mg/L of residual should be as lithium, so that the amount of lithium, chloride to be purchased is

$$3.48 \div 0.16 = 21.75 \text{ lb}, \qquad \text{say, 22 lb (10 kg) (costing approximately \$70)}$$

Make a solution that is approximately 20% (2.2 lb/gal). Provide five 4-gal capacity buckets. Place approximately 2 gal (15 L) of water in each of the five buckets, then add 4.4 lb of the chemical to each and slowly mix with a paddle. Heat will be generated during this dissolving process, but it will not be high enough to melt the plastic buckets.

Tracer Test for Chlorine Contact Tank In general, chlorine is added at the tank inlet by means of a diffuser pipe. Therefore, the tracer may be added through the same diffuser, if possible. If not, an additional diffuser may be installed next to the chlorine diffuser. Set the test conditions as follows:

Method	Step-dose method.
Tracer	Lithium chloride.
Level of tracer in sampled water	2 mg/L as lithium ion.
Tracer feeding	Lithium chloride solution is metered by a metering pump and is fed through the diffuser pipe.
Duration of tracer feeding	45 min of continuous feeding.

The amount of lithium chloride required is

$$(2 \times 8.34 \times 30) \times \frac{0.75}{24} = 15.64 \text{ lb } (7.1 \text{ kg})$$

Since lithium is 16.5% of lithium chlorine (100% purity), the actual amount of lithium chloride needed is

$$15.64 \div 0.16 = 97.7 \text{ lb}, \qquad \text{say, } 100 \text{ lb } (45.5 \text{ kg}) \text{ (cost approximately \$300)}$$

The capacity of the chemical feeder is calculated as follows. A 20% lithium chloride solution is to be prepared. Since 20% solution should contain 2.2 lb of the chemical per gallon of solution, 45.5 gal (172 L) of solution is prepared in a 55-gal (210-L) drum or other suitable container. The feeder capacity is determined to feed 45.5 gal of solution in 45 min.

$$45.5 \div \frac{45}{60} = 60.6 \text{ gph}, \qquad \text{say, 65-gph (246-L/h) capacity pump}$$

Figure A9-3 shows generalized tracer curves under certain physical conditions of the process unit as a reference.

NOTE:

The duration of tracer feeding depends on the size and shape of the tank, inlet and outlet conditions, and adequacy of internal baffling. Poorly designed tanks will require the tracer to be fed three to four times longer than the calculated mean detention time.

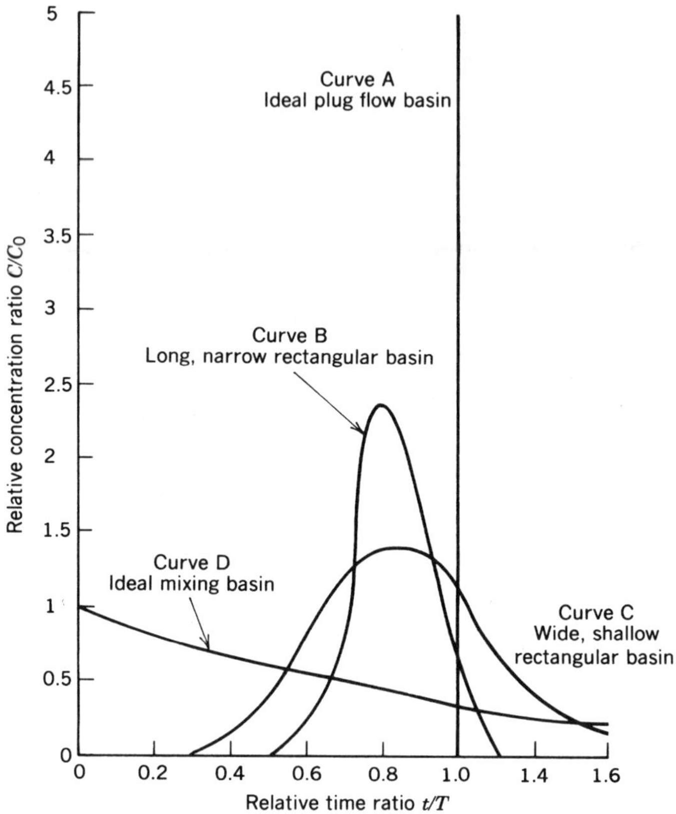

Figure A9-3 Generalized tracer curves under certain conditions (based on the slug-dose tracer test method).

APPENDIX 10

Recommended Jar Test Procedure

The following jar test procedure is recommended for use with the Phipps and Bird laboratory stirrer with regular 1-L beakers or 2-L square jars, but not applicable with a magnetic stirrer unit.

REGULAR JAR TEST PROCEDURE

Jar Test Procedure Place 1000 mL (1 L) or 2000 mL (2 L) of raw water in each beaker or jar, depending on the size of the containers. There should be a total of six beakers.

Flash Mixing To stimulate flash mixing, begin by initiating mixing for all six beakers at the highest speed of the jar tester (about 200 rpm). Then pipette a predetermined amount of alum solution to each beaker as quickly as possible. In general, the expected optimum alum or ferric dosage is 6 to 10 mg/L for good-quality raw water; thus a 0.1% coagulant solution should be prepared for the 1-L beaker and a 0.2 % solution for the 2-L jars. Addition of 1 mL of 0.1% solution to 1 L of sample water results in a 1 mg/L solution. Similarly, 1 mL of 0.2% solution added to 2-L of sample water equals 1 mg/L. When higher coagulant dosage is expected, 1% coagulant solution should be prepared for the 1-L beakers and 2% solution for the 2-L jars; in this case, 1 mL of coagulant added to the sample waters for both-size containers will result in 10 mg/L. Remember to adjust the mixing time of each beaker to compensate for the time delay that results from pipetting beakers 2 through 6.

NOTE: If an additional chemical such as lime or polymer is to be added, the chemical should be added separately from the alum since the two chemicals could immediately interreact. The proper procedure is first to add alum to all the beakers as described above, then repeat the flash mixing step using the second chemical.

Flocculation The flocculation process is simulated by applying the following *tapered mixing* immediately after flash mixing. Follow these instructions in the order given.

7.5 min at a mixing speed of 70 rpm	$G \times t = (7.5 \times 60) \times 60 = 2.7 \times 10^4$

7.5 min at a mixing speed of 40 rpm	$G \times t = (7.5 \times 60) \times 30 = 1.35 \times 10^4$
5 min at a mixing speed of 25 rpm	$G \times t = (5 \times 60) \times 15 = 4.5 \times 10^3$
Total of 20 min	$G \times t$ (approximately) $= 4.5 \times 10^4$

Figure A10-1(a) shows the relation between *G* values and mixing speed for regular 1-L cylindrical beaker. Figure A10-1(b) illustrates for the method of G value calibration and Figure A10-1(c) shows the relation between G value and mixing speed for a 2-L standard square jar as shown.

Settling Allow a standing time of 20 min for settling to occur. Avoid exposing the beakers to direct sunlight.

Sampling Carefully decant approximately 150 mL of the supernatant from each beaker. Avoid disturbing the settled particles and the scum on the surface of the water.

Measurements and Observations

Time Record the minimum time (in minutes) that is required to form pinpoint floc during the flocculation step.

Size At the end of the flocculation step, record the size of the (majority) floc by using the yard stick method shown in Figure A10-2.

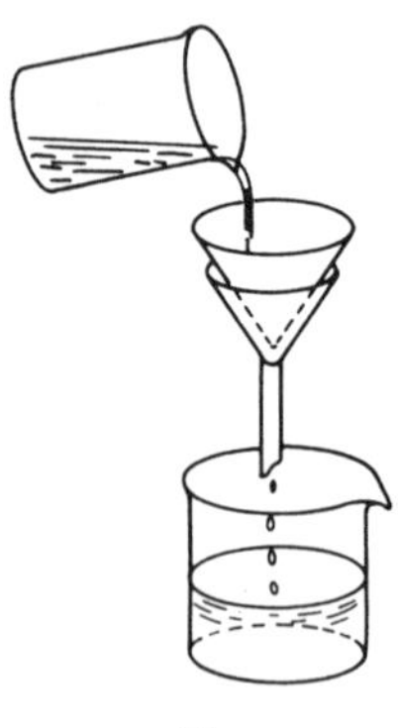

(*i*)

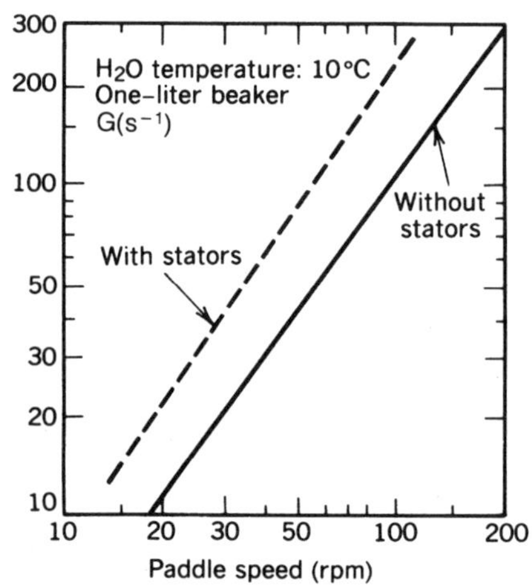

(*ii*)

Figure A10-1a Mixing speed versus *G* value and filterability test: (*i*) filterability test; (*ii*) velocity gradient versus mixing speed.

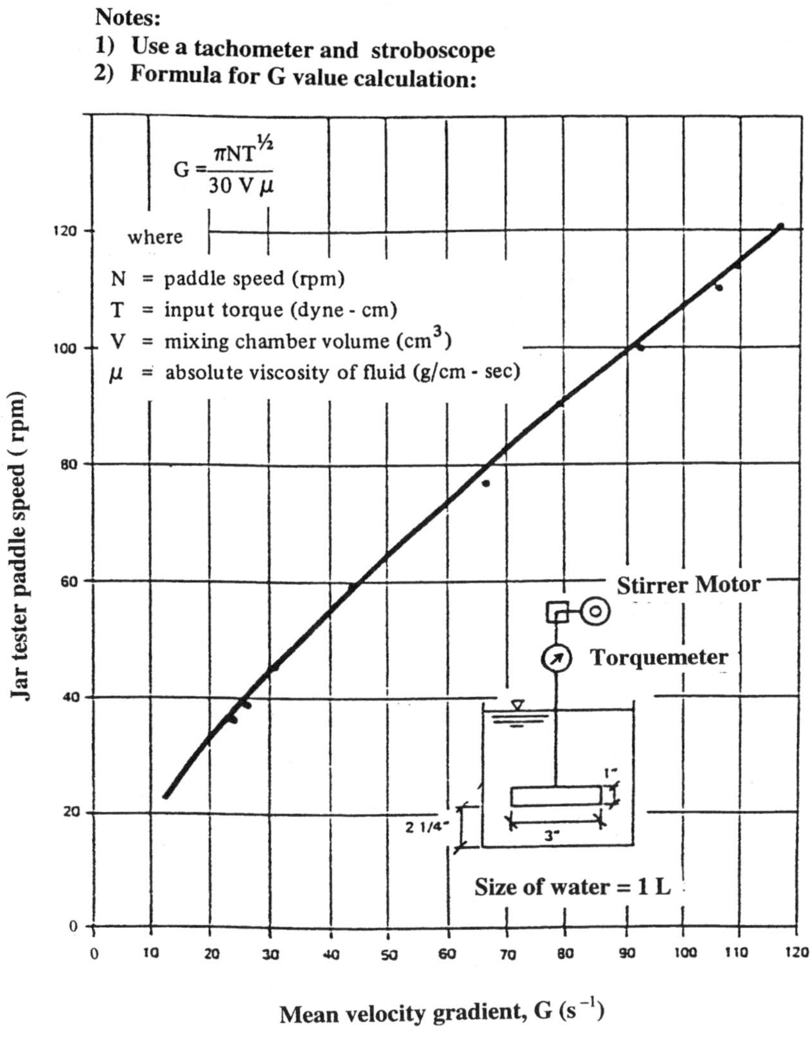

Figure A10-1b (Continued).

Water Quality Measure the following water quality characteristics for the finished water (the supernatant): water temperature, turbidity, and pH. Optionally, the color and alkalinity of the water may also be recorded. Refer to Table A10-1 for an example of the jar test data sheet.

Filterability Test The filterability of the finished water (supernatant) is determined by filtering 70 mL of the selected sample through Whatman No. 1 (approximately 10 μm pore size) or equivalent filter paper. This size of filter paper is chosen because the pore size of a regular sand filter bed is approximately 10 to 15 μm. The

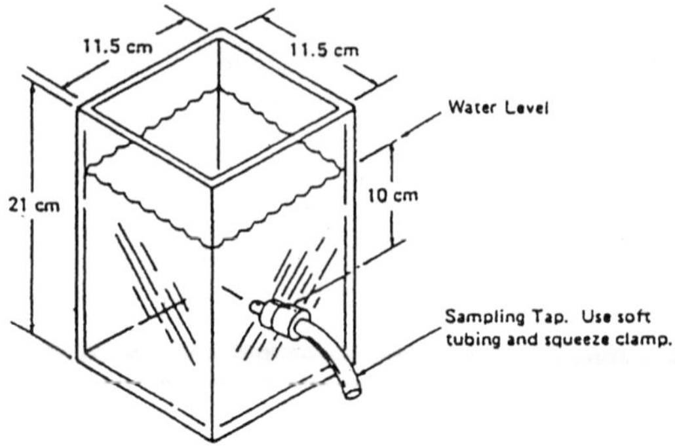

Two-liter jar for bench scale testing. Fabricate from plexiglass sheet.

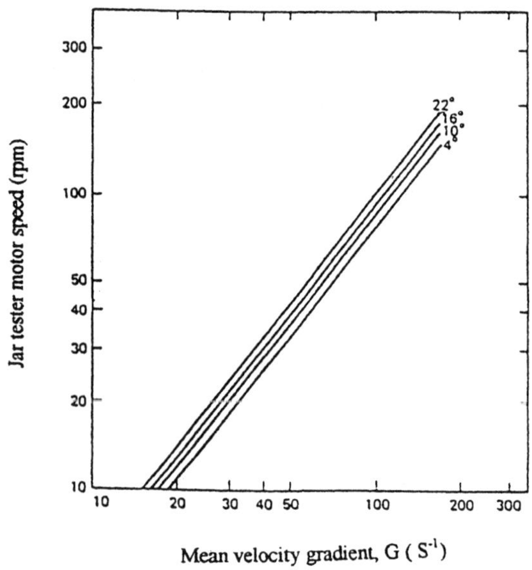

Figure A10-1c (Continued).

first 20 mL of filtered water should be wasted since it contains dust and fine fibers of the filter paper. Record the time required to filter 50 mL of the sample. Finally, measure the turbidity of the filtrate.

NOTE: By having stators in each beaker, both the mixing gradient and the velocity gradient are improved. Rectangular jars are preferred to regular beakers because their corners create more eddies and turbulence. However, rectangular jars are special-order items and are therefore more expensive. (See H. E. Hudson, Jr., *Water Clarification Process,* Van Nostrand Reinhold, New York, 1981.)

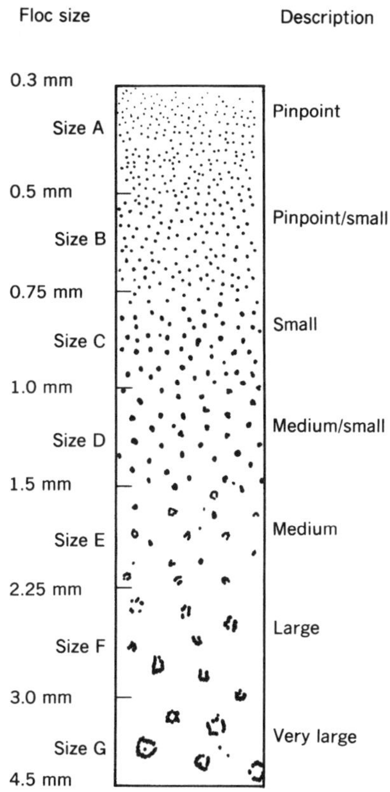

Figure A10-2 Floc size measuring scale.

Coagulation and Flocculation Control Tests

NOTE: These are not routine jar tests but basic water treatability evaluation tests.

Optimization of Coagulant Alum or ferric chloride should be chosen as a main coagulant. For economic reasons, the cationic polymers are usually considered to be a coagulant aid.

Proper Dosage The proper dosage of main coagulant is evaluated by a two-step process. First, a wide range of dosages is covered by adding 2.5, 5, 10, 20, 40, and 80 mg/L of a coagulant such as alum. The second step is a refinement of the first. For example, if the first test yields a good result at 20 mg/L of alum, the second step should evaluate the following dosages: 12.5, 15, 17.5, 20, 22.5, 25, and 27.5 mg/L.

Optimization of Cationic Polymer This should follow the same concept as optimization of alum dosage. However, the maximum cationic polymer dosage is gener-

TABLE A10-1 Jar Test Results

Jar Test Conditions

Rapid mixing: 100 rpm for 10 s ($G \times t = 1500$)
Flocculation: 60 rpm for 7.5 min
40 rpm for 7.5 min ($G \times t = 4.7 \times 10^4$)
20 rpm for 5.0 min
Settling time: 20 min
Filterability test: 50 mL of settled water to be filtered through Whatman No. 1 filter paper

	Chemicals		Floc		Settled Water		Filterability		
Beaker Number	Alum (mg/L)	(mg/L)	Time (min)	Size	Turbidity (NTU)	pH	Time (min)	Turbidity (NTU)	Remark
1	0	—	—	—	66	7.5	6.3	15	Raw water
2	15	—	8	C	13	7.2	8.4	4.4	Alum
3	20	—	2.5	C	7	7.1	5.3	2.3	
4	25	—	1.5	C	3.2	7.1	3.5	1.7	
5	30	—	1.5	C	2.5	6.9	3.3	1.4	
		$FeCl_3$							
6	—	6	3.5	C	11	7.1	8.1	3.8	$FeCl_3$
7	—	10	0.8	D	4	6.9	7.2	1.5	
8	—	14	0.5	D	2.7	6.7	5.1	1.2	
9	—	18	0.5	D	2	6.5	4.2	1	
		Chitosan							
10	—	1	2	C	12	7.4	6.3	1	Chitosan
11	—	2	1.5	C	6	7.4	—	—	
12	—	4	1.5	C	5.3	7.4	3.5	0.7	
13	—	8	1.5	C	4.9	7.3	4.5	0.6	
1	0	—	—	—	11	8.2	6.5	3.8	Raw water
2	2	—	—	—	—	—	—	—	Alum
3	5	—	7	A	6	8.1	6	1.2	
4	10	—	3.5	C	1.8	8.0	3.1	0.6	
		Cationic floc							
5	5	1	5	C	2.8	8.1	3.0	0.3	Alum and cationic polymer
6	10	1	2.8	D	1.3	8.0	2.8	0.23	
		$FeCl_3$							
7	—	5	6.5	B	4	8.0	5.6	1.2	$FeCl_3$
8	—	10	3.1	C	1.3	7.9	3	0.35	

ally limited to 5 mg/L. The suggested dosages are 0.25, 0.5, 1.2, and 4 mg/L. In cases where alum and cationic polymer are fed together, the following dosages may be tried:

Alum (mg/L)	6	12	20	6	12	20
Cationic polymers (mg/L)	0.25	0.25	0.25	0.5	0.5	0.5

Optimization of Mixing Energy and Time

A Phipps–Bird jar tester or equivalent unit is capable of adjusting G values of 15 to 100 s^{-1}. Mixing energy levels of 15, 30, 45, and 60 s^{-1} and mixing times of 10, 20, 30, and 40 min should be evaluated under the optimized coagulant dosage. As a rule of thumb, the target value is $G \times t = 5 \times 10^4$ (t in seconds), and the minimum mixing time is 15 min.

FLOC SETTLING VELOCITY MEASUREMENT

The jar test should be performed with 2-L beakers rather than 1-L beakers, under a particular set of conditions: whenever the floc settling velocity is related to the performance of the actual clarifier. When the flocculation process is complete, after the beginning of the settling period, samples of 50 mL or less are collected at 1, 2, 4, 8, and 16 min by pipetting or siphoning from a fixed depth of 10 cm from the original water level. The fixed depth represents settling velocities of 10, 5, 2, 2.5, 1.25, and 0.63 cm/min, respectively. A settling velocity profile curve is then derived by plotting the turbidity of the settled water samples (at the designated times) divided by the raw water turbidity, against the corresponding settling velocity, thus representing the ratio of raw water turbidity remaining as a percentage (see Figure A10-3). The settling velocities may be converted to the hydraulic surface loading of sedimentation tanks. For example, settling velocities of 2, 4, and 8 cm/min are approximately equivalent to 0.5, 1, and 2 gpm/ft^2 (1.2, 2.4, and 4.8 m/h).

EVALUATION OF THE CHEMICAL APPLICATION SEQUENCE

The supplemental chemicals should be evaluated under the optimized alum dosage and mixing conditions at the following times: when fed 2 min prior to the addition of alum, when fed at the same time as alum, and 5 to 10 min after alum flocculation has begun. Whenever the filter wash waste is recycled, the recycling point should also be evaluated both prior and after the alum application point.

EXAMPLE CALCULATIONS

The following calculation examples will help these who have not prepared chemical solutions for the jar test.

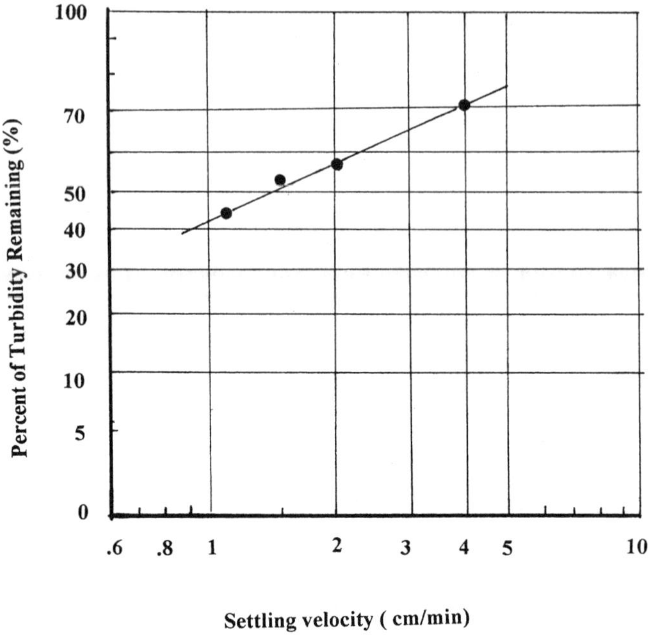

Figure A10-3 Floc settling velocity characteristic.

Solution Concentration

Example 1 An HCl solution has specific gravity of 1.19 and 37.23% of HCl by weight. Calculate the grams of anhydrous HCl in 5 mL of the solution.

Solution Weight of 5 mL HCl = 5 mL × 1.19 = 5.95 g. Thus the weight of anhydrous HCl in a 5-mL solution = 5.95 × 0.3723 = 2.22 g.

Example 2 A H_2SO_4 solution of specific gravity 1.84 and contains 98% H_2SO_4 by weight. Calculate the volume (mL) of the H_2SO_4 solution that contains 40 g of anhydrous H_2SO_4.

Solution 1 mL of the solution weighs 1.84 g and contains 0.98 × 1.84 = 1.8 g of anhydrous H_2SO_4. Thus 40 g of anhydrous H_2SO_4 is contained in 40 g ÷ 1.8 g/mL = 22.2 mL.

Example 3 Calculate the mL of NH_3 solution with specific gravity of 0.890 containing 31.7% NH_3 by weight that will be obtained by passing NH3 gas into 100 g of H_2O.

Solution Weight of 1 mL of the solution is 0.890 g. Weight of NH_3 in 1 mL of the solution = 0.317 × 0.89 g = 0.282 g; weight of H_2O in 1 mL of the solution = 0.890 − 0.282 g = 0.608 g. Thus 100 g of H_2O will make

$$\frac{100 \text{ g } H_2O}{0.608 \text{ g } H_2O/\text{mL solution}} = 164 \text{ mL solution}$$

Example 4 Calculate the weight of $Al_2(SO_4)_3 \cdot 18H_2O$ required to make 50 mL of an aqueous solution with a strength of 40 mg Al^{3+} per liter.

Solution Weight (g) of Al^{3+} in 50 mL of solution = 50 mL × 40 mg/mL = 2000 mg Al^{3+} (2 g). Atomic weight of Al = 27, molecular weight of alum is 666 g. Since 2 × 27 g = 54 g of Al^{3+} is contained in 666 g of alum, 1 g of Al^{3+} is contained in 666 g ÷ 54 g = 12.3 g of alum and 2 g of Al^{3+} is contained in 2 × 12.3 g = 24.7 g of alum. Thus weigh 24.6 g of the dry alum and place it in a volumetric flask. Then add distilled water to the 1000-mL line.

Dilution Problems

$$\text{Vol.}_{(1)} \times \text{concentration}_{(1)} = \text{Vol.}_{(2)} \times \text{concentration}_{(2)}$$

Example 1 A 50% NaOH solution has a specific gravity of 1.53, and a 25% NaOH solution has a specific gravity of 1.28. Calculate the amount of H_2O required to make a 25% solution from 50% solution.

Solution 1 mL of 50% solution weighs 1.53 g and has 0.5 × 1.53 = 0.765 g of NaOH in it. 1 mL of 25% solution weighs 1.28 g and has 0.25 × 1.28 = 0.32 g of NaOH in it. $V_1 \times C_1 = V_2 \times C_2$. Let X = volume after dilution (mL). Thus 1 mL × 0.765 g/mL = X × 0.32 g/mL. Therefore, X = 2.39 mL. Each mL of 50% NaOH solution must be diluted to a volume of 2.39 mL.

Example 2 Calculate the concentration of H_2SO_4 with a specific gravity of 1.84, containing 98% H_2SO_4 by weight, that must be diluted with H_2O to yield 100 mL of 20% H_2SO_4 solution.

Solution First, the concetration must be changed from a weight to a volume basis. Weight of H_2SO_4 per mL of 98% solution = 0.98 × 1.84 g/mL = 1.8 g/mL; Weight of H_2SO_4 per mL of 20% solution = 0.2 × 1.14 g/mL = 0.228 g/mL. Let Y = volume (mL) of 98% acid required for 100 mL of 20% acid.

$$V_1 \times C_1 = V_2\, C_2 \qquad 100\ \text{mL} \times 0.228\ \text{g/mL} = Y \times 1.8\ \text{g/mL}$$
$$Y = 12.7\ \text{mL}$$

Example 3 Calculate the volume of $N/2$ and $N/10$ HCl that must be mixed to give 2 L of $N/5$ HCl.

Solution Let x = liters of $N/2$ acid required. Thus $2x$ = liters of $N/10$ required. Gram-equivalent $N/5$ = gram-equivalent $N/2$ + gram-equivalent $N/10$. Therefore, 2 L × $\frac{1}{5}N = x$ L × $\frac{1}{2}N + 2x$ L × $\frac{1}{10}N$. Solving x gives x = 0.5 L. Therefore, 0.5 L of $N/2$ and 1.5 L of $N/10$ acids are required.

APPENDIX 11

Core Sampling of the Filter Bed

Preparation To obtain core samples, the engineer or plant operator should have a box of small plastic bags, such as sandwich bags, a non-water-soluble marker (permanent ink marker), and core sampling pipes similar to those illustrated in Figure A11-1. A minimum of 12 bags will be required per sampling point if samples are to be obtained before and after backwashing. The interior diameter of the pipe should be approximately 1.5 in. and lines should be sawed lightly on the outside of the pipe to indicate the depth of the core sample (Figure A11-1). The length of the pipe shown in the figure is for a bed depth of less than 4 ft (1.2 m). If the bed depth exceeds 4 ft [coarse, deep filter bed of 6 to 7 ft (1.8 to 2.1 m) deep], make the sampling pipe from two 4-ft pieces connected by a coupling. Arrangements should be made in advance with the chief plant operator to drain the filter water below the filter bed prior to sampling.

Core Sampling Technique First make sure that there is no standing water on the surface of the filter bed. Select three representative points in the filter box as the core sampling spots. Gently push the sampling pipe into the bed in a circular motion until the first marker line (1 in.) is aligned with the surface of the filter bed. Slowly pull the pipe up in a whirling motion, enlarging the size of the hole made by the sampling pipe. Next, gently blow the sample into a sampling bucket and transfer to a plastic bag. Label the bag "0–1 inch sample" with the indelible marker. Since a minimum of 200 mL is required, repeat the process if necessary; obtain subsequent samples next to the original hole.

Continue obtaining samples by carefully inserting the sampling pipe into the hole that was created by the first coring. Do not scrape medium off the sides of the hole, as this will contaminate the core sample. Carefully push the tube into the filter bed in the manner previously described until the second mark (2 in.) is aligned with the surface of the filter bed. Slowly remove the pipe while enlarging the hole with a circular motion. Blow

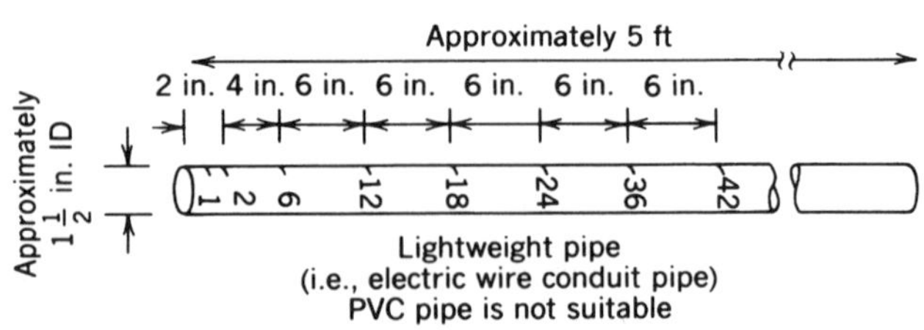

Figure A11-1 Core sampling pipe.

the sample into the bucket and transfer it to a plastic bag marked "1–2 inches." Repeat this process until the samples from the various depths are collected from the same spot.

When the sampling tube suddenly becomes difficult to push down, the top of the gravel bed has been reached. Operators and engineers will also know when the gravel bed has been reached because the gravel will not remain in the sampling pipe as the pipe is removed from the hole. Once the top of the gravel bed has been reached, measure and record this depth; this is the depth of the filter bed.

An alternative to the sampling procedure described above is to fabricate the sampling tube with two longitudinally cut pipes. The two pieces are held together by a metal band at the top and bottom portions of the pipe. This pipe is pushed into the bed in one continuous motion until the pipe hits the top of the gravel bed. The pipe is then removed and laid on a flat surface. The top half of the pipe is removed to expose the entire depth of the core sample. However, the sample is compressed in the depth direction so that no actual depth of the filter bed is shown. Also, fabrication of the sampler is not as easy as with the first procedure.

Sieve Analysis and Specific Gravity of the Various Samples These two qualities should be determined in accordance with the methods specified by Standard B-100 of AWWA (refer to Table A11-1) .

Sludge Retention Analysis Prepare 50 mL of test sample from each of the sample bags by lightly tamping the core samples into a 50-mL graduated cylinder until the volume of the sample becomes settled and is shown as 50 mL. In general, a 50-mL sample of anthracite coal weighs about 52.5 g and 72 g for 50-mL of sand. Transfer the 50-mL sample from the graduated cylinder into a 500-mL wide-mouth bottle or flask. Add 100 mL of tap water and shake vigorously with an up-and-down motion for 30 s. Drain the turbid water into another 500-mL beaker. Repeat this washing procedure four more times, using a total of 500 mL of water to wash out the sludge that is adhered to the filter medium. Measure and record the turbidity of the total 500 mL of wash water. Multiply the recorded turbidity by 2 so that the final tabulations for each depth will list the turbidity for 100 mL of sample instead of 50 mL of sample. Plot these results as the bed depth versus amount of sludge (turbidity), as in Figures 3.2.7-10 and 3.2.7-24.

Evaluation of the Results The three features that should be analyzed when evaluating the size characteristic of the filter bed are the size of the medium in the top 2 in. (5 cm), the medium-size profile across the entire depth of the bed, and the media interface of multimedia beds.

The first feature to be studied is the size of the medium in the upper 2 in. of the filter bed. This layer is composed of small grains. The effective size of these grains should be no smaller than 90% of the effective size specified for the medium type. If the medium is less than 80% of the specified size, this top layer should be scraped off to maintain a reasonable filter run length.

The second feature to be analyzed is the medium size profile across the entire depth of the filter bed. The size and uniformity coefficient of the medium in the upper 6 in. (15 cm) of the bed should be nearly equal. If the size of the medium becomes progressively larger toward the bottom of the filter bed, the chance of early turbidity breakthrough is increased. This problem may be minimized by specifying a uniformity coefficient that is less than 1.5, preferably 1.4.

TABLE A11-1 U.S. Standard Sieves

U.S. Standard Sieve Series No.	Sieve Opening mm	Sieve Opening (in.)	Sieve Wire Diameter (in.)
4	4.76	0.187	0.050
5	4.00	0.157	0.044
6	3.36	0.132	0.040
7	2.83	0.111	0.036
8	2.38	0.0937	0.0331
10	2.00	0.0787	0.0299
12	1.68	0.0661	0.0272
14	1.41	0.0555	0.0240
16	1.19	0.0469	0.0213
18	1.00	0.0394	0.0189
20	0.84	0.0331	0.0165
25	0.71	0.0280	0.0146
30	0.59	0.0232	0.0130
35	0.50	0.0197	0.0114
40	0.42	0.0165	0.0098
45	0.35	0.0138	0.0087
50	0.297	0.0117	0.0074
60	0.250	0.0098	0.0064
70	0.210	0.0083	0.0055
80	0.177	0.0070	0.0047
100	0.149	0.0059	0.0040
120	0.125	0.0049	0.0034
140	0.105	0.0041	0.0029
170	0.088	0.0035	0.0025
200	0.074	0.0029	0.0021
230	0.062	0.0024	0.0018
270	0.053	0.0021	0.0016
325	0.044	0.0017	0.0014

The last feature to be examined is the interface zone of dual- or trimedia filter beds. An interface zone over 6 in. will have a smaller void ratio and therefore promote accumulation of floc in the interface (Figure 3.2.7-24). This problem may be minimized or eliminated by specifying media with complementary characteristics (complementary with respect to fluidization during backwash) for each layer and by providing an effective auxiliary scouring system such as subsurface washing. Under these circumstances, air-scour wash is not necessarily the most effective system because the vigorous boiling action, generated by the compressed air, occurs only in the top 6 to 9 in. (15 to 22.5 cm) of the filter bed; thus, the air bubbles short-circuit to the surface of the bed with a "pulse-collapse" motion.

Cleanliness of the filter bed can be evaluated properly by the mud profile chart based on sludge retention analysis. Detailed discussions of this subject are given in Section 3.2.7, "Granular Medium Filtration," "Operation and Maintenance," and "Mud Deposition Profile."

APPENDIX 12

Venturi Flow Meter Selections*

CAPACITY TABLE - MGD
(Millions of Gallons per Day)

A thru E - Venturi Tube
B thru D - Venturi Insert Nozzle
A thru D - Dall Flow Tube

Size Throat	Range Tube Size - - Inches of Water: 20	30	40	60	80*	120	160	240	276	320	*Exact Diff. for Flows with 80" Range Tube
6A	----	----	0.23	0.28	0.32	0.40	0.45	0.56	0.60	0.65	78.23
B	----	0.32	0.37	0.45	0.52	0.65	0.72	0.90	0.95	1.00	83.32
C	----	0.52	0.60	0.72	0.85	1.00	1.20	1.40	1.50	1.70	81.02
D	----	0.85	0.98	1.20	1.40	1.70	2.00	2.40	2.50	2.80	84.35
E	----	1.10	1.20	1.50	1.70	2.20	2.40	2.90	3.00	3.40	77.02
8A	----	0.35	0.42	0.50	0.58	0.72	0.80	1.00	1.00	1.20	80.54
B	0.45	0.56	0.65	0.80	0.90	1.10	1.30	1.60	1.60	1.80	81.06
C	0.75	0.90	1.10	1.30	1.50	1.80	2.10	2.60	2.80	3.00	79.68
D	1.20	1.50	1.70	2.10	2.40	3.00	3.40	4.20	4.50	4.80	80.74
E	1.40	1.80	2.00	2.50	2.90	3.60	4.20	5.00	5.40	5.80	79.99
10A	0.45	0.55	0.65	0.80	0.90	1.10	1.30	1.60	1.70	1.80	79.69
B	0.75	0.90	1.00	1.30	1.50	1.80	2.10	2.60	2.70	3.00	82.60
C	1.20	1.50	1.70	2.10	2.40	3.00	3.40	4.20	4.50	4.80	78.05
D	2.00	2.40	2.80	3.40	4.00	4.80	5.50	6.80	7.20	8.00	86.08
E	2.40	2.90	3.40	4.20	5.00	5.80	6.60	8.50	8.80	9.50	83.71
12A	0.70	0.85	0.98	1.20	1.40	1.70	2.00	2.40	2.60	2.80	81.84
B	1.10	1.30	1.50	1.90	2.20	2.70	3.10	3.80	4.00	4.40	83.34
C	1.70	2.10	2.50	3.00	3.50	4.20	5.00	6.00	6.50	7.00	81.04
D	2.80	3.50	4.00	5.00	5.60	7.00	8.00	9.80	10.00	11.00	81.72
E	3.40	4.20	4.80	6.00	7.00	8.40	9.50	12.00	13.00	14.00	81.56
14A	0.90	1.10	1.30	1.60	1.80	2.30	2.60	3.20	3.40	3.70	76.01
B	1.40	1.80	2.00	2.50	2.90	3.50	4.00	5.00	5.20	5.80	83.70
C	2.40	2.90	3.40	4.20	4.80	5.80	6.80	8.50	8.80	9.50	81.95
D	3.80	4.80	5.50	6.80	7.50	9.50	11.00	13.00	14.00	15.00	78.14
E	4.80	5.60	6.60	8.00	9.50	12.00	13.00	16.00	17.00	20.00	78.62
16A	1.20	1.50	1.70	2.10	2.40	3.00	3.40	4.20	4.50	5.00	78.30
B	2.00	2.40	2.80	3.40	4.00	4.80	5.50	6.80	7.20	8.00	85.80
C	3.20	4.00	4.50	5.50	6.30	8.00	9.00	11.00	12.00	13.00	78.65
D	5.00	6.00	7.00	8.50	10.00	12.00	14.00	17.00	18.00	20.00	84.09
E	6.50	8.00	9.00	11.00	13.00	16.00	18.00	22.00	24.00	26.00	82.53
18A	1.50	1.80	2.10	2.60	3.00	3.70	4.20	5.20	5.60	6.00	80.05
B	2.40	2.90	3.30	4.20	4.80	5.80	6.80	8.00	8.50	9.50	85.42
C	3.80	4.80	5.50	6.80	8.00	9.50	11.00	13.00	14.00	16.00	85.25
D	6.50	8.00	9.00	11.00	13.00	16.00	18.00	22.00	23.00	25.00	86.83
E	8.00	9.50	11.00	13.00	16.00	19.00	22.00	26.00	29.00	33.00	81.83
20A	1.80	2.30	2.80	3.20	3.70	4.50	5.20	6.50	6.80	7.50	80.18
B	3.00	3.70	4.20	5.20	6.00	7.50	8.50	10.00	11.00	12.00	82.59
C	5.00	6.00	7.00	8.50	10.00	12.00	14.00	17.00	18.00	20.00	82.62
D	8.00	10.00	11.00	14.00	16.00	20.00	23.00	28.00	30.00	33.00	76.88
E	9.80	12.00	14.00	17.00	19.00	24.00	28.00	34.00	36.00	40.00	75.54
24A	2.70	3.30	3.70	4.50	5.40	6.50	7.50	9.50	9.80	11.00	82.60
B	4.30	5.40	6.20	7.50	8.50	10.00	12.00	15.00	16.00	17.00	79.66
C	7.00	8.50	9.80	12.00	14.00	17.00	20.00	24.00	26.00	28.00	81.31
D	11.00	14.00	16.00	19.00	22.00	27.00	32.00	40.00	40.00	45.00	79.49
E	13.00	17.00	19.00	23.00	26.00	32.00	38.00	48.00	50.00	54.00	73.44
30A	4.20	5.20	6.00	7.50	8.50	10.00	12.00	15.00	15.00	17.00	78.45
B	7.00	8.50	9.80	12.00	14.00	17.00	20.00	24.00	26.00	28.00	82.67
C	11.00	13.00	16.00	19.00	22.00	27.00	31.00	38.00	40.00	44.00	80.29
D	17.00	21.00	24.00	30.00	35.00	42.00	50.00	60.00	63.00	70.00	82.41
E	22.00	27.00	31.00	38.00	44.00	54.00	62.00	75.00	80.00	88.00	80.03
36A	6.50	8.00	9.00	11.00	13.00	16.00	18.00	22.00	24.00	25.60	83.53
B	9.80	12.00	14.00	17.00	20.00	24.00	28.00	34.00	36.00	40.00	82.77
C	16.00	20.00	23.00	28.00	32.00	40.00	45.00	56.00	60.00	65.00	80.30
D	25.00	31.00	36.00	44.00	51.00	62.00	72.00	88.00	90.00	100.00	84.39
E	31.00	38.00	44.00	54.00	62.00	80.00	88.00	110.00	110.00	130.00	76.63
42A	8.00	10.00	11.00	14.00	16.00	20.00	23.00	28.00	30.00	32.00	78.78
B	13.00	16.00	19.00	23.00	27.00	33.00	38.00	45.00	50.00	54.00	82.09
C	22.00	27.00	31.00	38.00	44.00	54.00	62.00	75.00	80.00	88.00	81.95
D	35.00	42.00	50.00	60.00	70.00	85.00	98.00	120.00	120.00	140.00	85.81
E	43.00	54.00	62.00	75.00	85.00	110.00	120.00	150.00	160.00	170.00	77.94

Source: Adapted from BIF Catalog.

CAPACITY TABLE - MGD
(Millions of Gallons per Day)

A thru E - Venturi Tube
B thru D - Venturi Insert Nozzle
A thru D - Dall Flow Tube

Size Throat	Range Tube Size - - Inches of Water: 20	30	40	60	80*	120	160	240	276	320	*Exact Diff. for Flows with 80" *Range Tub
48A	11.00	13.00	16.00	19.00	22.00	27.00	31.00	38.00	40.00	44.00	80.20
B	17.00	21.00	24.00	30.00	35.00	42.00	50.00	60.00	65.00	69.00	80.86
C	29.00	35.00	40.00	50.00	58.00	70.00	80.00	100.00	100.00	120.00	82.25
D	45.00	55.00	65.00	80.00	90.00	110.00	130.00	150.00	160.00	180.00	83.15
E	56.00	70.00	80.00	98.00	110.00	140.00	160.00	200.00	210.00	230.00	76.49
54A	14.00	17.00	20.00	24.00	28.00	34.00	38.00	48.00	51.00	56.00	81.10
B	22.00	27.00	31.00	38.00	44.00	54.00	62.00	75.00	80.00	88.00	79.78
C	36.00	45.00	51.00	63.00	72.00	90.00	100.00	130.00	130.00	150.00	78.99
D	58.00	70.00	80.00	98.00	110.00	140.00	160.00	200.00	210.00	230.00	77.54
E	70.00	85.00	98.00	120.00	140.00	170.00	200.00	240.00	260.00	280.00	81.45
60A	17.00	21.00	25.00	30.00	35.00	42.00	48.00	60.00	63.00	70.00	82.35
B	27.00	33.00	38.00	48.00	54.00	66.00	80.00	95.00	100.00	110.00	78.84
C	44.00	54.00	62.00	75.00	88.00	110.00	130.00	150.00	160.00	180.00	78.70
D	70.00	85.00	98.00	120.00	140.00	170.00	200.00	240.00	260.00	280.00	82.41
E	85.00	110.00	120.00	150.00	170.00	210.00	240.00	300.00	320.00	350.00	74.66
66A	21.00	25.00	29.00	36.00	42.00	51.00	60.00	72.00	75.00	85.00	80.99
B	33.00	40.00	48.00	58.00	66.00	80.00	95.00	110.00	120.00	130.00	80.44
C	54.00	66.00	75.00	95.00	110.00	130.00	150.00	190.00	200.00	220.00	83.10
D	80.00	100.00	110.00	140.00	160.00	200.00	230.00	280.00	300.00	320.00	78.69
E	100.00	130.00	150.00	180.00	210.00	250.00	290.00	360.00	380.00	420.00	82.10
72A	25.00	30.00	35.00	43.00	50.00	60.00	70.00	85.00	90.00	100.00	81.04
B	40.00	48.00	56.00	68.00	80.00	95.00	110.00	140.00	140.00	160.00	83.45
C	65.00	80.00	90.00	110.00	130.00	160.00	180.00	220.00	230.00	250.00	84.47
D	98.00	120.00	140.00	170.00	200.00	240.00	280.00	340.00	360.00	400.00	81.11
E	120.00	150.00	170.00	210.00	240.00	300.00	340.00	420.00	450.00	480.00	75.73
78A	29.00	35.00	42.00	50.00	58.00	70.00	85.00	100.00	110.00	120.00	79.18
B	48.00	58.00	65.00	80.00	95.00	110.00	130.00	160.00	170.00	180.00	83.48
C	75.00	95.00	110.00	130.00	150.00	180.00	210.00	260.00	280.00	300.00	80.06
D	120.00	140.00	160.00	200.00	230.00	280.00	330.00	400.00	430.00	480.00	77.88
E	140.00	180.00	200.00	250.00	290.00	350.00	400.00	500.00	540.00	580.00	79.96
84A	33.00	40.00	48.00	58.00	66.00	80.00	95.00	120.00	120.00	140.00	76.22
B	54.00	65.00	75.00	90.00	110.00	130.00	150.00	180.00	200.00	210.00	85.16
C	85.00	110.00	120.00	150.00	170.00	210.00	240.00	300.00	320.00	350.00	76.45
D	130.00	160.00	190.00	230.00	270.00	330.00	380.00	450.00	480.00	540.00	85.40
E	170.00	210.00	240.00	290.00	330.00	420.00	480.00	580.00	620.00	680.00	77.27
90A	38.00	48.00	54.00	66.00	75.00	95.00	110.00	130.00	140.00	150.00	76.62
B	65.00	80.00	90.00	110.00	130.00	160.00	180.00	220.00	240.00	250.00	83.44
C	98.00	120.00	140.00	170.00	200.00	240.00	280.00	340.00	360.00	400.00	80.30
D	160.00	190.00	220.00	270.00	310.00	380.00	440.00	540.00	580.00	620.00	79.82
E	200.00	240.00	280.00	340.00	400.00	480.00	560.00	680.00	720.00	800.00	81.65
96A	44.00	54.00	62.00	75.00	88.00	110.00	120.00	150.00	160.00	180.00	81.02
B	70.00	85.00	98.00	120.00	140.00	170.00	200.00	240.00	260.00	280.00	80.86
C	120.00	140.00	160.00	200.00	230.00	280.00	330.00	400.00	420.00	450.00	80.83
D	180.00	220.00	250.00	310.00	360.00	440.00	510.00	620.00	660.00	720.00	83.15
E	230.00	280.00	320.00	400.00	450.00	550.00	650.00	750.00	800.00	900.00	79.83
108A	55.00	68.00	80.00	95.00	110.00	140.00	160.00	190.00	200.00	220.00	79.03
B	85.00	110.00	120.00	150.00	170.00	210.00	240.00	300.00	320.00	350.00	76.18
C	140.00	180.00	200.00	250.00	290.00	350.00	420.00	500.00	540.00	580.00	81.42
D	230.00	280.00	320.00	400.00	450.00	550.00	650.00	750.00	800.00	900.00	81.11
E	280.00	350.00	400.00	500.00	580.00	700.00	800.00	980.00	1000.00	1100.00	82.79
120A	70.00	85.00	98.00	120.00	140.00	170.00	200.00	240.00	260.00	280.00	82.37
B	110.00	130.00	160.00	190.00	220.00	270.00	310.00	380.00	400.00	440.00	79.92
C	180.00	220.00	250.00	310.00	360.00	440.00	510.00	620.00	660.00	720.00	82.32
D	280.00	340.00	400.00	480.00	550.00	680.00	800.00	950.00	1000.00	1100.00	79.50
E	350.00	430.00	500.00	620.00	700.00	850.00	1000.00	1200.00	1300.00	1400.00	79.12

Water Treatment Chemicals and Preparation of Certain Percent Solutions

ALUMINUM SULFATE*

Properties

Chemical Formula............. $Al_2(SO_4)_3 \cdot xH_2O$ (approx. $14H_2O$)
Molecular Weight............... 594 (For $14H_2O$ product)
Appearance........................ White to cream color
Solubility, 32°F.................. 105 parts in 100 parts water
pH 1% Solution.................. Approximately 3.5

Bulk Density:		
	Powdered	38 to 45 pounds per cubic foot
	Ground	63 to 71 pounds per cubic foot
	Rice	57 to 61 pounds per cubic foot

Angle of Repose:		
	Powdered	65°
	Ground	43°
	Rice	38°

Grades

Commercial Dry...................... Powdered (97% thru 100 mesh)
Ground (90% thru 10 mesh)
Rice (100% thru 6 mesh)
(95% on 20 mesh)
Lump (90% thru 2.5 inch ring)

Commercial Liquid................. 8.3% Al_2O_3 Liquid

Iron-Free................................ Powdered
Ground
Lump

Iron-Free Liquid...................... 8.3% Al_2O_3 Liquid

	COMMERCIAL DRY	COMMERCIAL LIQUID	IRON-FREE DRY	IRON-FREE LIQUID
Total soluble Al_2O_3....	17.1 %	8.3 %	17.2 %	8.3 %
Free Al_2O_3	0.2 %	0.1 %	0.25 %	0.1 %
Total Fe (as Fe_2O_3)....	0.4 %	0.2 %	0.002%	0.004%
Actual Fe_2O_3	0.04%	0.03%	—	—
Insoluble in water......	0.05%	0.03%	0.01 %	0.005%

GRAVITY CONVERSIONS & WEIGHT EQUIVALENTS OF COMMERCIAL ALUMINUM SULFATE SOLUTIONS

These are typical values, varying slightly from one producing point to another because of variations in raw materials.

°Be at 60° F	Sp. Gr.	% Al_2O_3	% $Al_2(SO_4)_3$	Lbs./Gal.	% (Dry) 17% Alum	Pounds Dry Alum per Gallon	Grains Dry Alum per Gallon
32.16	1.2850	7.37	24.73	10.713	43.34	4.64	32,480
32.38	1.2875	7.42	24.90	10.734	43.63	4.68	32,760
32.63	1.2900	7.47	25.07	10.755	43.92	4.72	33,040
32.81	1.2925	7.52	25.24	10.776	44.22	4.77	33,390
33.03	1.2950	7.57	25.40	10.796	44.51	4.81	33,670
33.27	1.2975	7.61	25.54	10.817	44.75	4.84	33,880
33.46	1.3000	7.69	25.81	10.838	45.22	4.90	34,300
33.68	1.3025	7.74	25.97	10.859	45.51	4.94	34,580
33.89	1.3050	7.80	26.18	10.880	45.86	4.99	34,930
34.10	1.3075	7.86	26.38	10.901	46.22	5.04	35,280
34.31	1.3100	7.92	26.58	10.921	46.57	5.09	35,630
34.52	1.3125	7.97	26.75	10.942	46.86	5.13	35,910
34.73	1.3150	8.03	26.95	10.963	47.22	5.18	36,260
34.94	1.3175	8.08	27.11	10.984	47.51	5.22	36,540
35.15	1.3200	8.14	27.32	11.005	47.86	5.27	36,890
35.36	1.3225	8.19	27.48	11.026	48.16	5.31	37,170
35.57	1.3250	8.25	27.69	11.047	48.51	5.36	37,520
35.77	1.3275	8.30	27.85	11.067	48.80	5.40	37,800
35.98	1.3300	8.35	28.02	11.088	49.10	5.44	38,080
36.18	1.3325	8.40	28.19	11.109	49.39	5.49	38,430
36.39	1.3350	8.46	28.39	11.130	49.74	5.54	38,780
36.59	1.3375	8.52	28.59	11.151	50.10	5.59	39,130
36.79	1.3400	8.58	28.79	11.172	50.45	5.64	39,480
36.99	1.3425	8.64	28.99	11.192	50.80	5.69	39,830
37.19	1.3450	8.70	29.20	11.213	51.16	5.74	40,180
37.39	1.3475	8.75	29.36	11.234	51.45	5.78	40,460
37.59	1.3500	8.80	29.53	11.255	51.74	5.82	40,740
37.79	1.3525	8.85	29.70	11.276	52.04	5.87	41,090
37.99	1.3550	8.91	29.90	11.297	52.39	5.92	41,440
38.19	1.3575	8.97	30.10	11.317	52.74	5.97	41,790
38.38	1.3600	9.04	30.34	11.338	53.16	6.03	42,210

*Adapted from aluminum sulfate bulletins by Allied Chemical (1968) and Stauffer Chemical (1967).

CRYSTALLIZATION TEMPERATURE OF SULFURIC ACID AND ALUM

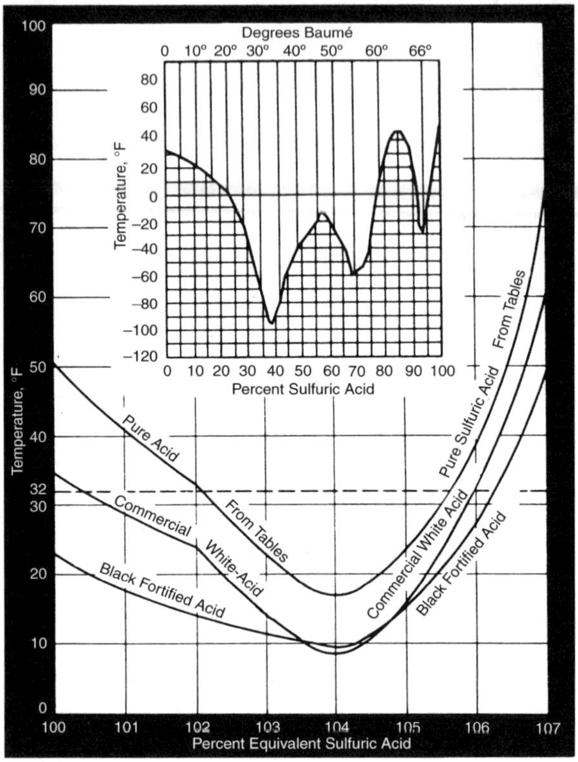

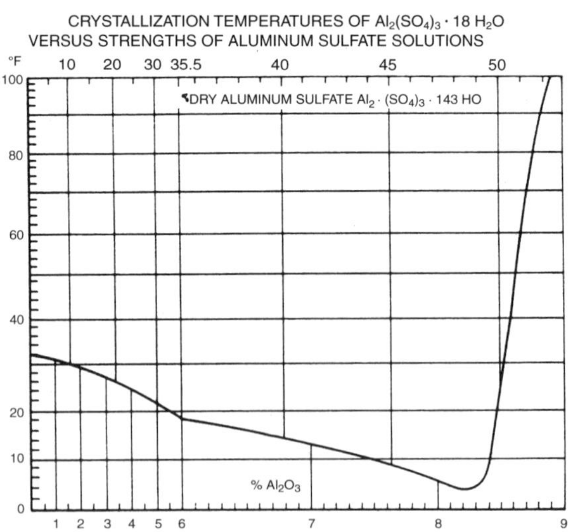

Figure A13-1 Crystallization temperature of sulfuric acid and alum.

FERRIC CHLORIDE*

PROPERTIES OF 42 BAUMÉ FERRIC CHLORIDE SOLUTION

Description of Material	Red brown
At Room Temperature	Solution
Formula	$FeCl_3$ in solution
Molecular Weight	162.22
$FeCl_3$	39%
Acidity as HCl	Less than 0.50%
Specific Gravity, 25°/4°C.	1.40
Melting Point	—6°C.
Boiling Point	109°C.
Viscosity, 25°C., Centistokes	6
Surface Tension, 25°C., dynes/cm.	60
Specific Heat, 0-98°C.	0.6

FERRIC CHLORIDE SOLUTIONS

Concentration-Sp. Gr. Data at 68-77° F.					Freezing Range °F.	
% $FeCl_3$	Sp. Gr.	° Bé	Lb. Sol. / Gal. Sol.	Lb. $FeCl_3$ / Gal. Sol.	Undisturbed	Agitated or Seeded
0	1.000	0.0	8.32	0.0	32	32
5	1.042	5.9	8.67	0.43	27	28
10	1.088	11.8	9.12	0.91	20	24
15	1.134	17.0	9.44	1.42	10	16
20	1.185	22.8	9.87	1.98	− 4	5
25	1.238	27.9	10.30	2.58	−22	−10
30	1.298	33.4	10.80	3.24	−50	−36
35	1.360	38.8	11.32	3.97	−42	−12
37	1.391	40.8	11.59	4.28	−15	15
38	1.402	41.8	11.69	4.44	− 5	21
39	1.415	42.6	11.78	4.59	4	23
40	1.430	43.7	11.90	4.77	10	25
41	1.445	44.8	12.03	4.93	15	28
42	1.459	45.5	12.14	4.98	20	31
43	1.472	46.6	12.24	5.27	24	35
44	1.485	47.4	12.36	5.44	28	40
45	1.500	48.6	12.48	5.62	32	45
46	1.514	49.3	12.60	5.79	35	50
47	1.530	50.3	12.73	5.98	39	58
48	1.545	51.1	12.86	6.17	43	68

*Adapted from Dow Chemical's ferric chloride bulletin (1965).

Density of Caustic Soda Solutions at 60°F (15.5°C)

% NaOH	% Na_2O	Specific Gravity 60/60°F.	Baumé Am. Std.	Twaddell 60°F.	Grams NaOH Per Liter	Pounds NaOH Per Gal.	Pounds NaOH Per Cu. Ft.	Pounds of Solution Per Gal.	Pounds of Solution Per Cu. Ft.
2	1.55	1.023	3.26	4.6	20.51	.17	1.28	8.53	63.80
4	3.10	1.045	6.24	9.0	41.65	.35	2.60	8.71	65.18
6	4.65	1.067	9.10	13.4	63.92	.53	3.99	8.90	66.55
8	6.20	1.090	11.97	18.0	87.15	.73	5.44	9.09	67.98
10	7.75	1.112	14.60	22.4	111.18	.93	6.94	9.27	69.36
12	9.30	1.134	17.13	26.8	136.01	1.13	8.49	9.45	70.74
14	10.85	1.156	19.57	31.2	161.64	1.36	10.09	9.63	72.10
16	12.40	1.178	21.91	35.6	188.40	1.57	11.76	9.82	73.47
18	13.95	1.201	24.27	40.2	215.31	1.80	13.44	10.01	74.91
20	15.50	1.223	26.44	44.6	244.47	2.04	15.26	10.20	76.28
22	17.05	1.245	28.53	49.0	273.62	2.28	17.08	10.38	77.65
24	18.60	1.267	30.56	53.4	303.74	2.53	18.96	10.56	79.02
26	20.15	1.289	32.51	57.8	334.82	2.79	20.90	10.75	80.39
28	21.69	1.310	34.31	62.0	366.54	3.06	22.88	10.92	81.70
30	23.24	1.332	36.14	66.4	399.22	3.33	24.92	11.11	83.08
32	24.79	1.353	37.83	70.6	432.54	3.61	27.00	11.28	84.39
34	26.34	1.374	39.47	74.8	466.82	3.89	29.14	11.46	85.71
36	27.89	1.394	40.98	78.8	501.43	4.18	31.30	11.62	86.95
38	29.44	1.415	42.53	83.0	537.15	4.48	33.53	11.80	88.25
40	30.99	1.434	43.88	86.8	573.20	4.78	35.78	11.96	89.45
42	32.54	1.454	45.28	90.8	610.20	5.09	38.09	12.12	90.70
44	34.09	1.473	46.56	94.6	647.53	5.40	40.42	12.28	91.87
46	35.64	1.492	47.82	98.4	685.82	5.72	42.81	12.44	93.06
48	37.19	1.511	49.04	102.2	724.74	6.04	45.24	12.60	94.24
50	38.74	1.530	50.23	106.0	764.47	6.38	47.72	12.76	95.43
52	40.29	1.549	51.39	109.8	804.84	6.71	50.24	12.92	96.62

Approximate Freezing Temperatures of Caustic Soda Solutions

% NaOH Solution	Approx. F.P. °F	% NaOH Solution	Approx. F.P. °F	% NaOH Solution	Approx. F.P. °F	% NaOH Solution	Approx. F.P. °F
2	30	22	−11	42	57	62	134
4	26	24	−2	44	51	64	140
6	22	26	+14	46	43	66	145
8	18	28	25	48	49	68	148
10	13	30	34	50	53	70	148
12	7	32	42	52	71	72	146
14	+1	34	51	54	88	73	144
16	−5	36	57	56	102	74	143
18	−13	38	60	58	116	74.2	140
20	−15	40	60	60	126	74.5	144

*Adapted from caustic soda bulletin by Stauffer Chemical (1965).

AQUA AMMONIA: NEUTRALIZATION GRADE*

CHEMICAL PROPERTIES

Property	Product Specification	Typical Properties
Ammonia Content	24.5% Min.	24.9%
Water Content	75.5% Max.	

PHYSICAL PROPERTIES

Chemical Formula	NH_4OH
Specific Gravity @ 60°F	0.910
Pounds Per Gallon @ 60°F	7.58
Color	Colorless

Vapor Pressure

°F	80	100	120	140
psig	-	0.9	8.3	19.3

Corrosivity: Corrosive to copper, copper alloy, aluminum and galvanized surfaces.

AQUA AMMONIA/26° BE: INDUSTRIAL GRADE*

CHEMICAL PROPERTIES

Property	Product Specification	Typical Properties [1]
Ammonia Content	29.4% Min.	29.6%
Carbon Dioxide		20.0 ppm Max.
Chlorides		0.5 ppm Max.
Phosphates		0.5 ppm Max.
Sodium		10.0 ppm Max.
Pyridine		Passes A.C.S. Test
Sulfur (As Sulfate)		3.0 ppm Max.
Iron		0.5 ppm Max.
Heavy Metals (As Lead)		1.0 ppm Max.
Organic		Passes A.C.S. Test
Residue After Ignition		5.0 ppm

PHYSICAL PROPERTIES

Chemical Formula	NH_4OH
Specific Gravity @ 60°F	0.897
Pounds Per Gallon @ 60°F	7.47
Color	Water White

Vapor Pressure

°F	50	80	100	120	140
psig	-	0	7.8	18.9	34.3

Corrosivity: Corrosive to copper, copper alloy, aluminum and galvanized surfaces.

*Adapted from UNOCAL 76 aqua ammonia bulletin (1984).

CHLORINE

Properties of Chlorine Gas

Symbol: Cl_2
Atomic wt.: 35.457
Atomic number: 17
Isotopes: 33, 34, 35, 36, 37, 38, 39
Density (see appendix) at 34°F and 1 atm: 0.2006 lb/ft^3
Specific gravity at 32°F and 1 atm: 2.482 (air = 1)
Liquefying point at 1 atm: −30.1°F (−34.5°C)
Viscosity (see Appendix) at 68°F: 0.01325 centipoise (approximately the same as saturated steam between 1 and 10 atm)
Specific heat at constant pressure of 1 atm and 59°F: 0.115 Btu/lb/°F
Specific heat at constant volume at 1 atm pressure and 59°F: 0.085 Btu/lb/°F
Thermal conductivity at 32°F: 0.0042 Btu/hr/ft^2/ft
Heat of reaction with NaOH: 626 Btu/lb Cl_2 gas
Solubility in water at 68°F and 1 atm: 7.29 g/L.

Combining Quantities

1 lb chlorine gas combines with

1.10 lb commercial hydrated lime (95% $Ca(OH)_2$

$$2Ca(OH)_2 + 2\ Cl_2 = Ca(OCl)_2 + Cl\ Cl_2 + 2H_2O$$

0.83 lb commercial quicklime (95% CaO)

$$2Ca + H_2O + Cl_2 = Ca(OCl)_2 + CaCl_2 + 2H_2O$$

1.13 lb caustic soda (100% NaOH)

$$2NaOH + Cl_2 = NaOCl + NaCl + H_2O$$

2.99 lb soda ash

$$2\ Na_2CO_3 + Cl_2 + H_2O = NaOCL + NaCl + 2NaHCO_3$$

Properties of Liquid Chlorine

Critical temperature	144°C; 291.2°F
Pretical pressure	1118.36 psia
Critical density	573 g/1; 35.77 lb/ft^3
Compressibility	0.0118% per unit vol per atm increase at 68°F
Density (see Appendix) at 32°F	91.67 lbs/ft^3
Specific gravity at 68°F	1.41 (water = 1)
Boiling point (liquefaction point) 1 atm	−34.5°C; −30.1°F
Freezing point	−100.98°C; −149.76°F
Viscosity (see Appendix) at 68°F	.345 centipoise (approx. 0.35 × water at 68°F)
1 volume liquid at 32°F and 1 atm	457.6 vol gas
1 lb liquid at 32°F and 1 atm	4.98 ft^3 gas
Specific heat	0.226 Btu/lb/°F
Latent heat of vaporization	123.8 Btu/lb at −29.3°F
Heat of fusion	41.2 Btu/lb at −150.7°F

Preparation of Certain Strength of Chemical Solutions for Water Treatment

The following table shows the number of pounds of chemical to add to 1 gal of water to make % solutions:

% Solution	lb/gal	% Solution	lb/gal	% Solution	lb/gal
0.1	0.008	3.0	0.258	20.0	2.20
0.2	0.017	5.0	0.440	25.0	2.76
0.5	0.042	6.0	0.553	30.0	3.56
1.0	0.084	10.0	0.927	40.0	5.52
2.0	0.170	15.0	1.473	50.0	8.33

CHLORINE GAS SCRUBBING SYSTEM

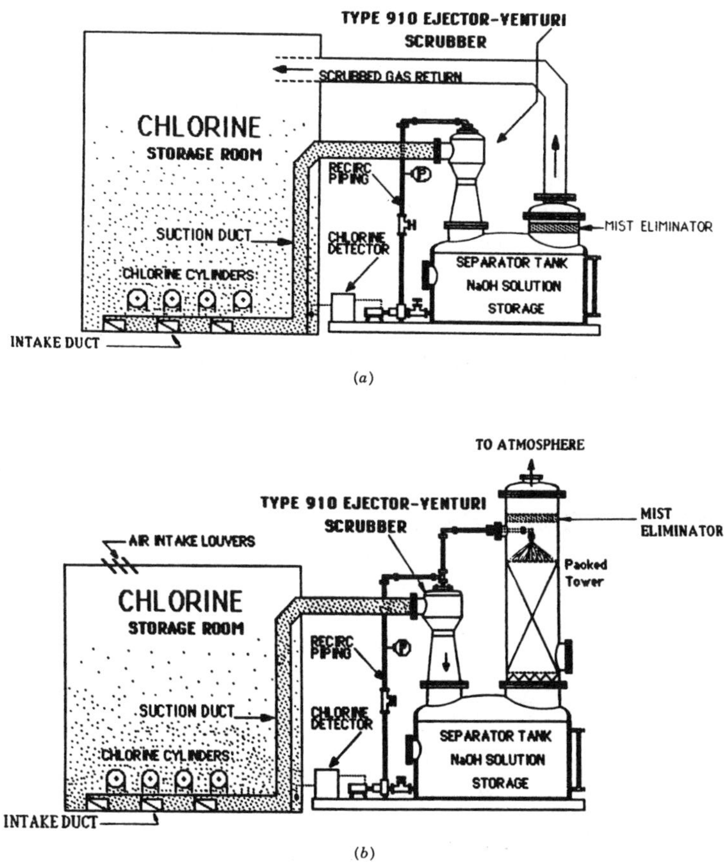

Figure A13-2. Typical emergency chlorine scrubbing systems: (*a*) recycle system; (*b*) once-through system. (Adapted from Environmental Systems Technology Corporation.)

APPENDIX 14

Hydraulics: Basic Data and Formulas

USEFUL CONVERSION FACTORS

1 million gallons per day (mgd) = 1.547 ft^3/s = 694.5 gpm
1 cubic foot per second (ft^3/s) = 448.8 gpm = 1.98 acre-foot/day
1 miner's inch = 1.5 ft^3/min
1 cubic meter per second (m^3/s) = 35.3 ft^3/s = 22.8 mgd
1 megaliter per day (ML/d) = 0.264 mgd
1 acre-foot per day = 0.326 mgd = 0.5 ft^3/s
1 cubic foot = 7.48 gal
1 pound per square inch (psi) = 0.07 kg/cm^2 = 6.9 Pa
1 horsepower = 7.46 kW
1 cubic foot of water = 62.4 lb
1 gallon of water = 8.34 lb
1 psi = 2.307 ft of water
Earth's gravitational acceleration g = 32.17 ft/s^2 = 9.806 m/s^2
Standard atmospheric pressure = 14.7 psi = 1013 millibars = 1 kg/cm^2 = 760 mm mercury

FUNDAMENTAL EQUATIONS

Equation of mass:

$$M = \frac{W}{g}$$

where

W = weight
g = gravitational acceleration

Manning's formula for open channel:

$$v = \frac{1.486}{n} R^{0.667} S^{0.5}$$

where

$$
\begin{aligned}
v &= \text{mean velocity (ft/s)} \\
R &= \text{hydraulic radius (ft) } (R = A/p) \\
S &= \text{hydraulic gradient} \\
n &= \text{coefficient of roughness (0.01 to 0.02)}
\end{aligned}
$$

From Manning's formula,

$$S = \frac{nv}{1.486R^{0.667}}$$

Chezy's formula for an open channel:

$$v = C(RS)^{0.5}$$

where

$$
\begin{aligned}
V, R, S &= \text{same as above} \\
C &= \text{coefficient}
\end{aligned}
$$

$$C = \frac{1.486}{n} R^{0.1667}$$

Darcy–Weisback formula for pipe line:

$$h_f = f\left(\frac{l}{d}\right)\left(\frac{v^2}{2g}\right)$$

where

$$
\begin{aligned}
h_f &= \text{friction loss (ft)} \\
l &= \text{length of pipeline (ft)} \\
d &= \text{inside diameter of pipe (ft)} \\
v &= \text{mean flow velocity (ft/s)} \\
g &= \text{acceleration of gravity (32.2 ft/s}^2\text{)} \\
f &= \text{friction coefficient (0.0002 to 0.002)}
\end{aligned}
$$

Hazen–William formula for pipeline:

$$v = 1.318CR^{0.63}S^{0.54}$$

where

$$
\begin{aligned}
v &= \text{mean velocity (ft/s)} \\
S &= \text{hydraulic gradient} \\
R &= \text{hydraulic radius (ft) } (R = A/p) \\
A &= \text{cross-sectional area (ft}^2\text{)} \\
P &= \text{wetted perimeter (ft)} \\
C &= \text{friction coefficient (80 to 140)}
\end{aligned}
$$

RELATION BETWEEN SURFACE CONDITIONS AND FRICTION COEFFICIENT

Condition	Friction Coefficient			Notes
	n (Manning's)	C (Hazen's)	f (Darcy's)	
Very smooth surface	0.010	140	0.0002	PVC pipe, clean cement-lined pipe
Fair condition surface	0.013	120	0.0012	Unlined pipe
Rough surface	0.016	100	0.0025	Rusted pipe

HEADLOSS IN OPEN CHANNEL

1. Sudden contraction (including inlet loss)

$$h = C_1 \frac{v_2 - v_1^2}{2g}$$

where

$$\begin{aligned} C_1 &= 0.5 \text{ for sharp-cornered entrance} \\ &= 0.25 \text{ for rounded-cornered entrance} \\ &= 0.05\text{–}0.1 \text{ for bell-mouthed entrance (use 0.1 for design)} \\ v_1 &= \text{flow velocity in upstream (ft/s)} \\ v_2 &= \text{flow velocity in downstream (ft/s)} \end{aligned}$$

2. Sudden enlargement (including outlet loss)

$$h = C_2 \frac{v_1^2 - v_2}{2g}$$

where

$$\begin{aligned} C_2 &= 1.0 \text{ for sudden enlargement} \\ &= 0.1 \text{ to } 0.2 \text{ for bell-mouthed outlet(use 0.2 for design)} \\ v_1 &= \text{flow velocity in upstream (ft/s)} \\ v_2 &= \text{flow velocity in downstream (ft/s)} \end{aligned}$$

3. Turns in the baffled channel (plus flow condition only)

$$h = K \frac{\nu^2}{2g}$$

where

$$\begin{aligned} K &= 1.7 \text{ for } 90^\circ \text{ turn} \\ &= 3.2 \text{ for } 180^\circ \text{ turn} \\ v &= \text{mean flow velocity (ft/s)} \end{aligned}$$

4. Flowing through bar screen

$$h = \frac{v^2 - v_a^2}{2g}\left(\frac{1}{0.7}\right)$$

where

v = flow velocity in-between bars ft/s
v_a = approaching flow velocity ft/s

5. Friction loss in open channel

$$h_f = \frac{2gn^2}{R^{0.333}}\frac{L}{R}\frac{v^2}{2g} \qquad R = \frac{A}{p}$$

where

L = length of the channel (ft)

ORIFICE

$$Q = Ca(2gh)^{0.5} \quad \text{or} \quad h = \frac{1}{2g}\left(\frac{Q}{Ca}\right)^2$$

where

Q = discharge rate (ft^3/s)
a = orifice area (ft^2)
C = coefficient, 0.65 to 0.8
h = headloss across the orifice (ft)

HEADLOSS FOR PIPE FLOW

Values of K for the basic formula: $h = K\dfrac{v^2}{2g}$

Item	K	Item	K
Butterfly valve (full open)	0.25 (>16 in. diameter	Return bend	
		Screwed ends	2.2
	0.35 (<12 in. diameter	Flanged ends	
		Regular	0.38
Gate valve (full open)	0.12 (>16 in. diameter)	Long radius	0.25
	0.20 (<12 in. diameter)	Regular Y branches (flow in the branches)	0.5–0.75
		Sudden contraction	
Plug valve (screwed end)	0.85	$d/D = \frac{1}{4}$	0.42
		$d/D = \frac{1}{2}$	0.35

Item	K	Item	K
Diaphragm valve	2.3	$d/D = \frac{3}{4}$	0.19
Check valve		*Note:* v to be used is in branches.	
Swing type	0.6–2.3	Sudden enlargement	
Silent check	2.5–3.3	$d/D = \frac{1}{4}$	0.92
Glove valve	4.0	$d/D = \frac{1}{2}$	0.35
Ball valve	0.15	$d/D = \frac{3}{4}$	0.19
Foot valve (excluding strainer)	0.8	*Note:* v is in smaller pipe.	
90° Elbow (flanged) ends)		Entrance loss	
Regular	0.2–0.3	Inward-projecting	0.75
Long radius	0.15–0.25	Sharp-cornered	0.5
90° Elbow (screwed ends)		Rounded	0.25
Short	0.9	Bell-mouthed	0.10
Medium	0.75	Outlet loss	
45° Elbow		Sudden enlargement	1.0
Regular (screwed)	0.3–0.45	Bell-mouthed	0.1–0.2
Long radius (flanged)	0.18–0.2	Tee (same sizes, same flow rate)	
22.5° Elbow (flanged)	0.10–0.12	90° turn	1.1–1.5
Miter bends (large size, welded ends)		Run of tee	0.2–0.5
90	0.25–0.36	Siphon	2.8
45	0.15–0.20	Increaser	0.25
22.5	0.07–0.12	*Note:* $h = 0.25 \dfrac{v_1^2 - v_2^2}{2g}$ where v_1 = velocity in smaller pipe, v_2 = velocity in larger pipe	
		Reducer	
		Bushing	1.4
		Regular	0.25
		Regular to large pipes (36 in.)	0.05
		Note: v is in the smaller pipe.	

WEIRS (FREE DISCHARGE CONDITION)

1. Sharp-crested rectangular weir

$$Q = 3.33LH^{1.5}$$

2. Broad-crested rectangular weir

$$Q = 3.09L(H + h_v)^{1.5}$$

3. 90° V-notched weir

$$Q = 2.54H^{2.5}$$

4. 60° V-notched weir

$$Q = 1.45H^{2.5}$$

5. 45° V-notched weir

$$Q = 1.07H^{2.5}$$

6. 22.5° V-notched weir

$$Q = 0.5H^{2.5}$$

where

Q = discharge rate (ft^3/s)
L = effective weir length (ft)
H = head (water height) over the crest (ft)
h_v = velocity head of approach (ft)

HYDRAULIC ELEMENTS FOR CIRCULAR PIPE

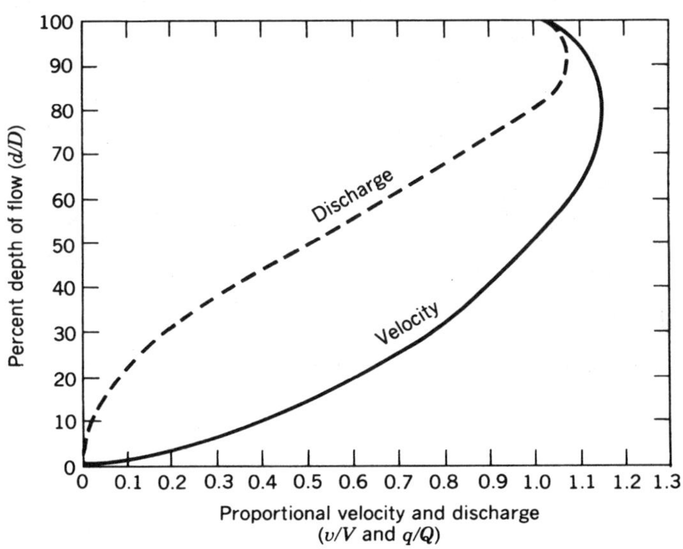

Figure A14-1 Proportional velocity and discharge for circular pipe.

Discharge Flow Rate Correction for Submerged Weirs

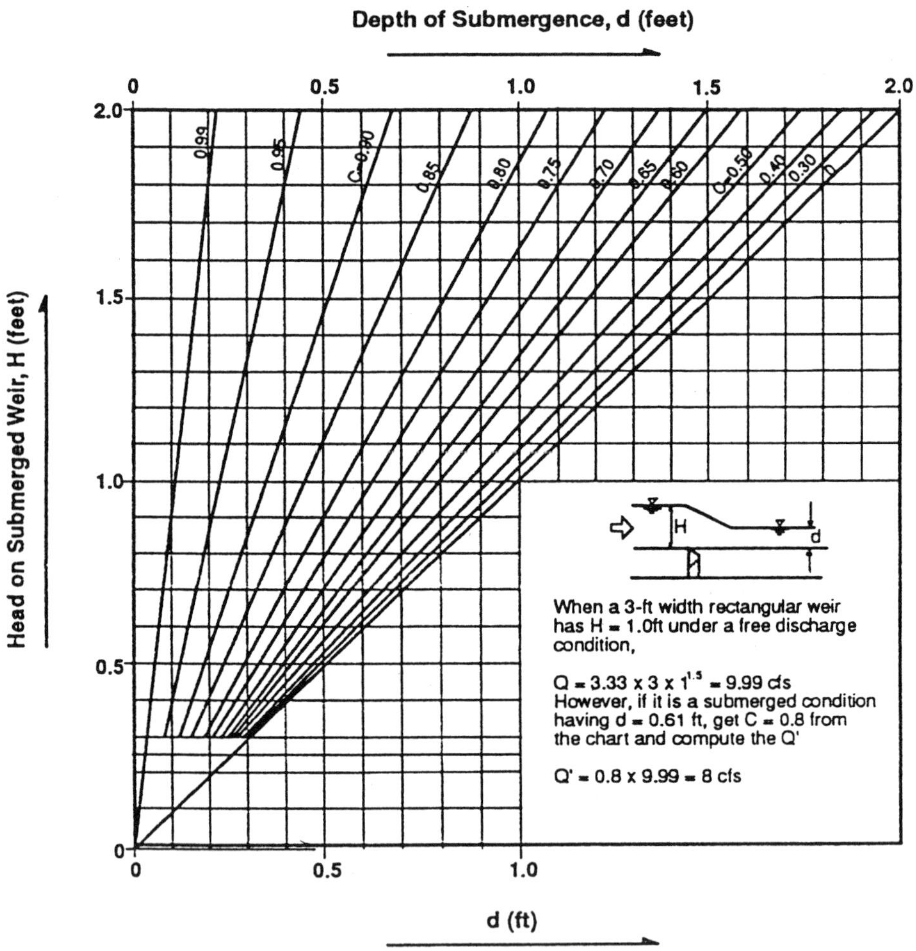

APPENDIX 16

Head–Discharge Curves of Telescoping Valves

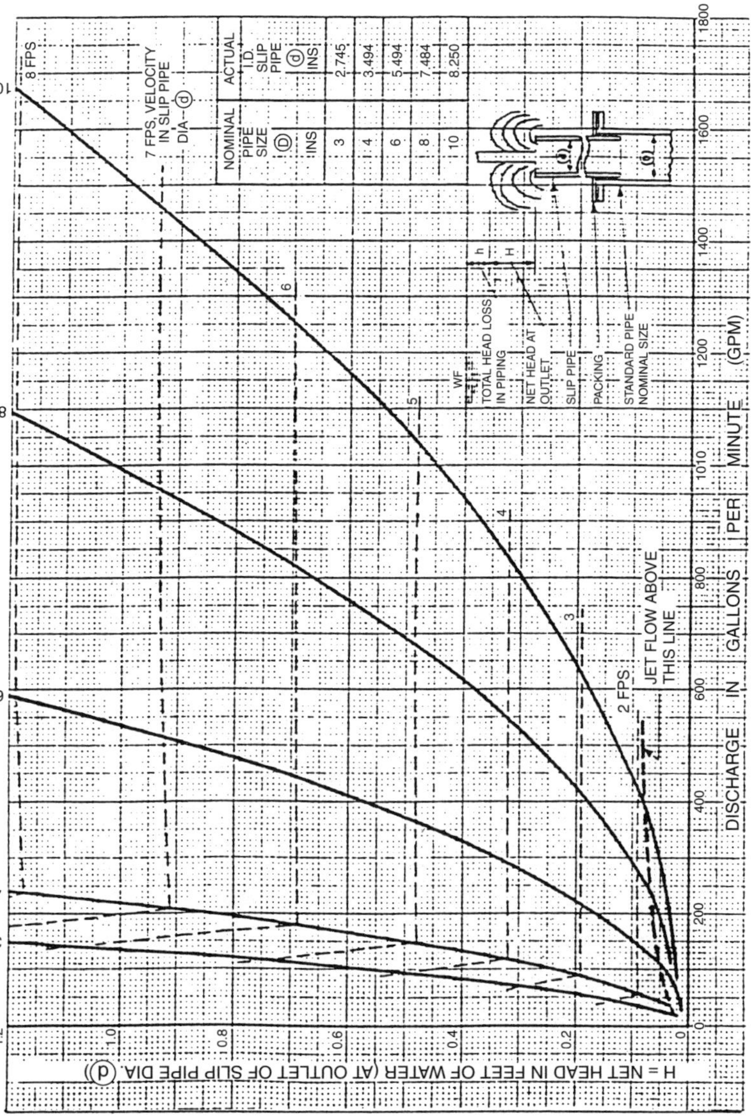

APPENDIX 17

Nomographical Determination of pH and Langelier Index

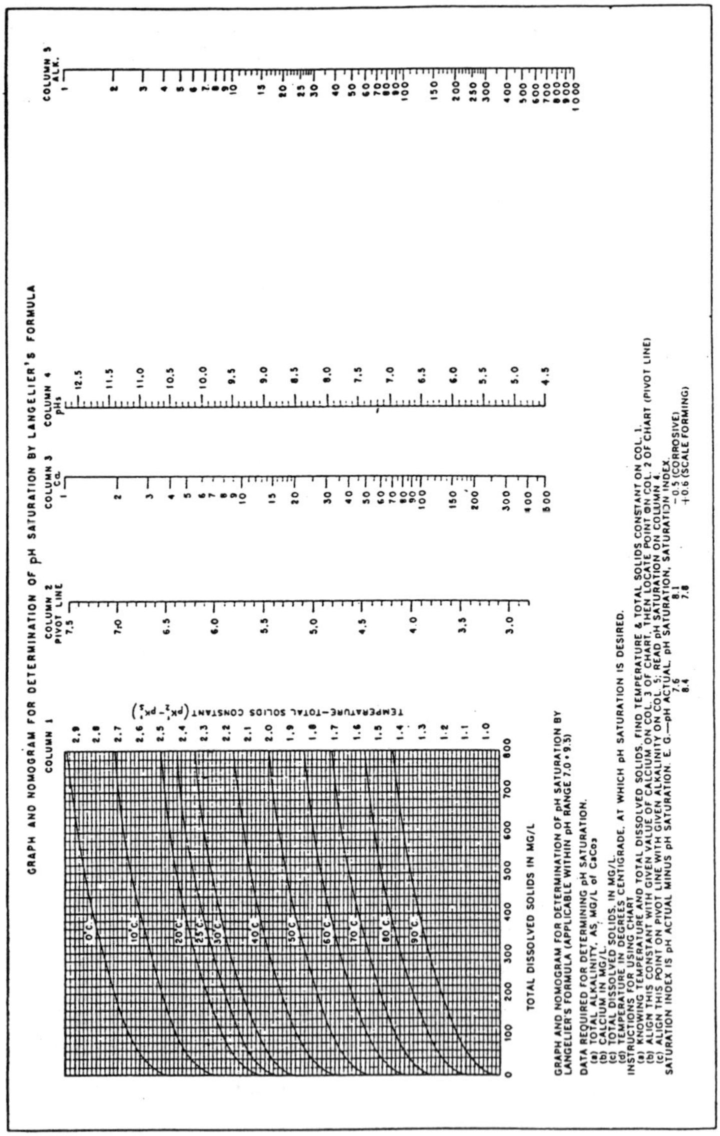

APPENDIX 18

Tertiary Filter Performance

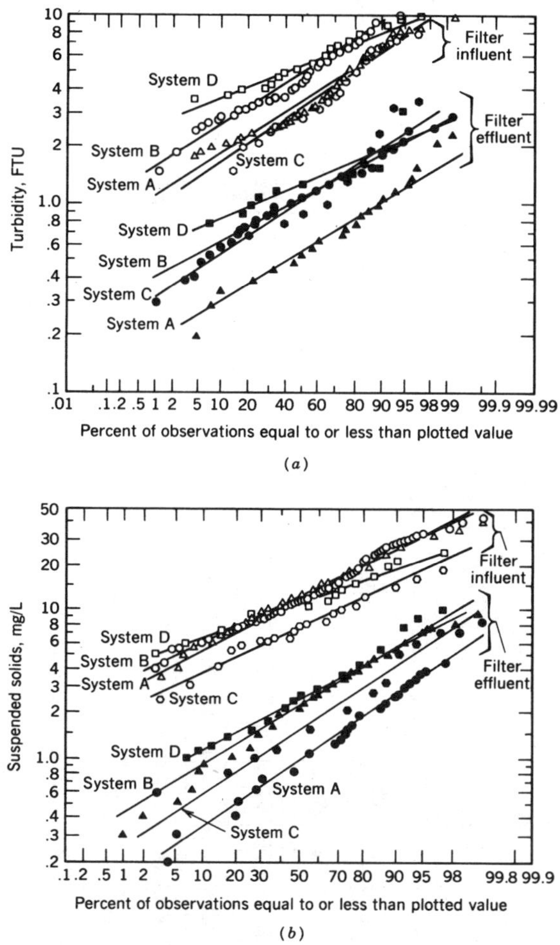

Figure A18-1 Frequency distributions of (*a*) filter influent and effluent turbidity and (*b*) filter influent and effluent suspended solids. System A: Flocculation, sedimentation, and filtration of secondary effluent with 50 mg/L alum and dual-media filter at 5 gpm/ft^2. System B: Coagulation and filtration (in-line filtration) of secondary effluent with 5 mg/L alum and dual-media filter at 5 gpm/ft^2. System C: Two-stage granular carbon adsorption process. Both beds have 10 min EBCT each. System D: Coagulation and filtration (in-line filtration) of nitrified secondary effluent with 5 mg/L alum and dual-media filter at 5 gpm/ft^2. (Adapted from Sanitation Districts of Los Angeles County Report.)

APPENDIX 19

Marston's Soil Coefficient for Pipe Trench Conditions

1. Trench condition:

$$W_d = C_d W B_d^2$$

where
W_d = fill load in pounds per linear foot of pipe
C_d = load coefficient (a function of H/B_d and $K\mu$)
W = unit weight of fill material in pounds per cu ft
B_d = width of trench at the level of the top of pipe, in ft.
H = backfill depth, ft

2. Embankment condition:

$$W_c = C_c W B_c^2$$

where
W_c = load per unit length of pipe
C_c = load coefficient (a function of H/B_c and $r_{sd}p$)
W = unit weight of fill material
B_c = outside diameter of the pipe, in feet

MARSTON SOIL COEFFICIENTS (C_d) FOR TRENCH CONDUITS

A = $K\mu$ = .1924 Granular materials without cohesion
B = $K\mu$ = .165 Maximum for sand and gravel
C = $K\mu$ = .150 Maximum for saturated top soil
D = $K\mu$ = .130 Ordinary maximum for clay
E = $K\mu$ = .110 Maximum for saturated clay

H/B_d	A	B	C	D	E
0.05	0.050	0.050	0.050	0.050	0.050
0.10	0.098	0.098	0.099	0.099	0.099
0.15	0.146	0.146	0.147	0.147	0.148
0.20	0.192	0.194	0.194	0.195	0.196
0.25	0.238	0.240	0.241	0.242	0.243
0.30	0.283	0.286	0.287	0.289	0.290
0.35	0.327	0.331	0.332	0.335	0.337
0.40	0.371	0.375	0.377	0.380	0.383
0.45	0.413	0.418	0.421	0.425	0.428
0.50	0.455	0.461	0.464	0.469	0.473
0.55	0.496	0.503	0.507	0.512	0.518
0.60	0.536	0.544	0.549	0.555	0.562
0.65	0.575	0.585	0.591	0.598	0.606
0.70	0.614	0.625	0.631	0.640	0.649
0.75	0.651	0.664	0.672	0.681	0.691
0.80	0.689	0.703	0.711	0.722	0.734
0.85	0.725	0.741	0.750	0.763	0.775
0.90	0.761	0.779	0.789	0.802	0.817
0.95	0.796	0.816	0.827	0.842	0.857
1.00	0.830	0.852	0.864	0.881	0.898
1.05	0.864	0.887	0.901	0.919	0.938
1.10	0.897	0.922	0.937	0.957	0.977
1.15	0.929	0.957	0.973	0.994	1.016
1.20	0.961	0.991	1.008	1.031	1.055
1.25	0.992	1.024	1.042	1.067	1.093
1.30	1.023	1.057	1.076	1.103	1.131
1.35	1.053	1.089	1.110	1.139	1.168
1.40	1.082	1.121	1.143	1.173	1.205
1.45	1.111	1.152	1.176	1.208	1.241
1.50	1.140	1.183	1.208	1.242	1.278
1.55	1.167	1.213	1.240	1.276	1.313
1.60	1.195	1.243	1.271	1.309	1.349
1.65	1.221	1.272	1.301	1.342	1.384
1.70	1.248	1.301	1.332	1.374	1.418
1.75	1.273	1.329	1.361	1.406	1.452
1.80	1.299	1.357	1.391	1.437	1.486
1.85	1.323	1.385	1.420	1.469	1.520
1.90	1.348	1.412	1.448	1.499	1.553
1.95	1.372	1.438	1.476	1.530	1.586
2.00	1.395	1.464	1.504	1.560	1.618
2.10	1.440	1.515	1.558	1.618	1.682
2.20	1.484	1.564	1.610	1.675	1.744
2.30	1.526	1.612	1.661	1.731	1.805
2.40	1.567	1.658	1.711	1.785	1.865
2.50	1.606	1.702	1.759	1.838	1.923
2.60	1.643	1.745	1.805	1.890	1.980
2.70	1.679	1.787	1.850	1.940	2.036
2.80	1.714	1.827	1.894	1.989	2.090
2.90	1.747	1.867	1.937	2.037	2.144

H/B_d	A	B	C	D	E
3.00	1.780	1.904	1.978	2.083	2.196
3.10	1.810	1.941	2.018	2.128	2.247
3.20	1.840	1.976	2.057	2.172	2.297
3.30	1.869	2.010	2.095	2.215	2.346
3.40	1.896	2.044	2.131	2.257	2.394
3.50	1.923	2.076	2.167	2.298	2.441
3.60	1.948	2.107	2.201	2.338	2.487
3.70	1.973	2.137	2.235	2.376	2.531
3.80	1.997	2.166	2.267	2.414	2.575
3.90	2.019	2.194	2.299	2.451	2.618
4.00	2.041	2.221	2.329	2.487	2.660
4.10	2.062	2.247	2.359	2.522	2.701
4.20	2.082	2.273	2.388	2.556	2.741
4.30	2.102	2.297	2.416	2.589	2.780
4.40	2.121	2.321	2.443	2.621	2.819
4.50	2.139	2.344	2.469	2.652	2.856
4.60	2.156	2.366	2.495	2.683	2.893
4.70	2.173	2.388	2.520	2.713	2.929
4.80	2.189	2.409	2.543	2.742	2.964
4.90	2.204	2.429	2.567	2.770	2.999
5.00	2.219	2.448	2.590	2.798	3.032
5.10	2.234	2.467	2.612	2.825	3.065
5.20	2.247	2.486	2.633	2.851	3.098
5.30	2.261	2.503	2.654	2.877	3.129
5.40	2.273	2.520	2.674	2.901	3.160
5.50	2.286	2.537	2.693	2.926	3.190
5.60	2.298	2.553	2.712	2.949	3.220
5.70	2.309	2.568	2.730	2.972	3.248
5.80	2.320	2.583	2.748	2.995	3.277
5.90	2.330	2.598	2.766	3.017	3.304
6.00	2.340	2.612	2.782	3.038	3.331
6.20	2.360	2.639	2.814	3.079	3.383
6.40	2.377	2.664	2.846	3.118	3.433
6.60	2.394	2.687	2.873	3.155	3.481
6.80	2.409	2.709	2.900	3.190	3.527
7.00	2.423	2.730	2.925	3.223	3.571
7.20	2.436	2.749	2.949	3.255	3.613
7.40	2.448	2.767	2.971	3.285	3.653
7.60	2.459	2.784	2.992	3.313	3.691
7.80	2.470	2.799	3.012	3.340	3.728
8.00	2.479	2.814	3.031	3.366	3.763
8.50	2.500	2.847	3.073	3.424	3.845
9.00	2.517	2.875	3.109	3.476	3.918
9.50	2.532	2.898	3.141	3.521	3.983
10.0	2.543	2.919	3.167	3.560	4.042
15.0	2.591	3.009	3.296	3.768	4.378
20.0	2.598	3.026	3.325	3.825	4.490
30.0	2.599	3.030	3.333	3.845	4.539
40.0	2.599	3.030	3.333	3.846	4.545

APPENDIX 20

Load Coefficient for Positive Projecting Embankment Condition

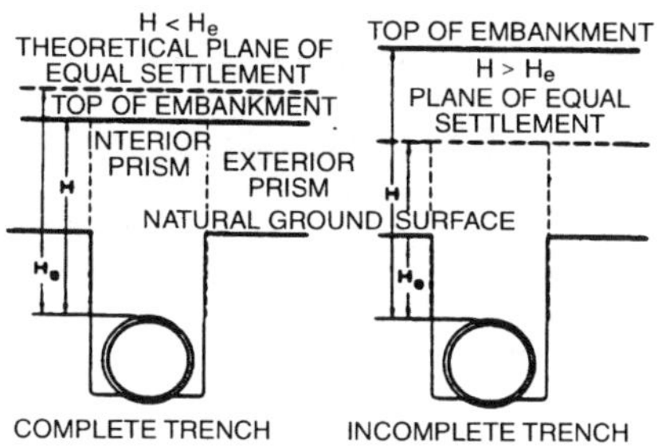

COMPLETE AND INCOMPLETE
TRENCH CONDITIONS

LOAD COEFFICIENT CHART FOR POSITIVE PROJECTING EMBANKMENT CONDITION

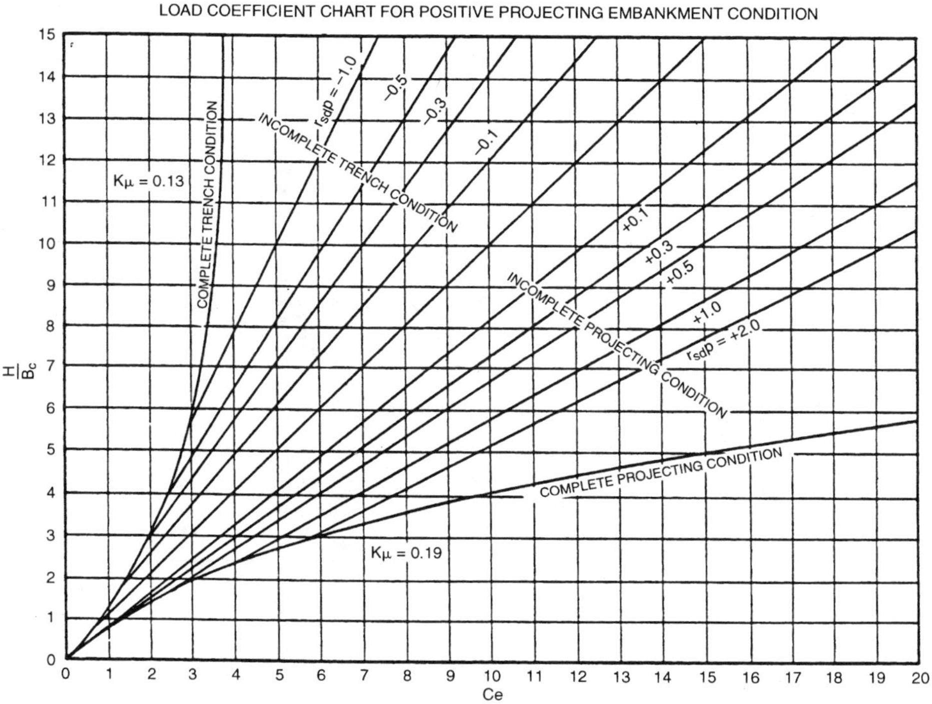

APPENDIX 21

USA Standard Sieves

Nominal Dimensions and Permissible Variations for Wire Cloth of Standard Test Sieves (U.S. Standard Series)

Sieve Designation		Nominal Sieve Opening	Permissible Variation of Average Opening from the Standard Sieve Designation	Maximum Opening Size for not More Than 5 percent of Openings	Maximum Individual Opening	Nominal Wire Diameter
Standard†	Alternative	*in.*				*mm*
125 mm	5 in.	5	±3.7 mm	130.0 mm	130.9 mm	8.0
106 mm	4.24 in.	4.24	±3.2 mm	110.2 mm	111.1 mm	6.40
100 mm	4 in.	4	±3.0 mm	104.0 mm	104.8 mm	6.30
90 mm	$3^1/_2$ in.	3.5	±2.7 mm	93.6 mm	94.4 mm	6.08
75 mm	3 in.	3	±2.2 mm	78.1 mm	78.7 mm	5.80
63 mm	$2^1/_2$ in.	2.5	±1.9 mm	65.6 mm	66.2 mm	5.50
53 mm	2.12 in.	2.12	±1.6 mm	55.2 mm	55.7 mm	5.15
50 mm	2 in.	2	±1.5 mm	52.1 mm	52.6 mm	5.05
45 mm	$1^3/_4$ in.	1.75	±1.4 mm	46.9 mm	47.4 mm	4.85
37.5 mm	$1^1/_2$ in.	1.5	±1.1 mm	39.1 mm	39.5 mm	4.59
31.5 mm	$1^1/_4$ in.	1.25	±1.0 mm	32.9 mm	33.2 mm	4.23
26.5 mm	1.06 in.	1.06	±0.8 mm	27.7 mm	28.0 mm	3.90
25.0 mm	1 in.	1	±0.8 mm	26.1 mm	26.4 mm	3.80
22.4 mm	$^7/_8$ in.	0.875	±0.7 mm	23.4 mm	23.7 mm	3.50
19.0 mm	$^3/_4$ in.	0.750	±0.6 mm	19.9 mm	20.1 mm	3.30
16.0 mm	$^5/_8$ in.	0.625	±0.5 mm	16.7 mm	17.0 mm	3.00
13.2 mm	0.530 in.	0.530	±0.41 mm	13.83 mm	14.05 mm	2.75
12.5 mm	$^1/_2$ in.	0.500	±0.39 mm	13.10 mm	13.31 mm	2.67
11.2 mm	$^7/_{16}$ in.	0.438	±0.35 mm	11.75 mm	11.94 mm	2.45
9.5 mm	$^3/_8$ in.	0.375	±0.30 mm	9.97 mm	10.16 mm	2.27
8.0 mm	$^5/_{16}$ in.	0.312	±0.25 mm	8.41 mm	8.58 mm	2.07
6.7 mm	0.265 in.	0.265	±0.21 mm	7.05 mm	7.20 mm	1.87
6.3 mm	$^1/_4$ in.	0.250	±0.20 mm	6.64 mm	6.78 mm	1.82
5.6 mm	No. $3^1/_2$	0.223	±0.18 mm	5.90 mm	6.04 mm	1.68
4.75 mm	No. 4	0.187	±0.15 mm	5.02 mm	5.14 mm	1.54
4.00 mm	No. 5	0.157	±0.13 mm	4.23 mm	4.35 mm	1.37
3.35 mm	No. 6	0.132	±0.11 mm	3.55 mm	3.66 mm	1.23
2.80 mm	No. 7	0.111	±0.095 mm	2.975 mm	3.070 mm	1.10
2.36 mm	No. 8	0.0937	±0.080 mm	2.515 mm	2.600 mm	1.00
2.00 mm	No. 10	0.0787	±0.070 mm	2.135 mm	2.215 mm	0.900
1.70 mm	No. 12	0.0661	±0.060 mm	1.820 mm	1.890 mm	0.810
1.40 mm	No. 14	0.0555	±0.050 mm	1.505 mm	1.565 mm	0.725
1.18 mm	No. 16	0.0469	±0.045 mm	1.270 mm	1.330 mm	0.650
1.00 mm	No. 18	0.0394	±0.040 mm	1.080 mm	1.135 mm	0.580
850 μm	No. 20	0.0331	±35 μm	925 μm	970 μm	0.510
710 μm	No. 25	0.0273	±30 μm	774 μm	815 μm	0.450
600 μm	No. 30	0.0234	±25 μm	660 μm	695 μm	0.390
500 μm	No. 35	0.0197	±20 μm	550 μm	585 μm	0.340
425 μm	No. 40	0.0165	±19 μm	471 μm	502 μm	0.290
355 μm	No. 45	0.0139	±16 μm	396 μm	425 μm	0.247

Standard Electric Motors and Their Characteristics

Synchronous and Full-Load Speeds of Standard AC Induction Motors

NUMBER OF POLES	60 CYCLE RPM		50 CYCLE RPM	
	SYNC.	F.L.	SYNC.	F.L.
2	3600	3500	3000	2900
4	1800	1770	1500	1450
6	1200	1170	1000	960
8	900	870	750	720
10	720	690	600	575
12	600	575	500	480
14	515	490	429	410
16	450	430	375	360
18	400	380	333	319
20	360	340	300	285
22	327	310	273	260
24	300	285	240	230
26	277	265	231	222
28	257	245	214	205
30	240	230	200	192

Full-Load Amperes for All Speeds and Frequencies

MOTOR HP	SINGLE-PHASE A-C		THREE-PHASE A-C INDUCTION TYPE SQUIRREL CAGE & WOUND ROTOR			DIRECT CURRENT	
	115 VOLTS	230 VOLTS*	230 VOLTS*	460 VOLTS	575 VOLTS	120 VOLTS	240 VOLTS
½	9.8	4.9	2.0	1.0	.8	5.2	2.6
¾	13.8	6.9	2.8	1.4	1.1	7.4	3.7
1	16	8	3.6	1.8	1.4	9.4	4.7
1½	20	10	5.2	2.6	2.1	13.2	6.6
2	24	12	6.8	3.4	2.7	17	8.5
3	34	17	9.6	4.8	3.9	25	12.2
5	56	28	15.2	7.6	6.1	40	20
7½	80	40	22	11	9	58	29
10	100	50	28	14	11	76	38
15			42	21	17	112	55
20			54	27	22	148	72
25			68	34	27	184	89
30			80	40	32	220	106
40			104	52	41	292	140
50			130	65	52	360	173
60			154	77	62	430	206
75			192	96	77	536	255
100			240	120	96		350
125			296	148	118		440
150			350	175	140		530
200			456	228	182		710
250			558	279	223		

*For full-load currents of 208- and 209-V motors, increase the corresponding 230- motor full-load current by 10 and 15% respectively.

APPENDIX 23

Geometrical Formulas

S = lateral or convex surface V = volume

SHAPE	FORMULAE
Parallelopiped 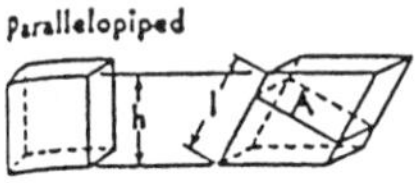	S = perimeter, P, perp. to sides x lat. length l : Pl : V = area of base, B, x perpendicular height, h : Bh. V = area of section, A, perp. to sides, x lat. length l : Al.
Prism right, or oblique, regular or irregular 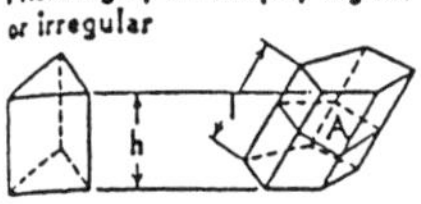	S = perimeter, P, perp. to sides x lat. length l : Pl : V = area of base, B, x perpendicular height, h : Bh. V = area of section, A, perp. to sides, x lat. length l : Al.
Cylinder, right or oblique, circular or elliptic etc. 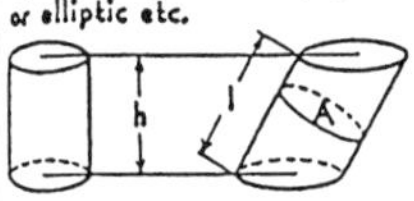	S = perimeter of base, P, x perp. height, h: Ph. S_1 = perimeter, P_1, perp., x lat. length, l : P_1 l. V = area of base, B, x perp. height, h: Bh. V = area of section, A, perp. to sides x lat. length l: Al.
Frustum of any prism or cylinder 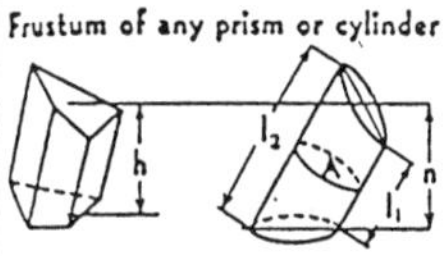	V = area of base, B, x perpendicular distance h, from base to centre of gravity of opposite face: Bh. for cylinder, $\frac{1}{2}A(l_1 + l_2)$
Pyramid or Cone, right and regular 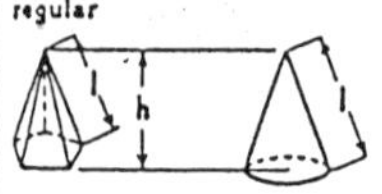	S = perimeter of base, P, x ½ slant height l : ½ Pl. V = area of base, B x ⅓ perpendicular ht., h : ⅓ Bh.
Pyramid or Cone, right or oblique, regular or irregular 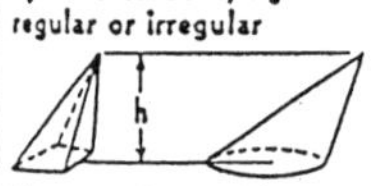	V = area of base, B, x ⅓ perp. height, h : ⅓ Bh. V = ⅓ vol. of prism or cylinder of same base & perp. height. V = ½ vol. of hemisphere of same base and perp. height.
Frustum of pyramid or cone, right and regular, parallel ends 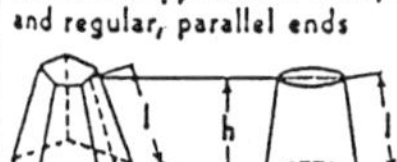	S = (sum of perimeter of base, P, and top, p) x ½ slant height l : ½ l (P + p). V = (sum of areas of base, B, and top, b + sq. root of their products) x ⅓ perp. height, h : $\frac{1}{3}h(B + b + \sqrt{Bb})$.
Frustum of any pyramid or cone, parallel ends 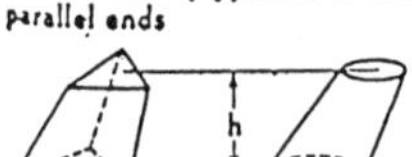	V = (sum of areas of base, B, and top, b_1 + sq. root of their products) x ⅓ perpendicular height, h : $\frac{1}{3}h(B + b + \sqrt{Bb})$.
Wedge, parallelogram face 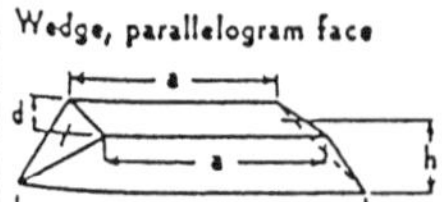	V = ⅙ (sum of three edges, a b a, x perpendicular height, h, x perpendicular width, d) : $\frac{1}{6}dh(2a + b)$.

S = lateral or convex surface V = volume

SHAPE	FORMULAE
Sphere 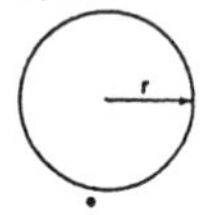	$S = 4\pi r^2 = \pi d^2 = 3.14159265\,d^2$. $V = \frac{4}{3}\pi r^3 = \frac{1}{6}\pi d^3 = 0.52359878\,d^3$. *Ellipse* 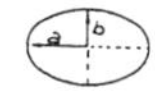 Area $= \pi ab$
Spherical Sector 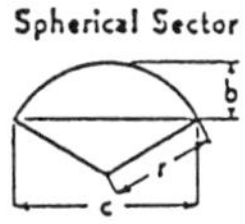	$S = \frac{1}{2}\pi r(4b + c)$. $V = \frac{2}{3}\pi r^2 b$.
Spherical Segment 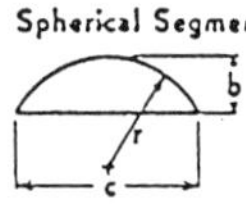	$S = 2\pi rb = \frac{1}{4}\pi(4b^2 + c^2)$. $V = \frac{1}{3}\pi b^2(3r - b) = \frac{1}{24}\pi b(3c^2 + 4b^2)$.
Spherical Zone 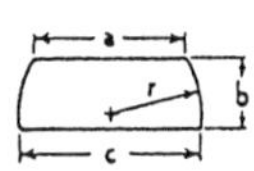	$S = 2\pi rb$. $V = \frac{1}{24}\pi b(3a^2 + 3c^2 + 4b^2)$. *Annulus* 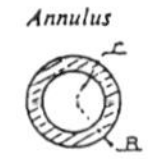 Area $= \pi(R^2 - r^2) = \pi(R - r)(R + r)$
Circular Ring 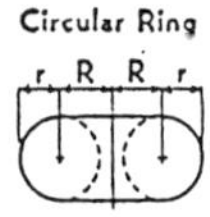	$S = 4\pi^2 Rr$. $V = 2\pi^2 Rr^2$.
Ungula of right, regular cylinder. 1. Base = segment, bab. 2. Base = half circle 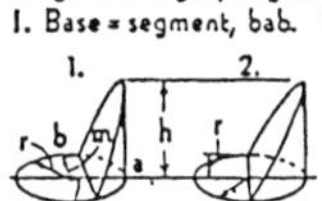	1. $S = (2rm - o \times \text{arc, bab})\frac{h}{r - o}$. $V = (\frac{2}{3}m^3 - o \times \text{area, bab})\frac{h}{r - o}$. 2. $S = 2rh$. $V = \frac{2}{3}r^2 h$.
1. Base = segment, cac. 2. Base = circle 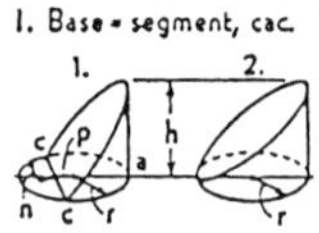	1. $S = (2rn + p \times \text{arc, cac})\frac{h}{r + p}$. $V = (\frac{2}{3}n^3 + p \times \text{area, cac})\frac{h}{r + p}$. 2. $S = \pi rh$. $V = \frac{1}{2}r^2\pi h$.
Ellipsoid 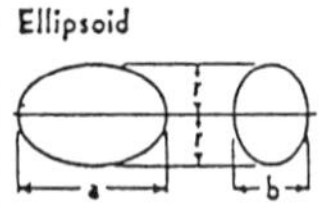	$V = \frac{1}{6}\pi rab$.
Paraboloid 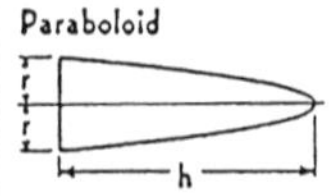	$V = \frac{1}{2}\pi r^2 h$. Ratio of corresponding volume of a Cone, Paraboloid, Sphere & Cylinder of equal height: ⅓, ½, ⅔, 1.

An Example of a Coarse Deep Bed Filter Design

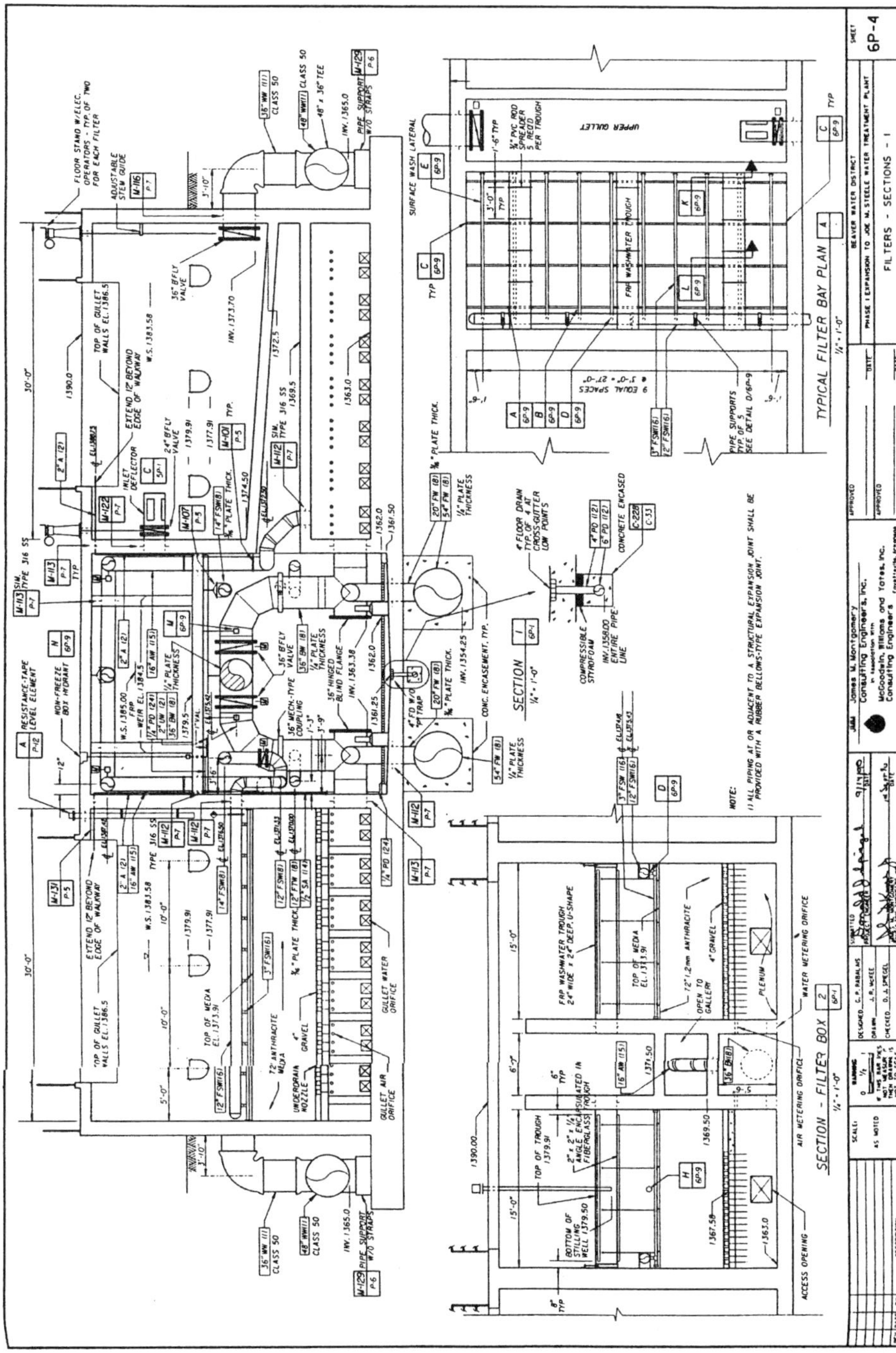

INDEX